Time-Varying Systems and Computations

Time-Varying Systems and Computations

by

PATRICK DEWILDE
and
ALLE-JAN VAN DER VEEN

DIMES, Delft University of Technology,
Delft, The Netherlands

SPRINGER-SCIENCE+BUSINESS MEDIA, B.V.

Library of Congress Cataloging-in-Publication Data

ISBN 978-1-4419-5045-1 ISBN 978-1-4757-2817-0 eBook)
DOI 10.1007/978-1-4757-2817-0

Printed on acid-free paper

Originally published by Kluwer Academic Publishers, Boston in 1998
Softcover reprint of the hardcover 1st edition 1998

Contents

Preface

Complex function theory and linear algebra provide much of the basic mathematics needed by engineers engaged in numerical computations, signal processing or control. The transfer function of a linear time invariant system is a function of the complex variable s or z and it is analytic in a large part of the complex plane. Many important properties of the system for which it is a transfer function are related to its analytic properties. On the other hand, engineers often encounter small and large matrices which describe (linear) maps between physically important quantities. In both cases similar mathematical and computational problems occur: operators, be they transfer functions or matrices, have to be simplified, approximated, decomposed and realized. Each field has developed theory and techniques to solve the main common problems encountered.

Yet, there is a large, mysterious gap between complex function theory and numerical linear algebra. For example, complex function theory has solved the problem to find analytic functions of minimal complexity and minimal supremum norm that approximate given values at strategic points in the complex plane. They serve *e.g.*, as optimal approximants for a desired behavior of a system to be designed. No similar approximation theory for matrices existed until recently, except for the case where the matrix is (very) close to singular. The relevant approximation theory in the complex plane is spectacular and has found a manifold of applications such as broadband matching, minimal sensitivity control, and the solution of differential game problems. A similar "linear algebra" result would without doubt be very desirable. Over the years we have discovered that a strong link between the two theories can indeed be developed.

To establish this link, one has to move away from the classical idiosyncrasies of the two basic theories, and develop a new and somewhat unusual paradigm, which, however, turns out to be quite natural and practical once one gets used to it. Classical matrix theory and linear algebra act on vectors and matrices. Very early in the development of these theories it was found beneficial to move from single scalar quantities and variables to vector representations. This has been an important lift in the level of abstraction, with great importance for physics and engineering, and also largely motivated by them. It has allowed for compact, insightful algebraic notations which have been adopted by a large variety of fields in which multidimensional objects interact with each other. Mechanics, electromagnetism, quantum mechanics, operations research, electrical network theory and signal processing all are fields which have been deeply influenced by vector and matrix calculus.

With the advent of powerful electronic computing, global vector or matrix-vector operations may even be viewed as atomic numerical operations. A vector computer can be programmed to execute them in parallel by a single instruction. A matrix-vector or matrix-matrix multiplication, a matrix inversion, and more complicated operations such as the calculation of matrix eigenvalues, can easily be conceived as simple sequences of such massive vector operations. In this book we will add another basic, vector-like quantity to the arsenal of objects handled by linear algebra. The new object represents a diagonal of a matrix or an operator. Thus, in addition to matrix operations acting on rows or columns, we shall consider elementary operations on diagonals. These, in fact, can be handled with the same ease by a vector or parallel computer, but they will have a very different algebraic interpretation.

What is to be gained by such an approach? In the course of the book, we develop a forceful argument that indeed, it allows us to solve several problems in linear algebra whose solutions were either unknown, or obscured by the traditional approach. In addition, the theory also creates its own class of new problems. The missing theoretical link is provided by system theory, in our case the theory of linear, time discrete and time varying dynamical systems. For example, we look at the meaning of "computational complexity" from a system theoretical point of view. Up to now, classical linear algebra had only a limited notion of complexity, restricted to either matrices that are sparse (most entries equal to zero), or matrices that are close to singular. The sparse structure is easily destroyed by algebraic operations: even the inverse of such a matrix is not sparse, and as a result, it seems that multiplication by this inverse is a full-complexity operation. This does not happen with a system theoretic "realization": it is straightforward to show that a minimal realization of the inverse has the same complexity as one for the original. In addition, system theory will allow us to derive a powerful approximation theory that maps a given matrix to a matrix of lowest computational complexity (in the system theoretical sense), given a certain tolerance.

System theory has already had a significant impact on linear algebra, mostly in the study of Toeplitz matrices and similar structured matrix problems. These are connected to time-invariant systems. Our approach in this book is complementary: we generalize to *time-varying* systems, which allows to describe any matrix. The structure in the matrix we are looking for is now less obvious, it is connected to the rank of certain strategic submatrices. In the course of the book, several classical results in the theory of time-varying systems are recovered: *e.g.,* we hit on the all-pervasive time-varying Riccati equation, the bounded real lemma and the related Kalman-Yakubovitch-Popov lemma. Still, we believe that we are placing the theory in the context of a new "paradigm", *i.e.,* a realm of problems and solutions with their own dynamics. Indeed, several results in time-varying system theory that were only known as abstract theory (such as proofs by Arveson of the existence of inner-outer factorizations) have now become explicit "constructive operator theory". Significant new results in this context are the time-varying Hankel-norm approximation theory, as well as the solution of several interpolation problems, leading to a generalization of the minimal sensitivity problem and optimal control theory.

An added value of the book is the very explicit link which it lays between numerical linear algebra and generalizations of analytic function theory on the open unit disc, as traditionally applied to transfer function calculus. The reader will discover the algebraic generalizations of a host of classical interpolation problems: Schur, Nevanlinna-Pick, Caratheodory-Fejer, Nehari, Schur-Takagi. These provide natural solutions to nice problems in algebra. Conversely, elementary methods in numerical analysis have interesting counterparts in system theory where their interpretation is non-trivial. E.g., we show that inner-outer factorization can be viewed as a generalization of QR factorization, and Hankel-norm model reduction can be used for efficient subspace estima tion and tracking.

We do not limit ourselves to finite matrices. The connection to system theory allows to derive meaningful results also for "infinite" matrices, or operators on a Hilbert space. From a linear algebra point of view, the results are perhaps uncanny: *e.g.*, the inverse of an upper triangular infinite matrix need not be upper triangular! The connection to time-varying systems gives insight into the mechanics of this: the inverse of an upper operator is upper if and only if the original system has a property which is called "outer". Even for linear algebra, infinite linear systems are useful: such systems occur *e.g.*, in the discretization of differential equations where the boundary condition is conveniently placed at infinity. Because the matrix entries become constant as we move away from the center of the matrix, it can still be described by a finite number of parameters. It has been amply demonstrated that such a procedure may lead to more accurate overall results. The downside of this generality is perhaps that, in order to obtain precise results, Hilbert space theory plays a major but sometimes also a mere technical role. (We summarize some basic notions of Hilbert space theory in an appendix.)

For whom is this book intended? We suggest the following.

- It can be used as a graduate course in linear time-varying system theory: all the main concepts and problems are there, and they are treated in a direct and streamlined manner. For this purpose we have been somewhat lengthy in developing the basic framework in the first chapters — our excuses to interested mathematicians!

- It can be used as a source of new problems in numerical linear algebra, with a concurrent new methodology to solve them. Several theories in the book scream for in-depth analysis, in particular the theory of optimal sensitivity reduction, the inversion theory for infinite systems of equations and the optimal matrix-operator approximation theory.

- It can also be used as an introductory course in a new topic: "theory of computational systems". Although the material presented here falls short of a comprehensive theory — the subject matter presently does not go far beyond linear problems and computations— we do think that there is already sufficient information to justify independent interest.

It is our hope that the algebraic system's community will find inspiration and motivation in the theory presented here. Although it has definite affinities to Arveson's "Nested Algebras" and Saeks and Feintuch's "Resolution Spaces", it does have a new flavor, mainly because of its direct link to numerical procedures via the treatment of diagonals as the new vectorial object.

Acknowledgments

Our gratitude goes to many fellow scientists and colleagues with whom we have had fruitful interactions, and we hope that we have given due reference to their contributions in the various chapters of this book. We do wish to single out two colleagues and friends who have contributed directly and materially to the realization of this book, in a scientific sense and by providing for hospitality and the environment in which the redactional work could progress. Harry Dym and Tom Kailath have been generous hosts to both of us, several times and for extended periods. Both had a decisive influence on the theory, providing insights and contributing key steps. We had intense exchanges of ideas and information with many of their students as well, too many to justify the compilation of a long list. Furthermore, we wish to thank students, colleagues and supporting staff in Delft, for the cordial atmosphere in which we were allowed to function. If one person in Delft has been outstanding in assisting us in almost any way one can think of (except scientifically), it is Mrs. Corrie Boers. Finally, Patrick Dewilde wants to voice his gratitude to his charming wife Anne for the continuous and warm support she has provided over the years.

1 INTRODUCTION

Two disciplines play a major role in this book. The first is linear algebra, which provides the setting for the derivation of efficient algorithms to do basic matrix calculations. The second is linear time-varying system theory, whose concepts we use to treat and analyze a large class of basic algorithms in a more general setting. In this introduction we explore important links between the two theories and show how linear time-varying system theory can be used to solve problems in linear algebra.

1.1 COMPUTATIONAL LINEAR ALGEBRA AND TIME-VARYING SYSTEMS

Concepts

As has been known for long, linear system theory and matrix algebra have a common footage. Indeed, if we represent a sampled signal by a vector, then a linear system—mapping an input signal to an output signal—has to be representable by a matrix. Of course, if the signals run from $t = -\infty$ to $t = +\infty$, then the matrix becomes infinite-dimensional and we rather speak of linear (Hilbert-space) operators instead. The connection between systems and matrices proves to be extremely fruitful.

Our first and foremost purpose in this book will be the "system"atic derivation of efficient algorithms for basic operations in linear algebra such as matrix multiplication, inversion and approximation, and their extension to operators which act on vectors of infinite dimensions yet have a finite numerical description. This endeavor will naturally lay in the intersection of linear algebra and system theory, a field that has been called *computational linear algebra*.

In most algorithms, the global, desired operation is decomposed into a sequence of local operations, each of which acts on a limited number of quantities (ultimately two). Intermediate variables are needed to connect the operations. The collection of these intermediate quantities at some point in the algorithmic sequence can be called the *state* of the algorithm at that point: it is what the algorithmic sequence has to remember from its past.

This point of view leads to the construction of a dynamical system that represents the algorithm and whose state equals the state of the computation at each point in the algorithmic sequence. In the case of the basic matrix operations mentioned above, the dynamical system will be linear. Although many matrix operations can be represented by some linear dynamical system, our interest is in matrices that possess a general kind of structure which results in a low dimensional state vector, and hence leads to efficient ("fast") algorithms: algorithms that exploit the structure. Structure in a matrix often has its origin in the physical map that it represents. Many problems in signal processing, finite element modeling, computational algebra and least-squares estimation produce structured matrices that can indeed be modeled by dynamical systems of low complexity. There are other very fruitful ways to represent and exploit structure in matrices. However, the time-varying system point of view produces so many results that it warrants an independent treatment.

Let us look in more detail at a linear transformation T which acts on some vector u,

$$u = \begin{bmatrix} u_0 & u_1 & u_2 & \cdots & u_n \end{bmatrix}$$

and yields an output vector $y = uT$. The vector u can just as well be viewed as an input sequence to a linear system which then produces the output sequence y. To this vector-matrix multiplication we can associate a network of computations that takes u and computes y. Intermediate quantities, *states*, are found as values on the internal edges of the network. Matrices with a sparse state structure have a computational network of low complexity so that using the network to compute y is more efficient than computing uT directly.

Consider *e.g.*, an upper triangular matrix T along with its inverse,

$$T = \begin{bmatrix} 1 & 1/2 & 1/6 & 1/24 \\ & 1 & 1/3 & 1/12 \\ & & 1 & 1/4 \\ & & & 1 \end{bmatrix}, \qquad T^{-1} = \begin{bmatrix} 1 & -1/2 & & \\ & 1 & -1/3 & \\ & & 1 & -1/4 \\ & & & 1 \end{bmatrix}. \tag{1.1}$$

The inverse of T is sparse, which is an indication of a sparse state structure. A computational network that models multiplication by T is depicted in figure 1.1(a). The reader can readily verify that this network does indeed compute $[y_1\ y_2\ y_3\ y_4] = [u_1\ u_2\ u_3\ u_4]T$ by trying the scheme on vectors of the form $[1\ 0\ 0\ 0]$ up to $[0\ 0\ 0\ 1]$. The computations in the network are split into sections, which we will call *stages*, where the k-th stage consumes u_k and produces y_k. At each point k the processor in the stage active at that point takes its input data u_k from the input sequence u and computes new

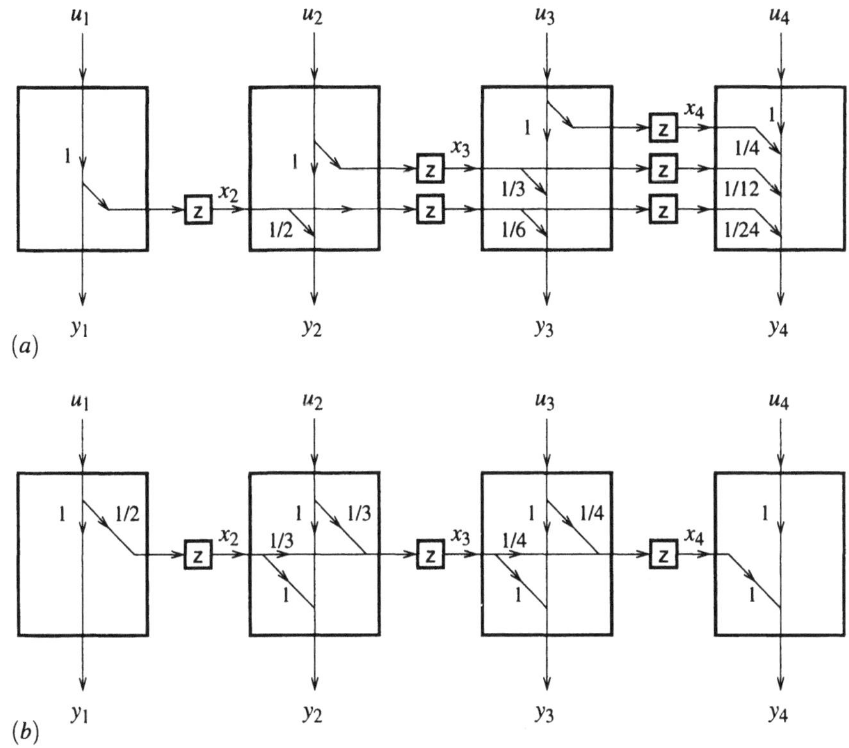

Figure 1.1. Computational networks corresponding to T. (a) Direct (trivial) realization, (b) minimal realization.

output data y_k which is part of the output sequence y generated by the system. The dependence of y_k on u_i $(i < k)$ introduces intermediate quantities x_k which we have called *states*, and which subsume the past history of the system as needed in future calculations. This state x_k is temporarily stored in registers indicated by the symbol z in the figure.[1] The complexity of the computational network is highly dependent on the number of states at each point. A non-trivial computational network to compute $y = uT$ which requires less states is shown in figure 1.1(b). The total number of (non trivial) multiplications in this network is 5, as compared to 6 in a direct computation using T. Although we have gained only one multiplication here, for a less moderate example, say an $(n \times n)$ upper triangular matrix with $n = 10000$ and $d \ll n$ states at each point, the number of multiplications in the network can be as low as $8dn$, instead of roughly $\frac{1}{2}n^2$ for a direct computation using T.

[1]This is a relic of an age-old tradition in signal processing which has little meaning in the present figure.

The (linear) computations in the network can be summarized by the following recursion, for $k = 1$ to n:

$$y = uT \quad \Leftrightarrow \quad \begin{array}{rcl} x_{k+1} & = & x_k A_k + u_k B_k \\ y_k & = & x_k C_k + u_k D_k \end{array} \tag{1.2}$$

or

$$[x_{k+1} \quad y_k] = [x_k \quad u_k]\mathbf{T}_k\,, \qquad \mathbf{T}_k = \begin{bmatrix} A_k & C_k \\ B_k & D_k \end{bmatrix}$$

in which x_k is the state vector at time k (taken to have d_k entries), A_k is a $d_k \times d_{k+1}$ (possibly non-square) matrix, B_k is a $1 \times d_{k+1}$ vector, C_k is a $d_k \times 1$ vector, and D_k is a scalar. More general computational networks may have any number of inputs and outputs, possibly also varying from stage to stage. In the example, we have a sequence of realization matrices

$$\mathbf{T}_1 = \begin{bmatrix} \cdot & \cdot \\ 1/2 & 1 \end{bmatrix}, \quad \mathbf{T}_2 = \begin{bmatrix} 1/3 & 1 \\ 1/3 & 1 \end{bmatrix}, \quad \mathbf{T}_3 = \begin{bmatrix} 1/4 & 1 \\ 1/4 & 1 \end{bmatrix}, \quad \mathbf{T}_4 = \begin{bmatrix} \cdot & 1 \\ \cdot & 1 \end{bmatrix},$$

where the '·' indicates entries that actually have dimension 0 (i.e. disappear) because the corresponding states do not exist. The recursion in equation (1.2) shows that it is a recursion for increasing values of k: the order of computations in the network is strictly from left to right, and we cannot compute y_k unless we know x_k, *i.e.*, until we have processed $u_1, \cdots, u_{k-1}$. Note that y_k does not depend on $u_{k+1}, \cdots, u_n$. This causality is a direct consequence of the fact that T has been chosen upper triangular, so that such an ordering of computations is indeed possible.

Time-varying systems

We obtain an obvious link with system theory when we regard T as the input-output map, alias the *transfer operator*, of a *non-stationary* causal linear system with input u and output $y = uT$. The i-th row of T then corresponds to the impulse response of the system when excited by an impulse at time instant i, that is, the output y caused by an input u with $u_k = \delta_{i-k}$, where δ_k is the Kronecker delta. The case where T has a Toeplitz structure then corresponds to a time-invariant system for which the response to an impulse at time $i+1$ is just the same as the response to an impulse at time i, shifted over one position. The computational network is called a state realization of T, and the number of states at each point in time is called the system order of the realization at that point. For time-invariant systems, the state realization can be chosen constant in time. For a time-varying system, the number of state variables need *not* be constant: it can increase and shrink. In this respect the time-varying realization theory is much richer, and we shall see in a later chapter that a time-varying number of states will enable the accuracy of some approximating computational network of T to be varied in time at will. If the network is regarded as the model of a physical time-varying system rather than a computational network, then the interpretation of a time-varying number of states is that the network contains switches that can switch on or off a certain part of the system and thus can make some states inaccessible for inputs or outputs at certain points in time.

Sparse computational models

If the number of state variables is relatively small, then the computation of the output sequence is efficient in comparison with a straight computation of $y = uT$. One example of an operator with a small number of states is the case where T is an upper triangular band matrix: $T_{ij} = 0$ for $j-i > p$. The state dimension is then equal to or smaller than $p-1$, since only $p-1$ of the previous input values have to be remembered at any point in the multiplication. However, the state model can be much more general, *e.g.*, if a banded matrix has an inverse, then this inverse is not bounded but is known to have a sparse state realization (of the same complexity) too, as we had in the example above. Moreover, this inversion can be easily carried out by local computations on the realization of T:[2] if $T^{-1} = S$, then $u = yS$ can be computed via

$$\left\{ \begin{array}{lcl} x_{k+1} &=& x_k A_k + u_k B_k \\ y_k &=& x_k C_k + u_k D_k \end{array} \right. \quad \Leftrightarrow \quad \left\{ \begin{array}{lcl} x_{k+1} &=& x_k(A_k - C_k D_k^{-1} B_k) + y_k D_k^{-1} B_k \\ u_k &=& -x_k C_k D_k^{-1} \qquad\qquad + y_k D_k^{-1} \end{array} \right.$$

hence S has a computational model given by

$$\mathbf{S}_k = \left[\begin{array}{cc} A_k - C_k D_k^{-1} B_k & -C_k D_k^{-1} \\ D_k^{-1} B_k & D_k^{-1} \end{array} \right] \tag{1.3}$$

Observe that the model for $S = T^{-1}$ is obtained in a *local* way from the model of T: $\mathbf{S}_k$ depends only on $\mathbf{T}_k$. Sums and products of matrices with sparse state structures have again sparse state structures with number of states at each point not larger than the sum of the number of states of its component systems, and computational networks built with these compositions (but not necessarily minimal ones) can easily be derived from those of its components.

In addition, a matrix T' that is not upper triangular can be split (or factored) into an upper triangular and a strictly lower triangular part, each of which can be separately modeled by a computational network. The computational model of the lower triangular part has a recursion that runs backward:

$$\begin{array}{lcl} x'_{k-1} &=& x'_k A'_k + u_k B'_k \\ y_k &=& x'_k C'_k + u_k D'_k \,. \end{array}$$

The model of the lower triangular part can be used to determine a model of a unitary upper matrix U which is such that U^*T is upper and has a sparse state structure. Thus, computational methods derived for upper matrices, such as the above inversion formula, can be generalized to matrices of mixed type [vdV95].

Besides matrix inversion, other matrix operations that can be computed efficiently using sparse computational models are for example the QR factorization (chapter 6) and the Cholesky factorization (chapter 13).

At this point, the reader may wonder for which class of matrices T there exists a sparse computational network (or state realization) that realizes the same multiplication operator. A general criterion will be derived in chapter 5, along with a recursive

[2]This applies to finite matrices only, for which the inverse of the matrix is automatically upper triangular again and D_k is square and invertible for all k. For infinite matrices (operators) and block matrices with non-uniform dimensions, the requirement is that T must be *outer*. See chapters 6 and 7.

$T =$ 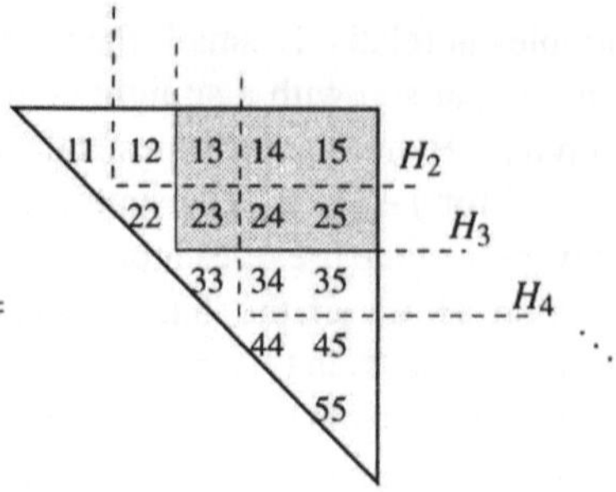

Figure 1.2. Hankel matrices are (mirrored) submatrices of T.

algorithm to determine such a network for a given matrix T. The criterion itself is not very complicated, but in order to specify it, we have to introduce an additional concept. For an upper triangular $(n \times n)$ matrix T, define matrices H_i $(1 \leq i \leq n)$, (which are mirrored submatrices of T), as

$$H_i = \begin{bmatrix} T_{i-1,i} & T_{i-1,i+1} & \cdots & T_{i-1,n} \\ T_{i-2,i} & T_{i-2,i+1} & & \vdots \\ \vdots & & \ddots & T_{2,n} \\ T_{1,i} & \cdots & T_{1,n-1} & T_{1,n} \end{bmatrix}$$

(see figure 1.2). The H_i are called (time-varying) Hankel matrices, as they have a Hankel structure (constant along anti-diagonals) if T has a Toeplitz structure.[3] In terms of the Hankel matrices, the criterion by which matrices with a sparse state structure can be detected is given by the following theorem, proven in chapter 5.

Theorem 1.1 *The number of states that are required at stage k in a minimal computational network of an upper triangular matrix T is equal to the rank of its k-th Hankel matrix H_k.*

Let's verify this statement for our example matrix (1.1). The Hankel matrices are

$$H_1 = [\cdot \cdot \cdot \cdot], \qquad H_2 = [1/2 \;\; 1/6 \;\; 1/24],$$

$$H_3 = \begin{bmatrix} 1/3 & 1/12 \\ 1/6 & 1/24 \end{bmatrix}, \qquad H_4 = \begin{bmatrix} 1/4 \\ 1/12 \\ 1/24 \end{bmatrix}.$$

[3]Warning: in the current context (arbitrary upper triangular matrices) the H_i do not have a Hankel structure and the predicate 'Hankel matrix' could lead to misinterpretations. The motivation for the use of this terminology can be found in system theory, where the H_i are related to an abstract operator H_T which is commonly called the Hankel operator. For time-invariant systems, H_T reduces to an operator with a matrix representation that has indeed a traditional Hankel structure.

Since $\text{rank}(H_1) = 0$, no states x_1 are necessary. One state is required for x_2 and one for x_4, because $\text{rank}(H_2) = \text{rank}(H_4) = 1$. Finally, also only one state is required for x_3, because $\text{rank}(H_3) = 1$. In fact, this is (for this example) the only non-trivial rank condition: if one of the entries in H_3 would have been different, then two states would have been necessary. In general, $\text{rank}(H_i) \le \min(i-1, n-i+1)$, and for a general upper triangular matrix T without state structure, a computational model indeed requires at most $\min(i-1, n-i+1)$ states for x_i. The statement is also readily verified for matrices with a band structure: if the band width of the matrix is equal to d, then the rank of each Hankel matrix is at most equal to d. As we have seen previously, the inverse of such a band matrix (if it exists) has again a low state structure, *i.e.*, the rank of the Hankel matrices of the inverse is again at most equal to d. For $d = 1$, such matrices have the form (after scaling of each row so that the main diagonal entries are equal to 1)

$$T = \begin{bmatrix} 1 & -a_1 & & \\ & 1 & -a_2 & \\ & & 1 & -a_3 \\ & & & 1 \end{bmatrix}, \qquad T^{-1} = \begin{bmatrix} 1 & a_1 & a_1 a_2 & a_1 a_2 a_3 \\ & 1 & a_2 & a_2 a_3 \\ & & 1 & a_3 \\ & & & 1 \end{bmatrix}$$

and it is seen that H_3 of T^{-1} is indeed of rank 1.

1.2 OBJECTIVES OF COMPUTATIONAL MODELING

Operations

With the preceding section as background material, we are now in a position to identify in more detail some of the objectives of computational modeling, as covered by this book. Many of the basic operations will assume that the given operators or matrices are upper triangular. Applications which involve other types of matrices will often require a transformation which converts the problem to a composition of upper (or lower) triangular matrices. For example, a vector-matrix multiplication with a general matrix can be written as the sum of two terms: the multiplication of the vector by the lower triangular part of the matrix, and the multiplication by the upper-triangular part. Efficient realizations for each will yield an efficient overall realization. In the case of matrix inversion, we would rather factor the matrix into a product of a lower and an upper triangular matrix, and treat the factors independently.

We will look at the class of matrices or operators for which the concept of a "sparse state structure" is *meaningful*, such that the typical matrix considered has a sequence of Hankel matrices that all have low rank (relative to the size of the matrix), or can be well approximated by a matrix that has that property.

Realization and cascade factorization. A central question treated in this book is *given a matrix (or operator), find a computational model* $\{\mathbf{T}_k\}_1^n$ *of minimal complexity*. Such a model could then *e.g.*, be used to efficiently compute multiplications of vectors by T. Often we want additional properties to be fulfilled, in particular we wish the computations to be *numerically stable*. One important strategy is derived from classical filter theory. It starts out by assuming T to be contractive (*i.e.*, $\|T\| \le 1$; if this is not the case, a normalization would pull the trick). Next, it subdivides the

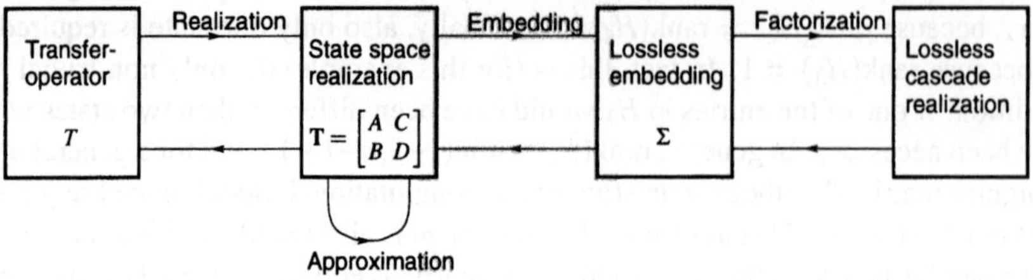

Figure 1.3. Objectives of computational modeling for matrix multiplication.

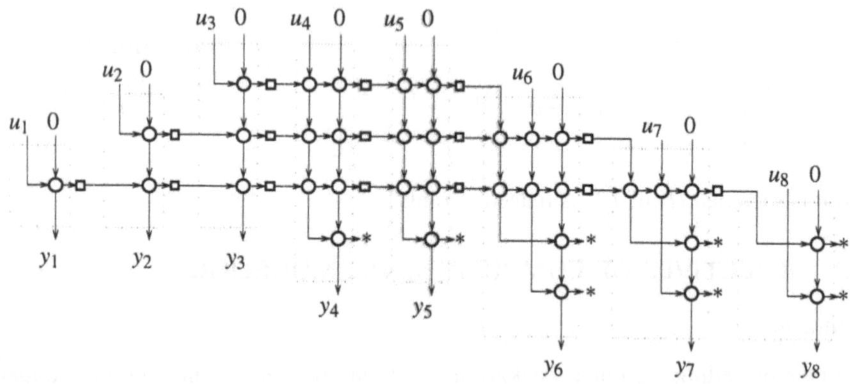

Figure 1.4. Cascade realization of a contractive 8×8 matrix T, with a maximum of 3 states at each point. The number of algebraic operations is minimal.

question in four subproblems, connected schematically as in figure 1.3: (1) realization of T by a suitable computational model, (2) embedding of this realization into a larger model that consists entirely of unitary (lossless) stages, (3) factorization of the stages of the embedding into a cascade of elementary (degree-1) lossless sections in an algebraically minimal fashion. One can show that this gives an algebraically minimal scheme to compute multiplications by T. At the same time, it is numerically stable because all elementary operations are bounded and cannot magnify intermediate errors due to noise or quantizations.

A possible minimal computational model for an example matrix T that corresponds to such a cascade realization is drawn in figure 1.4. In this figure, each circle indicates an elementary rotation of the form

$$[a_1 \quad b_1] \begin{bmatrix} \cos(\theta) & -\sin(\theta) \\ \sin(\theta) & \cos(\theta) \end{bmatrix} = [a_2 \quad b_2].$$

The precise form of the realization depends on whether the state dimension is constant, shrinks or grows. The realization can be divided into elementary *sections*, where each section describes how a single state entry of x_k is mapped to an entry of the "next state" vector x_{k+1}.

The cascade realization in figure 1.4 has a number of additional interesting properties. First, the number of operations to compute the next state and output is linear in the number of states at that point, rather than quadratic as would be the case for a general (non-factored) realization. Another is that the network is *pipelinable*, meaning that as soon as an operation has terminated it is ready to receive new data. This is interesting if the operation 'multiplication by T' is to be carried out on a collection of vectors u on a parallel computer or on a hardware implementation of the computational network. The property is a consequence of the fact that the signal flow in the network is strictly uni-directional: from top left to bottom right. Computations on a new vector u (a new u_k and a new x_k) can commence in the top-left part of the network, while computations on the previous u are still being carried out in the bottom-right part.

Approximation. It could very well be that the matrix that was originally given is known via a computational model of a very high order, *e.g.*, via a series expansion. Then intermediate in the above sequence of steps is (4) the approximation of a given realization of T by one that has the lowest possible complexity given an acceptable tolerance. For example, it could happen that the given matrix T is not of low complexity because numerical inaccuracies of the entries of T have increased the rank of the Hankel matrices of T, since the rank of a matrix is a very sensitive (ill-conditioned) parameter. But even if the given matrix T is known to be exact, an approximation by a reduced-order model could be appropriate, for example for design purposes in engineering, to capture the essential behavior of the model. With such a reduced-complexity model, the designer can more easily detect that certain features are not desired and can possibly predict the effects of certain changes in the design; an overly detailed model would rather mask these features.

While it is fairly well known in linear algebra how to obtain a (low-rank) approximant for certain norms to a matrix close to singular (*e.g.*, by use of the singular value decomposition (SVD)), such approximations are not necessarily appropriate for our purposes, because the approximant should be upper triangular again and have a lower system order than before. Moreover, the original operator may be far from singular. Because the minimal system order at each point is given by the rank of the Hankel matrix at that point, a possible approximation scheme is to approximate each Hankel operator by one that is of lower rank (this could be done using the SVD). The approximation error could then very well be defined in terms of the individual Hankel matrix approximations as the supremum over these approximations. Because the Hankel matrices have many entries in common, it is not immediately clear whether such an approximation scheme is feasible: replacing one Hankel matrix by one of lower rank in a certain norm might make it impossible for the other Hankel matrices to find an optimal approximant such that the part that they have in common with the original Hankel matrix will coincide with the original approximation. In other words: each individual local optimization might prevent a global optimum. The severity of this dilemma is mitigated by a proper choice of the error criterion. It is truly remarkable that this dilemma has a neat solution, and that this solution can be obtained in a closed form. The error for which a solution is obtained is measured in *Hankel norm:* it is the supremum over the spectral norm (the matrix 2-norm) of each individual Hankel matrix,

$$\| T \|_H = \sup_i \| H_i \| ,$$

and a generalization of the Hankel norm for time-invariant systems. In terms of the Hankel norm, the following theorem holds true and generalizes the model reduction techniques based on the Adamjan-Arov-Krein paper [AAK71] to time-varying systems:

Theorem 1.2 ([DvdV93]) *Let T be a strictly upper triangular matrix and let $\Gamma = \mathrm{diag}(\gamma_i)$ be a diagonal Hermitian matrix which parametrizes the acceptable approximation tolerance ($\gamma_i > 0$). Let H_k be the Hankel matrix of $\Gamma^{-1}T$ at stage k, and suppose that, for each k, none of the singular values of H_k are equal to 1. Then there exists a strictly upper triangular matrix T_a whose system order at stage k is equal to the number of singular values of H_k that are larger than 1, such that*

$$\| \Gamma^{-1}(T - T_a) \|_H \le 1 .$$

In fact, there is an algorithm that determines a state model for T_a directly from a model of T. Γ can be used to influence the local approximation error. For a uniform approximation, $\Gamma = \gamma I$, and hence $\|T - T_a\|_H \le \gamma$: the approximant is γ-close to T in Hankel norm, which implies in particular that the approximation error in each row or column of T is less than γ. If one of the γ_i is made larger than γ, then the error at the i-th row of T can become larger also, which might result in an approximant T_a that has fewer states. Hence Γ can be chosen to yield an approximant that is accurate at certain points but less tight at others, and whose complexity is minimal.

The realization problem is treated in chapter 5, the embedding problem is the subject of chapter 12, while the cascade factorization algorithm appears in chapter 14. The Hankel-norm approximation problem is solved in chapter 10.

QR factorization and matrix inversion. Direct methods to invert large matrices may give undesired "unstable" results. For example, suppose we try to invert

$$T = \begin{bmatrix} \ddots & \ddots & & & \\ & \boxed{1} & -2 & & \mathbf{0} \\ & & 1 & -2 & \\ & & & 1 & -2 \\ & \mathbf{0} & & & 1 & \ddots \\ & & & & & \ddots \end{bmatrix}$$

The inverse obtained by truncating the matrix to a large but finite size and inverting this part using standard linear algebra techniques produces

$$T^{-1} \stackrel{?}{=} \begin{bmatrix} \ddots & \vdots & & & \vdots & \\ & \boxed{1} & 2 & 4 & 8 & \cdots \\ & & 1 & 2 & 4 & \\ & & & 1 & 2 & \\ & \mathbf{0} & & & 1 & \cdots \\ & & & & & \ddots \end{bmatrix}$$

Clearly, this inverse is not bounded as we let the size of the submatrix grow. The true inverse is given by

$$T^{-1} = \begin{bmatrix} \ddots & \ddots & & & & \\ \cdots -1/2 & \boxed{0} & & & \mathbf{0} & \\ -1/4 & -1/2 & 0 & & & \\ -1/8 & -1/4 & -1/2 & 0 & & \\ \cdots -1/16 & -1/8 & -1/4 & -1/2 & \ddots & \\ \vdots & & & \vdots & & \ddots \end{bmatrix}.$$

Note that it is lower triangular, whereas the original is upper triangular. How could this have happened? We can obtain valuable insights in the mechanics of this effect by representing the matrix as a linear system for which it is the transfer operator:

$$T(z) = 1-2z \quad \Rightarrow \quad T^{-1}(z) = \frac{1}{1-2z} = \begin{cases} 1+2z+4z^2+\cdots \\ -\frac{1}{2}z^{-1} - \frac{1}{4}z^{-2} - \cdots. \end{cases}$$

Among other things, this will allow us to handle the instability by translating "unstable" into "anti-causal" yet bounded. In the above case, we see that $T^{-1}(z)$ has a pole inside the unit circle: it is not minimum phase and hence the causality reverses.

With time-varying systems, much more is possible. In general, we can conceptually have a time-varying number of zeros inside and outside the unit circle, —conceptually, because the notion of poles and zeros is not very well defined for time-varying systems. We can also have zeros that move from inside the circle to outside, or the other way around. This means that the inversion of infinite matrices is much more difficult, but also more interesting, than in the finite case.

The key to solving such inversion problems is to first compute a QR factorization, or "inner-outer factorization" in the system theoretical framework. This can be done using the realization of T as obtained earlier, hence can be done efficiently even on infinite-size matrices, and not surprisingly gives rise to time-varying Riccati equations. The inversion then reduces to inversion of each of the factors.

We derive the time-varying equivalent of the above example in chapter 7. Other factorizations, such as the Cholesky factorization, are discussed in chapter 13.

Interpolation and matrix completion. Several other topics are of interest as well. An important part of classical functional analysis and operator theory centers around solving constrained interpolation problems: *e.g.*, given "points" in the complex plane and "values" that a matrix-valued function should take in these points, construct an function that is constrained in norm and interpolates these values. In our present context, the functions are simply block-matrices or operators, the points are block diagonals, and the values are block diagonals as well. In chapter 9, we derive algebraic equivalents for very classical interpolation problems such as the Nevanlinna-Pick, Schur, Hermite-Fejer and Nudel'man problems. These problems are tightly connected to the optimal approximation problem discussed above. Lossless J-unitary matrices play a central role, and are discussed in chapter 8.

In linear system theory, interpolation problems have found application in the solution of robust control problems, as well as the minimal sensitivity problem: design a feedback such that a given system becomes stable and the worst-case energy amplification of a certain input to a certain output is smaller than a given bound. We treat only a single example of this: the solution of the four-block problem (section 9.7).

Finally, we consider the Nehari extension problem: for a given upper triangular matrix, try to find a lower-triangular extension such that the overall matrix has a norm bounded by a prescribed value (section 10.6). Again, the solution is governed by J-lossless matrices.

Operands

In the preceding section, the types of operations (realization, embedding, factorization, approximation, interpolation) that are considered in this book were introduced. We introduce now the types of operands to which these operations are applied. In principle, we work with bounded linear operators on Hilbert spaces of (vector) sequences. From an engineering point of view, such operators can be regarded as infinite-size matrices. The entries in turn can be block matrices. In general, they could even be operators, but we do not consider that case. There is no need for the block entries to have the same size: the only requirement is that all entries on a row of the operator have an equal number of rows, and all entries on a column of the operator have an equal number of columns, to ensure that all vector-matrix products are well defined. Consequently, the upper triangular matrices can have an "appearance" that is not upper triangular. For example, consider

$T =$ 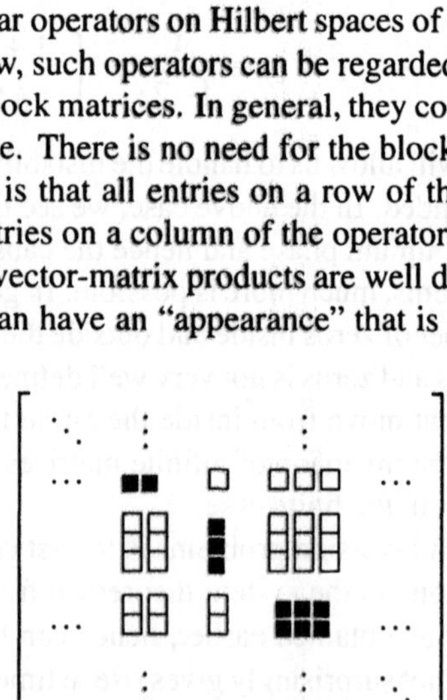 .

where in this case each box represents a complex number. The main diagonal is distinguished here by filled boxes.

We say that such an operator describes the input-output behavior of a linear time-varying system. The system is time invariant if the matrix representation of the operator is (block) Toeplitz: constant along diagonals. In general, we allow the upper triangular part to have an arbitrary structure, or even no structure at all. Special cases are periodically varying systems, which give block-Toeplitz operators, and systems that are time-invariant outside a finite interval in time, which give operators that are constant at the borders. A sequence on which the operator can be applied (the input of the system) is represented by a row vector whose entries are again finite-size vectors conforming to the block entries of the operator. This corresponds to a system with block inputs and block outputs. If the size of the block entries is not constant, then the system has a time-varying number of inputs and outputs, which corresponds physically to a system with switches that are used to switch on or off certain inputs and outputs at certain times. It is possible to model finite matrices this way, as was shown in the introduction. For finite matrices, there are no inputs and outputs before and after a certain interval in time.

A causal system corresponds to an operator whose matrix representation is upper triangular. We are interested in such systems because causality implies a computational direction: usually we can start calculations at the top-left end of the matrix and work towards the bottom-right end. Causality also introduces the notion of state. We allow the number of states to be time varying as well. This can be realized, for example, by switches that connect or disconnect parts of the system. The concept of a time-varying number of states allows the incorporation of a finer level of detail at certain intervals in time.

1.3 CONNECTIONS AND ALTERNATIVE APPROACHES

Low displacement rank

In recent times there has been quite an effort to study "structured matrices" in various guises. Besides sparse matrices (matrices with many zero entries) which fall within the context of our theory, two classical examples of structured matrices are the Toeplitz and Hankel matrices (matrices that are constant along diagonals or anti-diagonals). They represent the transfer operator of linear *time-invariant* (LTI) systems. The associated computational algorithms are well known. For example, for Toeplitz systems we have

- Schur recursions for LU and Cholesky factorization [Sch17, Kai86],
- Levinson recursions for the factorization of the inverse [Lev47],
- Gohberg/Semencul recursions for computing the inverse [GS72],
- Recursions for QR factorization [CKL87].

These algorithms have computational complexity of order $O(n^2)$ for matrices of size $(n \times n)$, as compared to $O(n^3)$ for algorithms that do not take the Toeplitz structure into account. Generalizations of the Toeplitz structure are obtained by considering matrices which have a *displacement structure* [KKM79, LK84, LK86, LK91]: matrices G for which there are (simple) matrices F_1, F_2 such that

$$G - F_1^* G F_2 \tag{1.4}$$

is of low rank, α say. This type of matrices occurs, *e.g.*, in stochastic adaptive prediction problems such as the covariance matrix of the received stochastic signal; the matrix is called of low displacement rank or α-stationary. Toeplitz matrices are a special case for which $F_1 = F_2$ are shift matrices Z and $\alpha = 2$. Related examples are block-Toeplitz and Toeplitz-block matrices, and, *e.g.*, the inverse of a Toeplitz matrix, which is itself not Toeplitz yet has a displacement rank of $\alpha = 2$. An overview of inversion and factorization algorithms for such matrices can be found in [Chu89]. Engineering applications are many, notably adaptive filtering [SK94]. In this book we do not consider low displacement matrices further (except sporadically in chapter 3, see section 3.6) and refer the reader to the extensive literature. Low displacement rank presupposes a structure that brings the operator under consideration "close to time-invariant". If an operator has that property, then it is very important to recognize and utilize it since it leads to efficient algorithms for almost any operation related to the operator. In addition, matrix-vector multiplication and inversion of a system of equations can then be done using an adaptation of the fast Fourier transform (FFT). It is possible to combine the properties of low-displacement matrices with a time-varying system theoretic approach, an account can be found in [Dew97].

Stability and control

The traditional focus of time-varying system theory has been control system design and related problems such as the stability of the system, optimal control, identification and modeling. On these topics there are quite a number of attractive textbooks and treatments, we mention [FS82, Kam79, Rug93]. Although some of the issues will appear in this book, they will not be our focus, which is essentially computational. We do give an extensive treatment of system identification—a central piece of theory—with the purpose of finding useful realizations for a linear operation. Reachability and observability spaces of a system will be omnipresent in many of our topics, such as in system approximation, algebraic operations on systems, embedding, and parametrization. The theory that we develop parallels the classical identification theory for time-varying systems, possibly in a more concrete way.

The notion of "uniform exponential stability" plays a central role in our theory as well. A linear computational scheme will have to be stable or it will not be usable. Many theorems are only valid under the condition of stability. However, conditions on stability of a system is not a great point of interest in the book, and we shall mostly assume them as a matter of course.

While this book was under redaction, Halanay and Ionescu published a book on linear time-varying discrete systems [HI94], using a framework very much like ours (and in fact partly inspired by it via publications in the mathematical literature). The contents of that book is very relevant to the work presented here, although the type of problems and their approach is often quite different. In the book of Halanay and Ionescu, basic concepts such as external factorization, inner-outer factorization, and J-inner embedding are related to the solution of specific types of (time varying) Riccati equations. We provide the derivation of the relevant Riccati equations as well, but systematically put them into a geometric context—the context provided by the reachability and observability operators of the system under consideration. On a number of other issues, the two books are unrelated. Halanay and Ionescu give an extensive treatment of optimal control and the related game theory. Although we treat some aspects of the former, *e.g.*, the four block problem, we do not consider the latter topic. On the other hand, since our focus is computational, we provide attractive algorithms such as "square root algorithms", parametrizations, and give an extensive treatment on model reduction and approximation. We have aimed at a textbook which could be used by engineering students with a good knowledge of linear algebra, but only a rudimentary knowledge of Hilbert space theory. We thought it remarkable that most essential properties could be approached from a relatively elementary point of view based on the geometry of reachability and observability spaces.

On the origin of this work

The *ansatz* for the computational modeling as studied in this book was a generalization of the Schur interpolation method to provide approximations of matrices to matrices with banded inverses, by Dewilde and Deprettere [DD87, DD88]. The motivation driving this research was the need to invert large matrices that occur in the finite element modeling of VLSI circuits [JD89, Nel89]. Subsequent research by Alpay, Dewilde, and Dym introduced an elegant diagonal notation by which the Schur interpolation method and similar such generalized, time-varying interpolation problems could be described [ADD90]. In these days, it became clear that the solution of many (time-invariant) interpolation problems can effectively be formulated in state space terms [BGR90]. The new diagonal notation was thus adopted and applied to the description of time-varying state space systems, resulting in a realization theory [vdVD91], orthogonal embedding theory with application to structural factorization [vdVD94a, vdVD93], and later an optimal Hankel-norm model reduction theory as well [DvdV93], and culminated in a workshop on the topic [DKV92], and the thesis of Van der Veen [vdV93b]. Subsequent work was on widening the algebraic aspects of the new theory [vdV96, GvdV96, vdV95], as well as H_∞ control aspects [Yu96, SV95, YSvdVD96, SV96].

The above provides the background for this book. In the mean time, there are many connections to parallel work by the "Amsterdam group" (Kaashoek, Gohberg, and coworkers) to interpolation and operator extension [GKW89, Woe89, GKW91, BGK92a], and to realization of time-varying systems [GKL92, BAGK94].

I REALIZATION

2 NOTATION AND PROPERTIES OF NON-UNIFORM SPACES

Time-varying linear systems can be compactly described by a recently developed notation in which the most important objects under consideration, sequences of vectors and the basic operators on them, are represented by simple symbols. Traditional time-varying system theory requires a clutter of indices to describe the precise interaction between signals and systems. The new notation helps to keep the number of indices in formulas at a minimum. Since in our case sequences of vectors may be of infinite length, we have to put them in a setting that can handle vectors of infinite dimensions. "Energy" also plays an important role, and since energy is measured by quadratic norms, we are naturally led to a Hilbert space setting, namely to Hilbert spaces of sequences of the ℓ_2-type. This should not be too big a step for engineers versed in finite vector space theory since most notions of Euclidean vector space theory carry over to Hilbert spaces. Additional care has to be exercised, however, with convergence of series and with properties of operators. The benefit of the approach is that matrix theory and system theory mesh in a natural way. To achieve that we must introduce a special additional flavor, namely that the dimensions of the entries of the vectors considered are not necessarily all equal.

The material covered in this chapter provides a minimal "working basis" for subsequent chapters. Additional properties and more advanced operator theoretic results are covered in chapter 4. A brief review of Hilbert space definitions and results which are relevant to later chapters can be found in Appendix A. In this work, we only need the space ℓ_2 of bounded sequences, subspaces thereof, and bounded operators on these spaces.

2.1 SPACES OF NON-UNIFORM DIMENSIONS

Non-uniform sequences

Let us consider (possibly infinite) sequences whose entries u_i are finite dimensional vectors:

$$u = [\cdots \quad u_{-1} \quad \boxed{u_0} \quad u_1 \quad u_2 \quad \cdots]. \tag{2.1}$$

Typically, we write such sequences out as *rows* of (row) vectors. We say that u represents a signal, where each component u_i is the value of the signal at time instant i. The square surrounding u_0 identifies it as the entry with index zero. If the u_i are scalar, then u is a one-channel signal. A more general situation is obtained by taking the u_i to be (row) vectors themselves, which makes u a multi-channel signal. It is not necessary that all u_i have equal dimensions: we allow for a *time-varying* number of channels, or equivalently, for non-uniform sequences. (Physically, such signals could be obtained by switches.) In order to specify such objects more precisely, we introduce the notion of *index sequences*.

Let $\{N_i \in \mathbb{N},\ i \in \mathbb{Z}\}$ be an indexed collection of natural numbers[1], such that $u_i \in \mathbb{C}^{N_i}$: N_i is the dimension of the vector u_i. The sequence N,

$$N = [N_i]_{-\infty}^{\infty} = [\cdots \quad N_{-1} \quad \boxed{N_0} \quad N_1 \quad N_2 \quad \cdots] \in \mathbb{N}^{\mathbb{Z}}$$

is called the *index sequence* of u. (The symbol $\mathbb{N}^{\mathbb{Z}}$ indicates the set (Cartesian product) of copies of $\mathbb{N}$ indexed by elements of $\mathbb{Z}$.) If we define $\mathcal{N}_i = \mathbb{C}^{N_i}$, then signals (2.1) live in the space of non-uniform sequences which is the Cartesian product of the $\mathcal{N}_i$:

$$\mathcal{N} = \cdots \times \mathcal{N}_{-1} \times \boxed{\mathcal{N}_0} \times \mathcal{N}_1 \times \mathcal{N}_2 \times \cdots =: \mathbb{C}^N,$$

Conversely, if $\mathcal{N} = \mathbb{C}^N$, then to retrieve the index sequence N from $\mathcal{N}$ we write

$$N = \#(\mathcal{N}).$$

A signal in $\mathcal{N}$ can be viewed as an infinite sequence that has a partitioning into finite dimensional components. Some of these components may have zero dimension ($N_i = 0$) to reflect the fact that no input signal is present at that point in time. In that case, we write $u_i = \cdot$, where '$\cdot$' is a marker or placeholder. Mathematically, '$\cdot$' can be viewed as the neutral (and only) element of the Hilbert space $\mathbb{C}^0$, the vector space of dimension zero. Formally, we must define some calculation rules with sequences or matrices that have blocks with dimension zero. Aside from obvious rules, the product of an "empty" matrix of dimension $m \times 0$ and an empty matrix of dimension $0 \times n$ is a matrix of dimension $m \times n$ with all elements equal to 0. All further rules of block matrix multiplication remain consistent. Using zero dimension indices, finite dimensional vectors are incorporated in the space of non-uniform sequences, by putting $N_i = 0$ for i outside a finite interval. We usually do not write these trailing markers if their presence is clear from the context or otherwise not relevant: this is consistent with the fact that for any set A,

[1] $\mathbb{Z}$ denotes the set of integers, $\mathbb{N}$ the non-negative integers $\{0, 1, \cdots\}$, and $\mathbb{C}$ the complex numbers.

$A \times \mathbb{C}^0 = A$. With abuse of notation, we will also usually identify $\mathbb{C}^0$ with the empty set $\emptyset$.

We say that a signal u as in (2.1) has finite energy if the sum

$$\sum_{i=-\infty}^{\infty} \| u_i \|_2^2$$

is finite. In that case we say that u belongs to $\ell_2^{\mathcal{N}}$, the space of (non-uniform) sequences in $\mathcal{N}$ with finite ℓ_2 norm. $\ell_2^{\mathcal{N}}$ is a 'Hilbert space', it is even a separable Hilbert space, which means that it has a countable basis. A Hilbert space of non-uniform sequences is of course isomorphic to a standard ℓ_2 Hilbert space, the non-uniformity provides an additional structure which has only system theoretical implications.

The inner product of two congruent (non-uniform) sequences f, g in $\mathcal{N}$ is defined in terms of the usual inner product of (row)-vectors in $\mathcal{N}_i$ as

$$(f, g) = \sum_i (f_i, g_i)$$

where $(f_i, g_i) = f_i g_i^*$ is equal to 0 if $N_i = 0$, by definition.[2] The corresponding norm is defined by

$$u = [u_i]_{-\infty}^{\infty} : \quad \| u \|_2^2 = (u, u) = \sum_{i=-\infty}^{\infty} \| u_i \|_2^2$$

so that $\| u \|_2^2$ represents the energy of the signal. $\ell_2^{\mathcal{N}}$ can thus be viewed as an ordinary separable Hilbert space of sequences on which a certain regrouping (of scalars into finite dimensional vectors) has been superimposed. Consequently, properties of Hilbert spaces carry over to the present context when this grouping is suppressed.

To illustrate some of the above, let $N = [\cdots\ 0\ 0\ \boxed{1}\ 3\ 2\ 0\ 0\ \cdots]$. The vector $u = [\boxed{6}, [3\ 2\ 1], [4\ 2]]$ is an element of the non-uniform sequence space $\mathcal{N} = \mathbb{C}^N$, suppressing entries with zero dimensions. The norm of u is given by $\| u \|_2 = [6^2 + (3^2 + 2^2 + 1^2) + (4^2 + 2^2)]^{1/2}$. We see that classical Euclidean vector space theory fits in easily.

Operators on non-uniform spaces

Let $\mathcal{M}$ and $\mathcal{N}$ be spaces of sequences corresponding to index sequences M, N. When we consider sequences in these spaces as signals, then a system that maps ingoing signals in $\mathcal{M}$ to outgoing signals in $\mathcal{N}$ is described by an operator from $\mathcal{M}$ to $\mathcal{N}$:

$$T : \mathcal{M} \rightarrow \mathcal{N}, \quad y = uT.$$

Following [ADD90, DD92], we adopt a convention of writing operators at the right of input sequences: $y = uT$. If for some reason there is confusion, we use brackets:

[2] * denotes the complex conjugate transpose of vectors or matrices, or the adjoint of operators.

'$T(u)$'. This unconventional notation is perhaps unnatural at first, but it does have advantages: signals correspond to row sequences, circuit diagrams read like the formulas, and the inverse scattering problem, which we shall treat extensively, appears more natural. Continuous applications of maps such as "$STU\cdots$" associate from left to right ($uSTU := ((uS)T)U$) and can often be interpreted as matrix products. Things get more complicated when S or T are maps defined on more complex objects than sequences. Notable examples are projection operators defined on spaces of operators, and the so-called Hankel operator which is introduced in the next chapter.

We denote by $\mathcal{X}(\mathcal{M},\mathcal{N})$ the space of *bounded* linear operators $\ell_2^{\mathcal{M}} \to \ell_2^{\mathcal{N}}$: an operator T is in $\mathcal{X}(\mathcal{M},\mathcal{N})$ if and only if for each $u \in \ell_2^{\mathcal{M}}$, the result $y = uT$ is in $\ell_2^{\mathcal{N}}$, and so that

$$\|T\| = \sup_{u\in\ell_2^{\mathcal{M}},u\neq 0} \frac{\|uT\|_2}{\|u\|_2}$$

is bounded. $\|T\|$ is called the induced operator norm of T. A bounded operator defined everywhere on separable Hilbert spaces admits a matrix representation which uniquely determines the operator [AG81]:

$$T = [T_{ij}]_{i,j=-\infty}^{\infty} = \begin{bmatrix} \ddots & & \vdots & & \iddots \\ & T_{-1,-1} & T_{-1,0} & T_{-1,1} & \\ \cdots & T_{0,-1} & \boxed{T_{00}} & T_{01} & \cdots \\ & T_{1,-1} & T_{10} & T_{11} & \\ \iddots & & \vdots & & \ddots \end{bmatrix} \tag{2.2}$$

(where the square identifies the 00-entry), so that it fits the usual vector-matrix multiplication rules. The block entry T_{ij} is an $M_i \times N_j$ matrix.

To identify the block-entries, rows and columns of T, it is convenient to have specific operators which construct a sequence from its entries. Following [ADD90], we define for a given space sequence $\mathcal{N}$, the operator π_k as

$$\cdot\pi_k: \quad \mathcal{N}_k \to \mathcal{N}: \quad a\pi_k = a[\cdots\ 0\ \ I_{\mathcal{N}_k}\ \ 0\ \cdots]. \tag{2.3}$$

Thus, π_k constructs a sequence out of an element of $\mathcal{N}_k$, by embedding it into a sequence which is otherwise zero (or empty, depending on the context). We define an "adjoint" to π_k as

$$\cdot\pi_k^*: \quad \mathcal{N} \to \mathcal{N}_k: \quad u_k = u\pi_k^*.$$

Thus, π_k^* retrieves the k-th (block) entry of a sequence.[3] We often implicitly use the facts that $\pi_k\pi_k^* = I_{\mathcal{N}_k}$ and $\sum_k \pi_k^*\pi_k = I_{\mathcal{N}}$, which is a "resolution of the identity". Clearly, both π_k and π_k^* have matrix representations. If an operator T with a congruent matrix representation is positioned to the right of π_k, then the (matrix or operator) product $\pi_k T$ makes sense and corresponds to taking the k-th row out of T. Similarly, $T\pi_k^*$ selects its k-th column.

[3]Properly speaking, the definition of an adjoint necessitates a Hilbert space context, but the operators do make obvious sense in a larger context as well.

The block entry T_{ij} of T is given by $T_{ij} = \pi_i T \pi_j^*$. With regard to (2.2), the operator $T_i = \pi_i T$ can be called the i-th (block) row of T, while $T\pi_j^*$ is the j-th column of T.

In $\mathcal{X}(\mathcal{M},\mathcal{N})$, we define the space of bounded upper operators

$$\mathcal{U}(\mathcal{M},\mathcal{N}) = \{T \in \mathcal{X}(\mathcal{M},\mathcal{N}) : T_{ij} = 0 \quad (i > j)\},$$

the space of bounded lower operators

$$\mathcal{L}(\mathcal{M},\mathcal{N}) = \{T \in \mathcal{X}(\mathcal{M},\mathcal{N}) : T_{ij} = 0 \quad (i < j)\},$$

and the space of bounded diagonal operators

$$\mathcal{D} = \mathcal{U} \cap \mathcal{L}.$$

As a matter of notational convenience, we often just write $\mathcal{X},\mathcal{U},\mathcal{L},\mathcal{D}$ when the underlying spaces are clear from the context or are of no particular relevance. For $A \in \mathcal{D}$, "A_i" serves as shorthand for the entry A_{ii}, and we write

$$A = \mathrm{diag}[\cdots A_{-1} \; \boxed{A_0} \; A_1 \; \cdots] = \mathrm{diag}[A_i].$$

$\mathcal{U}$, $\mathcal{L}$ and $\mathcal{D}$ satisfy the following elementary properties [ADD90]:

$$\begin{array}{rclcrcl} \mathcal{U}\cdot\mathcal{U} & \subset & \mathcal{U} & \qquad & \mathcal{L}^* & = & \mathcal{U} \\ \mathcal{L}\cdot\mathcal{L} & \subset & \mathcal{L} & & \mathcal{U}^* & = & \mathcal{L} \\ \mathcal{D}\cdot\mathcal{D} & \subset & \mathcal{D}. & & & & \end{array} \tag{2.4}$$

A link with classical linear time invariant (LTI) is established easily. In the time-invariant context, the sequences $\mathcal{M}$ and $\mathcal{N}$ are uniform, and the transfer operator behaves identically at each point in time: a shift of the input sequence over a few time slots produces still the same output sequence, but translated over the same shift. This translates to T having a *Toeplitz structure*: for all integers i, j and k, $T_{i,j} = T_{i+k,j+k}$, or, equivalently, all block entries on the same diagonal are equal. Toeplitz operators are often represented by their z-transform, which we define as follows. Denote by T_k the entry on the k-th diagonal (*i.e.*, $T_k = T_{i,i+k}$ for any i), and let

$$T(z) = \sum_{i=-\infty}^{+\infty} T_k z^k,$$

then $T(z)$ is called the matrix-valued transfer function associated to T. Note that this definition is purely formal, there is no guarantee that the series converges at any point of the complex plane. Occasionally, we will use a "meta-operator" $\mathcal{T}$ which associates a Toeplitz representation to a transfer function:

$$\mathcal{T}(T(z)) = \begin{bmatrix} \ddots & \ddots & \ddots & \ddots & \ddots & \ddots & \ddots \\ \ddots & T_{-1} & T_0 & T_1 & T_2 & T_3 & \ddots \\ \ddots & T_{-2} & T_{-1} & \boxed{T_0} & T_1 & T_2 & \ddots \\ \ddots & T_{-3} & T_{-2} & T_{-1} & T_0 & T_1 & \ddots \\ \ddots & \ddots & \ddots & \ddots & \ddots & \ddots & \ddots \end{bmatrix}.$$

Harmonic analysis on LTI systems will often provide interesting examples and counterexamples.

If $D \in \mathcal{D}$ and invertible, then $D^{-1} \in \mathcal{D}$, and $(D^{-1})_i = (D_i)^{-1}$ [ADD90]. However, unlike the situation for finite-size matrices on uniform sequences, the spaces $\mathcal{U}$ and $\mathcal{L}$ are not closed under inversion: if an upper operator $T \in \mathcal{U}$ is boundedly invertible, then the inverse is not necessarily upper. A simple example of this is given by the pair of Toeplitz operators

$$T = \begin{bmatrix} \ddots & \ddots & & \\ & \boxed{1} & -2 & \mathbf{0} \\ & & 1 & -2 \\ & \mathbf{0} & & 1 & \ddots \\ & & & & \ddots \end{bmatrix}, \qquad T^{-1} = \begin{bmatrix} \ddots & & & & \\ \ddots & \boxed{0} & & & \mathbf{0} \\ & -1/2 & 0 & & \\ & -1/4 & -1/2 & 0 & \\ \cdots & -1/8 & -1/4 & -1/2 & 0 \\ & \vdots & & \ddots & & \ddots \end{bmatrix}.$$

But also for finite-size matrices based on *non-uniform* space sequences, the same can happen. For example, let $T : \mathbb{C}^2 \times \mathbb{C}^1 \to \mathbb{C} \times \mathbb{C} \times \mathbb{C}$,

$$T = \begin{matrix} \\ \mathbb{C}^2 \{ \\ \\ \mathbb{C} \end{matrix} \overset{\begin{matrix}\mathbb{C} & \mathbb{C} & \mathbb{C}\end{matrix}}{\begin{bmatrix} \boxed{\begin{matrix}1 \\ 1/2\end{matrix}} & \begin{matrix}0 \\ 2\end{matrix} & \begin{matrix}0 \\ 0\end{matrix} \\ 0 & \underline{1/4} & 1 \end{bmatrix}}, \qquad T^{-1} = \begin{matrix} \mathbb{C} \\ \mathbb{C} \\ \mathbb{C} \end{matrix} \overset{\begin{matrix}\mathbb{C}^2 & & \mathbb{C}\end{matrix}}{\begin{bmatrix} \boxed{\begin{matrix}1 & 0\end{matrix}} & 0 \\ \begin{matrix}-1/4 & 1/2\end{matrix} & \underline{0} \\ \begin{matrix}1/16 & -1/8\end{matrix} & 1 \end{bmatrix}} \tag{2.5}$$

(the underscore identifies the position of the 0-th diagonal). When viewed as matrices without considering their structure, T^{-1} is of course just the matrix inverse of T. Mixed cases where the inverse has a lower and an upper part can also occur, and these inverses are not trivially computed, as they require a "dichotomy": a splitting of spaces into a part that determines the upper part and a part that gives the lower part. The topic will be investigated in chapter 7.

An important special case of upper operators with upper inverses is the following. An operator of the form $(I - X)$, where X is a bounded operator, has an inverse that is given by the series expansion (Neumann expansion)

$$(I - X)^{-1} = I + X + X^2 + \cdots \tag{2.6}$$

when the series converges in norm. It is known in operator theory that this will be the case when the geometric series $1 + \|X\| + \|X^2\| + \cdots$ converges, which occurs when the spectral radius $r(X)$ of X is smaller than 1:[4]

$$r(X) := \lim_{n \to \infty} \|X^n\|^{1/n} < 1 .$$

[4]For readers not familiar with the concept of spectral radius, we mention that for a finite matrix X, $r(X)$ is equal to the largest eigenvalue of X. In the context of operators, however, the spectrum is more complicated. See [AG81].

Proposition 2.1 *If $X \in \mathcal{U}$ and $r(X) < 1$, then $(I-X)^{-1}$ is given by (2.6) and is also in $\mathcal{U}$.*

It is known that the sequence $\| X^n \|^{1/n}$ converges when n goes to infinity (for an elementary proof, see [Yos71, p.212]). Also, $r(X) \le \| X \|$ because $\| X^n \|^{1/n} \le (\| X \|^n)^{1/n}$.

Hilbert-Schmidt operators

The *Hilbert-Schmidt norm* for objects in $\mathcal{X}(\mathcal{M},\mathcal{N})$ is defined as

$$\| A \|_{HS}^2 = \sum_{i,j} \|A_{ij}\|_{HS}^2 \qquad (A \in \mathcal{X}(\mathcal{M},\mathcal{N})),$$

where $\|A_{ij}\|_{HS}^2$ is, in turn, equal to the sum of the squared norms of the entries of A_{ij}.[5] For finite matrices, the Hilbert-Schmidt norm is usually called the Frobenius norm. The space in $\mathcal{X}(\mathcal{M},\mathcal{N})$ of operators which are bounded in Hilbert-Schmidt norm is given by

$$\mathcal{X}_2(\mathcal{M},\mathcal{N}) = \{A \in \mathcal{X}(\mathcal{M},\mathcal{N}) : \| A \|_{HS}^2 < \infty\}.$$

On $\mathcal{X}_2(\mathcal{M},\mathcal{N})$, the corresponding Hilbert-Schmidt inner product is

$$\langle A, B \rangle_{HS} = \mathrm{trace}(AB^*)$$

where the trace operator is a summation of the diagonal entries along the (block-)diagonal of AB^*. The Hilbert Schmidt norm satisfies $\| A \|_{HS}^2 = \langle A, A \rangle_{HS} = \mathrm{trace}(AA^*)$. $\mathcal{X}_2(\mathcal{M},\mathcal{N})$ is a Hilbert space for the Hilbert-Schmidt inner product (it becomes an ordinary Hilbert space of sequences if the entries A_{ij} are scalar and written as one sequence). Subspaces of $\mathcal{X}_2(\mathcal{M},\mathcal{N})$ are the spaces of upper, lower and diagonal Hilbert-Schmidt operators, respectively given by

$$\begin{aligned} \mathcal{U}_2 &= \mathcal{X}_2 \cap \mathcal{U} \\ \mathcal{L}_2 &= \mathcal{X}_2 \cap \mathcal{L} \\ \mathcal{D}_2 &= \mathcal{X}_2 \cap \mathcal{D}. \end{aligned} \tag{2.7}$$

We write $\mathbf{P}_{\mathcal{H}}$ for the orthogonal projection operator of $\mathcal{X}_2$ onto some subspace $\mathcal{H}$ of $\mathcal{X}_2$. We use an abbreviated notation for the following special projections:

$$\begin{aligned} &\mathbf{P} : \text{ the orthogonal projection of } \mathcal{X}_2 \text{ onto } \mathcal{U}_2 \\ &\mathbf{P}' : \text{ the orthogonal projection of } \mathcal{X}_2 \text{ onto } [\mathcal{X}_2 \ominus \mathcal{U}_2]\text{: } \mathbf{P}' = I - \mathbf{P} \\ &\mathbf{P}_0 : \text{ the orthogonal projection of } \mathcal{X}_2 \text{ onto } \mathcal{D}_2. \end{aligned} \tag{2.8}$$

The above projections are bounded operators on Hilbert-Schmidt spaces in the induced Hilbert-Schmidt operator norm. They can be generalized to operators on $\mathcal{X}$ on which $\mathbf{P}$, however, is not bounded (this is one of the reasons for introducing Hilbert-Schmidt

[5] If $A = [A_{ij}]$ is a doubly indexed collection of operators and is Hilbert-Schmidt summable: $\sum_{ij} \|A_{ij}\|^2 < \infty$, then A corresponds to a bounded operator in $\mathcal{X}$ automatically, since $\sum_j |\sum_i u_i A_{ij}|^2 \le \sum_j [\sum_i |u_i|^2 \cdot \sum_i \|A_{ij}\|^2] \le \sum_i |u_i|^2 \cdot \sum_{ij} \|A_{ij}\|^2 < \infty$.

spaces). This situation generalizes what already happens with Toeplitz operators; a few examples are given at the end of the section.

Elementary properties of $\mathbf{P}_0$ are

$$\begin{array}{ll} \mathbf{P}_0(D_1 X D_2) = D_1 \mathbf{P}_0(X) D_2 & (D_{1,2} \in \mathcal{D}, X \in \mathcal{X}), \\ [\mathbf{P}_0(X)]^* = \mathbf{P}_0(X^*) . & \end{array}$$

Operators in $\mathcal{X}_2$ satisfy the "two-sided ideal" properties: if $A \in \mathcal{X}_2$, and $B \in \mathcal{X}$ with dimensions such that the product AB is well defined, then $AB \in \mathcal{X}_2$. A similar result holds for BA if this product is well defined. A consequence is that operators in $\mathcal{X}$ can be thought of as maps from a Hilbert-Schmidt space $\mathcal{X}_2$ of the correct dimensions to another such space. We will use such spaces as generalized signal spaces ℓ_2.

2.2 SHIFTS AND DIAGONAL REPRESENTATIONS

Shift operators

For an index sequence $N = [\cdots\ N_{-1}\ \boxed{N_0}\ N_1\ \cdots]$, we denote the sequence right-shifted over k positions by $N^{(k)} = [\cdots\ N_{-k-1}\ \boxed{N_{-k}}\ N_{-k+1}\ \cdots]$. The corresponding right-shifted space sequence is denoted $\mathcal{N}^{(k)} = \mathbb{C}^{N^{(k)}}$. The right bilateral shift operator $Z = Z_{\mathcal{N}}$ on sequences $u \in \mathcal{N}$ is defined by $(uZ)_i = u_{i-1}$, *i.e.*,

$$[\cdots\ \boxed{u_0}\ \ u_1\ \ u_2\ \cdots]\, Z = [\cdots\ \boxed{u_{-1}}\ \ u_0\ \ u_1\ \cdots].$$

$Z_{\mathcal{N}}$ is an operator $\ell_2^{\mathcal{N}} \to \ell_2^{\mathcal{N}^{(1)}}$. It is readily checked from its definition that

$$Z_{ij} = \pi_i Z \pi_j^* = \begin{cases} I, & \text{if } j = i+1, \\ 0, & \text{otherwise,} \end{cases}$$

so that $Z \in \mathcal{U}$ and Z has a matrix representation

$$Z = \begin{bmatrix} \ddots & \ddots & & & & \\ & 0 & I_{N_{-1}\times N_{-1}} & & \mathbf{0} & \\ & & \boxed{0} & I_{N_0 \times N_0} & & \\ & & & 0 & I_{N_1\times N_1} & \\ & \mathbf{0} & & & 0 & \ddots \\ & & & & & \ddots \end{bmatrix}.$$

Z is unitary on $\ell_2^{\mathcal{N}}$: $ZZ^* = I$, $Z^*Z = I$, so that $Z^{-1} = Z^*$. The operator $Z^{[k]}$ denotes the k-times repeated application of Z:

$$Z^{[k]} = Z_{\mathcal{N}} Z_{\mathcal{N}^{(1)}} \cdots Z_{\mathcal{N}^{(k-1)}} .$$

Note that formally Z^k is not well defined because the dimensions in the multiplications do not match. Nonetheless, as a relaxation of notation we will in future sections often suppress dimension information in formulas and just write Z^k instead of $Z^{[k]}$.

Since $Z \in \mathcal{U}$, the properties in equation (2.4) specialize to [ADD90]

$$\begin{array}{lll} Z\mathcal{U} & \subset \mathcal{U} & \\ \mathcal{U}Z & \subset \mathcal{U} & \qquad \mathcal{L} \cap Z\mathcal{U} = 0 \\ Z^{-1}\mathcal{L} & \subset \mathcal{L} & \qquad \mathcal{L}Z^{-1} \cap \mathcal{U} = 0 \\ \mathcal{L}Z^{-1} & \subset \mathcal{L} & \end{array}$$

Similar properties hold for $\mathcal{U}_2$ and $\mathcal{L}_2$.

It is a fundamental fact (easy and proven in [ADD90]) that $\mathcal{U}_2 \perp \mathcal{L}_2 Z^{-1}$ and $\mathcal{U}_2 \perp Z^{-1}\mathcal{L}_2$, and that $\mathcal{X}_2$ admits an orthogonal decomposition

$$\mathcal{X}_2 = \mathcal{L}_2 Z^{-1} \oplus \mathcal{U}_2 = \mathcal{L}_2 Z^{-1} \oplus \mathcal{D}_2 \oplus \mathcal{U}_2 Z.$$

Previously (in equation (2.8)), we defined $\mathbf{P}'$ to be the orthogonal projection onto $[\mathcal{X}_2 \ominus \mathcal{U}_2]$. Hence $\mathbf{P}' = \mathbf{P}_{\mathcal{L}_2 Z^{-1}}$.

Diagonal shifts

Operators in $\mathcal{X}$ do not commute simply with the shift operator: let $T \in \mathcal{X}(\mathcal{M},\mathcal{N})$, and define $T^{(1)}$ by

$$Z_{\mathcal{M}} T^{(1)} = T Z_{\mathcal{N}},$$

that is, $T^{(1)} = Z^* T Z$, then $T^{(1)}$ is the operator T whose representation is shifted one position into the South-East direction: $(T^{(1)})_{i,j} = T_{i-1,j-1}$. If T commutes with the shift operator, $T^{(1)} = T$, then $T_{i,j} = T_{i-1,j-1}$ and T is a Toeplitz operator. More generally, the k-th diagonal shift of $T \in \mathcal{X}(\mathcal{M},\mathcal{N})$ into the southeast direction along the diagonals of T is defined by

$$T^{(k)} = (Z^{[k]})^* T Z^{[k]},$$

which is in $\mathcal{X}(\mathcal{M}^{(k)},\mathcal{N}^{(k)})$. Equivalently, $(T^{(k)})_{ij} = T_{i-k,j-k}$. The diagonal shift takes each of the spaces $\mathcal{L}$, $\mathcal{U}$ and $\mathcal{D}$ into themselves (albeit with shifted index sequences); it is readily verified that if $S, T \in \mathcal{X}$ such that the product ST is well defined, and

$$(ST)^{(k)} = S^{(k)} T^{(k)}, \qquad T^{(k+m)} = (T^{(k)})^{(m)}.$$

We will often run across products $(AZ)^n$, where $A \in \mathcal{X}(\mathcal{N},\mathcal{N}^{(-1)})$. These are evaluated as

$$\begin{array}{rcl} (AZ)^n & = & (AZ)(AZ)\cdots(AZ) \\ & = & Z^{[n]} A^{(n)} A^{(n-1)} \ldots A^{(1)} \\ & =: & Z^{[n]} A^{\{n\}} \end{array}$$

where $A^{\{n\}}$ is defined as

$$\begin{array}{rcl} A^{\{0\}} & = & I \\ A^{\{n\}} & = & A^{(n)} A^{\{n-1\}} = A^{(n)} A^{(n-1)} \ldots A^{(1)}. \end{array} \tag{2.9}$$

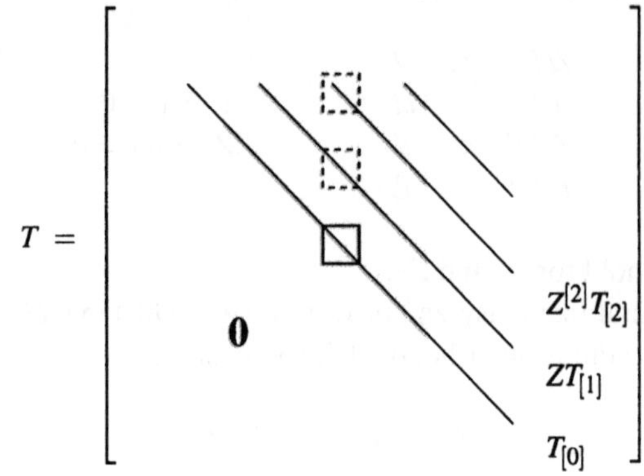

Figure 2.1. Diagonal decomposition of an operator $T \in \mathcal{U}$.

Diagonal representation

For $T \in \mathcal{X}(\mathcal{M},\mathcal{N})$, let $T_{[k]} \in \mathcal{D}(\mathcal{M}^{(k)},\mathcal{N})$ denote the k-th subdiagonal above the central (0-th) diagonal of T, defined as:

$$T_{[k]} = \mathbf{P}_0(Z^{-k}T),$$

so that $(T_{[k]})_i = T_{i-k,i}$. $T_{[k]}$ is a bounded operator because its entries are bounded by $\|T\|$ and $\|uT_{[k]}\| = \sup_i \|u_i T_{i-k,i}\| \le \|u\|\|T\|$. Based on a recursive use of the property $\mathcal{U} = \mathcal{D} \dotplus Z\mathcal{U}$, we see that, for $T \in \mathcal{U}$,

$$T - \sum_{k=0}^{n} Z^{[k]}T_{[k]} \in Z^{[n+1]}\mathcal{U}$$

so that T has a decomposition into a sum of shifted diagonals, at least formally (see figure 2.1). Although the collection $\{T_{[k]}\}_0^\infty$ uniquely specifies T, the sum does not necessarily converge to T for $n \to \infty$ in a uniform sense [ADD90]. However, for operators in $\mathcal{U}_2$ the sum does converge in the Hilbert-Schmidt norm, which provides another reason for the use of Hilbert-Schmidt spaces:

$$U \in \mathcal{X}_2: \qquad U = \sum_{-\infty}^{\infty} Z^{[k]}U_{[k]}, \qquad U_{[k]} = \mathbf{P}_0(Z^{[-k]}U).$$

Projections of operators onto $\mathcal{U}$ or $\mathcal{L}$

The projection of a bounded operator in $\mathcal{X}$ onto one in $\mathcal{U}$ or $\mathcal{L}$ may not lead to a bounded operator. This already happens for time-invariant systems, where it is known that projections of L_∞-functions of the unit circle onto their causal or anticausal parts may produce similar kinds of problems. The classical example (see appendix A) is the ideal low pass filter. Assume that $T(e^{i\theta})$ is real and specified by $T(e^{i\theta}) = 1$ for $-\frac{\pi}{2} \le \theta \le \frac{\pi}{2}$

Table 2.1. Glossary of notation.

$\mathcal{X}$	bounded operators	$\mathbf{P}_0$	proj. onto $\mathcal{D}$	$T_{[k]} = \mathbf{P}_0(Z^{-k}T)$
$\mathcal{U},\mathcal{L},\mathcal{D}$	b. upper, lower, diag.	$\mathbf{P}$	proj. onto $\mathcal{U}_2$	$T^{(k)} = Z^{[k]*}TZ^{[k]}$
$M = \#\mathcal{M} \in \mathbb{N}^{\mathbb{Z}}$	dimension sequence	$\mathbf{P}'$	proj. onto $\mathcal{L}_2 Z^{-1}$	$T^{\{k\}} = T^{(k)} \dots T^{(1)}$
$\mathcal{M} = \mathbb{C}^M$	sequence space			$T^{[k]} = T \cdots T^{(k-1)}$

and zero for other values of θ. We have $T_0 = \frac{1}{2}$ and for $k \neq 0$

$$T_k = \int_{-\pi}^{\pi} T(e^{i\theta})e^{-ik\theta}\frac{d\theta}{2\pi} = \int_{-\pi/2}^{\pi/2} e^{-ik\theta}\frac{d\theta}{2\pi} = \frac{1}{\pi k}\sin(\frac{\pi k}{2}).$$

Written out in matrix form, the corresponding transfer operator is given by

$$T = \frac{1}{\pi}\begin{bmatrix} \ddots & \ddots & \ddots & \ddots & \ddots & \ddots & \ddots & \ddots & \ddots & \ddots & \ddots \\ \ddots & 0 & 1 & \frac{\pi}{2} & 1 & 0 & -\frac{1}{3} & 0 & \frac{1}{5} & 0 & \ddots \\ \ddots & -\frac{1}{3} & 0 & 1 & \frac{\pi}{2} & 1 & 0 & -\frac{1}{3} & 0 & \frac{1}{5} & \ddots \\ \ddots & 0 & -\frac{1}{3} & 0 & 1 & \boxed{\frac{\pi}{2}} & 1 & 0 & -\frac{1}{3} & 0 & \ddots \\ \ddots & \frac{1}{5} & 0 & -\frac{1}{3} & 0 & 1 & \frac{\pi}{2} & 1 & 0 & -\frac{1}{3} & \ddots \\ \ddots & 0 & \frac{1}{5} & 0 & -\frac{1}{3} & 0 & 1 & \frac{\pi}{2} & 1 & 0 & \ddots \\ \ddots & \ddots & \ddots & \ddots & \ddots & \ddots & \ddots & \ddots & \ddots & \ddots & \ddots \end{bmatrix}$$

The projection $\mathbf{P}(T)$ is given by the series

$$\begin{aligned} \mathbf{P}(T)(z) &= \tfrac{1}{2} + \tfrac{1}{\pi}(z - \tfrac{1}{3}z^3 + \tfrac{1}{5}z^5 - \cdots) \\ &= \tfrac{1}{2} + \tfrac{1}{\pi}\arctan z \end{aligned}$$

since $\frac{d}{dz}(z - \frac{1}{3}z^3 + \frac{1}{5}z^5 + \cdots) = \frac{1}{1+z^2}$. It has an essential singularity at the points $z = \pm i$ on the unit circle ($i = \sqrt{-1}$), and hence neither belongs to L_∞ nor to H_∞, although it is analytic in the unit disc. Hence we see that $\mathbf{P}(T)(z)$ is unbounded in the operator norm, while $T(z)$ is perfectly bounded (the operator norm is equal to the $L_\infty(\mathbf{T})$-norm.)

We may expect similar problems with projection theory to upper and lower parts in time varying system theory also. In fact, we can construct simple examples from our knowledge of LTI theory. For example, the operator

$$T_1 = \frac{1}{\pi}\begin{bmatrix} \frac{\pi}{2} & 1 & 0 & -\frac{1}{3} & 0 & \frac{1}{5} & 0 & \cdots \\ 1 & \frac{\pi}{2} & 1 & 0 & -\frac{1}{3} & 0 & \frac{1}{5} & \cdots \\ 0 & 1 & \frac{\pi}{2} & 1 & 0 & -\frac{1}{3} & 0 & \cdots \\ -\frac{1}{3} & 0 & 1 & \frac{\pi}{2} & 1 & 0 & -\frac{1}{3} & \cdots \\ 0 & -\frac{1}{3} & 0 & 1 & \frac{\pi}{2} & 1 & 0 & \cdots \\ \frac{1}{5} & 0 & -\frac{1}{3} & 0 & 1 & \frac{\pi}{2} & 1 & \cdots \\ \ddots & \ddots & \ddots & \ddots & \ddots & \ddots & \ddots & \ddots \end{bmatrix}$$

is bounded as a sub-operator of T, but its projection to upper,

$$T_2 = \frac{1}{\pi}\begin{bmatrix} \frac{\pi}{2} & 1 & 0 & -\frac{1}{3} & 0 & \frac{1}{5} & 0 & \cdots \\ & \frac{\pi}{2} & 1 & 0 & -\frac{1}{3} & 0 & \frac{1}{5} & \cdots \\ & & \frac{\pi}{2} & 1 & 0 & -\frac{1}{3} & 0 & \cdots \\ & & & \frac{\pi}{2} & 1 & 0 & -\frac{1}{3} & \cdots \\ & & & & \frac{\pi}{2} & 1 & 0 & \cdots \\ \mathbf{0} & & & & & \frac{\pi}{2} & 1 & \cdots \\ & & & & & & \ddots & \ddots \end{bmatrix}$$

produces an unbounded operator. This can be shown directly from the properties of the series $[1, -\frac{1}{3}, \frac{1}{5}, \cdots]$, but the calculation would lead us too far astray here. The "logarithmic series" $[1, \frac{1}{2}, \frac{1}{3}, \frac{1}{4}, \cdots]$ and its subseries provide a wealth of additional examples well documented in the literature on harmonic analysis. We have reached the conclusion that *it is not true that the boundedness of T implies the boundedness of* $\mathbf{P}(T)$.

2.3 NOTES

The diagonal notation used in this book was originally introduced by Alpay and Dewilde in [AD90] (and subsequently in Alpay, Dewilde and Dym [ADD90]), who developed a generalization of the z-transform for upper non-commutative operators, called the W-transform, and investigated the interpolating properties of lossless time-varying systems represented by these operators. It has been refined a number of times to allow for sequences with non-uniform dimensions [vdVD91, DvdV93, vdV93b]. A time-continuous version was defined by Ball e.a. [BGK92b]. The basic mathematical properties were proven in [ADD90] and additional properties later in Dewilde and Dym [DD92].

There are a number of other approaches to describe time-varying systems. Starting in the 1950s [Zad50] (or even earlier), time-varying network theory and extensions of important system theoretic notions to the time-varying case have been discussed by many authors. While most of the early work is on continuous-time linear systems and differential equations with time-varying coefficients (see, *e.g.*, [Zad61] for a 1960 survey), discrete-time systems have gradually come into favor. There are some more recent approaches which are important, running in parallel with the time-varying state-space realization theory discussed later in chapters 3 and 5. These are presented in the monograph by Feintuch and Saeks [FS82], in which a Hilbert resolution space setting is taken, and in work by Kamen, Poolla and Khargonekar [KKP85, KP86, PK87], where time-varying systems are put into an algebraic framework of polynomial non-commutative rings. In the latter approach, a different kind of generalized z-transform is introduced. However, many of these results, in particular on controllability, detectability, stabilizability etc., have been discussed by many authors without using these specialized mathematical means, but rather by simply time indexing the state-space matrices $\{A, B, C, D\}$ and deriving expressions (iterations) in terms of these matrices. There is usually a one-to-one correspondence between these expressions and their equivalent in our notation.

3 TIME-VARYING STATE SPACE REALIZATIONS

Time-varying systems provide an especially fruitful point of view for the study of the properties of linear maps and operators acting on sequences of data vectors. The notation and preliminary results given in chapter 2 prepared the grounds for a realization theory of such systems. A linear operator may often be decomposed into a composition of *local* linear transformations in which intermediate data called states are generated for use in subsequent stages. This brings the theory of such transformations into the realm of linear dymamic system theory for discrete-time signals. The global transformation plays the role of input-output operator or *transfer operator*, while the decomposition can be interpreted as the *realization* of a computational scheme in which small local transformations are executed. Hence, methods from system theory can be used to yield schemes of minimal complexity, optimal approximations to systems of lower complexity, and so on.

The fact that there is a strong connection between system theory and linear algebra has long been known and exploited. For matrices with a Toeplitz or Hankel structure, this has resulted in fast matrix multiplications (via fast Fourier transforms), and Schur recursions for Cholesky factorizations. For the more general case of upper triangular matrices without such Toeplitz structure, the connection with systems theory becomes fruitful if we consider time-varying state realizations, and if we assume that the number of states in the realization is small compared to the size of the matrix.

The important first step in setting up a computational scheme for general upper matrices is to make the connection with system theory explicit, and in particular, to solve the *realization problem*. It is the problem of finding a decomposition of the original operator into a sequence of operations, each of which utilizes only partial data of the input sequence, generates intermediate quantities called *states*, and produces a part of the output. In doing so, we have made the implicit assumption in our computational model that the input data becomes available sequentially, and that the output data is generated sequentially as well. Since the original operator is assumed to be linear, the problem reduces to find, for a given upper triangular matrix T, a realization $\{A_k, B_k, C_k, D_k\}$ that has the given matrix as its input-output operator, *i.e.*, such that

$$[y_1 \quad y_2 \cdots y_n] = [u_1 \quad u_2 \cdots u_n]\,T \quad \Leftrightarrow \quad \left\{ \begin{array}{rcl} x_{k+1} & = & x_k A_k + u_k B_k \\ y_k & = & x_k C_k + u_k D_k \end{array} \right.$$

In the present chapter, we restrict the discussion to finite matrices and investigate state realizations and their relation with the matrices that they realize. We will discover how finite matrices are embedded in the more general framework of operators, consider some prime examples of finite matrices with a low number of states, and derive an algorithm for minimal state space realization of finite matrices. The more general case, the realization problem for operators on non-finite sequences of data, is deferred to chapter 5.

3.1 REALIZATIONS OF A TRANSFER OPERATOR

Transfer operator

Let $\ell_2^{\mathcal{M}}$ and $\ell_2^{\mathcal{N}}$ be two (non-uniform) spaces as defined in the previous chapter, and let the input-output behaviour of a linear time-varying and discrete time system be described by its *transfer operator* (input-output operator), which is an operator T which maps signals in $\ell_2^{\mathcal{M}}$ to signals in $\ell_2^{\mathcal{N}}$:

$$T : \ell_2^{\mathcal{M}} \to \ell_2^{\mathcal{N}} : \quad y = uT\,.$$

We call $\mathcal{M}$ the input space of the system, and $\mathcal{N}$ the output space.

We assume for the time being that T is bounded: it maps signals of bounded energy to other signals of bounded energy, with a uniform upper bound. Other spaces, such as ℓ_∞, could have been considered as signal spaces [Mur84], but ℓ_2 is mathematically more attractive. Many facts in operator theory are simplest for Hilbert spaces, and some facts, such as the existence of an adjoint operator, are dependent on the availability of an inner product. One could restrict the attention further and consider only input/output sequences with compact support: signals which are non-zero only on a finite number of time points. The argument for doing so is that most of the mathematical complications of the Hilbert Space context disappear, and since such sequences are dense in ℓ_2, the resulting system theory (save for the mathematical details) is closely related to the Hilbert space realization theory. This is the approach taken in the parallel time-varying system theory of [GKL92], and in a sense, the results are the same for finite matrices. In our case, however, we are interested mainly in problems of system approximation and numerical realization in which the ℓ_2 norm plays an essential role, so we keep to the Hilbert space setting.

At this point, let us introduce an important generalization of the ℓ_2-setting, which will be heavily used in subsequent chapters. Since time-varying systems may change at each point in time, we wish to consider collections of inputs and corresponding outputs which reveal characteristic properties of the system at each point in time, rather than just a single input and its corresponding output. Therefore, we wish to consider a type of input or output space more general than $\ell_2^{\mathcal{M}}$. We define this more general space so that it provides us with a *collection* of input and corresponding output time-sequences. An infinite collection of input sequences would fit in a (doubly infinite) matrix, with one row for each sequence, and each sequence with dimensions given by $\mathcal{M}$. The total matrix is formally a mapping from $\mathbb{C}^{\mathbb{Z}} := [\cdots\times\mathbb{C}\times\mathbb{C}\times\mathbb{C}\times\cdots]$ to $\mathcal{M}$. Here, $\mathbb{C}^{\mathbb{Z}}$ is just a sequence of copies of $\mathbb{C}$; the set $\mathbb{Z}$ contains the indices of the rows.

In the notation of section 2.1, we then define the Hilbert space

$$\mathcal{X}_2^{\mathcal{M}} := \mathcal{X}_2(\mathbb{C}^{\mathbb{Z}}, \mathcal{M}).$$

An element of this space can thus be viewed as an infinite collection of signal sequences from $\ell_2^{\mathcal{M}}$, stacked on top of each other, and such that the grand total energy of the collection is bounded.

Having collections of signals in one object allows us to apply a number of relevant input sequences to a system all at once, and collect the results in a similar collection of output sequences. There is no advantage in doing this with $\mathcal{X}_2^{\mathcal{M}}$ itself, but it is quite useful to act on certain subspaces of $\mathcal{X}_2^{\mathcal{M}}$, like $\mathcal{U}_2^{\mathcal{M}}$ (the space of "upper" signal collections). An element of $\mathcal{U}_2^{\mathcal{M}}$ is such that its i-th row is a signal in $\ell_2^{\mathcal{M}}$ which is identically zero before point i in time, for each i. The support of the signal on row i is completely in "the future", with respect to time point i. Since the systems we consider are time-varying, any analysis will have to take all time points i equally and separately into account, and this is precisely why it is useful to have the complete collection available in one object in $\mathcal{U}_2^{\mathcal{M}}$.

In a similar vein, elements of $\mathcal{D}_2^{\mathcal{M}}$ are signals that only have support at the "current point in time", for every point i, *i.e.*, it contains all impulses. Finally, $\mathcal{Z}^{-1}\mathcal{L}_2^{\mathcal{M}}$ is the collection of all signals with support "in the strict past".

A first use of the new notation for collections of signals is the following definition.

Definition 3.1 *A transfer operator T is causal if*

$$U \in \mathcal{U}_2 \qquad \Rightarrow \qquad Y = UT \in \mathcal{U}_2 .$$

Proposition 3.2 *T is causal if and only if it is an upper operator: $T \in \mathcal{U}$.*

An expression of causality in terms of ℓ_2-sequences is more elaborate, as it has to state that for all k and for all signals that are zero before point k, the corresponding response is also zero before point k.

The rows of T can be viewed as the impulse responses of the system. Indeed, in the single-input single-output case, and if T is a causal transfer operator, the response to the unit impulse at time i, $u = [\delta_i]_{-\infty}^{\infty}$, is $y = uT = [\cdots\ 0\ T_{ii}\ T_{i,i+1}\ T_{i,i+2}\ \cdots]$, precisely the i-th row of T. An obvious extension holds for general multi-dimensional sequences.

Realizations

Suppose that a transfer operator T is given. An important question is to know whether the corresponding system admits a *dynamical realization* in the form of a recursion on

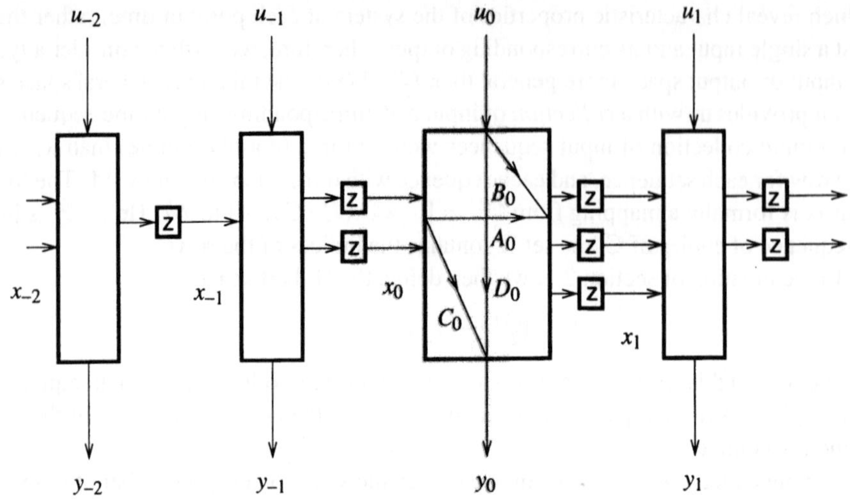

Figure 3.1. Time-varying state realization.

a sequence of states:

$$\begin{array}{rcll} x_{k+1} & = & x_k A_k + u_k B_k & \quad k = \cdots, -1, 0, 1, \cdots \\ y_k & = & x_k C_k + u_k D_k . & \end{array} \tag{3.1}$$

The expression states that the computation of y is performed as a sequence of stages, which are connected by intermediate quantities $\{x_k\}$, the states. The state at point k is data extracted from the input sequence u up to that point, such that knowledge of the state is sufficient to be able to compute future outputs without reference to the old input data. $\{A_k, B_k, C_k, D_k\}$ are called the *state realization matrices*. We require them to be uniformly bounded and to have finite dimensions, possibly varying with k. The state equations represent the structure of the computations as a sequence of operations, which is depicted in figure 3.1. In this figure, the symbols "z" stand for registers that store the values of the state variables when the computation goes from one point in time to the next. We often collect the matrices A_k, B_k, C_k, D_k into a single transition matrix, denoted by a boldface symbol, *e.g.*,

$$\mathbf{T}_k = \begin{bmatrix} A_k & C_k \\ B_k & D_k \end{bmatrix},$$

which allows to rewrite the state equations (3.1) as

$$[x_{k+1} \quad y_k] = [x_k \quad u_k] \mathbf{T}_k .$$

The realization automatically represents a causal operator: if $u_k = 0$ for all k less than some point k_0 in time, then $y_k = 0$ $(k < k_0)$.

Realizations of the type (3.1) can be rewritten in global operator form by assembling the matrices $\{A_k\}$, $\{B_k\}$ etc. as diagonal operators on spaces of sequences of appropriate dimensions:

$$A = \begin{bmatrix} \ddots & & \mathbf{0} \\ & A_k & \\ \mathbf{0} & & \ddots \end{bmatrix} \qquad C = \begin{bmatrix} \ddots & & \mathbf{0} \\ & C_k & \\ \mathbf{0} & & \ddots \end{bmatrix} \tag{3.2}$$

$$B = \begin{bmatrix} \ddots & & \mathbf{0} \\ & B_k & \\ \mathbf{0} & & \ddots \end{bmatrix} \qquad D = \begin{bmatrix} \ddots & & \mathbf{0} \\ & D_k & \\ \mathbf{0} & & \ddots \end{bmatrix}$$

Let $\ell_2^{\mathcal{M}}$ be the space of input sequences, $\ell_2^{\mathcal{N}}$ the space of output sequences, and let us define $\mathcal{B} = \cdots \oplus \mathcal{B}_0 \oplus \mathcal{B}_1 \oplus \cdots$ as the sequence of spaces to which the state belongs[1]. Then

$$\begin{array}{rclcl} u & = & [\cdots \ \boxed{u_0} \ u_1 \ u_2 \cdots] & \in & \ell_2^{\mathcal{M}} \\ y & = & [\cdots \ \boxed{y_0} \ y_1 \ y_2 \cdots] & \in & \ell_2^{\mathcal{N}} \\ x & = & [\cdots \ \boxed{x_0} \ x_1 \ x_2 \cdots] & \in & \mathcal{B} \\ xZ^{-1} & = & [\cdots \ \boxed{x_1} \ x_2 \ x_3 \cdots] & \in & \mathcal{B}^{(-1)}. \end{array}$$

The shift-operator Z was defined in section 2.1. Its inverse Z^{-1} shifts a sequence over one position to the left; by $\mathcal{B}^{(-1)}$ we denote the corresponding shifted space sequence. A discrete-time causal time-varying linear realization $\mathbf{T}$ consists of the set of four maps

$$\mathbf{T} = \begin{bmatrix} A & C \\ B & D \end{bmatrix}, \qquad \begin{array}{lcl} A & \in & \mathcal{D}(\mathcal{B},\mathcal{B}^{(-1)}), \\ B & \in & \mathcal{D}(\mathcal{M},\mathcal{B}^{(-1)}), \end{array} \quad \begin{array}{lcl} C & \in & \mathcal{D}(\mathcal{B},\mathcal{N}), \\ D & \in & \mathcal{D}(\mathcal{M},\mathcal{N}), \end{array} \tag{3.3}$$

which together represent the dynamical state equations

$$\begin{array}{rcl} xZ^{-1} & = & xA + uB \\ y & = & xC + uD. \end{array} \tag{3.4}$$

This definition constitutes the same set of time-varying state equations as in (3.1), but now written in an index-free form and acting on sequences. The state equations (3.1) are recovered by taking the k-th entry of each sequence and the corresponding k-th entry along the diagonal of each realization matrix. A difference between the equations (3.1) and (3.4) is that the former equations suggest a *recursion* which can be carried out to obtain the next state x_{k+1} and current output y_k from the current state x_k and input u_k, whereas the equations (3.4) are implicit conditions which some sequences u, x and y have to satisfy:

$$x(I - AZ) = uBZ. \tag{3.5}$$

If $(I - AZ)$ is boundedly invertible on the space $\ell_2^{\mathcal{B}}$ to which the state x belongs, then (3.5) has a solution

$$x = uBZ(I - AZ)^{-1}.$$

[1] We shall discuss the precise structure of $\mathcal{B}$ later on.

Substitution into the second equation of (3.4) leads to

$$y = u\left[D + BZ(I-AZ)^{-1}C\right],$$

so that the transfer operator corresponding to the state equations (3.4) is

$$T = D + BZ(I-AZ)^{-1}C.$$

Note the similarity of this expression for the transfer operator T and the familiar expression of the *transfer function* $T(z) = d + bz(1-az)^{-1}c$ for time-invariant systems with a time-invariant realization $\{a,b,c,d\}$, where a,b,c,d are matrices rather than diagonal operators with matrix entries.

However, even if the state sequences are Hilbert-Schmidt bounded (*i.e.,* they live in $\ell_2^{\mathcal{B}}$), $(I-AZ)^{-1}$ is not necessarily causal, as we showed by some examples in section 2.1. Only if $(I-AZ)^{-1} \in \mathcal{U}$ will the transfer $u \to x$ be causal. In contrast, the recursion (3.1) when started at some point in time, leads to a map which is always causal, but might be unbounded. In that case, (3.5) is not equivalent to (3.1).

According to proposition 2.1, $(I-AZ)$ has an inverse which is upper and given by the converging series

$$(I-AZ)^{-1} = I + AZ + (AZ)^2 + \cdots$$

if the spectral radius $\ell_A := r(AZ) < 1$. Since $r(AZ) = \lim_{n\to\infty}(AZ)^n$, and

$$(AZ)^n = AZAZ\cdots AZ = Z^{[n]}A^{(n)}A^{(n-1)}\cdots A^{(1)} = Z^{[n]}A^{\{n\}}$$

we find that

$$\ell_A = \lim_{k\to\infty} \|A^{\{k\}}\|^{1/k},$$

where $A^{\{n\}} := A^{(n)}\cdots A^{(1)}$ and $A^{(n)} := Z^{-n}AZ^n$ is a version of A, shifted downwards along the diagonal over n positions[2]. Note that $\ell_A < 1$ does *not* mean that $\|A\| < 1$. For example, the diagonal operator

$$A = \begin{bmatrix} \ddots & & & & & \\ & \frac{1}{2} & & & & \\ & & \frac{1}{2} & & & \\ & & & 1000 & & \\ & & & & \frac{1}{2} & \\ & & & & & \frac{1}{2} \\ & & & & & & \ddots \end{bmatrix}$$

has norm 1000 but $\ell_A = \frac{1}{2}$.

[2]There is a dual quantity to ℓ_A, namely the spectral radius of AZ^*, which equals $\lim_{n\to\infty} AA^{(1)}\cdots A^{(n-1)} = \lim_{n\to\infty} A^{[n]}$. Its value is not necessarily equal to ℓ_A. We shall not encounter it furtheron in this book.

Definition 3.3 *A realization* $\{A,B,C,D\}$ *is called* uniformly exponentially stable*(u.e. stable) if* $\ell_A < 1$.

If the transfer operator is a matrix of finite dimensions so that $\mathcal{B}$ has finite support, then the realization will always be u.e. stable. There are many definitions of stability in the control literature (*cf.* [SA68, AM69, AM81, AM92, Rug93]). Our definition is the only notion of stability that we use in the sequel.

If $\ell_A < 1$, then x is given by the series

$$\begin{aligned} x &= uBZ(I-AZ)^{-1} \\ &= uBZ + uBZ(AZ) + uBZ(AZ)^2 + \cdots \end{aligned} \tag{3.6}$$

which is convergent for any $u \in \ell_2^{\mathcal{M}}$. Clearly $x \in \ell_2^{\mathcal{B}}$, since the operator $BZ(I-AZ)^{-1}$ is bounded. Hence, if $\ell_A < 1$, the formal solution of the realization equations (3.4) for a given u equals the solution generated by the recursion (3.1), and

$$\begin{aligned} y &= uD + uBZC + uBZAZC + uBZ(AZ)^2C + \cdots \\ &= uD + uZB^{(1)}C + uZ^2B^{(2)}A^{(1)}C + uZ^3B^{(3)}A^{\{2\}}C + \cdots . \end{aligned} \tag{3.7}$$

If $\ell_A = 1$, then (3.6) may or may not converge to a sequence x with bounded entries, depending on u and B. Although the analysis of realizations for which $\ell_A = 1$ is certainly possible under suitable conditions, we shall usually limit our attention to the u.e. stable case. The analysis of ℓ_A to characterize u.e. stable ($\ell_A < 1$), marginally stable ($\ell_A = 1$) and unstable ($\ell_A > 1$) systems replaces the notion in LTI systems theory of poles (eigenvalues of A) that lie in, on, or outside the unit circle.

For the general case we can state the following definition (*cf.* equation (3.7)).

Definition 3.4 *A* 2×2 *matrix of block diagonals* $\mathbf{T}$ *is said to be a realization of a transfer operator* $T \in \mathcal{U}$ *if the diagonals* $T_{[k]} = \mathbf{P}_0(Z^{-k}T)$ *of* T *equal the diagonal expansion (3.7):*

$$T_{[k]} = \begin{cases} 0, & k<0, \\ D, & k=0, \\ B^{(k)}A^{\{k-1\}}C, & k>0. \end{cases} \tag{3.8}$$

Equivalently, the entries T_{ij} of T are given by

$$T_{ij} = \begin{cases} 0, & i>j \\ D_i, & i=j \\ B_iA_{i+1}\cdots A_{j-1}C_j, & i<j, \end{cases} \tag{3.9}$$

and it follows that the transfer operator which corresponds to the realization $\{A,B,C,D\}$ has the matrix representation

$$T = \begin{bmatrix} \ddots & \vdots & & & \vdots & \\ & D_{-1} & B_{-1}C_0 & B_{-1}A_0C_1 & B_{-1}A_0A_1C_2 & \cdots \\ & & \boxed{D_0} & B_0C_1 & B_0A_1C_2 & \\ & & & D_1 & B_1C_2 & \\ & \mathbf{0} & & & D_2 & \cdots \\ & & & & & \ddots \end{bmatrix}. \tag{3.10}$$

Definition 3.5 *Let $T \in \mathcal{U}$. An operator $T \in \mathcal{U}$ is said to be* locally finite *if it has a state realization whose state space sequence $\mathcal{B}$ is such that each $\mathcal{B}_k$ has finite dimension. The order of the realization is the index sequence $\#(\mathcal{B})$ of $\mathcal{B}$.*

The concept of locally finite operators is a generalization of rational transfer functions to the context of time-varying systems.

Realizations on $\mathcal{X}_2$

We can extend the realization (3.4) further by considering generalized inputs U in $\mathcal{X}_2^{\mathcal{M}}$ and outputs Y in $\mathcal{X}_2^{\mathcal{N}}$:

$$\begin{aligned} XZ^{-1} &= XA + UB \\ Y &= XC + UD. \end{aligned} \tag{3.11}$$

If $\ell_A < 1$, then $X = UBZ(I - AZ)^{-1}$, so that $X \in \mathcal{X}_2^{\mathcal{B}}$. The classical realization (3.4) may be recovered by selecting corresponding rows in U, Y and X. Indeed, we can interpret the rows of $U \in \mathcal{X}_2^{\mathcal{M}}$ as a *collection* of input sequences $u \in \ell_2^{\mathcal{M}}$, applied simultaneous to the system. Likewise, $Y \in \ell_2^{\mathcal{N}}$ contains the corresponding output sequences $y \in \ell_2^{\mathcal{N}}$. This interpretation will be pursued at length in the following chapters.

A recursive description for the realization (3.11) is a generalization of (3.1), and is obtained by selecting the k-th diagonal of U, Y, and X in (3.11):

$$\begin{aligned} X_{[k+1]}^{(-1)} &= X_{[k]}A + U_{[k]}B \\ Y_{[k]} &= X_{[k]}C + U_{[k]}D. \end{aligned} \tag{3.12}$$

Note that the k-th diagonal of XZ^{-1} is $X_{[k+1]}^{(-1)}$, which contains a diagonal shift. The same remarks on the relation between this recursive realization and the equations (3.11) as made earlier on the ℓ_2-realizations are in order here. Starting with chapter 5, we will heavily use this type of realizations, where we act on sequences of diagonals rather than scalars.

State transformations

Two realizations $\{A, B, C, D\}$ and $\{A', B', C', D'\}$ are called *equivalent* if they realize the same transfer operator T,

$$\begin{aligned} D &= D' \\ B^{(k)}A^{\{k-1\}}C &= B'^{(k)}A'^{\{k-1\}}C' \qquad (\text{all } k \geq 0). \end{aligned} \tag{3.13}$$

Given a realization of an operator $T \in \mathcal{U}$, it is straightforward to generate other realizations that are equivalent to it. For a boundedly invertible diagonal operator R (sometimes called a Lyapunov transformation), inserting $x = x'R$ in the state equations (3.11) leads to

$$\left\{\begin{array}{lcl} x'RZ^{-1} & = & x'RA + uB \\ y & = & x'RC + uD \end{array}\right.$$

$$\Leftrightarrow \left\{\begin{array}{lcl} x'Z^{-1}R^{(-1)} & = & x'RA + uB \\ y & = & x'RC + uD \end{array}\right.$$

$$\Leftrightarrow \left\{\begin{array}{lcl} x'Z^{-1} & = & x'RAR^{-(-1)} + uBR^{-(-1)} \\ y & = & x'RC + uD \end{array}\right.$$

$$\Leftrightarrow \left\{\begin{array}{lcl} x'Z^{-1} & = & x'A' + uB' \\ y & = & x'C' + uD'. \end{array}\right.$$

Proposition 3.6 *Let $R \in \mathcal{D}(\mathcal{B},\mathcal{B})$ be boundedly invertible in $\mathcal{D}$. If $\{A,B,C,D\}$ is a realization of a system with transfer operator T, then an equivalent realization is given by $\{A',B',C',D'\}$, where*[3]

$$\begin{bmatrix} A' & C' \\ B' & D' \end{bmatrix} = \begin{bmatrix} R & \\ & I \end{bmatrix} \begin{bmatrix} A & C \\ B & D \end{bmatrix} \begin{bmatrix} \left[R^{(-1)}\right]^{-1} & \\ & I \end{bmatrix}. \tag{3.14}$$

In addition, the spectral radii of AZ and $A'Z$ are the same: $\ell_A = \ell_{A'}$.

PROOF We have already $D = D'$, and

$$\begin{aligned} & B'^{(k)}A'^{\{k-1\}}C' \\ & = B^{(k)}R^{-(k-1)} \cdot R^{(k-1)}A^{\{k-1\}}R^{-(k-2)} \cdot R^{(k-2)}A^{\{k-2\}}R^{-(k-3)} \cdots R^{(1)}A^{(1)}R^{-1} \cdot RC \\ & = B^{(k)}A^{\{k-1\}}C. \end{aligned}$$

Stability is preserved under the transformation:

$$\begin{array}{lcl} \ell_{RAR^{-(-1)}} & = & \lim_{n\to\infty} \| (RAR^{-(-1)}Z)^n \|^{1/n} \\ & = & \lim_{n\to\infty} \| (RAZR^{-1})^n \|^{1/n} \\ & = & \lim_{n\to\infty} \| R(AZ)^n R^{-1} \|^{1/n} \\ & \leq & \lim_{n\to\infty} \| R \|^{1/n} \cdot \| (AZ)^n \|^{1/n} \cdot \| R^{-1} \|^{1/n} \; = \; \ell_A \end{array} \tag{3.15}$$

since $\| R \|^{1/n} \to 1$ and $\| R^{-1} \|^{1/n} \to 1$. Because $\ell_A \leq \ell_{RAR^{-(-1)}}$ can be proven in the same way, it follows that $\ell_A = \ell_{RAR^{-(-1)}}$. □

If the realizations $\{A,B,C,D\}$ and $\{A',B',C',D'\}$ are related by (3.14) using bounded R with bounded R^{-1}, then we call them Lyapunov equivalent.

3.2 SPECIAL CLASSES OF TIME-VARYING SYSTEMS

In this section, we examine the behavior of certain interesting subclasses of systems. Since it takes an infinite amount of data and time to describe a general time-varying

[3] In future equations, we write, for shorthand, $R^{-(-1)} := [R^{(-1)}]^{-1}$.

system, it pays to consider special classes of operators in which computations can be carried out in finite time. Interesting classes are (1) finite matrices, (2) periodically varying systems, (3) systems which are initially time-invariant or periodic, then start to change, and become again time-invariant or periodic after some finite period (time-invariant or periodic at the borders), (4) systems that are quasi-periodic with a given law of quasi-periodicity, and (5) systems with low displacement rank [KKM79]. Sometimes we can even treat the general case with finite computations, especially when we are interested only in the behavior of the system in a finite window of time.

Finite matrices

Matrices of finite size can be embedded in the general framework in several ways. For example, if the input space sequence $\mathcal{M} = \cdots \oplus \mathcal{M}_{-1} \oplus \mathcal{M}_0 \oplus \mathcal{M}_1 \oplus \cdots$ has $\mathcal{M}_i = \emptyset$ for i outside a finite interval, $[1,n]$ say, and if the output space sequence $\mathcal{N}$ has $\mathcal{N}_i = \emptyset$ also for i outside $[1,n]$, then $T \in \mathcal{U}(\mathcal{M},\mathcal{N})$ is an upper triangular $n \times n$ (block) matrix:

$$T = \begin{bmatrix} \cdot & \cdot & \cdot & \cdot & & \cdot & \cdot & \cdot \\ & \cdot & \cdot & \cdot & & \cdot & \cdot & \cdot \\ & & T_{11} & T_{12} & \cdots & T_{1n} & \cdot & \cdot \\ & & & T_{22} & \cdots & T_{2n} & \cdot & \cdot \\ & & & & \ddots & \vdots & & \vdots \\ & & & & & T_{nn} & \cdot & \cdot \\ & & & & & & \cdot & \cdot \\ & & & & & & & \cdot \end{bmatrix} \equiv \begin{bmatrix} T_{11} & T_{12} & \cdots & T_{1n} \\ & T_{22} & \cdots & T_{2n} \\ & & \ddots & \vdots \\ & & & T_{nn} \end{bmatrix}$$

where "·" stands for an entry in which one or both dimensions are zero. We can choose the sequence of state spaces $\mathcal{B}$ to have zero dimensions outside the index interval $[2,n]$ in this case, so that computations start and end with zero dimensional state vectors. Doing so yields computational networks in the form described in chapter 1. The finite matrices form an important subclass of the bounded operators, because (*i*) initial conditions are known precisely (a vanishing state vector) (*ii*) computations are finite, so that boundedness and convergence are not an issue (these issues become important again, of course, for very large matrices). In particular, $\ell_A = 0$ always.

By taking the dimensions of the non-empty $\mathcal{M}_i$ non-uniform, block-matrices are special cases of finite matrices, and sometimes, matrices that are not upper triangular in the ordinary sense, are block-upper, *i.e.*, in $\mathcal{U}(\mathcal{M},\mathcal{N})$, where $\mathcal{M}$ and $\mathcal{N}$ are chosen appropriately. An example is given in figure 3.2(a). An extreme representative of a block-upper matrix is obtained by taking

$$\begin{array}{rcl} \mathcal{M} & = & \cdots \oplus \emptyset \oplus \mathcal{M}_1 \oplus \emptyset \oplus \emptyset \cdots \\ \mathcal{N} & = & \cdots \oplus \emptyset \oplus \emptyset \oplus \mathcal{N}_2 \oplus \emptyset \cdots \end{array}$$

so that a matrix $T \in \mathcal{U}$ has the form

$$T = \begin{bmatrix} \cdot & \cdot & \cdot \\ \cdot & T_{12} & \cdot \\ & \cdot & \cdot \\ & & \cdot \end{bmatrix} \equiv [T_{12}] \,,$$

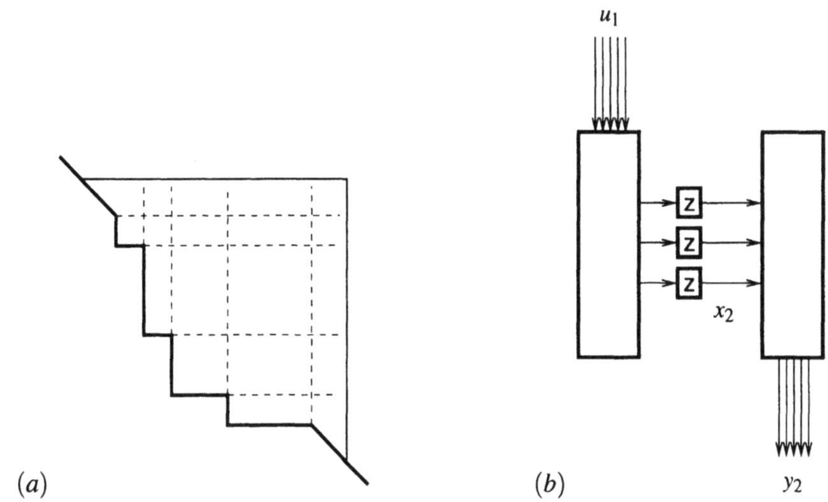

Figure 3.2. (a) A block-upper matrix; (b) a state realization of a special case of a block-upper matrix, $T = [T_{12}]$.

that is, $T = T_{12}$ is just any matrix of any size. Figure 3.2(b) depicts the time-varying state realization of such a system. Inputs are only present at time 1, and outputs are only generated at time 2. The number of states that are needed in going from time 1 to time 2 is, for a minimal realization, equal to the rank of T_{12}, as we will see in the next section. There are applications in low-rank matrix approximation theory that use this degenerate view of a matrix [vdV96].

Time-invariant on the borders

A second important subclass of time-varying systems are systems for which the state realization matrices $\{A_k, B_k, C_k, D_k\}$ are time-invariant for k outside a finite time interval, again say $[1, n]$. This class properly contains the finite matrix case. The structure resulting from such realizations is depicted in figure 3.3. Computations on such systems can typically be split in a time-invariant part, for which methods of classical system theory can be used, and a time-varying part, which will typically involve recursions starting from initial values provided by the time-invariant part. Boundedness often reduces to a time-invariant issue. For example, ℓ_A is equal to $\max(r(A_0), r(A_{n+1}))$, solely governed by the stability of the time-invariant parts.

Periodic systems

A third subclass is the class of periodically varying systems. If a system has a period n, then it can be viewed as a time-invariant system T with block entries $T_{ij} = T_{i-j}$ of size $n \times n$: T is a block Toeplitz operator. The realization matrices $\{\underline{A}, \underline{B}, \underline{C}, \underline{D}\}$ of this

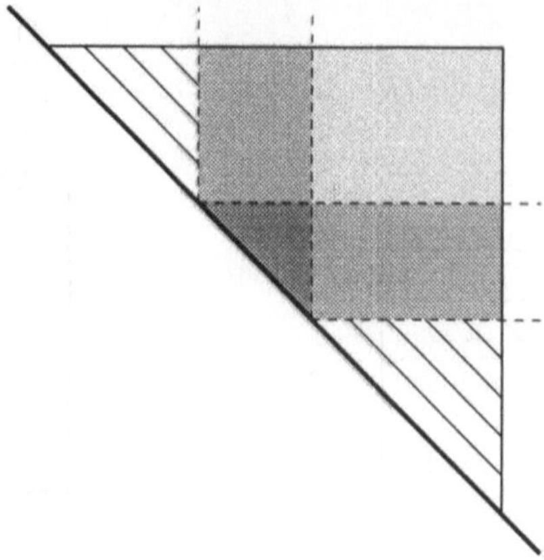

Figure 3.3. Transfer operator of a system that is time-invariant on the borders. Only the shaded areas are non-Toeplitz.

block operator are given in terms of the time-varying $\{A_k, B_k, C_k, D_k\}$ as

$$\underline{A} = A_1 A_2 \cdots A_n, \qquad \underline{C} = \begin{bmatrix} C_1 & A_1 C_2 & A_1 A_2 C_3 & \cdots & A_1 \cdots A_{n-1} C_n \end{bmatrix}$$

$$\underline{B} = \begin{bmatrix} B_1 A_2 A_3 \cdots A_n \\ B_2 A_3 \cdots A_n \\ \vdots \\ B_n \end{bmatrix} \qquad \underline{D} = \begin{bmatrix} D_1 & B_1 C_2 & B_1 A_2 C_3 & \cdots & B_1 A_2 \cdots A_{n-1} C_n \\ & D_2 & B_2 C_3 & & B_2 A_3 \cdots A_{n-1} C_n \\ & & \ddots & & \vdots \\ & & & & D_n \end{bmatrix}$$

Computations involving such operators can often be done by taking it as a block time-invariant system, which will provide exact initial conditions at the beginning of each period. These time-invariant solutions can be computed in classical ways. However, if the period is large, this may not be attractive, in which case the general methods may be more appealing. The classical Floquet theorem for time continuous systems which states that there is a periodic state transformation which transforms a periodic state transition matrix $A(t)$ into a time invariant one does not have a simple time discrete counterpart. A generic counterexample is a system with period 2 and A matrices given by

$$A_1 = \begin{bmatrix} 0 & 1 \\ 0 & 0 \end{bmatrix}, \qquad A_2 = \begin{bmatrix} 1 & 0 \\ 0 & 1 \end{bmatrix},$$

and suppose that there would exist state transformation matrices R_1 and R_2 such that $R_1 A_1 R_2^{-1} = R_2 A_2 R_1^{-1} =: \alpha$, then we should have $(R_1^{-1} \alpha R_1)^2 = \left[\begin{smallmatrix} 0 & 1 \\ 0 & 0 \end{smallmatrix}\right]$ which is impossible since this latter matrix has no square root.

Systems of low displacement rank

An important class of structured matrices is formed by matrices of low "displacement rank", and was extensively studied by Kailath and his many students and co-workers

[KKM79, KS95] (see in particular the theses of Lev-Ari [Lev83] and Sayed [Say92]). Although the brunch of this book is devoted to another kind of structure, namely low-dimensional state realizations, we give here a short introduction to low displacement matrices, and at the end of this chapter, we give a theory which combines results of the two theories, namely low complexity parametrizations for systems that are at the same time of low displacement type *and* of low systems rank.

Let R be an $n \times n$ positive definite matrix, and let us define the $n \times n$ "displacement matrix", alias restricted shift operator

$$\sigma = \begin{bmatrix} 0 & & & \mathbf{0} \\ 1 & \ddots & & \\ & \ddots & \ddots & \\ \mathbf{0} & & 1 & 0 \end{bmatrix}.$$

This matrix is similar to the reverse shift operator Z^*, but unlike Z^*, it truncates the shifted sequence and introduces a zero, so that it is not an invertible operator. The displacement of R is defined as $R - \sigma R \sigma^*$ [KKM79]. We assume that it has inertia (p, o, q), which means that there exist matrices

$$G = [g_0 \;\; g_1 \;\; \cdots \;\; g_{n-1}], \qquad J = \begin{bmatrix} I_p & \\ & -I_q \end{bmatrix}$$

of dimensions $(p+q) \times n$ and $(p+q) \times (p+q)$ respectively, such that

$$R - \sigma R \sigma^* \;=\; G^* J G \;=\; \begin{bmatrix} g_0^* \\ g_1^* \\ \vdots \\ g_{n-1}^* \end{bmatrix} J [g_0 \;\; \cdots \;\; g_{n-1}],$$

in which G has full row rank. It is convenient to split each entry g_k according to the inertia formula:

$$g_k = \begin{matrix} p \\ q \end{matrix} \begin{bmatrix} g_{k1} \\ g_{k2} \end{bmatrix}.$$

$\alpha := p + q$ is called the *displacement rank*. If α is small compared to n, then R is said to be of low displacement rank. Clearly, R is parametrized by the entries of G. Important calculations on R such as the determination of its Cholesky factorization $R = L^* L$ or that of its inverse can be done efficiently on G rather than on R itself. In fact, this is not limited to positive definite matrices R, but can be generalized to any matrix T with (block matrix) entries. In the sequel we shall just look at additive and multiplicative decompositions of a positive definite matrix $R = L^* L = \frac{1}{2}(F + F^*)$, because that covers the most important applications.

The two notions, low displacement rank and low system order are not related to each other. We know of systems that score high for one and low for the other. The prototype example of a low displacement rank matrix is a Toeplitz matrix, which has displacement rank one or two. Such a matrix may not have a useful low dimensional state space realization. For example, the LTI transfer function $T(z) = z + \frac{1}{2}z^2 + \frac{1}{3}z^3 + \cdots$ is of course of low displacement rank but no finite dimensional state space realization is capable of reproducing the decay $\frac{1}{n}$, characteristic of a transfer function of logarithmic type.

Uniformly exponentially stable systems

Finally, a large class of systems for which precise and finite calculations are possible is the class of u.e. stable systems: systems that have a realization for which $\ell_A < 1$. Recursions on such systems are typically convergent, that is, independent of the precise initial value at point k as $k \to -\infty$. This means that it is possible to limit attention to a finite time-interval, and to obtain arbitrarily accurate initial values for this interval by performing a finite recursion on data outside the interval, starting with initial values set to 0.

For example, if in a computation for $k > 1$, an initial state x_1 is required, then this latter value can be approximated to arbitrary precision using a finite sequence of past input samples and system matrices, since

$$x_1 = x_{-n}A_{-n}\cdots A_0 + \begin{bmatrix} u_{-n} & u_{-n+1} & \cdots & u_0 \end{bmatrix} \begin{bmatrix} B_{-n}A_{-n+1}A_{-n+2}\cdots A_0 \\ B_{-n+1}A_{-n+2}\cdots A_0 \\ \vdots \\ B_0 \end{bmatrix}. \tag{3.16}$$

If the system is u.e. stable, then $\|A_{-n}\cdots A_0\|$ can be made arbitrarily small by choosing n large enough. Neglect of the first term in (3.16) then gives an accurate approximation for x_1. The same approximation would of course be obtained by choosing $x_{-n} = 0$ if that were possible, and computing x_1 via the state recursion.

3.3 EXAMPLES AND EXTENSIONS

Using the connection of a matrix or operator in $\mathcal{U}$ and its realization as visualized in equation (3.10), we study some simple classes of matrices and their corresponding realizations, as well as some simple operations on matrices such as sums and products, a special case of matrix inversion, and extensions to more general realization frameworks.

Banded matrices

One of the easiest examples of a matrix for which it is possible to write down a realization directly is the case of a banded matrix. Thus let $T \in \mathcal{U}$ be given by

$$T = \begin{bmatrix} T_{11} & T_{12} & T_{1,d} & 0 & 0 \\ & T_{22} & T_{2,d} & T_{2,d+1} & 0 \\ & & \ddots & \ddots & T_{n-d+1,n} \\ & & & T_{n-1,n-1} & T_{n-1,n} \\ \mathbf{0} & & & & T_{n,n} \end{bmatrix}.$$

The width of the band is in this case equal to d. A trivial realization for T requires up to $d-1$ states per stage:

$$\mathbf{T}_1 = \left[\begin{array}{c|c} \cdot & \cdot \\ \hline 1 & T_{11} \end{array}\right] \qquad \mathbf{T}_k = \left[\begin{array}{cccc|c} 0 & & & & T_{k-d+1,k} \\ 1 & \ddots & \mathbf{0} & & T_{k-d+2,k} \\ & \ddots & \ddots & & \vdots \\ \mathbf{0} & & 1 & 0 & T_{k-1,k} \\ \hline 0 & \cdots & 0 & 1 & T_{k,k} \end{array}\right]$$

$$\mathbf{T}_2 = \left[\begin{array}{cc|c} 1 & 0 & T_{12} \\ \hline 0 & 1 & T_{22} \end{array}\right] \qquad \mathbf{T}_{n-1} = \left[\begin{array}{cccc|c} 0 & & & & T_{n-d,n-1} \\ 1 & \ddots & \mathbf{0} & & T_{n-d+1,n-1} \\ & \ddots & \ddots & & \vdots \\ \mathbf{0} & & 1 & 0 & T_{n-2,n-1} \\ \hline 0 & \cdots & 0 & 1 & T_{n-1,n-1} \end{array}\right]$$

$$\mathbf{T}_3 = \left[\begin{array}{ccc|c} 1 & 0 & 0 & T_{13} \\ 0 & 1 & 0 & T_{23} \\ \hline 0 & 0 & 1 & T_{33} \end{array}\right] \qquad \mathbf{T}_n = \left[\begin{array}{c|c} \cdot & T_{n-d+1,n} \\ \cdot & \vdots \\ \cdot & T_{n-1,n} \\ \hline \cdot & T_{n,n} \end{array}\right].$$

This is not necessarily a minimal realization: there might exist realizations with a smaller number of states, depending on the precise values of T_{ij}. Even for general banded matrices, the number of states in the last d stages can be made smaller than presented here, although this will introduce some irregularities in the structure of these sections.

Other matrices for which one can obtain realizations directly are matrices with a staircase band structure, and band matrices with some spurious entries in the upper right hand corner. The latter type of matrix arises in finite difference modeling of one dimensional differential equations with periodic boundary conditions (figure 3.4). In simple cases, the non-zero entries of the matrix are just $+1$ and -1 (for a first-order differential equation), and the matrix has a Toeplitz structure (constant along the diagonals). A three-diagonal matrix occurs with simple discretizations of second-order ODEs. Nonuniform spacing of the discretization points leads to banded matrices without the Toeplitz structure.

Sum of two realizations

Let $T_1, T_2 \in \mathcal{U}(\mathcal{M},\mathcal{N})$ be two transfer operators, with realizations $\{A_1,B_1,C_1,D_1\}$ and $\{A_2,B_2,C_2,D_2\}$, respectively. Then the sum of these two operators, $T = T_1 + T_2$, has a realization given directly in terms of these two realizations as

$$\left[\begin{array}{c|c} A & C \\ \hline B & D \end{array}\right] = \left[\begin{array}{cc|c} A_1 & 0 & C_1 \\ 0 & A_2 & C_2 \\ \hline B_1 & B_2 & D_1 + D_2 \end{array}\right]$$

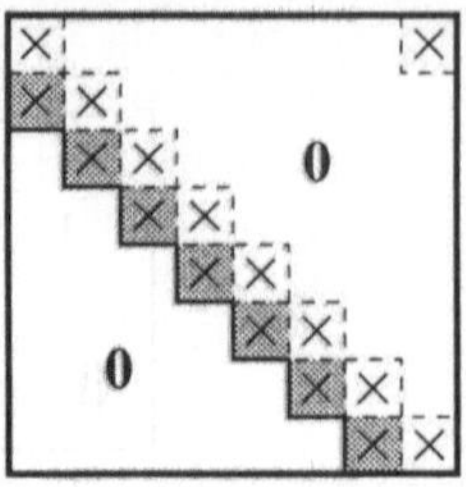

Figure 3.4. A block-upper matrix with a maximal state dimension of 2. The main diagonal is shaded; $\mathcal{N}_1 = \emptyset$ and $\mathcal{M}_{n+1} = \emptyset$. This type of matrix arises after discretization in certain 1-D finite difference modeling problems with periodic boundary conditions.

The state dimension sequence of this realization is equal to the sum of the state dimension sequences of T_1 and T_2. Note, however, that this realization is not necessarily minimal: there might exist a realization of T whose state dimension sequence is smaller (for reduction of the state space to minimal dimensions, see the next section).

Product of two realizations

The product of $T_1 \in \mathcal{U}(\mathcal{M},\mathcal{N}_1)$ and $T_2 \in \mathcal{U}(\mathcal{N}_1,\mathcal{N}_2)$ can also be obtained using realizations: if T_1 has a realization $\{A_1,B_1,C_1,D_1\}$ and T_2 has a realization $\{A_2,B_2,C_2,D_2\}$, then $T = T_1T_2$ has a realization given by

$$\left[\begin{array}{c|c} A & C \\ \hline B & D \end{array}\right] = \left[\begin{array}{cc|c} A_1 & & C_1 \\ & I & 0 \\ \hline B_1 & 0 & D_1 \end{array}\right]\left[\begin{array}{cc|c} I & & 0 \\ & A_2 & C_2 \\ \hline 0 & B_2 & D_2 \end{array}\right] = \left[\begin{array}{cc|c} A_1 & C_1B_2 & C_1D_2 \\ 0 & A_2 & C_2 \\ \hline B_1 & D_1B_2 & D_1D_2 \end{array}\right] \tag{3.17}$$

Again, the state dimensions sequences of T_1 and T_2 add up to the state dimension sequence of T, and again, this realization is not necessarily minimal: there might exist realizations of T that have smaller state dimension sequences.

Realization of an upper inverse

Let $T \in \mathcal{U}$ be an invertible operator or matrix, and suppose that it is known that $T^{-1} \in \mathcal{U}$ is also upper, then it is straightforward to derive a realization for T^{-1}. From $T^{-1}T = I$ and $TT^{-1} = I$, we obtain that $D = T_{[0]}$ must be invertible, and

$$\left\{\begin{array}{rcl} xZ^{-1} & = & xA + uB \\ y & = & xC + uD \end{array}\right. \quad\Leftrightarrow\quad \left\{\begin{array}{rcll} xZ^{-1} & = & x(A - CD^{-1}B) & +\; yD^{-1}B \\ u & = & -xCD^{-1} & +\; yD^{-1}. \end{array}\right.$$

Hence, $S = T^{-1}$ has a realization

$$\mathbf{S} = \left[\begin{array}{cc} A - CD^{-1}B & -CD^{-1} \\ D^{-1}B & D^{-1} \end{array}\right] = \left[\begin{array}{cc} A & \\ & 0 \end{array}\right] + \left[\begin{array}{c} -C \\ I \end{array}\right] D^{-1}[B \;\; I]. \tag{3.18}$$

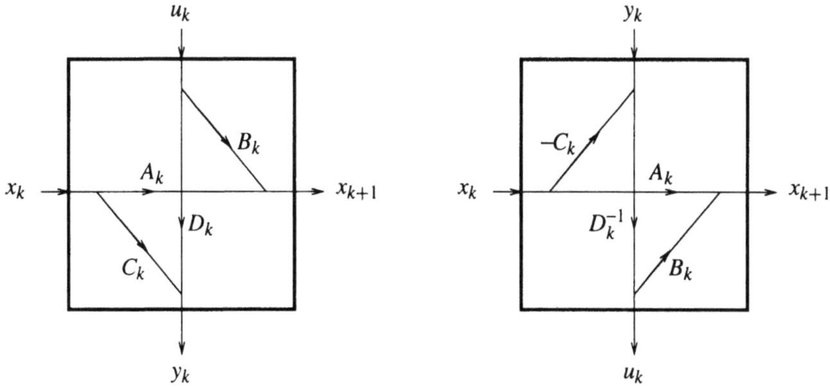

Figure 3.5. Simple local computational scheme for inverting a system

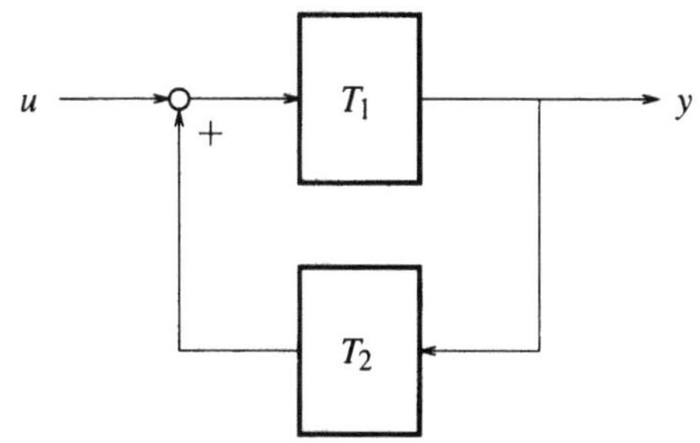

Figure 3.6. Feedback configuration.

As shown in figure 3.5, the latter factorization allows to apply the inverse in an efficient manner: only D_k^{-1} has to be computed.

$A - CD^{-1}B$ has the same dimensions as A, so that the state dimension of the realization of the inverse is at each point equal to the state dimension of T at that point. We will see in section 13.2 (proposition 13.2) that, under some assumptions on the realization of T, $\ell_A < 1 \Leftrightarrow \ell_{A-CD^{-1}B} < 1$, so that for u.e. stable realizations, the realization of the inverse is u.e. stable, too.

The above is only valid if T^{-1} is upper. Not every operator in $\mathcal{U}$ is invertible in $\mathcal{U}$; the condition is that T must be *outer*, a notion that we will define in chapter 6. Some examples of matrices that are not outer have been given in chapter 2; in particular, block-upper matrices of which the entries on the main diagonal are not square give rise to inverses that need not be upper, but have a lower triangular part, too. General matrix/operator inversion is studied in chapter 7.

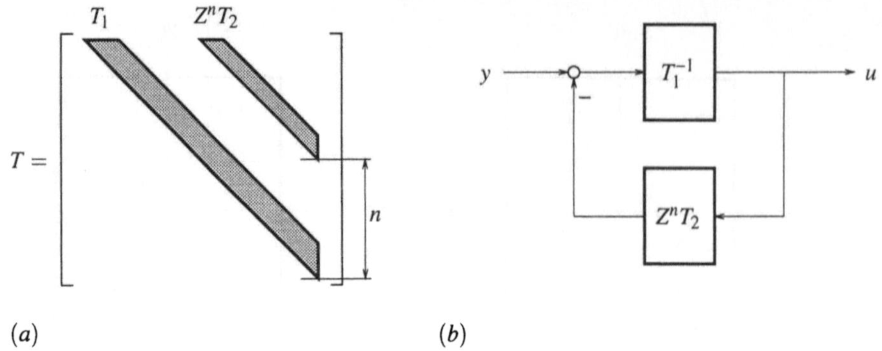

Figure 3.7. (a) Multiband matrix, (b) feedback structure that models T^{-1}.

Feedback

Suppose that two systems $T_1 \in \mathcal{U}$ and $T_2 \in \mathcal{U}$ are connected in a feedback configuration (figure 3.6). If the resulting transfer operator $T = T_1(I - T_2T_1)^{-1}$ is bounded, then a realization of T can directly be written down in terms of realizations of T_1 and T_2, using $x = [x_1 \;\; x_2]$ as the state vector:

$$\mathbf{T} = \left[\begin{array}{cc|c} A_1 & C_1B_2 & 0 \\ 0 & A_2 & \\ \hline 0 & & 0 \end{array}\right] + \begin{bmatrix} C_1D_2 \\ C_2 \\ I \end{bmatrix} (I - D_1D_2)^{-1} [B_1 \quad D_1B_2 \quad D_1] .$$

Feedback configurations arise in the inversion of a sum of two operators. An example is given in the following subsection.

Extension to multi-band matrices

Let $T = T_1 + Z^nT_2$, where T_1 and T_2 are band matrices and $n > 0$. Then T is a multi-band matrix (figure 3.7(a)). A realization of T has the following structure:

$$\mathbf{T} = \left[\begin{array}{cccc|c} A_2 & & & & C_2 \\ B_2 & & & & D_2 \\ & I_{n-1} & & & 0 \\ & & 0 & A_1 & C_1 \\ \hline & & 1 & B_1 & D_1 \end{array}\right] .$$

(If T_1 and T_2 are not SISO, then the identity matrices must have sufficiently large dimensions.) If n is large, then the state dimension of $\mathbf{T}$ is not small, but it has a sparse structure, so that multiplications are still efficient. It is sometimes possible to keep this sparse structure during operations on T. Consider e.g. the inverse of a multi-band matrix, which (if it exists as a bounded upper matrix) is full: $S = T^{-1} = (T_1 + Z^nT_2)^{-1} = (I + T_1^{-1}Z^nT_2)^{-1}T_1^{-1}$. S can be interpreted as a feedback model (figure 3.7(b)). Its realization still has a sparse structure with complexity essentially independent of n:

$$
\begin{aligned}
\mathbf{S} &= \left[\begin{array}{cccc|c}
A_2 & & -C_2 D_1^{-1} & -C_2 D_1^{-1} B_1 & -C_2 D_1^{-1} \\
B_2 & & -D_2 D_1^{-1} & -D_2 D_1^{-1} B_1 & -D_2 D_1^{-1} \\
 & I_{n-1} & & & 0 \\
 & & -C_1 D_1^{-1} & A_1 - C_1 D_1^{-1} B_1 & -C_1 D_1^{-1} \\
\hline
 & & D_1^{-1} & D_1^{-1} B_1 & D_1^{-1}
\end{array}\right] \\
&= \left[\begin{array}{cccc|c}
A_2 & & & & \\
B_2 & & & & \\
 & I_{n-1} & & & \\
 & & 0 & A_1 & \\
\hline
 & & & & 0
\end{array}\right] + \begin{bmatrix} -C_2 \\ -D_2 \\ 0 \\ -C_1 \\ 1 \end{bmatrix} D_1^{-1} \begin{bmatrix} 0 & 0 & 1 & B_1 & 1 \end{bmatrix}.
\end{aligned}
$$

Using the latter factorization, multiplication by T^{-1} can be performed just as efficiently as multiplication by T. Other operations such as QR-factorization lead to a complexity that is essentially linear in n.

Systems of mixed causality; general matrices

Throughout this book, many basic properties assume that the transfer operator T is upper triangular, so that the corresponding forward state recursions are stable. This, however, does not mean that all results are limited to this case: it is possible to fit general matrices or operators in $\mathcal{X}$ into the time-varying systems framework. Viewing such operators as the sum or product of an upper and a lower triangular matrix (provided each of these parts on its own is bounded), it is possible to determine realizations of $T \in \mathcal{X}$ as the sum or product of a forward-running and a backward-running set of state equations. The computation of a factorization of such operators into a product of a lower triangular unitary matrix and an upper triangular matrix is in fact a (partial) QR factorization. We will see in chapter 6 that, given realizations of the upper and the lower triangular part, it can be computed using state space matrices only.

To extend our framework to this more general situation, let $\{\mathbf{T}_k\}_1^n$, $\{\mathbf{T}'_k\}_1^n$ be a series of matrices with block entries

$$
\mathbf{T}_k = \begin{bmatrix} A_k & C_k \\ B_k & D_k \end{bmatrix}, \qquad \mathbf{T}'_k = \begin{bmatrix} A'_k & C'_k \\ B'_k & 0 \end{bmatrix}, \qquad k = 1, \cdots, n,
$$

and consider the time-varying forward and backward state recursions, for $k = 1, \cdots, n$,

$$
(\mathbf{T}) \left\{ \begin{array}{rcl} x_{k+1} &=& x_k A_k + u_k B_k \\ y_k &=& x_k C_k + u_k D_k \end{array} \right. \qquad (\mathbf{T}') \left\{ \begin{array}{rcl} x'_{k-1} &=& x'_k A'_k + u_k B'_k \\ y'_k &=& x'_k C'_k \end{array} \right. \tag{3.19}
$$

using the initial values

$$
x_1 = [\cdot], \quad x'_n = [\cdot],
$$

and let the output of the system be the sum of the forward and backward recursions:

$$
z_k = y_k + y'_k .
$$

The intermediate quantities in the recursion are x_k, the forward state, and x'_k, the backward state. The relation between $u = [u_1, u_2, \cdots, u_n]$ and $z = [z_1, z_2, \cdots, z_n]$, as generated by the given state recursions, is

$$z = u \begin{bmatrix} D_1 & B_1C_2 & B_1A_2C_3 & B_1A_2A_3C_4 & \cdots & \\ B'_2C'_1 & D_2 & B_2C_3 & B_2A_3C_4 & & \vdots \\ B'_3A'_2C'_1 & B'_3C'_2 & D_3 & B_3C_4 & & \\ B'_4A'_3A'_2C'_1 & B'_4A'_3C'_2 & B'_4C'_3 & D_4 & \cdots & B_{n-2}A_{n-1}C_n \\ \vdots & & & \vdots & \ddots & B_{n-1}C_n \\ & & \cdots & B'_nA'_{n-1}C'_{n-2} & B'_nC'_{n-1} & D_n \end{bmatrix}.$$

As shown in the next section, any finite matrix can be written in this form. The recursions (3.19) can be used to compute a vector-matrix multiplication $z = uT$ efficiently, provided the matrix T is specified in terms of its realization and the state dimensions are relatively small in comparison with the size of the matrix. Accordingly, we say that matrices $\{\mathbf{T}_k\}_1^n$, $\{\mathbf{T}'_k\}_1^n$ form a time-varying realization of mixed causality for a matrix $T \in \mathcal{X}$, if the block entries of T are given by

$$T_{ij} = \begin{cases} D_i, & i = j, \\ B_iA_{i+1}\cdots A_{j-1}C_j, & i < j, \\ B'_iA'_{i-1}\cdots A'_{j+1}C'_j, & i > j. \end{cases}$$

3.4 REALIZATION THEORY FOR FINITE MATRICES

An important part of chapter 5 is concerned with the *realization problem*: the problem to determine a realization $\{A, B, C, D\}$ for a given operator $T \in \mathcal{U}$. In this chapter, we give a solution for the case where T is a finite (block)-upper triangular matrix, rather than a more general operator acting on infinite sequences. The proof becomes simple and direct since difficulties with convergence and boundedness are avoided. For clarity of exposition, we use expressions with indices rather than diagonals.

Realization algorithm for upper triangular matrices

Let us assume that we are given a finite upper triangular matrix T, as a special case of a bounded operator. Assume that $\{A_k, B_k, C_k, D_k\}$ defines a not yet known time-varying state realization for T, which specifies T via the time-varying state equations (3.1):

$$\begin{aligned} x_{k+1} &= x_kA_k + u_kB_k \\ y_k &= x_kC_k + u_kD_k \end{aligned}$$

We look for properties of this realization that enable us to derive it from T, *i.e.*, to find a realization given the transfer specification. According to definition 3.4, the entries T_{ij} of T can be expressed in terms of $\{A_k, B_k, C_k, D_k\}$ as (see (3.9))

$$T_{ij} = \begin{cases} 0, & i > j \\ D_i, & i = j \\ B_iA_{i+1}\cdots A_{j-1}C_j, & i < j. \end{cases} \tag{3.20}$$

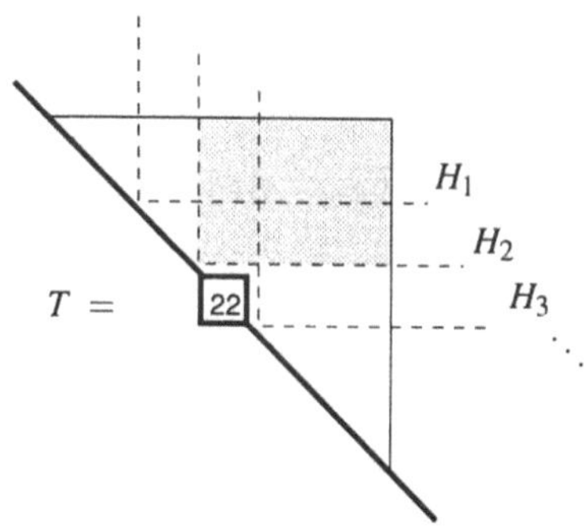

Figure 3.8. Hankel matrices are submatrices of T. H_2 is shaded.

We assume the T_{ij} are known; the problem is to find $\{A_k, B_k, C_k, D_k\}$. There is no ambiguity about D_k: $D_k = T_{k,k}$. The main realization problem is to determine suitable $\{A_k, B_k, C_k\}$. Because state transformations are allowed, these matrices are not unique.

The key to the solution of the realization problem is the analysis of certain submatrices H_k of T. Define

$$H_k = \begin{bmatrix} T_{k-1,k} & T_{k-1,k+1} & T_{k-1,k+2} & \cdots \\ T_{k-2,k} & T_{k-2,k+1} & & \\ T_{k-3,k} & & \ddots & \\ \vdots & & & \end{bmatrix}. \tag{3.21}$$

We call this matrix a *time-varying Hankel operator, at point k*, since it would have the special structure known as "Hankel matrix" in the time-invariant case, namely that the elements on the antidiagonals are equal: $(H_k)_{ij} = (H_k)_{\ell m}$ if $i+j = \ell+m$. In the time-varying case the collection of matrices $\{H_k\}$ still has a special structure, as we will see soon. The entries of H_k are taken from the submatrix of the matrix T above and to the right of entry $T_{k,k}$, as depicted in figure 3.8. For finite matrices T, H_k is a finite $(k-1)\times(n-k)$ matrix. We have (traditionally) reversed the ordering of rows of H_k in comparison to the ordering of rows of T, because we will allow for operators ("infinite matrices") later on, and we wish to have the infinite sides of semi-infinite matrices at the bottom and right. If $\{A_k, B_k, C_k, D_k\}$ is a realization of T then substitution of (3.20) into (3.21) produces

$$H_k = \begin{bmatrix} B_{k-1}C_k & B_{k-1}A_kC_{k+1} & B_{k-1}A_kA_{k+1}C_{k+2} & \cdots \\ B_{k-2}A_{k-1}C_k & B_{k-2}A_{k-1}A_kC_{k+1} & & \\ B_{k-3}A_{k-2}A_{k-1}C_k & & \ddots & \\ \vdots & & & \end{bmatrix}.$$

A first observation is that H_k has a factorization due to the regular structure of its entries, as

$$H_k = \begin{bmatrix} B_{k-1} \\ B_{k-2}A_{k-1} \\ B_{k-3}A_{k-2}A_{k-1} \\ \vdots \end{bmatrix} [C_k \quad A_kC_{k+1} \quad A_kA_{k+1}C_{k+2} \quad \cdots] =: \mathcal{C}_k\mathcal{O}_k\,, \qquad (3.22)$$

where we have defined

$$\mathcal{C}_k = \begin{bmatrix} B_{k-1} \\ B_{k-2}A_{k-1} \\ B_{k-3}A_{k-2}A_{k-1} \\ \vdots \end{bmatrix}, \qquad \mathcal{O}_k = [C_k \quad A_kC_{k+1} \quad A_kA_{k+1}C_{k+2} \quad \cdots]\,. \qquad (3.23)$$

$\mathcal{C}_k$ is called the *reachability matrix* at point k, while $\mathcal{O}_k$ is called the *observability matrix* at point k. We explain the reason for this terminology later.

If A_k has size $d_k \times d_{k+1}$, then from the factorization (3.22) it follows that the rank of H_k is less than or equal to d_k, the number of columns of $\mathcal{C}_k$ and rows of $\mathcal{O}_k$. A realization will be *minimal* if the rank of H_k is equal to d_k, for all k. Obviously, no realizations exist for which the state dimension is smaller.

A second observation that follows from the factorization is a *shift-invariance property*. Let $H_k^{\leftarrow}$ be the matrix defined by removing the first column from H_k (which also can be viewed as H_k shifted one notch to the left with the first column chopped off), then we obtain the factorization

$$H_k^{\leftarrow} = \begin{bmatrix} B_{k-1} \\ B_{k-2}A_{k-1} \\ B_{k-3}A_{k-2}A_{k-1} \\ \vdots \end{bmatrix} \cdot A_k \cdot [C_{k+1} \quad A_{k+1}C_{k+2} \quad A_{k+1}A_{k+2}C_{k+3} \quad \cdots] = \mathcal{C}_kA_k\mathcal{O}_{k+1}\,.$$

The underlying property is $\mathcal{O}_k^{\leftarrow} = A_k\mathcal{O}_{k+1}$. Shifting upward in a dual way, we have $H_{k+1}^{\uparrow} = \mathcal{C}_kA_k\mathcal{O}_{k+1}$, and $\mathcal{C}_{k+1}^{\uparrow} = \mathcal{C}_kA_k$. The shift-invariance properties allow us to determine the A_k from the $\mathcal{O}_k$ or the $\mathcal{C}_k$. If the factorization (3.22) is minimal, then the columns of $\mathcal{C}_k$ are linearly independent as well as the rows of $\mathcal{O}_k$, so that $\mathcal{C}_k^*\mathcal{C}_k > 0$ and $\mathcal{O}_k\mathcal{O}_k^* > 0$. These matrices are of full rank, $\mathcal{C}_k$ has a left inverse given by $\mathcal{C}_k^{\dagger} = (\mathcal{C}_k^*\mathcal{C}_k)^{-1}\mathcal{C}_k^*$ while $\mathcal{O}_k$ has a right inverse given by $\mathcal{O}_k^{\dagger} = \mathcal{O}_k^*(\mathcal{O}_k\mathcal{O}_k^*)^{-1}$, and we can solve for A_k:

$$A_k = \mathcal{O}_k^{\leftarrow}\mathcal{O}_{k+1}^{\dagger} = \mathcal{C}_k^{\dagger}\mathcal{C}_{k+1}^{\uparrow}\,. \qquad (3.24)$$

From the definitions of $\mathcal{C}_k$ and $\mathcal{O}_k$, we also have

$$\begin{aligned} B_k &= [\text{first row of } \mathcal{C}_{k+1}]\,, \\ C_k &= [\text{first column of } \mathcal{O}_k]\,. \end{aligned} \qquad (3.25)$$

Hence, once all H_k have been factored into $H_k = \mathcal{C}_k\mathcal{O}_k$, the constituting $\{A_k, B_k, C_k\}$ can be derived from the structure in $\mathcal{C}_k$ and $\mathcal{O}_k$.

Theorem 3.7 *Let T be an upper triangular matrix, and let d_k be the rank of its Hankel matrices H_k. For each k, let $H_k = \mathcal{C}_k \mathcal{O}_k$, where $\mathcal{C}_k, \mathcal{O}_k$ are rank-d_k factors. Then $\{A_k, B_k, C_k, D_k\}$ is a minimal realization for T, where A_k, B_k, C_k are given by (3.24) and (3.25), and $D_k = T_{k,k}$.*

This theorem parallels the famous Kronecker realization theorem [Kro90] in the present setting.

PROOF For $k = 1, \cdots, n$, let $H_k = \mathcal{C}_k \mathcal{O}_k$ be a minimal factorization of H_k, and let us choose $A_k = \mathcal{O}_k^{\leftarrow} \mathcal{O}_{k+1}^{\dagger}$ and B_k, C_k as in (3.25). We must show that $T_{ij} = B_i A_{i+1} \cdots A_{j-1} C_j$ $(i < j)$. As all these elements are entries of some H_k, this is equivalent to showing that $\mathcal{C}_k$ and $\mathcal{O}_k$ are given by equation (3.23).

We will first prove that $\mathcal{O}_k^{\leftarrow} = A_k \mathcal{O}_{k+1}$, where $A_k = \mathcal{O}_k^{\leftarrow} \mathcal{O}_{k+1}^{\dagger}$. Note that

$$H_{k+1}^{\uparrow} = H_k^{\leftarrow}, \tag{3.26}$$

i.e., $H_k^{\leftarrow}$ is obtained by removing the top row of H_{k+1}. Hence, the row span of $H_k^{\leftarrow}$ is contained in that of H_{k+1}. Because the factorizations $H_k = \mathcal{C}_k \mathcal{O}_k$ and $H_{k+1} = \mathcal{C}_{k+1} \mathcal{O}_{k+1}$ are minimal, these row spans are equal to the row spans of $\mathcal{O}_k^{\leftarrow}$ and $\mathcal{O}_{k+1}$. It follows that there exist matrices A_k such that $O_k^{\leftarrow} = A_k \mathcal{O}_{k+1}$. One solution is $A_k = \mathcal{O}_k^{\leftarrow} \mathcal{O}_{k+1}^{\dagger}$.

Substituting the given factorizations into (3.26) yields $\mathcal{C}_{k+1}^{\uparrow} \mathcal{O}_{k+1} = \mathcal{C}_k \mathcal{O}_k^{\leftarrow} = \mathcal{C}_k A_k \mathcal{O}_{k+1}$, so that, as $\mathcal{O}_{k+1}$ is right-invertible, $\mathcal{C}_{k+1}^{\uparrow} = \mathcal{C}_k A_k$.

We will now derive the expression for $\mathcal{O}_k$. By the definition of C_k, and because $\mathcal{O}_k^{\leftarrow} = A_k \mathcal{O}_{k+1}$, we have

$$\begin{aligned} \mathcal{O}_k &= [C_k \quad \mathcal{O}_k^{\leftarrow}] \\ &= [C_k \quad A_k \mathcal{O}_{k+1}]. \end{aligned} \tag{3.27}$$

Recursion on k now gives the required expression (3.23) for $\mathcal{O}_k$. The expression for $\mathcal{C}_k$ is similarly derived from the definition of B_k and $\mathcal{C}_{k+1}^{\uparrow} = \mathcal{C}_k A_k$, which yields

$$\begin{aligned} \mathcal{C}_{k+1} &= \begin{bmatrix} B_k \\ \mathcal{C}_{k+1}^{\uparrow} \end{bmatrix} \\ &= \begin{bmatrix} B_k \\ \mathcal{C}_k A_k \end{bmatrix}. \end{aligned}$$

Expanding $\mathcal{O}_{k+1}$ and $\mathcal{C}_k$ recursively produces (3.23), and thus by definition of H_k the required values of T_{ij}. □

The realization algorithm is shown in figure 3.9. It is natural that $d_1 = 0$ and $d_{n+1} = 0$, so that a minimal realization starts and ends with a zero number of states. The algorithm is reminiscent of the principal component identification method of system theory [Kun78]. Some numerical issues are discussed later in this section.

Corollary 3.8 *If, for some k, $\mathcal{C}_k^* \mathcal{C}_k = I$ and $\mathcal{C}_{k+1}^* \mathcal{C}_{k+1} = I$, then $A_k^* A_k + B_k^* B_k = I$. If, for some k, $\mathcal{O}_k \mathcal{O}_k^* = I$ and $\mathcal{O}_{k+1} \mathcal{O}_{k+1}^* = I$, then $C_k C_k^* + A_k A_k^* = I$.*

PROOF The second claim follows from the first equation in (3.27) by taking the square of this expression, and using the fact that $\mathcal{O}_k^{\leftarrow} = A_k \mathcal{O}_{k+1}$. The first claim follows dually. □

In: T (an upper triangular $n \times n$ matrix)
Out: $\{\mathbf{T}_k\}$ (a minimal realization)

$$
\begin{array}{l}
\mathcal{C}_1 = [\cdot], \mathcal{O}_1 = [\cdot] \\
\text{for } k = 1, \cdots, n \\
\left[\begin{array}{lcl}
d_{k+1} & = & \text{rank}(H_{k+1}) \\
H_{k+1} & =: & \mathcal{C}_{k+1}\mathcal{O}_{k+1} \quad \text{(take any minimal factorization)} \\
A_k & = & [0 \;\; \mathcal{C}_k^{\dagger}]\mathcal{C}_{k+1} \\
B_k & = & [\text{first row of } \mathcal{C}_{k+1}] \\
C_k & = & [\text{first column of } \mathcal{O}_k] \\
D_k & = & T_{k,k}
\end{array} \right. \\
\text{end}
\end{array}
$$

Figure 3.9. The realization algorithm. The factorization $H_k = \mathcal{C}_k \mathcal{O}_k$ can be obtained from a QR-factorization or an SVD.

Realizations for which $\mathcal{C}_k^* \mathcal{C}_k = I$ for all k are said to be in input normal form, whereas realizations for which $\mathcal{O}_k \mathcal{O}_k^* = I$ for all k are in output normal form. E.g., the trivial realization for banded matrices, discussed in section 3.3, has $\mathcal{C}_k = I$, and gives a realization in input normal form, although not necessarily minimal.

Numerical example

As an example of the realization theorem and the algorithm in figure 3.9, let the transfer matrix be given by

$$
T = \left[\begin{array}{ccc|ccc}
1 & .800 & .200 & .050 & .013 & .003 \\
0 & .900 & .600 & .240 & .096 & .038 \\
0 & 0 & .800 & .500 & .250 & .125 \\
\hline
0 & 0 & 0 & .700 & .400 & .240 \\
0 & 0 & 0 & 0 & .600 & .300 \\
0 & 0 & 0 & 0 & 0 & .500
\end{array} \right] \tag{3.28}
$$

The position of the Hankel matrix H_4 is indicated (recall that this submatrix must be mirrored to obtain H_4). A stable numerical way to obtain the minimal rank factorization of H_k as $H_k = \mathcal{C}_k \mathcal{O}_k$ is by computing its singular value decomposition (SVD) [GV89]. The SVDs of the Hankel matrices are computed as $H_k = \hat{U}_k \hat{\Sigma}_k \hat{V}_k^*$, where

$$
\begin{array}{lcl}
H_1 & = & [\cdot] \\
H_2 & = & [\; .800 \;\; .200 \;\; .050 \;\; .013 \;\; .003 \;] = 1 \cdot 0.826 \cdot [.968 \;\; .242 \;\; .061 \;\; .015 \;\; .004] \\
H_3 & = & \left[\begin{array}{cccc} .600 & .240 & .096 & .038 \\ .200 & .050 & .013 & .003 \end{array} \right]
\end{array}
$$

$$= \begin{bmatrix} .955 & .298 \\ .298 & -.955 \end{bmatrix} \begin{bmatrix} .685 & 0 \\ 0 & .033 \end{bmatrix} \begin{bmatrix} .922 & .356 & .139 & .055 \\ -.374 & .729 & .511 & .259 \end{bmatrix}$$

$$H_4 = \begin{bmatrix} .500 & .250 & .125 \\ .240 & .096 & .038 \\ .050 & .013 & .003 \end{bmatrix}$$

$$= \begin{bmatrix} .908 & .405 & .112 \\ .412 & -.808 & -.420 \\ .080 & -.428 & .901 \end{bmatrix} \begin{bmatrix} .631 & 0 & 0 \\ 0 & .029 & 0 \\ 0 & 0 & .001 \end{bmatrix} \begin{bmatrix} .882 & .424 & .205 \\ -.448 & .622 & .642 \\ .145 & -.658 & .739 \end{bmatrix}$$

etcetera. In the above, columns and rows that correspond to zero singular values have been omitted. The non-zero singular values of the Hankel operators of T are

	H_1	H_2	H_3	H_4	H_5	H_6
σ_1		.826	.685	.631	.553	.406
σ_2			.033	.029	.023	
σ_3				.001		

Hence T has a state-space realization which grows from zero states ($k = 1$) to a maximum of 3 states ($k = 4$), and then shrinks back to 0 states ($k > 6$). Small singular values represent states that are not very important. We apply the realization algorithm, using the factorizations $H_k = \mathcal{C}_k \mathcal{O}_k = (\hat{U}_k)(\hat{\Sigma}_k \hat{V}_k^*)$. This yields as time-varying state realization for T the collection $\{\mathbf{T}_k\}_1^6$,

$$\mathbf{T}_1 = \left[\begin{array}{c|c} \cdot & \cdot \\ \hline 1.000 & 1.000 \end{array}\right] \qquad \mathbf{T}_4 = \left[\begin{array}{cc|c} .536 & -.810 & .557 \\ .045 & .308 & -.013 \\ -.000 & .040 & .000 \\ \hline .843 & .498 & .700 \end{array}\right]$$

$$\mathbf{T}_2 = \left[\begin{array}{cc|c} .298 & -.955 & .800 \\ \hline .955 & .298 & .900 \end{array}\right] \qquad \mathbf{T}_5 = \left[\begin{array}{c|c} .671 & .481 \\ .051 & -.012 \\ \hline .739 & .600 \end{array}\right]$$

$$\mathbf{T}_3 = \left[\begin{array}{ccc|c} .417 & -.899 & -.133 & .632 \\ .047 & .167 & -.985 & -.012 \\ \hline .908 & .405 & .112 & .800 \end{array}\right] \qquad \mathbf{T}_6 = \left[\begin{array}{c|c} \cdot & .406 \\ \hline \cdot & .500 \end{array}\right].$$

As is seen from the table of singular values, H_4 is close to a singular matrix, and hence one expects that T can be approximated by a matrix close to it such that only two states are needed. That this is indeed possible will be shown in chapter 10.

System-theoretic interpretation

In the previous section, we have noted two properties of the Hankel matrices: their ranks are equal to the minimal system order at each point in time, and they satisfy a shift-invariance property. These properties have a fundamental system-theoretical nature, which we briefly explain now. We go into more details in chapter 5.

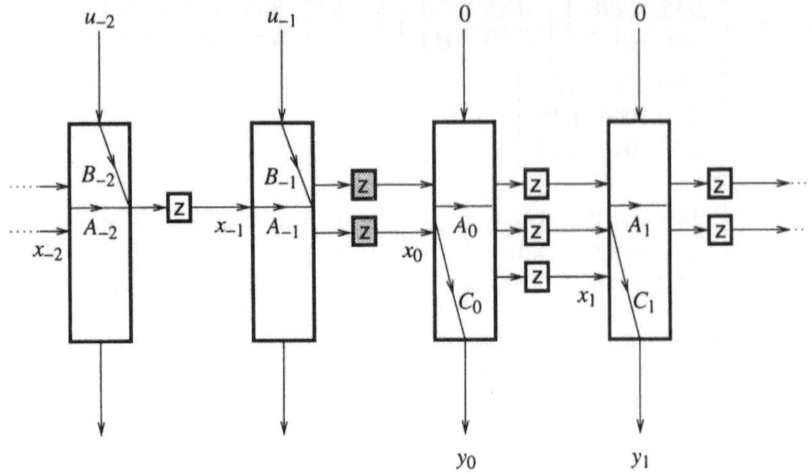

Figure 3.10. Principle of the identification of a time-varying state-space model. In this diagram, the current time is $k = 0$. All possible inputs with non-zero values up to time $k = -1$ (the past) are applied, and the corresponding output sequences are recorded from time $k = 0$ on (the future). Thus, only part of T is used: H_0, the Hankel operator at instant $k = 0$. The rank of the Hankel operator determines the state dimension at that point.

Let T be a given input-output operator. Denote a certain time instant as "current time", say point i. Apply an input sequence $u \in \ell_2$ to the system which is arbitrary up to $k = i-1$ and equal to 0 from $k = i$ on. We say that such an input has support in "the past", with respect to time $k = i$. The corresponding output sequence $y = uT$ is taken into consideration only from time $k = i$ on, *i.e.*, we record only the "future" part of y. See figure 3.10. The following two observations form the cornerstone of realization theory. Let $y_{f(i)}$ denote the half-sided sequence $y_{f(i)} = [y_i \; y_{i+1} \; \cdots] \in \ell_2^+$, and likewise define $u_{p(i)} = [u_{i-1} \; u_{i-2} \; \cdots] \in \ell_2^-$. The future output sequence is dependent only on x_i:

$$\begin{aligned} y_{f(i)} = [y_i \quad y_{i+1} \quad \cdots] &= x_i [C_i \quad A_i C_{i+1} \quad A_i A_{i+1} C_{i+2} \quad \cdots] \\ &= x_i \mathcal{O}_i . \end{aligned}$$

Hence upon applying all possible inputs that are zero from $k = i$ on, the corresponding possible outputs $y_{f(i)}$ are restricted by the finite dimension of x_i to a subspace of small dimensions in ℓ_2^+ (in the example: two dimensions). This subspace is called the natural *output state space*, or space of natural responses, at time $k = i$. Of course, if we select another point in time as current time, then a similar property holds, mutatis mutandis.

A second observation is almost trivial. If we stop the input at $k = i-1$, but now only record the output from $k = i+1$ on, then we reach a subset of the subspace $\{y_{f(i+1)}\}$. This subset is again a subspace, now of the form

$$\{ x_i A_i [C_{i+1} \; A_{i+1} C_{i+2} \; A_{i+1} A_{i+2} C_{i+3} \; \cdots] : x_i \in \mathbb{C}^{d_i} \} . \qquad (3.29)$$

A refinement of this observation leads to the mathematical concept of *shift invariance*: the subspace (3.29) is equal to the output state space at time i after the application of a shift, and this subspace is contained in the next output state space, at time $i+1$. The appearance of A_i in this expression enables us to identify it.

Write $u_{p(i)} = [u_{i-1} \; u_{i-2} \; u_{i-3} \; \cdots]$. Then from the relation $y = uT$ follows $y_{f(i)} = u_{p(i)}H_i$, where

$$H_i = \begin{bmatrix} T_{i-1,i} & T_{i-1,i+1} & T_{i-1,i+2} & \cdots \\ T_{i-2,i} & T_{i-2,i+1} & & \\ T_{i-3,0} & & \ddots & \\ \vdots & & & \end{bmatrix}.$$

Repeating the same exercise for all the signal pairs $u_{p(k)}$, $y_{f(k)}$, we obtain a sequence of operators H_k, which are precisely the Hankel matrices defined in (3.21). In the time-invariant case, where T has a Toeplitz structure, the construction yields H_k which are all the same and do indeed possess a Hankel structure (constant along anti-diagonals). Although we have lost the traditional anti-diagonal Hankel structure in the time-varying case, we have retained two important properties: the rank property and a shift-invariance property.

With regard to the rank property: suppose that we have a factorization of H_k: $H_k = \mathcal{C}_k\mathcal{O}_k$. Then the multiplication $y_{f(k)} = u_{p(k)}H_k$ can be split into two stages using an intermediate quantity x_k which is precisely the state at time k:

$$\begin{aligned} x_k &= u_{p(k)}\mathcal{C}_k \\ y_{f(k)} &= x_k\mathcal{O}_k. \end{aligned}$$

This factorization is typical of any state realization: the future output $y_{f(k)}$ is not directly computed, but uses an intermediate quantity x_k. From the decomposition $H_k = \mathcal{C}_k\mathcal{O}_k$, it is directly inferred that the rank of H_k determines the minimal dimensions of $\mathcal{C}_k$ and $\mathcal{O}_k$. If the decomposition is *minimal*, that is, if $\mathcal{C}_k$ and $\mathcal{O}_k$ are full-rank factors ($\mathcal{C}_k^*\mathcal{C}_k > 0$, $\mathcal{O}_k\mathcal{O}_k^* > 0$), then the dimension of the state space of the realization corresponding to $\mathcal{C}_k$ and $\mathcal{O}_k$ is equal to rank(H_k). If all $\mathcal{C}_k$ satisfy $\mathcal{C}_k^*\mathcal{C}_k > 0$, then we call the resulting realization *reachable*, and if all $\mathcal{O}_k$ satisfy $\mathcal{O}_k\mathcal{O}_k^* > 0$, then we call the realization *observable*. Hence, if the realization is both reachable and observable, it is minimal. The reason for this nomenclature is that if a realization is reachable (at point k), any state x_k can be reached using some $u_{p(k)}$: it suffices to take $u_{p(k)} = x_k\mathcal{C}_k^\dagger$, where $\mathcal{C}_k^\dagger = (\mathcal{C}_k^*\mathcal{C}_k)^{-1}\mathcal{C}_k^*$. Similarly, if a realization is observable, then from an observed output $y_{f(k)}$, and assuming $u_{f(k)} = 0$, the state x_k can be retrieved as $x_k = y_{f(k)}\mathcal{O}_k^\dagger$, where $\mathcal{O}_k^\dagger = \mathcal{O}_k^*(\mathcal{O}_k\mathcal{O}_k^*)^{-1}$.

In chapter 5, we elaborate on the concepts of reachability, observability, and input/output state spaces. This plays a fundamental role throughout the remainder of this book. It is possible to define them in an index-free notation using diagonal operators, and this will prove valuable in derivations later on.

Numerical issues

The key part of the realization algorithm is to obtain bases $\mathcal{C}_k$ and $\mathcal{O}_k$ for the column space and row space of each Hankel matrix H_k of T. The singular value decomposition

(SVD) [GV89] is a robust tool for doing this. It is a decomposition of H_k into factors U_k, Σ_k, V_k, where U_k and V_k are unitary matrices whose columns contain the left and right singular vectors of H_k, and Σ_k is a diagonal matrix with positive entries (the singular values of H_k) on the diagonal. The integer d_k is set equal to the number of non-zero singular values of H_k, and the first d_k columns of U_k and V_k constitute basis vectors for the column spans of H_k and H_k^*.

Figure 3.9 only gives an algorithmic outline of the realization procedure. Because H_{k+1} has a large overlap with H_k, an efficient SVD updating algorithm can be devised that takes this structure into account. Other decompositions from linear algebra that identify subspaces can be used instead. In theory a QR factorization of the H_k should work, although this is not advisable in practice because a QR factorization is not rank revealing: the addition of a small amount of noise on the entries of T will make all Hankel matrices have full rank, thus producing a realization of high order. Decompositions that can be used instead of QR are rank revealing QR [Fos86, Cha87, BS92], and the URV decomposition [Ste92], which is equivalent to SVD but computationally less demanding.

Note that, based on the singular values of H_k, a reduced order model can be obtained by omitting some vectors in $\mathcal{C}_k$ and $\mathcal{O}_k$, in particular those that correspond to small singular values. For time-invariant systems, this technique leads to a so-called balanced model reduction. Although widely used for time-invariant systems, this is for time-varying systems in fact a "heuristic" model reduction theory, because the modeling error norm is not known. (For LTI systems, a potentially large upper bound on the modeling error is given by the sum of the truncated singular values [Glo84].) A precise approximation theory results if the tolerance on the error is given in terms of the *Hankel norm*, which is the subject of chapter 10. The approximation algorithm in that chapter is in fact a competitor for the rank revealing QR method.

Computational issues

We mention some other issues related to theorem 3.7 and the corresponding realization algorithm, which are of some importance for a practical implementation of the algorithm.

Let T be a given upper triangular matrix, and consider its sequence of Hankel matrices $\{H_k\}$, where H_k has rank d_k. If for each H_k a submatrix $\hat{H}_k$ is known such that $\text{rank}(\hat{H}_k) = d_k$ also, then it is possible to determine a realization of T based on factorizations of the $\hat{H}_k$ rather than factorizations of H_k. This generalization of the time-invariant analog [Kal65] is useful since it can yield considerable computational savings if the $\hat{H}_k$ have small dimensions in comparison with H_k. A remaining practical problem is how to obtain the $\hat{H}_k$ in an efficient way, because, unlike the time-invariant case, T need not be diagonally dominant even if its Hankel matrices have low rank, so that the $\hat{H}_k$ can still be matrices of large size. A trivial example of the latter is provided by taking T to be an $n \times n$ matrix consisting of zeros, except for the $(1,n)$-entry.

In this section, we use the matrix $\pi_r := [I_r \;\; 0 \;\; 0 \;\; \cdots \;]$ to select the first r rows of a matrix at its right. We use, as before, the notation $H_k^{\leftarrow}$ to denote H_k with its first column deleted, and let $\dagger$ denote the generalized (left or right) inverse of a matrix. The following result (and proof) can be found in [GKL92].

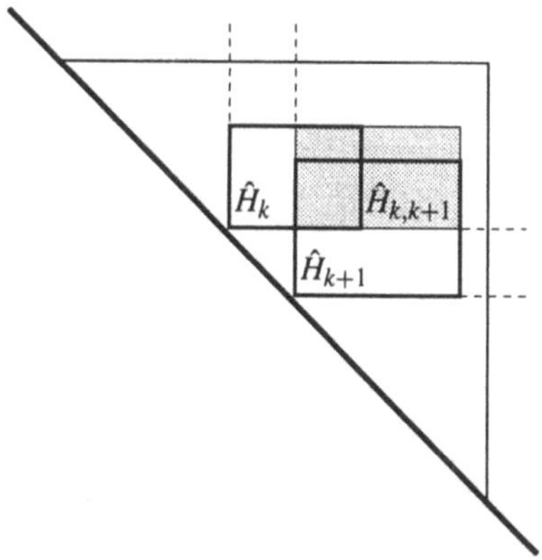

Figure 3.11. Relation between $\hat{H}_k$ and $\hat{H}_{k+1}$.

Theorem 3.9 *Let T be an upper triangular matrix with Hankel matrices H_k having rank d_k. For each k, suppose that the numbers $r(k)$ and $c(k)$ are such that the submatrices $\hat{H}_k = \pi_{r(k)} H_k \pi^*_{c(k)}$ have rank d_k. Let $\hat{H}_k = \hat{\mathcal{C}}_k \hat{\mathcal{O}}_k$ be a factorization of $\hat{H}_k$ into minimal rank factors. Then a realization of T is given by*

$$\begin{array}{rclrcl} \hat{A}_k &=& \hat{\mathcal{C}}_k^\dagger \hat{H}_{k,k+1} \hat{\mathcal{O}}_k^\dagger , & \hat{C}_k &=& \hat{\mathcal{O}}_k \pi_1^* , \\ \hat{B}_k &=& \pi_1 \hat{\mathcal{C}}_k , & \hat{D}_k &=& T_{k,k} , \end{array}$$

where $\hat{H}_{k,k+1} = \pi_{r(k)} H_k^{\leftarrow} \pi^*_{c(k+1)}$.

PROOF A diagram of the relations between $\hat{H}_k$, $\hat{H}_{k+1}$ and $\hat{H}_{k,k+1}$ is provided in figure 3.11. The proof consists of two parts. We first verify that the full size Hankel matrix H_k has a minimal factorization into rank d_k factors $\mathcal{C}_k$ and $\mathcal{O}_k$ such that

$$\hat{\mathcal{C}}_k = \pi_{r(k)} \mathcal{C}_k , \qquad \hat{\mathcal{O}}_k = \mathcal{O}_k \pi^*_{c(k)} , \tag{3.30}$$

i.e., , certain extensions of $\hat{\mathcal{C}}_k$ and $\hat{\mathcal{O}}_k$. Indeed, let $H_k = \tilde{\mathcal{C}}_k \tilde{\mathcal{O}}_k$ be any minimal factorization, then $\hat{H}_k = \pi_{r(k)} H_k \pi^*_{c(k)} = (\pi_{r(k)} \tilde{\mathcal{C}}_k)(\tilde{\mathcal{O}}_k \pi^*_{c(k)})$. Because $\text{rank}(\hat{H}_k) = d_k$ also, it follows that $\pi_{r(k)} \tilde{\mathcal{C}}_k$ and $\tilde{\mathcal{O}}_k \pi^*_{c(k)}$ are full rank factors of $\hat{H}_k$, so that these are related to the given factorization $\hat{H}_k = \hat{\mathcal{C}}_k \hat{\mathcal{O}}_k$ as $\hat{\mathcal{C}}_k = (\pi_{r(k)} \tilde{\mathcal{C}}_k) R_k$ and $\hat{\mathcal{O}}_k = R_k^{-1} (\tilde{\mathcal{O}}_k \pi^*_{c(k)})$, where R_k is an invertible state transformation. Putting $\mathcal{C}_k = \tilde{\mathcal{C}}_k R_k$ and $\mathcal{O}_k = R_k^{-1} \tilde{\mathcal{O}}_k$ gives (3.30).

The second step is to verify that $\{\hat{A}_k, \hat{B}_k, \hat{C}_k, \hat{D}_k\}$ is a realization of T. This is done by proving that it is precisely equal to the realization based on the full-size factors $\mathcal{C}_k$ and $\mathcal{O}_k$. The main issue is to show that $A_k = \mathcal{C}_k^\dagger H_k^{\leftarrow} \mathcal{O}_{k+1}^\dagger$ is equal to $\hat{A}_k$. Expressions for these generalized inverses are

$$\begin{array}{rcl} \mathcal{C}_k^\dagger \mathcal{C}_k &=& \hat{\mathcal{C}}_k^\dagger \pi_{r(k)} \mathcal{C}_k \\ \mathcal{O}_k \mathcal{O}_k^\dagger &=& \mathcal{O}_k \pi^*_{c(k)} \hat{\mathcal{O}}_k^\dagger \end{array}$$

because $\mathcal{C}_k^\dagger \mathcal{C}_k = I_{d_k} = \hat{\mathcal{C}}_k^\dagger \hat{\mathcal{C}}_k = \hat{\mathcal{C}}_k^\dagger \pi_{r(k)} \mathcal{C}_k$, and likewise for $\mathcal{O}_k^\dagger$. Hence

$$\begin{aligned} A_k &= \mathcal{C}_k^\dagger \mathcal{C}_k A_k \mathcal{O}_{k+1} \mathcal{O}_{k+1}^\dagger \\ &= \hat{\mathcal{C}}_k^\dagger \pi_{r(k)} \mathcal{C}_k A_k \mathcal{O}_{k+1} \pi_{c(k+1)}^* \hat{\mathcal{O}}_{k+1}^\dagger \\ &= \hat{\mathcal{C}}_k^\dagger \pi_{r(k)} H_k^{\leftarrow} \pi_{c(k+1)}^* \hat{\mathcal{O}}_{k+1}^\dagger \\ &= \hat{\mathcal{C}}_k^\dagger \hat{H}_{k,k+1} \hat{\mathcal{O}}_{k+1}^\dagger \;=\; \hat{A}_k\,. \end{aligned}$$

With less effort, it follows that $B_k = \pi_1 \mathcal{C}_k = \pi_1 \pi_{r(k)} \mathcal{C}_k = \pi_1 \hat{\mathcal{C}}_k = \hat{B}_k$, and likewise $C_k = \hat{C}_k$. □

The theorem shows that even for Hankel matrices with infinite dimensions we can find a realization, as long as we are sure that the finite Hankel matrices have their rank equal to the actual system order at that point. Unlike for time-invariant systems, we can never be sure that a finite size Hankel matrix has indeed the maximal rank without making further assumptions on the matrix. Hence, without making assumptions on the matrix, it is not really possible to work with finite size Hankel matrices and obtain exact state space models. An approximate realization algorithm is discussed in chapter 10.

3.5 IDENTIFICATION FROM INPUT-OUTPUT DATA

Theorem 3.7 and the realization algorithm assume knowledge of the input-output operator T. This is equivalent to assuming that the time-varying impulse response of the system is known. Many applications, however, provide only input-output data, *i.e.*, pairs of input sequences $u \in \ell_2^{\mathcal{M}}$ with corresponding outputs $y \in \ell_2^{\mathcal{N}}$. We will need several such pairs. In that case, these rows can be stacked into matrices U and Y. Since we have

$$Y = UT$$

it follows that if U has a left inverse $U^\dagger$ such that $U^\dagger U = I$, then T can be computed as $T = U^\dagger Y$, and from T we can obtain the realization as in theorem 3.7 or 3.9. Thus, system identification from input-output data is, at this level, not much different from the realization problem with known impulse response data.

In the time-invariant case, it suffices to have a single input-output pair (u,y), since other independent pairs can be found simply by time-shifting the sequences, exploiting the time-invariance of the system. For time-invariant systems, the condition on u so that U has a left inverse is called *persistently exciting*.

For time-varying systems, of course, we cannot generate multiple input-output pairs from a single one, and we really need a collection of input-output pairs. Whether this can be realized in practice depends on the application: *e.g.*, we might have multiple copies of the system to obtain input-output pairs that span the same period in time. Even so, we have the problem that with a finite collection of input-output pairs we can only estimate the part of the system that has actually been excited.

Let's again consider the finite matrix case, with a time window running from 1 to n. There are two possibilities. If the system starts and stops with zero state dimensions, then T is an $n \times n$ matrix, and we need at least n independent input-output pairs (u,y), stacked into $n \times n$ matrices U and Y, to proceed as indicated above. However, a more general case occurs if the input-output pairs have been *observed* over a finite time interval of length n, but in actuality span a much larger period. In that case, the initial state x_1 need not be zero, so that we obtain

$$y = x_1 \mathcal{O}_1 + uT.$$

Here, T is a finite $n \times n$ matrix which is a submatrix of the actual much larger input-output operator, spanning only the window of interest. The above equation can be derived in several ways, *e.g.*, by using linearity: the output y is the sum of the effect of the input with zero initial state, and of the initial state with zero input. With a stack of inputs and outputs collected as rows in U and Y, we obtain

$$Y = X_1 \mathcal{O}_1 + UT,$$

where X_1 is a column vector containing the initial states. Let's assume that we have N such finite input-output pairs, and no knowledge of X_1. Our objective is to construct a system T of minimal complexity that is consistent with the given data. The main idea for doing this is to get rid of the influence of the term $X_1 \mathcal{O}_1$ by using the causality of T.

The first step is to employ a QR factorization to reduce the data matrices, *i.e.*, to compute a unitary $N \times N$ matrix Q such that

$$Q^*[U \quad Y] = \begin{bmatrix} R_{11} & R_{12} \\ 0 & R_{22} \\ 0 & 0 \end{bmatrix} =: [U' \quad Y'].$$

Since the system is linear, the premultiplication by Q^* can be viewed as simply generating new input-output pairs, consisting of linear combinations of the old pairs. Thus, we have new pairs $[U', Y']$ for which $Y' = X_1' \mathcal{O}_1 + U'T$. We can go further and premultiply the top block row by R_{11}^{-1} (assuming the inputs were chosen such that it is invertible), which produces a new pair $[U'', Y'']$ where $U'' = [\begin{smallmatrix} I \\ 0 \end{smallmatrix}]$. Dropping the quotes for readability, we can say that after these steps we have a data matrix Y such that

$$Y = \begin{bmatrix} Y_1 \\ \vdots \\ Y_N \end{bmatrix} = X_1 \mathcal{O}_1 + \begin{bmatrix} T \\ 0 \end{bmatrix} \tag{3.31}$$

where T is upper triangular.

The second term on the right has only n nonzero rows. Thus,

$$\begin{bmatrix} Y_{n+1} \\ \vdots \\ Y_N \end{bmatrix} = \begin{bmatrix} (X_1)_{n+1} \\ \vdots \\ (X_1)_N \end{bmatrix} \mathcal{O}_1.$$

This allows us to identify a basis for $\mathcal{O}_1$: it is the row span of this part of the data. The initial state dimension d_1 follows as well, as the rank of this submatrix. We need at least $n + d_1$ independent input sequences u to do this.

The first part of the data specifies

$$\begin{bmatrix} Y_{1,1} & \cdots & Y_{1,n} \\ \vdots & & \vdots \\ Y_{n,1} & \cdots & Y_{n,n} \end{bmatrix} = \begin{bmatrix} (X_1)_1 \\ \vdots \\ (X_1)_n \end{bmatrix} \mathcal{O}_1 + \begin{bmatrix} T_{1,1} & \cdots & \cdots & T_{1,n} \\ & T_{2,2} & \cdots & T_{2,n} \\ & & \ddots & \vdots \\ \mathbf{0} & & & T_{n,n} \end{bmatrix}.$$

With $\mathcal{O}_1$ known, the next objective is to estimate the above first part of X_1, so that this term can be subtracted. Here, we have to use the fact that T is upper triangular: we can select submatrices where T is zero. Thus

$$[Y_{k,1} \ \cdots \ Y_{k,k}] = (X_1)_k[(\mathcal{O}_1)_1 \ \cdots \ (\mathcal{O}_1)_k], \qquad (k=2,\cdots,n). \tag{3.32}$$

The last factor on the right has d_1 rows. Thus, only for $k \geq d_1$ can $(X_1)_k$ be consistently estimated as

$$(X_1)_k = [Y_{k,1} \ \cdots \ Y_{k,k}][(\mathcal{O}_1)_1 \ \cdots \ (\mathcal{O}_1)_k]^\dagger, \qquad (k=d_1,\cdots,n). \tag{3.33}$$

The first d_1 states cannot be recovered, but we can choose something reasonable so that (3.32) holds, *e.g.*, by using the same equation (3.33). This will provide a realization that is consistent with the given data, although the initial $[B_k, D_k]$ (for $k = 1,\cdots,d_1-1$) will be some arbitrary fit.

At this point, we have recovered the initial state X_1. Thus, the term $X_1\mathcal{O}_1$ can be subtracted from Y, which leaves an estimate for T. The realization of T can now be obtained as before using the realization procedure in theorem 3.7.

The above procedure is only intended as a framework. There are several points for improvement:

1. $\mathcal{O}_1$ is estimated from the zero block in (3.31). However, also the zero submatrices in the lower triangular part of T should be used for this, as well as the nonzero (Hankel) submatrices of the upper triangular part of T.

2. The pseudo-inverses in (3.33) are nested and of increasing dimensions. This could be exploited in a computationally efficient implementation.

3. In practical situations, Y is perturbed by noise. A major point of interest is to devise an identification procedure that is asymptotically consistent even in this case.

Similar algorithms for identification of time-varying systems from input-output ensembles have been proposed in [VY95, Yu96]. Those algorithms differ in the last step, where they try to remove the influence of X_1 by projection onto the complement of $\mathcal{O}_1$, and shifts thereof. This was inspired by recent subspace-based identification schemes in the time-invariant domain [MMVV89, VD92a, VD92b, Ver94, Ove95, Vib94]. Some applications to actual time-varying systems (the dynamics of a human joint) can be found in [KH90, KKMH91, YV93].

3.6 REALIZATION THEORY FOR MATRICES OF LOW DISPLACEMENT RANK

In section 3.2 we discussed systems of low displacement rank: matrices R for which the displacement $R-\sigma R\sigma^*$ has low rank. We continue the discussion on systems with low displacement rank by deriving a realization theory for such systems, in case they also have a low state dimension. We consider the finite-size matrix case, which necessarily leads to time-varying realizations. The objective is to use the structure provided by the low displacement rank to derive a recursive rule for computing the state space matrices of the next time instant in terms of the current one.

Thus let R be a positive definite matrix of size $n\times n$, possibly with block matrix entries which we take square of fixed dimensions. Let us write $R=[r_{ij}]$ as

$$R = L^*L = \frac{1}{2}(F+F^*)$$

in which L and F are upper triangular matrices. Hence,

$$F = \begin{bmatrix} r_{00} & 2r_{01} & \cdots & 2r_{0,n-1} \\ & r_{11} & \ddots & \vdots \\ & & \ddots & 2r_{n-2,n-1} \\ \mathbf{0} & & & r_{n-1,n-1} \end{bmatrix}.$$

Recall the definition of the restricted backward shift operator

$$\sigma = \begin{bmatrix} 0 & & & \mathbf{0} \\ 1 & \ddots & & \\ & \ddots & \ddots & \\ \mathbf{0} & & 1 & 0 \end{bmatrix}.$$

We will assume that R has a displacement structure:

$$R-\sigma R\sigma^* = G^*JG = \begin{bmatrix} g_0^* \\ \vdots \\ g_{n-1}^* \end{bmatrix} J[g_0 \ \cdots \ g_{n-1}].$$

where J has p positive and q negative signature entries, and G has $p+q$ rows.

Realization of the additive component F

Let us write for simplicity $R-\sigma R\sigma^* =: X$, then it is easy to see that R can be recovered from X via the formula

$$R = X + \sigma X\sigma^* + \cdots + \sigma^{n-1}X(\sigma^*)^{n-1}.$$

The contribution of each term to the Hankel operators for F is straightforward to evaluate. Indeed, consider the Hankel operator $H_k(F)$ for F ($k=1,\cdots,n-1$). The contributions of the individual terms to $H_k(F)$ are

$$
\begin{aligned}
H_k(X) &= \begin{bmatrix} g_{k-1}^* \\ \vdots \\ g_0^* \end{bmatrix} J \begin{bmatrix} g_k & \cdots & g_{n-1} \end{bmatrix} \\
H_k(\sigma X \sigma^*) &= \begin{bmatrix} g_{k-2}^* \\ \vdots \\ g_0^* \\ 0 \end{bmatrix} J \begin{bmatrix} g_{k-1} & \cdots & g_{n-2} \end{bmatrix} \\
H_k(\sigma^{k-1} X (\sigma^*)^{k-1}) &= \begin{bmatrix} g_0^* \\ 0 \\ \vdots \\ 0 \end{bmatrix} J \begin{bmatrix} g_1 & \cdots & g_{n-k} \end{bmatrix}.
\end{aligned}
$$

Putting these terms together and using the outer product representation of a matrix, we obtain

$$
H_k(F) = 2 \begin{bmatrix} g_0^*J & g_1^*J & \ddots & g_{k-1}^*J \\ & g_0^*J & \ddots & \ddots \\ & & \ddots & g_1^*J \\ \mathbf{0} & & & g_0^*J \end{bmatrix} \begin{bmatrix} g_1 & g_2 & \cdot^{\cdot^{\cdot}} & g_{n-k} \\ g_2 & \cdot^{\cdot^{\cdot}} & \cdot^{\cdot^{\cdot}} & g_{n-k+1} \\ \cdot^{\cdot^{\cdot}} & & & \cdot^{\cdot^{\cdot}} \\ g_k & \cdot^{\cdot^{\cdot}} & \cdot^{\cdot^{\cdot}} & g_{n-1} \end{bmatrix}, \tag{3.34}
$$

which is of the form Toeplitz matrix times Hankel matrix. From (3.34) we conclude that

$$
\operatorname{rank}(H_k(F)) \le \operatorname{rank} \begin{bmatrix} g_1 & g_2 & \cdot^{\cdot^{\cdot}} & g_{n-k} \\ g_2 & \cdot^{\cdot^{\cdot}} & \cdot^{\cdot^{\cdot}} & g_{n-k+1} \\ \cdot^{\cdot^{\cdot}} & & & \cdot^{\cdot^{\cdot}} \\ g_k & \cdot^{\cdot^{\cdot}} & \cdot^{\cdot^{\cdot}} & g_{n-1} \end{bmatrix},
$$

where the latter matrix is a submatrix of the semi-infinite Hankel operator for the LTI system

$$
g_0 + g_1 z + g_2 z^2 + g_3 z^3 + \cdots .
$$

A (standard) realization for this LTI system can be used as a starting point for the realization of F. Assuming that the rank of the global Hankel operator is δ, so that we need a state space of dimension δ, we find matrices α, β, γ of dimensions $\delta \times \delta, (p+q) \times \delta, \delta \times 1$ such that

$$
g_i = \beta \alpha^{i-1} \gamma. \tag{3.35}
$$

The k-th Hankel matrix for the series $\{g_i\}$ as needed in (3.34) is then

$$
\begin{bmatrix} g_1 & \cdots & g_{n-k} \\ \vdots & & \vdots \\ g_k & \cdots & g_{n-1} \end{bmatrix} = \begin{bmatrix} \beta \\ \beta\alpha \\ \vdots \\ \beta\alpha^{k-1} \end{bmatrix} [\gamma \;\; \alpha\gamma \;\; \cdots \;\; \alpha^{n-k-1}\gamma].
$$

Thus, the k-th Hankel matrix for F is

$$H_k(F) = 2\begin{bmatrix} g_0^*J & g_1^*J & \ddots & g_{k-1}^*J \\ & g_0^*J & \ddots & \ddots \\ & & \ddots & g_1^*J \\ \mathbf{0} & & & g_0^*J \end{bmatrix} \begin{bmatrix} \beta \\ \beta\alpha \\ \vdots \\ \beta\alpha^{k-1} \end{bmatrix} [\gamma \quad \alpha\gamma \ \cdots \ \alpha^{n-k-1}\gamma]. \qquad (3.36)$$

Clearly, this is a factorization of the form $H_k(F) = \mathcal{C}_k\mathcal{O}_k$, and a realization for F can have a time-invariant $A_k = \alpha$ and $C_k = \gamma$. The $\{B_k, D_k\}$-part of the realization will be time-varying. B_{k-1} is found from the first row of the first factor,

$$\begin{aligned} B_{k-1} &= 2(g_0^*J\beta + g_1^*J\beta\alpha + \cdots + g_{k-1}^*J\beta\alpha^{k-1}) \\ &= 2g_0^*J\beta + 2\gamma^*[\beta^*J\beta\alpha + \cdots + (\alpha^*)^{k-2}\beta^*J\beta\alpha^{k-1}]. \end{aligned}$$

Let us define

$$\Lambda_k := \beta^*J\beta + \cdots + (\alpha^*)^{k-1}\beta^*J\beta\alpha^{k-1},$$

then Λ_k satisfies the recursive Lyapunov equation[4]

$$\Lambda_k = \beta^*J\beta + \alpha^*\Lambda_{k-1}\alpha,$$

and B_k can easily be computed from Λ_k via

$$B_k = 2(g_0^*J\beta + \gamma^*\Lambda_k\alpha).$$

Similarly,

$$D_k = g_k^*Jg_k + \cdots + g_0^*Jg_0$$

which satisfies the recursion

$$D_0 = g_0^*Jg_0; \qquad D_k = g_k^*Jg_k + D_{k-1}, \quad (k \geq 1).$$

This constitutes a low rank recursive realization for F. The algorithm is summarized in figure 3.12. The realization is not necessarily (locally) minimal: for this, it should at least start and end with zero state dimensions. If the dimension n of R grows while R stays bounded, then $|\alpha| < 1$ and the scheme converges gradually to a time invariant realization, since Λ_k, B_k, D_k converge as $k \to \infty$.

A realization for the multiplicative Cholesky factor L

We had before

$$R = \frac{1}{2}(F + F^*) = L^*L.$$

[4]Such equations are discussed in extenso later in section 5.3.

In: A generator G and signature J such that $R - \sigma R \sigma^* = G^* J G$
Out: A realization for $F \in \mathcal{U}$ such that $R = \frac{1}{2}(F + F^*)$.

Find a realization $\{\alpha, \beta, \gamma\}$ from the LTI system (3.35)
$A_0 = \cdot$ (size $0 \times \delta$)
$C_0 = \cdot$ (size 0×1)
$\Lambda_0 = 0$ (size $\delta \times \delta$)
$D_0 = g_0^* J g_0$
for $k = 1, \cdots, n-1$
  $A_k = \alpha$, $C_k = \gamma$
  $\Lambda_k = \beta^* J \beta + \alpha^* \Lambda_{k-1} \alpha$
  $B_{k-1} = 2(g_0^* J \beta + \gamma^* \Lambda_{k-1} \alpha)$
  $D_k = g_k^* J g_k + D_{k-1}$
end

Figure 3.12. Realization algorithm for the additive component F of a positive definite matrix R of low displacement rank.

We will try to find a realization for L of the form (in diagonal notation)

$$L = D_L + B_L Z(I - AZ)^{-1} C, \tag{3.37}$$

where we keep $A_k = \alpha$ and $C_k = \gamma$ from before, and compute new $B_{L,k}$, $D_{L,k}$ from the realization $\{A_k, B_k, C_k, D_k\}$ of F of the preceding section. We should then have

$$\begin{aligned} L^* L = D_L^* D_L &+ C^*(I - Z^* A^*)^{-1} Z^* B_L^* D_L + D_L^* B_L Z(I - AZ)^{-1} C \\ &+ C^*(I - Z^* A^*)^{-1} Z^* B_L^* B_L Z(I - AZ)^{-1} C. \end{aligned}$$

The last term in this expression is quadratic. It can be subjected to a partial fraction expansion, which in this generalized context leads to

$$\begin{aligned} (I - Z^* A^*)^{-1} (B_L^* B_L)^{(1)} (I - AZ)^{-1} &= (I - Z^* A^*)^{-1} M + M(I - AZ)^{-1} - M \\ &= M + MAZ(I - AZ)^{-1} + (I - Z^* A^*)^{-1} Z^* A^* M \end{aligned}$$

where the block diagonal matrix M satisfies the equation

$$M^{(-1)} = B_L^* B_L + A^* M A. \tag{3.38}$$

This equation in diagonals is actually again a recursive Lyapunov-Stein equation, and it has indeed a (unique) solution which can be computed recursively, provided that at each step $B_{L,k}^* B_{L,k}$ is known. Indeed, with initial point $M_0 = 0$, the expansion of (3.38) into its diagonal entries leads to

$$M_{k+1} = B_{L,k}^* B_{L,k} + A_k^* M_k A_k \qquad (k = 0, 1, \cdots).$$

When we introduce the partial fraction decomposition in the equation for $L^* L$ above, and identify the strictly upper triangular, diagonal and strictly lower triangular parts,

In: A generator G and signature J such that $R-\sigma R\sigma^* = G^*JG$
Out: A realization for $L \in \mathcal{U}$ such that $R = L^*L$.

α, γ from the LTI system (3.35).
B_0, D_0 from the algorithm in figure 3.12
$D_{L,0} = [\frac{1}{2}(D_0 + D_0^*)]^{1/2}$
$B_{L,0} = \frac{1}{2} D_{L,0}^{-*} B_0$
$M_1 = B_{L,0}^* B_{L,0}$
for $i = 1, \cdots, n-1$
    $A_i = \alpha,\ C_i = \gamma$
    B_i, D_i from the algorithm in figure 3.12
    $D_{L,i} = [\frac{1}{2}(D_i + D_i^*) - C^* M_i C]^{1/2}$
    $B_{L,i} = D_{L,i}^{-*}[\frac{1}{2}B_i - C^* M_i A]$
    $M_{i+1} = B_{L,i}^* B_{L,i} + A^* M_i A$
end

Figure 3.13. Realization of a Cholesky factor L of R

viz.

$$\begin{array}{rcl} L^*L & = & D_L^*D_L + C^*MC + (D_L^*B_L + C^*MA)(I-AZ)^{-1}C + [*] \\ \frac{1}{2}(F+F^*) & = & \frac{1}{2}(D + D^*) + \frac{1}{2}BZ(I-AZ)^{-1}C + [*], \end{array}$$

then we see that the block diagonal matrices B_L and D_L must satisfy the set of equations

$$\left\{ \begin{array}{rcl} \frac{1}{2}(D+D^*) & = & D_L^*D_L + C^*MC \\ \frac{1}{2}B & = & D_L^*B_L + C^*MA \\ M^{(-1)} & = & B_L^*B_L + A^*MA\,. \end{array} \right. \tag{3.39}$$

This set of equations clearly leads to a recursive algorithm, at least if they are consistently solvable (which we have to show) — see the algorithm in figure 3.13.

One may think that a solution must exist, almost by construction (since the starting point of the recursion is well known — $M_0 = 0$), but there is reasonable doubt that at some point k, the equation for $D_{L,k}$,

$$D_{L,k}^* D_{L,k} = \tfrac{1}{2}(D_k + D_k^*) - C_k^* M_k C_k$$

cannot be satisfied because the second member is possibly not positive definite. It is instructive to show that this cannot happen. We construct the proof by looking at the Cholesky factorization of R in a "Crout-Doolittle" fashion — the classical "LU factorization" or Gauss elimination method to solve a system of linear equations, see [Ste77]. The general Crout-Doolittle method (without pivoting) consists in the recursive construction of a tableau for the lower/upper factorization of a general matrix T, but it applies almost without modification to the Cholesky decomposition of a (strictly) positive matrix R. In this case, we recursively construct the factorization

$$\begin{bmatrix} r_{00} & r_{01} & r_{02} & \cdots \\ r_{10} & r_{11} & r_{12} & \cdots \\ r_{20} & r_{21} & r_{22} & \cdots \\ \vdots & \vdots & \vdots & \ddots \end{bmatrix} = \begin{bmatrix} l_{00}^* & & & \mathbf{0} \\ l_{01}^* & l_{11}^* & & \\ l_{02}^* & l_{12}^* & l_{22}^* & \\ \vdots & \vdots & \vdots & \ddots \end{bmatrix} \begin{bmatrix} l_{00} & l_{01} & l_{02} & \cdots \\ & l_{11} & l_{12} & \cdots \\ & & l_{22} & \cdots \\ \mathbf{0} & & & \ddots \end{bmatrix}$$

by peeling off the first row and column of R:

$$R = \begin{bmatrix} l_{00}^* \\ l_{01}^* \\ \vdots \end{bmatrix} [l_{00} \quad l_{01} \quad \cdots] + \begin{bmatrix} 0 & 0 \quad \cdots \\ 0 & \\ \vdots & R' \end{bmatrix}.$$

It is clear that

$$[l_{00} \quad l_{01} \quad \cdots] = r_{00}^{-*/2} [r_{00} \quad r_{01} \quad \cdots].$$

In this way, the first column and row of L^* and L are computed; the procedure is then repeated on the smaller matrix R' to find the next columns and rows, etcetera. The entries of L are thus recursively determined by

- Step 0: $$\begin{aligned} l_{00} &= r_{00}^{1/2} \\ l_{0j} &= l_{00}^{-*} r_{0j} \qquad (j = 1, 2, \cdots), \end{aligned}$$

- Step i: $$\begin{aligned} l_{ii} &= (r_{ii} - \Sigma_{k=0}^{i-1} l_{ki}^* l_{ki})^{1/2} \\ l_{ij} &= l_{ii}^{-*}(r_{ij} - \Sigma_{k=0}^{i-1} l_{ki}^* l_{kj}) \qquad (j = i+1, i+2, \cdots). \end{aligned}$$

A standard proof (see *e.g.*, lemma 12.2 later in the book) shows that, for finite matrices

$$R > 0 \quad \Leftrightarrow \quad \left\{ \begin{array}{lcl} r_{00} & > & 0 \\ R' & > & 0. \end{array} \right.$$

This can be used to derive the central property in the algorithm: the pivot is strictly positive definite whenever R is, so that its square root can be taken, in the case of scalar as well as matrix block entries.

In our case we have, thanks to the realization for F,

$$R = \begin{bmatrix} \frac{1}{2}[D_0 + D_0^*] & \frac{1}{2}B_0\gamma & \frac{1}{2}B_0\alpha\gamma & \frac{1}{2}B_0\alpha^2\gamma & \cdots \\ \frac{1}{2}\gamma^* B_0^* & \frac{1}{2}[D_1 + D_1^*] & \frac{1}{2}B_1\gamma & \frac{1}{2}B_1\alpha\gamma & \cdots \\ \frac{1}{2}\gamma^*\alpha^* B_0^* & \frac{1}{2}\gamma^* B_1^* & \frac{1}{2}[D_2 + D_2^*] & \frac{1}{2}B_2\gamma & \cdots \\ \vdots & \vdots & \vdots & \vdots & \ddots \end{bmatrix}.$$

We now show that the recursions that solve (3.39) in effect generate the Crout-Doolittle recursion.

- Step 0: $M_0 = \cdot$, $\frac{1}{2}[D_0 + D_0^*] = D_{L,0}^* D_{L,0}$, $\frac{1}{2}B_0 = D_{L,0}^* B_{L,0}$ are of course solvable for $D_{L,0}$ and $B_{L,0}$, and produce the first row of L as

$$[l_{00} \quad l_{01} \quad \cdots] = [D_{L,0} \quad B_{L,0}\gamma \quad B_{L,0}\alpha\gamma \quad B_{L,0}\alpha^2\gamma \cdots].$$

- Step i: let us assume that the first i rows (*i.e.*, with indices $0,\cdots,i-1$) of L are known and satisfy
$$l_{ki} = B_{L,k}\alpha^{k-i-1}\gamma \qquad (k<i).$$
We have to show for the i-th row that $D^*_{L,i}D_{L,i}$ is well defined (*i.e.*, the expression for it is positive definite), and also that the rest of the row with index i is correct. The Crout-Doolittle scheme applies and, thanks to the induction hypothesis, it says that
$$r_{ii}-\sum_{k=0}^{i-1} l^*_{ki}l_{ki} = \tfrac{1}{2}[D_i+D^*_i]-\Sigma_{k=0}^{i-1}\gamma^*[\alpha^*]^{i-1-k}B^*_{L,k}B_{L,k}\alpha^{i-k-1}\gamma$$
is positive definite. The recursion for M on the other hand gives an expression for the sum:
$$M_i = \sum_{k=0}^{i-1}[\alpha^*]^{i-1-k}B^*_{L,k}B_{L,k}\alpha^{i-k-1}$$
so that the formula is in fact
$$r_{ii}-\sum_{k=0}^{i-1} l^*_{ki}l_{ki} = \tfrac{1}{2}[D_i+D^*_i]-\gamma^*M_i\gamma,$$
which is hence positive-definite and can be factored as $D^*_{L,i}D_{L,i}$. This also gives $D_{L,i}=l_{ii}$. A further identification with the Crout-Doolittle scheme produces, for $j>i$,
$$\begin{aligned} l_{ij} &= D^{-*}_{L,i}\{\tfrac{1}{2}B_i\alpha^{j-i-1}\gamma-\Sigma_{k=0}^{i-1}(B_{L,k}\alpha^{i-1-k}\gamma)^*(B_{L,k}\alpha^{j-1-k}\gamma)\} \\ &= D^{-*}_{L,i}[\tfrac{1}{2}B_i-\gamma^*M_i\alpha]\alpha^{j-i-1}\gamma. \end{aligned}$$
Since we defined $B_{L,i}$ to satisfy
$$\tfrac{1}{2}B_i = D^*_{L,i}B_{L,i}+\gamma^*M_i\alpha$$
we find $l_{ij}=B_{L,i}\alpha^{i-j+1}\gamma$, which is the posed realization for L. We may conclude that the scheme given by (3.38) always produces a strictly positive-definite expression for
$$\tfrac{1}{2}[D_i+D^*_i]-\gamma^*M_i\gamma,$$
when the original R is strictly positive definite.

This concludes the proof of the existence of the realization for L as given by the algorithm.

Discussion

A legitimate question is of course, "what do we gain in complexity reduction of calculations when we apply the realization theory to a system of low displacement rank?" A meaningful answer to such a question requires an understanding of what we mean by 'calculations'. We consider two cases: calculations aiming at the construction of a model for the system or its inverse, and calculations that aim at applying the model to an input. Both the low displacement rank and the low Hankel degree will contribute in both cases — and in the expected manner. The low displacement rank allows for realizations of F and L in which only the matrices B_k and D_k vary from one point to next using simple update equations, depending only on the actual g_k, which is itself only dependent on time-invariant data. If the Hankel rank is δ and the original system has scalar inputs and outputs, then the complexity of the realization $\{\alpha, \beta, \gamma\}$ is of the order of $(p+q)\delta$ (see chapter 14 for further discussions on complexity of the time varying state space model). Hence we end up with a parameter update scheme which can be very efficient, depending on the precise values of the three parameters. The computational efficiency of a vector-matrix products realized by a state space computational scheme is directly proportional to the size of the state δ. As for the inverse (say of L), we also face two types of computations: updates and the application of the computing scheme to inputs. Again, the update of the realization matrices for the inverse is a purely local matter dependent only on the actual g_k or their realization, via the formulas given by (3.18). The computations can be restricted to the computation of D_k^{-1} only.

Of course, the usage of a time varying systems model for computation precludes the utilization of the FFT as complexity reducing engine. An FFT scheme, however, requires a complete shuffle of the data, either at the input side, or in the course of the computations. It is (in some variations) the computational scheme that uses the smallest number of multiplications and additions possible, but at the cost of maximal shuffling of data. It also does not utilize the fact that relevant impulse responses can have a lot of structure or can be approximated with very little data. In selective applications, accuracy will suffer. This is the reason why in many signal processing applications, filtering algorithms are the preferred mode of implementation, although they coexist with the FFT. Even intermediate forms are possible, utilized in subband or multiresolution coding schemes, in which some shuffling of data takes place, combined with classical filtering. Therefore, a clear-cut statement concerning the advantage of one or the other is hard to make outside a specific application domain. This holds true even for the Toeplitz case. Here, Hankel realization theory reduces to the classical LTI realization theory, and the displacement rank may be just 1, so that vector-matrix multiplication reduces to a single FFT. The relative computational efficiency then pitches the system's degree δ against the logarithm of the time sequence — $\ln n$, which might appear to be to the advantage of the latter. But then, not all items in the complexity calculation have been included! E.g., the 'pipeline' complexity of the FFT is again n against δ, which may be very disadvantageous in concrete cases. And if selective accuracy is included in the considerations, then the length of the FFT and the wordlength to be used may just be impractical.

4 DIAGONAL ALGEBRA

In the theory of discrete time systems, there are two classes of "most elementary" operators, namely instantaneous or non-dynamic operators which affect only the current input and leave the state undisturbed, and "simple shifts" (unit delays). In our notation, the first class corresponds to diagonal transfer operators (block diagonal matrices, elements of $\mathcal{D}$ or matrices for which only the main diagonal is non-zero), whereas simple shifts are represented by Z: a matrix whose only non zero block-entries are identity matrices on the first off-diagonal. With these two basic components, we can set up a "diagonal algebra" which yields expressions that look like those of classical time-invariant system theory. Many results from that theory carry over straightforwardly as well: the notation is not just cosmetically interesting.

In chapter 3, we have looked at systems T that map input sequences in $\ell_2^{\mathcal{M}}$ to output sequences in $\ell_2^{\mathcal{N}}$, and we have briefly considered the use of stackings of such sequences into stacked spaces $\mathcal{X}_2^{\mathcal{M}} := \mathcal{X}_2(\mathbb{C}^{\mathbb{Z}}, \mathcal{M})$. Interestingly, such a generalized input sequence can be brought into the elementary scheme of algebra of diagonals, by viewing an element of $\mathcal{X}_2(\mathbb{C}^{\mathbb{Z}}, \mathcal{M})$ simply as a (row) sequence of *diagonals*. Based on this idea, we set up a non-commutative algebra in which diagonals play the role of scalars and the Hilbert space of ℓ_2-sequences becomes a Hilbert space *module* of sequences of diagonals (*cf.* [GH77]). In the same way, the scalar Hilbert space inner product translates to a diagonal inner product in the Hilbert space module. The idea of using such a diagonal algebra originated in the papers of Alpay, Dewilde and Dym [ADD90]. We omit the (standard) proof that an *algebra* is obtained, and confine ourselves to proving the properties that we actually need.

In this chapter, we introduce the necessary algebraic background and concepts, so that we can focus on the system theoretical consequences in chapter 5 and further.

4.1 SEQUENCES OF DIAGONALS

Collections of signals

Let $\mathbb{C}^{\mathbb{Z}}$ denote a doubly infinite sequence of one dimensional copies of the complex plane $\mathbb{C}$, and let us consider $\mathcal{X}_2(\mathbb{C}^{\mathbb{Z}},\mathcal{M})$ for some input space sequence $\mathcal{M}$. If $U \in \mathcal{X}_2(\mathbb{C}^{\mathbb{Z}},\mathcal{M})$, then each row in U is a sequence in $\ell_2^{\mathcal{M}}$. An operator T mapping $\ell_2^{\mathcal{M}}$ to $\ell_2^{\mathcal{N}}$ can easily be extended to an operator mapping $\mathcal{X}_2(\mathbb{C}^{\mathbb{Z}},\mathcal{M})$ to $\mathcal{X}_2(\mathbb{C}^{\mathbb{Z}},\mathcal{N})$: just let T act on individual rows of U, in agreement with the matrix representation of UT. In this way, T is upgraded to an operator which maps one Hilbert-Schmidt space to another, with the same norm $\|T\|$. Note that T is quite a special operator on $\mathcal{X}_2$-spaces: it is described by a matrix representation with only two indices.

For simplicity of notation, we will write from now on

$$\begin{aligned} \mathcal{X}_2^{\mathcal{M}} &= \mathcal{X}_2(\mathbb{C}^{\mathbb{Z}},\mathcal{M}), \\ \mathcal{L}_2^{\mathcal{M}} &= \mathcal{L}_2(\mathbb{C}^{\mathbb{Z}},\mathcal{M}), \\ \mathcal{U}_2^{\mathcal{M}} &= \mathcal{U}_2(\mathbb{C}^{\mathbb{Z}},\mathcal{M}). \end{aligned} \tag{4.1}$$

Also, we will often simply write $\mathcal{X}_2$ instead of $\mathcal{X}_2^{\mathcal{M}}$ if the precise structure of $\mathcal{M}$ is not particularly relevant to the argument.

A second way to represent an element of $\mathcal{X}_2$ was indicated in section 2.1:

$$U \in \mathcal{X}_2: \qquad U = \sum_{-\infty}^{\infty} Z^{[k]} U_{[k]}, \qquad U_{[k]} = \mathbf{P}_0(Z^{[-k]}U).$$

Thus, U can also be viewed as a sequence of diagonals. Applying U to T, it is seen that T acts on U as it would act on an ℓ_2-sequence:

$$Y = UT = (\cdots + Z^{[-1]}U_{[-1]} + U_{[0]} + ZU_{[1]} + Z^{[2]}U_{[2]} + \cdots)T.$$

If we introduce a diagonal expansion for T as well, we can work out the expression for a diagonal $Y_{[n]}$ of the result:

$$\begin{aligned} Y_{[n]} &= \mathbf{P}_0(Z^{[-n]}UT) \\ &= \mathbf{P}_0\left(Z^{[-n]}\textstyle\sum_k (Z^{[k]}U_{[k]}\sum_i Z^{[i]}T_{[i]})\right) \\ &= \textstyle\sum_k Z^{[k-n]}U_{[k]}Z^{[n-k]}T_{[n-k]} \\ &= \textstyle\sum_k U_{[k]}^{(n-k)}T_{[n-k]}. \end{aligned}$$

This expression plays the role of convolution in the present diagonal setting. If $T \in \mathcal{U}$, then it is a causal transfer operator, and the summation runs from $k = -\infty$ to $k = n$.

D-invariance and snapshots

Since elements of an $\mathcal{X}_2$ space have natural matrix representations, general operators mapping an $\mathcal{X}_2$ space to an $\mathcal{X}_2$ space require a *tensor* representation with four indices. It turns out that most operators that we use throughout this book have a special structure, called *left D-invariance*, which allows them to be specified by at most three indices. Singling out one of these indices as a parameter (often the index corresponding to time evolution), we obtain a representation of the operator as a sequence of matrices which we can aptly name a sequence of "snapshots". We say that an operator $T : \mathcal{X}_2 \to \mathcal{X}_2$ is *left D-invariant* if for all $D \in \mathcal{D}$, $U \in \mathcal{X}_2$,

$$D(UT) = (DU)T .$$

An operator which is left D-invariant maps the rows of U independently of each other: the k-th row of $Y = UT$ depends only on the k-th row of U. To see this, it suffices to take for U an operator in $\mathcal{X}_2$ which has zero rows except possibly for the k-th row. Let D be a diagonal operator which is zero except for the k-th diagonal entry, which is taken equal to I. Then $DY = D(UT) = (DU)T = UT = Y$, which implies that Y has zero rows except for the k-th row. This can also be checked in the more formal π_k-notation of chapter 2, equation (2.3), in which $\pi_k U = u_k$, the k-th row of U; $\pi_k^*(\pi_k U) = U$, and $D = \pi_k^* \pi_k$.

Definition 4.1 *Let $T : \mathcal{X}_2 \to \mathcal{X}_2$ be a left D-invariant operator. Then T_k,*

$$T_k : \quad \ell_2 \to \ell_2 : \quad u \mapsto uT_k = \pi_k([\pi_k^* u]T) .$$

is called a snapshot of T at point k.

Note that $\pi_k^* u$ is an operator U in $\mathcal{X}_2$, whose k-th row is equal to u, and which is zero otherwise. Hence, this operator has the correct dimensions as left argument for T. Because T is left D-invariant, the resulting operator $Y = UT$ also has zero rows except for the k-th row. Denote this row by $y \in \ell_2$, then applying the definition, we obtain that $Y = UT \Leftrightarrow y = uT_k$.

More in general, we can take any $U \in \mathcal{X}_2$, break it apart into its rows $u_k = \pi_k U \in \ell_2$, apply T_k to u_k, for each k, and assemble the resulting rows $y_k = u_k T_k$ into $Y = \sum \pi_k^* y_k$. By D-invariance, the result is equal to $Y = UT$. This proves the following proposition.

Proposition 4.2 *Let $T : \mathcal{X}_2 \to \mathcal{X}_2$ be a left D-invariant operator. Then, for all $U \in \mathcal{X}_2$,*

$$Y = UT \quad \Leftrightarrow \quad y_k = u_k T_k \quad (\text{all } k).$$

Hence, the collection of snapshots $\{T_k\}$ forms a complete description of T.

Examples

As an example, consider the projection operator $\mathbf{P}$, which projects $\mathcal{X}_2$ onto $\mathcal{U}_2$. It is easily verified that $\mathbf{P}$ is a left D-invariant operator: $\mathbf{P}(DU) = D\mathbf{P}(U)$ for all $D \in \mathcal{D}$. Hence $\mathbf{P}$ has a collection of snapshots $\{\mathbf{P}_k\}$. Applying the definition, we obtain that the snapshots $\mathbf{P}_k$ are

$$\mathbf{P}_k = \begin{bmatrix} \ddots & \ddots & & & \ddots \\ \ddots & & & & \\ & \ddots & 0 & 0 & & \\ & & 0 & \underline{1} & 0 & \\ & & & 0 & 1 & \ddots \\ & \ddots & & & \ddots & \ddots \end{bmatrix} \tag{4.2}$$

where the underlined entry of $\mathbf{P}_k$ is at the (k,k)-th position. For a sequence $u \in \ell_2$, all entries in $u\mathbf{P}_k$ with index smaller than k are zero, while the other entries remain equal to the originals in the sequence u. The collection of operators $\{\mathbf{P}_k\}$ is *nested*: $\mathbf{P}_{k+1} \prec \mathbf{P}_k$ for an obvious definition of the ordering relation "$\prec$". This property can be used to describe time-varying systems [FS82].

A second, trivial example is formed by the transfer operators $T: \ell_2^{\mathcal{M}} \to \ell_2^{\mathcal{N}}$ upgraded to $T: \mathcal{X}_2(\mathbb{C}^{\mathbb{Z}}, \mathcal{M}) \to \mathcal{X}_2(\mathbb{C}^{\mathbb{Z}}, \mathcal{N})$. All its snapshots are the same, and equal to T.

A more elaborate example is the following. Let $M = N = [\cdots\ 0\ \boxed{1}\ 1\ 1\ 1\ 0\ \cdots]$ and $T: \ell_2^{\mathcal{M}} \to \ell_2^{\mathcal{N}}$ be given by

$$T = \begin{bmatrix} \boxed{1} & t_{01} & t_{02} & t_{03} \\ & 1 & t_{12} & t_{13} \\ & & 1 & t_{23} \\ & & & 1 \end{bmatrix}$$

T is an operator $\mathcal{X}_2 \to \mathcal{X}_2$ as well. Consider the operator $H_T: \mathcal{X}_2 \to \mathcal{X}_2: H_T = \mathbf{P}(\mathbf{P}'(\cdot)T)$, *i.e.*, an argument to H_T is first projected onto $\mathcal{L}_2 Z^{-1}$, subsequently multiplied by T, then projected onto $\mathcal{U}_2$. H_T is left D-invariant. Its non-zero snapshots are given by

$$H_1 = \begin{bmatrix} \boxed{0} & t_{01} & t_{02} & t_{03} \\ & 0 & 0 & 0 \\ & & 0 & 0 \\ & & & 0 \end{bmatrix}, \quad H_2 = \begin{bmatrix} \boxed{0} & 0 & t_{02} & t_{03} \\ & 0 & t_{12} & t_{13} \\ & & 0 & 0 \\ & & & 0 \end{bmatrix},$$
$$H_3 = \begin{bmatrix} \boxed{0} & 0 & 0 & t_{03} \\ & 0 & 0 & t_{13} \\ & & 0 & t_{23} \\ & & & 0 \end{bmatrix}. \tag{4.3}$$

4.2 THE DIAGONAL ALGEBRA OF $\mathcal{X}_2$

Diagonal inner product

An operator $U \in \mathcal{X}_2^{\mathcal{M}}$ consists of rows $U_i = \pi_i U \in \ell_2^{\mathcal{M}}$ such that $U = \sum_i \pi_i^* U_i$. $\mathcal{X}_2^{\mathcal{M}}$ is the direct orthogonal sum of its subspaces $\pi_i^* \pi_i \mathcal{X}_2^{\mathcal{M}}$, each of which is isomorphic to $\ell_2^{\mathcal{M}}$. If T is an operator $\ell_2^{\mathcal{M}} \to \ell_2^{\mathcal{N}}$, extended to $\mathcal{X}_2^{\mathcal{M}} \to \mathcal{X}_2^{\mathcal{N}}$, then T is left D-invariant, and the rows of U act as independent input sequences to T. Consequently, the norm of an operator T on ℓ_2 is also equal to

$$\|T\| = \sup_{U \in \mathcal{X}_2^{\mathcal{M}}} \frac{\|UT\|_{HS}}{\|U\|_{HS}}$$

In the space $\mathcal{X}_2^{\mathcal{M}}$, we define the *diagonal inner product* as [ADD90]

$$\{A,B\} := \mathbf{P}_0(AB^*) \qquad (A, B \in \mathcal{X}_2^{\mathcal{M}}). \tag{4.4}$$

This inner product takes diagonal values and plays a similar role as the scalar inner product in Hilbert spaces.[1] Some properties are $\{A,B\} \in \mathcal{D}_2(\mathcal{M},\mathcal{M})$, and $\langle A,B\rangle_{HS} = \text{trace}\{A,B\}$. The i-th entry of $\{A,B\}$ on the diagonal is equal to the ordinary inner product of ℓ_2-sequences (A_i,B_i):

$$\{A,B\} = \text{diag}[\,(A_i,B_i)\,]_{-\infty}^{\infty},$$

where $A_i = \pi_i A$ and $B_i = \pi_i B$ are the i-th rows of A and B, respectively. In particular, we have that

$$A=0 \quad \Leftrightarrow \quad \langle A,A\rangle_{HS}=0 \quad \Leftrightarrow \quad \{A,A\}=0, \tag{4.5}$$

$$\langle DA,B\rangle_{HS}=0 \quad (\text{all } D \in \mathcal{D}) \quad \Leftrightarrow \quad \{A,B\}=0. \tag{4.6}$$

Positive and contractive operators

A Hermitian operator A in $\mathcal{X}(\mathcal{M},\mathcal{M})$ is *positive*, $A \geq 0$, if for all $u \in \ell_2^{\mathcal{M}}$,

$$(uA,u) \geq 0.$$

We say that A is *strictly positive*,[2] notation $A \gg 0$, if there is an $\varepsilon > 0$ such that, for all u in $\ell_2^{\mathcal{M}}$,

$$(uA,u) \geq \varepsilon(u,u).$$

It is known that a positive operator $A \in \mathcal{X}$ is strictly positive if and only if A is boundedly invertible in $\mathcal{X}$. The above definitions can be formulated in terms of the diagonal inner product, as follows.

Proposition 4.3 *Let $A \in \mathcal{X}(\mathcal{M},\mathcal{M})$ be a bounded Hermitian operator.*

$$\begin{array}{lll} A \geq 0 & \Leftrightarrow & \text{for all } U \in \mathcal{X}_2^{\mathcal{M}}: \quad \{UA,U\} \geq 0, \\ A \gg 0 & \Leftrightarrow & \exists\, \varepsilon>0: \quad \text{for all } U \in \mathcal{X}_2^{\mathcal{M}}: \{UA,U\} \geq \varepsilon\{U,U\}. \end{array}$$

PROOF A Hermitian diagonal operator is positive if and only if all its diagonal entries are positive. Since the diagonal inner product is a diagonal of ordinary inner products: $\{UA,U\} = \text{diag}[\,(U_iA,U_i)\,]_{-\infty}^{\infty}$, where $U_i = \pi_i U$ is the i-th row of U, we have that for all $U \in \mathcal{X}_2^{\mathcal{M}}$,

$$\begin{array}{rll} \{UA,U\} \geq 0 & \Leftrightarrow & (U_iA,U_i) \geq 0 \quad (\text{all } i) \\ & \Leftrightarrow & \text{for all } V \in \mathcal{X}_2^{\mathcal{M}}: \langle VA,V\rangle_{HS} = \sum_i (V_iA,V_i) \geq 0. \end{array}$$

A similar reasoning applies to the second part of the proposition. □

Let T be an operator in $\mathcal{X}(\mathcal{M},\mathcal{N})$. T is said to be *contractive* if $y = uT \Rightarrow \|y\| \leq \|u\|$, that is, if $(uT,uT) \leq (u,u)$ for all $u \in \ell_2^{\mathcal{M}}$. T is *strictly contractive* if there is $\varepsilon > 0$ such that $(uT,uT) \leq (1-\varepsilon)(u,u)$ for all $u \in \ell_2^{\mathcal{M}}$. Hence T is contractive, respectively strictly contractive, if

$$I - TT^* \geq 0, \qquad \text{resp.} \qquad I - TT^* \gg 0.$$

[1] The diagonal inner product does not evaluate to a scalar and hence it is not an inner product in the usual Hilbert space theory, but rather in a Hilbert space module sense.

[2] More precisely, uniformly strictly positive.

Left D-invariant subspaces

Consider subspaces (*i.e.*, closed linear manifolds) of the Hilbert space $\mathcal{X}_2^{\mathcal{M}}$ with the standard Hilbert-Schmidt inner product and which satisfy the additional property of *left D-invariance*: $\mathcal{H}$ in $\mathcal{X}_2^{\mathcal{M}}$ is said to be left D-invariant if $F \in \mathcal{H} \Rightarrow DF \in \mathcal{H}$ for any diagonal $D \in \mathcal{D}(\mathbb{C}^{\mathbb{Z}}, \mathbb{C}^{\mathbb{Z}})$, *i.e.*,

$$\mathcal{D}\mathcal{H} \subset \mathcal{H}.$$

A left D-invariant subspace has the property that it falls apart naturally into a stack of independent *slices*: just as $\mathcal{X}_2^{\mathcal{M}} = \cdots \ell_2^{\mathcal{M}} \times \ell_2^{\mathcal{M}} \times \cdots$ earlier, we can write

$$\mathcal{H} = \cdots \times \mathcal{H}_0 \times \mathcal{H}_1 \times \cdots \tag{4.7}$$

where each $\mathcal{H}_i = \pi_i \mathcal{H}$ is a subspace in $\ell_2^{\mathcal{M}}$. Indeed, if $F \in \mathcal{H}$, then $DF \in \mathcal{H}$, and by taking D equal to $[D_i = 1, D_k = 0\ (k \neq i)]$, that is, $D = \pi_i^* \pi_i \in \mathcal{D}$, it follows that $\pi_i^* \pi_i F = \pi_i^* F_i \in \mathcal{H}$, hence $(\cdots \times \mathcal{H}_0 \times \mathcal{H}_1 \times \cdots) \subset \mathcal{H}$ and $F \in \mathcal{H} \Rightarrow F_i \in \mathcal{H}_i$. Conversely, $F \in \mathcal{X}_2^{\mathcal{M}}, F_i \in \mathcal{H}_i \Rightarrow F \in \mathcal{H}$ by definition of the $\mathcal{H}_i$'s. The D-invariance property implies that the $\mathcal{H}_i$ are "uncoupled": the fact that an element F of $\mathcal{H}$ has a component F_i in $\mathcal{H}_i$ does not pose conditions on other rows of F. A closely related alternative to the description (4.7) is provided by the following lemma:

Lemma 4.4 *Let $\mathcal{H} \in \mathcal{X}_2^{\mathcal{M}}$ be a left D-invariant subspace, and let $\mathcal{H}_i = \pi_i \mathcal{H} \in \ell_2^{\mathcal{M}}$. The spaces $\pi_i^* \mathcal{H}_i$ are subspaces of $\mathcal{H}$ which are pairwise orthogonal and together span $\mathcal{H}$:*

$$\mathcal{H} = \cdots \oplus \pi_0^* \mathcal{H}_0 \oplus \pi_1^* \mathcal{H}_1 \oplus \cdots.$$

PROOF An element of $\pi_i^* \mathcal{H}_i$ has all its rows equal to zero, except possibly the i-th row. $\pi_i^* \mathcal{H}_i$ is a subspace of $\mathcal{H}$ because $\pi_i^* \mathcal{H}_i = \pi_i^* \pi_i \mathcal{H} = D\mathcal{H} \subset \mathcal{H}$, where $D = \pi_i^* \pi_i \in \mathcal{D}$. $\pi_i^* \mathcal{H}_i$ is orthogonal to $\pi_j^* \mathcal{H}_j$ if $i \neq j$ because, for $F_i \in \mathcal{H}_i, F_j \in \mathcal{H}_j$, we have that $\langle \pi_i^* F_i, \pi_j^* F_j \rangle_{HS} = \text{trace}\ \pi_i^* (F_i, F_j) \pi_j$ and $\text{trace}\ \pi_i^* A \pi_j = 0\ (i \neq j)$ for all A of appropriate dimensions. The collection $\{\pi_i^* \mathcal{H}_i\}$ spans $\mathcal{H}$ because $\sum_i \pi_i^* \pi_i = I$. □

Let $\mathcal{H}$ be a left D invariant subspace in $\mathcal{X}_2^{\mathcal{M}}$. Each of its slices $\mathcal{H}_i$ is a subspace in the Hilbert space $\ell_2^{\mathcal{M}}$. Let N_i be the dimension of the subspace $\mathcal{H}_i$. If each of these dimensions is finite then we say that $\mathcal{H}$ is of *locally finite dimension*. Note that the dimension of $\mathcal{H}$ is equal to the sum of all N_i, and $\mathcal{H}$ can be a finite or infinite dimensional subspace in $\mathcal{X}_2^{\mathcal{M}}$. The index sequence $N = [N_i]_{-\infty}^{\infty}$ is called the (left) dimension sequence of the left D-invariant subspace $\mathcal{H}$, and we write

$$N = \text{sdim}\,(\mathcal{H}).$$

The orthogonal complement of a subspace $\mathcal{H}$ in $\mathcal{X}_2^{\mathcal{M}}$ is

$$\mathcal{H}^{\perp} = \{F \in \mathcal{X}_2 : \langle F, G \rangle_{HS} = 0, \text{ all } G \in \mathcal{H}\}.$$

Since $\mathcal{X}_2^{\mathcal{M}}$ is a Hilbert space, $\mathcal{H}^{\perp}$ is a subspace, and $\mathcal{H} \oplus \mathcal{H}^{\perp} = \mathcal{X}_2^{\mathcal{M}}$.

Proposition 4.5 *If $\mathcal{H}$ is a left D invariant subspace in $\mathcal{X}_2^{\mathcal{M}}$, then $\mathcal{H}^{\perp}$ is also left D invariant, and*

$$\mathcal{H}^{\perp} = \{F \in \mathcal{X}_2^{\mathcal{M}} : \{F, G\} = 0, \text{ all } G \in \mathcal{H}\}.$$

PROOF A straightforward proof uses (4.6) twice. Let $F \in \mathcal{H}^\perp$, $G \in \mathcal{H}$, then the D-invariance property of $\mathcal{H}$ implies

$$\langle F, DG\rangle_{HS} = 0 \quad (\text{all } D \in \mathcal{D}) \quad \Leftrightarrow \quad \{F, G\} = 0 \quad \Leftrightarrow \quad \langle DF, G\rangle_{HS} = 0 \quad (\text{all } D \in \mathcal{D})$$

so that $DF \in \mathcal{H}^\perp$. □

Consequently, $\mathcal{H}^\perp$ also falls apart into subspaces $(\mathcal{H}^\perp)_i$, and it is easy to show that $(\mathcal{H}^\perp)_i = (\mathcal{H}_i)^\perp$, so that the orthogonal complement of a left D-invariant subspace $\mathcal{H}$ consists of the complement of its slices $\mathcal{H}_i$.

We list some more straightforward properties of D-invariant subspaces.

- If $\mathcal{A}$ and $\mathcal{B}$ are left D-invariant subspaces, then so are $\mathbf{P}_{\mathcal{A}}(\mathcal{B})$ and $\mathbf{P}_{\mathcal{A}^\perp}(\mathcal{B})$, the projections of $\mathcal{B}$ onto $\mathcal{A}$ and $\mathcal{A}^\perp$, respectively.
- If $\mathcal{A}$ or $\mathcal{B}$ is locally finite, then so is $\mathbf{P}_{\mathcal{A}}(\mathcal{B})$.
- If two linearly independent subspaces $\mathcal{A}$ and $\mathcal{B}$ of $\mathcal{X}_2^{\mathcal{M}}$ are locally finite, then so is their direct sum $\mathcal{A} \dotplus \mathcal{B}$.
- If $\mathcal{A}$ is a left D-invariant subspace and $B \in \mathcal{X}$ is a bounded linear operator, then $\overline{\mathcal{A}B}$ is also a left D-invariant subspace, with

$$\text{sdim}\,(\overline{\mathcal{A}B}) \leq \text{sdim}\,(\mathcal{A})\,. \tag{4.8}$$

(The overbar denotes closure.)

4.3 SLICED BASES AND PROJECTIONS IN $\mathcal{X}_2$

Sliced bases of locally finite subspaces

Let $\mathcal{H}$ be a left D-invariant subspace of $\mathcal{X}_2^{\mathcal{M}}$. Since $\mathcal{X}_2^{\mathcal{M}}$ is separable in the Hilbert-Schmidt metric, $\mathcal{H}$ has an orthonormal basis. We have seen that $\mathcal{H}$ falls apart into slices $\mathcal{H}_i = \pi_i \mathcal{H}$, which are subspaces in $\ell_2^{\mathcal{M}}$. If each of these subspaces has finite dimension (N_i, say), then $\mathcal{H}$ is by definition locally finite. In this section, we consider special basis representations for such subspaces which are consistent with the sliced structure.

Let $\mathcal{H}_i$ have an orthonormal basis $\{(q_i)_1, \cdots, (q_i)_{N_i}\}$, with each $(q_i)_j \in \ell_2^{\mathcal{M}}$. Because of lemma 4.4, an orthonormal basis of $\mathcal{H}$ is the set $\{\pi_i^*(q_i)_j\}$ $(j = 1, \cdots, N_i,\ i = -\infty, \cdots, \infty)$. It is notationally convenient to collect the set of $(q_i)_j$ into one operator $\mathbf{Q}$. This is done in two steps.

- Stack $\{(q_i)_j\}_{j=1..N_i}$ as one operator $\mathbf{Q}_i \in [\mathbb{C}^{N_i} \to \ell_2^{\mathcal{M}}]$. Note that $\Lambda_i = \mathbf{Q}_i \mathbf{Q}_i^*$ is well defined, it is the Gram matrix of the basis of $\mathcal{H}_i$. In the current situation, the basis is orthonormal and $\Lambda_i = I$. The subspace $\mathcal{H}_i$ is generated by the basis operator $\mathbf{Q}_i$ in the sense that $\mathcal{H}_i = \mathbb{C}^{N_i}\mathbf{Q}_i$: it consists of all linear combinations of the $(q_i)_j$.
- Stack the $\mathbf{Q}_i$ further as one operator

$$\mathbf{Q} = \sum_i \pi_i^* \mathbf{Q}_i \tag{4.9}$$

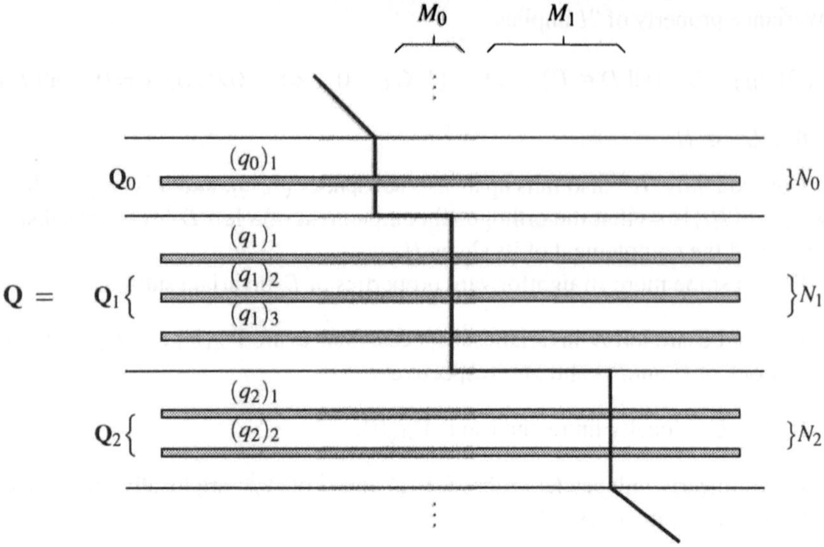

Figure 4.1. Basis representation $\mathbf{Q}$ of some subspace in $\mathcal{X}_2$.

with rows $\pi_i \mathbf{Q} = \mathbf{Q}_i$. See figure 4.1. Usually, we have $\mathcal{H} \subset \mathcal{U}_2$ or $\mathcal{H} \subset \mathcal{L}_2 Z^{-1}$, which is signified by the diagonal staircase line in the figure (for these subspaces, the basis vectors are zero at the left, resp. right of the line).

We call $\mathbf{Q}$ an (orthonormal) sliced basis representation of the given basis of $\mathcal{H}$. A number of properties of such a basis operator are listed below.

Proposition 4.6 *Let $\mathcal{H}$ be a locally finite D-invariant subspace in $\mathcal{X}_2^{\mathcal{M}}$, with* sdim $(\mathcal{H}) = N$, *and let $\mathbf{Q}$ be an orthonormal sliced basis representation for $\mathcal{H}$. Let $\mathcal{N} = \mathbb{C}^N$. Then any $F \in \mathcal{H}$ can be uniquely written as*

$$F = D_F \mathbf{Q},$$

for a certain $D_F \in \mathcal{D}_2^{\mathcal{N}}$. In particular, $\mathbf{Q}$ is bounded on $\mathcal{D}_2^{\mathcal{N}}$ and generates $\mathcal{H}$ via

$$\mathcal{H} = \mathcal{D}_2^{\mathcal{N}} \mathbf{Q}.$$

PROOF Let us start from the orthonormal basis $\{(q_i)_1, \cdots, (q_i)_{N_i}\}$ of each $\mathcal{H}_i$. Because $\{\pi_i^*(q_i)_j\}$ $(j = 1, \cdots, N_i,\ i = -\infty, \cdots, \infty)$ is a basis of $\mathcal{H}$, any $F \in \mathcal{H}$ can be written as the linear combination of the basis sequences

$$F = \sum_{i,j} (\alpha_i)_j \cdot \pi_i^* (q_i)_j , \tag{4.10}$$

where the coefficients $(\alpha_i)_j$ are uniquely determined by F and $\sum_{ij} |(\alpha_i)_j|^2 = \| F \|_{HS}^2 < \infty$. Using $\mathbf{Q}_i$, equation (4.10) becomes

$$F = \sum_i \alpha_i \cdot \pi_i^* \mathbf{Q}_i , \tag{4.11}$$

where $\alpha_i = [(\alpha_i)_1, \cdots, (\alpha_i)_{N_i}] \in \mathbb{C}^{1\times N_i}$ satisfies $\sum_i \|\alpha_i\|_2^2 < \infty$. In terms of $\mathbf{Q}$, equation (4.11) in turn becomes

$$F = D_F \mathbf{Q}, \qquad D_F = \mathrm{diag}[\alpha_i]_{-\infty}^{\infty} \in \mathcal{D}_2^{\mathcal{N}}, \tag{4.12}$$

so that $\mathcal{H} = \mathcal{D}_2^{\mathcal{N}}\mathbf{Q}$. The expression $\mathcal{H} = \mathcal{D}_2^{\mathcal{N}}\mathbf{Q}$ shows that $\mathbf{Q}$ is bounded as a $[\mathcal{D}_2 \to \mathcal{X}_2]$ operator. □

Example

Let $M = N = [\cdots\ 0\ \boxed{1}\ 1\ 1\ 1\ 0\ \cdots]$ and $T : \ell_2^{\mathcal{M}} \to \ell_2^{\mathcal{N}}$ be given by

$$T = \begin{bmatrix} \boxed{1} & \alpha_1 & \alpha_1\alpha_2 & \alpha_1\alpha_2\alpha_3 \\ & 1 & \alpha_2 & \alpha_2\alpha_3 \\ & & 1 & \alpha_3 \\ & & & 1 \end{bmatrix}$$

Consider $\mathcal{H} = \mathbf{P}(\mathcal{L}_2 Z^{-1} T)$. (This type of subspace will be frequently used in the following chapters.) $\mathcal{H}$ is a left D-invariant subspace. To obtain a basis representation for $\mathcal{H}$, we first look at its slices $\mathcal{H}_i$. Note that, by D-invariance, $\pi_i^* \mathcal{H}_i = \mathbf{P}(\pi_i^* \pi_i \mathcal{L}_2 Z^{-1} T)$, and that $\pi_i \mathcal{L}_2 Z^{-1}$ is the subspace in $\ell_2^{\mathcal{M}}$ consisting of sequences that are zero from entry i on. After multiplying with T, the resulting sequences are acted upon by $\mathbf{P}(\pi_i^* \cdot)$, whose action can also be described as setting al entries before point i equal to zero, and embedding the result in $\mathcal{X}_2^{\mathcal{N}}$. It is clear that only $\mathcal{H}_1, \cdots, \mathcal{H}_3$ are nonzero. These subspaces are given by

$$\begin{aligned} \mathcal{H}_1 &= \mathrm{row}(H_1), & H_1 &= \begin{bmatrix} \alpha_1 & \alpha_1\alpha_2 & \alpha_1\alpha_2\alpha_3 \end{bmatrix} \\ \mathcal{H}_2 &= \mathrm{row}(H_2), & H_2 &= \begin{bmatrix} \alpha_1\alpha_2 & \alpha_1\alpha_2\alpha_3 \\ \alpha_2 & \alpha_2\alpha_3 \end{bmatrix} \\ \mathcal{H}_3 &= \mathrm{row}(H_3), & H_3 &= \begin{bmatrix} \alpha_1\alpha_2\alpha_3 \\ \alpha_2\alpha_3 \\ \alpha_3 \end{bmatrix} \end{aligned}$$

The connection with the snapshots of H_T in (4.3) is not coincidental, and will be worked out in chapter 5. Assuming $\alpha_i \neq 0$, we thus have sdim $(\mathcal{H}) = [\cdots\ \boxed{0}\ 1\ 1\ 1\ 0\ \cdots]$, and an unnormalized basis for $\mathcal{H}$ is given by

$$\mathbf{Q} = \left[\begin{array}{cccc} \boxed{\cdot} & \cdot & \cdot & \cdot \\ \hline & 1 & \alpha_2 & \alpha_2\alpha_3 \\ \hline & & 1 & \alpha_3 \\ \hline & & & 1 \end{array}\right],$$

suppressing the remaining empty dimensions.

Boundedness and computing rules for a sliced basis representation

$\mathbf{Q}$ can be viewed as an operator from (a domain in) $\mathcal{X}_2^{\mathcal{N}}$ to $\mathcal{X}_2^{\mathcal{M}}$, but it is not necessarily a bounded operator. A simple example of an unbounded $\mathbf{Q}$ is obtained by taking $\mathbf{Q}_i = [\cdots 0 \ \boxed{1} \ 0 \cdots]$ (all i), so that

$$u\mathbf{Q} = [\cdots u_{-1} \quad u_0 \quad u_1 \cdots] \begin{bmatrix} & \vdots & \\ \mathbf{0} & 1 & \mathbf{0} \\ & \vdots & \end{bmatrix} = [\cdots 0 \quad \sum_{-\infty}^{\infty} u_i \quad 0 \cdots],$$

which can be infinite since an ℓ_2-sequence need not be summable (as is demonstrated by the sequence $[1, \frac{1}{2}, \frac{1}{3}, \cdots]$). Although it is usually enough to consider $\mathbf{Q}$ with domain restricted to $\mathcal{D}_2$, sometimes we need properties which seem to involve a more general domain, and we derive such properties below. (A reader not interested in these details can continue with proposition 4.7.)

To start, note that along with $\mathbf{Q}$, operators $D\mathbf{Q}$ and $\mathbf{Q}X$ ($D \in \mathcal{D}, X \in \mathcal{X}$) are also bounded $[\mathcal{D}_2 \to \mathcal{X}_2]$ operators since $D\mathcal{D}_2 \in \mathcal{D}_2$, $\mathcal{X}_2 X \in \mathcal{X}_2$. The domain of definition of $\mathbf{Q}$ can be extended: for example, the application of $\mathbf{Q}$ on elements of the type $\mathcal{D}_2 Z$ is consistently defined via

$$D(Z\mathbf{Q}) = Z(D^{(1)}\mathbf{Q}), \tag{4.13}$$

and can be consistently extended $\mathbf{Q}$ (though not necessarily in a bounded fashion), to all finite sums of terms of the type $DZ^{[k]}\mathbf{Q}$. Hence $\mathbf{Q}$ is densely defined on $\mathcal{X}_2^{\mathcal{N}}$ by extension.

We have defined, see (2.8), the operator $\mathbf{P}_0$ on $\mathcal{X}_2$ as the projection onto $\mathcal{D}_2$. We have already extended $\mathbf{P}_0$ to operators in $\mathcal{X}$: $\mathbf{P}_0(X) = \operatorname{diag}[X_{ii}] \in \mathcal{D}$, where $X_{ii} = \pi_i X \pi_i^*$ is bounded for each i. $\mathbf{P}_0$ can also be extended to unbounded operators that are bounded as $[\mathcal{D}_2 \to \mathcal{X}_2]$ operators: because $\pi_i^*\pi_i \in \mathcal{D}_2$ and hence $\pi_i^*\pi_i\mathbf{Q} \in \mathcal{X}_2$, $\mathbf{Q}_{ii} = \pi_i\mathbf{Q}\pi_i^* = \pi_i(\pi_i^*\pi_i\mathbf{Q})\pi_i^*$ is uniformly bounded over i. Thus $\mathbf{P}_0(\mathbf{Q}) = \operatorname{diag}(\mathbf{Q}_{ii})$ is well defined and bounded: $\mathbf{P}_0(\mathbf{Q}) \in \mathcal{D}$. The extension satisfies the usual homogeneity rule for $\mathbf{P}_0$: if $D_{1,2} \in \mathcal{D}$, then $\mathbf{P}_0(D_1\mathbf{Q}D_2) = D_1\mathbf{P}_0(\mathbf{Q})D_2$.

As a bounded operator $\mathbf{Q} : \mathcal{D}_2^{\mathcal{N}} \to \mathcal{X}_2^{\mathcal{M}}$, $\mathbf{Q}$ has a bounded adjoint: $\mathbf{Q}^a : \mathcal{X}_2^{\mathcal{M}} \to \mathcal{D}_2^{\mathcal{N}}$. But also as a (possibly unbounded) operator $[\mathcal{X}_2^{\mathcal{N}} \to \mathcal{X}_2^{\mathcal{M}}]$, $\mathbf{Q}$ has an (unbounded) adjoint $\mathbf{Q}^* : \mathcal{X}_2^{\mathcal{M}} \to \mathcal{X}_2^{\mathcal{N}}$, see [AG81, §44]. It is defined as follows: let $\operatorname{dom}(\mathbf{Q})$ be the domain of $\mathbf{Q}$ in $\mathcal{X}_2^{\mathcal{N}}$. The domain of $\mathbf{Q}^*$ consists of all elements $G \in \mathcal{X}_2^{\mathcal{M}}$ for which there is a $F' \in \mathcal{X}_2^{\mathcal{N}}$ such that for every $F \in \operatorname{dom}(\mathbf{Q})$,

$$\langle F\mathbf{Q}, G\rangle_{HS} = \langle F, F'\rangle_{HS}, \tag{4.14}$$

and we write $F' = G\mathbf{Q}^*$. The existence of $\mathbf{Q}^*$ implies symmetrically: if $F \in \operatorname{dom}(\mathbf{Q})$ then for all $G \in \operatorname{dom}(\mathbf{Q}^*)$ and $F' = G\mathbf{Q}^* \in \mathcal{X}_2^{\mathcal{N}}$ we have that $\langle F\mathbf{Q}, G\rangle_{HS} = \langle F, F\mathbf{Q}^*\rangle_{HS}$. Restricting F to $\mathcal{D}_2 \subset \operatorname{dom}(\mathbf{Q})$ on which it is a bounded operator, and since then it is true (for any F') that

$$\langle F, F'\rangle_{HS} = \langle F, \mathbf{P}_0(F')\rangle_{HS},$$

we have $\langle F\mathbf{Q}, G\rangle_{HS} = \langle F, \mathbf{P}_0(G\mathbf{Q}^*)\rangle_{HS}$, and hence $\mathbf{P}_0(\cdot\,\mathbf{Q}^*)$ is the adjoint operator of [$\mathbf{Q}$ restricted to $\mathcal{D}_2^{\mathcal{N}}$]. Since the latter operator is bounded, its adjoint is a bounded $[\mathcal{X}_2^{\mathcal{M}} \to$

$\mathcal{D}_2^{\mathcal{N}}]$ operator. The $\mathbf{Q}^a$ alluded to earlier is in fact given by $\mathbf{Q}^a = \mathbf{P}_0(\cdot\mathbf{Q}^*)$, so that $\mathbf{Q}^*$ can be viewed as a natural extension of $\mathbf{Q}^a$.

As a corollary, $\mathbf{P}_0(\cdot\mathbf{Q}\mathbf{Q}^*)$ is a bounded $[\mathcal{D}_2^{\mathcal{N}} \to \mathcal{D}_2^{\mathcal{N}}]$ operator, hence

$$\Lambda_{\mathbf{Q}} := \mathbf{P}_0(\mathbf{Q}\mathbf{Q}^*) \in \mathcal{D}(\mathcal{N},\mathcal{N})$$

is well defined by the extension of the domain of $\mathbf{P}_0$ discussed earlier. The operator $\Lambda_{\mathbf{Q}}$ is the Gram operator of the basis $\{(\pi_i^* q_i)_j\}$ of $\mathcal{H}$. It is a diagonal operator whose entries $\Lambda_i = \mathbf{Q}_i\mathbf{Q}_i^*$ contain the Gram matrices of the bases of the subspaces $\mathcal{H}_i$ of $\mathcal{H}$. Because these bases have been chosen orthonormal, $\Lambda_{\mathbf{Q}} = I$.

Finally, using the definition (4.13), the adjoint of $\cdot(Z\mathbf{Q})$ restricted to $\mathcal{D}_2$ is formally equal to $\mathbf{P}_0(Z^{-1}\cdot\mathbf{Q}^*)^{(-1)}$: let $D \in \mathcal{D}_2$, $X \in \mathcal{X}_2$, then

$$\begin{aligned} \{DZ\mathbf{Q},X\} &= \{ZD^{(1)}\mathbf{Q},X\} \\ &= \{D^{(1)}\mathbf{Q},Z^{-1}X\}^{(-1)} \\ &= \{D^{(1)},\mathbf{P}_0(Z^{-1}X\mathbf{Q}^*)\}^{(-1)} \\ &= \{D,\mathbf{P}_0(Z^{-1}X\mathbf{Q}^*)^{(-1)}\}. \end{aligned} \tag{4.15}$$

The computing rules on unbounded basis operators introduced so far are sufficient for our purposes. The importance of such basis representations is illustrated by the following proposition.

Proposition 4.7 *Let $\mathcal{H}$ be a locally finite D-invariant subspace in $\mathcal{X}_2^{\mathcal{M}}$, and let $\mathbf{Q}$ be a sliced basis representation of $\mathcal{H}$. Let $F \in \mathcal{H}$, then*

$$F = \mathbf{P}_0(F\mathbf{Q}^*)\,\mathbf{Q} = \{F,\mathbf{Q}\}\mathbf{Q}.$$

PROOF Let $N = \text{sdim } \mathcal{H}$ and $\mathcal{N} = \mathbb{C}^N$. According to (4.12), any element F of $\mathcal{H}$ has a representation $F = D_F\,\mathbf{Q}$ in terms of $\mathbf{Q}$, where $D_F \in \mathcal{D}_2^{\mathcal{N}}$. The diagonal of coefficients D_F is obtained as

$$D_F = \mathbf{P}_0(F\mathbf{Q}^*).$$

Since $F \in \mathcal{X}_2^{\mathcal{M}}$, we have indeed that $D_F \in \mathcal{D}_2^{\mathcal{N}}$. □

Non-orthogonal bases of locally finite subspaces

The preceding discussion can be generalized to non-orthonormal bases. Again, let $\mathcal{H}$ be a locally finite left D-invariant subspace in $\mathcal{X}_2^{\mathcal{M}}$. $\mathcal{H}$ falls apart into subspaces $\mathcal{H}_i = \pi_i\mathcal{H}$ with finite dimensions N_i. For each i, let $\{(f_i)_1,\cdots,(f_i)_{N_i}\}$ be a complete system of vectors whose Gram matrix $\Lambda_i = [((f_i)_j,(f_i)_k)]_{j,k=1}^{N_i}$ is bounded and boundedly invertible. The total collection $\{\pi_i^*(f_i)_j\}$ ($j = 1,\cdots,N_i$, all i) is called a *Riesz basis* of $\mathcal{H}$. The condition on Λ is equivalent to demanding that it be strictly positive: $\Lambda \gg 0$. We call such a basis a *strong sliced basis*. For such a strong sliced basis, we can construct operators $\mathbf{F}_i$ and stack them in an operator $\mathbf{F}$ in the same way as before. We obtain similar results: $\mathbf{F}$ generates $\mathcal{H}$ via

$$\mathcal{H} = \mathcal{D}_2^{\mathcal{N}}\mathbf{F},$$

it may be an unbounded operator, densely defined on $\mathcal{X}_2^{\mathcal{N}}$, but it is bounded as a $[\mathcal{D}_2^{\mathcal{N}} \to \mathcal{X}_2^{\mathcal{M}}]$ operator, and its adjoint $\mathbf{F}^*$ exists in $\mathcal{X}_2$, which in general may be unbounded as well. The operator $\mathbf{P}_0(\cdot\,\mathbf{F}^*) : \mathcal{X}_2^{\mathcal{M}} \to \mathcal{D}_2^{\mathcal{N}}$ is well defined and bounded, and is the adjoint of $\mathbf{F}$ with domain restricted to $\mathcal{D}_2^{\mathcal{N}}$. Consequently, the operator $\Lambda_{\mathbf{F}} = \mathbf{P}_0(\mathbf{F}\mathbf{F}^*)$ is in $\mathcal{D}(\mathcal{N},\mathcal{N})$, and is equal to the Gram operator Λ of the chosen basis:

$$\Lambda_{\mathbf{F}} = \mathbf{P}_0(\mathbf{F}\mathbf{F}^*) = \operatorname{diag}[\Lambda_i]\,.$$

If $\mathbf{Q}$ is an orthonormal sliced basis representation of $\mathcal{H}$, then $\mathbf{F}$ can be expressed in terms of $\mathbf{Q}$:

$$\mathbf{F} = R\mathbf{Q}\,, \qquad R \in \mathcal{D}(\mathcal{N},\mathcal{N})\,,$$

where R is given explicitly as $R = \mathbf{P}_0(\mathbf{F}\mathbf{Q}^*)$.

If $\mathbf{F}$ is a given strong sliced basis representation, then it can be orthonormalized by factoring $\Lambda_{\mathbf{F}}$ into invertible factors R as

$$\Lambda_{\mathbf{F}} = \mathbf{P}_0(\mathbf{F}\mathbf{F}^*) =: RR^*\,.$$

Since $\Lambda_{\mathbf{F}} \gg 0$, this is always possible. The orthonormal sliced basis representation $\mathbf{Q}$ is given by $\mathbf{Q} = R^{-1}\mathbf{F}$; indeed

$$\Lambda_{\mathbf{Q}} = \mathbf{P}_0(R^{-1}\mathbf{F}\mathbf{F}^*R^{-*}) = R^{-1}\mathbf{P}_0(\mathbf{F}\mathbf{F}^*)R^{-*} = I\,.$$

Orthogonal projection onto subspaces

Using the sliced representation for left D-invariant subspaces, we now turn our attention to the projection onto subspaces. We shall need the following proposition.

Proposition 4.8 *Let $\mathcal{H}$ be a locally finite left D-invariant subspace in $\mathcal{X}_2^{\mathcal{M}}$, and let $\mathbf{Q}$ be an orthonormal sliced basis representation of $\mathcal{H}$, then (for $X \in \mathcal{X}_2^{\mathcal{M}}$),*

$$X \perp \mathcal{H} \qquad \Leftrightarrow \qquad \mathbf{P}_0(X\mathbf{Q}^*) = 0\,.$$

PROOF Any Y in $\mathcal{H}$ can be written as $Y = D\mathbf{Q}$, for some $D \in \mathcal{D}_2$. Then $X \perp Y \Leftrightarrow \{X,Y\} = \mathbf{P}_0(XY^*) = 0$, and $\mathbf{P}_0(XY^*) = \mathbf{P}_0(X\mathbf{Q}^*D^*) = \mathbf{P}_0(X\mathbf{Q}^*)D^*$. Letting Y range over $\mathcal{H}$, this expression is zero for all D in $\mathcal{D}_2$, and it follows that $\mathbf{P}_0(X\mathbf{Q}^*) = 0$. □

Let $\mathcal{H}$ be a subspace in $\mathcal{X}_2^{\mathcal{M}}$. Then $\mathcal{X}_2^{\mathcal{M}} = \mathcal{H} \oplus \mathcal{H}^{\perp}$, so that every $X \in \mathcal{X}_2^{\mathcal{M}}$ can be written (uniquely) as $X = X_1 + X_2$, where $X_1 \in \mathcal{H}$ and $X_2 \in \mathcal{H}^{\perp}$. The operator of (orthogonal) projection onto $\mathcal{H}$ is defined as $\mathbf{P}_{\mathcal{H}}(X) = X_1$.

Theorem 4.9 *Let $\mathcal{H}$ be a locally finite left D-invariant subspace in $\mathcal{X}_2^{\mathcal{M}}$, and let $\mathbf{Q}$ be a sliced orthonormal basis representation of $\mathcal{H}$. The orthogonal projection of any $X \in \mathcal{X}_2^{\mathcal{M}}$ onto $\mathcal{H}$ is given by*

$$\mathbf{P}_{\mathcal{H}}(X) = \mathbf{P}_0(X\mathbf{Q}^*)\mathbf{Q}\,. \tag{4.16}$$

PROOF Let $X = X_1 + X_2$, where $X_1 = \mathbf{P}_{\mathcal{H}}(X) \in \mathcal{H}$ and $X_2 \in \mathcal{H}^{\perp}$. Then

$$\begin{aligned} \mathbf{P}_0(X\mathbf{Q}^*)\,\mathbf{Q} &= \mathbf{P}_0((X_1+X_2)\mathbf{Q}^*)\,\mathbf{Q} \\ &= \mathbf{P}_0(X_1\mathbf{Q}^*)\,\mathbf{Q} + \mathbf{P}_0(X_2\mathbf{Q}^*)\,\mathbf{Q} \\ &= \mathbf{P}_0(X_1\mathbf{Q}^*)\,\mathbf{Q} && \text{[prop. 4.8]} \\ &= X_1 && \text{[prop. 4.7]} \end{aligned}$$

Hence $\mathbf{P}_{\mathcal{H}}(X) = \mathbf{P}(X\mathbf{Q}^*)\,\mathbf{Q}$. □

Corollary 4.10 *Let $\mathcal{H}$ be a locally finite left D-invariant subspace in $\mathcal{X}_2^{\mathcal{M}}$, and let* $\mathbf{F}$ *be a strong sliced basis representation of $\mathcal{H}$. The orthogonal projection of $X \in \mathcal{X}_2^{\mathcal{M}}$ onto $\mathcal{H}$ is given by*

$$\mathbf{P}_{\mathcal{H}}(X) = \mathbf{P}_0(X\mathbf{F}^*)\,\Lambda_{\mathbf{F}}^{-1}\,\mathbf{F}. \qquad (4.17)$$

PROOF If $\mathbf{F}$ is a strong sliced basis representation generating $\mathcal{H}$, then $\mathbf{F} = R\mathbf{Q}$, where $\mathbf{Q}$ is an orthonormal basis representation and $R \in \mathcal{D}$ is any boundedly invertible factor of $\Lambda_{\mathbf{F}} = RR^*$. Inserting $\mathbf{Q} = R^{-1}\mathbf{F}$ in (4.16), the result is obtained. □

Equation (4.17) generalizes the classical projection formula to the present diagonal algebra context. As in the classical use, an operator $\mathbf{P}$ defined everywhere on $\mathcal{X}_2$ is an orthogonal projector if and only if it is idempotent and Hermitian: $\mathbf{PP} = \mathbf{P}$, $\mathbf{P}^* = \mathbf{P}$. These properties are readily verified for the definition in (4.16):

$\mathbf{P}_{\mathcal{H}}$ is idempotent since

$$\begin{aligned} \mathbf{P}_{\mathcal{H}}\left(\mathbf{P}_{\mathcal{H}}(X)\right) &= \mathbf{P}_0\left(\mathbf{P}_0(X\mathbf{Q}^*)\cdot\mathbf{Q}\mathbf{Q}^*\right)\cdot\mathbf{Q} = \\ & \mathbf{P}_0(X\mathbf{Q}^*)\,\mathbf{P}_0(\mathbf{Q}\mathbf{Q}^*)\cdot\mathbf{Q} = \mathbf{P}_0(X\mathbf{Q}^*)\cdot\mathbf{Q} = \mathbf{P}_{\mathcal{H}}(X). \end{aligned}$$

$\mathbf{P}_{\mathcal{H}}$ is Hermitian if $\{\mathbf{P}_{\mathcal{H}}(A), B\} = \{A, \mathbf{P}_{\mathcal{H}}(B)\}$ for all $A, B \in \mathcal{X}_2$. Expanding the first term yields

$$\{\mathbf{P}_{\mathcal{H}}(A), B\} = \mathbf{P}_0\left(\mathbf{P}_0(A\mathbf{Q}^*)\cdot\mathbf{Q}B^*\right) = \mathbf{P}_0(A\mathbf{Q}^*)\,\mathbf{P}_0(\mathbf{Q}B^*).$$

The second term is equal to

$$\begin{aligned} \{A, \mathbf{P}_{\mathcal{H}}(B)\} &= \mathbf{P}_0(A\,[\mathbf{P}_0(B\mathbf{Q}^*)\cdot\mathbf{Q}]^*) \\ &= \mathbf{P}_0\,(A\mathbf{Q}^*\,\mathbf{P}_0(\mathbf{Q}B^*)) \\ &= \mathbf{P}_0(A\mathbf{Q}^*)\,\mathbf{P}_0(\mathbf{Q}B^*). \end{aligned}$$

Hence $\mathbf{P}_{\mathcal{H}}$ is Hermitian.

5 OPERATOR REALIZATION THEORY

The realization problem for time-varying systems is to find a (minimal) state space description for the input-output operator of a time-varying system, solely based on the collection of time-varying impulse responses. An important role in its solution is played by the Hankel operator, which is a restriction or suboperator of the input-output operator. It maps input signals with support in the "past" to output signals restricted to the future. Its relevance to the realization theory of time-invariant systems has been known since the early 1960s and resulted in Ho and Kalman's canonical realization algorithm in 1966 [HK66]. The fundamental properties that enable one to derive a realization are not the linearity or time invariance of the system (although these properties greatly simplify the problem), but rather its causality and the existence of a factorization of the Hankel operator into a surjective and an injective part [KFA70]. Thus, the problem of realization was brought into the algebraic context of the characterization of the Hankel operator. The algorithm derived by Ho and Kalman does not require knowledge of these invariant factors but uses the underlying structure to find the state representation of the system.

In section 3.4 we studied realization theory for the finite matrix case, and we introduced the Hankel operator as a sequence of matrices $\{H_k\}$. A direct extension of this approach to operators ("infinite matrices") is grosso modo possible but faces additional difficulties with boundedness and convergence. The operator case is nonetheless interesting: it allows us to treat large classes of matrices and operators which correspond *e.g.*, to systems that are initially time-invariant (matrices that are partially Toeplitz) or are periodically varying, and allows us also to analyze very large matrices for which only the behavior on a finite submatrix is of interest (*viz.* section 3.2). The purpose of this chapter is to extend the finite matrix approach of section 3.4 to an operator-theoretic setting. To this end we introduce concepts such as input and output state spaces, and basis representations for these spaces. These are fundamental ingredients for our realization theory and are used throughout the remaining chapters of the book. In addition, we formulate the reachability and observability operator as single (diagonal) operators rather than as an indexed collection of matrices, and connect these operators to the basis representations of the input/output state spaces. The index-free notation proves to be extremely valuable in subsequent chapters: it enables short proofs of theorems that would be burdensome otherwise. In fact, many proofs are almost carbon copies of those for the time-invariant case, with the difference that the shift-operator Z does not commute with most other operators: $AZ \neq ZA$.

5.1 THE HANKEL OPERATOR

In section 3.4, we have introduced sequences of Hankel matrices $\{H_k\}$ as submatrices of a given upper triangular matrix T. We move now to a more formal approach, which allows us to represent this sequence by a single operator H_T, which we will call the Hankel operator. It is now necessary to work not on single input sequences, but on collections of them, namely one for each point in time. In chapter 2 we have introduced the spaces $\mathcal{X}_2^{\mathcal{M}}$ as generalized input space and $\mathcal{X}_2^{\mathcal{N}}$ as generalized output space, and have indicated how the transfer operator acts between them. We define the Hankel operator as acting between subspaces of these spaces.

Definition of the Hankel operator

Using the projection operators $\mathbf{P}$ and $\mathbf{P}'$ defined in (2.8) in chapter 2, define the *past part of a signal* $U \in \mathcal{X}_2$ as its projection onto $\mathcal{L}_2Z^{-1}$: $U_p = \mathbf{P}'(U)$, and its *future part* as its projection onto $\mathcal{U}_2$: $U_f = \mathbf{P}(U)$, so that $U = U_p + U_f$. The same definitions apply to the past and future part of an output Y. For an operator $T \in \mathcal{U}$, mapping $\mathcal{X}_2^{\mathcal{M}}$ into $\mathcal{X}_2^{\mathcal{N}}$, the action of T onto $U \in \mathcal{X}_2^{\mathcal{M}}$ can then be split into three parts:

$$Y = UT \quad \Leftrightarrow \quad \left\{ \begin{array}{rcl} Y_p & = & U_pK_T \\ Y_f & = & U_pH_T + U_fE_T \end{array} \right. \tag{5.1}$$

where

$$\begin{array}{llll} K_T: & \mathcal{L}_2Z^{-1} \to \mathcal{L}_2Z^{-1}: & & U_pK_T = \mathbf{P}'(U_pT) \\ H_T: & \mathcal{L}_2Z^{-1} \to \mathcal{U}_2: & & U_pH_T = \mathbf{P}(U_pT) \\ E_T: & \mathcal{U}_2 \to \mathcal{U}_2: & & U_fE_T = \mathbf{P}(U_fT) = U_fT\,. \end{array} \tag{5.2}$$

Note that there is no transfer from U_f to Y_p, due to causality. Since T is a bounded operator and the projections are contractions on $\mathcal{X}_2$, these operators are also bounded. The operator $H_T = \mathbf{P}(\cdot\, T)\big|_{\mathcal{L}_2Z^{-1}}$ is called the Hankel operator of T: it is the map of inputs in $\mathcal{L}_2Z^{-1}$ to the part in $\mathcal{U}_2$ of the corresponding outputs. See figure 5.1(b). The operators K_T and E_T do not have special names attached to them; they will occasionally be used in later chapters.

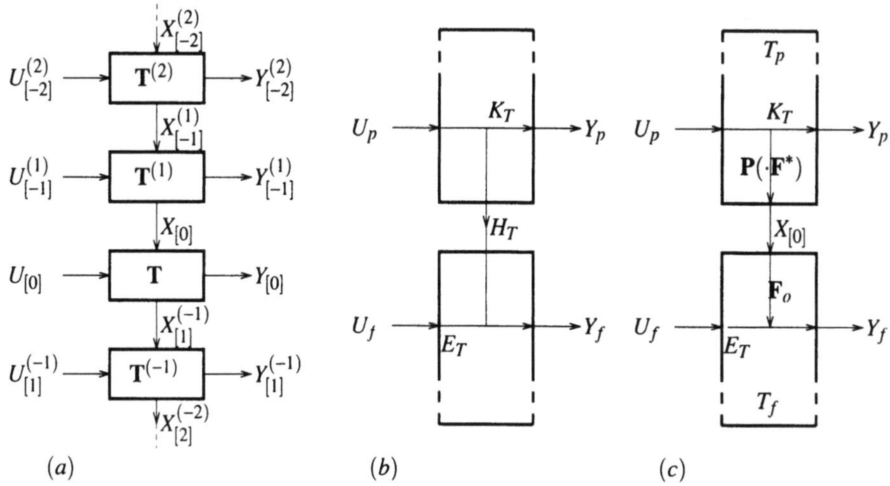

Figure 5.1. (a) realization $\mathbf{T}$ of T, (b) splitting into past and future signals, (c) representation by T_p and T_f, using the factorization of H_T in (5.17).

Definition of state spaces

In the study of the Hankel operator, the range and kernel of H_T and its adjoint $\cdot H_T^* = \mathbf{P}'(\cdot T^*)\big|_{\mathcal{U}_2}$ play a major role. Neither $\text{ran}(H_T)$ nor $\text{ran}(H_T^*)$ have to be closed. We will use these subspaces throughout the remainder of the book, and therefore we attach specific symbols to them. Some preliminary properties are derived at this point.

Let the subspaces $\mathcal{H}$ and $\mathcal{K}$ in $\mathcal{L}_2Z^{-1}$ be defined as

$$\begin{array}{ll} \text{input null space:} & \mathcal{K} = \ker(\cdot H_T) = \{U \in \mathcal{L}_2Z^{-1} : \mathbf{P}(UT) = 0\} \\ \text{input state space:} & \mathcal{H} = \text{ran}(\cdot H_T^*) = \mathbf{P}'(\mathcal{U}_2T^*)\,. \end{array} \tag{5.3}$$

$\mathcal{K}$, as a kernel, is always a closed subspace. It is readily verified that these subspaces are left D-invariant; *e.g.*, $\mathcal{K}$ is left D-invariant since for $D \in \mathcal{D}$, $\mathbf{P}(UT) = 0 \Rightarrow \mathbf{P}(DUT) = D\mathbf{P}(UT) = 0$. The kernel of a linear operator defines equivalence classes: we say that an input $U_1 \in \mathcal{L}_2Z^{-1}$ in the *past* is *Nerode equivalent* to $U_2 \in \mathcal{L}_2Z^{-1}$ if and only if they have the same *future* outputs: $\mathbf{P}(U_1T) = \mathbf{P}(U_2T)$. Consequently, $\mathbf{P}[(U_1 - U_2)T] = 0$, hence U_1 is Nerode equivalent to U_2 if $U_1 - U_2 \in \mathcal{K}$. This means that, as far as the future part of the output signal is concerned, there is no distinction between U_1 and U_2: for the purpose of computing Y_f, a collection of Nerode-equivalent signals may be represented by a single one of them. The idea underlying state realizations is that the selected signal, in turn, will be represented by a state variable in $\mathcal{D}_2$. Signals that are Nerode-equivalent are mapped to the same state.

The kernel of an operator and the closure of the range of its adjoint span the whole space on which they are defined (*cf.* (A.7)). Hence, we have that $\ker(\cdot H_T)$ is orthogonal to $\overline{\text{ran}}(\cdot H_T^*)$ and $\ker(H_T) \oplus \overline{\text{ran}}(H_T^*) = \mathcal{L}_2Z^{-1}$, or

$$\overline{\mathcal{H}} \oplus \mathcal{K} = \mathcal{L}_2Z^{-1}\,. \tag{5.4}$$

In a dual way, we define the

$$\begin{array}{ll} \text{output state space:} & \mathcal{H}_o = \operatorname{ran}(\cdot H_T) = \mathbf{P}(\mathcal{L}_2 \mathcal{Z}^{-1} T) \\ \text{output null space:} & \mathcal{K}_o = \ker(\cdot H_T^*) = \{Y \in \mathcal{U}_2 : \mathbf{P}'(YT^*) = 0\}. \end{array} \tag{5.5}$$

$\mathcal{H}_o$, as the range of H_T, is the left D-invariant manifold containing the projections onto $\mathcal{U}_2$ of all outputs of the system that can be generated from inputs in $\mathcal{L}_2\mathcal{Z}^{-1}$. $\mathcal{K}_o$ is its complement in $\mathcal{U}_2$:

$$\overline{\mathcal{H}}_o \oplus \mathcal{K}_o = \mathcal{U}_2. \tag{5.6}$$

The input and output null and state spaces satisfy the following relations:

$$\begin{array}{ll} \mathbf{P}(\mathcal{K}T) = 0, & \mathcal{H}_o = \overline{\mathcal{H}} H_T = \mathbf{P}(\overline{\mathcal{H}} T) \\ \mathbf{P}'(\mathcal{K}_o T^*) = 0, & \mathcal{H} = \overline{\mathcal{H}}_o H_T^* = \mathbf{P}'(\overline{\mathcal{H}}_o T^*). \end{array} \tag{5.7}$$

(The two equations on the right follow from inserting (5.4) and (5.6) into the definitions of $\mathcal{H}$ and $\mathcal{H}_o$, and using the two equations on the left.) These relations ensure that $\mathcal{H}$ and $\mathcal{H}_o$ have the same dimension sequences:

Proposition 5.1 *If $\overline{\mathcal{H}}$ and $\overline{\mathcal{H}}_o$ are locally finite subspaces, then*

$$\operatorname{sdim}(\overline{\mathcal{H}}) = \operatorname{sdim}(\overline{\mathcal{H}}_o).$$

PROOF Apply equation (4.8) to (5.7): $\overline{\mathcal{H}}_o = \overline{\overline{\mathcal{H}} H_T}$ and $\overline{\mathcal{H}} = \overline{\overline{\mathcal{H}}_o H_T^*}$. This yields that $\operatorname{sdim}(\overline{\mathcal{H}}_o) \le \operatorname{sdim}(\overline{\mathcal{H}})$ and $\operatorname{sdim}(\overline{\mathcal{H}}) \le \operatorname{sdim}(\overline{\mathcal{H}}_o)$. □

We will see later in this chapter that the sequence of dimensions of the state spaces is equal to the minimal system order of T.

Connection of H_T with $\{H_k\}$

A consequence of the fact that H_T is a left D-invariant operator is that the spaces $\mathcal{H}$, $\mathcal{K}$, $\mathcal{H}_o$, and $\mathcal{K}_o$ are left D-invariant; *e.g.*,

$$Y_f = U_p H_T = \mathbf{P}(U_p T) \quad \Rightarrow \quad DY_f = D\mathbf{P}(U_p T) = \mathbf{P}(DU_p T) = (DU_p) H_T.$$

As indicated in section 4.1, the operator H_T (and likewise, K_T and E_T, which are also left D-invariant operators) can be viewed as an indexed sequence of *snapshots*. According to definition 4.1, the snapshot H_k is obtained via

$$\forall U \in \mathcal{L}_2\mathcal{Z}^{-1}: \qquad (\pi_k^* U_k) H_T = \pi_k^* (U_k H_k) \tag{5.8}$$

where $U_k = \pi_k U$ is the k-th row of U. (The operator π_k was defined in (2.3) and projects elements of $\mathcal{X}_2$ onto rows in ℓ_2.) Hence H_k is an operator such that

$$Y = UH_T \quad \Leftrightarrow \quad Y_k = U_k H_k \quad (\text{all } k).$$

Since $U \in \mathcal{L}_2\mathcal{Z}^{-1}$, U_k is a sequence which has zero entries from its k-th entry on. Likewise, $Y = UH_T \in \mathcal{U}_2$ has rows Y_k which have zero entries before time k. Thus, H_k has the matrix representation

$$H_k = \left[\begin{array}{c|cccc} \mathbf{0} & \vdots & & & \\ & T_{k-3,k} & & \iddots & \\ & T_{k-2,k} & T_{k-2,k+1} & & \\ & T_{k-1,k} & T_{k-1,k+1} & T_{k-1,k+2} & \cdots \\ \hline \mathbf{0} & & \mathbf{0} & & \end{array}\right]. \tag{5.9}$$

This yields the definition of H_k of the previous section, equation (3.21), save for an (isomorphic) mirroring operation.

If we write $\mathcal{H}_k = \pi_k \mathcal{H}$ as the k-th *slice* of the D-invariant subspace $\mathcal{H}$ (as we have done in section 4.2), and similarly for $\mathcal{H}_o$, then

$$\mathcal{H}_k = \operatorname{ran}(\cdot H_k^*), \qquad (\mathcal{H}_o)_k = \operatorname{ran}(\cdot H_k). \tag{5.10}$$

5.2 REACHABILITY AND OBSERVABILITY OPERATORS

Factorization of H_T

If a u.e. stable realization $\{A,B,C,D\}$ of T is given, then from (3.23) it follows that each H_k can be factored as $H_k = \mathcal{C}_k\mathcal{O}_k$. An obvious question that emerges at this point is whether the operator H_T admits a similar factorization. The answer should be affirmative, of course, in view of the close connection between H_T and its snapshots H_k. The key is the identification of the state as an intermediate quantity through which the input-output map factors.

Recall from equation (3.11) the state equations that describe the mapping $Y = UT$ based on input and output spaces of $\mathcal{X}_2$-type:

$$\left\{\begin{array}{lcl} XZ^{-1} & = & XA + UB \\ Y & = & XC + UD \end{array}\right. \tag{5.11}$$

The state X can be written in terms of its diagonals as

$$X = \sum_{k=-\infty}^{\infty} Z^k X_{[k]}, \qquad X_{[k]} = \mathbf{P}_0(Z^{-k}X),$$

and likewise for U and Y. The recursive description of (5.11) is (*viz.* (3.12))

$$\begin{array}{rcl} X_{[k+1]}^{(-1)} & = & X_{[k]}A + U_{[k]}B \\ Y_{[k]} & = & X_{[k]}C + U_{[k]}D \end{array} \tag{5.12}$$

If $\ell_A < 1$ and $U \in \mathcal{X}_2^{\mathcal{M}}$, then (5.11) can be solved for X which leads to

$$X = UBZ(I-AZ)^{-1}.$$

Specializing to the 0-th diagonal (considering all the "presents" at once) produces

$$X_{[0]} = \mathbf{P}_0(U_p BZ(I-AZ)^{-1}). \tag{5.13}$$

Note that U_f does not play a role because $BZ(I-AZ)^{-1} \in \mathcal{U}Z$. Now assume $U = U_p$ (*i.e.*, $U_f = 0$). Equation (5.11) then implies $\mathbf{P}(XZ^{-1}) = \mathbf{P}(X)A$; $Y_f = \mathbf{P}(X)C$, so that the "future state" $X_f = \mathbf{P}(X)$ satisfies

$$\mathbf{P}(X_f Z^{-1}) = X_f A.$$

Since $\mathbf{P}(X_f Z^{-1}) = X_f Z^{-1} - X_{[0]}Z^{-1}$, we have $X_f = X_{[0]} + X_f AZ$, or

$$\begin{aligned} X_f &= X_{[0]}(I-AZ)^{-1} \\ Y_f &= X_f C = X_{[0]}(I-AZ)^{-1}C. \end{aligned} \tag{5.14}$$

Equations (5.13) and (5.14) represent a factorization of the Hankel operator. It is illustrated in figure 5.1(a) and (b): H_T is the transfer of U_p to Y_f for $U_f = 0$. The past input U_p determines the state $X_{[0]}$, which then determines the future output Y_f, provided $U_f = 0$. We reserve special symbols for the main operators in this development. Let $\mathbf{F}$ and $\mathbf{F}_o$ be defined by

$$\begin{aligned} \mathbf{F}^* &= BZ(I-AZ)^{-1} &= BZ + BZAZ + BZ(AZ)^2 + \cdots \\ \mathbf{F}_o &= (I-AZ)^{-1}C &= C + AZC + (AZ)^2C + \cdots. \end{aligned} \tag{5.15}$$

$\mathbf{F}$ is strictly lower, $\mathbf{F}_o$ is upper, and they satisfy the equations

$$\begin{aligned} \mathbf{F}^* &= \mathbf{F}^* AZ + BZ \\ \mathbf{F}_o &= AZ\mathbf{F}_o + C. \end{aligned} \tag{5.16}$$

The above derivation is valid for $\ell_A < 1$. If ℓ_A is not strictly less than 1, then $\mathbf{F}$ and $\mathbf{F}_o$ are not necessarily bounded and have to be used with care. We restrict our attention to the u.e. stable case ($\ell_A < 1$) at this point, although generalizations will be needed later on.

The following theorem summarizes the previous development. $Y_f = U_p H_T$ can be computed as $Y_f = \mathbf{P}_0(U_p\mathbf{F}^*)\mathbf{F}_o$, so that H_T has a factorization into a product of two operators: $Y_f = X_{[0]}\mathbf{F}_o$, where $X_{[0]} = \mathbf{P}_0(U_p\mathbf{F}^*)$.

Theorem 5.2 *Let $T \in \mathcal{U}$, and let $\{A,B,C,D\}$ be a u.e. stable locally finite realization of T. Let $\mathbf{F}$ and $\mathbf{F}_o$ be as given in (5.15). Then H_T has a factorization*

$$\cdot H_T = \mathbf{P}_0(\cdot\mathbf{F}^*)\mathbf{F}_o. \tag{5.17}$$

The factorization of H_T is equivalent to the factorization $H_k = \mathcal{C}_k\mathcal{O}_k$ in (3.23). Indeed, taking snapshots of $\mathbf{P}_0(\cdot\mathbf{F}^*)$ and $\mathbf{F}_o$ produces $\mathcal{C}_k$ and $\mathcal{O}_k$. In view of (5.1) and the factorization of the Hankel operator, the computation of $Y = UT$ can be written as the composition of two operators T_p and T_f, using an intermediate quantity $X_{[0]}$, the state.

$$\left\{ \begin{array}{l} [X_{[0]} \;\; Y_p] = U_p T_p \\ Y_f = [X_{[0]} \;\; U_f] T_f \end{array} \right. \quad \text{where} \quad \left\{ \begin{array}{lcl} T_p &=& [\mathbf{P}_0(\cdot\mathbf{F}^*) \;\; K_T] \\ T_f &=& \begin{bmatrix} \mathbf{F}_o \\ E_T \end{bmatrix}. \end{array} \right. \tag{5.18}$$

We interpret T_p as the "past" input-output operator and T_f as the "future" input-output operator. Past and future are only connected via $X_{[0]}$. See figure 5.1(c).

Definitions of reachability and observability

An important property that the factorization $H_T = \mathbf{P}_0(\cdot\,\mathbf{F}^*)\,\mathbf{F}_o$ can possess is (local) minimality, since that implies the minimality of the dimension sequence $\#\mathcal{B}$ of $X_{[0]}$ and thus the minimality of the realization. Let $\{A,B,C,D\}$ be a u.e. stable locally finite realization of T where $A \in \mathcal{D}(\mathcal{B},\mathcal{B}^{(-1)})$. With $\mathbf{F}$ and $\mathbf{F}_o$ as in equation (5.15), we define the

$$\begin{array}{ll}\text{reachability operator:} & \mathbf{P}_0(\cdot\,\mathbf{F}^*)\big|_{\mathcal{L}_2\mathcal{Z}^{-1}} \\ \text{observability operator:} & \cdot\,\mathbf{F}_o\big|_{\mathcal{D}_2^{\mathcal{B}}}\end{array}$$

Reachability, observability and minimality are defined as properties of the ranges of these operators, as follows.

Definition 5.3 *A realization is reachable if* $\mathbf{P}_0(\mathcal{L}_2\mathcal{Z}^{-1}\,\mathbf{F}^*)$ *is dense in* $\mathcal{D}_2^{\mathcal{B}}$, *and uniformly reachable if* $\mathbf{P}_0(\mathcal{L}_2\mathcal{Z}^{-1}\,\mathbf{F}^*) = \mathcal{D}_2^{\mathcal{B}}$.

A realization is observable if $\mathbf{P}_0(\mathcal{U}_2\mathbf{F}_o^*)$ *is dense in* $\mathcal{D}_2^{\mathcal{B}}$, *and uniformly observable if* $\mathbf{P}_0(\mathcal{U}_2\mathbf{F}_o^*) = \mathcal{D}_2^{\mathcal{B}}$.

A realization is said to be minimal if it is both reachable and observable.

Using the fact that the closure of the range of an operator and the kernel of its adjoint are complementary subspaces, we immediately obtain the following proposition.

Proposition 5.4 *A realization is reachable if and only if the operator* $\cdot\,\mathbf{F}\big|_{\mathcal{D}_2}$ *is one-to-one:* $D\mathbf{F} = 0 \Rightarrow D = 0$ *(all* $D \in \mathcal{D}_2$*). If the realization is uniformly reachable, then* $\mathcal{D}_2\mathbf{F}$ *is a closed subspace.*

A realization is observable if and only if the operator $\cdot\,\mathbf{F}_o\big|_{\mathcal{D}_2}$ *is one-to-one:* $D\mathbf{F}_o = 0 \Rightarrow D = 0$ *(all* $D \in \mathcal{D}_2$*). If the realization is uniformly observable, then* $\mathcal{D}_2\mathbf{F}_o$ *is closed.*

PROOF If $\mathbf{P}_0(\cdot\,\mathbf{F}^*)$ is regarded as an operator from $\mathcal{L}_2\mathcal{Z}^{-1} \to \mathcal{D}_2^{\mathcal{B}}$, then its adjoint operator is $\cdot\,\mathbf{F}$ with domain restricted to $\mathcal{D}_2^{\mathcal{B}}$. In view of (A.7), we obtain the decomposition $\mathcal{D}_2^{\mathcal{B}} = \overline{\text{ran}}\,\mathbf{P}_0(\cdot\,\mathbf{F}^*) \oplus \ker(\cdot\,\mathbf{F}\big|_{\mathcal{D}_2})$. The range of $\mathbf{P}_0(\cdot\,\mathbf{F}^*)$ is dense in $\mathcal{D}_2$ if and only if $\ker(\cdot\,\mathbf{F}\big|_{\mathcal{D}_2}) = 0$, *i.e.*, $\cdot\,\mathbf{F}\big|_{\mathcal{D}_2}$ is one-to-one. Finally, the range of an operator is closed if and only if the range of its adjoint is closed. □

Let us now recall the definitions of the input and output state spaces (equations (5.3) and (5.5)):

$$\mathcal{H} = \text{ran}(\cdot\,H_T^*)\,, \qquad \mathcal{H}_o = \text{ran}(\cdot\,H_T)\,.$$

Using the fact that H_T has a factorization $H_T = \mathbf{P}_0(\cdot\,\mathbf{F}^*)\mathbf{F}_o$, we can prove the following relations between these spaces and $\mathbf{F}$, $\mathbf{F}_o$.

Proposition 5.5 *Let* $\{A,B,C,D\}$ *be a u.e. stable locally finite realization of* T, *with* $A \in \mathcal{D}(\mathcal{B},\mathcal{B}^{(-1)})$, *and let* $\mathbf{F}$ *and* $\mathbf{F}_o$ *be the associated reachability and observability op-*

erators. Then $\mathcal{H}_o \subset \mathcal{D}_2^{\mathcal{B}}\mathbf{F}_o$ and $\mathcal{H} \subset \mathcal{D}_2^{\mathcal{B}}\mathbf{F}$, and we have the following implications:

$$\begin{array}{rcl} \textit{reachability} & \Rightarrow & \overline{\mathcal{H}_o} = \overline{\mathcal{D}_2^{\mathcal{B}}\mathbf{F}_o} \\ \textit{uniform reachability} & \Rightarrow & \mathcal{H}_o = \mathcal{D}_2^{\mathcal{B}}\mathbf{F}_o \\ \textit{observability} & \Rightarrow & \overline{\mathcal{H}} = \overline{\mathcal{D}_2^{\mathcal{B}}\mathbf{F}} \\ \textit{uniform observability} & \Rightarrow & \mathcal{H} = \mathcal{D}_2^{\mathcal{B}}\mathbf{F}. \end{array}$$

PROOF Since $H_T = \mathbf{P}_0(\cdot\mathbf{F}^*)\mathbf{F}_o$, it follows that $\mathcal{H}_o = \operatorname{ran} H_T = \mathbf{P}_0(\mathcal{L}_2 Z^{-1}\mathbf{F}^*)\mathbf{F}_o \subset \mathcal{D}_2^{\mathcal{B}}\mathbf{F}_o$. If the realization is uniformly reachable, then $\mathbf{P}_0(\mathcal{L}_2 Z^{-1}\mathbf{F}^*) = \mathcal{D}_2^{\mathcal{B}}$, so that, indeed, $\mathcal{H}_o = \mathcal{D}_2^{\mathcal{B}}\mathbf{F}_o$. We also have $\mathcal{K}_o = \ker H_T^* = \ker \mathbf{P}_0(\cdot\mathbf{F}_o^*)\mathbf{F}|_{\mathcal{U}_2}$. If the realization is reachable, then $\mathbf{F}$ is one-to-one and $\mathcal{K}_o = \ker \mathbf{P}_0(\cdot\mathbf{F}_o^*)|_{\mathcal{U}_2}$, with complement $\overline{\mathcal{H}_o} = \overline{\operatorname{ran}}(\cdot\mathbf{F}_o|_{\mathcal{D}_2}) = \overline{\mathcal{D}_2^{\mathcal{B}}\mathbf{F}_o}$. The remaining statements are proven in the same manner.

□

Proposition 5.6 *If a realization of T is both uniformly reachable and uniformly observable, then the range of H_T: $\mathcal{H}_o = \mathcal{D}_2^{\mathcal{B}}\mathbf{F}_o$, and the range of H_T^*: $\mathcal{H} = \mathcal{D}_2^{\mathcal{B}}\mathbf{F}$, are closed subspaces.*

Conversely, let $\mathcal{H}_o$ and $\mathcal{H}$ be closed subspaces. If the realization is reachable and uniformly observable, then it is uniformly reachable. Likewise, if the realization is observable and uniformly reachable, then it is uniformly observable.

PROOF The first part of the proposition follows immediately from proposition 5.5: since the realization is uniformly reachable, $\mathcal{H}_o = \mathcal{D}_2^{\mathcal{B}}\mathbf{F}_o$. Because the realization is uniformly observable, proposition 5.4 asserts that $\mathcal{D}_2^{\mathcal{B}}\mathbf{F}_o$ is a closed subspace, and hence $\mathcal{H}_o = \operatorname{ran} H_T$ is a closed subspace.

We now prove the second part. According to proposition 5.5, uniform observability implies $\mathcal{H} = \mathcal{D}_2\mathbf{F}$ is closed, so that the range of the adjoint of $\cdot\mathbf{F}|_{\mathcal{D}_2}$ is closed, too: $\mathbf{P}_0(\mathcal{L}_2 Z^{-1}\mathbf{F}^*)$ is closed. Reachability means, by definition, $\overline{\mathbf{P}_0(\mathcal{L}_2 Z^{-1}\mathbf{F}^*)} = \mathcal{D}_2$, hence $\mathbf{P}_0(\mathcal{L}_2 Z^{-1}\mathbf{F}^*) = \mathcal{D}_2$: the realization is uniformly reachable. □

Propositions 5.4 and 5.5 have a direct corollary, which is part of a Kronecker-type theorem for time-varying systems. The second part appears as theorem 5.19 in the next section.

Corollary 5.7 (Kronecker-type thm, I) *Let $T \in \mathcal{U}$ be a locally finite transfer operator which has a u.e. stable realization with state dimension sequence $\mathcal{B}$. If the realization is minimal, then $\#\mathcal{B} = \operatorname{sdim} \mathcal{H} = \operatorname{sdim} \mathcal{H}_o$.*

PROOF The given realization defines $\mathbf{F}$ and $\mathbf{F}_o$ by equations (5.15). Reachability implies $\overline{\mathcal{H}_o} = \overline{\mathcal{D}_2^{\mathcal{B}}\mathbf{F}_o}$. Observability implies that $\mathbf{F}_o$ is one-to-one, hence $\operatorname{sdim} \mathcal{D}_2^{\mathcal{B}}\mathbf{F}_o = \operatorname{sdim} \mathcal{D}_2^{\mathcal{B}} = \#\mathcal{B}$. □

The corollary can be stated at the local level as well: if the realization is minimal and the k-th slice $\pi_k\mathcal{H}_o = (\mathcal{H}_o)_k$ of $\mathcal{H}_o$ has a dimension d_k, then d_k is equal to the state dimension of the realization at point k. Hence, we recover part of the realization theorem for finite matrices (theorem 3.7). It is also true that $(\mathcal{H}_o)_k$ is equal to the range of H_k, the k-th snapshot of the Hankel operator, as we have seen in (5.10), and that d_k is equal to the rank of H_k.

Most of the remainder of this chapter is concerned with a proof of the converse of the corollary, *i.e.*, to show that if sdim $\mathcal{H} =$ sdim $\mathcal{H}_o = [\cdots\ d_0\ d_1\ d_2\ \cdots]$ is a uniformly bounded sequence of dimensions, where $d_k = \operatorname{rank} H_k$, then there exist realizations of T with d_k equal to the system order at point k. We call the sequence the *minimal system order* of T. The actual construction of such minimal realizations is the subject of section 5.4, where the converse of corollary 5.7 appears as theorem 5.19.

Computation rules for $\mathbf{F}$ *and* $\mathbf{F}_o$

Equations (5.16) will often be used in the following form.

$$\begin{aligned} \mathbf{P}_0(Z^{-1}\cdot\mathbf{F}^*)^{(-1)} &= \mathbf{P}_0(\cdot[\mathbf{F}^*A+B]) \\ Z\mathbf{F} &= A^*\mathbf{F}+B^* \end{aligned} \qquad (5.19)$$

$$\begin{aligned} \mathbf{F}_o &= C+AZ\mathbf{F}_o \\ \mathbf{P}_0(\cdot\mathbf{F}_o^*) &= \mathbf{P}_0(Z^{-1}\cdot\mathbf{F}_o^*)^{(-1)}A^*+\mathbf{P}_0(\cdot C^*) \end{aligned} \qquad (5.20)$$

$$\begin{aligned} T &= D+\mathbf{F}^*C \\ T &= D+BZ\mathbf{F}_o\,. \end{aligned} \qquad (5.21)$$

5.3 REACHABILITY AND OBSERVABILITY GRAMIANS

Whether a given realization is reachable and observable is important: it determines whether the realization is minimal. Some state space operations to be dealt with in following chapters can only be carried out on realizations that are reachable and/or observable. However, the form in which reachability and observability properties have been presented so far (as range conditions) does not give a straightforward method to determine these properties for a given realization. The purpose of this section is to make these properties more concrete.

Proposition 5.5 states that if a realization is reachable and observable, then the input and output state spaces are given by $\mathcal{H} = \mathcal{D}_2\mathbf{F}$, $\mathcal{H}_o = \mathcal{D}_2\mathbf{F}_o$. Hence $\mathbf{F}$ and $\mathbf{F}_o$ can be viewed as basis representations that generate these subspaces. In view of this, we define the Gramians of these bases as

$$\begin{aligned} &\text{reachability Gramian:} && \Lambda_{\mathbf{F}} = \mathbf{P}_0(\mathbf{F}\mathbf{F}^*) \\ &\text{observability Gramian:} && \Lambda_{\mathbf{F}_o} = \mathbf{P}_0(\mathbf{F}_o\mathbf{F}_o^*)\,. \end{aligned}$$

$\Lambda_{\mathbf{F}}$ and $\Lambda_{\mathbf{F}_o}$ are bounded diagonal operators, see section 4.3.

Proposition 5.8 *A realization is reachable if and only if* $\Lambda_{\mathbf{F}} > 0$, *and uniformly reachable if and only if* $\Lambda_{\mathbf{F}} \gg 0$.

A realization is observable if and only if $\Lambda_{\mathbf{F}_o} > 0$, *and uniformly observable if and only if* $\Lambda_{\mathbf{F}_o} \gg 0$.

PROOF In terms of diagonal inner products, $D\mathbf{F} = 0 \Leftrightarrow \{D\mathbf{F}, D\mathbf{F}\} = 0$, and $\{D\mathbf{F}, D\mathbf{F}\} = \mathbf{P}_0(D\mathbf{F}\mathbf{F}^*D^*) = D\mathbf{P}_0(\mathbf{F}\mathbf{F}^*)D^*$. With proposition 5.4, this implies that the realization is reachable if and only if the Gram operator $\Lambda_\mathbf{F} = \mathbf{P}_0(\mathbf{F}\mathbf{F}^*) > 0$. Proof of the proposition on uniformity also works by transforming to the local level. If $\Lambda_\mathbf{F} \gg 0$ then Λ^{-1} exists and is bounded. In that case, let D_n be a sequence in $\mathcal{D}_2$ such that $U_n := D_n\mathbf{F} \to U$ for some $U \in \mathcal{L}_2Z^{-1}$. Then $\mathbf{P}_0(D_n\mathbf{F}\mathbf{F}^*) = \mathbf{P}_0(U_n\mathbf{F}^*)$ so that the sequence $D_n = \mathbf{P}_0(U_n\mathbf{F}^*)\Lambda_\mathbf{F}^{-1}$ is bounded and converges to a diagonal D for which $U = D\mathbf{F}$. This shows closure. Conversely, if the map is closed, then by standard Hilbert space arguments $\Lambda_\mathbf{F}$ must be boundedly invertible and since it is positive already, it then must be strictly positive definite as well.[1] □

The reachability and observability Gramians will play an important role in many of the topics of the remaining chapters, because it is often possible to compute them recursively.

Lyapunov equations

Proposition 5.9 *Let $\{A,B,C,D\}$ be a u.e. stable realization, and let the operator $\mathbf{F}$ be given by equation (5.15), with Gramian $\Lambda_\mathbf{F} = \mathbf{P}_0(\mathbf{F}\mathbf{F}^*)$.*

$\Lambda_\mathbf{F}$ satisfies the equation

$$\Lambda_\mathbf{F}^{(-1)} = B^*B + A^*\Lambda_\mathbf{F}A\,. \qquad (5.22)$$

PROOF Using equation (5.19),

$$\begin{aligned}
\Lambda_\mathbf{F}^{(-1)} &= \mathbf{P}_0(Z^{-1}[Z\mathbf{F}]\mathbf{F}^*)^{(-1)} \\
&= \mathbf{P}_0([B^* + A^*\mathbf{F}]\,[B + \mathbf{F}^*A]) \\
&= \mathbf{P}_0(B^*B) + \mathbf{P}_0(A^*\mathbf{F}\mathbf{F}^*A) + \mathbf{P}_0(B^*\mathbf{F}^*A) + \mathbf{P}_0(A^*\mathbf{F}B) \\
&= B^*B + A^*\mathbf{P}_0(\mathbf{F}\mathbf{F}^*)A + 0 + 0\,.
\end{aligned}$$

□

Equations of the type

$$M^{(-1)} = A^*MA + B^*B, \qquad M \in \mathcal{D}(\mathcal{B},\mathcal{B}) \qquad (5.23)$$

are known as Lyapunov or Lyapunov-Stein equations. If $\ell_A < 1$, then it is easy to verify by substitution that the equation has a solution given by

$$M = \sum_{k=0}^{\infty} (A^{\{k\}})^*(B^*B)^{(k+1)}A^{\{k\}}\,, \qquad (5.24)$$

where $A^{\{k\}} = A^{(k)} \cdots A^{(1)}$ for $k \geq 1$ and $A^{\{0\}} = I$. If $\ell_A < 1$, then the summation converges and the solution is unique: if Λ is another solution, then

$$\begin{aligned}
(M-\Lambda)^{(-1)} &= A^*(M-\Lambda)A \\
\Rightarrow \quad M-\Lambda &= (A^{\{k\}})^*(M-\Lambda)^{(k)}A^{\{k\}}
\end{aligned}$$

[1] This is a corollary to the closed graph theorem, see [Rud66] p. 122.

and $\ell_A < 1$ implies $A^{\{k\}} \to 0$ so that $\Lambda = M$. If $\ell_A = 1$, then the Lyapunov equation does not necessarily have a unique solution. For example, if $A = I$ and $B = 0$, then the resulting equation is $M^{(-1)} = M$ so that any M which is Toeplitz and diagonal will do, whereas $\Lambda_{\mathbf{F}} = \mathbf{P}_0(\mathbf{F}\mathbf{F}^*) = 0$ in this example.

We obtain the dual to proposition 5.9 in a similar way.

Proposition 5.10 *Let $\{A,B,C,D\}$ be a u.e. stable realization, and let the operator $\mathbf{F}_o$ be given by equation (5.15), with Gramian $\Lambda_{\mathbf{F}_o} = \mathbf{P}_0(\mathbf{F}_o\mathbf{F}_o^*)$. Then $\Lambda_{\mathbf{F}_o}$ satisfies the (dual) Lyapunov equation*

$$\Lambda_{\mathbf{F}_o} = CC^* + A\Lambda_{\mathbf{F}_o}^{(-1)}A^*. \tag{5.25}$$

Again, if $\ell_A < 1$, then the solution to the equation $Q = CC^* + AQ^{(-1)}A^*$ is unique and equal to $\Lambda_{\mathbf{F}_o}$.

For a given realization, the Lyapunov equation is computable: by taking the k-th entry of every diagonal in this equation, we obtain the recursion

$$M_{k+1} = A_k^* M_k A_k + B_k^* B_k, \qquad k = \cdots,-1,0,1,\cdots, \tag{5.26}$$

and M_k can be computed, for $k = \cdots,-1,0,1,\cdots$, provided we have an appropriate initial point for the recursion. Exact initial points can be obtained in most cases of interest, as follows.

- If the realization is a realization for a finite $n\times n$ matrix, then we can assume that the realization $\{A_k,B_k,C_k,D_k\}$ starts with a zero number of states at time 1, say. An exact initial value is then $M_1 = [\cdot]$, a matrix with zero dimensions.
- For systems which are time invariant before some point in time ($k = 1$, say), an exact initial value can be computed analytically from the time-invariant algebraic equation that holds before time $k = 1$:

$$M_0 = A_0^* M_0 A_0 + B_0^* B_0.$$

The solution to this equation follows from an eigenvalue decomposition (Schur decomposition) of A_0, or by using Kronecker products, see [HJ89].

- If the system is periodically time-varying, then it can be viewed as a time-invariant system T with block entries $T_{ij} = T_{i-j}$ of size $n\times n$: T is a block Toeplitz operator. As discussed in section 3.2, we can assume that the realization is periodical, too, in which case we can replace it by a block realization $\{\underline{A},\underline{B},\underline{C},\underline{D}\}$ that is time-invariant. The Lyapunov equation can be solved for this time-invariant system, although this is not really attractive if the period is large.
- Finally, if we have a time-varying realization for which $\ell_A < 1$, then, as we have shown before, the Lyapunov recursion is strongly convergent. In that case, M_k at some point k is independent of the precise initialization of the recursion at $k \approx -\infty$, say. Hence it is possible to limit attention to a finite time-interval, and to obtain arbitrarily accurate initial values for this interval by performing a finite recursion on data outside the interval, starting with initial values set to 0. For the Lyapunov recursion example, M_1

can be determined as

$$\begin{aligned} M_1 &= A_0^* M_0 A_0 + B_0^* B_0 \\ &= A_0^* A_{-1}^* M_{-1} A_{-1} A_0 + B_0^* B_0 + A_0^* B_{-1}^* B_{-1} A_0 \\ &= A_0^* \cdots A_{-n}^* M_{-n} A_{-n} \cdots A_0 + \\ &\quad + \left\{ B_0^* B_0 + A_0^* B_{-1}^* B_{-1} A_0 + \sum_{i=2}^{n} A_0^* \cdots A_{-i+1}^* B_{-i}^* B_{-i} A_{-i+1} \cdots A_0 \right\}. \end{aligned}$$

If the system is u.e. stable, then $\|A_{-n} \cdots A_0\|$ can be made arbitrarily small by choosing n large enough. Neglecting for this n the first term gives an approximation of M_1. The same approximation would have been obtained by choosing $M_{-n} = 0$, and computing M_1 via the recursion (5.26).

Normalized realizations

Lyapunov equations arise in the *normalization* of a given realization. Suppose that we are given a u.e. stable minimal realization $\{A,B,C,D\}$ of some locally finite operator $T \in \mathcal{U}$. The objective is to find a similar realization $\{A',B',C',D\}$ which is in input normal form, *i.e.*, for which $\Lambda_{\mathbf{F}'} = I$. In view of (5.22), such a realization satisfies $A'^*A' + B'^*B' = I$. Let $\mathbf{F}$ and $\mathbf{F}_o$ be the reachability and observability operators of $\mathbf{T}$ as in (5.15), and define $\mathbf{F}'$ and $\mathbf{F}'_o$ likewise for $\mathbf{T}'$. If R is a state transformation that brings $\mathbf{T}$ into $\mathbf{T}'$ according to (3.14), then $\mathbf{F} = R^*\mathbf{F}'$ and $R\mathbf{F}_o = \mathbf{F}'_o$, and the corresponding Gram operators satisfy

$$\begin{aligned} \Lambda_{\mathbf{F}} &= R^* \Lambda_{\mathbf{F}'} R \\ \Lambda_{\mathbf{F}'_o} &= R \Lambda_{\mathbf{F}_o} R^* . \end{aligned} \tag{5.27}$$

The first equation gives

$$\Lambda_{\mathbf{F}} = R^* R,$$

so that the required state transformation R is a factor of $\Lambda_{\mathbf{F}}$. R is boundedly invertible if and only if $\Lambda_{\mathbf{F}}$ is uniformly positive, that is, if the given realization is uniformly reachable. If $\ell_A < 1$, then R is obtained by solving the Lyapunov equation (5.23) for M, followed by solving the factorization $M = R^*R$. Another way to arrive at the Lyapunov equation directly is by inserting the relations $A' = RAR^{-(-1)}$ and $B' = BR^{-(-1)}$ into the normalization condition $A'^*A' + B'^*B' = I$, and putting $M = R^*R$. Likewise, a realization in output normal form (for which $\Lambda_{\mathbf{F}'_o} = I$ so that $A'A'^* + C'C'^* = I$) is obtained by factoring $\Lambda_{\mathbf{F}_o} = R^{-1}R^{-*}$, and we see that the given realization must be uniformly observable. Again, if $\ell_A < 1$, then R can be obtained by solving the Lyapunov equation $Q = CC^* + AQ^{(-1)}A^*$ for Q after which R is obtained as a factor of Q^{-1}. The Lyapunov equation is directly obtained by inserting the relations $A' = RAR^{-(-1)}$ and $C' = RC$ into the condition $A'A'^* + C'C'^* = I$.

Equivalent minimal realizations

Reachability and observability Gramians can also be used to compute equivalent minimal realizations from realizations that are not reachable and/or not observable. Suppose that we are given a u.e. stable realization $\{A,B,C,D\}$ of a locally finite operator

$T \in \mathcal{U}$, and let us assume that it is not in reachable form. To transform it into a canonical form, let $\Lambda_{\mathbf{F}} \in \mathcal{D}$ be the reachability Gramian of the given realization of T. Since $\Lambda_{\mathbf{F}} \geq 0$, it has a factorization

$$\Lambda_{\mathbf{F}} = R^* \begin{bmatrix} \Lambda_{11} & \\ & 0 \end{bmatrix} R, \qquad R = \begin{bmatrix} R_1 \\ R_2 \end{bmatrix}, \tag{5.28}$$

where $\Lambda_{11} > 0$ and R is an invertible operator in $\mathcal{D}$ (*e.g.*, R can be chosen unitary). Note that the range of Λ_{11} is not necessarily closed: it need not be *uniformly* positive. In case R is unitary and has the indicated block decomposition, then $\operatorname{ran}(\cdot R_2) = \ker(\cdot \Lambda_{\mathbf{F}})$, $\operatorname{ran}(\cdot R_1) = \overline{\operatorname{ran}}(\cdot \Lambda_{\mathbf{F}})$. Applying R as state transformation to $\mathbf{T}$ leads to a realization $\mathbf{T}' = \{A', B', C', D\}$ given by

$$\begin{bmatrix} A' & C' \\ B' & D \end{bmatrix} = \begin{bmatrix} R & \\ & I \end{bmatrix} \begin{bmatrix} A & C \\ B & D \end{bmatrix} \begin{bmatrix} R^{-(-1)} & \\ & I \end{bmatrix}.$$

$\Lambda_{\mathbf{F}'} := \left[\begin{smallmatrix} \Lambda_{11} & \\ & 0 \end{smallmatrix}\right]$ is the reachability Gramian of $\mathbf{T}'$, and satisfies the Lyapunov equation $\Lambda_{\mathbf{F}'}^{(-1)} = A'^* \Lambda_{\mathbf{F}'} A' + B'^* B'$. Partition A', B', C' conformably to the partitioning of R. Then

$$A' = \begin{bmatrix} A'_{11} & 0 \\ A'_{21} & A'_{22} \end{bmatrix}, \qquad B' = [B'_1 \quad 0], \qquad C' = \begin{bmatrix} C'_1 \\ C'_2 \end{bmatrix}, \qquad D' = D, \tag{5.29}$$

because the Lyapunov equation leads, in particular, to $0 = B_2'^* B_2' + A_{12}'^* \Lambda_{11} A'_{12}$, so that $B_2 = 0$ and $A_{12} = 0$ since $\Lambda_{11} > 0$. It follows that $\{A'_{11}, B'_1, C'_1, D\}$ is a (smaller) realization of T which is reachable, with reachability Gramian equal to Λ_{11}.

Similarly, a realization which is not observable can be transformed into the canonical form

$$A' = \begin{bmatrix} A'_{11} & A'_{12} \\ 0 & A'_{22} \end{bmatrix}, \qquad B' = [B'_1 \quad B'_2], \qquad C' = \begin{bmatrix} C'_1 \\ 0 \end{bmatrix}, \qquad D' = D,$$

by computing a factorization of the observability Gramian $\Lambda_{\mathbf{F}_o}$ as

$$\Lambda_{\mathbf{F}_o} = R^{-1} \begin{bmatrix} (\Lambda_o)_{11} & \\ & 0 \end{bmatrix} R^{-*},$$

and now $\{A'_{11}, B'_1, C'_1, D\}$ form an equivalent realization with observable states.

Realizations that are neither reachable nor observable can be transformed into a minimal realization by applying both transformations in succession, as follows. In the first step the reduction of the reachability Gramian yields

$$A' = \begin{bmatrix} A'_1 & 0 \\ A'_{21} & A'_2 \end{bmatrix}, \qquad B' = [B'_1 \quad 0], \qquad C' = \begin{bmatrix} C'_1 \\ C'_2 \end{bmatrix}.$$

Reducing the systems $\{A'_1, B'_1, C'_1\}$ and $\{A'_2, 0, C'_2\}$ separately produces state transformations R'_1 and R'_2 such that

$$\{R'_1 A'_1 R_1'^{-(-1)}, B'_1 R_1'^{-(-1)}, R'_1 C'_1\} = \left\{ \begin{bmatrix} A''_{11} & A''_{12} \\ 0 & A''_{22} \end{bmatrix}, [B''_1 \quad B''_2], \begin{bmatrix} C''_1 \\ 0 \end{bmatrix} \right\}$$

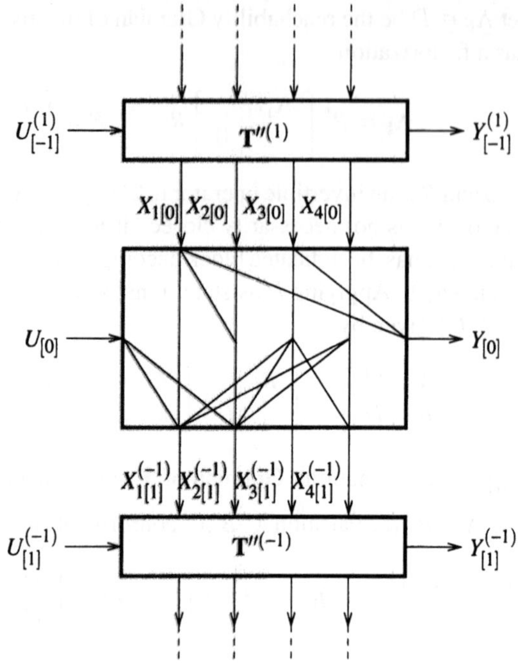

Figure 5.2. Splitting of the state space into four parts. Only state X_1 is useful, *i.e.*, both reachable and observable.

and

$$\{R'_2 A'_2 R_2^{\prime -(-1)}, B'_2 R_2^{\prime -(-1)}, R'_2 C'_2\} = \left\{ \begin{bmatrix} A''_{33} & A''_{34} \\ 0 & A''_{44} \end{bmatrix}, [0\ \ 0], \begin{bmatrix} C''_2 \\ 0 \end{bmatrix} \right\}.$$

If we now apply the transformation $\begin{bmatrix} R'_1 & \\ & R'_2 \end{bmatrix}$ to the primed system, we obtain

$$\mathbf{T}'' = \begin{bmatrix} A'' & C'' \\ B'' & D'' \end{bmatrix} = \left[\begin{array}{cc|cc|c} A''_{11} & A''_{12} & 0 & 0 & C''_1 \\ 0 & A''_{22} & 0 & 0 & 0 \\ \hline A''_{31} & A''_{32} & A''_{33} & A''_{34} & C''_2 \\ A''_{41} & A''_{42} & 0 & A''_{44} & 0 \\ \hline B''_1 & B''_2 & 0 & 0 & D \end{array} \right].$$

The structure of this realization is shown in figure 5.2. The state space is split into four subspaces. Only state X_1 is useful. States X_2 and X_4 are not observable, states X_3 and X_4 get no excitation and hence remain zero if they were initially zero (X_3 is observable if some "deus ex machina" has put a non-zero value there). Non-minimal realizations should be avoided from numerical and computational points of view: they lead to extra, unnecessary operations on the data. Physically, the spurious states can be

In: $\mathbf{T} = (A,B,C,D)$ (a locally finite realization)
Out: $\mathbf{T}' = (A',B',C',D')$ (a) an equivalent reachable realization
(b) an equivalent observable realization

(a) Initialize $\hat{Q}_1$
for $k = 1,2,\cdots$

Compute $\hat{Q}_{k+1}$ from an LQ factorization:

$$\left[\begin{array}{c|c} \hat{Q}_k A_k & \hat{Q}_k C_k \\ \hline B_k & D_k \end{array}\right] =: \left[\begin{array}{cc|c} A'_k & 0 & C'_k \\ \hline B'_k & 0 & D'_k \end{array}\right] \left[\begin{array}{c|c} \hat{Q}_{k+1} & 0 \\ * & \\ \hline 0 & I \end{array}\right]$$

end

(b) Initialize $\hat{Q}_{n+1}$
for $k = n, n-1, \cdots$

Compute $\hat{Q}_k$ from a QR factorization:

$$\left[\begin{array}{c|c} A_k\hat{Q}_{k+1} & C_k \\ \hline B_k\hat{Q}_{k+1} & D_k \end{array}\right] =: \left[\begin{array}{cc|c} \hat{Q}_k & * & 0 \\ \hline 0 & 0 & I \end{array}\right] \left[\begin{array}{c|c} A'_k & C'_k \\ 0 & 0 \\ \hline B'_k & D'_k \end{array}\right]$$

end

Figure 5.3. Algorithm to bring a realization into (a) reachable form, (b) observable form.

a source of noise and instability, and hence we usually wish to retain only the minimal part:

$$\hat{\mathbf{T}}'' = \begin{bmatrix} A''_{11} & C''_{11} \\ B''_{11} & D'' \end{bmatrix}.$$

A practical algorithm to bring a realization into reachable form uses the observation that in (5.29) we are only interested in A'_{11}, B'_1, C'_1, D, and want to reject the remaining blocks. Thus, if $R = \left[\begin{smallmatrix} R_1 \\ R_2 \end{smallmatrix}\right]$ and $R^{-1} = [(R^{-1})_1 \; (R^{-1})_2]$, then we only need R_1 and $(R^{-1})_1$. It is convenient to use a *unitary* state transformation $Q = \left[\begin{smallmatrix} Q_1 \\ Q_2 \end{smallmatrix}\right]$ in place of R, because $Q^{-1} = Q^* = [Q_1^* \; Q_2^*]$. Hence, we need only retain Q_1 and do not have to invert any matrix. The resulting algorithm is summarized in figure 5.3(a), where Q_1 is called $\hat{Q}$. The algorithm can often be combined with other forward recursions, to ensure reachability "on the fly". Since the state transformation matrices are not inverted, the algorithm could in principle be applied to a realization for which the reachability operator does not have closed range, provided that the QR factorization algorithm used is reliable for nearly singular matrices. In that case, the resulting realization is reachable, but not uniformly reachable.

The initialization of the algorithm depends on the situation at hand. For a finite matrix T, A_1 starts with 0 states, so that $\hat{Q}_1 = [\cdot]$. For systems that are time-invariant before time $k = 1$, $\hat{Q}_1 = \hat{Q}_0$ is derived from the solution of the time-invariant Lyapunov equation

$$\Lambda_0 = A_0^* \Lambda_0 A_0 + B_0^* B_0\,, \quad \Lambda_0 =: Q_0^* \begin{bmatrix} \Lambda_{11} & \\ & 0 \end{bmatrix} Q_0 = \hat{Q}_0^* \Lambda_{11} \hat{Q}_0\,.$$

The solution of the Lyapunov equation for other cases of interest is discussed in the beginning of this section.

5.4 ABSTRACT REALIZATION THEORY

In the preceding section, we assumed knowledge of a u.e. stable realization $\{A,B,C,D\}$ of an operator T. We will now investigate how such realizations can be derived. This is done by the analysis of H_T and its characteristic subspaces, $\mathcal{H}$ and $\mathcal{H}_o$. We show how a shift-invariance property of these spaces, along with the choice of a basis in either one of them, produces minimal realizations which are either in "input normal form" (or in "canonical controller form") or in "output normal form" (canonical observer form). In all four cases, realizations with $\ell_A \le 1$ are obtained.

Shift-invariance properties

Recall the definitions of the input state space $\mathcal{H}$ and the input null space $\mathcal{K}$ in equation (5.3):

$$\begin{aligned} \mathcal{H} &= \operatorname{ran}(\cdot H_T^*) &&= \mathbf{P}'(\mathcal{U}_2 T^*) \\ \mathcal{K} &= \ker(\cdot H_T) &&= \{U \in \mathcal{L}_2 Z^{-1} : \mathbf{P}(UT) = 0\}\,. \end{aligned}$$

$\mathcal{K}$ satisfies the *shift-invariance property*

$$Z^{-1}\mathcal{K} \subset \mathcal{K}\,. \qquad (5.30)$$

Indeed, if $U \in \mathcal{K}$, then $\mathbf{P}(UT) = 0$, hence $UT \in \mathcal{L}_2 Z^{-1}$ and thus $Z^{-1}UT \in \mathcal{L}_2 Z^{-1}$, too. But this means that $\mathbf{P}(Z^{-1}UT) = 0$ so that $Z^{-1}U \in \mathcal{K}$.

The shift-invariance property of $\mathcal{K}$ implies a shift-invariance property of its complement $\mathcal{H}$. We will use it in the following form.

Lemma 5.11 *Let* $\mathbf{A}(\cdot) := \mathbf{P}_{\mathcal{H}}(Z^{-1}\cdot)$. *Then*
(*a*) $\mathbf{A}(\mathbf{P}_{\mathcal{K}}(U)) = 0$ *for all* $U \in \mathcal{X}_2$.
(*b*) *On* $\mathcal{L}_2 Z^{-1}$, $\mathbf{A}^n(\cdot) = \mathbf{P}_{\mathcal{H}}(Z^{-n}\cdot)$, *for* $n > 0$.

PROOF (*a*) is a consequence of $\overline{\mathcal{H}} \perp \mathcal{K}$ and $Z^{-1}\mathcal{K} \subset \mathcal{K}$, so that $\overline{\mathcal{H}} \perp Z^{-1}\mathcal{K}$.
(*b*) For any $U \in \mathcal{L}_2 Z^{-1}$,

$$\begin{aligned} \mathbf{P}_{\mathcal{H}}[Z^{-1}\mathbf{P}_{\mathcal{H}}(Z^{-1}U)] &= \mathbf{P}_{\mathcal{H}}[Z^{-1}\mathbf{P}_{\mathcal{H}}(Z^{-1}U) + Z^{-1}\mathbf{P}_{\mathcal{K}}(Z^{-1}U)] \\ &= \mathbf{P}_{\mathcal{H}}(Z^{-2}U)\,. \end{aligned}$$

The result for $n > 2$ follows by induction. □

Canonical controller operator realizations

Let T be a given bounded linear causal time-varying system transfer operator in $\mathcal{U}$, and let $\mathcal{H}, \mathcal{H}_o, \mathcal{K}$ and $\mathcal{K}_o$ be its input-output state and null spaces, respectively. Then,

$$\text{for } U \in \mathcal{L}_2 Z^{-1}: \qquad \mathbf{P}(UT) = \mathbf{P}[\mathbf{P}_{\mathcal{H}}(U)T].$$

This property is related to Nerode equivalence: as far as the "future output" $\mathbf{P}(UT)$ of T is concerned, inputs $U \in \mathcal{L}_2 Z^{-1}$ are equivalent to their projection $\mathbf{P}_{\mathcal{H}}(U)$ onto $\overline{\mathcal{H}}$. It follows immediately that the Hankel operator $H_T = \mathbf{P}(\cdot T)|_{\mathcal{L}_2 Z^{-1}}$ can be factored:

$$\cdot H_T = \mathbf{P}[\mathbf{P}_{\mathcal{H}}(\cdot)T]. \tag{5.31}$$

Introducing the "state" $\mathbf{X}_0$, this becomes more clearly visible: for $U_p \in \mathcal{L}_2 Z^{-1}$,

$$Y_f = U_p H_T \quad \Leftrightarrow \quad \left\{ \begin{array}{rcl} \mathbf{X}_0 & = & \mathbf{P}_{\mathcal{H}}(U_p) \\ Y_f & = & \mathbf{P}(\mathbf{X}_0 T). \end{array} \right.$$

More in general, for $U \in \mathcal{X}_2$ and any $k \in \mathbb{Z}$, and with $U_{[k]} = \mathbf{P}_0(Z^{-k}U)$ equal to the k-th diagonal of U,

$$Y = UT \quad \Leftrightarrow \quad \left\{ \begin{array}{rcl} \mathbf{X}_k & = & \mathbf{P}_{\mathcal{H}}(Z^{-k}U) \\ Y_{[k]} & = & \mathbf{P}_0(\mathbf{X}_k T) + U_{[k]} T_{[0]} \end{array} \right. \tag{5.32}$$

where we have introduced a state $\mathbf{X}_k$, and used (i) $(Z^{-k}U)T = Z^{-k}Y$, (ii) by causality, $Y_{[k]}$ does not depend on $U_{[j]}$ for $j > k$.

The shift-invariance property in lemma 5.11 directly gives a recursion for $\mathbf{X}_k$. Together, these lead to an operator state space model in a form that is already familiar in a number of other contexts (see *e.g.*, [KFA70, Fuh76, Hel74, FS82, You86]).

Theorem 5.12 *Let $T \in \mathcal{U}(\mathcal{M}, \mathcal{N})$ be a given transfer operator with input state space $\mathcal{H}$. Define bounded operators $\mathbf{A}, \mathbf{B}, \mathbf{C}, \mathbf{D}$ by*

$$\begin{array}{llll} \mathbf{A}: & \overline{\mathcal{H}} \to \overline{\mathcal{H}} & \mathbf{C}: & \overline{\mathcal{H}} \to \mathcal{D}_2^{\mathcal{N}} \\ \mathbf{B}: & \mathcal{D}_2^{\mathcal{M}} \to \overline{\mathcal{H}} & \mathbf{D}: & \mathcal{D}_2^{\mathcal{M}} \to \mathcal{D}_2^{\mathcal{N}} \end{array} \qquad \left[\begin{array}{cc} \mathbf{A} & \mathbf{C} \\ \mathbf{B} & \mathbf{D} \end{array} \right] = \left[\begin{array}{cc} \mathbf{P}_{\mathcal{H}}(Z^{-1}\cdot) & \mathbf{P}_0(\cdot T) \\ \mathbf{P}_{\mathcal{H}}(Z^{-1}\cdot) & \mathbf{P}_0(\cdot T) \end{array} \right]$$

Then, (1) the (uniformly bounded) sequence $\{\mathbf{X}_k\}$ defined by $\mathbf{X}_k = \mathbf{P}_{\overline{\mathcal{H}}}(Z^{-k}U)$ for $U \in \mathcal{X}_2^{\mathcal{M}}$ satisfies

$$\mathbf{X}_{k+1} = \mathbf{X}_k \mathbf{A} + U_{[k]} \mathbf{B}. \tag{5.33}$$

(2) If $Y = UT$ and $U \in \mathcal{X}_2^{\mathcal{M}}$, then $Y_{[k]}$ satisfies

$$Y_{[k]} = \mathbf{X}_k \mathbf{C} + U_{[k]} \mathbf{D}.$$

(3) The spectral radius $r(\mathbf{A})$ satisfies $r(\mathbf{A}) \le 1$. If $r(\mathbf{A}) < 1$, then $\{\mathbf{X}_k\} \in \overline{\mathcal{H}}$ is the only uniformly bounded sequence in $\overline{\mathcal{H}}$ which satisfies the recursion (5.33).

PROOF The proof goes in three steps.
(1) $\mathbf{X}_k = \mathbf{P}_{\overline{\mathcal{H}}}(Z^{-k}U)$ *satisfies (5.33):*

$$\begin{aligned}\mathbf{X}_{k+1} &= \mathbf{P}_{\mathcal{H}}(Z^{-k-1}U) = \mathbf{P}_{\mathcal{H}}(Z^{-1}(Z^{-k}U)) \\ &= \mathbf{P}_{\mathcal{H}}(Z^{-1}(\mathbf{P}_{\mathcal{H}}(Z^{-k}U) + \mathbf{P}_{\mathcal{K}}(Z^{-k}U) + \mathbf{P}(Z^{-k}U))) \\ &= \mathbf{P}_{\mathcal{H}}(Z^{-1}\mathbf{X}_k) + 0 + \mathbf{P}_{\mathcal{H}}(Z^{-1}\mathbf{P}_0(Z^{-k}U)) \\ &= \mathbf{X}_k\mathbf{A} + U_{[k]}\mathbf{B}.\end{aligned}$$

(2) *The output equation is also satisfied:* with $\mathbf{X}_k = \mathbf{P}_{\mathcal{H}}(Z^{-k}U)$, equation (5.32) ensures that

$$Y = UT \quad \Leftrightarrow \quad Y_{[k]} = \mathbf{P}_0(\mathbf{X}_k T) + \mathbf{P}_0(U_{[k]}T) = \mathbf{X}_k\mathbf{C} + U_{[k]}\mathbf{D} \quad \text{(all } k\text{)}.$$

(3) *If $r(\mathbf{A}) < 1$, then $\{\mathbf{X}_k\}$ is unique:* suppose there is another sequence $\{\mathbf{X}'_k\}$ which satisfies 5.33, then for $\mathbf{X}''_k := \mathbf{X}_k - \mathbf{X}'_k$ and $n > 0$ we have

$$\mathbf{X}''_k = \mathbf{P}_{\overline{\mathcal{H}}}(Z^{-n}\mathbf{X}''_{k-n}) = \mathbf{X}''_{k-n}\mathbf{A}^n.$$

Hence, $\|\mathbf{X}''_k\| \le \|\mathbf{X}''_{k-n}\|\|\mathbf{A}^n\|$. Let M be an upper bound on $\{\|\mathbf{X}''_k\|\}$, and $r(\mathbf{A}) < \rho < 1$, then for n large enough we have (by definition of the spectral radius) that $\|\mathbf{A}^n\| < \rho^n$. Hence,

$$\|\mathbf{X}''_k\| < M\rho^n$$

for large enough n. Since n can be arbitrarily large, it follows that $\|\mathbf{X}''_k\| = 0$, and $\mathbf{X}_k$ must be unique. □

Without the uniform bound on the sequence $\{\|\mathbf{X}_k\|\}$, uniqueness of the sequence cannot be assured. This already occurs in the LTI context, and example of this is given at the end of the chapter.

The realization which we have obtained has its state in $\overline{\mathcal{H}}$, and may therefore be called a canonical controller state realization. Alternatively, we can choose the state operator in the output state space, in which case we obtain a canonical observer realization. This will be derived in a subsequent section.

Canonical controller realization

Although the state-space description in the form of operator recursions as in theorem 5.12 is the core of any state realization, it is not very useful for our purposes yet. If we assume the state space to be of locally finite dimension, then by choosing a sliced orthonormal basis representation $\mathbf{Q}$ in $\overline{\mathcal{H}}$ such that $\overline{\mathcal{H}} = \mathcal{D}_2\mathbf{Q}$ (*viz.* section 4.3), we can derive concrete matrix representations for the abstract operators $\mathbf{A}$, $\mathbf{B}$ in terms of $\mathbf{Q}$, and produce state space descriptions based on diagonal operators A,B,C,D. Thus let

$$\mathbf{X}_k = X_{[k]}\mathbf{Q}, \qquad X_{[k]} \in \mathcal{D}_2.$$

Using the projection formula, $\mathbf{P}_{\mathcal{H}}(\cdot) = \mathbf{P}_0(\cdot\,\mathbf{Q}^*)\mathbf{Q}$ (thm. 4.9), gives $\mathbf{X}_k = \mathbf{P}_{\mathcal{H}}(Z^{-k}U) = \mathbf{P}_0(Z^{-k}U\mathbf{Q}^*)\mathbf{Q}$, so that

$$X_{[k]} = \mathbf{P}_0(Z^{-k}U\mathbf{Q}^*).$$

Also, $\mathbf{P}(\mathbf{X}_k T) = X_{[k]}\mathbf{P}(\mathbf{Q}T)$. The factorization of the Hankel operator in (5.31) thus becomes

$$\cdot H_T = \mathbf{P}_0(\cdot\,\mathbf{Q}^*)\mathbf{P}(\mathbf{Q}T). \tag{5.34}$$

Comparing (5.34) with the factorization of H_T obtained in section 5.1, we see that the reachability operator of a realization based on this factorization is $\mathbf{P}_0(\cdot\,\mathbf{Q}^*)$. As $\Lambda_\mathbf{Q} = \mathbf{P}_0(\mathbf{Q}\mathbf{Q}^*) = I$, such a realization will be uniformly reachable, and even be in input normal form. The operator $\mathbf{F}_o := \mathbf{P}(\mathbf{Q}T)$ is the observability operator. $\mathbf{F}_o$ is one-to-one, because

$$\begin{aligned} D\mathbf{F}_o = 0 \quad &\Leftrightarrow \quad \mathbf{P}(D\mathbf{Q}T) = 0 \\ &\Leftrightarrow \quad D\mathbf{Q} \in \mathcal{K} \\ &\Rightarrow \quad D = 0, \end{aligned}$$

since $\mathbf{Q}$ forms a sliced basis of the orthogonal complement of $\mathcal{K}$. Hence a realization based on this factorization is observable and minimal.

The following theorem is the main theorem of this section: it gives an explicit realization $\{A,B,C,D\}$ in terms of the basis $\mathbf{Q}$ of the input state space.

Theorem 5.13 (canonical controller realization) *Let $T \in \mathcal{U}(\mathcal{M},\mathcal{N})$ be a given transfer operator with input state space $\mathcal{H}$ of locally finite dimensions. Let $\mathbf{Q}$ be a sliced orthonormal basis representation of $\overline{\mathcal{H}}$: $\overline{\mathcal{H}} = \mathcal{D}_2^{\mathcal{B}}\mathbf{Q}$, $\Lambda_\mathbf{Q} = I$, where $\mathcal{B}$ is defined by* sdim $\mathcal{H} = \#\mathcal{B}$. *Define*

$$\begin{array}{ll} A \in \mathcal{D}(\mathcal{B},\mathcal{B}^{(-1)}) & C \in \mathcal{D}(\mathcal{B},\mathcal{N}) \\ B \in \mathcal{D}(\mathcal{M},\mathcal{B}^{(-1)}) & D \in \mathcal{D}(\mathcal{M},\mathcal{N}) \end{array} \qquad \begin{bmatrix} A & C \\ B & D \end{bmatrix} = \begin{bmatrix} \mathbf{P}_0(Z^{-1}\mathbf{Q}\mathbf{Q}^*)^{(-1)} & \mathbf{P}_0(\mathbf{Q}T) \\ \mathbf{P}_0(Z^{-1}\mathbf{Q}^*)^{(-1)} & \mathbf{P}_0(T) \end{bmatrix}$$

Then, for $U \in \mathcal{X}_2^{\mathcal{M}}$, $Y \in \mathcal{X}_2^{\mathcal{N}}$, there exists a uniformly bounded sequence of states $X_{[k]} \in \mathcal{D}_2$, $k = -\infty,\cdots,\infty$ such that

$$Y = UT \quad \Rightarrow \quad \left\{ \begin{array}{rcl} X_{[k+1]}^{(-1)} &=& X_{[k]}A + U_{[k]}B \\ Y_{[k]} &=& X_{[k]}C + U_{[k]}D \qquad (\text{all } k)\,. \end{array} \right. \tag{5.35}$$

This realization is observable and uniformly reachable (hence minimal), in input normal form, and $\ell_A \le 1$.

If $\ell_A < 1$, then there is a unique uniformly bounded solution $\{X_{[k]}\}$. In that case, the operator X given by $X = \sum_k Z^k X_{[k]}$ is in $\mathcal{X}_2^{\mathcal{B}}$, and (5.35) is equivalent to

$$\left\{ \begin{array}{rcl} XZ^{-1} &=& XA + UB \\ Y &=& XC + UD. \end{array} \right.$$

(At the end of this chapter we give an example to show that the qualification "uniformly bounded" in the uniqueness assertion is needed.)

PROOF Starting from the operator realization in theorem 5.12, define $X_{[k]} := \mathbf{P}_0(\mathbf{X}_k\mathbf{Q}^*)$, so that $\mathbf{X}_k = X_{[k]}\mathbf{Q}$. Then

$$\begin{array}{rclcll} \mathbf{X}_{k+1} = X_{[k+1]}\mathbf{Q} &=& \mathbf{X}_k\mathbf{A} &+& U_{[k]}\mathbf{B} & \\ &=& \mathbf{P}_\mathcal{H}(Z^{-1}\mathbf{X}_k) &+& \mathbf{P}_\mathcal{H}(Z^{-1}U_{[k]}) & \\ &=& \mathbf{P}_0(Z^{-1}\mathbf{X}_k\mathbf{Q}^*)\mathbf{Q} &+& \mathbf{P}_0(Z^{-1}U_{[k]}\,\mathbf{Q}^*)\mathbf{Q} & [\text{thm. } 4.9] \\ &=& \mathbf{P}_0(Z^{-1}X_{[k]}\mathbf{Q}\mathbf{Q}^*)\mathbf{Q} &+& \mathbf{P}_0(Z^{-1}U_{[k]}\,\mathbf{Q}^*)\mathbf{Q} & \\ &=& X_{[k]}^{(1)}\mathbf{P}_0(Z^{-1}\mathbf{Q}\mathbf{Q}^*)\mathbf{Q} &+& U_{[k]}^{(1)}\mathbf{P}_0(Z^{-1}\mathbf{Q}^*)\mathbf{Q}, & \end{array}$$

that is, $X_{[k+1]}^{(-1)} = X_{[k]}\mathbf{P}_0(Z^{-1}\mathbf{Q}\mathbf{Q}^*)^{(-1)} + U_{[k]}\mathbf{P}_0(Z^{-1}\mathbf{Q}^*)^{(-1)}$.

In the same way,

$$\begin{aligned} Y_{[k]} &= \mathbf{X}_k\mathbf{C} + U_{[k]}\mathbf{D} \\ &= \mathbf{P}_0(\mathbf{X}_k T) + \mathbf{P}_0(U_{[k]}T) \\ &= X_{[k]}\mathbf{P}_0(\mathbf{Q}T) + U_{[k]}\mathbf{P}_0(T)\,. \end{aligned}$$

The definition of A is connected to the definition of $\mathbf{A}$ via the chosen basis $\mathbf{Q}$ as

$$(D\mathbf{Q})\mathbf{A} = D^{(1)}A^{(1)}\mathbf{Q} \qquad (\text{any } D \in \mathcal{D}_2^{\mathcal{B}})\,.$$

Recursive application gives

$$(D\mathbf{Q})\mathbf{A}^n = D^{(n)}A^{\{n\}}\mathbf{Q} \qquad (\text{any } D \in \mathcal{D}_2^{\mathcal{B}})\,, \tag{5.36}$$

where $A^{\{n\}} = A^{(n)}\cdots A^{(1)}$. Hence $\|A^{\{n\}}\| = \|\mathbf{A}^n\| = \|\mathbf{P}_{\mathcal{H}}(Z^{-n}\cdot)\|$, so that $\ell_A = r(\mathbf{A}) \leq 1$. □

Equation (5.36) shows that $\mathbf{Q}$ and A are closely connected. In particular, since (5.36) is valid for any $D \in \mathcal{D}_2^{\mathcal{B}}$, we can derive

$$\mathbf{P}_0(Z^{-n}\mathbf{Q}\mathbf{Q}^*) = A^{\{n\}}\,. \tag{5.37}$$

Similarly, we can show that

$$\mathbf{P}_0(Z^{-n}\mathbf{Q}^*) = B^{(n)}A^{\{n-1\}}\,. \tag{5.38}$$

In view of these relations, it comes as no surprise that we can relate the stability properties of A to the boundedness of $\mathbf{Q}$, as is shown in the following proposition.

Proposition 5.14 *Let $\mathbf{Q}$ be a sliced orthonormal basis representation of $\mathcal{H}$, and let the operators $\mathbf{A}$ and A be as given in theorems 5.12 and 5.13. Then $\mathbf{Q}$ is bounded on $\mathcal{X}_2$ if and only if $\ell_A < 1$.*

PROOF The proof is technical and relegated to an appendix at the end of the chapter. □

The formulation in theorem 5.13 has carefully avoided to state that $X \in \mathcal{X}_2$. Clearly $\mathbf{X}_k = X_{[k]}\mathbf{Q} = \mathbf{P}_{\mathcal{H}}(Z^{-k}U)$ produces an $X = \sum_k Z^k X_{[k]}$ which satisfies $XZ^{-1} = XA + UB$. However, X is not guaranteed to be HS-bounded, unless $\ell_A < 1$, since then $X = UBZ(I-AZ)^{-1}$ with $U \in \mathcal{X}_2$ and $BZ(I-AZ)^{-1} \in \mathcal{X}$. An alternative expression for X follows from

$$X_{[k]} = \mathbf{P}_0(Z^{-k}U\mathbf{Q}^*)\,.$$

Hence we can write $X = U\mathbf{Q}^*$. Again this does not guarantee that $X \in \mathcal{X}_2$, since $\cdot\mathbf{Q}^*$ is not necessarily a bounded operator on $\mathcal{U}_2$, although the expression is well defined as a collection of inner products. (See section 4.3 for a discussion on this.)

If $\ell_A = 1$, then an equation like $XZ^{-1} = XA$ may have non-zero solutions with uniformly bounded norms $\{\|X_{[k]}\|\}$. E.g., if $A = I$, it will suffice to take for all $\{i,j\}$, $X_{i,j} = X_{i,j+1}$. A solution with uniformly bounded $\{\|X_{[k]}\|\}$ will not be unique!

Related realizations can be derived if a different, possibly non-orthogonal basis in $\overline{\mathcal{H}}$ is chosen. For canonical results, we have to require that this alternative basis is a strong (sliced) basis, *i.e.*, has a uniformly positive Gramian (*viz.* the definitions in section 4.3). The realization that is obtained in this case is linked to the realization based on $\mathbf{Q}$ via an invertible state transformation.

Theorem 5.15 *Let $T \in \mathcal{U}(\mathcal{M},\mathcal{N})$ be a given transfer operator with input state space $\mathcal{H}$ of locally finite dimensions.*

If $\mathbf{F}$ is a strong sliced basis representation of $\mathcal{H}$, $\mathcal{H} = \mathcal{D}_2^{\mathcal{B}}\mathbf{F}$, such that $\Lambda_{\mathbf{F}}$ is bounded and $\Lambda_{\mathbf{F}} = \mathbf{P}_0(\mathbf{F}\mathbf{F}^*) \gg 0$, then T has a state realization

$$\begin{bmatrix} A & C \\ B & D \end{bmatrix} = \begin{bmatrix} \Lambda_{\mathbf{F}}^{-1}\mathbf{P}_0(Z^{-1}\mathbf{F}\mathbf{F}^*)^{(-1)} & \Lambda_{\mathbf{F}}^{-1}\mathbf{P}_0(\mathbf{F}T) \\ \mathbf{P}_0(Z^{-1}\mathbf{F}^*)^{(-1)} & \mathbf{P}_0(T) \end{bmatrix}$$

$$\begin{array}{ll} A \in \mathcal{D}(\mathcal{B},\mathcal{B}^{(-1)}) & C \in \mathcal{D}(\mathcal{B},\mathcal{N}) \\ B \in \mathcal{D}(\mathcal{M},\mathcal{B}^{(-1)}) & D \in \mathcal{D}(\mathcal{M},\mathcal{N}). \end{array}$$

This realization is uniformly reachable, observable, and has $\ell_A \le 1$.

PROOF The realization follows from theorem 5.12 in the same way as the realization in theorem 5.13 was derived, but now with the projector onto $\overline{\mathcal{H}}$ written in terms of $\mathbf{F}$: $\mathbf{P}_{\mathcal{H}}(\cdot) = \mathbf{P}_0(\cdot\,\mathbf{F}^*)\,\Lambda_{\mathbf{F}}^{-1}\,\mathbf{F}$ (*viz.* equation (4.17)), and the choice of $X_{[k]} = \mathbf{P}_0(\mathbf{X}_k\mathbf{F}^*)$ so that $\mathbf{X}_k = X_{[k]}\Lambda_{\mathbf{F}}^{-1}\mathbf{F}$. (The rest of the proof is straightforward and omitted.) □

When $\mathbf{F}$ is written in terms of a sliced orthonormal basis representation $\mathbf{Q}$ of $\overline{\mathcal{H}}$,

$$\begin{aligned} \mathbf{F} &= R^*\mathbf{Q} \\ \Lambda_{\mathbf{F}} &= \mathbf{P}_0(\mathbf{F}\mathbf{F}^*) = R^*R \end{aligned}$$

(where $R \in \mathcal{D}(\mathcal{B},\mathcal{B})$ is a boundedly invertible factor of $\Lambda_{\mathbf{F}}$), then the above realization based on $\mathbf{F}$ can be "normalized" to obtain the realization based on $\mathbf{Q}$ via a state transformation $X \to X'R$, where X' is a state in the realization based on $\mathbf{Q}$. This provides another way to derive theorem 5.15 from theorem 5.13.

The realization based on the basis representation $\mathbf{F}$ of $\mathcal{H}$ provides a factorization of H_T into

$$\cdot H_T \;=\; \mathbf{P}_0(\cdot\,\mathbf{F}^*)\,\Lambda_{\mathbf{F}}^{-1}\mathbf{P}(\mathbf{F}T). \tag{5.39}$$

The realization is uniformly reachable by construction: the reachability operator is given by $\mathbf{P}_0(\cdot\,\mathbf{F}^*)$, with Gramian $\Lambda_{\mathbf{F}} \gg 0$. The observability operator is $\mathbf{F}_o = \Lambda_{\mathbf{F}}^{-1}\mathbf{P}(\mathbf{F}T)$. The fact that $\mathbf{F}_o$ is one-to-one on $\mathcal{D}_2^{\mathcal{B}}$ is proven in the same way as done for the realization based on $\mathbf{Q}$, and hence the realization is observable and minimal.

Numerical example

To illustrate some of the above with a numerical example, consider again the transfer matrix T given in equation (3.28). The range of the Hankel operator H_T is given locally by the row spaces of the Hankel matrices $\{H_k\}$, and likewise for the range of H_T. Basis vectors for these ranges are given in turn by the V- and U-matrices of the SVDs of the $\{H_k\}$ that have already been computed in section 3.4. Hence, for example,

$$\begin{aligned} \mathcal{C}_1 &= [\cdot] \\ \mathcal{C}_2 &= [1] \\ \mathcal{C}_3 &= \begin{bmatrix} .955 & .298 \\ .298 & -.955 \end{bmatrix} \end{aligned}$$

etcetera. The operator $\mathbf{F} = \mathbf{Q}$ as used in the present section is obtained by stacking these matrices into one upper operator. This gives

$$\mathbf{F} = \left[\begin{array}{cccccccc} & & \vdots & & & & \vdots & \\ \hline \cdots & \cdot & \boxed{\cdot} & \cdot & \cdot & \cdot & \cdot & \cdots \\ \hline \cdots & \cdot & 1 & 0 & 0 & 0 & & \\ \hline \cdots & \cdot & .298 & .955 & 0 & 0 & & \\ \cdots & \cdot & -.955 & .298 & 0 & 0 & & \\ \hline \cdots & \cdot & 0.080 & 0.412 & 0.908 & 0 & & \\ \cdots & \cdot & -0.428 & -0.808 & 0.405 & 0 & & \\ \cdots & \cdot & 0.901 & -0.420 & 0.112 & 0 & & \\ \hline & & \vdots & & & & \vdots & \end{array}\right]. \tag{5.40}$$

In fact, the i-th row of $\mathbf{F}$ is given by the entries of $\mathcal{C}_i^*$ (after permutation of the rows of $\mathcal{C}_i$ since the definitions of H_k in chapter 3 and this chapter differ in that respect). It is readily verified that $\mathbf{F}$ satisfies $\mathbf{P}_0(\mathbf{F}\mathbf{F}^*) = I$. It can also be shown that theorem 5.13 applied to $\mathbf{F} = \mathbf{Q}$ gives the same realization as the realization algorithm in chapter 3.

Canonical observer realizations

In the previous section, we have defined the state $\mathbf{X}_k$ at point k to be the projection of the "past input" with respect to point k, $U_{p(k)} := \mathbf{P}'(Z^{-k}U)$, onto the input state space $\overline{\mathcal{H}}$. If we select a sliced orthonormal basis or another sliced strong basis of $\mathcal{H}$, we obtain a canonical realization which we called a controller realization because the state is defined via an orthogonal projection of the input data. Dually, we can derive realizations based on a definition of state at the output side of the system. In that case, we obtain canonical realizations in observer form (the state is observed at the output). To this end, we define the operator state $\mathbf{X}_k$ to be the projection of the past input, after transformation by T, onto the output state space $\overline{\mathcal{H}}_o$:

$$\mathbf{X}_k = \mathbf{P}(U_{p(k)}T) \in \mathcal{H}_o. \tag{5.41}$$

The procedure gives rise to an almost trivial factorization of the Hankel operator: for $U_p \in \mathcal{L}_2 Z^{-1}$,

$$Y_f = U_p H_T \quad \Leftrightarrow \quad \left\{ \begin{aligned} \mathbf{X}_0 &= \mathbf{P}(U_p T) \\ Y_f &= \mathbf{X}_0 \end{aligned} \right. \tag{5.42}$$

which generalizes to

$$Y = UT \quad \Leftrightarrow \quad \begin{cases} \mathbf{X}_k &= \mathbf{P}(U_{p(k)}T) \\ Y_{[k]} &= \mathbf{P}_0(\mathbf{X}_k) + U_{[k]}T_{[0]} \,. \end{cases} \tag{5.43}$$

The shift-invariance property from which the state recursions are derived is a consequence of the identity $\mathbf{P}(Z^{-1}\mathbf{P}(\cdot)) = \mathbf{P}(Z^{-1}\cdot)$, from which it follows that

$$[\mathbf{P}(Z^{-1}\cdot)]^n = \mathbf{P}(Z^{-n}\cdot) \tag{5.44}$$

and also that the output state space $\mathcal{H}_o = \mathbf{P}(\mathcal{L}_2 Z^{-1} T)$ is restricted shift-invariant: with $U \in \mathcal{L}_2 Z^{-1} \Rightarrow \mathbf{P}(Z^{-1}\mathbf{P}(UT)) = \mathbf{P}(Z^{-1}UT)$, and hence

$$\mathbf{P}(Z^{-1}\mathcal{H}_o) \subset \mathcal{H}_o \,. \tag{5.45}$$

Theorem 5.16 *Let $T \in \mathcal{U}(\mathcal{M},\mathcal{N})$ be a given transfer operator with output state space $\mathcal{H}_o$. Define bounded operators $\mathbf{A},\mathbf{B},\mathbf{C},\mathbf{D}$ as*

$$\begin{array}{ll} \mathbf{A}: \; \overline{\mathcal{H}}_o \to \overline{\mathcal{H}}_o & \mathbf{C}: \; \overline{\mathcal{H}}_o \to \mathcal{D}_2^{\mathcal{N}} \\ \mathbf{B}: \; \mathcal{D}_2^{\mathcal{M}} \to \overline{\mathcal{H}}_o & \mathbf{D}: \; \mathcal{D}_2^{\mathcal{M}} \to \mathcal{D}_2^{\mathcal{N}} \end{array} \qquad \begin{bmatrix} \mathbf{A} & \mathbf{C} \\ \mathbf{B} & \mathbf{D} \end{bmatrix} = \begin{bmatrix} \mathbf{P}(Z^{-1}\cdot) & \mathbf{P}_0(\cdot) \\ \mathbf{P}(Z^{-1}\cdot T) & \mathbf{P}_0(\cdot T) \end{bmatrix}$$

Then, for $U \in \mathcal{X}_2^{\mathcal{M}}$, $Y \in \mathcal{X}_2^{\mathcal{N}}$, the sequence $\{\mathbf{X}_k = \mathbf{P}(U_{p(k)}T)\}$ is uniformly bounded, and satisfies

$$\begin{cases} \mathbf{X}_{k+1} &= \mathbf{X}_k\mathbf{A} + U_{[k]}\mathbf{B} \\ Y_{[k]} &= \mathbf{X}_k\mathbf{C} + U_{[k]}\mathbf{D} \qquad (\text{all } k). \end{cases} \tag{5.46}$$

$\mathbf{A}$ has spectral radius $r(\mathbf{A}) \le 1$. If $r(\mathbf{A}) < 1$, then there there is only one uniformly bounded solution for (5.46) and it is given by $\{\mathbf{X}_k\}$.

PROOF We first show that, for $\cdot\mathbf{A} = \mathbf{P}(Z^{-1}\cdot)$, $\cdot\mathbf{B} = \mathbf{P}(Z^{-1}\cdot T)$, it follows that $\mathbf{X}_k = \mathbf{P}(U_{p(k)}T)$ satisfies

$$\mathbf{X}_{k+1} = \mathbf{X}_k\mathbf{A} + U_{[k]}\mathbf{B}.$$

From (5.44), we have that $\cdot\mathbf{A}^n = \mathbf{P}(Z^{-n}\cdot)$, and $\cdot\mathbf{B}\mathbf{A}^{n-1} = \mathbf{P}(Z^{-n}\cdot T)$. Also, with $\mathbf{X}_k = \mathbf{P}(U_{p(k)}T)$, equation (5.43) directly gives

$$Y_{[k]} = \mathbf{P}_0(\mathbf{X}_k) + U_{[k]}T_{[0]} = \mathbf{X}_k\mathbf{C} + U_{[k]}\mathbf{D},$$

for $\mathbf{C} = \mathbf{P}_0(\cdot)$ and $\mathbf{D} = \mathbf{P}_0(\cdot T)$. Uniqueness is shown in a similar fashion as in theorem 5.12. □

Note that, if $\mathbf{X}_k \in \overline{\mathcal{H}}_o$, then $\mathbf{X}_k\mathbf{A} = \mathbf{P}(Z^{-1}\mathbf{X}_k) \in \overline{\mathcal{H}}_o$ as required, because of the shift-invariance property of $\overline{\mathcal{H}}_o$ (equation 5.45)).

A realization is obtained by chosing a strong sliced basis in $\overline{\mathcal{H}}_o$. Assume the system to be locally finite, and let $\mathbf{G}$ be an orthonormal sliced basis representation: $\overline{\mathcal{H}}_o = \mathcal{D}_2\mathbf{G}$, $\Lambda_{\mathbf{G}} = I$. Then

$$\mathbf{X}_k = X_{[k]}\mathbf{G}, \qquad X_{[k]} = \mathbf{P}_0(\mathbf{X}_k\mathbf{G}^*).$$

In particular, $\mathbf{X}_0 = \mathbf{P}(U_pT) = \mathbf{P}_{\mathcal{H}_o}(U_pT) = \mathbf{P}_0((U_pT)\mathbf{G}^*)\mathbf{G}$, so that $X_{[0]} = \mathbf{P}_0(U_pT\mathbf{G}^*)$. Hence, the factorization of H_T in equation (5.42) becomes

$$\cdot H_T = \mathbf{P}_0(\cdot T\mathbf{G}^*)\,\mathbf{G}.$$

A realization based on this factorization has observability operator $\mathbf{G}$, with observability Gramian $\Lambda_\mathbf{G} = I$, and reachability operator $\mathbf{P}_0(\cdot T\mathbf{G}^*) =: \mathbf{P}_0(\cdot \mathbf{F}^*)$, where $\mathbf{F} = \mathbf{P}'(\mathbf{G}T^*)$. Its kernel $\ker(\cdot\mathbf{F})\big|_{\mathcal{D}_2} = 0$, because, for any $D \in \mathcal{D}_2^{\mathcal{B}}$,

$$\begin{array}{rcl} D\mathbf{F} = 0 & \Leftrightarrow & \mathbf{P}'(D\mathbf{G}T^*) = 0 \\ & \Leftrightarrow & D\mathbf{G} \in \mathcal{K}_o \\ & \Rightarrow & D = 0. \end{array}$$

Hence, the realization is reachable, but not necessarily uniformly.

Theorem 5.17 (canonical observer realization) *Let $T \in \mathcal{U}(\mathcal{M},\mathcal{N})$ be a given transfer operator with output state space $\mathcal{H}_o$ of locally finite dimensions. Let $\mathbf{G}$ be an orthonormal sliced basis representation of $\overline{\mathcal{H}}_o$: $\overline{\mathcal{H}}_o = \mathcal{D}_2^{\mathcal{B}}\mathbf{G}$, $\Lambda_\mathbf{G} = I$.*

$$\begin{array}{ll} A \in \mathcal{D}(\mathcal{B},\mathcal{B}^{(-1)}) & C \in \mathcal{D}(\mathcal{B},\mathcal{N}) \\ B \in \mathcal{D}(\mathcal{M},\mathcal{B}^{(-1)}) & D \in \mathcal{D}(\mathcal{M},\mathcal{N}) \end{array} \qquad \begin{bmatrix} A & C \\ B & D \end{bmatrix} = \begin{bmatrix} \mathbf{P}_0(Z^{-1}\mathbf{G}\mathbf{G}^*)^{(-1)} & \mathbf{P}_0(\mathbf{G}) \\ \mathbf{P}_0(Z^{-1}T\mathbf{G}^*)^{(-1)} & \mathbf{P}_0(T) \end{bmatrix}$$

Then, for $U \in \mathcal{X}_2^{\mathcal{M}}$, $Y \in \mathcal{X}_2^{\mathcal{N}}$, there exists a uniformly bounded sequence of states $X_{[k]} \in \mathcal{D}_2, k = -\infty,\cdots,\infty$ such that

$$Y = UT \quad \Rightarrow \quad \left\{ \begin{array}{rcl} X_{[k+1]}^{(-1)} & = & X_{[k]}A + U_{[k]}B \\ Y_{[k]} & = & X_{[k]}C + U_{[k]}D \end{array} \right. \quad (\text{all } k). \tag{5.47}$$

The realization is reachable and uniformly observable (hence minimal), in output normal form, and has $\ell_A \leq 1$. If $\ell_A < 1$, then (5.47) has a unique uniformly bounded solution for which $X = \sum_k Z^k X_{[k]}$ is in $\mathcal{X}_2^{\mathcal{B}}$.

PROOF For a given $\mathbf{X}_k$ in $\overline{\mathcal{H}}_o$, put $\mathbf{X}_k = X_{[k]}\mathbf{G}$, with $X_{[k]} \in \mathcal{D}_2^{\mathcal{B}}$. Then

$$\begin{array}{rcccl} \mathbf{X}_{k+1} = X_{[k+1]}\mathbf{G} & = & \mathbf{P}(Z^{-1}\mathbf{X}_k) & + & \mathbf{P}(Z^{-1}U_{[k]}T) \\ & = & \mathbf{P}_{\mathcal{H}_o}(Z^{-1}\mathbf{X}_k) & + & \mathbf{P}_{\mathcal{H}_o}(Z^{-1}U_{[k]}T) \\ & = & \mathbf{P}_{\mathcal{H}_o}(Z^{-1}X_{[k]}\mathbf{G}) & + & \mathbf{P}_{\mathcal{H}_o}(Z^{-1}U_{[k]}T) \\ & = & \mathbf{P}_0(Z^{-1}X_{[k]}\mathbf{G}\mathbf{G}^*)\mathbf{G} & + & \mathbf{P}_0(Z^{-1}U_{[k]}T\mathbf{G}^*)\mathbf{G} \\ & = & X_{[k]}^{(1)}\mathbf{P}_0(Z^{-1}\mathbf{G}\mathbf{G}^*)\mathbf{G} & + & U_{[k]}^{(1)}\mathbf{P}_0(Z^{-1}T\mathbf{G}^*)\mathbf{G}. \end{array}$$

Hence $A = \mathbf{P}_0(Z^{-1}\mathbf{G}\mathbf{G}^*)^{(-1)}$ and $B = \mathbf{P}_0(Z^{-1}T\mathbf{G}^*)^{(-1)}$. In the same way,

$$\mathbf{P}_0(\mathbf{X}_k) = \mathbf{P}_0(X_{[k]}\mathbf{G}) = X_{[k]}\,\mathbf{P}_0(\mathbf{G}),$$

hence $C = \mathbf{P}_0(\mathbf{G})$. The fact that the realization is minimal follows from the minimality of the corresponding factorization of H_T. The state variables are uniformly bounded:

$$\|X_{[k]}\| \leq \|U\|\,\|T\|.$$

Finally, uniqueness is proven in the same way as in theorem 5.13. □

We can generalize the canonical observer realization if we allow a non-orthonormal sliced basis $\mathbf{F}_o$ for $\overline{\mathcal{H}}_o$.

Theorem 5.18 *Let $T \in \mathcal{U}(\mathcal{M},\mathcal{N})$ be a given transfer operator with output state space $\mathcal{H}_o$ of locally finite dimensions.*

If $\mathbf{F}_o$ is a strong sliced basis representation of $\overline{\mathcal{H}}_o$: $\overline{\mathcal{H}}_o = \mathcal{D}_2^{\mathcal{B}}\mathbf{F}_o$, such that $\Lambda_{\mathbf{F}_o}$ is bounded and $\Lambda_{\mathbf{F}_o} = \mathbf{P}_0(\mathbf{F}_o\mathbf{F}_o^*) \gg 0$, then T has a state realization

$$\begin{bmatrix} A & C \\ B & D \end{bmatrix} = \begin{bmatrix} \mathbf{P}_0(Z^{-1}\mathbf{F}_o\mathbf{F}_o^*)^{(-1)}\Lambda_{\mathbf{F}_o}^{-(-1)} & \mathbf{P}_0(\mathbf{F}_o) \\ \mathbf{P}_0(Z^{-1}T\mathbf{F}_o^*)^{(-1)}\Lambda_{\mathbf{F}_o}^{-(-1)} & \mathbf{P}_0(T) \end{bmatrix}$$

$$\begin{array}{ll} A \in \mathcal{D}(\mathcal{B},\mathcal{B}^{(-1)}) & C \in \mathcal{D}(\mathcal{B},\mathcal{N}) \\ B \in \mathcal{D}(\mathcal{M},\mathcal{B}^{(-1)}) & D \in \mathcal{D}(\mathcal{M},\mathcal{N})\,. \end{array}$$

This realization is reachable, uniformly observable, and has $\ell_A \leq 1$.

PROOF The proof follows from theorem 5.17 and can be derived by taking a state transformation $X = X'R$, such that $\mathbf{F}_o = R\mathbf{G}$ for a sliced orthogonal basis $\mathbf{G}$.

□

The factorization of H_T corresponding to this realization is

$$H_T = \mathbf{P}_0(\cdot T\mathbf{F}_o^*)\Lambda_{\mathbf{F}_o}^{-1}\mathbf{F}_o\,. \tag{5.48}$$

The realization is uniformly observable by construction: the observability Gramian is $\Lambda_{\mathbf{F}_o} \gg 0$. The reachability operator is given by $\cdot\mathbf{F}^* = \mathbf{P}_{Z\mathcal{U}}(\cdot T\mathbf{F}_o^*)\Lambda_{\mathbf{F}_o}^{-1}$; the fact that $\mathbf{F}$ is one-to-one on $\mathcal{D}_2^{\mathcal{B}}$ is proven in the same way as before, and hence the realization is reachable and minimal.

Realization theorem for operators

The preceding theorems, along with proposition 5.6, amount to a converse of corollary 5.7:

Theorem 5.19 (Kronecker-type thm, II) *Let $T \in \mathcal{U}$ be a locally finite operator, and let $\mathcal{H}$ and $\mathcal{H}_o$ be respectively the corange and range of its Hankel operator $\cdot H_T$. Then there exist minimal realizations $\{A,B,C,D\}$ for T for which $\ell_A \leq 1$ and for which the state is observable and uniformly reachable.*

Dually, there exist minimal realizations for which $\ell_A \leq 1$ and for which the state is reachable and uniformly observable. Uniformly minimal realizations with $\ell_A \leq 1$ exist if and only if the range of H_T is closed.

As mentioned before, we are primarily interested in cases where $\ell_A < 1$: u.e. stable realizations. Such a realization occurs if the basis for the subspaces $\mathcal{H}$ or $\mathcal{H}_o$ from which the realization is constructed generates a bounded operator $\mathbf{Q}$ or $\mathbf{G}$. It is possible that a given $T \in \mathcal{U}$ has subspaces $\mathcal{H}$ and $\mathcal{H}_o$ that do not have such bounded basis operators, so that it does not have a u.e. stable realization. An example is given later in this section.

Because the canonical controller and observer realizations both provide a factorization of the Hankel operator H_T, there is a connection between the two representations.

Theorem 5.20 *Given a bounded system transfer operator $T \in \mathcal{U}$ with locally finite dimensional state spaces $\overline{\mathcal{H}}$ and $\overline{\mathcal{H}}_o$, let* $\mathbf{F}$ *be the representation of a strong sliced basis in $\overline{\mathcal{H}}$. Put*

$$\mathbf{F}_o = \Lambda_{\mathbf{F}}^{-1}\mathbf{P}(\mathbf{F}T)$$

and suppose that $\mathbf{F}_o$ represents a strong sliced basis ($\Lambda_{\mathbf{F}_o} \gg 0$). Then the canonical realization based on $\mathbf{F}$ *(theorem 5.15) is identical to the canonical realization based on $\mathbf{F}_o$ (theorem 5.18).*

PROOF The factorizations of H_T in equations (5.39) and (5.48) are

$$H_T = \mathbf{P}_0(\cdot\mathbf{F}^*)\Lambda_{\mathbf{F}}^{-1}\mathbf{P}(\mathbf{F}T) = \mathbf{P}_0(\cdot T\mathbf{F}_o^*)\Lambda_{\mathbf{F}_o}^{-1}\mathbf{F}_o\,.$$

The realization corresponding to the first factorization has $\mathbf{P}_0(\cdot\mathbf{F}^*)$ as its reachability operator and $\Lambda_{\mathbf{F}}^{-1}\mathbf{P}(\mathbf{F}T)$ as its observability operator; the realization corresponding to the second factorization has $\mathbf{P}_0(\cdot T\mathbf{F}_o^*)\Lambda_{\mathbf{F}_o}$ as its reachability operator and $\mathbf{F}_o$ as its observability operator. If we take $\mathbf{F}_o = \Lambda_{\mathbf{F}}^{-1}\mathbf{P}(\mathbf{F}T)$, then the two realizations have the same observability operator. As the realizations are observable, we must have that $\mathbf{P}_0(\cdot\mathbf{F}^*) = \mathbf{P}_0(\cdot T\mathbf{F}_o^*)\Lambda_{\mathbf{F}_o}$, so that they also have the same reachability operator. The result follows by noting that two realizations that have the same reachability operator must have the same $\{A,B\}$-pair, and two realizations that have the same observability operator must have the same $\{A,C\}$-pair. □

SVD-based realizations and balanced realizations

We obtain bases $\mathbf{Q}$ and $\mathbf{G}$ in a generic way via a singular value decomposition of the snapshots of H_T. Let $T \in \mathcal{U}$ be locally finite. Then there exist $\mathbf{Q}$, $\mathbf{G}$, $\hat{\Sigma}$ such that

$$\cdot H_T = \mathbf{P}_0(\cdot\mathbf{Q}^*)\hat{\Sigma}\mathbf{G} \qquad \text{with} \qquad \left\{\begin{array}{rcll} \mathcal{D}_2^{\mathcal{B}}\mathbf{Q} &=& \overline{\mathcal{H}}, & \Lambda_{\mathbf{Q}} = I \\ \mathcal{D}_2^{\mathcal{B}}\mathbf{G} &=& \overline{\mathcal{H}}_o, & \Lambda_{\mathbf{G}} = I \\ \hat{\Sigma} &\in& \mathcal{D}(\mathcal{B},\mathcal{B}), & \hat{\Sigma}^* = \hat{\Sigma}. \end{array}\right. \tag{5.49}$$

in which, moreover, each $\hat{\Sigma}$ is diagonal and has non-negative entries in decreasing order. We produce this factorization of H_T by computing the singular value decomposition of its snapshots H_k (as in section 3.4), putting the singular vectors whose span is the range of H_k^* and H_k into $\mathbf{Q}_k$ and $\mathbf{G}_k$, and putting the non-zero singular values into $\hat{\Sigma}_k$. Then $\mathbf{Q}$, $\mathbf{G}$ are obtained by stacking the $\mathbf{Q}_i$ and $\mathbf{G}_i$ (like was done in equations (4.9)), and setting $\hat{\Sigma} = \mathrm{diag}[\hat{\Sigma}_k]_{-\infty}^{\infty}$. Since $\|H_k\| = \|\hat{\Sigma}_k\|$, also $\|H_T\| = \|\hat{\Sigma}\|$. The ensuing factorizations corresponding to the canonical realizations we derived earlier in this section are

$$\begin{array}{rcll} H_T &=& [\mathbf{P}_0(\cdot\mathbf{Q}^*)]\,[\hat{\Sigma}\mathbf{G}] = \mathbf{P}_0(\cdot\mathbf{F}^*)\,\mathbf{F}_o\,, & (\mathbf{F} = \mathbf{Q},\ \mathbf{F}_o = \hat{\Sigma}\mathbf{G}) \\ &=& [\mathbf{P}_0(\cdot\mathbf{Q}^*)\hat{\Sigma}]\,\mathbf{G} = \mathbf{P}_0(\cdot\mathbf{F}'^*)\,\mathbf{F}'_o\,, & (\mathbf{F}' = \hat{\Sigma}\mathbf{Q},\ \mathbf{F}'_o = \mathbf{G})\,. \end{array}$$

The factorization of H_T on the first line corresponds to a canonical controller realization on $\mathbf{Q}$ for which $\Lambda_{\mathbf{F}_o} = \hat{\Sigma}^2$, while the second factorization corresponds to a canonical observer realization based on $\mathbf{G}$ and has $\Lambda_{\mathbf{F}'} = \hat{\Sigma}^2$. The actual construction of the realization based on $\mathbf{G}$, according to theorem 5.17, can be done along the lines of algorithm 3.9 in section 3.4.

A realization is said to be *balanced* if its reachability Gramian is equal to its observability Gramian, and if all diagonal entries of $\Lambda_{\mathbf{F}} = \Lambda_{\mathbf{F}_o}$ are diagonal matrices. A realization based on the SVD factorization

$$H_T = [\mathbf{P}_0(\cdot\mathbf{Q}^*)\hat{\Sigma}^{1/2}]\,[\hat{\Sigma}^{1/2}\mathbf{G}]$$

is balanced: $\Lambda_{\mathbf{F}} = \hat{\Sigma}^{1/2}$ and $\Lambda_{\mathbf{F}_o} = \hat{\Sigma}^{1/2}$.

Proposition 5.21 *Let $T \in \mathcal{U}$ be a locally finite operator, and let its Hankel operator H_T have an SVD-based factorization given by (5.49). $\mathcal{H}$ and $\mathcal{H}_o$ are closed subspaces if and only if $\hat{\Sigma}$ is boundedly invertible, and a realization of T which is uniformly reachable and uniformly observable exists if and only if this condition holds.*

PROOF Consider the SVD-based factorization of H_T in terms of (5.49). A realization based on $\mathbf{Q}$ is uniformly reachable, and because $\mathbf{F}_o = \hat{\Sigma}\mathbf{G}$, the observability Gramian is $\Lambda_{\mathbf{F}_o} = \hat{\Sigma}^2$. Hence the realization is observable. It is uniformly observable, $\hat{\Sigma}^2 \gg 0$, if and only if $\hat{\Sigma}^{-1}$ is bounded. According to proposition 5.6, this occurs if and only if $\mathcal{H}$ and $\mathcal{H}_o$ are both closed subspaces. Proposition 5.6 already implied that any other realization can be both uniformly reachable and uniformly observable if and only if these subspaces are closed. □

Anomalies

Some anomalies noted in the previous sections are

1. the basis representations $\mathbf{Q}$, $\mathbf{G}$ of $\mathcal{H}$ and $\mathcal{H}_o$ can be unbounded operators, which occurs if and only if $\ell_A = 1$ (proposition 5.14),
2. H_T, H_T^* can have ranges $\mathcal{H}_o$, $\mathcal{H}$ which are not closed, which occurs if $\hat{\Sigma}$ in proposition 5.21 is not boundedly invertible.

We show by some examples that these phenomena are unconnected. An example that shows that it is not true that $\mathbf{Q}$ and $\mathbf{G}$ bounded implies that $\hat{\Sigma}^{-1}$ is bounded, is provided by

$$T = \begin{bmatrix} 0 & 1/2 & & & \mathbf{0} \\ & \boxed{0} & 1/4 & & \\ & & 0 & 1/8 & \\ \mathbf{0} & & & \ddots & \ddots \end{bmatrix}.$$

$\mathbf{Q}$, $\mathbf{G}$ and $\hat{\Sigma}$ are given by

$$\mathbf{Q} = \begin{bmatrix} 1 & \boxed{0} & & \mathbf{0} \\ & 1 & 0 & \\ \mathbf{0} & & 1 & 0 \\ & & & \ddots & \ddots \end{bmatrix}, \quad \mathbf{G} = \begin{bmatrix} \boxed{1} & & \mathbf{0} \\ & 1 & \\ \mathbf{0} & & 1 \\ & & & \ddots \end{bmatrix}, \quad \hat{\Sigma} = \begin{bmatrix} \boxed{1/2} & & \mathbf{0} \\ & 1/4 & \\ \mathbf{0} & & 1/8 \\ & & & \ddots \end{bmatrix}.$$

$\mathbf{Q}$ and $\mathbf{G}$ are bounded, but $\hat{\Sigma}^{-1}$ is unbounded. A realization based on $\mathbf{Q}$ yields $A_k = 0$, $B_k = 1$, $C_k = 2^{-k-1}$, $D_k = 0$ $(k \geq 0)$. Indeed, the realization is not uniformly observable.

It is also not true that $\hat{\Sigma}^{-1}$ bounded implies that $\mathbf{Q}$, $\mathbf{G}$ are bounded. An example is obtained by considering inner operators (operators T which are both unitary and upper). As shown in chapter 6, such operators have Hankel matrices H_k that are isometries, so that $\hat{\Sigma} = I_{\mathcal{B}}$. We also show in that chapter that a unitary realization $\mathbf{T} = \{A, B, C, D\}$ realizes a unitary operator T. It is, however, possible to construct a sequence of unitary matrices $\mathbf{T}_k$ such that $\ell_A = 1$, a trivial example being

$$\mathbf{T}_k = \begin{bmatrix} c_k & s_k \\ -s_k^* & c_k^* \end{bmatrix}, \qquad c_k^* c_k + s_k^* s_k = 1\,,$$

where $c_k \to 1$ for $k \to \infty$. With $\ell_A = 1$, $\mathbf{Q}$ and $\mathbf{G}$ are unbounded.

Hence there is no connection between the properties $\ell_A < 1$ ($\mathbf{Q}$ and $\mathbf{G}$ bounded) and the fact that $\mathcal{H}$ and $\mathcal{H}_o$ are closed subspaces ($\hat{\Sigma}$ boundedly invertible).

As a pathological example in which some of the above-mentioned aspects occur, consider the operator

$$T = \begin{bmatrix} \boxed{0} & 1 & \frac{1}{2} & \frac{1}{4} & \frac{1}{8} & \frac{1}{16} & \cdots \\ & 0 & \frac{1}{2} & \frac{1}{4} & \frac{1}{8} & \frac{1}{16} & \cdots \\ & & 0 & \frac{1}{4} & \frac{1}{8} & \frac{1}{16} & \cdots \\ & & & 0 & \vdots & & \ddots \end{bmatrix}.$$

T is a bounded operator: it is equal to a diagonal scaling of the bounded LTI system $z(1 - \frac{1}{2}z)^{-1}$. One possible (SVD-based) factorization of its Hankel operators H_k is

$$H_k = \sigma_k \cdot \frac{1}{\sqrt{k}} \left.\begin{bmatrix} 1 \\ 1 \\ \vdots \\ 1 \\ 0 \\ \vdots \end{bmatrix}\right\} k \cdot \frac{1}{p} \begin{bmatrix} 1 & \frac{1}{2} & \frac{1}{4} & \cdots \end{bmatrix} \qquad (k > 0)\,,$$

where $\sigma_k = \frac{p\sqrt{k}}{2^{k-1}}$ and p is equal to the norm of the vector $[1\ \frac{1}{2}\ \frac{1}{4}\ \cdots]$. Each Hankel matrix H_k has only one singular value unequal to 0, and $\sigma_k \to 0$ if $k \to \infty$, hence $\hat{\Sigma}$ is not boundedly invertible. $\mathbf{Q}$ and $\mathbf{G}$ follow from the above decomposition as

$$\mathbf{Q} = \begin{bmatrix} \boxed{\cdot} & & & & & \\ 1 & 0 & & & & \\ \frac{1}{\sqrt{2}} & \frac{1}{\sqrt{2}} & 0 & & & \\ \frac{1}{\sqrt{3}} & \frac{1}{\sqrt{3}} & \frac{1}{\sqrt{3}} & 0 & & \\ \frac{1}{\sqrt{4}} & \frac{1}{\sqrt{4}} & \frac{1}{\sqrt{4}} & \frac{1}{\sqrt{4}} & 0 & \\ & \vdots & & & & \ddots \end{bmatrix} \qquad \mathbf{G} = \begin{bmatrix} \boxed{\cdot} & \frac{1}{p} & \frac{1}{2p} & \frac{1}{4p} & \frac{1}{8p} & \cdots \\ & & \frac{1}{p} & \frac{1}{2p} & \frac{1}{4p} & \cdots \\ & & & \frac{1}{p} & \frac{1}{2p} & \cdots \\ & & & & \ddots & \end{bmatrix}.$$

G is bounded, but **Q** is unbounded, which can be seen, *e.g.*, from the fact that the norms of its columns are unbounded. A realization based on **G** has

$$A_k = \tfrac{1}{2}, \qquad C_k = \tfrac{1}{p}, \qquad (k>0), \tag{5.50}$$

so that $\ell_A = \frac{1}{2}$, but $B_k = \frac{p}{2^{k-1}} \to 0$ $(k \to \infty)$ and the realization is not uniformly reachable. A realization based on **Q** is

$$\begin{aligned} A_k &= \frac{1}{\sqrt{k}}\frac{k}{\sqrt{k+1}} = \frac{\sqrt{k}}{\sqrt{k+1}} \to 1 \quad (k\to\infty) \\ B_k &= \frac{1}{\sqrt{k+1}} \to 0 \end{aligned} \tag{5.51}$$

and indeed $\ell_A = 1$, which was to be expected as **Q** is unbounded.

On the uniqueness of the solution of canonical state equations

As indicated in theorem 5.12, the state sequence $\{X_{[k]}\}$ has to be uniformly bounded or else it cannot be assured to be unique. Some insight in this point might be gained by looking at a time-invariant example, for which we can easily exhibit the non-uniqueness alluded to. Let us take the simple case for which the transfer function is given by $1/(1-az)$, with $0<|a|<1$. We study it in the LTI domain and then translate to our LTV formalism. With scalar inputs and outputs, the input and output $\mathcal{X}_2$ spaces are analogous to L_2 of the unit circle **T** of the complex plane, which we indicate by $L_2(\mathbf{T})$. "Future" inputs and outputs are in $H_2(\mathbf{T})$, the subspace of $L_2(\mathbf{T})$ of functions whose Fourier coefficients are zero for strictly negative indices, while inputs and outputs that belong to the strict past are in $H_2^\perp$, the orthogonal complement of H_2 in $L_2(\mathbf{T})$. In our example, the relevant state spaces and null spaces are known to be[2]

$$\begin{aligned} \hat{\mathcal{H}} &= \left\{ \frac{dz^{-1}}{1-\bar{a}z^{-1}} : d \in \mathbb{C} \right\} \\ \hat{\mathcal{K}} &= H_2^\perp \cdot \left[\frac{z^{-1}-a}{1-\bar{a}z^{-1}} \right] \\ \hat{\mathcal{H}}_o &= \left\{ \frac{d}{1-az} : d \in \mathbb{C} \right\} \\ \hat{\mathcal{K}}_o &= H_2 \cdot \left[\frac{z-\bar{a}}{1-az} \right]. \end{aligned} \tag{5.52}$$

The "hatted" spaces indicate the analogs of the spaces defined earlier, but now in the time-invariant Fourier transform context. (A formal correspondence of spaces can be set up, but here we just present it intuitively.) Translated to our time-varying formalism, the functions of z become Toeplitz operators. This can be viewed as replacing the scalar z in the series expansion of the function by the shift operator Z. For brevity, recall the transformation operator $\mathcal{T}$:

$$f(z) = \cdots z^{-1} f_{-1} + f_0 + z f_1 + \cdots \quad \Leftrightarrow \quad \mathcal{T}(f(z)) = \cdots Z^{-1} f_{-1} + f_0 + Z f_1 + \cdots.$$

[2]For example, $u(z) \in \hat{\mathcal{K}} \Leftrightarrow u(z)\cdot(1/(1-az)) \in H_2^\perp$, which is the case if and only if $u(z)$ can be written as $u_1(z)\cdot \frac{z^{-1}-a}{1-\bar{a}z^{-1}}$ with $u_1(z) \in H_2^\perp$. This follows directly from the fact that $\frac{z^{-1}-a}{1-\bar{a}z^{-1}} \cdot \frac{1}{1-az} = \frac{z^{-1}}{1-\bar{a}z^{-1}}$.

We find that

$$\begin{aligned}
\mathcal{H} &= \{D\mathcal{T}(\tfrac{dz^{-1}}{1-\bar{a}z^{-1}}) : D \in \mathcal{D}_2\} \\
\mathcal{K} &= \mathcal{L}_2 Z^{-1} \cdot \mathcal{T}\left[\tfrac{z^{-1}-a}{1-\bar{a}z^{-1}}\right] \\
\mathcal{H}_o &= \{D\mathcal{T}(\tfrac{d}{1-\bar{a}z}) : D \in \mathcal{D}_2\} \\
\mathcal{K}_o &= \mathcal{U}_2\mathcal{T}\left[\tfrac{z-\bar{a}}{1-az}\right].
\end{aligned} \tag{5.53}$$

The orthogonal space decompositions $H_2^{\perp} = \hat{\mathcal{K}} \oplus \hat{\mathcal{H}}$ and $H_2 = \hat{\mathcal{H}}_o \oplus \hat{\mathcal{K}}_o$ carry over to $\mathcal{L}_2 Z^{-1} = \mathcal{K} \oplus \overline{\mathcal{H}}$ and $\mathcal{U}_2 = \overline{\mathcal{H}}_o \oplus \mathcal{K}_o$ — the proofs for the time-invariant case can be deduced from those for the LTV case. Let

$$E := \mathcal{T}(\frac{z^{-1}}{1-\bar{a}z^{-1}})$$

and let us define

$$\mathbf{X}_k := a^k E. \tag{5.54}$$

Since $z^{-1}a^k \frac{z^{-1}}{1-\bar{a}z^{-1}} = a^k \left(\frac{(z^{-1}-a)z^{-1}}{1-\bar{a}z^{-1}} + a \cdot \frac{z^{-1}}{1-\bar{a}z^{-1}} \right)$ and $\frac{(z^{-1}-a)z^{-1}}{1-\bar{a}z^{-1}} \perp \hat{\mathcal{H}}$, we have

$$\mathbf{P}_{\hat{\mathcal{H}}}(z^{-1}a^k \frac{z^{-1}}{1-\bar{a}z^{-1}}) = a^{k+1} \frac{z^{-1}}{1-\bar{a}z^{-1}}.$$

Hence,

$$\begin{aligned}
\mathbf{P}_{\mathcal{H}}(Z^{-1}a^k E) &= \mathbf{P}_{\mathcal{H}}(Z^{-1}a^k \mathcal{T}(\tfrac{z^{-1}}{1-\bar{a}z^{-1}})) \\
&= \mathbf{P}_{\mathcal{H}}(a^k \mathcal{T}(\tfrac{z^{-2}}{1-\bar{a}z^{-1}})) \\
&= a^k \mathcal{T}(\mathbf{P}_{\hat{\mathcal{H}}}(\tfrac{z^{-2}}{1-\bar{a}z^{-1}})) \\
&= a^{k+1} E,
\end{aligned}$$

so that

$$\mathbf{X}_k \mathbf{A} = \mathbf{P}_{\mathcal{H}}(a^k Z^{-1} E) = \mathbf{X}_{k+1}. \tag{5.55}$$

The sequence $\mathbf{X}_k$ is unbounded for $k \to -\infty$, but it does satisfy $\mathbf{X}_k \mathbf{A} = \mathbf{X}_{k+1}$. When $r(\mathbf{A}) = 1$, obviously no uniqueness statement can be made, *e.g.*, when $a = 1$, the sequence just exhibited would be a (uniformly) bounded solution of the autonomous system.

5.5 NOTES

The concept of state originated as an abstraction of computer memory in automaton theory [Ner58]. It entered system theory in the late 1950s when the connection with first-order differential equations became clear. During the 1960s, much effort was put into the construction of state models for continuous-time LTI and LTV systems specified by their impulse response $H(t,\tau)$, such that $y(t) = \int H(t,\tau)u(\tau)d\tau$. Among the initial results was the proof that realizability is equivalent to the separability of the impulse response matrix into $H(t,\tau) = \Psi(t)\Theta(\tau)$. However, the effective construction of this factorization was difficult, and even not always possible, and the direct realizations that were produced were not always asymptotically stable [Kam79]. For LTI systems, state-space realization synthesis began with the work of Kalman and his co-workers

[Kal63, HK66, KFA70], Gilbert [Gil63] and Youla [You66]. The use of the Hankel matrix, which does not require a separable form of the impulse response matrix, resulted in the Ho-Kalman algorithm [HK66], which was independently obtained by Youla and Tissi [YT66]. In the 1970s, a new tool carried over from linear algebra into the world of system theory: the singular value decomposition. With this tool, a numerically robust way became available to compute the factorization of the Hankel matrix. The SVD was incorporated into the realization algorithm by Moore in 1978 (see [Moo79, Moo81]) in the context of continuous-time systems for the purpose of *balancing* the realization. There are closely related papers by Zeiger and McEwen [ZM74] and by Pernebo and Silverman [PS79]. It was realized at that time that a balanced realization can be approximated very straightforwardly, and the resulting combination (reported by Kung in 1978 [Kun78] for discrete-time systems) gave rise to a class of robust *identification* algorithms, called Principal Component identification techniques.

For continuous-time time-varying systems with a constant system order, a realization theory was developed by Silverman and Meadows [SM66, SA68, SM69]. Reachability and stability issues were treated also in [AM69]. Kamen extended Kalman's algebraic module theory to incorporate a continuous-time pure delay operator [Kam75, Kam76a], and considered the realization by state-space models of systems $Ay(t) = Bu(t)$, where A and B are matrix polynomials in the differential operator p and unit delay operator d. For time-varying systems, these results could be extended by using a non-commutative ring of polynomials [Kam76b].

The development of discrete-time realization theory for LTV systems started in the 1970s with the work of Weiss [Wei72] and Evans [Eva72]. The concepts of reachability, observability and minimality were defined (see also [AM81]), but the realization theory was limited to state dimensions of constant rank. An algebraic approach was followed by Kamen, Khargonekar, and Poolla [KH79, KKP85, PK87], who defined time-varying systems via modules of non-commutative rings of polynomials acting on signals in $\ell_\infty(\mathbb{Z})$. Many definitions and results in [KKP85] can be translated directly into the diagonal algebra considered in this book: instead of Z, two operators z and σ are used, where σ is a time-shift operator on sequences, and z is an algebraic symbol. The description of objects using z and σ is equivalent to our description of diagonals and polynomials (in Z) of diagonals. The aspect of varying state dimensions was first published in Van der Veen and Dewilde [vdVD91]. A similar realization theory for lower triangular block matrices was presented by Gohberg, Kaashoek and Lerer [GKL92], in which operators on $\ell_\infty(\mathbb{Z})$ were considered. Many ingredients (*e.g.*, the definition of the Hankel operator and its factorization) can also be found in Halanay and Ionescu [HI94].

In a parallel development, mathematicians and "fundamental" engineers considered state-space theory for operators on a Hilbert space. Besides the mathematical elegance, Hilbert spaces seemed necessary to incorporate infinite dimensional systems in a state space theory. Such systems arise in a natural way in the time-continuous context of systems which contain "pure delays", *e.g.*, networks with lossless transmission lines. Scattering theory for such networks was developed by Phillips and Lax [LP67], but without using state-space theory. Connections between the fields of Hilbert space operator theory (in particular the work of Sz.-Nagy and Foias [SNF70]) and network synthesis were made by Livsic in 1965 in Russia and with other viewpoints by Dewilde [Dew76], Helton [Hel72, Hel74, Hel76] and Fuhrmann [Fuh74, Fuh75, Fuh76, Fuh81] in the West. These efforts put the algebraic realization theory of Kalman into the Hardy space context of shift-invariant subspaces à la Helson [Hel64], Beurling-Lax representations of such subspaces by inner functions [Beu49, Lax59], and coprime factorizations. More recently, additional results on this type of realization theory (the existence of balanced realizations for infinite-dimensional discrete-time systems) have been obtained by Young [You86]. These ideas and results on infinite-dimensional realization theory of operators in Hilbert space are fundamental to the time-varying realization theory as treated in this chapter, and to a number of results in the chapters to come.

Finally, one different but related approach to the time-varying realizations of operators in Hilbert space is the work of Feintuch and Saeks [FS82]. Their theory is based on a Hilbert space resolution of the identity in terms of a nested series of projectors that endow the abstract Hilbert space with a time structure. The projectors are projectors of sequences onto the past, with respect to each point k in time. With the projectors, one can define various types of causality, and the theory provides operators with a state structure via a factorization of the Hankel operator, which is also defined in terms of the projections. Many of the issues mentioned in the present chapter are also discussed in the book [FS82], but in a different language.

Appendix 5.A: Proof of proposition 5.14

PROOF of proposition 5.14. Because we know already that $\ell_A \le 1$, the proof that $\mathbf{Q}$ is bounded if and only if $\ell_A < 1$ can consist of the two steps,

1. $\ell_A = 1 \Rightarrow$ the operator $[I + AZ + (AZ)^2 + \cdots]$ is unbounded on $\mathcal{D}_2^{\mathcal{B}}$,
2. $\mathbf{Q}$ bounded $\Rightarrow$ the operator $[I + AZ + (AZ)^2 + \cdots]$ is bounded on $\mathcal{D}_2^{\mathcal{B}}$.

Proof of step 1. By definition, $\ell_A = r(AZ) = \lim_{n\to\infty} \| (AZ)^n \|^{1/n}$. We already know that $\|AZ\| \le 1$. Suppose that for some finite n we have $\| (AZ)^n \|^{1/n} < 1$. Then also $\| (AZ)^n \| < 1$, so that $\| (AZ)^{2n} \| \le \| (AZ)^n \|^2 < 1$ and $\| (AZ)^{2n} \|^{1/2n} \le \| (AZ)^n \|^{1/n} < 1$. It follows that

$$\begin{aligned} \ell_A = 1 \quad &\Rightarrow \quad \| (AZ)^n \| = 1 \quad (\text{ for all } n) \\ &\Rightarrow \quad \sup_{D: \|D\|_{HS}=1} \|D(AZ)^n\|_{HS} = 1 \quad (\text{ for all } n). \end{aligned} \tag{5.A.1}$$

Because $\|AZ\| \le 1$ implies $\|D(AZ)^{n-1}\|_{HS} \ge \|D(AZ)^n\|_{HS}$ for any $D \in \mathcal{D}_2$, we have from (5.A.1) that

$$\sup_{D:\|D\|_{HS}=1} \sum_{k=0}^{n} \|D(AZ)^k\|_{HS}^2 \ge \sup_{D:\|D\|_{HS}=1} n\|D(AZ)^n\|_{HS}^2 = n. \tag{5.A.2}$$

This follows from the following reasoning: for any $0 < \varepsilon < 1$ and any n, choose D such that $\|D\|_{HS} = 1$ and $\|D(AZ)^n\|_{HS}^2 \ge 1 - \varepsilon/n$, then for all $k \le n$ we have $\|D(AZ)^k\|_{HS}^2 \ge 1 - \varepsilon/n$, and hence

$$\sup_{D:\|D\|_{HS}=1} \sum_{k=1}^{n} \|D(AZ)^k\|_{HS}^2 \ge n - \varepsilon.$$

Since ε was arbitrary, (5.A.2) follows.

Now since, for any n,

$$\sup_{D:\|D\|_{HS}=1} \|D\left[I+AZ+(AZ)^2+\cdots\right]\|_{HS}^2 \geq \sup_{D:\|D\|_{HS}=1} \sum_{k=0}^{n} \|D(AZ)^k\|_{HS}^2,$$

it follows from (5.A.2) by taking the limit for $n \to \infty$, that the left-hand side of this expression is equal to infinity. This proves that $[I+AZ+(AZ)^2+\cdots]$ is unbounded on $\mathcal{D}_2$.

Proof of step 2. We first remark that equation (5.36), along with $\mathbf{A}^n = \mathbf{P}_{\mathcal{H}}(Z^{-n}\cdot)$ (lemma 5.11) and $\mathbf{P}_{\mathcal{H}}(\cdot) = \mathbf{P}_0(\cdot\mathbf{Q}^*)\mathbf{Q}$ (theorem 4.9), results in the expression

$$A^{\{n\}} = \mathbf{P}_0(Z^{-n}\mathbf{Q}\mathbf{Q}^*) \qquad (n \geq 0). \tag{5.A.3}$$

If $\mathbf{Q}$ is a bounded operator, then the operator $\mathbf{P}(\cdot\mathbf{Q}\mathbf{Q}^*)$ acting on $D \in \mathcal{D}_2$ is bounded and in $\mathcal{U}_2$. But, using (5.A.3), $\mathbf{P}(D\mathbf{Q}\mathbf{Q}^*)$ can be evaluated as

$$\begin{aligned}
\mathbf{P}(D\mathbf{Q}\mathbf{Q}^*) &= \textstyle\sum_0^\infty Z^n \mathbf{P}_0(Z^{-n}D\mathbf{Q}\mathbf{Q}^*) \\
&= \textstyle\sum_0^\infty Z^n D^{(n)} \mathbf{P}_0(Z^{-n}\mathbf{Q}\mathbf{Q}^*) \\
&= \textstyle\sum_0^\infty Z^n D^{(n)} A^{\{n\}} \\
&= D \textstyle\sum_0^\infty (AZ)^n \\
&= D\left[I+AZ+(AZ)^2+\cdots\right].
\end{aligned}$$

Hence $\mathbf{Q}$ bounded implies that $[I+AZ+(AZ)^2+\cdots]$ is bounded on $\mathcal{D}_2$. □

6 ISOMETRIC AND INNER OPERATORS

Lossless systems play an important role in the class of linear systems. They are causal systems which "conserve energy". If energy is measured as the square of a quadratic norm $\|\cdot\|$, a lossless system transforms an input signal u with bounded energy $\|u\|$ to an output signal $y = uT$ which contains the same total energy: $\|u\| = \|y\|$. In filter theory, scalar lossless systems are also known as allpass filters, with a flat amplitude spectrum but a variable phase. They have many interesting properties. One is that any passive rational filter may be realized as the partial response of a lossless filter. Another property is that lossless systems may be implemented in a locally lossless way as well, by using a state space realization in which every section is itself lossless. Such realizations do not amplify noise introduced at any point in the system, and they can be made robust with respect to parameter deviations as well.

The most elementary algebraic expression of losslessness is the orthogonal or Jacobi rotation (in which ϕ is an angle):

$$\begin{bmatrix} \cos\phi & -\sin\phi \\ \sin\phi & \cos\phi \end{bmatrix}.$$

It plays a central role in many algorithms for linear algebra, *e.g.*, for computing QR factorizations and the singular value decomposition. The Jacobi rotation is a building block for more general classes of matrices or operators called isometric, unitary and inner. We study their main system theoretic properties in this and the following chapter. The embedding of passive systems into lossless systems is the topic of chapter 12, and the implementation of a pointwise lossless realization by a cascade of Jacobi rotations is in chapter 14.

Conservation of energy between input and output ($\|u\| = \|y\|$) requires only that the corresponding operator $V : y = uV$ is isometric ($VV^* = I$). V does not have to be causal to be isometric. The class of causal isometric operators characterizes DZ-invariant subspaces in $\mathcal{U}_2$: subspaces that are both left D-invariant and invariant under shifts. This is the content of a general version of the Beurling-Lax theorem, possibly due to Arveson [Arv75], which will play a central role in this chapter. We rederive it in our context (the proof is illuminating and parallels the classical proof), and use it to characterize the main system theoretical input and output spaces. It is a non-trivial question, and one of fierce controversy in the literature, to identify the class of causal, isometric operators that can be embedded into a causal unitary operator. A useful characterization is given in this chapter.

We say that a transfer operator V is inner if $V \in \mathcal{U}$ satisfies both $VV^* = I$ and $V^*V = I$. We first show that if an operator is inner and locally finite, then it admits a realization $\left[\begin{smallmatrix} A & C \\ B & D \end{smallmatrix}\right]$ which is unitary. Conversely, if a realization is unitary and has $\ell_A < 1$, then the corresponding transfer operator is inner. With this background, we look at certain standard factorizations of transfer operators T. The first factorization that we consider is what we call the *external* factorization: a factorization of the type

$$T = \Delta^* V$$

where V is inner and $\Delta \in \mathcal{U}$. (In the literature the term *inner-coprime* is often used, which we shall reserve for the case where the factorization is minimal.) Such a factorization exists if the output nullspace $\mathcal{K}_o(T)$ of T can be represented as $\mathcal{K}_o(T) = \mathcal{U}_2 V$, where V is an inner function, which will imply that $\mathcal{H}_o(V) = \overline{\mathcal{H}_o(T)}$. Because of this property, inner operators play an important role in the derivation of reduced-order models discussed in chapter 10. The factorization can be derived in two ways: via a constructive proof using realizations, but also via the generalized Beurling-Lax theorem. Finally, we utilize the external factorization to give a general embedding theorem which characterizes the set of causal operators that have a (minimal) unitary extension. A similar factorization, the inner-outer factorization, is treated in chapter 7.

6.1 REALIZATION OF INNER OPERATORS

Definitions

An operator $V \in \mathcal{X}$ is called an *isometry* if $VV^* = I$, a *(co-)isometry* if $V^*V = I$, and *unitary* if both $VV^* = I$ and $V^*V = I$, or $V^{-1} = V^*$. Equivalently, an operator is an isometry if its domain and range are closed subspaces in $\mathcal{X}_2$ and if inner products are conserved: for $F, G \in \mathcal{X}_2$, $\langle FV, GV \rangle_{HS} = \langle F, G \rangle_{HS}$, or $\{FV, GV\} = \{F, G\}$ in the diagonal inner product notation. We shall say that an operator is *inner* if it is unitary and upper. Systems described by isometric or inner operators satisfy an energy conservation property: let $U, Y \in \mathcal{X}_2$,

$$\begin{array}{llll} \text{if } VV^* = I \text{ then} & Y = UV & \Rightarrow & \|Y\|_{HS} = \|U\|_{HS} \\ \text{if } V^*V = I \text{ then} & Y = UV^* & \Rightarrow & \|Y\|_{HS} = \|U\|_{HS}. \end{array}$$

Another elementary property is that they leave orthogonality intact:

$$\begin{array}{lll} \text{if } VV^* = I & \text{then} & X \perp Y \Leftrightarrow XV \perp YV \\ \text{if } V^*V = I & \text{then} & X \perp Y \Leftrightarrow XV^* \perp YV^* . \end{array}$$

If V is an isometry, then it maps closed sets into closed sets: since distances between elements of the set are preserved, $x_n \to x \Rightarrow x_n V \to xV$.

For finite matrices (operators in $\mathcal{U}(\mathcal{M},\mathcal{N})$ with index sequences that vanish outside a finite interval), the notion of inner is particularly tied to how the index sequences of $\mathcal{M}$ and $\mathcal{N}$ run. If they are all pointwise scalar and equal, then an inner matrix will trivially be a diagonal: non-trivial inner matrices are possible only when the dimensions of $\mathcal{M}$ and $\mathcal{N}$ are varying. This is because an upper triangular and unitary matrix with scalar entries is necessarily diagonal. However, many more types of matrices qualify as upper in our formalism. E.g., with the proper choice of input and output sequences, a unitary matrix of the form

$$\begin{bmatrix} \star & \star & \star & \star & \star \\ \boxed{\star} & \star & \star & \star & \star \\ 0 & \underline{\star} & \star & \star & \star \\ 0 & 0 & \underline{\star} & \star & \star \\ 0 & 0 & 0 & \underline{\star} & \star \end{bmatrix}$$

may be considered upper and thus inner.

Let $\mathbf{V} = \begin{bmatrix} A & C \\ B & D \end{bmatrix}$ be a realization operator. The realization is called unitary if $\mathbf{V}\mathbf{V}^* = I$ and $\mathbf{V}^*\mathbf{V} = I$.

The purpose of this section is to show that if V is a locally finite inner operator, then it has a realization that is unitary. Conversely, a u.e. stable unitary realization corresponds to an inner operator. There are various ways to prove these properties. For example, we can start with a realization of V in input normal form, *i.e.*, $A^*A + B^*B = I$. To have a unitary realization $\mathbf{V}$, it suffices to show that $VV^* = V^*V = I$ implies that there exist C and D such that $A^*C + B^*D = 0$, $C^*C + D^*D = I$. However, a direct proof of this is not so easy. We propose a more elegant indirect proof, which gives valuable insight into the geometrical properties of the underlying state spaces as well.

State-space properties of inner operators

For a transfer operator $T \in \mathcal{U}$, we have defined the input/output state and null spaces in chapter 5 in terms of the ranges and kernels of the Hankel operator H_T and its adjoint (equations (5.3), (5.5)):

$$\begin{array}{lllll} \mathcal{K}(T) & = & \ker(H_T) & = & \{U \in \mathcal{L}_2 Z^{-1} : \mathbf{P}(UT) = 0\} \\ \mathcal{H}(T) & = & \operatorname{ran}(H_T^*) & = & \mathbf{P}'(\mathcal{U}_2 T^*) \\ \mathcal{H}_o(T) & = & \operatorname{ran}(H_T) & = & \mathbf{P}(\mathcal{L}_2 Z^{-1} T) \\ \mathcal{K}_o(T) & = & \ker(H_T^*) & = & \{Y \in \mathcal{U}_2 : \mathbf{P}'(YT^*) = 0\} . \end{array}$$

These subspaces provide decompositions of $\mathcal{L}_2 Z^{-1}$ and $\mathcal{U}_2$ as

$$\begin{array}{lllll} \overline{\mathcal{H}}(T) & \oplus & \mathcal{K}(T) & = & \mathcal{L}_2 Z^{-1} \\ \overline{\mathcal{H}}_o(T) & \oplus & \mathcal{K}_o(T) & = & \mathcal{U}_2 . \end{array}$$

For inner operators V, the null spaces take on a more specific structure.

Proposition 6.1 *Let $V \in \mathcal{U}$ be an inner operator. Then*

$$\begin{array}{rclcrcl} \mathcal{K}(V) & = & \mathcal{L}_2 Z^{-1} V^*, & & \mathcal{H}(V) & = & \mathcal{L}_2 Z^{-1} \ominus \mathcal{L}_2 Z^{-1} V^* \\ \mathcal{K}_o(V) & = & \mathcal{U}_2 V, & & \mathcal{H}_o(V) & = & \mathcal{U}_2 \ominus \mathcal{U}_2 V. \end{array}$$

$\mathcal{H}$ and $\mathcal{H}_o$ are closed subspaces. In addition, $\mathcal{H}_o = \mathcal{H} V$, $\mathcal{H} = \mathcal{H}_o V^*$.

PROOF Since $\mathcal{X}_2 V \subset \mathcal{X}_2$ and $VV^* = I$, $\mathcal{X}_2 \subset \mathcal{X}_2 V^* \subset \mathcal{X}_2$ and hence $\mathcal{X}_2 = \mathcal{X}_2 V^* = (\mathcal{L}_2 Z^{-1} \oplus \mathcal{U}_2) V^*$. Because $V^* V = I$, $\mathcal{L}_2 Z^{-1} V^* \perp \mathcal{U}_2 V^*$, so that

$$\mathcal{X}_2 = \mathcal{L}_2 Z^{-1} V^* \oplus \mathcal{U}_2 V^*. \tag{6.1}$$

Both $\mathcal{L}_2 Z^{-1} V^*$ and $\mathcal{U}_2 V^*$ are closed subspaces, and because $V \in \mathcal{U}$, $\mathcal{L}_2 Z^{-1} V^* \subset \mathcal{L}_2 Z^{-1}$. Projecting equation (6.1) onto $\mathcal{L}_2 Z^{-1}$ produces

$$\mathcal{H} = \mathbf{P}'(\mathcal{U}_2 V^*) = \mathcal{L}_2 Z^{-1} \ominus \mathcal{L}_2 Z^{-1} V^*.$$

As an orthogonal complement, this is a closed subspace, so that $\mathcal{H}$ is closed. Hence

$$\mathcal{L}_2 Z^{-1} = \mathcal{L}_2 Z^{-1} V^* \oplus \mathcal{H}, \tag{6.2}$$

so that $\mathcal{K} = \mathcal{L}_2 Z^{-1} V^*$. From (6.2), it also follows immediately that

$$\mathcal{L}_2 Z^{-1} \oplus \mathcal{H} V = \mathcal{L}_2 Z^{-1} V.$$

Hence $\mathcal{H} V \subset \mathcal{U}_2$, and $\mathcal{H} V = \mathbf{P}(\mathcal{L}_2 Z^{-1} V) = \mathcal{H}_o$. The remaining results are obtained by dual arguments. □

For general transfer operators T, we had already that $\mathcal{H}_o = \mathbf{P}(\mathcal{H} T)$. Thus, inner operators are special in the sense that they map their input state space fully into the output state space, without the intervention of a projection. Likewise, the Hankel operator of V, H_V, satisfies $\cdot H_V = \cdot V$ on $\mathcal{H}$. Since $\cdot H_V = 0$ on $\mathcal{K}$, we see that H_V is an isometry. In the locally finite case, the non-zero singular values of its snapshots are all equal to 1: in the SVD-based factorization $H_V = \mathbf{P}_0(\cdot \mathbf{Q}^*) \hat{\Sigma} \mathbf{G}$ of equation (5.49), we have $\hat{\Sigma} = I$.

Unitary realizations

We now show that (*i*) if a locally finite operator V is inner, then it has a unitary realization $\mathbf{V}$ (which is obtained by a canonical realization based on $\mathbf{Q}$ or $\mathbf{G}$); and conversely, (*ii*) if $\mathbf{V}$ is a unitary realization with $\ell_A < 1$, then the corresponding operator V is inner. The case $\ell_A = 1$ is much more complicated and deferred to sections 6.3 and especially 6.4.

We start with a lemma which is actually a corollary of proposition 6.1.

Lemma 6.2 *Let $V \in \mathcal{U}$ be a locally finite inner operator. If $\mathbf{Q}$ is a sliced orthonormal basis representation of the input state space $\mathcal{H}$ of V, then $\mathbf{G} = \mathbf{Q} V$ is a sliced orthonormal basis representation of its output state space $\mathcal{H}_o$, and the canonical controller* *realization based on* $\mathbf{Q}$ *(theorem 5.13) is equal to the canonical observer realization based on* $\mathbf{G}$ *(theorem 5.17).*

PROOF According to theorem 5.20, a sliced basis of $\mathcal{H}_o$ is obtained as $\mathbf{F}_o = \mathbf{P}(\mathbf{Q}V)$. Because $\mathcal{H}_o = \mathcal{H}V$, it follows that $\mathbf{F}_o = \mathbf{P}(\mathbf{Q}V) = \mathbf{Q}V = \mathbf{G}$. $\mathbf{G}$ is an orthonormal basis of $\mathcal{H}_o$, because $\Lambda_\mathbf{G} = \mathbf{P}_0(\mathbf{Q}VV^*\mathbf{Q}^*) = \Lambda_\mathbf{Q} = I$. The canonical realizations are obtained from theorems 5.13 and 5.17, respectively, as

$$\mathbf{V} = \begin{bmatrix} \mathbf{P}_0(Z^{-1}\mathbf{Q}\mathbf{Q}^*)^{(-1)} & \mathbf{P}_0(\mathbf{Q}V) \\ \mathbf{P}_0(Z^{-1}\mathbf{Q}^*)^{(-1)} & \mathbf{P}_0(V) \end{bmatrix} \text{ and } \mathbf{V}' = \begin{bmatrix} \mathbf{P}_0(Z^{-1}\mathbf{G}\mathbf{G}^*)^{(-1)} & \mathbf{P}_0(\mathbf{G}) \\ \mathbf{P}_0(Z^{-1}V\mathbf{G}^*)^{(-1)} & \mathbf{P}_0(V) \end{bmatrix}. \tag{6.3}$$

The fact that both realizations are equal follows directly by inserting $\mathbf{G} = \mathbf{Q}V$. □

Theorem 6.3 *Let $V \in \mathcal{U}$ be a locally finite inner operator. Then V has a realization* $\mathbf{V}$ *which is unitary and both uniformly reachable and uniformly observable.*

PROOF Let $\mathbf{Q}$ be an orthonormal basis representation for $\mathcal{H}(V)$, and let $\mathbf{V}$ be given by the canonical controller realization (6.3). This realization satisfies the properties (5.19)–(5.21):

$$Z\mathbf{Q} = A^*\mathbf{Q} + B^*, \qquad V = D + \mathbf{Q}^*C. \tag{6.4}$$

We set out to prove that $\mathbf{V}^*\mathbf{V} = I$, *i.e.*,

$$A^*A + B^*B = I, \qquad C^*C + D^*D = I, \qquad A^*C + B^*D = 0.$$

$A^*A + B^*B = I$ follows from the fact that $\mathbf{Q}$ is an orthonormal basis: $\Lambda_\mathbf{Q} = I$, which satisfies the Lyapunov equation (5.22). To show that $C^*C + D^*D = I$, use equation (6.4) and the fact that $\mathbf{Q}$ is strictly lower:

$$\begin{array}{lll} \mathbf{P}_0(V^*V) = I & \Rightarrow & \mathbf{P}_0([D^* + C^*\mathbf{Q}][D + \mathbf{Q}^*C]) \\ & & = D^*D + C^*\mathbf{P}_0(\mathbf{Q}\mathbf{Q}^*)C + D^*\mathbf{P}_0(\mathbf{Q}^*)C + C^*\mathbf{P}_0(\mathbf{Q})D \\ & & = D^*D + C^*C + 0 + 0 \quad = I. \end{array}$$

$A^*C + B^*D = 0$ follows from lemma 6.2: $\mathbf{G} = \mathbf{Q}V$ spans $\mathcal{H}_o(V)$, hence $\mathbf{G} \in \mathcal{U}$ so that $\mathbf{P}_0(Z\mathbf{Q}V) = \mathbf{P}_0(Z\mathbf{G}) = 0$. With equation (6.4), we obtain

$$\begin{array}{lll} \mathbf{P}_0(Z\mathbf{Q}V) = 0 & \Rightarrow & \mathbf{P}_0([B^* + A^*\mathbf{Q}][D + \mathbf{Q}^*C]) \\ & & = B^*D + A^*\mathbf{P}_0(\mathbf{Q}\mathbf{Q}^*)C \\ & & = B^*D + A^*C \quad = 0. \end{array}$$

Hence $\mathbf{V}^*\mathbf{V} = I$. Dually, we find in the same way that $\mathbf{V}'$ in (6.3) satisfies $\mathbf{V}'\mathbf{V}'^* = I$. Since $\mathbf{V} = \mathbf{V}'$ if $\mathbf{G} = \mathbf{Q}V$ (lemma 6.2), it follows that $\mathbf{V}$ is unitary. □

The converse of this theorem is in general true only if, in addition, $\ell_A < 1$: in that case, a unitary realization corresponds to an inner operator. If $\ell_A = 1$, then additional assumptions on the reachability and observability of the realization must be made. The latter case is deferred to theorem 6.12 in the next section.

Theorem 6.4 *Let* $\mathbf{V} = \left[\begin{smallmatrix} A & C \\ B & D \end{smallmatrix}\right]$ *be a state realization of a locally finite operator* $V \in \mathcal{U}$. *If* $\ell_A < 1$, *then* $\mathbf{V}$ *unitary implies that* V *is inner.*

PROOF If $\ell_A < 1$, then $(I-AZ)^{-1}$ is bounded, so that we can write

$$\begin{aligned} I-V^*V &= I-[D+BZ(I-AZ)^{-1}C]^*\,[D+BZ(I-AZ)^{-1}C] \\ &= I-D^*D - C^*(I-Z^*A^*)^{-1}Z^*B^*D - D^*BZ(I-AZ)^{-1}C \\ &\quad -C^*(I-Z^*A^*)^{-1}Z^*B^*BZ(I-AZ)^{-1}C \\ &= I-D^*D + C^*(I-Z^*A^*)^{-1}Z^*A^*C + C^*AZ(I-AZ)^{-1}C + \\ &\quad -C^*(I-Z^*A^*)^{-1}Z^*\{I-A^*A\}Z(I-AZ)^{-1}C \\ &= C^*C + C^*(I-Z^*A^*)^{-1}\{Z^*A^* + AZ - I - Z^*A^*AZ\}(I-AZ)^{-1}C \end{aligned}$$

since $B^*D = -A^*C$, $B^*B = I-A^*A$ and $I-D^*D = C^*C$, and hence

$$\begin{aligned} I-V^*V &= C^*(I-Z^*A^*)^{-1}\{(I-Z^*A^*)(I-AZ) + \\ &\quad + Z^*A^* + AZ - I - Z^*A^*AZ\}(I-AZ)^{-1}C \\ &= 0. \end{aligned}$$

$I-VV^* = 0$ is verified by an analogous procedure. □

A slightly more general version of this, not using normalized realizations, is given by the following corollary, where M is the reachability Gramian of the given realization, and Q its observability Gramian. (A comparable result can be found in [HI94, §2.5].)

Corollary 6.5 *Let* $T \in \mathcal{U}$ *be a locally finite input-output operator with u.e. stable state realization* $\mathbf{T}$. *Then*

$$\begin{array}{llll} \exists M \in \mathcal{D}: & \mathbf{T}^* \begin{bmatrix} M & \\ & I \end{bmatrix} \mathbf{T} = \begin{bmatrix} M^{(-1)} & \\ & I \end{bmatrix} & \Rightarrow & T^*T = I \\ \exists Q \in \mathcal{D}: & \mathbf{T} \begin{bmatrix} Q^{(-1)} & \\ & I \end{bmatrix} \mathbf{T}^* = \begin{bmatrix} Q & \\ & I \end{bmatrix} & \Rightarrow & TT^* = I. \end{array}$$

Conversely, if $T^*T = I$ *and* $TT^* = I$, *and the realization is uniformly reachable or observable, then the left-hand sides are satisfied with* $M = Q^{-1}$.

6.2 EXTERNAL FACTORIZATION

Definition

Let $T \in \mathcal{U}$ be some transfer operator. We call an *external factorization* a factorization of the form

$$T = \Delta^* V,$$

where $\Delta = VT^* \in \mathcal{U}$ and $V \in \mathcal{U}$ is an inner operator. If the factorization is such that V is an inner transfer operator of smallest possible local degree such that $\Delta = VT^*$ is upper, then we call the factorization *inner coprime*. We show that if T has a locally finite state space and a uniformly observable realization for which $\ell_A < 1$, then such factorizations exist. They can readily be computed from a state realization. If V has the same output state space as T, then the factorization is minimal. The minimal inner factor will be unique except for a left unitary diagonal factor.

To obtain a better understanding of the external (inner-coprime) factorization, consider the scalar time-invariant case. Let

$$T = \frac{z-\alpha^*}{1-\beta z}, \qquad |\alpha|, |\beta| < 1.$$

Then T has an inner-coprime factorization as

$$T = \Delta^* V = \frac{z-\alpha^*}{z-\beta^*} \cdot \frac{z-\beta^*}{1-\beta z}, \qquad \Delta = \frac{1-\alpha z}{1-\beta z}, \qquad V = \frac{z-\beta^*}{1-\beta z}.$$

Hence the poles of T are collected in the inner factor V. These poles also appear as poles of Δ, unless they are matched by complementary zeros of T.

The existence of external and inner-coprime factorizations has great system theoretical importance. Aside from the fact that it plays a key role in important practical questions such as the design of low-sensitivity controllers, it is directly related to the existence of a meaningful state-space representation. Since it is also a matter of controversy in the literature, we devote a few words to introduce the question; for the time-invariant case deeper treatments can be found in [Dew76] and [Fuh81].

In the case of a single-input, single-output time-invariant system, the existence of inner-coprime factorizations is equivalent to the existence of non-trivial system null-spaces $\mathcal{K}(T)$ and $\mathcal{K}_o(T)$. The Fourier transform of $\mathcal{K}_o(T)$ is a subspace of H_2, the space of Fourier transforms of one-sided ℓ_2-sequences whose support is the non-negative integers. $\mathcal{K}_o(T)$ has a special property: it is z-invariant: $z \cdot \mathcal{K}_o(T) \subset \mathcal{K}_o(T)$. Dually, $\mathcal{K}(T)$ is a z^{-1}-invariant subspace of the orthogonal complement $H_2^\perp$ of H_2 which represents past inputs.

Beurling's celebrated theorem [Hel64, Hof62] states that $\mathcal{K}_o(T)$ *is either trivial* (= $\{0\}$) *or there exists an inner function* $\phi_o(z)$ *such that* $\mathcal{K}_o(T) = \phi_o(z) H_2$ (in this simple context, "inner" means that $\phi_o(z)$ is analytic in the open unit disc of the complex plane, and that $|\phi_o(e^{i\theta})| = 1$ almost everywhere on the unit circle; in other words, ϕ_o is a pure phase function). Dually, either $\mathcal{K}(T) = \{0\}$ or there exists an inner $\phi(z)$ such that $\mathcal{K}(T) = \phi^*(z) H_2^\perp$. In the first case, the null-space is trivial and the system remembers its full past. In that case there is no meaningful state space description: the state is equivalent to the whole input sequence, and the state space description is nothing but the input-output description. In the second case, the null-space is very large, and each state stands for an input collection isomorphic to H_2. One can say that the system forgets almost everything from its past. There is no in-between: once a system forgets one input, it will forget an infinity of them.

In [Dew76] such systems have been called "roomy". Their transfer functions can be characterized by an analytical property, they are "pseudo-meromorphically continuable", see the work of Helton [Hel74]. It turns out that roominess is a necessary and sufficient condition for the lossless (inner) embedding of a causal contractive transfer function. This fact was discovered independently by Arov and Dewilde around 1971. For multi-dimensional systems the situation is more complex, but the property that a lossless embedding exists if and only if the causal contractive system is roomy still applies. We shall find many of these properties back in the time-varying case. Again, external and coprime factorization play a major role. Similar time-varying coprime factorizations have also been reported in [PK87, DS92, RPK92].

Derivation

The following simple observation is crucial in the computation of the inner factor of an external factorization.

Proposition 6.6 *Let be given operators $T \in \mathcal{U}$ and $V \in \mathcal{U}$. Then $\Delta := VT^*$ is upper if and only if $\mathcal{U}_2 V \subset \mathcal{K}_o(T)$.*

PROOF $\Delta \in \mathcal{U} \Leftrightarrow \mathbf{P}'(\mathcal{U}_2\Delta) = 0$. Substitution of $\Delta = VT^*$ produces, if $\mathcal{U}_2 V \subset \mathcal{K}_o(T)$,

$$\begin{array}{rcl} \mathbf{P}'(\mathcal{U}_2\Delta) & = & \mathbf{P}'(\mathcal{U}_2 VT^*) \\ & \subset & \mathbf{P}'(\mathcal{K}_o(T)T^*) \\ & = & 0. \end{array}$$

This produces the "if" statement. The converse follows from the property $\mathcal{K}_o(T) = \{u \in \mathcal{U}_2 : \mathbf{P}'(uT^*) = 0\}$, hence $\mathbf{P}'[(uV)T^*] = \mathbf{P}'(u\Delta) = 0$, and $uV \in \mathcal{K}_o(T)$. □

$\mathcal{K}_o(T)$ is the largest subspace in $\mathcal{U}_2$ which remains upper under mapping by T^*. It follows that a system V with lowest state dimensions such that $\Delta = VT^* \in \mathcal{U}$ is obtained if $\mathcal{U}_2 V = \mathcal{K}_o(T)$, since the larger the nullspace, the smaller the state dimension. We shall make this observation more precise soon.

If V is inner, then from proposition 6.1, we have that $\mathcal{K}_o(V) = \mathcal{U}_2 V$, which provides the following additional result.

Corollary 6.7 *If V is inner, then $\Delta = VT^*$ is upper if and only if $\mathcal{H}_o(T) \subset \mathcal{H}_o(V)$.*

The next step in the construction of the external factorization is the calculation of an operator V such that $\mathcal{H}_o(V) = \overline{\mathcal{H}_o(T)}$. This can be done in a state-space context, directly on a realization of T. Let T be a locally finite operator in $\mathcal{U}$. We start from a realization of T in output normal form, *i.e.*, such that

$$AA^* + CC^* = I, \tag{6.5}$$

which means that at each point k in time the equation $A_k A_k^* + C_k C_k^* = I$ is satisfied. Such a realization is obtained from a canonical observer realization (*viz.* theorem 5.17), or by normalizing any uniformly observable realization (section 5.3). We assume that $T \in \mathcal{U}(\mathcal{M},\mathcal{N})$, with state-space sequence $\mathcal{B}$, so that $A \in \mathcal{D}(\mathcal{B},\mathcal{B}^{(-1)})$. For each time instant k, we augment the state transition matrices $[A_k \;\; C_k]$ of T with as many extra rows as needed to yield a unitary (hence square) matrix $\mathbf{V}_k$:

$$\mathbf{V}_k = \begin{array}{c} \\ \mathcal{B}_k \\ (\mathcal{M}_V)_k \end{array} \!\!\begin{array}{c} \begin{array}{cc} \mathcal{B}_{k+1} & \mathcal{N}_k \end{array} \\ \begin{bmatrix} A_k & C_k \\ (B_V)_k & (D_V)_k \end{bmatrix} \end{array}. \tag{6.6}$$

The added rows introduce a space $(\mathcal{M}_V)_k$ with dimensions satisfying $\#\mathcal{B}_k + \#(\mathcal{M}_V)_k = \#\mathcal{B}_{k+1} + \#\mathcal{N}_k$. Since $[A_k \;\; C_k]$ must have full row rank to enable $A_kA_k^* + C_kC_k^* = I$, it follows that $\#\mathcal{B}_{k+1} + \#\mathcal{N}_k \geq \#\mathcal{B}_k$, hence $\#(\mathcal{M}_V)_k \geq 0$. Assemble the individual matrices $\{A_k, (B_V)_k, C_k, (D_V)_k\}$ into diagonal operators $\{A, B_V, C, D_V\}$, and define V by taking the corresponding operator $\mathbf{V}$ as a state-space realization for V. By theorem 6.4, V is inner if also $\ell_A < 1$, and because $\mathbf{T}$ and $\mathbf{V}$ have the same (A,C)-matrices, $\mathcal{H}_o(V) = \overline{\mathcal{H}_o(T)}$, as required to make $\Delta \in \mathcal{U}$.

Although the construction is the same whether $\ell_A < 1$ or $\ell_A = 1$, the proof that it yields an external factorization is less elementary (and only conditionally true) for the case $\ell_A = 1$, so the latter case is omitted in the following theorem.

Theorem 6.8 *Let T be a locally finite operator in $\mathcal{U}$. If T has a realization which is uniformly observable and for which $\ell_A < 1$, then there exists an inner operator V such that*

$$T = \Delta^* V$$

where $\Delta = VT^* \in \mathcal{U}$.

PROOF Under the present conditions on T, it has a minimal realization $\mathbf{T}$ which is in output normal form and has $\ell_A < 1$. Then the above construction gives a unitary realization $\mathbf{V}$. Since this realization has $\ell_A < 1$, theorem 6.4 ensures that $\mathbf{V}$ is a minimal realization and that the corresponding operator V is inner. By construction $\mathcal{H}_o(V) = \overline{\mathcal{H}_o(T)}$, so that application of corollary 6.7 shows that $\Delta := VT^*$ is upper. Because V is inner, this implies that $T = \Delta^* U$. □

The fact that $\Delta = VT^*$ is upper can also be verified by a direct computation of Δ. Let's assume for generality that the realization for T has observability Gramian $Q \gg 0$. Then the corresponding unnormalized realization $\mathbf{V} = \left[\begin{smallmatrix} A & C \\ B_V & D_V \end{smallmatrix}\right]$ satisfies in particular the relations $AQ^{(-1)}A^* + CC^* = Q$, $B_VQ^{(-1)}A^* + D_VC^* = 0$, and it follows that

$$\begin{aligned}
\Delta = VT^* &= \left[D_V + B_VZ(I-AZ)^{-1}C\right]\left[D^* + C^*(I-Z^*A^*)^{-1}Z^*B^*\right] \\
&= \left[D_V + B_VZ(I-AZ)^{-1}C\right]D^* + \underline{D_VC^*}(I-Z^*A^*)^{-1}Z^*B^* + \\
&\quad + B_VZ(I-AZ)^{-1}\underline{CC^*}(I-Z^*A^*)^{-1}Z^*B^* \\
&= \left[D_V + B_VZ(I-AZ)^{-1}C\right]D^* - B_VQ^{(-1)}A^*(I-Z^*A^*)^{-1}Z^*B^* + \\
&\quad + B_VZ(I-AZ)^{-1}(Q-AQ^{(-1)}A^*)(I-Z^*A^*)^{-1}Z^*B^*.
\end{aligned}$$

Now, we make use of the relation

$$\begin{aligned}
&Z(I-AZ)^{-1}(Q-AQ^{(-1)}A^*)(I-Z^*A^*)^{-1}Z^* \\
&\quad = (I-ZA)^{-1}Z(Q-AQ^{(-1)}A^*)(Z-A^*)^{-1} \\
&\quad = (I-ZA)^{-1}Q^{(-1)} + Q^{(-1)}A^*(Z-A^*)^{-1} \\
&= Q^{(-1)} + Z(I-AZ)^{-1}AQ^{(-1)} + Q^{(-1)}A^*(I-Z^*A^*)^{-1}Z^*
\end{aligned}$$

where the second step is easily verified by pre- and postmultiplying with $(I-ZA)$ and $(Z-A^*)$, respectively. Plugging this relation into the expression for Δ, it is seen that the anti-causal parts of the expression cancel, and we obtain

$$\Delta = D_VD^* + B_VQ^{(-1)}B^* + B_VZ(I-AZ)^{-1}(AQ^{(-1)}B^* + CD^*).$$

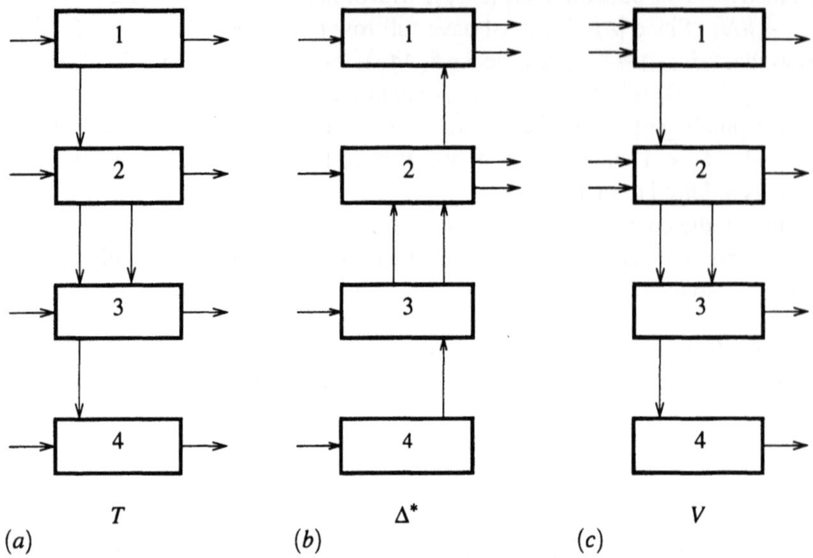

Figure 6.1. External factorization: (a) The structure of a state realization for an example T, (b) the structure of the corresponding Δ^* and (c) inner factor V such that $T = \Delta^* V$.

In summary, if $\ell_A < 1$, $AQ^{(-1)}A^* + CC^* = Q \gg 0$, then T has an external factorization $T = \Delta^* V$ with realizations of the form

$$\mathbf{T} = \begin{bmatrix} A & C \\ B & D \end{bmatrix} \Rightarrow \mathbf{V} = \begin{bmatrix} A & C \\ B_V & D_V \end{bmatrix}, \quad \boldsymbol{\Delta} = \begin{bmatrix} A & AQ^{(-1)}B^* + CD^* \\ B_V & B_V Q^{(-1)} B^* + D_V D^* \end{bmatrix}. \tag{6.7}$$

This realization is not necessarily minimal: if, for example, T is itself inner, then $B = B_V$ and $D = D_V$, so that $C_\Delta = 0$, and the realization for Δ is not observable.

A dual result is a factorization $T = U\Delta^*$ with realizations of the form

$$\mathbf{T} = \begin{bmatrix} A & C \\ B & D \end{bmatrix} \Rightarrow \mathbf{U} = \begin{bmatrix} A & C_U \\ B & D_U \end{bmatrix}, \quad \boldsymbol{\Delta} = \begin{bmatrix} A & C_U \\ C^*MA + D^*B & C^*MC_U + D^*D_U \end{bmatrix} \tag{6.8}$$

which is valid for $\ell_A < 1$, $A^*MA + B^*B = M^{(-1)} \gg 0$.

Because the A_k are not necessarily square matrices, the dimension of the state space may vary in time. A consequence of this is that the number of inputs of V varies in time for an inner V with minimal state dimension. The varying number of inputs of V are of course matched by a varying number of outputs of Δ^*. Figure 6.1 illustrates this point.

Algorithm

If we do not assume that the realization for T is in output normal form, then the recursion to normalize $\mathbf{T}$ and the complementation to compute $\mathbf{V}$ and $\boldsymbol{\Delta}$ can conveniently be combined into a single "QR iteration":

Proposition 6.9 *Under conditions of theorem 6.8, let (A,B,C,D) be any uniformly observable realization of T. Denote realizations of V and Δ by*

$$\mathbf{V} = \left[\begin{array}{cc} A_V & C_V \\ B_V & D_V \end{array}\right], \quad \mathbf{\Delta} = \left[\begin{array}{cc} A_\Delta & C_\Delta \\ B_\Delta & D_\Delta \end{array}\right].$$

Then **V** *and* **Δ** *such that $T = \Delta^* V$ follow (backward) recursively from the LQ factorizations*

$$\left[\begin{array}{cc} A_k R_{k+1} & C_k \\ \hline I & 0 \\ B_k R_{k+1} & D_k \end{array}\right] =: \left[\begin{array}{cc} R_k & 0 \\ \hline \multicolumn{2}{c}{} \\ \multicolumn{2}{c}{\mathbf{\Delta}_k^*} \end{array}\right] \mathbf{V}_k, \qquad k = \cdots, n, n-1, n-2, \cdots \qquad (6.9)$$

where $\mathbf{V}_k$ is unitary and $R_k : d_k \times d_k$ is a recursively determined square matrix.

Dually, let (A,B,C,D) be any uniformly reachable realization of T. Realizations **V** *and* **Δ** *such that $T = V\Delta^*$ follow recursively from the QR factorizations*

$$\left[\begin{array}{c|cc} R_k A_k & I & R_k C_k \\ B_k & 0 & D_k \end{array}\right] =: \mathbf{V}_k \left[\begin{array}{c|c} R_{k+1} & \mathbf{\Delta}_k \\ 0 & \end{array}\right], \qquad k = \cdots, n, n+1, n+2, \cdots.$$

PROOF Postmultiplying (6.9) with its transpose removes $\mathbf{V}_k$ and produces the equation

$$A_k (R_{k+1} R_{k+1}^*) A_k^* + C_k C_k^* = (R_k R_k^*).$$

Hence R_k is the square root of the solution of the Lyapunov equation associated to (A,C) (*viz.* (5.25)), and a state transformation by R will bring **T** into output normal form, as discussed in section 5.3. Working out (6.9) gives the equations

$$[R^{-1} A R^{(-1)} \quad R^{-1} C] \left[\begin{array}{cc} A_V^* & B_V^* \\ C_V^* & D_V^* \end{array}\right] = [I \quad 0]$$

$$\mathbf{\Delta}^* = \left[\begin{array}{cc} I & 0 \\ BR^{(-1)} & D \end{array}\right] \left[\begin{array}{cc} A_V^* & B_V^* \\ C_V^* & D_V^* \end{array}\right] = \left[\begin{array}{cc} A_V^* & B_V^* \\ (BR^{(-1)}) A_V^* + D C_V^* & (BR^{(-1)}) B_V^* + D D_V^* \end{array}\right].$$

After taking the state transformation by R into account, these are precisely the defining equations (6.6) for **V** and (6.7) for **Δ**. □

Both recursions require an initial R_n (for some adequate n). Since R is the square root of the solution of a Lyapunov equation, it may be initialized in a similar way as in section 5.3. In particular,

1. if T is a finite $n \times n$ matrix, we can start with $R_n = [\cdot]$,
2. if T is Toeplitz starting from some point n in time, then we can initialize (6.9) by taking R_n to be the solution of the time-invariant Lyapunov equation

$$A_n Q A_n^* + C_n C_n^* = Q, \qquad Q =: R_n^* R_n.$$

3. We already had to assume $\ell_A < 1$ to guarantee the existence of the external factorization. The Lyapunov equation is strongly convergent for $\ell_A < 1$, hence even if we start with an imprecise initial $\hat{R}_n$, it will converge towards the true solution ($\hat{R}_k \to R_k$). Thus, we may start with any invertible R_n, *e.g.*, $R_n = I$. Here, n should be sufficiently far away from the interval in which the external factorization is of interest.

Remarks

One remaining issue with the external factorization is to explain why (and when) it can be called inner coprime. Two upper operators T_1 and T_2 are called (left inner) coprime if they do not have a common, non-trivial left inner factor [Dew76], *i.e.*, if

$$\begin{aligned} T_1 &= WT_1' \\ T_2 &= WT_2' \end{aligned}$$

(where $T_{1,2}' \in \mathcal{U}$ and W is inner) implies $W \in \mathcal{D}$. With this definition of inner coprimeness, it is possible to show that Δ and V in the factorization $T = \Delta^* V$ are inner coprime if $\mathcal{K}_o(T) = \mathcal{U}_2 V = \mathcal{K}_o(V)$. Indeed, suppose that they have a common left inner factor W, then $T = \Delta_1^* V_1$, where

$$\begin{aligned} \Delta_1 &= W^*\Delta \in \mathcal{U} \\ V_1 &= W^*V \in \mathcal{U}. \end{aligned}$$

On the one hand, $\mathcal{U}_2 V = \mathcal{U}_2 W V_1 \subset \mathcal{U}_2 V_1$. On the other, $\Delta_1 \in \mathcal{U} \Rightarrow \mathcal{U}_2\Delta_1 = \mathcal{U}_2[V_1 T^*] = [\mathcal{U}_2 V_1]T^* \subset \mathcal{U}_2$, hence $\mathcal{U}_2 V_1 \subset \mathcal{U}_2 V$, since $\mathcal{U}_2 V = \mathcal{K}_o(T)$ is the largest subspace in $\mathcal{U}_2$ that is mapped by T^* to $\mathcal{U}_2$. Combining both observations gives $\mathcal{U}_2 V_1 = \mathcal{U}_2 V$, so that V_1 is equal to V, up to a left diagonal unitary factor.

6.3 STATE-SPACE PROPERTIES OF ISOMETRIC SYSTEMS

In section 6.1, we derived a number of state space properties of inner systems. In preparation of a treatment on inner-outer factorizations in chapter 7, it is necessary to consider also the state space properties of *isometric* operators. It will turn out that an inner-outer factorization with an inner operator as defined earlier is not always possible, even in the locally finite case.

The equivalent of proposition 6.1 for isometric operators is more complicated:

Proposition 6.10 *Let $V \in \mathcal{U}$. Then*

$$VV^* = I \quad \Rightarrow \quad \left\{ \begin{array}{rcl} \mathcal{K}_o &=& \mathcal{U}_2 V \oplus \ker(\cdot V^*|_{\mathcal{U}_2}), \\ \mathcal{H} &=& \overline{\mathcal{H}_o V^*} \\ \overline{\mathcal{U}_2 V^*} &=& \mathcal{U}_2 \oplus \overline{\mathcal{H}} \\ \multicolumn{3}{l}{\ker(\cdot V^*|_{\mathcal{X}_2}) = \{0\} \quad \Rightarrow \quad V \text{ is inner}} \end{array} \right.$$

$$V^*V = I \quad \Rightarrow \quad \left\{ \begin{array}{rcl} \mathcal{K} &=& \mathcal{L}_2 Z^{-1} V^* \oplus \ker(\cdot V|_{\mathcal{L}_2 Z^{-1}}). \\ \mathcal{H}_o &=& \overline{\mathcal{H} V} \\ \overline{\mathcal{L}_2 Z^{-1} V} &=& \mathcal{L}_2 Z^{-1} \oplus \overline{\mathcal{H}_o} \\ \multicolumn{3}{l}{\ker(\cdot V|_{\mathcal{X}_2}) = \{0\} \quad \Rightarrow \quad V \text{ is inner}} \end{array} \right.$$

PROOF Let $VV^* = I$. Because V is an isometry, the subspace $\mathcal{X}_2 V = \mathrm{ran}(V)$ is closed. Because $\mathcal{X}_2 V = \mathcal{L}_2 Z^{-1} V \oplus \mathcal{U}_2 V$, both $\mathcal{U}_2 V$ and $\mathcal{L}_2 Z^{-1} V$ are closed subspaces.

$\mathcal{U}_2 V \subset \mathcal{K}_o$, because $\mathbf{P}'([\mathcal{U}_2 V]V^*) = 0$. The remaining subspace $\mathcal{K}_o \ominus \mathcal{U}_2 V$ consists of elements

$$\begin{aligned} \mathcal{K}_o \ominus \mathcal{U}_2 V &= \{X \in \mathcal{U}_2 : \mathbf{P}'(XV^*) = 0 \wedge \mathbf{P}(XV^*) = 0\} \\ &= \{X \in \mathcal{U}_2 : XV^* = 0\} \\ &= \ker(\cdot V^*|_{\mathcal{U}_2}). \end{aligned}$$

Hence $\mathcal{K}_o = \mathcal{U}_2 V \oplus \ker(\cdot V^*|_{\mathcal{U}_2})$.

To show that $\mathcal{H} = \overline{\mathcal{H}_o V^*}$, take $U \in \mathcal{L}_2 Z^{-1}$. Then $UV = U_1 + Y$, where $U_1 \in \mathcal{L}_2 Z^{-1}$ and $Y = \mathbf{P}(UV) \in \mathcal{H}_o \subset \mathcal{U}_2$. All of $\mathcal{H}_o$ can be reached by Y if U ranges over $\mathcal{L}_2 Z^{-1}$. Multiplication by V^* gives $U = U_1 V^* + YV^*$, and since $V^* \in \mathcal{L}$, it follows that $YV^* \in \mathcal{L}_2 Z^{-1}$, and this is true for all $Y \in \mathcal{H}_o$. Hence $\mathcal{H}_o V^* \subset \mathcal{L}_2 Z^{-1}$ and also

$$\overline{\mathcal{H}_o V^*} \subset \mathcal{L}_2 Z^{-1}.$$

Since $\mathcal{H} = \mathbf{P}'(\mathcal{U}_2 V^*) = \mathbf{P}'(\overline{\mathcal{H}_o V^*})$, we obtain $\mathcal{H} = \overline{\mathcal{H}_o V^*}$. Thirdly, the expressions for $\mathcal{H}_o$ and $\mathcal{K}_o$ combined give

$$\mathcal{U}_2 = \overline{\mathcal{H}_o} \oplus \mathcal{U}_2 V \oplus \ker(\cdot V^*|_{\mathcal{U}_2})$$

hence $\overline{\mathcal{U}_2 V^*} = \overline{\mathcal{H}_o V^*} \dot{+} \mathcal{U}_2$. Because $\overline{\mathcal{H}_o V^*} = \mathcal{H} \in \mathcal{L}_2 Z^{-1}$, the two components are actually orthogonal. Finally, since V is an isometry, its range is closed, and if $\ker(\cdot V^*|_{\mathcal{X}_2}) = \{0\}$ then that range is actually $\mathcal{X}_2$. Hence V has a left inverse, which must be equal to the right inverse V^*, $V^*V = I$ and V is inner.

Dual results hold in case $V^*V = I$. □

The spaces $\ker(\cdot V^*|_{\mathcal{U}_2})$ and $\ker(\cdot V^*|_{\mathcal{X}_2})$ are fundamentally different: because the inputs are restricted to $\mathcal{U}_2$, the first can be the zero space while the other contains nonzero elements — this fact will be of great importance for the inversion theory of the next chapter. A dual remark holds for $\cdot V$.

Isometric realizations

Theorems 6.3 and 6.4 on the realizations of inner operators have specializations to isometric operators and realizations. For later use, we now consider the case $\ell_A = 1$ as well, which complicates the proof of theorem 6.12.

Theorem 6.11 *Let $V \in \mathcal{U}$ be a locally finite operator. Then*

$V^*V = I$	$\Rightarrow$	*The canonical controller realization* $\mathbf{V}$ *of* V *satisfies* $\mathbf{V}^*\mathbf{V} = I$ *and is observable and uniformly reachable.*
$VV^* = I$	$\Rightarrow$	*The canonical observer realization* $\mathbf{V}$ *of* V *satisfies* $\mathbf{V}\mathbf{V}^* = I$ *and is reachable and uniformly observable.*

PROOF The proof is the same as the proof of theorem 6.3. □

Theorem 6.12 *Let $\mathbf{V} = \left[\begin{smallmatrix} A & C \\ B & D \end{smallmatrix}\right]$ be a state realization of a locally finite operator $V \in \mathcal{U}$. Let $\Lambda_{\mathbf{F}}$ and $\Lambda_{\mathbf{F}_o}$ be the reachability and the observability Gramians of the given realization. If $\ell_A < 1$, then*

$$\begin{array}{lcl} \mathbf{V}^*\mathbf{V} = I & \Rightarrow & V^*V = I, \quad \Lambda_{\mathbf{F}} = I, \\ \mathbf{V}\mathbf{V}^* = I & \Rightarrow & VV^* = I, \quad \Lambda_{\mathbf{F}_o} = I. \end{array} \qquad (6.10)$$

If $\ell_A \le 1$, then

$$\begin{array}{lcl} \mathbf{V}^*\mathbf{V} = I, \quad \Lambda_{\mathbf{F}} = I & \Rightarrow & V^*V = I, \\ \mathbf{V}\mathbf{V}^* = I, \quad \Lambda_{\mathbf{F}_o} = I & \Rightarrow & VV^* = I. \end{array}$$

PROOF If $\ell_A < 1$, then $\mathbf{V}^*\mathbf{V} = I$ implies *a.o.* $A^*A + B^*B = I$. This expression can be compared with the Lyapunov equation for $\mathbf{F}$: $A^*\Lambda_\mathbf{F}A + B^*B = \Lambda_\mathbf{F}^{(-1)}$. Since $\ell_A < 1$, the equation has a unique solution, which must be $\Lambda_\mathbf{F} = I$. A dual result holds for $\Lambda_{\mathbf{F}_o}$ in case $\mathbf{V}\mathbf{V}^* = I$. In contrast, if $\ell_A = 1$, then $\Lambda_\mathbf{F}$ cannot be uniquely determined: we cannot conclude uniform reachability from $A^*A + B^*B = I$. Hence, in that case we have to put this as a requirement.

Assume $\mathbf{V}^*\mathbf{V} = I$ and $\Lambda_\mathbf{F} = I$. Since it is an orthonormal basis, we write $\mathbf{Q}$ for $\mathbf{F}$ from now on. Equations (6.4) hold:

$$\begin{aligned} \mathbf{P}_0(\cdot V) &= \mathbf{P}_0(\cdot[D + \mathbf{Q}^*C]) \\ V^* &= D^* + C^*\mathbf{Q}. \end{aligned}$$

To show $V^*V = I$, we show that $\mathbf{P}_0(Z^{-n}V^*V)$ is $= I$ for $n = 0$, and $= 0$ otherwise. For $n = 0$:

$$\begin{aligned} \mathbf{P}_0(V^*V) &= \mathbf{P}_0([D^* + C^*\mathbf{Q}]\,[D + \mathbf{Q}^*C]) \\ &= \mathbf{P}_0(D^*D) + \mathbf{P}_0(D^*\mathbf{Q}^*C) + \mathbf{P}_0(C^*\mathbf{Q}D) + \mathbf{P}_0(C^*\mathbf{Q}\mathbf{Q}^*C) \\ &= D^*D + C^*C = I. \end{aligned}$$

For $n > 0$,

$$\begin{aligned} &\mathbf{P}_0(Z^{-n}V^*V) \\ &\quad = \mathbf{P}_0(Z^{-n}[D^* + C^*\mathbf{Q}]\,[D + \mathbf{Q}^*C]) \\ &\quad = \mathbf{P}_0(Z^{-n}D^*D) + \mathbf{P}_0(Z^{-n}D^*\mathbf{Q}^*C) + \mathbf{P}_0(Z^{-n}C^*\mathbf{Q}D) + \mathbf{P}_0(Z^{-n}C^*\mathbf{Q}\mathbf{Q}^*C). \end{aligned}$$

Using equations (5.37) and (5.38), *viz.*

$$\begin{aligned} \mathbf{P}_0(Z^{-n}\mathbf{Q}\mathbf{Q}^*) &= A^{\{n\}} && (n \geq 0) \\ \mathbf{P}_0(Z^{-n}\mathbf{Q}^*) &= B^{(n)}A^{\{n-1\}} && (n > 0), \end{aligned}$$

gives

$$\begin{aligned} \mathbf{P}_0(Z^{-n}V^*V) &= 0 + 0 + D^{*(n)}B^{(n)}A^{\{n-1\}}C + C^{*(n)}A^{\{n\}}C \\ &= [D^*B + C^*A]^{(n)}A^{\{n-1\}}C \\ &= 0. \end{aligned}$$

Taking adjoints shows that $\mathbf{P}_0(Z^{-n}V^*V) = 0$ for $n < 0$, too. Hence $V^*V = I$.
The fact $[\mathbf{V}\mathbf{V}^* = I,\ \Lambda_{\mathbf{F}_o} = I] \Rightarrow VV^* = I$ can be shown in a dual way. □

Theorem 6.12 has an interpretation in terms of conservation of energy. Let $\mathbf{V}$ be a realization for some bounded operator, such that $\mathbf{V}\mathbf{V}^* = I$. With $[X_{[k+1]}^{(-1)}\ Y_{[k]}] = [X_{[k]}\ U_{[k]}]\mathbf{V}$, this property ensures that, for each k,

$$\|\,[X_{[k+1]}^{(-1)}\ Y_{[k]}]\,\|_{HS}^2 = \|\,[X_{[k]}\ U_{[k]}]\,\|_{HS}^2. \qquad (6.11)$$

Summing this equation over all k yields

$$\|\,Y\,\|_{HS}^2 + \|\,X\,\|_{HS}^2 = \|\,U\,\|_{HS}^2 + \|\,X\,\|_{HS}^2.$$

If $\ell_A < 1$, then $X \in \mathcal{X}_2$ so that $\|\,X\,\|_{HS}^2 < \infty$, and it follows that $\|\,Y\,\|_{HS} = \|\,U\,\|_{HS}$, so that $VV^* = I$. In the case where $\ell_A = 1$, $\|\,X\,\|_{HS}^2$ can be unbounded: energy can remain in the state $X_{[k]}$ for $k \to \infty$, so that the system is not lossless. If the realization has observability Gramian equal to I, this can in fact not occur, but observability cannot be determined from $AA^* + CC^* = I$ if $\ell_A = 1$.

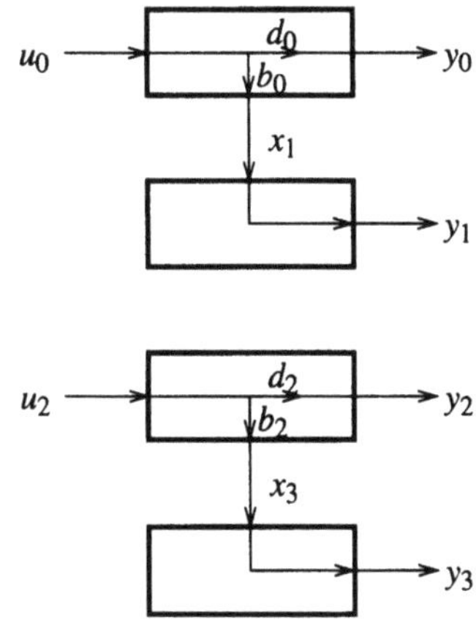

Figure 6.2. A simple isometric system.

Example

As an example, let $V \in \mathcal{U}(\mathcal{M},\mathcal{N})$ be given by

$$V = \begin{bmatrix} \boxed{d_0} & b_0 & 0 & 0 & 0 & 0 & \cdots \\ \cdot & \vdots & \cdot & \cdot & \cdot & \cdot & \cdots \\ 0 & 0 & \underline{d_2} & b_2 & 0 & 0 & \cdots \\ \cdot & \cdot & \cdot & \vdots & \cdot & \cdot & \cdots \\ 0 & 0 & 0 & 0 & \underline{d_4} & b_4 & \cdots \\ & & & & & & \ddots \end{bmatrix} \qquad \begin{array}{lcl} \#\mathcal{M} & = & [1\ 0\ 1\ 0\ 1\ 0 \cdots] \\ \#\mathcal{N} & = & [1\ 1\ 1\ 1\ 1\ 1\ \cdots] \\ \#\mathcal{B} & = & [0\ 1\ 0\ 1\ 0\ 1\ \cdots], \end{array}$$

where $d_k^2 + b_k^2 = 1$ (the underlined entries form the main diagonal, the '$\cdot$' denotes an entry with zero dimensions). V is an isometry: $VV^* = I$. It has an isometric realization, $\mathbf{VV}^* = I$, given by

$$\mathbf{V}_k = \left[\begin{array}{c|c} \cdot & \cdot \\ \hline b_k & d_k \end{array}\right] \quad (\text{even } k), \qquad \mathbf{V}_k = \left[\begin{array}{c|c} \cdot & 1 \\ \hline \cdot & \cdot \end{array}\right] \quad (\text{odd } k).$$

See figure 6.2. Let $b_k \to 0$, for $i \to \infty$. Then the output state space $\mathcal{H}_o(V) = \mathbf{P}(\mathcal{L}_2 Z^{-1} V)$ is not a closed subspace: it is the range of the Hankel operator H_V of V, with snapshots

$$(H_V)_k = 0 \quad (\text{even } k), \qquad (H_V)_k = \begin{bmatrix} b_{k-1} & 0 & \cdots \\ 0 & 0 & \\ \vdots & & \ddots \end{bmatrix} \quad (\text{odd } k).$$

The row range of $(H_V)_k$ determines the k-th slice of $\mathcal{H}_o(V)$. For odd k, the Hankel matrix has rank 1, but the range of the whole collection is not closed because $b_k \to 0$ but never becomes equal to 0.

In this example, V can be extended to an inner operator W, by adding extra inputs. This is straightforwardly done by completing each realization matrix $\mathbf{V}_k$ to a unitary matrix $\mathbf{W}_k$, which yields

$$\mathbf{W}_k = \left[\begin{array}{c|c} \cdot & \cdot \\ \hline b_k & d_k \\ -d_k & b_k \end{array}\right] \quad (\text{even } k), \qquad \mathbf{W}_k = \left[\begin{array}{c|c} \cdot & 1 \\ \hline \cdot & \cdot \end{array}\right] \quad (\text{odd } k),$$

$$W = \begin{bmatrix} \boxed{d_0} & b_0 & 0 & 0 & 0 & \cdots \\ b_0 & -d_0 & 0 & 0 & 0 & \cdots \\ \cdot & \vdots & \cdot & \cdot & \cdot & \cdots \\ 0 & 0 & \underline{d_2} & b_2 & 0 & \cdots \\ 0 & 0 & \underline{b_2} & -d_2 & 0 & \cdots \\ & & & & \ddots & \end{bmatrix} \qquad \begin{array}{rcl} \#\mathcal{M}_W & = & [2\ 0\ 2\ 0\ 2\ 0 \cdots] \\ \#\mathcal{N}_W & = & [1\ 1\ 1\ 1\ 1\ 1\ \cdots] \\ \#\mathcal{B}_W & = & [0\ 1\ 0\ 1\ 0\ 1\ \cdots]. \end{array}$$

W satisfies $WW^* = I_{\mathcal{M}_W}$ and $W^*W = I_{\mathcal{N}_W}$. Its output state space is closed, and it is the closure of the output state space of V: $\mathcal{H}_o(W) = \overline{\mathcal{H}}_o(V)$. Indeed, the snapshots of the Hankel operator of W are given by

$$(H_W)_k = 0 \quad (\text{even } k), \qquad (H_W)_k = \begin{bmatrix} b_{k-1} & 0 & \cdots \\ -d_{k-1} & 0 & \\ 0 & 0 & \\ \vdots & & \ddots \end{bmatrix} \quad (\text{odd } k),$$

and each odd Hankel operator snapshot has one nonzero singular value, equal to 1.

Not every isometric transfer operator can be embedded in an inner one, although every isometric realization can be completed to a unitary one. A counterexample is given in the next section.

6.4 BEURLING-LAX LIKE THEOREM

The existence of the external factorization was shown to depend on the construction of an inner operator V such that $\mathcal{U}_2 V$ is equal to some specified subspace $\mathcal{K}_o(T)$, the output null space of the system T. There is, however, a general result, which states that any subspace[1] $\mathcal{K}_0$ which is left D-invariant and Z-invariant (*i.e.*, such that $Z\mathcal{K}_0 \subset \mathcal{K}_0$) is of the form $\mathcal{U}_2 V$, for some isometric operator V. Such a theorem is known in the Hardy space setting as a Beurling-Lax theorem [Beu49, Lax59, Hel64]. It not only provides the external factorization in the locally finite case, but other factorizations as well, such as the inner-outer factorization in section 7.2.

[1] The index 0 in $\mathcal{K}_0$ will get meaning later in this section, when we consider a nested sequence of spaces $\mathcal{K}_n$ constructed from $\mathcal{K}_0$.

From the next theorem, it follows that the input space $\mathcal{M}$ of $V \in \mathcal{U}(\mathcal{M},\mathcal{N})$ satisfying $\mathcal{K}_0 = \mathcal{U}_2^{\mathcal{M}} V$ is of locally finite dimension only if $\mathcal{K}_0 \ominus Z\mathcal{K}_0$ is a locally finite subspace. Although $\mathcal{M}$ will be locally finite in the application to inner-outer factorization, we will prove theorem 6.13 for the more general situation. This calls for an extension of some of the definitions in chapter 2, to include operators with matrix representations whose entries are again operators. The extensions are straightforward (see [DD92]).

Theorem 6.13 (BeurlingLax-like) *All DZ-invariant subspaces $\mathcal{K}_0$ in $\mathcal{U}_2^{\mathcal{N}}$ have the form $\mathcal{K}_0 = \mathcal{U}_2^{\mathcal{M}} V$, where $V \in \mathcal{U}(\mathcal{M},\mathcal{N})$ is an isometry ($VV^* = I$). V is uniquely defined except for a right diagonal unitary factor.*

PROOF Let $\mathcal{R}_0 = \mathcal{K}_0 \ominus Z\mathcal{K}_0$. This is a D-invariant subspace in $\mathcal{U}_2^{\mathcal{N}}$. We can assume that it is non-empty, for else $\mathcal{K}_0 = Z\mathcal{K}_0 = Z^n\mathcal{K}_0$ for all $n \geq 0$. In that case, note that $X \in \mathcal{U}_2 \Rightarrow \lim_{n\to\infty} \mathbf{P}(Z^{-n}X) = 0$, so that in particular, for $X \in \mathcal{K}_0 \subset \mathcal{U}_2$, we have $Z^{-n}X \in \mathcal{K}_0$, and $\lim_{n\to\infty} \mathbf{P}(Z^{-n}X) = \lim_{n\to\infty} Z^{-n}X = 0$. This implies that $\mathcal{K}_0 = \{0\}$, so that there is nothing to prove when $\mathcal{R}_0$ is empty.

Now define $\mathcal{R}_n = Z^n\mathcal{K}_0 \ominus Z^{n+1}\mathcal{K}_0$. Then $\mathcal{R}_n = Z^n\mathcal{R}_0$, and $\mathcal{K}_0 \subset \mathcal{R}_0 \oplus \mathcal{R}_1 \oplus \mathcal{R}_2 \oplus \cdots$. In fact $\mathcal{K}_0 = \mathcal{R}_0 \oplus \mathcal{R}_1 \oplus \mathcal{R}_2 \oplus \cdots$, for suppose that $f \in \mathcal{K}_0$ and $f \perp \mathcal{R}_0 \oplus \mathcal{R}_1 \oplus \cdots$, then it follows that $f \in Z^n\mathcal{K}_0 \subset Z^n\mathcal{U}_2$ for all $n \geq 1$, and hence $f = 0$.

Suppose sdim $\mathcal{R}_0 = M$, and define the sequence of Hilbert spaces $\mathcal{M}$ to have entries $\mathcal{M}_k = \mathbb{C}^{M_k}$ ($\mathcal{M}_k = \ell_2$ if $M_k = \infty$).[2] Then there exist isometries $V_k : \mathcal{M}_k \to (\mathcal{R}_0)_k$ such that $(\mathcal{R}_0)_k = \mathcal{M}_k V_k$. Let V be the operator whose k-th block-row is equal to V_k. Stacking the V_k into one operator V, we obtain an orthonormal basis representation of $\mathcal{R}_0$, as in chapter 4, such that

$$\mathcal{R}_0 = \mathcal{D}_2^{\mathcal{M}} V, \qquad \mathbf{P}_0(VV^*) = I.$$

It follows that $\mathcal{R}_n = \mathcal{D}_2 Z^n V$, and because $\mathcal{R}_i \perp \mathcal{R}_j$ ($i \neq j$), that $D_1 Z^n V \perp D_2 V$ ($n \geq 1$) for all $D_{1,2} \in \mathcal{D}_2$, *i.e.*,

$$\begin{aligned} \mathbf{P}_0(Z^n VV^*) &= 0 \\ \mathbf{P}_0(VV^*Z^{-n}) &= 0 \end{aligned}$$

so that $VV^* = I$: V is an isometry. The orthogonal collection $\{\mathcal{D}_2 Z^n V\}_{n=0}^{\infty} \in \mathcal{K}_0$, and together spans the space $\mathcal{U}_2 V$. Hence $\mathcal{K}_0 = \{\mathcal{D}_2 Z^n V\}_0^{\infty} = \mathcal{U}_2 V$.

The uniqueness follows easily by retracing the steps and showing that any characteristic V actually defines an orthonormal basis for the "wandering subspace" $\mathcal{R}_0$.

□

The above proof is in the style of Helson [Hel64, §VI.3] for the time-invariant Hardy space setting. This proof was in turn based on Beurling's work [Beu49] for the scalar (SISO) case and Lax [Lax59] for the extension to vector valued functions.

[2]Let N be the index sequence corresponding to $\mathcal{N}$, with entries N_i. It follows that the dimension sequence M has entries $M_i < N_i + N_{i+1} + \cdots$. Although M_i can be infinite, an orthonormal basis for $(\mathcal{R}_0)_i = \pi_i \mathcal{R}_0$ is still *countable*, and the construction of an orthonormal basis representation of $\mathcal{R}_0$ can be done as explained in the proof of the theorem.

Doubly shift-invariant subspaces

Theorem 6.13 is instrumental in completing the description of isometric operators given in proposition 6.10. In that proposition, it was found that V is inner if $VV^* = I$ and $\ker(\cdot V^*|_{\mathcal{X}_2}) = \{0\}$. A remaining issue is to give conditions in state space terms under which V is actually inner, or can be extended/embedded into an inner transfer operator. As we already know, a *sufficient* condition is that $\ell_A < 1$. A precise condition involves the notion of "doubly shift-invariant subspaces".

For time-invariant systems, V will be inner if and only if the corresponding output null space $\mathcal{K}_o(V)$ is "full range" [Hel64].[3] Systems T for which $\mathcal{K}_o(T)$ is full range are called "roomy" in [Dew76]. Time invariant systems of finite degree are roomy: if $\mathcal{H}_o(T)$ is finite dimensional, then its complement $\mathcal{K}_o(T)$ is automatically full range. In the time-varying setting this turns out *not* to be true. To show this, we start out with a study of the geometry of the state spaces of an isometry.

If V is inner, then $\mathcal{K}_o(V) = \mathcal{U}_2 V$ and $\mathcal{H}_o(V) = \mathcal{U}_2 \ominus \mathcal{U}_2 V$. If V is an isometry, then the structure of the orthogonal complement of $\overline{\mathcal{H}}_o(V)$ is more involved. Let $\mathcal{K}_o = \mathcal{U}_2 V$ and $\mathcal{K}'_o = \ker(\cdot V^*|_{\mathcal{U}_2}) = \{X \in \mathcal{U}_2 : XV^* = 0\}$, then, by proposition 6.10,

$$\mathcal{U}_2 = \overline{\mathcal{H}}_o(V) \oplus \mathcal{K}'_o \oplus \mathcal{K}_o. \tag{6.12}$$

However, the condition $\mathcal{K}'_o = \{0\}$ does not entail $\{X \in \mathcal{X}_2 : XV^* = 0\} = \{0\}$, so that $\mathcal{K}'_o = \{0\}$ does not imply that V is inner (an elementary example is given in chapter 7). The space $\mathcal{K}'_o$, if non-empty, can be absorbed in an isometric embedding of V, with output state space $\mathcal{H}_o(V)$ and output null space $\mathcal{U}_2 V \oplus \mathcal{K}'_o$. The result is not necessarily an inner operator, but one which has a unitary realization, which makes it "almost" inner but not quite. Indeed, there might be elements in $\ker(\cdot V^*|_{\mathcal{X}_2})$ that are not in $\mathcal{K}'_o$ and hence cannot be absorbed. This "defect space" will be shown to satisfy a double shift invariance property.

Let $\mathcal{K}_0 = \mathcal{K}_o = \mathcal{U}_2 V$ and $\mathcal{K}_n = \mathbf{P}(Z^{-n}\mathcal{K}_0)$. Define $\mathcal{H}_0 = \mathcal{U}_2 \ominus \mathcal{K}_0$, and, for $n > 0$, $\mathcal{H}_n = \mathbf{P}(Z^{-n}\mathcal{H}_0)$.

Proposition 6.14 *With the definitions given above and for $n \geq 0$,*

$$\begin{aligned} \mathcal{K}_n &\subset \mathcal{K}_{n+1} \\ \mathcal{K}'_o &= \mathcal{U}_2 \ominus \textstyle\bigcup_0^\infty \mathcal{K}_n . \end{aligned}$$

PROOF Because $Z\mathcal{K}_0 \subset \mathcal{K}_0$, it follows that $\mathcal{K}_0 = \mathbf{P}(Z^{-1}Z\mathcal{K}_0) \subset \mathbf{P}(Z^{-1}\mathcal{K}_0) = \mathcal{K}_1$. Repeating the argument gives $\mathcal{K}_n \subset \mathcal{K}_{n+1}$. Let $X \in \mathcal{U}_2$. Then, and because $\mathcal{K}_0 = \mathcal{U}_2 V$,

$$\begin{aligned} X \in \mathcal{K}'_o \;&\Leftrightarrow\; XV^* = 0 \\ &\Leftrightarrow\; \mathbf{P}_0(XV^*Z^n) = 0 \qquad (\text{all } n \in \mathbb{Z}) \\ &\Leftrightarrow\; X \perp \mathbf{P}(Z^{-n}\mathcal{K}_0) \qquad (\text{all } n \geq 0) \\ &\Leftrightarrow\; X \perp \textstyle\bigcup_0^\infty \mathbf{P}(Z^{-n}\mathcal{K}_0) \qquad (\text{all } n \geq 0). \end{aligned}$$

[3] The notion of full range refers to the space spanned by z-transforms of functions of $\mathcal{K}_o$ at each point of the unit circle in the complex plane (a so-called "analytic range function").

□

This property can also be given in terms of $\mathcal{H}_n$:

Corollary 6.15 *With the definitions given above and for $n \geq 0$,*

$$\begin{array}{rcl} \mathcal{H}_n & = & \mathcal{U}_2 \ominus \mathcal{K}_n\,, \\ \mathcal{H}_{n+1} & \subset & \mathcal{H}_n\,, \\ \mathcal{K}'_o & = & \bigcap_0^\infty \mathcal{H}_n\,. \end{array}$$

PROOF

$$\begin{array}{rcl} X \in \mathcal{U}_2 \ominus \mathcal{K}_n & \Leftrightarrow & X \in \mathcal{U}_2\,, \quad X \perp \mathcal{K}_n \\ & \Leftrightarrow & X \in \mathcal{U}_2\,, \quad Z^n X \perp \mathcal{K}_0 \\ & \Leftrightarrow & X \in \mathcal{U}_2\,, \quad Z^n X \in \mathcal{H}_0 \\ & \Leftrightarrow & X \in \mathcal{U}_2\,, \quad X \in Z^{-n}\mathcal{H}_0 \\ & \Leftrightarrow & X \in \mathbf{P}(Z^{-n}\mathcal{H}_0) \quad = \quad \mathcal{H}_n\,. \end{array}$$

Hence $\mathcal{H}_n = \mathcal{U}_2 \ominus \mathcal{K}_n$. The remaining issues are a corollary of proposition 6.14. □

Proposition 6.16 *$\mathcal{K}'_o$ is a doubly shift-invariant subspace in $\mathcal{U}_2$: $Z\mathcal{K}'_o \subset \mathcal{K}'_o$, $\mathbf{P}(Z^{-1}\mathcal{K}'_o) \subset \mathcal{K}'_o$.*

PROOF $Z\mathcal{K}'_o \subset \mathcal{K}'_o$, because

$$\begin{array}{rcl} Z\mathcal{K}'_o & = & \{ZX : X \in \mathcal{K}'_o\} \\ & = & \{ZX : X \in \mathcal{U}_2 \wedge XV^* = 0\} \\ & = & \{ZX : X \in \mathcal{U}_2 \wedge ZXV^* = 0\} \\ & = & \{Y \in Z\mathcal{U}_2 : YV^* = 0\} \\ & \subset & \mathcal{K}'_o\,. \end{array}$$

But also $\mathbf{P}(Z^{-1}\mathcal{K}'_o) \subset \mathcal{K}'_o$, because $\mathbf{P}(Z^{-1}\mathcal{H}_n) = \mathcal{H}_{n+1} \subset \mathcal{H}_n$, and $\mathcal{H}_n = \bigcap_{k=0}^n \mathcal{H}_n$. Hence $\mathbf{P}(Z^{-1} \bigcap_{k=0}^n \mathcal{H}_k) = \mathcal{H}_{n+1} \subset \bigcap_{k=0}^n \mathcal{H}_k$. Letting $n \to \infty$ yields $\mathbf{P}(Z^{-1}\mathcal{K}'_o) \subset \mathcal{K}'_o$. □

An important corollary of the preceding discussion is that an isometric transfer operator $V \in \mathcal{U}$ for which $\mathcal{K}'_o(V) \neq \{0\}$ admits a completion by another isometric U into a larger isometric operator $W = \left[\begin{smallmatrix} U \\ V \end{smallmatrix}\right]$ for which $\mathcal{K}'_o(W) = \{0\}$ and which has a unitary realization. W is "almost" inner, since from theorem 6.4 we know it has to be inner if $\ell_A < 1$. The existence of U follows from the fact that the kernel $\mathcal{K}'_o = \ker(\cdot\, V^*|_{\mathcal{U}_2})$ is shift-invariant (proposition 6.16), so that, according to theorem 6.13, it can be written as $\mathcal{K}'_o = \mathcal{U}_2 U$:

Proposition 6.17 *If $V \in \mathcal{U}(\mathcal{M},\mathcal{N})$ is a locally finite isometry ($VV^* = I$), then there exists an isometry $U \in \mathcal{U}(\mathcal{M}_U,\mathcal{N})$ such that $\ker(\cdot\, V^*|_{\mathcal{U}_2^{\mathcal{N}}}) = \mathcal{U}_2^{\mathcal{M}_U} U$. The operator*

$$W = \begin{bmatrix} U \\ V \end{bmatrix}$$

is again isometric, now with $\mathcal{K}'_o(W) = \{0\}$, and it has a unitary realization. Conversely, if V is isometric and has a unitary realization, then the corresponding $\mathcal{K}'_o = \{0\}$.

PROOF If V is an isometry, then (proposition 6.10)

$$\mathcal{U}_2^{\mathcal{N}} = \overline{\mathcal{H}}_o(V) \oplus \ker(\cdot V^*|_{\mathcal{U}_2^{\mathcal{N}}}) \oplus \mathcal{U}_2^{\mathcal{M}} V, \tag{6.13}$$

where $\mathcal{K}'_o := \ker(\cdot V^*|_{\mathcal{U}_2})$ is left DZ-invariant. According to theorem 6.13 there exists an isometry $U \in \mathcal{U}(\mathcal{M}_U,\mathcal{N})$ such that $\mathcal{K}'_o = \mathcal{U}_2^{\mathcal{M}_U} U$. To conclude that $WW^* = I$, it remains to show that $UV^* = 0$, which is true because $\mathcal{U}_2 V \perp \mathcal{U}_2 U$. Hence $\mathcal{U}_2 W = \mathcal{U}_2 U \oplus \mathcal{U}_2 V$, and since $\overline{\mathcal{H}}_o(W) \supset \overline{\mathcal{H}}_o(V)$, we must have (from equation (6.13)) that $\overline{\mathcal{H}}_o(W) = \overline{\mathcal{H}}_o(V)$ and $\mathcal{K}'_o(W) = \ker(\cdot W^*|_{\mathcal{U}_2}) = \{0\}$.

By theorem 6.11, V has a canonical observer realization

$$\mathbf{V} = \begin{bmatrix} A_V & B_V \\ C_V & D_V \end{bmatrix}$$

which has observability Gramian $\Lambda_{\mathbf{F}_o} = I$ and satisfies $\mathbf{V}\mathbf{V}^* = I$. Since U and W constructed above are isometric as well, and have the same output state space as V, their canonical observer realizations $\mathbf{U}$ and $\mathbf{W}$ have the same A_V, C_V and are also isometric. Hence, we must have that

$$\mathbf{W} = \begin{bmatrix} A_V & B_V \\ C_V & D_V \\ B_U & D_U \end{bmatrix}. \tag{6.14}$$

By theorem 6.12, it then follows that $\mathbf{W}$ is the realization of an operator W such that $WW^* = I$. If at this point $\mathbf{W}$ would not be unitary then this can only be because its local realizations $\mathbf{W}_k$ would not be square matrices (since they have finite size). In that case, $\mathbf{W}$ can be extended to a unitary matrix, but after application of theorem 6.12, it would follow that W is not yet an isometry, because its extension is. From this contradiction, it follows that $\mathbf{W}$ must be unitary.

The same argument also proves the converse statement in the theorem. □

It is relatively easy to construct an isometric transfer function V for which $\mathcal{K}'_o := \ker(\cdot V^*|_{\mathcal{U}_2}) = \{0\}$ but not $\mathcal{K}''_o := \ker(\cdot V^*|_{\mathcal{X}_2}) = \{0\}$, and we do so in chapter 7. This shows that W in the last proposition is not necessarily inner. We already know from proposition 6.10 that W will be inner if $\ker(\cdot W^*|_{\mathcal{X}_2}) = \{0\}$. In addition, from theorem 6.12, we can conclude that W is inner if the realization of V has $\ell_A < 1$, or in case $\ell_A = 1$, if the realization of W is both uniformly observable and uniformly controllable. For this it is necessary that the input and output state space of W are closed. Problems can be expected if this does not hold for V.

$\mathcal{K}''_o$ is doubly shift invariant on all of $\mathcal{X}_2$ ($Z\mathcal{K}''_o \subset \mathcal{K}''_o$ and $Z^{-1}\mathcal{K}''_o \subset \mathcal{K}''_o$). This fact is very important and characterizes this subspace. We explore the matter a little further in the following proposition; additional results will be proven in section 7.5.

Proposition 6.18 *Assume that V is an isometry for which $\mathcal{K}'_o := \ker(\cdot V^*|_{\mathcal{U}_2}) = \{0\}$, but $\mathcal{K}''_o := \ker(\cdot V^*|_{\mathcal{X}_2}) \neq \{0\}$. Then $\mathbf{P}(\mathcal{K}''_o) \subset \overline{\mathcal{H}}_o(V)$. Moreover, let A be defined by the canonical observer realization of V. Then $\ell_A = 1$.*

PROOF By assumption, $\mathcal{K}''_o$ contains non-zero members, and clearly, it forms a left D,Z,Z^{-1} (*i.e.*, doubly shift invariant) subspace of $\mathcal{X}_2$. Let $y_o \in \mathcal{K}''_o$ and $y = \mathbf{P}(y_o)$, and consider the diagonal inner product $\{y, uV\}$ for an arbitrary $u \in \mathcal{U}_2$:

$$\begin{aligned} \{y, uV\} &= \mathbf{P}_0(yV^*u^*) \\ &= \mathbf{P}_0([y_o - (y_o - y)]V^*u^*) \\ &= -\mathbf{P}_0((y_o - y)V^*u^*) = 0, \end{aligned}$$

since $(y_o - y) \in \mathcal{L}_2 Z^{-1}$. Hence $y \in \overline{\mathcal{H}}_o(V) = \mathcal{U}_2 \ominus \mathcal{U}_2 V$.

Furthermore, an output normal realization based on an orthonormal basis representation $\mathbf{G}$ of $\mathcal{H}_o$ will produce an A-matrix for which $yA^{(1)} = \mathbf{P}(Z^{-1}y)$, or more generally, $yA^{\{n\}} = \mathbf{P}(Z^{-n}y)$, see section 5.4. To show that $\ell_A = 1$, we have to show that $\lim_{n\to\infty} \|A^{\{n\}}\|^{\frac{1}{n}} = 1$. This we do by showing that for all $n > 0$, $\|A^{\{n\}}\| = 1$. Pick an $n \geq 1$ and an arbitrary $\varepsilon < 1$. If $y_o \in \mathcal{K}_o''$, then it is also true for any k that $Z^k y_o \in \mathcal{K}_o''$, due to double shift invariance. By taking y_o to be a unit-norm member of $\mathcal{K}_o''$ shifted far enough to the right, we can guarantee that the corresponding $y = \mathbf{P}y_o$ as well as $\mathbf{P}(Z^{-n}y_o)$ have their norm as close to 1 as we wish. Thus, given ε choose a $y_o \in \mathcal{K}_o''$ and $y = \mathbf{P}(y_o)$ such that (1) $\|y\| = 1$, (2) $\|y_o - y\| < \varepsilon/3$ and (3) $\|Z^{-n}y_o - \mathbf{P}(Z^{-n}y_o)\| < \varepsilon/3$. It follows that

$$\begin{aligned} \|\mathbf{P}(Z^{-n}y) - Z^{-n}y\| &\leq \|\mathbf{P}(Z^{-n}(y - y_o))\| + \|\mathbf{P}(Z^{-n}y_o) - Z^{-n}y_o\| + \|Z^{-n}y_o - Z^{-n}y\| \\ &< \varepsilon, \end{aligned}$$

and since $\|Z^{-n}y\| = \|y\| = 1$,

$$\|\mathbf{P}(Z^{-n}y)\| > 1 - \varepsilon.$$

Since $A^{\{n\}}$ maps y on $\mathbf{P}(Z^{-n}y)$, it must be that $\|A^{\{n\}}\| > 1 - \varepsilon$, and since ε was arbitrary to start with, $\|A^{\{n\}}\| = 1$ and $\ell_A = 1$. □

The second part of this theorem should not come as a surprise in view of the inner realization theorem 6.4, for if ℓ_A had been less than one, then V would have been inner, and $\mathcal{K}_o'' = \{0\}$ (this observation amounts to an indirect proof of the property).

Embedding through external factorization

Suppose that an isometric transfer operator $V \in \mathcal{U}$ (with $VV^* = I$) is given, and assume that it has a right coprime factorization

$$V = \Delta^* U$$

with $\Delta \in \mathcal{U}$ and U inner. We show that Δ is actually diagonal, and there exists a unitary diagonal U_d such that

$$\Delta^* = [I \quad 0]\, U_d, \qquad V = [I \quad 0]\,(U_d U).$$

The latter expression is an alternative right coprime factorization for V.

The property is easy to prove from the previous theory, in particular proposition 6.18. An alternative proof follows from an adaptation of the LTI theory in [Dew76], and we give a sketch of how it would work in the present context.

If Δ and U form a right coprime pair, then there exist sequences of transfer operators $M_n \in \mathcal{U}$ and $N_n \in \mathcal{U}$ such that

$$\lim_{n\to\infty} (UM_n + \Delta N_n) = I$$

(only a weak limit is needed, the sequences of operators do not necessarily converge individually). Now, $VV^* = I$ implies $\Delta^*\Delta = I$, and hence

$$\lim_{n\to\infty} (\Delta^* UM_n + N_n) = \Delta^* .$$

It follows that

$$\Delta^* = \lim_{n\to\infty} (VM_n + N_n) \in \mathcal{U} ,$$

and hence Δ must be diagonal, and isometric. A further reduction brings it to the form $[I \;\; 0]U_d$. The embedding theory will be given a further extensive treatment in chapter 12.

6.5 EXAMPLE

As an example of the use of inner-coprime factorizations, consider a mixed-causality operator $T \in \mathcal{X}$ with a decomposition $T = T_{\mathcal{L}} + T_{\mathcal{U}}$ where $T_{\mathcal{L}} \in \mathcal{L}$ and $T_{\mathcal{U}} \in \mathcal{ZU}$. Our objective is to compute a QR factorization of T,

$$Q^*T = R, \qquad Q \text{ unitary}, \quad R \in \mathcal{U} .$$

Note that $Q^*(T_{\mathcal{L}} + T_{\mathcal{U}}) = Q^*T_{\mathcal{L}} + Q^*T_{\mathcal{U}}$. Suppose that we compute an inner-coprime factorization of $T_{\mathcal{L}}^*$:

$$T_{\mathcal{L}}^* = \Delta^* V , \qquad \Delta \in \mathcal{U}, \quad V \in \mathcal{U}, \text{inner.}$$

Then $VT_{\mathcal{L}} = \Delta \in \mathcal{U}$, and also $VT_{\mathcal{U}} \in \mathcal{U}$ since both factors are upper. The QR factorization is thus given by

$$Q = V^* \in \mathcal{L}, \qquad R = \Delta + VT_{\mathcal{U}} \in \mathcal{U} .$$

In view of theorem 6.8, the factorization is possible if $T_{\mathcal{L}}^*$ has a locally finite realization that is uniformly observable and u.e. stable.

The factorization can be computed by the algorithm in proposition 6.9; this is worked out in detail in chapter 7. For a simple numerical example, which is amenable to direct calculations, take

$$T = \left[\begin{array}{ccc|ccccc}
\ddots & & & & & & & \\
 & 1 & & & & \mathbf{0} & & \\
 & & 1 & & & & & \\ \hline
 & & & \boxed{1} & & & & \\
 & & & 1/2 & 1 & & & \\
 & & & 1/4 & 1/2 & 1 & & \\
 & \mathbf{0} & & 1/8 & 1/4 & 1/2 & 1 & \\
 & & & \vdots & & & \vdots & \ddots
\end{array}\right] .$$

The inner-coprime factorization $T = Q^*R$ is

$$
Q^* = \left[\begin{array}{cc|ccccc}
\ddots & & & & & & \\
 & 1 & & & \mathbf{0} & & \\
 & & 1 & & & & \\
\hline
 & & \boxed{[1\ 0]} & & & & \\
 & & [0\ \tfrac{1}{2}\sqrt{3}] & 1/2 & & & \\
 & & [0\ \tfrac{1}{4}\sqrt{3}] & -3/4 & 1/2 & & \\
 & \mathbf{0} & [0\ \tfrac{1}{8}\sqrt{3}] & -3/8 & -3/4 & 1/2 & \\
 & & \vdots & & & \vdots & \ddots
\end{array}\right]
$$

$$
R = \left[\begin{array}{cc|ccccc}
\ddots & & & & & & \\
 & 1 & & & \mathbf{0} & & \\
 & & 1 & & & & \\
\hline
 & & \begin{bmatrix} 1 \\ \tfrac{1}{3}\sqrt{3} \end{bmatrix} & \begin{bmatrix} 0 \\ \tfrac{2}{3}\sqrt{3} \end{bmatrix} & \begin{bmatrix} 0 \\ \tfrac{1}{3}\sqrt{3} \end{bmatrix} & \begin{bmatrix} 0 \\ \tfrac{1}{6}\sqrt{3} \end{bmatrix} & \cdots \\
 & & & 0 & -1 & -\tfrac{1}{2} & -\tfrac{1}{4} \ \cdots \\
 & & & & 0 & -1 & -\tfrac{1}{2} \\
 & \mathbf{0} & & & & 0 & -1 \ \cdots \\
 & & & & & & \ddots
\end{array}\right].
$$

It is not hard to verify this by direct multiplications: Q is unitary and $T = Q^*R$, but obviously, this factorization is not trivially obtained. It has been computed by Matlab using the state space algorithm in proposition 6.9. Note that the number of inputs of Q and R is not constant: it is equal to 2 at time $k = 0$.

7 INNER-OUTER FACTORIZATION AND OPERATOR INVERSION

Direct methods to invert large matrices may give undesired "unstable" results. We can obtain valuable insights into the mechanics of this effect by representing the matrix as a time-varying system for which it is the transfer operator. Among other things, this will allow us to handle the instability by translating "unstable" into "anti-causal" yet bounded.

A central role in doing that is played by inner-outer factorizations of the relevant transfer operator: yet another consequence of the Beurling-Lax like theory of the previous chapter. The inner parts of the operator capture the part of the operator that causes the instability in the inverse, while the outer part can be straightforwardly inverted. The theory of inner-outer factorization may appear to be complex at first, but numerically it simply amounts to the computation of a sequence of QR factorization steps on the state space description of the original transfer operator, much like the computation of the external factorization of chapter 6. In fact, the inner-outer factorization provides a QR factorization of an upper operator, which is interesting only for infinite operators or finite matrices that are singular or have nonsquare blocks.

This chapter will be mostly motivated by the inversion problem, but we also cover important theoretical grounds on inner-outer factorization, since this forms the basis for the inversion algorithms. The chapter is concluded by a brief investigation of the "zero structure" of a transfer operator. For this we could analyze the pole structure of the inverse operator, but only if it exists. The inner-outer factorization provides precisely the same information without this complication, and we study its limit behavior in a specific case.

7.1 INTRODUCTION

The inversion of large structured matrices is a delicate problem which often arises in finite element modeling applications, or (implicitly) in non-stationary inverse filtering problems in signal processing. To stress the fact that these matrices might be fairly large and even so large that ordinary linear algebra techniques might fail, we allow them to have infinite size, *i.e.*, they are operators on the space of ℓ_2-sequences. To set the scene, consider the infinite Toeplitz matrix

$$T = \begin{bmatrix} \ddots & \ddots & & & \\ & \boxed{1} & -1/2 & & \mathbf{0} \\ & & 1 & -1/2 & \\ & & & 1 & -1/2 \\ & \mathbf{0} & & & 1 & \ddots \\ & & & & & \ddots \end{bmatrix}. \tag{7.1}$$

The inverse of T is given by

$$T^{-1} = \begin{bmatrix} \ddots & \vdots & & & \vdots & \\ & \boxed{1} & 1/2 & 1/4 & 1/8 & \cdots \\ & & 1 & 1/2 & 1/4 & \\ & & & 1 & 1/2 & \\ & \mathbf{0} & & & 1 & \cdots \\ & & & & & \ddots \end{bmatrix},$$

as is readily verified: $TT^{-1} = I$, $T^{-1}T = I$. One way to guess T^{-1} in this case is to restrict T to a (sufficiently large) finite matrix and invert that matrix. For example,

$$\begin{bmatrix} 1 & -1/2 & 0 \\ 0 & 1 & -1/2 \\ 0 & 0 & 1 \end{bmatrix}^{-1} = \begin{bmatrix} 1 & 1/2 & 1/4 \\ 0 & 1 & 1/2 \\ 0 & 0 & 1 \end{bmatrix}$$

already gives an indication. In general, however, this method will not give correct results. Another way to obtain T^{-1}, perhaps more appealing to engineers, goes via the z-transform:

$$\begin{aligned} T(z) &= 1 - \tfrac{1}{2}z \\ \Rightarrow \quad T^{-1}(z) &= \frac{1}{1-\frac{1}{2}z} = 1 + \tfrac{1}{2}z + \tfrac{1}{4}z^2 + \cdots . \end{aligned}$$

The expansion is valid for $|z| \leq 2$.

What happens if we now take

$$T = \begin{bmatrix} \ddots & \ddots & & & \\ & \boxed{1} & -2 & & \mathbf{0} \\ & & 1 & -2 & \\ & & & 1 & -2 \\ & \mathbf{0} & & & 1 & \ddots \\ & & & & & \ddots \end{bmatrix} \tag{7.2}$$

and treat it in the same way? The method "restricting to finite" would yield

$$T^{-1} \overset{?}{=} \begin{bmatrix} \ddots & \vdots & & & \vdots & \\ & \boxed{1} & 2 & 4 & 8 & \cdots \\ & & 1 & 2 & 4 & \\ & & & 1 & 2 & \\ & \mathbf{0} & & & 1 & \cdots \\ & & & & \vdots & \ddots \end{bmatrix}.$$

In transfer function parlance, that would correspond to writing

$$T^{-1}(z) = 1 + 2z + 4z^2 + \cdots.$$

Thus, T^{-1} is *unbounded*, and the series expansion for $T^{-1}(z)$ is not even valid for $|z| < 1$. The correct, bounded inverse is e.g. obtained via

$$\begin{aligned} T^{-1}(z) = \frac{1}{1-2z} &= \frac{-\frac{1}{2}z^{-1}}{1-\frac{1}{2}z^{-1}} \\ &= -\tfrac{1}{2}z^{-1} - \tfrac{1}{4}z^{-2} - \cdots \end{aligned}$$

$$\Rightarrow \quad T^{-1} = \begin{bmatrix} \ddots & \ddots & & & & \\ \cdots & -1/2 & \boxed{0} & & \mathbf{0} & \\ & -1/4 & -1/2 & 0 & & \\ & -1/8 & -1/4 & -1/2 & 0 & \\ \cdots & -1/16 & -1/8 & -1/4 & -1/2 & \ddots \\ & \vdots & & & \vdots & \ddots \end{bmatrix}. \tag{7.3}$$

Again, it is readily verified that $TT^{-1} = I$, $T^{-1}T = I$. This inverse is *bounded* but not causal. We see that the inverse of an upper operator need not be upper. In the light of finite dimensional linear algebra, this seems to be a strange result. An intuitive explanation is that, because the matrix is so large, the location of the main diagonal is not clear: a shift of the whole matrix over one (or a few) positions should be allowed and should only give a similar (reverse) shift in the inverse. For example, T^{-1} can be guessed from finite matrix calculus if one shifts the origin over one position:

$$\begin{bmatrix} -2 & 0 & 0 \\ 1 & -2 & 0 \\ 0 & 1 & -2 \end{bmatrix}^{-1} = \begin{bmatrix} -1/2 & 0 & 0 \\ -1/4 & -1/2 & 0 \\ -1/8 & -1/4 & -1/2 \end{bmatrix}.$$

A better explanation is to say that $T(z)$ is not an *outer function*, that is to say it is non-minimum phase, and hence $T^{-1}(z)$ is not causal, which translates to a lower triangular matrix representation (we shall make this more precise soon).

The above example gives a very elementary insight in how system theory (in this case the z-transform) can help with the bounded inversion of large matrices. The examples so far were cast in a time-invariant framework: all matrices were Toeplitz. We now go beyond this and consider general matrices and their connection to *time-varying* systems.

A simple illustrative example is provided by the combination of the above two cases:

$$T = \left[\begin{array}{ccc|ccccc} \ddots & \ddots & & & & & & \\ & 1 & -1/2 & & & & \mathbf{0} & \\ & & 1 & -1/2 & & & & \\ \hline & & & \boxed{1} & -2 & & & \\ & & & & 1 & -2 & & \\ & & & & & 1 & -2 & \\ & \mathbf{0} & & & & & 1 & \ddots \\ & & & & & & & \ddots \end{array}\right]. \tag{7.4}$$

Is T^{-1} upper? But then it would be unbounded:

$$T^{-1} \stackrel{?}{=} \left[\begin{array}{ccc|ccccc} \ddots & \vdots & & \vdots & & & \vdots & \\ & 1 & 1/2 & 1/4 & 1/2 & 1 & 2 & \cdots \\ & & 1 & 1/2 & 1 & 2 & 4 & \\ \hline & & & \boxed{1} & 2 & 4 & 8 & \cdots \\ & & & & 1 & 2 & 4 & \\ & \mathbf{0} & & & & 1 & 2 & \\ & & & & & & 1 & \cdots \\ & & & & & & & \ddots \end{array}\right].$$

Something similar happens if we opt for a lower triangular matrix representation. A bounded T^{-1} (if it exists!) will most likely be a combination of upper (the top-left corner) and lower (the bottom-right corner), and some unknown interaction in the center: something like

$$T^{-1} \stackrel{?}{=}$$

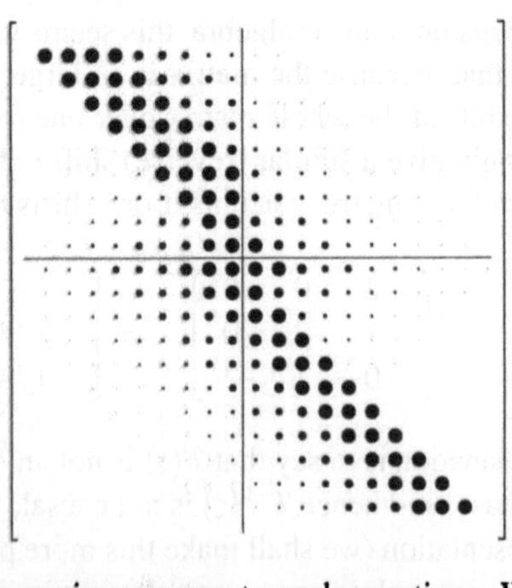

(7.5)

The purpose of this chapter is to give precise answers to such questions. We shall see that, in fact, T is not directly invertible, although it has a closed range. The form given in the figure is a Moore-Penrose pseudoinverse.

There are several potential applications of the theory in this chapter:

1. *Time-varying filter inversion*: T in (7.4) could represent an adaptive FIR filter, with a zero that moves from $z = 2$ to $z = 1/2$. Think *e.g.*, of an adaptive channel estimate that has to be inverted to retrieve the input signal from an observed output signal [TD95]. As the example shows, a direct inversion might lead to unstable results.

2. *Finite element matrix inversion*: Finite element matrices are often very large, and hence the effects observed above might play a role. Presently, stability of the inverse is ensured by careful selection of boundary conditions: the borders of the matrix are chosen such that its inverse (as determined by finite linear algebra) is well behaved. Time-varying techniques might give additional insight. Under stability assumptions, it is even envisioned that one might do without explicit boundary conditions: extend T to an infinite matrix, which is constant (Toeplitz) towards $(-\infty,-\infty)$ and $(+\infty,+\infty)$. LTI systems theory gives explicit starting points for inversion recursions. It is even possible to "zoom in" on a selected part of T^{-1}, without computing all of it.

3. System inversion also plays a role in control, *e.g.*, the manipulation of a flexible robot-arm [BL93].

4. Matrices and operators for which we already have a state realization with a low number of states can be inverted efficiently in this way. This is the topic of section 7.3. A prime example of such operators is the band matrix: this corresponds to a time-varying FIR filter. See chapter 3 for more examples.

7.2 INNER-OUTER FACTORIZATIONS

For rational time-invariant single-input single-output systems, the *inner-outer factorization* is a factorization of an analytical (causal) transfer function $T(z)$ into the product of an inner and an outer transfer function: $T(z) = V(z)T_o(z)$. The inner factor $V(z)$ has its poles outside the unit disc and has modulus 1 on the unit circle, whereas the outer factor $T_o(z)$ and its inverse are analytical in the open unit disc. Such functions are called *minimum phase* in engineering. For example, (with $|\alpha|,|\beta| < 1$)

$$z\frac{z-\alpha^*}{1-\beta z} = z\frac{z-\alpha^*}{1-\alpha z}\cdot\frac{1-\alpha z}{1-\beta z}.$$

The resulting outer factor is such that its inverse is again a stable system, provided it has no zeros on the unit circle. For multi-input multi-output systems, the definition of the outer factor is more complicated (see *e.g.*, Halmos [Hel64]) and takes the form of a range condition: $T_o(z)$ is outer if $T_o(z)H_m^2 = H_n^2$, where H_m^2 is the Hardy space of analytical m-dimensional vector-valued functions. Because matrix multiplication is not commutative, there is now a distinction between left and right outer factors. We shall see that generalizations of these definitions to the time-varying context are fairly straightforward.

An operator $T_\ell \in \mathcal{U}$ is said to be *left outer* if

$$\overline{\mathcal{U}_2 T_\ell} = \mathcal{U}_2. \tag{7.6}$$

Other definitions are possible;[1] this definition is such that $\overline{\text{ran}}(\cdot T_\ell) = \overline{\mathcal{X}_2 T_\ell} = \mathcal{X}_2$, so that $\ker(\cdot T_\ell^*) = \{0\}$ and T_ℓ has an algebraic left inverse (which can be unbounded if $\mathcal{X}_2 T_\ell$ is not closed).

A factorization of an operator $T \in \mathcal{U}$ into

$$T = T_\ell V, \qquad T_\ell \text{ left outer}, \quad VV^* = I$$

(V inner if possible) is called an outer-inner factorization. This factorization can be obtained from the Beurling-Lax type theorem 6.13 by taking a different definition of $\mathcal{K}_0$ than was the case in the external factorization (where we took $\mathcal{K}_0$ equal to the output null space $\mathcal{K}_o(T)$). Note that the closure in (7.6) is necessary: for example, the system $T = I - Z$ has inner factor $V = I$ and of necessity an outer factor $T_\ell = I - Z$. T_ℓ is not boundedly invertible, and $\mathcal{U}_2 T_\ell$ is only dense in $\mathcal{U}_2$. This happens when the range of T is not a closed subspace. The time-invariant equivalent of this example is $T(z) = 1 - z$, which has a zero on the unit circle. Here also, $V(z) = 1$, and $T_\ell(z) = T(z)$ is not boundedly invertible. Also note that if T is not an invertible operator, then it is not possible to obtain an inner factor: V can only be isometric since we have chosen T_ℓ to be (left) invertible.

Dually, we define $T_r \in \mathcal{U}$ to be *right outer* if

$$\overline{\mathcal{L}_2 Z^{-1} T_r^*} = \mathcal{L}_2 Z^{-1}. \tag{7.7}$$

The corresponding inner-outer factorization is

$$T = VT_r, \qquad V^*V = I, \quad T_r \text{ right outer}. \tag{7.8}$$

The two factorizations can be combined to obtain

$$T = URV, \qquad U^*U = I, \quad VV^* = I, \quad R \text{ left and right outer}.$$

This is similar to a *complete orthogonal decomposition* in linear algebra [GV89]. R is algebraically invertible in $\mathcal{U}$, and R^{-1} is bounded if the range of T is closed. $T^\dagger := V^* R^{-1} U^*$ will be a Moore-Penrose pseudo-inverse for T and it will be bounded if R^{-1} is. If U and V are both inner, then $T^\dagger$ will be the inverse of T. If T has a state space realization then there will be state space realizations for U, V, R and R^{-1} as well, but the last one may be unbounded.

Theorem 7.1 (inner-outer factorization) *Let $T \in \mathcal{U}(\mathcal{M},\mathcal{N})$. Then T has a factorization (outer-inner factorization)*

$$T = T_\ell V,$$

where $V \in \mathcal{U}(\mathcal{M}_V,\mathcal{N})$ is an isometry ($VV^* = I$), $T_\ell \in \mathcal{U}(\mathcal{M},\mathcal{M}_V)$ is left outer, and $\#(\mathcal{M}_V) \le \#(\mathcal{M})$. V is inner if and only if $\ker(\cdot T^*) = \{0\}$.

[1] See *e.g.*, Arveson [Arv75], who, translated to our notation, requires that $\mathcal{U}_2 T_\ell$ is dense in $\mathbf{P}(\mathcal{X}_2 T_\ell)$ and that the projection operator onto the range of T_ℓ is diagonal.

Dually, T has a factorization (inner-outer factorization)

$$T = VT_r$$

where $V \in \mathcal{U}(\mathcal{M},\mathcal{N}_V)$ is a coisometry ($V^*V = I$), $T_r \in \mathcal{U}(\mathcal{N}_V,\mathcal{N})$ is right outer, and $\mathcal{N}_V \subset \mathcal{N}$. V is inner if and only if $\ker(\cdot T) = \{0\}$.

PROOF Let $\mathcal{K}_0 = \overline{\mathcal{U}_2 T}$. Then $\mathcal{K}_0$ is a D-invariant subspace which is shift-invariant: $Z\mathcal{K}_0 \subset \mathcal{K}_0$. According to theorem 6.13, there is a space sequence $\mathcal{M}_V$ and an isometric operator $V \in \mathcal{U}(\mathcal{M}_V,\mathcal{N})$, $VV^* = I$, such that $\overline{\mathcal{U}_2^{\mathcal{M}} T} = \mathcal{U}_2^{\mathcal{M}_V} V$.

Define $T_\ell = TV^*$. Then $T_\ell \in \mathcal{U}(\mathcal{M},\mathcal{M}_V)$ and $\overline{\mathcal{U}_2^{\mathcal{M}} T_\ell} = \overline{\mathcal{U}_2 TV^*} = \overline{\overline{\mathcal{U}_2 T} V^*} = \overline{\mathcal{U}_2 VV^*} = \mathcal{U}_2^{\mathcal{M}_V}$, so that T_ℓ is left outer. It remains to prove that $T = T_\ell V$, *i.e.*, $T = TV^*V$. This is immediate if V is inner. If V is not inner, then it follows from the fact that $\cdot V^*V$ is an orthogonal projection onto the range of $\cdot V$. Indeed, since $\overline{\mathcal{U}_2 T} = \mathcal{U}_2 V$, also $\overline{\mathcal{X}_2 T} = \mathcal{X}_2 V$: the closure of the range of T is equal to the range of V, and T is not changed by the projection. Since $\overline{\operatorname{ran}}(\cdot T) \oplus \ker(\cdot T^*) = \mathcal{X}_2$, we must have $\ker(\cdot T^*) = \ker(\cdot V^*)$. By proposition 6.10, V is inner if and only if the latter space is empty.

By construction, $\overline{\mathcal{U}_2^{\mathcal{M}} T} = \mathcal{D}_2^{\mathcal{M}_V} V \oplus Z\mathcal{U}_2^{\mathcal{M}_V} V$ with $\mathcal{M}_V$ of minimal dimensions. Hence, $\mathbf{P}_0(\overline{\mathcal{U}_2^{\mathcal{M}} TV^*}) = \mathbf{P}_0(\overline{\mathcal{U}_2^{\mathcal{M}} T_\ell}) = \mathcal{D}_2^{\mathcal{M}} \mathbf{P}_0(T_\ell) = \mathcal{D}_2^{\mathcal{M}_V}$ and it follows that $\#(\mathcal{M}_V) \leq \#(\mathcal{M})$ (since the dimension of the range of a matrix cannot exceed the dimension of the domain).

□

Thus, the isometric factor V of the outer-inner factorization is defined by the property $\overline{\mathcal{U}_2 T} = \mathcal{U}_2 V$. If $\overline{\operatorname{ran}}\, T$ is not all of $\mathcal{X}_2^{\mathcal{M}}$, then V is not full range either, so that it is not inner. One can define a factorization based on the extension of V to an inner operator $W = \left[\begin{smallmatrix} U \\ V \end{smallmatrix}\right]$, if such an embedding exists; see proposition 6.17 in the previous chapter. This then gives a factorization $T = T_\ell W$ for which

$$\begin{aligned} \overline{\mathcal{U}_2^{\mathcal{M}} T_\ell} &= \overline{\mathcal{U}_2 T} W^* \\ &= \mathcal{U}_2 V W^* \\ &= \mathcal{U}_2^{\mathcal{M}_V} [0 \quad I] \subset \mathcal{U}_2^{\mathcal{M}_W}, \end{aligned}$$

so that T_ℓ is upper but not precisely outer:[2] it reaches only a subset of $\mathcal{U}_2^{\mathcal{M}_W}$. This is the best we can hope for, in view of the fact that T is not "full range".

[2] T_ℓ is outer according to Arveson's definition [Arv75].

Example

Let

$$T = \left[\begin{array}{ccc|ccccc}
\ddots & \ddots & & & & & & \\
& 1 & -1/2 & & & \mathbf{0} & & \\
& & 1 & -1/2 & & & & \\ \hline
& & & \boxed{1} & -2 & & & \\
& & & & 1 & -2 & & \\
& \mathbf{0} & & & & 1 & -2 & \\
& & & & & & 1 & \ddots \\
& & & & & & & \ddots
\end{array}\right]. \tag{7.9}$$

T is a prototype time-varying system: for negative times it coincides with a minimal phase time invariant system, while for positive times, the system has switched to a time invariant behavior which now has a zero at $z = 1/2$ and has thus become "maximum phase".

T has a right inverse obtained from the time-invariant behavior and given by

$$T_i = \left[\begin{array}{cccc|cccc}
\ddots & \ddots & \ddots & \vdots & & & & \\
& 1 & 1/2 & 1/4 & & & & \\
\mathbf{0} & & 1 & 1/2 & & \mathbf{0} & & \\
& & & 1 & & & & \\ \hline
& & & & \boxed{0} & & & \\
& & & & -1/2 & 0 & & \mathbf{0} \\
& \mathbf{0} & & & -1/4 & -1/2 & 0 & \\
& & & & \vdots & \ddots & \ddots & \ddots
\end{array}\right]. \tag{7.10}$$

Hence T itself is right outer: the inner-outer factorization is $T = I \cdot T_r$. Since the right inverse T_i is a bounded operator, the range of T is closed as well.[3] It is not hard to see that $Tx = 0$ for $x = [\cdots \ \frac{1}{4} \ \frac{1}{2} \ \boxed{1} \ \frac{1}{2} \ \frac{1}{4} \ \cdots]^T$. Hence the columns of T are not linearly independent (although looking at T one would have guessed differently!): $\ker(\cdot T^*|_{\mathcal{X}_2}) \neq \{0\}$. Thus, T is not invertible, and the right inverse is not a left inverse. In addition, the inverse displayed in (7.10) is not the Moore-Penrose inverse, see further in this chapter.

We can try to construct the outer-inner factorization for T from the property that $\mathcal{K}_o(V) = \mathcal{U}_2 V = \overline{\mathcal{U}_2 T}$. As explored later in this chapter, for this we should look for the largest sliced upper basis $\mathbf{F}_o$ satisfying $\mathbf{P}(\mathbf{F}_o T^*) = 0$, which will then be a basis for the output state space $\mathcal{H}_o(V)$, the orthogonal complement of $\mathcal{K}_o(V)$. By inspection, we

[3] Suppose that in the range of T, there is a sequence $y_n \to y$, then there is a sequence u_n in the domain of T such that $y_n = u_n T$. Since $TT_i = I$ we have that $u_n = u_n T T_i = y_n T_i \to y T_i$ is a convergent series. It follows that $y = uT$ since in turn $y = \lim_{n\to\infty} u_n T = uT$.

obtain

$$\mathbf{F}_o = \left[\begin{array}{cccc|ccccc} \ddots & \vdots & \vdots & \vdots & \vdots & \vdots & \vdots & \vdots & \\ & 1/8 & 1/4 & 1/2 & 1 & 1/2 & 1/4 & 1/8 & \cdots \\ & & 1/4 & 1/2 & 1 & 1/2 & 1/4 & 1/8 & \cdots \\ & & & 1/2 & 1 & 1/2 & 1/4 & 1/8 & \cdots \\ \hline & & & & \boxed{1} & 1/2 & 1/4 & 1/8 & \cdots \\ & & & & & 1 & 1/2 & 1/4 & \ddots \\ & & \mathbf{0} & & & & 1 & 1/2 & \ddots \\ & & & & & & & 1 & \ddots \\ & & & & & & & & \ddots \end{array}\right]. \qquad (7.11)$$

(Note that the basis is not a bounded operator in $\mathcal{X}$, but $\mathcal{D}_2\mathbf{F}_o \subset \mathcal{U}_2$.) Based on $\mathbf{F}_o$, we can construct a realization for V by normalizing $\mathbf{F}_o$ to an orthonormal basis representation $\mathbf{G}$, defining $A_V = \mathbf{P}_0(\mathcal{Z}^{-1}\mathbf{G}\mathbf{G}^*)^{(-1)}$, $C_V = \mathbf{P}_0(\mathbf{G})$, as in the canonical observer realization in chapter 5, theorem 5.17, followed by pointwise completion of (A_V, C_V) to a square unitary realization. The result is shown in (7.34) at the end of the chapter, where the outer-inner factorization will be obtained in a more structured way.

As remarked above, $\ker(\cdot T^*|_{\mathcal{X}_2}) \neq \{0\}$. Hence, by theorem 7.1, the V-operator in the outer-inner factorization which we just constructed cannot be inner and is only isometric. This is tightly connected to the existence of a doubly shift invariant subspace, and this illustrates the discussion of section 6.4. The space $\mathcal{K}_o'' := \ker(\cdot V^*|_{\mathcal{X}_2})$ as used in that section is equal to $\ker(\cdot T^*|_{\mathcal{X}_2})$, and clearly it is a left D-invariant and doubly shift invariant subspace. An (unnormalized) sliced basis for this subspace is

$$\begin{bmatrix} & \vdots & \vdots & \vdots & \vdots & \vdots & \\ \cdots & 1/4 & 1/2 & 1 & 1/2 & 1/4 & \cdots \\ \cdots & 1/4 & 1/2 & \boxed{1} & 1/2 & 1/4 & \cdots \\ \cdots & 1/4 & 1/2 & 1 & 1/2 & 1/4 & \cdots \\ & \vdots & \vdots & \vdots & \vdots & \vdots & \end{bmatrix}.$$

The fact that the slices consist of look-alike basis vectors is characteristic for the (left) double shift invariance of $\mathcal{K}_o''$: obviously, shifting the rows up or down gives the same result. As announced in the proof of proposition 6.18, the projection of $\mathcal{K}_o''$ onto $\mathcal{U}_2$ will be in $\mathcal{D}_2\mathbf{F}_o$; in the present case we obtain even all of $\mathcal{D}_2\mathbf{F}_o$ in this way. The derivation in section 7.4 (equation (7.34)) will show that although V has a unitary realization, it has $\ell_A = 1$ and it is not an inner operator, in accordance to proposition 6.18. This matches with the fact that $\ker(\cdot V^*) = \mathcal{K}_o''$ is not zero and $\overline{\text{ran}}(\cdot V)$ is not the whole output space. Finally, it is not hard to see by direct calculation that $\mathcal{K}_o' := \ker(\cdot V^*|_{\mathcal{U}_2}) = \{0\}$.

Computation of the inner-outer factorization $T = VT_r$

In this section, we choose to work with the inner-outer factorization of T, as in (7.8): $T = VT_r$ where T_r is right outer: $\overline{\mathcal{L}_2\mathcal{Z}^{-1}T_r^*} = \mathcal{L}_2\mathcal{Z}^{-1}$, and the left inner (isometric) factor

V satisfies $V^*V = I$ and is obtained by setting $\mathcal{K}(V) := \mathcal{L}_2 Z^{-1} V^*$ equal to $\overline{\mathcal{L}_2 Z^{-1} T^*}$. For this factorization,

$$\mathcal{K}' := \ker(\cdot V|_{\mathcal{L}_2 Z^{-1}}) = \ker(\cdot T|_{\mathcal{L}_2 Z^{-1}}).$$

On the one hand,

$$\mathcal{L}_2 Z^{-1} \ominus \mathcal{L}_2 Z^{-1} V^* = \overline{\mathcal{H}}(V) \oplus \mathcal{K}',$$

and on the other (for $K_T = \mathbf{P}'(\cdot T|_{\mathcal{L}_2 Z^{-1}})$),

$$\mathcal{L}_2 Z^{-1} = \ker(\cdot K_T) \oplus \overline{\operatorname{ran}}(\cdot K_T^*) = \{u \in \mathcal{L}_2 Z^{-1} : uT \in \mathcal{U}_2\} \oplus \overline{\mathcal{L}_2 Z^{-1} T^*},$$

so that, with $\mathcal{L}_2 Z^{-1} V^* = \overline{\mathcal{L}_2 Z^{-1} T^*}$,

$$\overline{\mathcal{H}}(V) \oplus \mathcal{K}' = \{u \in \mathcal{L}_2 Z^{-1} : uT \in \mathcal{U}_2\}.$$

Thus we see that $\overline{\mathcal{H}}(V)$ is the *largest* subspace in $\mathcal{L}_2 Z^{-1}$ for which $\overline{\mathcal{H}}(V) K_T = \{0\}$ and which is orthogonal to $\mathcal{K}'$. This property provides a way to compute the inner-outer factorization. Note that if $\overline{\mathcal{H}}(V)$ is too small, then $\mathcal{L}_2 Z^{-1} V^* \supset \mathcal{L}_2 Z^{-1} T^*$, *i.e.*, $\mathcal{L}_2 Z^{-1} T^* \subset \mathcal{L}_2 Z^{-1} T^* V \subset \mathcal{L}_2 Z^{-1}$. In that case, V^*T is not outer, although the range might have improved on T itself. This defines a hierarchy of partial solutions. In terms of subspaces, the maximal solution is unique.

Let $\mathbf{Q}$ be a sliced orthonormal basis representation of $\overline{\mathcal{H}}(V)$: $\overline{\mathcal{H}}(V) = \mathcal{D}_2 \mathbf{Q} \in \mathcal{L}_2 Z^{-1}$, and let $\mathbf{F}_o$ be a sliced basis representation of $\overline{\mathcal{H}}_o(T)$, or more generally, for a subspace in $\mathcal{U}_2$ containing $\overline{\mathcal{H}}_o(T)$. The fact that $\overline{\mathcal{H}}(V) K_T = 0$ translates to the condition $\mathbf{Q}T \in \mathcal{U}$. Because $\overline{\mathcal{H}}(V) T \subset \overline{\mathcal{H}}_o(T)$, we must have that $\mathbf{Q}T = Y\mathbf{F}_o$ for some bounded diagonal operator Y, which will play an instrumental role in the derivation of a state realization for V. It remains to implement the condition $\overline{\mathcal{H}}(V) \perp \mathcal{K}'$. Suppose that $\mathbf{Q}$ has a component in $\mathcal{K}'$, so that $D\mathbf{Q} \in \mathcal{K}'$, for some $D \in \mathcal{D}_2$. Then, since $\mathcal{K}' = \ker(\cdot T|_{\mathcal{L}_2 Z^{-1}})$,

$$D\mathbf{Q} \in \mathcal{K}' \quad \Leftrightarrow \quad D\mathbf{Q}T = DY\mathbf{F}_o = 0 \quad \Leftrightarrow \quad D \in \ker(\cdot Y) \tag{7.12}$$

(since $\mathbf{F}_o$ is assumed to represent a basis). Hence $\overline{\mathcal{H}}(V) = \mathcal{D}_2 \mathbf{Q}$ can be described as the *largest* subspace of type $\mathcal{D}_2 \mathbf{Q}$ for which $\mathbf{Q}T = Y\mathbf{F}_o$ with $\ker(\cdot Y) = \{0\}$.

If $\mathcal{B}$ is the space sequence of the state of the given realization for T, and $\mathcal{B}_V$ is the space sequence of the state of the realization for V, then $Y \in \mathcal{D}(\mathcal{B}_V, \mathcal{B})$. The condition $\ker(\cdot Y) = \{0\}$ implies that $\mathcal{B}_V \subset \mathcal{B}$ (pointwise), so that the state dimension of V is at each point in time less than or equal to the state dimension of T at that point (the condition forces each diagonal component Y_{kk} of Y to be a square or "wide" matrix, the number of columns is equal to or larger than the number of rows).

In the following theorems, we shall say that (A, B) is a realization for some sliced basis $\mathbf{Q}$ in $Z^{-1}\mathcal{L}_2^{\mathcal{B}}$, if $\mathbf{Q}^* = BZ(I-AZ)^{-1}$, provided $\ell_A < 1$. If $\ell_A = 1$, then we have to be more prudent and say that the k-th diagonal $\mathbf{P}_0(Z^{-k}\mathbf{Q}^*)$ of $\mathbf{Q}^*$ matches $B^{(k)}A^{\{k-1\}}$, for each $k \geq 0$. Dually, if $\mathbf{G}$ is a sliced basis in $\mathcal{U}_2$, then we shall say that (A, C) is a realization for it, if $\mathbf{G} = (I-AZ)^{-1}C$ (for $\ell_A < 1$), and in general if the k-th diagonal $\mathbf{P}_0(Z^{-k}\mathbf{G})$ of $\mathbf{G}$ matches $A^{\{k-1\}}C$.

Lemma 7.2 *Let $T \in \mathcal{U}$ be a locally finite input-output operator, and suppose that $\mathbf{T} = \{A,B,C,D\}$ is an observable and u.e. stable realization of T. Also let (A_V, B_V) be a realization for some orthonormal basis $\mathbf{Q}$ in $\mathcal{L}_2 Z^{-1}$. Then*

$$\mathcal{D}_2 \mathbf{Q}T \in \mathcal{U}_2 \quad \Leftrightarrow \quad \exists\, Y \in \mathcal{D}(\mathcal{B}_V, \mathcal{B}) : \quad \left\{ \begin{array}{llcl} (a) & A_V^* Y A + B_V^* B & = & Y^{(-1)} \\ (b) & A_V^* Y C + B_V^* D & = & 0 \\ (c) & \ker(\cdot Y) & = & \{0\}. \end{array} \right.$$

Y is unique, and bounded: $Y^*Y \le \Lambda_{\mathbf{F}}$, where $\Lambda_{\mathbf{F}}$ is the reachability Gramian of $\mathbf{T}$.

PROOF Let $\mathbf{F}^* = BZ(I-AZ)^{-1}$ and $\mathbf{F}_o = (I-AZ)^{-1}C$. We use in this proof the relations (*cf.* (5.19)–(5.21))

$$T = D + \mathbf{F}^* C, \quad Z\mathbf{F} = A^*\mathbf{F} + B^*, \quad Z\mathbf{Q} = A_V^*\mathbf{Q} + B_V^*.$$

We first show that Y: $\mathbf{P}(\mathbf{Q}T) = Y\mathbf{F}_o \Rightarrow (a)$. Recall that the Hankel operator associated to T is $H_T = \mathbf{P}(\cdot T)|_{\mathcal{L}_2 Z^{-1}} = \mathbf{P}_0(\cdot\, \mathbf{F}^*)\mathbf{F}_o$ (*cf.* theorem 5.2). Hence $\mathbf{P}(\mathbf{Q}T) = \mathbf{P}_0(\mathbf{Q}\mathbf{F}^*)\mathbf{F}_o$, and if the realization is observable, this implies that Y is unique and given by

$$Y = \mathbf{P}_0(\mathbf{Q}\mathbf{F}^*) = \mathbf{P}_0\left[(Z - A_V^*)^{-1} B_V^* B (Z^* - A)^{-1}\right]. \tag{7.13}$$

Y is well defined because $\ell_A < 1$ so that the summation is convergent in norm. Furthermore,

$$\begin{aligned} Y^{(-1)} &= \mathbf{P}_0(Z\mathbf{Q}\mathbf{F}^* Z^*) \\ &= \mathbf{P}_0([A_V^*\mathbf{Q} + B_V^*][\mathbf{F}^* A + B]) \\ &= A_V^* Y A + B_V^* B, \end{aligned}$$

hence (a) holds. Next we show that $(a) \Rightarrow Y = \mathbf{P}_0(\mathbf{Q}\mathbf{F}_o)$. Since $\ell_A < 1$, this equation has a unique and bounded solution, which is seen from an expansion of the equation into a summation similar as in (5.24). Necessarily, the solution satisfies $\mathbf{P}(\mathbf{Q}T) = Y\mathbf{F}_o$. Hence also $(a) \Rightarrow \mathbf{P}(\mathbf{Q}T) = Y\mathbf{F}_o$.

Let Y be the solution of (a). Now, to derive the equivalence of (b) with $\mathcal{D}_2\mathbf{Q}T \in \mathcal{U}_2$, we use the fact that $\mathcal{D}_2\mathbf{Q}T \in \mathcal{U}_2 \Leftrightarrow \mathbf{P}_0(Z^n\mathbf{Q}T) = 0$ for all $n > 0$. Recurring:

$$\begin{aligned} n = 1: \qquad \mathbf{P}_0(Z\mathbf{Q}T) &= \mathbf{P}_0([A_V^*\mathbf{Q} + B_V^*][D + \mathbf{F}^* C]) \\ &= A_V^* \mathbf{P}_0(\mathbf{Q}\mathbf{F}^*) C + B_V^* D + 0 + 0 \\ &= A_V^* Y C + B_V^* D. \end{aligned}$$

Hence $\mathbf{P}_0(Z\mathbf{Q}T) = 0 \Leftrightarrow A_V^* Y C + B_V^* D = 0$. For $n > 1$, assume $\mathbf{P}_0(Z^{n-1}\mathbf{Q}T) = 0$. Then

$$\begin{aligned} \mathbf{P}_0(Z^n\mathbf{Q}T) &= \mathbf{P}_0(Z^{n-1}[Z\mathbf{Q}T]) \\ &= \mathbf{P}_0(Z^{n-1}[A_V^*\mathbf{Q}]T) + \mathbf{P}_0(Z^{n-1}B_V^* T) \\ &= A_V^{*(n-1)} \mathbf{P}_0(Z^{n-1}\mathbf{Q}T) + B_V^{*(n-1)} \mathbf{P}_0(Z^{n-1}T) \\ &= 0 + 0. \end{aligned}$$

Hence (b) is both necessary and sufficient for the condition $\mathcal{D}_2\mathbf{Q}T \in \mathcal{U}_2$ to be satisfied.

The bound on Y^*Y follows from the observation that

$$\begin{aligned} Y^*Y &= \mathbf{P}_0(\mathbf{FQ}^*)\mathbf{P}_0(\mathbf{QF}^*) \\ &= \mathbf{P}_0(\mathbf{P}_0(\mathbf{FQ}^*)\mathbf{QF}^*) \\ &= \mathbf{P}_0(\mathbf{P}_{\mathcal{H}}(\mathbf{F})\mathbf{F}^*)\,, \qquad \mathbf{P}_{\mathcal{H}}(\cdot) = \mathbf{P}_0(\cdot\,\mathbf{Q}^*)\mathbf{Q} \\ &\le \mathbf{P}_0(\mathbf{FF}^*) = \Lambda_{\mathbf{F}} \end{aligned}$$

where we have used the fact that $\mathbf{P}_{\mathcal{H}}(\cdot)$ is an orthogonal projector (onto $\mathcal{D}_2\mathbf{Q}$, *i.e.*, the input state space of V), *viz.* theorem 4.9. □

The computation of Y amounts to a generalized partial fraction splitting of expression (7.13). The quadratic term can in this case be split in linear terms because there is an automatic "dichotomy": half of the expression lays in (an extension of) $\mathcal{L}$ and the other half in $\mathcal{U}$. The uniqueness of Y is of course dependent on the choice of $\mathbf{Q}$ —any sliced DZ-invariant subspace of $\mathcal{D}_2\mathbf{Q}$ would provide a solution for Y as well, but one that has smaller dimensions.

Proposition 7.3 *Let $T \in \mathcal{U}$ be a locally finite input-output operator, and suppose that $\mathbf{T} = \{A,B,C,D\}$ is an observable and u.e. stable realization of T.*

Also let $W = [U \;\; V] \in \mathcal{U}$ be isometric ($W^*W = I$) and have a unitary realization $\mathbf{W} = \left[\begin{smallmatrix} A_V & C_U & C_V \\ B_V & D_U & D_V \end{smallmatrix}\right]$ with state dimension $\mathcal{B}_V$.

Let $T_r \in \mathcal{U}$ have a (not necessarily minimal) realization (A, B_{T_r}, C, D_{T_r}). Then the following statements are equivalent:

1. *$V^*T = T_r$ is right outer and $U^*T = 0$,*
2. *$(\mathbf{W}, Y, B_{T_r}, D_{T_r})$ is a solution of*

$$\mathbf{W}^* \begin{bmatrix} YA & YC \\ B & D \end{bmatrix} = \begin{bmatrix} Y^{(-1)} & 0 \\ 0 & 0 \\ B_{T_r} & D_{T_r} \end{bmatrix} \qquad (7.14)$$
$$\ker(\cdot\, Y) = \{0\}$$
$$\ker(\cdot\, D_{T_r}) = \{0\}$$

where $\mathbf{W}$ is unitary and the state dimension $\#\mathcal{B}_V$ is maximal among all possible solutions.

The "maximal solution" Y is bounded and unique up to a left diagonal unitary factor.

PROOF Equation (7.14), written out in full, reads

$$\begin{array}{llcl@{\qquad}llcl} (a) & A_V^*YA + B_V^*B &=& Y^{(-1)} & (e) & C_V^*YC + D_V^*D &=& D_{T_r} \\ (b) & A_V^*YC + B_V^*D &=& 0 & (f) & C_V^*YA + D_V^*B &=& B_{T_r} \\ (c) & C_U^*YA + D_U^*B &=& 0 & (g) & \ker(\cdot\, Y) &=& \{0\} \\ (d) & C_U^*YC + D_U^*D &=& 0 & (h) & \ker(\cdot\, D_{T_r}) &=& \{0\}\,. \end{array} \qquad (7.15)$$

($\Rightarrow$) Suppose that $W = [U \;\; V]$ is isometric with a unitary realization $\mathbf{W}$ and such that V^*T is right outer, and $U^*T = 0$. Let $\mathbf{Q}$ be the sliced orthonormal basis of $\mathcal{H}(W)$ corresponding to the realization $\mathbf{W}$. $W^*T \in \mathcal{U}$, so that, by lemma 7.2, there is $Y \in \mathcal{D}$, given by $(a),(b)$, such that $\mathbf{Q}T = Y\mathbf{F}_o$. Also, since $T = D + BZ\mathbf{F}_o$,

$$\begin{aligned} U^*T &= [D_U^* + C_U^*\mathbf{Q}]T \\ &= D_U^*T + C_U^*\mathbf{Q}T \\ &= D_U^*[D + BZ\mathbf{F}_o] + C_U^* Y\mathbf{F}_o \\ &= [D_U^*D + C_U^*YC] + [D_U^*B + C_U^*YA]Z\mathbf{F}_o . \end{aligned} \tag{7.16}$$

Hence

$$U^*T = 0 \quad \Leftrightarrow \quad \left\{ \begin{array}{lcl} C_U^*YA + D_U^*B &=& 0 \\ C_U^*YC + D_U^*D &=& 0 \end{array} \right. \tag{7.17}$$

which proves (c) and (d). Much as in (7.16),

$$T_r = V^*T \;=\; [C_V^*YC + D_V^*D] \;+\; [C_V^*YA + D_V^*B]Z\mathbf{F}_o .$$

A realization of T_r is thus given by (A, B_{T_r}, C, D_{T_r}), with B_{T_r}, D_{T_r} given by (e), (f). The condition that $T_r = V^*T$ is right outer implies that $\ker(\cdot V|_{\mathcal{L}_2Z^{-1}}) = \ker(\cdot T|_{\mathcal{L}_2Z^{-1}})$ so that by (7.12), $\ker(\cdot Y) = \{0\}$. Finally, (h) holds, for else T_r cannot be right outer.

($\Leftarrow$) Suppose we have a solution of (7.15). Let $\mathbf{Q}$ be the orthonormal sliced basis generated by (A_V, B_V). By lemma 7.2, (a) and (b) imply that $\mathbf{Q}T \in \mathcal{U}$. $\mathbf{W}$ is a unitary completion of $[A_V \; B_V]$ and $\mathcal{H}(W) = \mathcal{D}_2\mathbf{Q}$. Hence $\mathcal{H}(W)T \in \mathcal{U}_2$, *i.e.*, $\mathcal{H}(W) \subset \{u \in \mathcal{L}_2Z^{-1} : uT \in \mathcal{U}_2\}$. Condition (g) implies that $\mathcal{H}(W) \perp \ker(\cdot T|_{\mathcal{L}_2Z^{-1}})$. Since (A_V, B_V) are of largest possible dimension, it follows that

$$\mathcal{H}(W) = \{u \in \mathcal{L}_2Z^{-1} : uT \in \mathcal{U}_2\} \ominus \ker(\cdot T|_{\mathcal{L}_2Z^{-1}})$$

(The existence of a W such that equality is obtained follows from the existence of the inner-outer factorization for the locally finite case.) Since W has a unitary realization, we have

$$\mathcal{L}_2Z^{-1} = \mathcal{H}(W) \oplus \mathcal{L}_2Z^{-1}W^*$$

We also have, for $K_T = \mathbf{P}'(\cdot T|_{\mathcal{L}_2Z^{-1}})$, that

$$\mathcal{L}_2Z^{-1} = \ker(\cdot K_T) \oplus \overline{\operatorname{ran}}(\cdot K_T^*) = \{u \in \mathcal{L}_2Z^{-1} : uT \in \mathcal{U}_2\} \oplus \overline{\mathcal{L}_2Z^{-1}T^*}$$

Hence $\mathcal{L}_2Z^{-1}W^* = \ker(\cdot T|_{\mathcal{L}_2Z^{-1}}) \oplus \overline{\mathcal{L}_2Z^{-1}T^*}$.

We now look at the decomposition $W = [U \;\; V]$. From (c), (d) and equation (7.17) it follows that $U^*T = 0$, and from (h) that U is the largest operator with $\mathcal{H}(U) = \mathcal{H}(W)$ to do so. Hence $\mathcal{L}_2Z^{-1}U^* = \ker(\cdot T|_{\mathcal{L}_2Z^{-1}})$, so that $\mathcal{L}_2Z^{-1}V^* = \overline{\mathcal{L}_2Z^{-1}T^*}$. This implies that T_r is outer.

The bound on Y follows from lemma 7.2 ($Y^*Y \le \Lambda_{\mathbf{F}}$), and its uniqueness from the fact that the basis $\mathbf{Q}$ of $\mathcal{H}(W)$ is unique up to a unitary diagonal state transformation.

□

Proposition 7.3 directly leads to an algorithm to compute the inner-outer factorization recursively (see figure 7.1). The main step in the algorithm is a QL (unitary-lower) factorization. Given Y_k, this produces all necessary state space matrices at point k, and Y_{k+1} for the next step. Because both Y_{k+1} and $D_{T_r,k}$ have full row rank, the dimensions

In:	$\{\mathbf{T}_k\}$	(an observable realization of T)
Out:	$\{\mathbf{V}_k\}, \{(\mathbf{T}_r)_k\}$	(realizations of the isometric and right outer factor)

Initialize Y_1

for $k = 1, 2, \cdots$

Compute a QL factorization: $\mathbf{W}'$ unitary such that

$$\begin{bmatrix} Y_k A_k & Y_k C_k \\ B_k & D_k \end{bmatrix} =: \mathbf{W}'_k \begin{bmatrix} 0 & 0 \\ Y_{k+1} & 0 \\ B_{T_r,k} & D_{T_r,k} \end{bmatrix}, \qquad \begin{matrix} \ker(\cdot Y_{k+1}) = 0, \\ \ker(\cdot D_{T_r,k}) = 0 \end{matrix}$$

$$\mathbf{W}'_k =: \begin{bmatrix} C_{U,k} & A_{V,k} & C_{V,k} \\ D_{U,k} & B_{V,k} & D_{V,k} \end{bmatrix}$$

$$\mathbf{V}_k := \begin{bmatrix} A_{V,k} & C_{V,k} \\ B_{V,k} & D_{V,k} \end{bmatrix}$$

$$\mathbf{T}_{r,k} := \begin{bmatrix} A_k & C_k \\ B_{T_r,k} & D_{T_r,k} \end{bmatrix}$$

end

Figure 7.1. Inner-outer factorization algorithm for $T = VT_r$

of the QL factorization are unique, and the factorization itself is unique up to block-diagonal unitary factors acting on columns of $\mathbf{W}$ and rows of Y_{k+1} and $[B_{T_r,k} \; D_{T_r,k}]$, corresponding to unitary state space transformations on $\mathbf{W}$ and unitary left diagonal factors on T_r and $\mathbf{V}^*$. Both transformations are admissible.

The main issue left to be discussed concerns the initialization of Y. Note that once Y_1 is fixed, the remainder of the recursion is determined. Hence, the choice of Y_1 has to ensure that we end up with the *maximal* solution that is required to obtain outer factors. In this respect, note that if D is invertible, then $Y = [\cdot]$ is always a solution, but perhaps not the maximal solution.

Nonetheless, for *finite* $n \times n$ *block matrices*, we may simply set $Y_1 = [\cdot]$, assuming $\mathbf{T}$ is a realization that starts with zero states. Interesting situations can in this case only occur if D does not have full row rank.

For systems which are *time-invariant* before $k = 1$, the recursion becomes an equation, in fact leading to an eigenvalue problem. In terms of Y (or rather, Y^*Y), this equation is an algebraic Riccati equation, and its solution will be discussed in the next subsection.

For systems which are *periodic,* the corresponding LTI Riccati equation of the enclosing LTI system has to be solved as well, which is not attractive if the period is large. In terms of Riccati equations, an alternative solution for periodical systems was proposed in [HL94], in which an iteration over a chain of QZ decompositions is computed. This has numerical advantages, as only orthogonal transformations are used, and no products of A-matrices have to be evaluated. The method also allows to do a preprocessing with state transformations on the realization, in particular to transform all A-matrices into upper Hessenberg form, and thus have faster convergence of the QZ steps.

Alternatively (and in fact not much differently), we can act as in the *unstructured case with* $\ell_A < 1$: start with a random $\hat{Y}_1$ of full rank d_1, where d_1 is the state dimension of T at time $k = 1$. For example, set $\hat{Y}_1 = (\Lambda_{\mathbf{F}}^{1/2})_1$: the reachability Gramian of the realization of T. We have to show that the resulting sequence $\{\hat{Y}\}_k$, $k = 1,2,\cdots$ will converge down to the true maximal-size solution Y_k. The analysis of this is deferred to section 13.4, where a similar Riccati equation is investigated. (Note that if the rank of the initial Y_1 is too small, it will usually converge towards a non-maximal solution of the equations: T_r will not be outer.) The speed of convergence is only linear.

Riccati equation

In the time-invariant setting, it is well known that the outer factor T_r of T can be written in closed form in terms of the original state matrices $\{A,B,C,D\}$ of T and only one unknown intermediate quantity, M say, which is the solution of a Riccati equation with $\{A,B,C,D\}$ as parameters. One way to obtain the Riccati equation is by looking at a spectral factorization of the squared relation $T^*T = T_r^*T_r$. Riccati equations can be solved recursively; efficient solution methods for the recursive version are the *square-root algorithms*, in which extra intermediate quantities are introduced to avoid the computation of inverses and square roots. In fact, algorithm (7.1) to compute the realization for T_r is precisely the square-root algorithm. We show in this section how the corresponding Riccati equation is derived.

Theorem 7.4 *Let $T \in \mathcal{U}$ be a locally finite transfer operator, let $\mathbf{T} = \{A,B,C,D\}$ be an observable realization of T, and assume $\ell_A < 1$. Then the Riccati equation*

$$M^{(-1)} = A^*MA + B^*B - \left[A^*MC + B^*D\right](D^*D + C^*MC)^{\dagger}\left[D^*B + C^*MA\right] \qquad (7.18)$$

has a solution $M \in \mathcal{D}$, $M \geq 0$ of (pointwise) maximal rank.[4] *The maximal solution is unique and bounded: $M \leq \Lambda_{\mathbf{F}}$.*

Define D_{T_r} to be a minimal full range factor ($\ker(\cdot D_{T_r}) = \{0\}$) of $D_{T_r}^* D_{T_r} = D^*D + C^*MC$. Then a realization of the right outer factor T_r of T so that $T_r = V^*T$ is given by

$$\mathbf{T}_r = \begin{bmatrix} A & C \\ D_{T_r}^{\dagger *}(C^*MA + D^*B) & D_{T_r} \end{bmatrix}.$$

PROOF We start from proposition 7.3, in particular equation (7.14). Premultiplying this equation with its Hermitian transpose, using $\mathbf{W}^*\mathbf{W} = I$, and denoting $M := Y^*Y$

[4] $(\cdot)^{\dagger}$ denotes the operator pseudo-inverse [BR76]. Although it is non-unique, the definition of $M^{(-1)}$ is. In practice, it is advantageous to choose the unique least-squares pseudo-inverse for $(\cdot)^{\dagger}$.

produces

$$\left\{\begin{array}{lcl} M^{(-1)} & = & A^*MA+B^*B-B_{T_r}^*B_{T_r} \\ D_{T_r}^*D_{T_r} & = & C^*MC+D^*D \\ D_{T_r}^*B_{T_r} & = & C^*MA+D^*B \end{array}\right.$$

$\Leftrightarrow$

$$\left\{\begin{array}{lcl} M^{(-1)} & = & A^*MA+B^*B-[A^*MC+B^*D]\,(D^*D+C^*MC)^\dagger\,[D^*B+C^*MA] \\ D_{T_r} & = & Q(C^*MC+D^*D)^{1/2}, \quad (Q\in\mathcal{D} \text{ isometric, s.t. } \ker(\cdot D_{T_r})=\{0\}) \\ B_{T_r} & = & D_{T_r}^{\dagger *}[C^*MA+D^*B] \end{array}\right.$$

(The right inverse $D_{T_r}^{\dagger}$ need not be bounded, which happens if the range of D_{T_r} is not closed. However, the product $D_{T_r}^{\dagger *}[C^*MA+D^*B]$ is bounded, which can be motivated from the fact that B_{T_r} is the same as in proposition 7.3, but can also proven directly. We omit the details.) □

Equation (7.18) is a time-varying Riccati equation. It is a (generalized) quadratic equation which often arises in problems involving spectral factorizations, or Cholesky factorizations, once the state equations are substituted for the operator. We will encounter it several more times, *e.g.*, in the solution of the time-varying lossless embedding problem (chapter 12), and in the spectral factorization problem discussed in chapter 13. The algebraic Riccati equation has a rich history; a list of contributions and contributors can be found in [Nic92, BLW91, LR95].

Necessary and sufficient conditions for the existence of positive semidefinite, and "stabilizing" (or outer) solutions for the LTI Riccati equation were proven by Wonham [Won68], Kucera [Kuc72], and Molinari [Mol75], but under the assumption that $D_{T_r}^*D_{T_r} > 0$.

By taking the k-th entry of each diagonal in equation (7.18), we obtain the recursion

$$\begin{array}{rcl} M_{k+1} & = & A_k^*M_kA_k+B_k^*B_k- \\ & & -\left[A_k^*M_kC_k+B_k^*D_k\right](D_k^*D_k+C_k^*M_kC_k)^\dagger\left[D_k^*B_k+C_k^*M_kA_k\right]. \end{array} \tag{7.19}$$

Initial conditions for the recursion can be obtained for our usual list of special cases.

1. When T starts with zero states at some point k_0 in time, then $M_{k_0} = [\cdot]$. If T is time invariant before k_0, then M_{k_0} is given by a time-invariant Riccati equation.

2. The exact solution of (7.19) for the LTI case can be computed in several ways. In comparison with standard solutions based on connections with the linear-quadratic optimal control problem, complications arise because in our problem we can neither assume D^*D nor A to be invertible. In that case, the solution is not given directly in terms of the eigenvalues and eigenvectors of a Hamiltonian matrix, but of a matrix pencil. Pencil techniques were perhaps first introduced in [PLS80], to avoid the inversion of A. To avoid inversion of D^*D as well, the pencil matrices have to be extended, *e.g.*, as done in [LR95, §15.2], which we follow here. Suppose $A: d\times d$ and $D: m\times n$. The realization is assumed to be observable. Define

$$F_e = \begin{bmatrix} I_d & 0 & 0 \\ 0 & A^* & 0 \\ 0 & -C^* & 0 \end{bmatrix}, \qquad G_e = \begin{bmatrix} A & 0 & C \\ -B^*B & I_d & -B^*D \\ D^*B & 0 & D^*D \end{bmatrix}$$

Compute the solutions of the pencil $\lambda F_e - G_e$, preferably via the QZ decomposition [GV89]: find matrices Q, Z (unitary), R_F, R_G (upper triangular), and V (the generalized eigenvectors) such that

$$\begin{aligned} QF_eZ &= R_F \\ QG_eZ &= R_G \end{aligned}$$

$$F_eV\,\mathrm{diag}(R_G) = G_eV\,\mathrm{diag}(R_F).$$

Let V' contain the columns of V for which $|(R_F)_{ii}| < |(R_G)_{ii}|$ (*i.e.*, the eigenvectors of the eigenvalues inside the unit circle), and partition V' into

$$V' = \begin{matrix} d \\ d \\ n \end{matrix} \begin{bmatrix} V_1 \\ V_2 \\ V_3 \end{bmatrix}.$$

It is shown in [LR95] for the case where D^*D is invertible and where there are no zeros on the unit circle (the pencil is regular), that V' has d columns, that V_1 is invertible, and that $V_2V_1^{-1}$ is Hermitian, positive semidefinite, and in fact the maximal solution to the LTI Riccati equation. Thus, $M = V_2V_1^{-1}$ is the solution of the LTI Riccati equation that gives the outer factor. It seems possible to extend the method to the more general case where D^*D is not invertible, and to allow zeros on the unit circle. This is still an open research area, and additional conditions (on reachability) seem to be in order.

3. Periodic Riccati equations were studied in [KN79, BCN88, dS91, Nic92, BGD92, HL94]. Necessary and sufficient conditions for convergence of the periodic Riccati recursion to the maximal solution from any initial point $\hat{M}_0$ which satisfies $\hat{M}_0 \geq \varepsilon I_d$ or $\hat{M}_0 \geq M_0$ were established in [dS91] (assuming $D^*D > 0$). An interesting method to find the periodic solution is described in [HL94]. Instead of directly following the Riccati recursion, the method is based on a cyclic (period p) QZ factorization of the pencil

$$E_k = \begin{bmatrix} A_k^\times & 0 \\ B^*D(D^*D)^{-1}D^*B - B_k^*B_k & 0 \end{bmatrix}, \qquad F_k = \begin{bmatrix} I & C_k(D_k^*D_k)^{-1}C_k^* \\ 0 & A_k^{\times *} \end{bmatrix}$$

$$A_k^\times := A_k - C_k(D_k^*D_k)^{-1}D_k^*B_k,$$

(again assuming D^*D to be invertible) into a chain

$$\begin{aligned} Q_1^*E_1Z_2 &= R_{E_1}, & Q_1^*F_1Z_1 &= R_{F_1} \\ Q_2^*E_2Z_3 &= R_{E_2}, & Q_2^*F_2Z_2 &= R_{F_2} \\ &\vdots & &\vdots \\ Q_p^*E_pZ_{p+1} &= R_{E_p}, & Q_p^*F_pZ_p &= R_{F_p} \\ Z_1 = Z_{p+1} & & & \end{aligned}$$

where Q_k, Z_k are all unitary, and R_{E_k}, R_{F_k} are all upper triangular. The periodic condition is that $Z_1 = Z_{p+1}$. The decomposition is basically obtained by first reducing all E_k and F_k to triangular or Hessenberg forms, and following the suggested iteration a number of times, starting with some Z_1. In general, convergence is linear. Once the decomposition is found, the generalized eigenvalues are given by the product $\Lambda = \text{diag}(R_{E_1})\text{diag}(R_{F_1})^{-1}\cdots\text{diag}(R_{E_p})\text{diag}(R_{F_p1})^{-1}$, provided the inverses exist, or else by a more complicated expression (see [HL94]). The decomposition can be made sorted in such a way that the first d "eigenvalues" λ_k are smaller than 1, in which case the solution of the periodic Riccati equation is obtained as $M_k = Z_{21,k}Z_{11,k}^{-1}$, where

$$Z_k = \begin{bmatrix} Z_{11,k} & Z_{12,k} \\ Z_{21,k} & Z_{22,k} \end{bmatrix}.$$

Computation of the outer-inner factorization $T = T_\ell V$

For completeness and future reference, we present at this point also the derivation of the outer-inner factorization $T = T_\ell V$ in which T_ℓ is left outer and V is isometric, $VV^* = I$. It is of course completely analogous to that of the inner-outer factorization in the preceding section. This time, V is defined via the property

$$\mathcal{U}_2 V = \overline{\mathcal{U}_2 T} \tag{7.20}$$

in accordance to the generalized Beurling-Lax theory of chapter 6. The further elaboration given there shows that we have the decomposition

$$\mathcal{U}_2 \;=\; \mathcal{H}_o(V) \oplus \mathcal{U}_2 V \oplus \ker(\cdot V^*|_{\mathcal{U}_2}) \tag{7.21}$$

in which it also holds that $\ker(\cdot V^*|_{\mathcal{U}_2}) = \ker(\cdot T^*|_{\mathcal{U}_2})$. This space is a defect space or kernel for T^* which we characterize here as a causal sliced space[5] annihilated by T^*. However, the kernel or defect space of T^* may be larger, it is indeed possible that $\cdot T^*|_{\mathcal{U}_2}$ is strictly smaller than $\cdot T^*|_{\mathcal{X}_2}$. In that case, there is a component in the defect space which is intrinsically non-causal, and V is isometric but not inner. This aspect is investigated later in section 7.5.

Let us define $\mathbf{G}$ to be a sliced orthonormal basis for $\mathcal{H}_o(V)$, and let the corresponding realization be given by the observability pair (A_V, C_V). $\mathbf{G}$ is not necessarily bounded in operator norm, and we only assume $\ell_{A_V} \leq 1$. We have, by definition, that

$$\mathbf{G} = C_V + A_V Z \mathbf{G}$$

and that a causal, again not necessarily bounded realization for V is given by

$$V = D_V + B_V Z \mathbf{G}$$

[5] In classical analytical function theory, valid when T is LTI, it would correspond to an "analytical range space" in the sense of Helson [Hel64], *i.e.*, a range space with a basis consisting of functions which are uniformly bounded and analytic in the unit disc of the complex plane, corresponding to causal and bounded transfer functions. He also shows that the (left or right) nullspace of a rational transfer matrix is an analytic range space. The property does not hold in general for non-rational transfer functions.

in which an additional pair (B_V, D_V) makes

$$\begin{bmatrix} A_V & C_V \\ B_V & D_V \end{bmatrix}$$

isometric and of the appropriate dimensions (see further).

Let $\mathbf{F}_T$ be a minimal basis for the input state space of T, and $\mathbf{F}_{oT}$ the corresponding basis of the output state space, satisfying the realization equations (5.19)–(5.21), *viz.*

$$\begin{array}{rclcrcl} \mathbf{F}_{oT} &=& AZ\mathbf{F}_{oT}+C & \qquad & \mathbf{F}_T^* Z^* &=& \mathbf{F}_T^* A + B \\ T &=& BZ\mathbf{F}_{oT}+D & & T &=& \mathbf{F}_T^* C + D\,. \end{array}$$

The main property that follows from (7.20) and the orthogonal decomposition (7.21) is that

$$\mathbf{G}T^* = Y\mathbf{F}_T \tag{7.22}$$

for some bounded diagonal operator Y with $\ker(\cdot Y) = \{0\}$. In particular, $\mathbf{G}T^*$ is anticausal, and since the adjoint Hankel operator for T can be expressed in terms of $\mathbf{F}_T$ and $\mathbf{F}_{oT}$ as

$$\cdot H_T^* = \mathbf{P}_0(\cdot \mathbf{F}_{oT}^*)\mathbf{F}_T$$

we find

$$Y = \mathbf{P}_0(\mathbf{G}\mathbf{F}_{oT}^*) \tag{7.23}$$

which proves in particular its boundedness, and the fact that its sequence of row dimensions are pointwise less than that of $\mathbf{F}_{oT}^*$. Inserting the expressions for $\mathbf{G}$ and $\mathbf{F}_{oT}$ gives a recursion for Y in terms of the state space matrices for T and V:

$$\begin{array}{rcl} Y &=& \mathbf{P}_0\{[C_V + A_V Z\mathbf{G}][C^* + \mathbf{F}_{oT}^* Z^* A^*]\} \\ &=& C_V C^* + A_V Y^{(-1)} A^*\,. \end{array}$$

The definition of all the quantities involved is found by working out the two relations

$$\begin{array}{rcl} T_\ell^* &=& VT^* \\ Y\mathbf{F}_T &=& \mathbf{G}T^*\,. \end{array}$$

In detail,

$$\begin{array}{rcl} T_o^* &=& VT^* = [D_V + B_V Z\mathbf{G}]T^* \\ &=& D_V T^* + B_V Z\mathbf{G}T^* = D_V T^* + B_V ZY\mathbf{F}_T \\ &=& D_V[D^* + C^*\mathbf{F}_T] + B_V Y^{(-1)} Z\mathbf{F}_T\,, \end{array}$$

and since $Z\mathbf{F}_T = B^* + A^*\mathbf{F}_T$, we find

$$\begin{array}{rcl} T_o^* &=& D_V[D^* + C^*\mathbf{F}_T] + B_V Y^{(-1)}[B^* + A^*\mathbf{F}_T] \\ &=& [D_V D^* + B_V Y^{(-1)} B^*] + [D_V C^* + B_V Y^{(-1)} A^*]\mathbf{F}_T \\ &=:& D_\ell^* + C_\ell^* \mathbf{F}_T\,, \end{array}$$

which shows that T_ℓ inherits A, B from T, and has C_ℓ, D_ℓ as shown. From $\mathbf{G}T^* = Y\mathbf{F}_T$ we obtain

$$\begin{array}{rcl} \mathbf{G}T^* &=& [C_V + A_V Z\mathbf{G}]T^* = C_V T^* + A_V Z\mathbf{G}T^* \\ &=& C_V T^* + A_V Y^{(-1)} Z\mathbf{F}_T \\ &=& C_V[D^* + C^*\mathbf{F}_T] + A_V Y^{(-1)}[B^* + A^*\mathbf{F}_T] \\ &=& C_V D^* + A_V Y^{(-1)} B^* + [C_V C^* + A_V Y^{(-1)} A^*]\mathbf{F}_T \end{array}$$

showing that $C_V D^* + A_V Y^{(-1)} B^* = 0$ in addition to the recursion for Y which we know already.

Finally, the space $\ker(\cdot T^*|_{\mathcal{U}_2})$ is also characterized by a causal isometry U such that its closure equals $\mathcal{U}_2 U$, for which $UT^* = 0$ (and $UV^* = 0$ as well). Working out the relation $UT^* = 0$ as before for VT^* leads to $D_U D^* + B_U Y^{(-1)} B^* = 0$ and $D_U C^* + B_U Y^{(-1)} A^* = 0$, while U inherits the observality space $\mathcal{H}_o(V)$. Putting it all together gives the recursion

$$\begin{bmatrix} A_V & C_V \\ B_V & D_V \\ B_U & D_U \end{bmatrix} \begin{bmatrix} Y^{(-1)} A^* & Y^{(-1)} B^* \\ C^* & D^* \end{bmatrix} = \begin{bmatrix} Y & 0 \\ C_\ell^* & D_\ell^* \\ 0 & 0 \end{bmatrix}. \tag{7.24}$$

where the matrix on the far left is unitary. These relations, together with the additional properties

$$\begin{aligned} \ker(\cdot Y) &= \{0\} \\ \ker(\cdot D_\ell^*) &= \{0\} \end{aligned}$$

fully characterize the unknown quantities, up to orthogonal equivalences. These are the square root equations for the outer-inner factorization. As before, the corresponding Riccati equation can be obtained by eliminating the leftmost unitary matrix in (7.24) via premultiplication with the complex conjugates.

Geometric "innovation" interpretation

Outer-inner factorization hinges on the determination of an orthonormal basis of maximal dimensions for the space $\mathcal{R}_0 := \mathcal{K}_0 \ominus Z\mathcal{K}_0$ with $\mathcal{K}_0 = \overline{\mathcal{U}_2 T}$ —see the Beurling-Lax type theorem 6.13. Such a basis admits a geometric interpretation as an "innovations" sequence and can —in principle— be calculated using a Levinson or Schur algorithm, as sketched below.

By theorem 6.13, there exists a $V \in \mathcal{U}$, $VV^* = I$, such that $\mathcal{R}_0 = \mathcal{D}_2 V$ (indeed, V is the right inner factor of T). Since $\mathcal{R}_0 \subset \mathcal{K}_0$ there must be diagonal operators $G_i \in \mathcal{D}$ such that

$$V = \sum_{i=0}^{\infty} G_i \cdot Z^i T.$$

Formally, we can write $G = \sum_{i=0}^{\infty} G_i Z^i$ (although the sum need not converge to a bounded operator), so that $V = GT$. In fact, if $T = T_\ell V$ is an outer-inner factorization in which T_ℓ is boundedly invertible, then $G = T_\ell^{-1}$. Let us assume that this is the case.

The orthogonality condition $V \perp Z\mathcal{K}_0$ can be written as $\forall F \in \mathcal{K}_0 : \mathbf{P}_0(VF^*Z^{-1}) = 0$. In particular, for all $n \geq 1$: $\mathbf{P}_0(VT^*Z^{-n}) = 0$, *i.e.*, $\mathbf{P}_0(GTT^*Z^{-n}) = 0$. If we do a diagonal expansion of this, we obtain with $C_n = \mathbf{P}_0(TT^*Z^{-n})$,

$$[G_0 \;\; G_1 \;\; \cdots] \begin{bmatrix} C_0 & C_1^* & C_2^* & \cdots \\ C_1 & C_0^{(-1)} & C_1^{*(-1)} & \ddots \\ C_2 & C_1^{(-1)} & C_0^{(-2)} & \ddots \\ \vdots & \vdots & \ddots & \ddots \end{bmatrix} = [G_0^{-*} \;\; 0 \;\; \cdots]. \tag{7.25}$$

Here, all equations but the first express the orthogonalities. The first equation is obtained from the fact that $T_\ell^{-1}(TT^*)T_\ell^{-*} = I \Leftrightarrow G(TT^*) = G^{-*} \in \mathcal{L}$, specialized to the main diagonal. The fact that T_ℓ is outer also implies that $(G^{-*})_0 = (G_0)^{-*}$.

V can be interpreted as the normalized innovations of T with respect to ZT, Z^2T, etc. As a consequence, G can be found through a limiting procedure based on successive partial (normalized) innovations. In particular, if G_n is the n-th innovation, we could expect that G_n will converge to G when $n \to \infty$. G_n in turn can be found through a recursive algorithm known as Schur's algorithm. We do not pursue the matter further.

7.3 OPERATOR INVERSION

The strategy for the inversion of an operator $T \in \mathcal{X}$ is to determine the following factorizations:

$$
\begin{array}{rcllll}
T & = & Q^*R & \text{[Inner-coprime]:} & Q \text{ inner} \\
R & = & UR_r & \text{[Inner-outer]:} & U^*U = I, \\
 & & & & R_r \text{ right outer} \\
R_r & = & R_{\ell r}V & \text{[Outer-inner]:} & VV^* = I, \\
 & & & & R_{\ell r} \text{ left outer}
\end{array}
$$

$$
\text{(all factors in } \mathcal{U}), \qquad (7.26)
$$

so that $T = Q^*UR_{\ell r}V$. The final factor, $R_{\ell r}$, is upper and both left and right outer, hence invertible in $\mathcal{U}$, and its inverse is easily obtained. T is not necessarily invertible: U and V are isometries, and might not be unitary. In any case, T has a Moore-Penrose (pseudo-)inverse

$$
T^\dagger = V^*R_{\ell r}^{-1}U^*Q,
$$

and T is invertible with $T^{-1} = T^\dagger$ if U and V are both unitary. The inverse is thus specified as a lower-upper-lower-upper factorization. The factors may be multiplied to obtain an explicit matrix representation of $T^\dagger$, but because each of them will be known by its state representation, it is computationally efficient to keep it in factored form. In this section we consider the connection of matrix inversion with state representations in detail.

Time-varying state realizations of mixed causality

Let $\{\mathbf{T}_k\}$, $\{\mathbf{T}'_k\}$ be series of matrices with block entries

$$
\mathbf{T}_k = \begin{bmatrix} A_k & C_k \\ B_k & D_k \end{bmatrix}, \qquad \mathbf{T}'_k = \begin{bmatrix} A'_k & C'_k \\ B'_k & 0 \end{bmatrix},
$$

and consider the time-varying forward and backward *state recursions*,

$$
(\mathbf{T}) \quad \left\{ \begin{array}{rcl} x_{k+1} & = & x_kA_k + u_kB_k \\ y_k & = & x_kC_k + u_kD_k \end{array} \right.
$$

$$
(\mathbf{T}') \quad \left\{ \begin{array}{rcl} x'_{k-1} & = & x'_kA'_k + u_kB'_k \\ y'_k & = & x'_kC'_k \end{array} \right.
$$

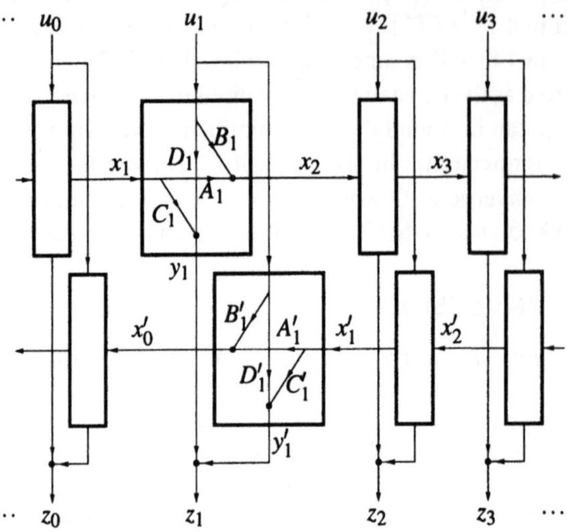

Figure 7.2. State realization which models the multiplication $z = uT$.

$$z_k = y_k + y'_k.$$

See figure 7.2. The recursion maps input sequences $[u_k]$ to output sequences $[y_k]$, $[y'_k]$ and finally $[z_k]$. The intermediate quantities in the recursion are x_k, the forward state, and x'_k, the backward state. The matrices $\{A_k, B_k, C_k, D_k, A'_k, B'_k, C'_k\}$ must have compatible dimensions in order for the multiplications to make sense, but they need not be square or have constant dimensions. Zero dimensions are also allowed. The relation between input $u = [\cdots\; u_1\; u_2\; \cdots]$ and output $z = [\cdots\; z_1\; z_2\; \cdots]$, as generated by the above state recursions, is

$$z = uT: \qquad T = \begin{bmatrix} \ddots & \vdots & & & \vdots & \\ \cdots & D_1 & B_1C_2 & B_1A_2C_3 & B_1A_2A_3C_4 \cdots & \\ & B'_2C'_1 & D_2 & B_2C_3 & B_2A_3C_4 & \\ & B'_3A'_2C'_1 & B'_3C'_2 & D_3 & B_3C_4 & \\ \cdots & & B'_4A'_3C'_2 & B'_4C'_3 & D_4 & \cdots \\ & \vdots & & & \vdots & \ddots \end{bmatrix}$$

so that the state recursions can be used to compute a vector-matrix multiplication $z = uT$, where the matrix T is of the above form. Accordingly, we will say that a matrix T has a (time-varying) *state realization* if there exist matrices $\{\mathbf{T}_k\}$, $\{\mathbf{T}'_k\}$ such that the block entries of $T = [T_{ij}]$ are given by

$$T_{ij} = \begin{cases} D_i, & i = j, \\ B_iA_{i+1}\cdots A_{j-1}C_j, & i < j, \\ B'_iA'_{i-1}\cdots A'_{j+1}C'_j, & i > j. \end{cases} \tag{7.27}$$

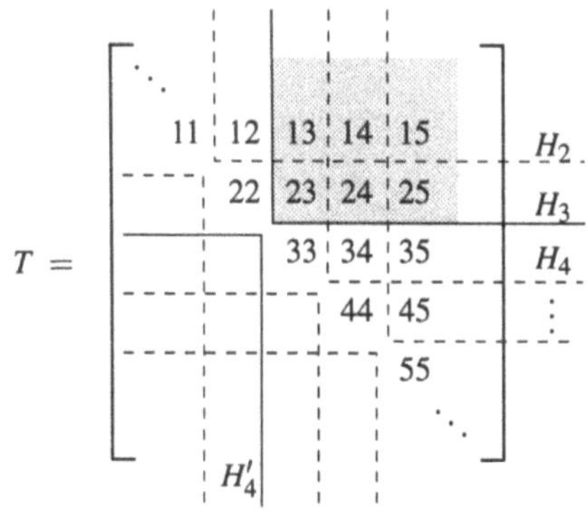

Figure 7.3. Hankel matrices are submatrices of T. H_3 is shaded.

The upper triangular part of T is generated by the forward state recursions $\{\mathbf{T}_k\}$, the lower triangular part by the backward state recursions $\{\mathbf{T}'_k\}$. To have nicely converging expressions in (7.27), we always require realizations to be *exponentially stable*, in the sense that

$$\begin{cases} \ell_A = \lim_{n\to\infty} \sup_i \|A_{i+1}\cdots A_{i+n}\|^{\frac{1}{n}} < 1\,, \\ \ell_{A'} = \lim_{n\to\infty} \sup_i \|A'_{i-1}\cdots A'_{i-n}\|^{\frac{1}{n}} < 1\,. \end{cases}$$

The computation of a vector-matrix product using the state equations is more efficient than a direct multiplication if, for all k, the dimensions of x_k and x'_k are relatively small compared to the matrix size. If this dimension is, on average, equal to d, and T is an $n\times n$ matrix, then a vector-matrix multiplication has complexity $\mathcal{O}(d^2 n)$ (this can be reduced further to $\mathcal{O}(dn)$ by considering minimal parametrizations of the realization, *viz.* [vdVD93, Dew95]) and chapter 14), and a matrix inversion has complexity $\mathcal{O}(d^2 n)$ rather than $\mathcal{O}(n^3)$.

Computation of a state realization

Computation of a minimal state realization for a given matrix or operator T was the topic of chapters 3 and 5. We summarize the main points, and generalize to the operators of mixed causality that we have here.

Minimal realizations are connected to time-varying Hankel matrices, in the present case

$$H_k = \begin{bmatrix} T_{k-1,k} & T_{k-1,k+1} & \cdots \\ T_{k-2,k} & T_{k-2,k+1} & \\ \vdots & & \ddots \end{bmatrix}, \qquad H'_k = \begin{bmatrix} T_{k,k-1} & T_{k,k-2} & \cdots \\ T_{k+1,k-1} & T_{k+1,k-2} & \\ \vdots & & \ddots \end{bmatrix}. \quad (7.28)$$

See figure 7.3. When we substitute the realization equations (7.27) into (7.28), we obtain that H_k (and also H'_k) have structured factorizations of the form

$$
\begin{aligned}
H_k &= \begin{bmatrix} B_{k-1}C_k & B_{k-1}A_kC_{k+1} & \cdots \\ B_{k-2}A_{k-1}C_k & B_{k-2}A_{k-1}A_kC_{k+1} & \\ B_{k-3}A_{k-2}A_{k-1}C_k & & \ddots \\ \vdots & & \end{bmatrix} \\
&= \begin{bmatrix} B_{k-1} \\ B_{k-2}A_{k-1} \\ B_{k-3}A_{k-2}A_{k-1} \\ \vdots \end{bmatrix} [C_k \;\; A_kC_{k+1} \;\; A_kA_{k+1}C_{k+2} \cdots] = \mathcal{C}_k\mathcal{O}_k\,.
\end{aligned}
$$

The rank of the factorization of H_k is (at most) equal to the state dimension d_k at time k, and similarly for H'_k and d'_k. Conversely, the structure of this factorization can be used to derive realizations from it.

Theorem 7.5 *Let $T \in \mathcal{X}$, and define $d_k = \operatorname{rank}(H_k)$, $d'_k = \operatorname{rank}(H'_k)$. If all d_k, d'_k are finite, then there are (marginally) exponentially stable time-varying state realizations that realize T. The minimal dimension of x_k and x'_k of any state realization of T is equal to d_k and d'_k, respectively.*

Hence, the state dimensions of the realization (which determines the computational complexity of multiplications and inversions using state realizations) are equal to the ranks of the Hankel matrices. These ranks are not necessarily the same for all k, so that the number of states may be time-varying.

Minimal state realizations are obtained from minimal factorizations of the H_k and H'_k. In principle, the following algorithm from section 3.4 is suitable. Let $H_k = Q_kR_k$ be a QR factorization of H_k, where Q_k is an isometry ($Q_k^*Q_k = I_{d_k}$), and R_k has full row rank d_k. Likewise, let $H'_k = Q'_kR'_k$. Then a realization of T is given by

$$
\mathbf{T}: \begin{array}{rcl} A_k &=& [0 \;\; Q_k^*]Q_{k+1} \\ B_k &=& (Q_{k+1})(1,:) \\ C_k &=& R_k(:,1) \\ D_k &=& T_{k,k} \end{array}
\qquad
\mathbf{T}': \begin{array}{rcl} A'_k &=& [0 \;\; Q'^*_{k+1}]Q'_k \\ B'_k &=& Q'_k(1,:) \\ C'_k &=& R'_{k+1}(:,1) \\ D'_k &=& 0. \end{array}
$$

(For a matrix X, the notation $X(1,:)$ denotes the first row of X, and $X(:,1)$ the first column.) Important refinements are possible. For example, it is not necessary to act on the infinite size matrix H_k: it is sufficient to consider a principal submatrix that has rank d_k (theorem 3.9). Also note that H_k and H_{k+1} have many entries in common, which can be exploited by considering updating algorithms for the QR factorizations.

State complexity of the inverse

Suppose that T is an invertible matrix or operator with a state realization of low complexity. Under some regularity conditions, it is straightforward to prove that the inverse has a state realization of the same complexity.

Proposition 7.6 *Let $T \in \mathcal{X}$ be an invertible operator with finite dimensional Hankel matrices $(H_T)_k$ and $(H'_T)_k$, defined by (7.28). Put $d_k := \text{rank}\,(H_T)_k$ and $d'_k := \text{rank}(H'_T)_k$.*
If, for each k, at least one of the submatrices $[T_{ij}]_{i,j=-\infty}^{k-1}$ or $[T_{ij}]_{i,j=k}^{\infty}$ is invertible, then $S = T^{-1}$ has Hankel matrices with the same ranks: $\text{rank}\,(H_S)_k = d_k$ and $\text{rank}(H'_S)_k = d'_k$.

PROOF We will use Schur's inversion lemma. In general, let A, B, C, D be matrices or operators such that A and D are square, and A is invertible, then

$$\begin{bmatrix} A & B \\ C & D \end{bmatrix} = \begin{bmatrix} I & 0 \\ CA^{-1} & I \end{bmatrix} \begin{bmatrix} A & 0 \\ 0 & D-CA^{-1}B \end{bmatrix} \begin{bmatrix} I & A^{-1}B \\ 0 & I \end{bmatrix}.$$

If in addition the inverse of this block matrix exists, then $D^{\times} := D - CA^{-1}B$ is invertible and the inverse of the block matrix is given by

$$\begin{aligned} \begin{bmatrix} A' & B' \\ C' & D' \end{bmatrix} &= \begin{bmatrix} I & -A^{-1}B \\ 0 & I \end{bmatrix} \begin{bmatrix} A^{-1} & 0 \\ 0 & (D^{\times})^{-1} \end{bmatrix} \begin{bmatrix} I & 0 \\ -CA^{-1} & I \end{bmatrix} \\ &= \begin{bmatrix} (*) & -A^{-1}B(D^{\times})^{-1} \\ -(D^{\times})^{-1}CA^{-1} & (D^{\times})^{-1} \end{bmatrix}. \end{aligned}$$

In particular, D' is invertible, $\text{rank}\,B' = \text{rank}\,B$, $\text{rank}\,C' = \text{rank}\,C$. The proposition follows if $\left[\begin{smallmatrix} A & B \\ C & D \end{smallmatrix}\right]$ is taken to be a partioning of T, such that $B = (H_T)_k$ and $C = (H'_T)_k$.

□

Outer inversion

If a matrix or operator is block upper and has an inverse which is again block upper (*i.e.*, the corresponding time-varying system is both left and right outer), then it is straightforward to derive a state realization of the inverse.

Proposition 7.7 *Let $T \in \mathcal{U}$ be invertible and left and right outer, so that $S := T^{-1} \in \mathcal{U}$. If T has a state realization $\mathbf{T} = \{A_k, B_k, C_k, D_k\}$, then a realization of S is given by*

$$\mathbf{S}_k = \begin{bmatrix} A_k - C_k D_k^{-1} B_k & -C_k D_k^{-1} \\ D_k^{-1} B_k & D_k^{-1} \end{bmatrix}.$$

PROOF From $T^{-1}T = I$ and $TT^{-1} = I$, and the fact that T^{-1} is upper, we obtain that all $D_k = T_{k,k}$ must be invertible. Using this, we rewrite the state equations:

$$\begin{aligned} &\left\{ \begin{array}{lcl} xZ^{-1} &=& xA + uB \\ y &=& xC + uD \end{array} \right. \\ \Leftrightarrow \quad &\left\{ \begin{array}{lcll} xZ^{-1} &=& x(A - CD^{-1}B) &+\; yD^{-1}B \\ u &=& -xCD^{-1} &+\; yD^{-1}. \end{array} \right. \end{aligned}$$

The second set of state equations generates the inverse mapping $y \to u$, so that it must be a realization of T^{-1}. The remaining part of the proof is to show that $\{A_k - C_k D_k^{-1} B_k\}$ is a *stable* state operator. The proof of this is omitted, but it is essentially a consequence of the fact that T is outer invertible and hence has a bounded upper inverse. See also proposition 13.2.

□

Note that the realization of the inverse is obtained locally: it is, at point k, only dependent on the realization of the given matrix at point k. Hence, it is quite easy to compute the inverse of an operator once we know that it is left and right outer.

Inner-coprime factorization

In order to use the above inversion proposition on a matrix T which is not block upper, we compute a kind of QR factorization of T as $T = Q\Delta$, where Q is block lower and unitary, and Δ is block upper. Since Q is unitary, its inverse is equal to its Hermitian transpose and can trivially be obtained. We first consider the special case where T is lower triangular. This case is related to the inner-coprime factorization in section 6.2.

Proposition 7.8 *(a) Suppose that $T \in \mathcal{L}$ has an exponentially stable finite dimensional state realization $\mathbf{T}' = \{A'_k, B'_k, C'_k, D'_k\}$, with $A'_k : d'_k \times d'_{k-1}$. Then T has a factorization $T = Q^*R$, where $Q \in \mathcal{U}$ is inner and $R \in \mathcal{U}$.*

(b) Denote realizations of Q and R by

$$\mathbf{Q}_k = \begin{bmatrix} (A_Q)_k & (C_Q)_k \\ (B_Q)_k & (D_Q)_k \end{bmatrix}, \quad \mathbf{R}_k = \begin{bmatrix} (A_R)_k & (C_R)_k \\ (B_R)_k & (D_R)_k \end{bmatrix}.$$

Then $\mathbf{Q}_k$ and $\mathbf{R}_k$ follow recursively from the QR factorization

$$\left[\begin{array}{c|cc} Y_k A'_k & I & Y_k C'_k \\ B'_k & 0 & D'_k \end{array}\right] = \mathbf{Q}_k^* \left[\begin{array}{c|c} Y_{k-1} & \mathbf{R}_k \\ 0 & \end{array}\right] \tag{7.29}$$

where $Y_k : d'_k \times d'_k$ is a square matrix.

The state operators of $\mathbf{Q}$ and $\mathbf{R}$ are the same: $(A_Q)_k = (A_R)_k$, and they are related to A'^*_k via a state transformation. The resulting number of inputs of Q and R may be time-varying. In particular, Q can be a block matrix whose entries are matrices, even if T itself has scalar entries.

Equation (7.29) is a recursion: for a given initial matrix Y_{k_0}, we can compute $\mathbf{Q}_{k_0}$, $\mathbf{R}_{k_0}$, and Y_{k_0-1}. Hence we obtain the state realization matrices for Q and R in turn for $k = k_0-1, k_0-2, \cdots$. All we need is a correct initial value for the recursion. Exact initial values can be computed in the case of systems that are LTI for large k ($Y^*_{k_0} Y_{k_0}$ satisfies a Lyapunov equation), or periodically varying, or that have zero state dimensions for $k > k_0$. However, even if this is not the case, we can obtain Q and R to any precision we like by starting the recursion with any (invertible) initial value, such as $\tilde{Y}_{k_0} = I$. The assumption that T has an exponentially stable realization implies that $\tilde{Y}_k \to Y_k$ ($k \to -\infty$), the correct value for Y. Convergence is monotonic, and the speed of convergence is depending on the "amount of stability" of the A'_k.

The more general case ($T \in \mathcal{X}$) is a corollary of the above proposition. Split $T = T_\mathcal{L} + T_\mathcal{U}$, with $T_\mathcal{L} \in \mathcal{L}$ and $T_\mathcal{U} \in \mathcal{ZU}$ (strictly upper). The above inner-coprime factorization, applied to $T_\mathcal{L}$, gives $T_\mathcal{L} = Q^*R$. Then T has a factorization $T = Q^*(R + QT_\mathcal{U}) =: Q^*\Delta$, where $\Delta \in \mathcal{U}$. The realization for Q is only dependent on $T_\mathcal{L}$, and follows from the recursion (7.29). A realization for Δ is obtained by multiplying Q with $T_\mathcal{U}$, and adding R. These operations can be done in state space. Using the fact that $A_Q = A_R$

and $B_Q = B_R$, we obtain

$$\mathbf{\Delta}_k = \left[\begin{array}{cc|c} (A_Q)_k & (C_Q)_k B_k & (C_R)_k \\ 0 & A_k & C_k \\ \hline (B_Q)_k & (D_Q)_k B_k & (D_R)_k \end{array}\right].$$

Inner-outer factorization

Let $T \in \mathcal{U}$, with exponentially stable finite dimensional realization $\mathbf{T} = \{A_k, B_k, C_k, D_k\}$, where $A_k : d_k \times d_{k+1}$, $A'_k : d'_k \times d'_{k-1}$. The inner-outer factorization $T = UT_r$, where $U^*U = I$ and T_r is right outer, can be computed recursively, as follows. Suppose that, at point k, we know the matrix Y_k. Compute the following QR factorization:

$$\begin{array}{c} \\ m_k \\ (d_Y)_k \end{array} \begin{array}{c} \begin{array}{cc} n_k & d_{k+1} \end{array} \\ \begin{bmatrix} D_k & B_k \\ Y_k C_k & Y_k A_k \end{bmatrix} \end{array} =: \mathbf{W}_k \begin{array}{c} \\ (m_r)_k \\ (d_Y)_{k+1} \\ \\ \end{array} \begin{array}{c} \begin{array}{cc} n_k & d_{k+1} \end{array} \\ \begin{bmatrix} (D_r)_k & (B_r)_k \\ 0 & Y_{k+1} \\ 0 & 0 \end{bmatrix} \end{array} \tag{7.30}$$

where $\mathbf{W}_k$ is unitary, and the partitioning of the factors at the right hand side of (7.30) is such that $(D_r)_k$ and Y_{k+1} both have full row rank. This also defines the dimensions $(m_r)_k$ and $(d_Y)_{k+1}$. Since the factorization produces Y_{k+1}, we can perform the QR factorization (7.30) in turn for $k+1, k+2, \cdots$.

Theorem 7.1 in section 7.2 claimed that this recursion determines the inner-outer factorization. $\mathbf{W}_k$ has a partitioning as

$$\mathbf{W}_k = \begin{array}{c} \\ m_k \\ (d_Y)_k \end{array} \begin{array}{c} \begin{array}{ccc} (m_r)_k & (d_Y)_{k+1} & \end{array} \\ \begin{bmatrix} (D_U)_k & (B_U)_k & * \\ (C_U)_k & (A_U)_k & * \end{bmatrix} \end{array}.$$

It turns out that $\mathbf{U} = \{(A_U)_k, (B_U)_k, (C_U)_k, (D_U)_k\}$ is a realization of U, and $\mathbf{T}_r = \{A_k, (B_r)_k, C_k, (D_r)_k\}$ is a realization of T_r.

In [vdV93a], the inner-outer factorization was solved using a time-varying Riccati equation (see also [Nic92]). The above recursive QR factorization is a square-root variant of it. Correct initial points for the recursion can be obtained in a similar way as for the inner-coprime factorization. If T is Toeplitz for $k < k_0$, then Y_{k_0} can be computed from the underlying time-invariant Riccati equation (viz. section 7.2) which is retrieved upon squaring of (7.30), thus eliminating $\mathbf{W}_k$. As is well known, this calls for the solution of an eigenvalue problem. Similar results hold for the case where T is periodically varying before $k < k_0$, or has zero state dimensions ($d_k = 0, k < k_0$). But, as for the inner-coprime factorization, we can in fact take any invertible starting value, such as $\tilde{Y}_{k_0} = I$, and perform the recursion: because of the assumed stability of A, $\tilde{Y}_k \to Y_k$. In a sense, we are using the familiar QR-iteration [GV89] for computing eigenvalues! (Open question is how the shifted QR iteration fits in this framework.)

The outer-inner factorization $T = T_\ell V$ ($VV^* = I$, T_ℓ left outer) is computed similarly, now by the backward recursive LQ factorization

$$\begin{array}{c} \\ m_k \\ d_k \end{array} \begin{array}{c} \begin{array}{cc} n_k & (d_Y)_k \end{array} \\ \begin{bmatrix} D_k & B_k Y_k \\ C_k & A_k Y_k \end{bmatrix} \end{array} =: \begin{array}{c} \\ m_k \\ d_k \end{array} \begin{array}{c} \begin{array}{ccc} (n_\ell)_k & (d_Y)_{k-1} & \end{array} \\ \begin{bmatrix} (D_\ell)_k & 0 & 0 \\ (C_\ell)_k & Y_{k-1} & 0 \end{bmatrix} \end{array} \mathbf{W}_k. \tag{7.31}$$

The partitioning is such that $(D_\ell)_k$ and Y_{k-1} have full column rank. $\mathbf{W}_k$ is unitary and has a partitioning as

$$\mathbf{W}_k = \begin{matrix} \\ (n_\ell)_k \\ (d_Y)_{k-1} \\ \\ \end{matrix} \begin{matrix} n_k & (d_Y)_k \\ \left[\begin{matrix} (D_V)_k \\ (C_V)_k \\ * \end{matrix}\right. & \left.\begin{matrix} (B_V)_k \\ (A_V)_k \\ * \end{matrix}\right] \end{matrix} .$$

Realizations of the factors are given by

$$\begin{aligned} \mathbf{V} &= \{(A_V)_k, (B_V)_k, (C_V)_k, (D_V)_k\} \\ \mathbf{T}_\ell &= \{A_k, B_k, (C_\ell)_k, (D_\ell)_k\}. \end{aligned}$$

An example of the outer-inner factorization is given in section 7.4.

Inversion

At this point, we have obtained state space versions of all operators in the factorization $T = Q^*UR_{\ell r}V$ of equation (7.26): Q is obtained by the backward inner-coprime factorization of section 7.3, U by the forward inner-outer QR recursion in equation (7.30), and V by the backward outer-inner LQ recursion in equation (7.31). We also have obtained a state space expression for the inverse of the outer factor $R_{\ell r}$, viz. section 7.3. The realizations of the (pseudo-)inverses of the inner (isometric) factors are obtained simply via transposition: *e.g.*, the realization for V^* is anti-causal and given by $\{(A_V)_k^*, (C_V)_k^*, (B_V)_k^*, (D_V)_k^*\}$. The pseudo-inverse of T is given by $T^\dagger = V^*R_{\ell r}^{-1}U^*Q$.

It is possible to obtain a single set of state matrices for $T^\dagger$, by using formulas for the multiplication and addition of realizations. This is complicated to some extent because of the alternating upper-lower nature of the factors. Moreover, it is often not necessary to obtain a single realization: matrix-vector multiplication is carried out more efficiently on a factored representation than on a closed-form realization. This is because for a closed-form representation, the number of multiplications per point in time is roughly equal to the square of the sum of the state dimensions of all factors, whereas in the factored form it is equal to the sum of the square of these dimensions. See also section 14.

7.4 EXAMPLES

We illustrate the preceding sections with some examples of the inner-outer factorization algorithm on finite (4×4) matrices and on a simple infinite matrix. In the finite matrix case, interesting things can occur only when T is singular or when the dimensions of T are not uniform.

Finite size matrices

1. Using algorithm 7.1 on

$$T = \begin{bmatrix} \underline{0} & 1 & 4 & 6 \\ 0 & \underline{0} & 2 & 5 \\ 0 & 0 & \underline{0} & 3 \\ 0 & 0 & 0 & \underline{0} \end{bmatrix}$$

(the underlined entries form the 0-th diagonal) yields an almost trivial left isometric factor V or left inner factor W:

$$V = \begin{bmatrix} \vdots & 1 & 0 & 0 \\ \cdot & \underline{0} & 1 & 0 \\ \cdot & 0 & \underline{0} & 1 \\ \cdot & 0 & 0 & \underline{0} \end{bmatrix} \qquad W = \begin{bmatrix} \vdots & 1 & 0 & 0 & 0 \\ \cdot & \underline{0} & 1 & 0 & 0 \\ \cdot & 0 & \underline{0} & 1 & 0 \\ \cdot & 0 & 0 & \underline{0} & \underline{1} \end{bmatrix} \qquad \begin{array}{lcl} \#\mathcal{M}_W & = & [1\ 1\ 1\ 1] \\ \#\mathcal{N}_W & = & [0\ 1\ 1\ 2] \\ \#\mathcal{B}_W & = & [0\ 1\ 1\ 1] \end{array}$$

It is seen that V is not inner, because T is singular. W is the inner extension of V. The only effect of W is a redefinition of time intervals: W acts as a shift operator. $T_r = W^*T$ is

$$W^*T = \begin{bmatrix} \vdots & \cdot & \cdot & \cdot \\ 0 & \underline{1} & 4 & 6 \\ 0 & 0 & \underline{2} & 5 \\ 0 & 0 & 0 & \underline{3} \\ 0 & 0 & 0 & \underline{0} \end{bmatrix} \qquad \begin{array}{lcl} \#\mathcal{M}_{T_r} & = & [0\ 1\ 1\ 2] \\ \#\mathcal{N}_{T_r} & = & [1\ 1\ 1\ 1]. \end{array}$$

The multiplication by W^* has shifted the rows of T downwards. This is possible: the result T_r is still upper. V^*T is equal to W^*T with its last row removed.

2. Take

$$T = \begin{bmatrix} \underline{0} & 1 & 4 & 6 \\ 0 & \underline{1} & 2 & 5 \\ 0 & 0 & \underline{1} & 3 \\ 0 & 0 & 0 & \underline{1} \end{bmatrix} \qquad \begin{array}{lcl} \#\mathcal{M} & = & [1\ 1\ 1\ 1] \\ \#\mathcal{N} & = & [1\ 1\ 1\ 1] \\ \#\mathcal{B} & = & [0\ 1\ 2\ 1]. \end{array}$$

Hence T is again singular, but now a simple shift will not suffice. The algorithm computes W as

$$W = \begin{bmatrix} \vdots & -0.707 & 0.577 & 0.367 & 0.180 \\ \cdot & \underline{-0.707} & -0.577 & -0.367 & -0.180 \\ \cdot & 0 & \underline{0.577} & -0.733 & -0.359 \\ \cdot & 0 & 0 & \underline{-0.440} & \underline{0.898} \end{bmatrix} \qquad \begin{array}{lcl} \#\mathcal{M}_W & = & [1\ 1\ 1\ 1] \\ \#\mathcal{N}_W & = & [0\ 1\ 1\ 2] \\ \#\mathcal{B}_W & = & [0\ 1\ 1\ 1] \end{array}$$

$$T_r = W^*T = \begin{bmatrix} \vdots & \cdot & \cdot & \cdot \\ 0 & \underline{-1.414} & -4.243 & -7.778 \\ 0 & 0 & \underline{1.732} & 2.309 \\ 0 & 0 & 0 & \underline{-2.273} \\ 0 & 0 & 0 & \underline{0} \end{bmatrix} \qquad \begin{array}{lcl} \#\mathcal{M}_{T_r} & = & [0\ 1\ 1\ 2] \\ \#\mathcal{N}_{T_r} & = & [1\ 1\ 1\ 1]. \end{array}$$

V is equal to W with its last column removed, so that $T_r = V^*T$ is equal to the above T_r with its last row removed.

3. In the previous examples, we considered only systems T with a constant number of inputs and outputs (equal to 1), for which $V \neq I$ only if T is singular. However, a non-identical V can also occur if the number of inputs and outputs of T varies in

time. Thus consider

$$T = \begin{bmatrix} \underline{1.000} & 0.500 & 0.250 & 0.125 \\ \underline{1.000} & 0.300 & 0.100 & 0.027 \\ 0 & \underline{1.000} & 0.500 & 0.250 \\ 0 & 0 & \underline{1.000} & 0.300 \\ \cdot & \cdot & \cdot & \vdots \end{bmatrix} \qquad \begin{array}{lcl} \#\mathcal{M} & = & [2\ 1\ 1\ 0] \\ \#\mathcal{N} & = & [1\ 1\ 1\ 1] \\ \#\mathcal{B} & = & [0\ 1\ 2\ 1] \end{array}$$

$$V = \begin{bmatrix} \underline{-0.707} & 0.099 & 0.025 & -0.699 \\ \underline{-0.707} & -0.099 & -0.025 & 0.699 \\ 0 & \underline{0.990} & -0.005 & 0.139 \\ 0 & 0 & \underline{0.999} & 0.035 \\ \cdot & \cdot & \cdot & \vdots \end{bmatrix} \qquad \begin{array}{lcl} \#\mathcal{M}_V & = & [2\ 1\ 1\ 0] \\ \#\mathcal{N}_V & = & [1\ 1\ 1\ 1] \\ \#\mathcal{B}_V & = & [0\ 1\ 1\ 1]. \end{array}$$

In this case, V is itself inner. The outer factor T_r follows as

$$T_r = V^*T = \begin{bmatrix} \underline{-1.414} & -0.565 & -0.247 & -0.107 \\ 0 & \underline{1.010} & 0.509 & 0.257 \\ 0 & 0 & \underline{1.001} & 0.301 \\ 0 & 0 & 0 & \underline{-0.023} \end{bmatrix} \qquad \begin{array}{lcl} \#\mathcal{M}_{T_r} & = & [1\ 1\ 1\ 1] \\ \#\mathcal{N}_{T_r} & = & [1\ 1\ 1\ 1]. \end{array}$$

Infinite-size matrices

A simple doubly infinite example which can be computed by hand is

$$T = \left[\begin{array}{cccc|cccc} \ddots & & & & & & & \\ & 1 & & & & & \mathbf{0} & \\ & & 1 & & & & & \\ & & & 1 & 1 & & & \\ \hline & & & & \boxed{0} & 1 & & \\ & & & & & 0 & 1 & \\ & \mathbf{0} & & & & & 0 & 1 \\ & & & & & & & \ddots \quad \ddots \end{array}\right]. \tag{7.32}$$

In this example, T does not have full row span: $[\cdots\ 0\ 0\ \boxed{1}\ 0\ 0\ \cdots]$ is not contained in it, and $\ker(\cdot T^*) \neq \{0\}$. The outer-inner factorization is

$$T = T_\ell V$$
$$= \left[\begin{array}{ccc|ccc} \ddots & & & & & \\ & 1 & & & \mathbf{0} & \\ & & 1 & & & \\ & & & \sqrt{2} & & \\ \hline & & & 1 & & \\ & & & & 1 & \\ & \mathbf{0} & & & & 1 \\ & & & & & \ddots \end{array}\right] \left[\begin{array}{ccc|ccc} \ddots & & & & & \\ & 1 & & & \mathbf{0} & \\ & & 1 & & & \\ & & \frac{1}{\sqrt{2}} & \frac{1}{\sqrt{2}} & & \\ \hline & & & 0 & 1 & \\ & & & & 0 & 1 \\ & \mathbf{0} & & & & 0 \quad \ddots \\ & & & & & \ddots \end{array}\right].$$

T_ℓ obviously has a left inverse T_ℓ^{-1} which is upper (it is even diagonal and a right inverse in this case). V is only an isometry: $VV^* = I$, but $V^*V \neq I$, consistent with theorem 7.1. The inner-outer factorization is

$$
\begin{aligned}
T &= UT_r \\
&= \left[\begin{array}{cccc|cccc}
\ddots & & & & \cdot & & & \\
 & 1 & & & \cdot & & \mathbf{0} & \\
 & & 1 & & \cdot & & & \\
 & & & 1 & \cdot & & & \\
\hline
 & & & & \boxdot & 1 & & \\
 & & & & \cdot & 0 & 1 & \\
 & \mathbf{0} & & & \cdot & & 0 & \ddots \\
 & & & & \cdot & & & \ddots
\end{array}\right]
\left[\begin{array}{cccc|cccc}
\ddots & & & & & & & \\
 & 1 & & & & & & \\
 & & 1 & & & & & \\
 & & & 1 & 1 & & & \\
\hline
\cdot & \cdot & \cdot & \cdot & \boxdot & \cdot & \cdot & \cdot \\
 & & & & & 1 & & \\
 & & & & & & 1 & \\
 & & & & & & & \ddots
\end{array}\right].
\end{aligned}
$$

U has a column with zero horizontal dimension (signified by '$\cdot$'), but $U^*U = I$ nonetheless. T_r has a right inverse T_r^{-1} which is upper,

$$
T_r^{-1} = \left[\begin{array}{cccc|cccc}
\ddots & & & & \cdot & & & \\
 & 1 & & & \cdot & & & \\
 & & 1 & & \cdot & & & \\
 & & & 1 & \cdot & & & \\
\hline
 & & & & \boxdot & & & \\
 & & & & \cdot & 1 & & \\
 & & & & \cdot & & 1 & \\
 & & & & \cdot & & & \ddots
\end{array}\right],
$$

but $T_r^{-1}T_r \neq I$: it is not a left inverse. If our purpose is the inversion of T, then it is clear in this case that T only has a right inverse. The outer-inner factorization is useful for computing this inverse: it is equal to $V^*T_\ell^{-1}$.

An interesting observation from these examples is that the inner-outer factorization of finite matrices T is equal to the QR factorization of T when it is considered as an ordinary matrix without block entries. In combination with the external factorization, this observation can be used to efficiently compute the QR factorization of a general block matrix (mixed upper-lower) if both its upper and its lower parts have state realizations of low dimensions. Let X be such a matrix, then first compute U such that $T = UX$ is upper (U follows from an external factorization of $\mathbf{P}(X^*) =: \Delta^*U$), and subsequently compute the inner-outer factorization of T as $T = VT_r$. Then the QR factorization of X follows as $X = (U^*V)T_r$. Note that if the square-root algorithm is used, then the global QR factorization of X is replaced by local QR factorizations of state-space matrices.

Matrix inversion example

As a last example for this section, consider again T from equation (7.4). A realization for T is straightforward to obtain, since it is a banded matrix:

$$k = -\infty, \cdots, 0: \qquad \mathbf{T}_k = \left[\begin{array}{c|c} 0 & -1/2 \\ \hline 1 & 1 \end{array}\right]$$

$$k = 1, \cdots, \infty: \qquad \mathbf{T}_k = \left[\begin{array}{c|c} 0 & -2 \\ \hline 1 & 1 \end{array}\right].$$

T is already upper, so an inner-coprime factorization is not necessary. We pointed out before that the inner-outer factorization of T is $T = I \cdot T$. This is because the initial point of the recursion (7.30), given by the LTI solution of the inner-outer factorization of the top-left block of T, produces $(d_Y)_0 = 0$, and hence all subsequent Y_k's have zero dimensions. Consequently, T is immediately seen to be right outer by itself.

The next step is to compute the outer-inner factorization of the right outer factor, *i.e.*, of T. An initial point for the recursion (7.31) is obtained as $Y_k = \sqrt{3}$, $k \geq 1$. It requires the solution of an LTI Riccati equation to find it (this equation is the dual of (7.19) specialized to LTI, and its solution can be found using the pencil technique described below that equation), but it is easy to verify that it is a stationary solution of (7.31) for $k \geq 1$: it satisfies the equation

$$\underbrace{\begin{bmatrix} 1 & \sqrt{3} \\ -2 & 0 \end{bmatrix}}_{\begin{bmatrix} D & BY \\ C & AY \end{bmatrix}} = \underbrace{\begin{bmatrix} 2 & 0 & \cdot \\ -1 & \sqrt{3} & \cdot \end{bmatrix}}_{\begin{bmatrix} D_\ell & 0 & 0 \\ C_\ell & Y & 0 \end{bmatrix}} \underbrace{\begin{bmatrix} \frac{1}{2} & \frac{1}{2}\sqrt{3} \\ -\frac{1}{2}\sqrt{3} & \frac{1}{2} \\ \cdot & \cdot \end{bmatrix}}_{\mathbf{W}_k} \tag{7.33}$$

(where '$\cdot$' denotes zero dimensions). Alternatively, we can start the recursion with $\tilde{Y}_{20} = 1$, say, and obtain $\tilde{Y}_0 = 1.7321 \cdots \approx \sqrt{3}$. Equation (7.33) also shows that the realization of the outer factor has $(D_\ell)_k = 2$ and $(C_\ell)_k = -1$, for $k \geq 0$. Continuing with the recursion gives us

$$(\mathbf{T}_\ell)_1 = \left[\begin{array}{c|c} 0 & -1 \\ \hline 1 & 2 \end{array}\right], \qquad \mathbf{V}_1 = \left[\begin{array}{c|c} 0.5 & -0.866 \\ \hline 0.866 & 0.5 \end{array}\right]$$

$$Y_0 = 1.732,$$

$$(\mathbf{T}_\ell)_0 = \left[\begin{array}{c|c} 0 & -0.25 \\ \hline 1 & 2 \end{array}\right], \qquad \mathbf{V}_0 = \left[\begin{array}{c|c} 0.5 & -0.866 \\ \hline 0.866 & 0.5 \end{array}\right]$$

$$Y_{-1} = 0.433,$$

$$(\mathbf{T}_\ell)_{-1} = \left[\begin{array}{c|c} 0 & -0.459 \\ \hline 1 & 1.090 \end{array}\right], \qquad \mathbf{V}_{-1} = \left[\begin{array}{c|c} 0.918 & -0.397 \\ \hline 0.397 & 0.918 \end{array}\right]$$

$$Y_{-2} = 0.199,$$

$$(\mathbf{T}_\ell)_{-2} = \left[\begin{array}{c|c} 0 & -0.490 \\ \hline 1 & 1.020 \end{array}\right], \qquad \mathbf{V}_{-2} = \left[\begin{array}{c|c} 0.981 & -0.195 \\ \hline 0.195 & 0.981 \end{array}\right]$$

$$Y_{-3} = 0.097,$$

$$(\mathbf{T}_\ell)_{-3} = \left[\begin{array}{c|c} 0 & -0.498 \\ \hline 1 & 1.005 \end{array}\right], \qquad \mathbf{V}_{-3} = \left[\begin{array}{c|c} 0.995 & -0.097 \\ \hline 0.097 & 0.995 \end{array}\right]$$

$$Y_{-4} = 0.049,$$

$$(\mathbf{T}_\ell)_{-4} = \left[\begin{array}{c|c} 0 & -0.499 \\ \hline 1 & 1.001 \end{array}\right], \qquad \mathbf{V}_{-4} = \left[\begin{array}{c|c} 0.999 & -0.048 \\ \hline 0.048 & 0.999 \end{array}\right]$$

$$Y_{-5} = 0.024,$$

$$(\mathbf{T}_\ell)_{-5} = \left[\begin{array}{c|c} 0 & -0.500 \\ \hline 1 & 1.000 \end{array}\right], \qquad \mathbf{V}_{-5} = \left[\begin{array}{c|c} 1.000 & -0.024 \\ \hline 0.024 & 1.000 \end{array}\right]$$

$$Y_{-6} = 0.012.$$

Thus, Y_k tends towards zero as $k \to -\infty$, and at the same time, $\mathbf{V}_k$ tends towards the identity matrix. For this reason, $\ell_A = 1$, and $\mathbf{V}$ is not uniformly reachable. The latter is seen from the definition of the snapshots of the reachability operator $\mathcal{C}_k$ in (3.23), *viz.*

$$\mathcal{C}_k = \begin{bmatrix} B_{k-1} \\ B_{k-2}A_{k-1} \\ B_{k-3}A_{k-2}A_{k-1} \\ \vdots \end{bmatrix}, \qquad \mathcal{O}_k = [C_k \quad A_k C_{k+1} \quad A_k A_{k+1} C_{k+2} \quad \cdots].$$

Since $\mathcal{C}_k$ only looks at A_n for a decreasing sequence $n = k, k-1, k-2, \cdots$, and $B_n \to 0$, we can make $\|\mathcal{C}_k\|_{HS}$ arbitrarily small for k sufficiently close to $-\infty$. The implication is that, although all $\mathbf{V}_k$ are unitary, the corresponding operator V is not inner, and T is not right invertible. Note that we do have that V is isometric: $VV^* = I$, because $\mathbf{V}$ is uniformly observable ($\mathcal{O}_k$ looks at A_n for increasing values of n). All this is consistent with theorem 7.1: since $\ker(\cdot T^*) \neq \{0\}$, V cannot be inner. The fact that $\ell_A = 1$ is consistent with proposition 6.18.

We could continue with $\mathbf{V}$ as defined above, but in practice, we settle for an approximation. At a certain point, (say around $k = -10$, but actually depending on the desired accuracy),[6] we will decide on $d_{Y,k-1} = 0$, after which the number of states in $\mathbf{V}_k$ will be reduced to zero as well:

$$\begin{aligned} \mathbf{V}_{-9} &= \left[\begin{array}{c|c} 1.000 & -0.000 \\ \hline 0.000 & 1.000 \end{array}\right] \\ \mathbf{V}_{-10} &= \left[\begin{array}{c|c} \cdot & \cdot \\ \hline 0.000 & 1.000 \end{array}\right] \\ \mathbf{V}_{-11} &= \left[\begin{array}{c|c} \cdot & \cdot \\ \hline \cdot & 1.000 \end{array}\right]. \end{aligned}$$

[6] A decent approximation theory is found in chapter 10. The convergence of Y_k and its decomposition into a full-rank and a singular part is studied in section 7.5.

This brings us back to the LTI solution for this part of T. It is seen from $\mathbf{V}_{-10}$ that it is not unitary at this point in time: only $\mathbf{V}_{-10}\mathbf{V}_{-10}^* = I$ holds, but $\mathbf{V}_{-10}^*\mathbf{V}_{-10} \neq I$. Consequently, $VV^* = I$ but $V^*V \neq I$, as we had without the approximation. Now it is clear that V is not unitary but only isometric, and hence T is only left invertible. The situation is not unlike T in (7.32), but less pronounced.

The outer-inner factorization of T is thus

$$T_\ell = \left[\begin{array}{cccccc|cccc}
\ddots & \ddots & & & & & & & & \\
 & 1 & -0.5 & & & & & & \mathbf{0} & \\
 & & 1 & -0.5 & & & & & & \\
 & & & 1 & -0.49 & & & & & \\
 & & & & 1 & -0.46 & & & & \\
 & & & & & 1.09 & -0.25 & & & \\
\hline
 & & & & & & 2 & -1 & & \\
 & & & & & & & 2 & -1 & \\
 & \mathbf{0} & & & & & & & 2 & \ddots \\
 & & & & & & & & & \ddots
\end{array}\right] \qquad (7.34)$$

$$V = \left[\begin{array}{cccccc|cccc}
\ddots & \vdots & & & & & & & \vdots & \\
 & 1 & -0.00 & -0.00 & -0.00 & -0.01 & -0.02 & -0.01 & -0.00 & \cdots \\
 & & 1 & -0.00 & -0.01 & -0.02 & -0.04 & -0.02 & -0.01 & \\
 & & & 1 & -0.02 & -0.04 & -0.08 & -0.04 & -0.02 & \\
 & & & & 0.98 & -0.08 & -0.15 & -0.08 & -0.04 & \\
 & & & & & 0.92 & -0.34 & -0.17 & -0.09 & \\
\hline
 & & & & & & 0.5 & -0.75 & -0.37 & \\
 & & & & & & & 0.5 & -0.75 & \\
 & \mathbf{0} & & & & & & & 0.5 & \cdots \\
 & & & & & & & & & \ddots
\end{array}\right].$$

The (left) inverse of T is

$$T^\dagger = V^*T_\ell = \left[\begin{array}{ccccc|cccc}
\ddots & \vdots & & & & & & & \vdots \\
\cdots & 1.00 & 0.49 & 0.24 & 0.10 & 0.01 & 0.01 & 0.00 & 0.00\cdots \\
 & -0.01 & 0.99 & 0.48 & 0.20 & 0.03 & 0.01 & 0.01 & 0.00 \\
 & -0.01 & -0.02 & 0.95 & 0.40 & 0.05 & 0.03 & 0.01 & 0.01 \\
 & -0.02 & -0.05 & -0.10 & 0.80 & 0.10 & 0.05 & 0.03 & 0.01 \\
\hline
 & -0.05 & -0.10 & -0.20 & -0.40 & 0.20 & 0.10 & 0.05 & 0.03 \\
 & -0.02 & -0.05 & -0.10 & -0.20 & -0.40 & 0.05 & 0.03 & 0.01 \\
 & -0.01 & -0.02 & -0.05 & -0.10 & -0.20 & -0.47 & 0.01 & 0.01 \\
\cdots & -0.01 & -0.01 & -0.02 & -0.05 & -0.10 & -0.24 & -0.49 & 0.00\cdots \\
 & \vdots & & & & & & & \vdots\ddots
\end{array}\right].$$

It has indeed the structure which we announced in equation (7.5): it is Toeplitz towards $(-\infty, -\infty)$ and $(+\infty, +\infty)$, and equal to the solution of the LTI subsystems of T in those regions. In addition, there is some limited interaction in the center which glues the two solutions together. All entries are nicely bounded.

7.5 ZERO STRUCTURE AND ITS LIMITING BEHAVIOR

In this section we study the zero-structure of a transfer operator further for as far as it is relevant to system inversion. Here, the term "zero-structure" relates to the system dynamics of the inverse system. The latter is often captured through a left or right external factorization. In particular, we know from chapter 6 that a locally finite, uniformly exponentially stable system has a left and a right external factorization with inner functions that characterize (share) the system dynamics of the original system and are obtained through unitary completions, respectively of a coisometric pair $[{}^{A}_{B}]$ in a canonical input normal form or an isometric pair $[A \;\; C]$ in a canonical output normal form of the original system. If the system T has a causal and uniformly exponentially stable inverse T^{-1}, then external factorizations on T^{-1} would provide similar kind of information on T^{-1}, and it would be logical to call that the "zero structure" of T. For more general T we cannot work on T^{-1}, since it does not exist, but inner-outer type factorizations come to the rescue. The structural description, however, turns out to be considerably more complicated, and interestingly so, not in the least because certain effects occur in the time-varying case that have no equivalent in the LTI case. In particular we can expect to encounter quite simple isometric transfer functions with unitary realizations *and which are not inner*, while studying the inner-outer factorization of even quite simple isometric transfer functions. We shall see that such transfer functions have a non-trivial kernel or *defect space* which plays an important role in the zero structure of T.

The theory presented in this section allows us to make statements on the inversion of upper transfer operators which are uniformly locally finite, and very precise ones on systems which have an LTI realization for very small (toward $-\infty$) and very large time indices (toward ∞), while changing in between.

The exploration of the zero-structure of a transfer operator T starts out via an investigation of its outer-inner factorization $T = T_\ell V$ in which T_ℓ is left outer and V is isometric, $VV^* = I$, see the end of section 7.2. V is defined via the property

$$\mathcal{U}_2 V = \overline{\mathcal{U}_2 T}, \tag{7.35}$$

and we have the decomposition

$$\mathcal{U}_2 = \mathcal{H}_o(V) \oplus \mathcal{U}_2 V \oplus \ker(\cdot V^*|_{\mathcal{U}_2}) \tag{7.36}$$

where also $\ker(\cdot V^*|_{\mathcal{U}_2}) = \ker(\cdot T^*|_{\mathcal{U}_2})$. It is a sliced upper space, characterized via the extended Beurling-Lax theorem by a causal isometric operator U such that $\ker(\cdot V^*|_{\mathcal{U}_2}) = U\mathcal{U}_2$.

However, the nullspace of T^* may be larger: it is indeed possible that $\cdot T^*|_{\mathcal{U}_2}$ is strictly smaller than $\cdot T^*|_{\mathcal{X}_2}$. In that case, there is a component in the nullspace which is intrinsically non-causal, and which could be termed a (right-) "defect" space for T. Its investigation is the topic of this section, and it is connected to the doubly invariant subspaces mentioned in section 6.4. It may even be larger than (with '$\bigvee$' indicating a sum of subspaces)

$$\bigvee_{n=0}^{-\infty} Z^n \ker(\cdot V^*)|_{\mathcal{U}_2},$$

which may be zero while $\ker(\cdot T^*)|_{\mathcal{X}_2}$ is not.

Let us define[7]

$$\mathcal{K}''_o = \ker(\cdot T^*)|_{\mathcal{X}_2} \ominus \bigvee_{n=0}^{-\infty} Z^n \ker(\cdot V^*)|_{\mathcal{U}_2}. \tag{7.37}$$

From proposition 6.18 applied to the extension $W = \left[\begin{smallmatrix} V \\ U \end{smallmatrix}\right]$ we know that $\mathbf{P}(\mathcal{K}''_o) \subset \mathcal{H}_o(V)$. Our first theorem asserts that this space has finite dimensional slices when the original system T is uniformly locally finite. Since $\mathcal{K}''_o$ is doubly Z-invariant ($Z\mathcal{K}''_o \subset \mathcal{K}''_o$ and $Z^*\mathcal{K}''_o \subset \mathcal{K}''_o$), the k-th slice $\pi_k\mathcal{K}''_o$ will have the same basis for each integer k. We denote by $\mathbf{g}''$ this common basis of each $\pi_k\mathcal{K}''_o$. $\pi_k\mathcal{K}''_o$ is a subspace of $\pi_k\mathcal{X}_2$, which is itself isomorphic to $\ell_2((-\infty,\infty))$ — for ease of discussion and notation we just identify these two spaces.

Let $\mathbf{P}_k$ be the k-th snapshot of the projection operator $\mathbf{P}$, *i.e.*, a diagonal matrix with $[\mathbf{P}_k]_{i,i} = 0$ for $i < k$ and I for $i \geq k$, as defined in equation (4.2). We know already that for all k, the rows of $\mathbf{g}''\mathbf{P}_k$ are contained in $\pi_k\mathcal{H}_o(V)$. This observation leads to the following theorem.

Theorem 7.9 *Suppose that T is uniformly locally finite with the upper bound on the dimension given by some integer δ, then $\mathbf{g}''$ is finite dimensional.*

PROOF From proposition 6.18, we have that for all k

$$\mathbf{g}''\mathbf{P}_k = (\pi_k\mathcal{K}''_o)\mathbf{P}_k \subset \pi_k\mathcal{H}_o.$$

For the purpose of establishing a contradiction, suppose now that $\pi_k\mathcal{K}''_o$ would not be of finite dimension. Let, for $\ell > \delta$, $\{f_1, \cdots, f_\ell\}$ be orthonormal basis vectors of $\pi_k\mathcal{K}''_o$ (they are basis vectors of any slice of $\mathcal{K}''_o$, since all the slices are equal). Let ε be a positive number much smaller than 1, and choose k close enough to $-\infty$ so that

$$\forall f_n \in \{f_1, \cdots, f_\ell\}: \qquad \|f_n\mathbf{P}_k - f_n\| < \varepsilon.$$

Then $\{f_n\mathbf{P}_k\}$ also form a basis, which for small ε is almost orthonormal, and they are contained in $\pi_k\mathcal{H}_o$. This contradicts the assumption that the dimension of $\pi_k\mathcal{H}_o$ is less than δ. Hence, $\pi_k\mathcal{K}''_o$ cannot have more than δ basis vectors. □

Locally finite systems with compact support

In the remainder of this section we specialize to the case where

(A) the system T has a u.e. stable realization and becomes an LTI system represented by $T_{-\infty}$ when the time index $k \to -\infty$. We assume moreover that we know an initial point Y_{k_0} of the backward recursion (7.24) which governs the computation of the outer-inner factorization (this would *e.g.*, be the case if the system would also be LTI for $k \to +\infty$);

[7] Note: this definition is a generalization of our earlier use of $\mathcal{K}''_o$ in proposition 6.18, since now we do not assume that $\ker(\cdot V^*)|_{\mathcal{U}_2} = \{0\}$. The extension $W = \left[\begin{smallmatrix} V \\ U \end{smallmatrix}\right]$ absorbs this subspace.

(B) $\ker(\cdot T^*)|_{\mathcal{U}_2} = \{0\}$;

(C) $\ker(\cdot T^*_{-\infty})|_{\mathcal{U}_2} = \{0\}$, too.

(The more general case can be done just as well but leads to a considerably more detailed and technical development which we wish to avoid here).

Let $T_{-\infty} = [T_{-\infty}]_\ell V_{-\infty}$ be the outer-inner factorization of $T_{-\infty}$ and let $\mathcal{H}_o(V_{-\infty})$ be the corresponding observability space. We try to find the relation between $\pi_k \mathcal{H}_o(V)$ and $\pi_k \mathcal{H}_o(V_{-\infty})$ when $k \to -\infty$. Let $\mathbf{G}$ be an orthonormal sliced basis representation of $\mathcal{H}_o(V)$. The defining properties for $f \in \pi_k \mathcal{H}_o(V)$ can be formulated as

(1) $f = f\mathbf{P}_k$ — causality;

(2) $f \perp \operatorname{ran}(\cdot \mathbf{P}_k T)$.

(2) is a direct consequence of the relation $\mathbf{G}T^* \in \mathcal{L}_2 Z^*$, and sufficient as defining relation for $\mathbf{G}$ because of hypothesis (B). Likewise, we have that $f' \in \pi_k \mathcal{H}_o(V_{-\infty})$ if and only if $f' = f'\mathbf{P}_k$ *and* $f' \perp \operatorname{span}(\mathbf{P}_k T_{-\infty})$, this time because of hypothesis (C).

Because of proposition 6.18, we know that the space

$$\mathcal{H}_{2k} := \pi_k \mathbf{P}(\mathcal{K}''_o)$$

is contained in $\pi_k \mathcal{H}_o(V)$. We also know from theorem 7.9 that the space $\pi_k \mathcal{K}''_o$ is finite dimensional and independent of k. Hence, in the limit for very small k, the space $\pi_k \mathcal{K}''_o$ itself is contained in $\pi_k \mathcal{H}_o(V)$. The question is: what else is in $\pi_k \mathcal{H}_o(V)$? A strong candidate is $\pi_k \mathcal{H}_o(V_{-\infty})$, or at least a space close to it, since for small k, T is behaving like $T_{-\infty}$. We claim that the space $\pi_k \mathcal{H}_o(V_{-\infty})$ is in fact *nearly* orthogonal to the set of row vectors $\{\mathbf{P}_k T\}$. The near orthogonality will become better for smaller k.

Based on that fact it seems but a small step to look for a subspace close to $\pi_k \mathcal{H}_o(V_{-\infty})$ as orthogonal complement for $\pi_k \mathbf{P}(\mathcal{K}''_o)$ in $\pi_k \mathcal{H}_o(V)$ for $k \to -\infty$. Let ε be a positive number much smaller than 1. We say that two subspaces $\mathcal{S}_1$ and $\mathcal{S}_2$ are ε-close to each other (and we write $\mathcal{S}_1 \overset{\varepsilon}{\approx} \mathcal{S}_2$) if, for the respective orthonormal projectors $\mathbf{P}_{\mathcal{S}_1}$ and $\mathbf{P}_{\mathcal{S}_2}$, $\|\mathbf{P}_{\mathcal{S}_1} - \mathbf{P}_{\mathcal{S}_2}\| \le \varepsilon$. In particular, $\mathcal{S}_1$ and $\mathcal{S}_2$ must then necessarily have the same dimensions, and the maximum angle between them must be of the order of ε.

The near orthogonality of $\pi_k \mathcal{H}_o(V_{-\infty})$ on the vectors which define the orthogonal complement of $\pi_k \mathcal{H}_o(V)$ does not guarantee the existence of a subspace in $\pi_k \mathcal{H}_o(V)$ that is actually orthogonal to the space defined by those vectors, namely $\operatorname{ran}(\cdot \mathbf{P}_k T)$. For example, in three dimensional space, $[0,0,1]$ will become nearly orthogonal on the collection $[1,0,0]$, $[0,1,0]$, $[0,0,\varepsilon]$ when $\varepsilon \to 0$. but the span of the latter three vectors remains the whole space. It turns out that $\pi_k \mathcal{H}_o(V)$ *may* contain a subspace, which we will call $\mathcal{H}_{1k}$,

$$\mathcal{H}_{1k} := \pi_k \mathcal{H}_o(V) \ominus \pi_k \mathbf{P}(\mathcal{K}''_o),$$

that is ε-close to $\pi_k \mathcal{H}_o(V_{-\infty})$, and orthogonal to $\pi_k \mathcal{K}''_o$. $\mathcal{H}_{1k}$ may have any dimension between zero and the dimension of $\pi_k \mathcal{H}_o(V_{-\infty})$, depending on the actual data, but cannot be larger. In the following theorem we show that there can be *nothing else* in $\pi_k \mathcal{H}_o(V)$, and this fact will allow us to study the convergence of the recursion for Y in equation (7.24).

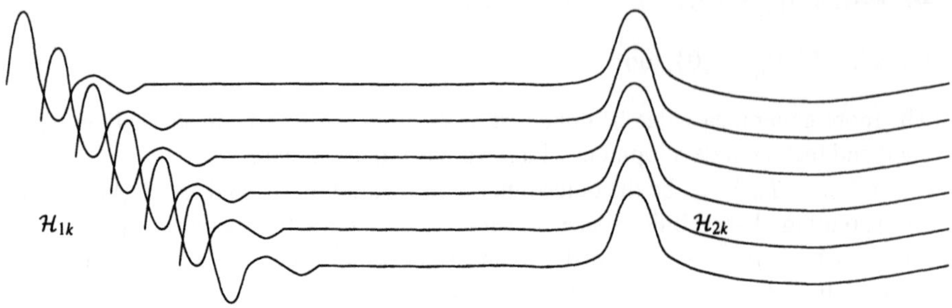

Figure 7.4. The evolution of the left zero structure when $k \to -\infty$.

Figure 7.4 depicts the situation. In our standard example T, equation (7.9), we see that $\mathcal{H}_{1k}$ is actually empty (as expected since $\mathcal{H}_o(V_{-\infty}) = \{0\}$), while a basis vector for $\pi_k \mathbf{P}(\mathcal{K}_o'')$ is given by

$$[\cdots\ 0\ 0\ 2^k\ \cdots\ \tfrac{1}{2}\ \boxed{1}\ \tfrac{1}{2}\ \tfrac{1}{4}\ \cdots].$$

We give another, opposite example at the end of this section in which both $\mathcal{H}_{1k}$ and $\mathcal{H}_{2k}$ are trivial, while $\pi_k \mathcal{H}_o(V_{-\infty})$ is not.

Theorem 7.10 *Let T be a transfer operator satisfying the hypotheses (A)–(C) above and ε a positive number much smaller than 1. Then there exists an integer k_1 which is such that for all $k < k_1$, $\pi_k \mathcal{H}_o(V) = \mathcal{H}_{1k} \oplus \mathcal{H}_{2k}$, where $\mathcal{H}_{1k} \overset{\varepsilon}{\subset} \pi_k \mathcal{H}_o(V_{-\infty})$ and $\mathcal{H}_{2k} := \pi_k \mathbf{P}(\mathcal{K}_o'') \overset{\varepsilon}{\approx} \pi_k \mathcal{K}_o''$. In particular,*

$$dim(\pi_k \mathcal{H}_o(V)) \leq dim(\pi_k \mathcal{H}_o(V_{-\infty})) + dim(\pi_k \mathcal{K}_o''),$$

for all k. $\mathcal{H}_{1k}$ is dependent on k only in the sense that all $\mathcal{H}_{1k} Z^{-k}$ are ε-close to each other, while alle H_{2k} are ε-close to each other.

PROOF The proof is based on the construction of the index k_1 in the $-\infty$ LTI zone chosen small enough so that there exists a space $\mathcal{H}_{1k} \subset \mathcal{H}_o(V)$ nearly contained in $\mathcal{H}_o(V_{-\infty})$ and orthogonal to $\mathcal{H}_{2k} = \pi_k \mathbf{P}(\mathcal{K}_o'')$, and such that $\mathcal{H}_o(V) = \mathcal{H}_{1k} \oplus \mathcal{H}_{2k}$. For that purpose we select a collection of four integers $k_1 < \cdots < k_4$ as follows (see figure 7.5).

1. let k_4 be such that T has an LTI realization for all $k < k_4$;

2. choose $k_3 < k_4$ such that for all $k < k_3$, T is essentially equal to $T_{-\infty}$, meaning that $\|\pi_k T - \pi_k T_{-\infty}\| < \varepsilon$, *and* $\mathcal{K}_o''$ is essentially concentrated on $[k_3, \infty)$, meaning that for all $f \in \pi_k \mathcal{K}_o''$, $\|f - f\mathbf{P}_{k_3}\| < \varepsilon$. Such a k_3 can always be found, because of proposition 6.18 and the assumption that T has a u.e. stable realization;

3. choose next an interval-size K which essentially supports (that is within ε) the functions in $\pi_0 \mathcal{H}_o(V_{-\infty})$. We have, under the assumption of local finiteness, that for any

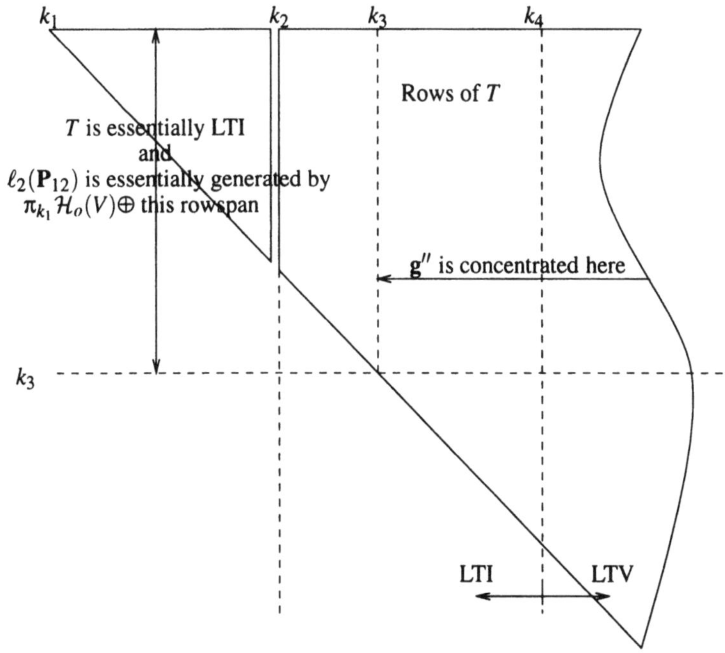

Figure 7.5. The definition of the various indices relevant to the proof of theorem 7.10.

k, the $\pi_k\mathcal{H}_o(V_{-\infty})$ are finite dimensional subspaces of $\ell_2([k,\infty))$, which are, moreover, shifted versions of each other in the sense that

$$\pi_\ell\mathcal{H}_o(V_{-\infty}) = \pi_k\mathcal{H}_o(V_{-\infty})Z^{[\ell-k]}.$$

We pick K large enough to insure that the constituent functions of $\pi_0\mathcal{H}_o(V_{-\infty})$ essentially vanish outside the interval $[0,K)$. Because of time-invariance such an interval can be used at other time points than 0 as well.

4. Then choose $k_1 < k_3 - K$ such that $\ell_2([k_1,\infty))(\mathbf{P}_{k_1} - \mathbf{P}_{(k_1+K)})$ (which is isomorphic to $\ell_2([k_1,k_1+K)))$ is essentially generated by $\mathrm{ran}(\cdot(\mathbf{P}_{k_1} - \mathbf{P}_{k_3})T_{-\infty}) \oplus \mathcal{H}_o(V_{-\infty})$. Such a K can be found because of the hypothesis that $\ker(\cdot T_{-\infty}^*)\big|_{\mathcal{U}_2} = \{0\}$ so that actually

$$\ell_2([k_1,\infty)) = \pi_{k_1}\mathcal{H}_o(V_{-\infty}) \oplus \overline{\mathrm{ran}}(\cdot(\mathbf{P}_{k_1}T_{-\infty}))\,, \tag{7.38}$$

itself a specialization of the relation $\mathcal{U}_2 = \mathcal{H}_o(V_{-\infty}) \oplus \overline{(\mathcal{U}_2 T)}$. If we take the span in the second member of (7.38) large enough, we can nearly generate any given finite dimensional subspace of $\ell_2([k_1,\infty))$, and in particular $\ell_2([k_1,\infty))(\mathbf{P}_{k_1} - \mathbf{P}_{k_1+K})$.

5. Put $k_2 = k_1 + K$.

The proof now runs in steps.

- We start out by remarking that $\pi_{k_1}\mathcal{H}_o(V_{-\infty})$ is nearly orthogonal to the vectors in $\{\mathbf{P}_{k_1}T\}$. The reason is that if $f \in \pi_{k_1}\mathcal{H}_o(V_{-\infty})$ with $\|f\| = 1$, then because of the choice of k_1 and k_2, f is essentially (*i.e.*, within norm ε) supported on the interval $[k_1,k_2)$. Since now $\{\mathbf{P}_{k_1}T_{-\infty}\mathbf{P}_{12}\} = \{\mathbf{P}_{k_1}T\mathbf{P}_{12}\}$, we have that $f \stackrel{\varepsilon}{\perp} \{\mathbf{P}_{k_1}T\}$.

- Now we show that there is nothing more in $\pi_{k_1}\mathcal{H}_o(V)$ than the orthogonal sum of a space[8] $\mathcal{H}_1$, which is essentially contained in $\pi_{k_1}\mathcal{H}_o(V_{-\infty})$ and hence essentially supported on the interval $[k_1,k_2)$, and the space $\mathcal{H}_2 = \pi_{k_1}\mathbf{P}(\mathcal{K}_o'') = (\pi_{k_1}\mathcal{K}_o'')\mathbf{P}_{k_1}$ which is ε-close to $\pi_{k_1}\mathcal{K}_o''$ and whose functions are (uniformly) essentially supported on the interval $[k_3,\infty)$. We know already that $\mathcal{H}_2$ is ε-orthogonal to the row vectors in $\{T\mathbf{P}_{k_2}\}$, by definition of $\mathcal{K}_o''$ and k_3. A converse statement is true as well: if $f \in \ell_2([k_2,\infty))$ is such that $\|f\| = 1$ *and* $\forall g : g \in \text{ran}(\cdot T\mathbf{P}_{k_2})$ with $\|g\| = 1, |(f,g)| < \varepsilon$, then f is ε-close to $\pi_{k_1}\mathcal{K}_o''$, for the extension $[\cdots\ 0\ \ f]$ of f from $\ell_2([k_2,\infty))$ to $\ell_2((-\infty,\infty))$ will be nearly orthogonal to $\text{ran}(\cdot T)$, and hence be ε-close to $\mathcal{K}_o''$, which is by definition the orthogonal complement of $\text{ran}(\cdot T)$, under the assumption that $\ker(\cdot T^*)|_{\mathcal{U}_2} = \{0\}$.

  Suppose now that $f \in \pi_{k_1}\mathcal{H}_0(V)$, $\|f\| = 1$ and that f is nearly orthogonal on $\mathcal{H}_o(V_{-\infty})$. We show that f is then nearly contained in $\mathcal{H}_2$. This we do by showing that f is essentially supported on the interval $[k_2,\infty)$ (*i.e.*, $\|f - f\mathbf{P}_{k_2}\| < \varepsilon$), since we have just shown that it will then nearly belong to $\mathcal{H}_2$. But $\ell_2([k_1,\infty))\,(\mathbf{P}_{k_1} - \mathbf{P}_{k_2})$ is ε-close to $\pi_{k_1}\mathcal{H}_o(V_{-\infty}) \oplus \text{ran}(\cdot(\mathbf{P}_{k_1} - \mathbf{P}_{k_3})T_{-\infty})$ by construction, and since $\text{ran}(\cdot(\mathbf{P}_{k_1} - \mathbf{P}_{k_3})T_{-\infty})$ is ε-close to $\text{ran}(\cdot(\mathbf{P}_{k_1} - \mathbf{P}_{k_3})T)$, also by construction, we have that f is ε-orthogonal to $\ell_2([k_1,\infty))\,(\mathbf{P}_{k_1} - \mathbf{P}_{k_2})$. It follows that the support of f is essentially on $[k_2,\infty)$ and hence that f is nearly in $\mathcal{H}_2$. It follows that any $f \in \mathcal{H}_o(V)$ can be decomposed as $f = f_1 + f_2$ with $f_1 \stackrel{\varepsilon}{\in} \mathcal{H}_o(V_{-\infty})$ and $f_2 \in \mathcal{H}_2$, and we have that

$$\pi_{k_1}\mathcal{H}_o(V) \stackrel{\varepsilon}{\subset} \pi_{k_1}\mathcal{H}_o(V_{-\infty}) + \mathcal{H}_2,$$

  in which the two spaces on the right hand side are nearly orthogonal. Let now $\mathcal{H}_1 = \pi_{k_1}\mathcal{H}_o(V) \ominus \mathcal{H}_2$, then it follows that $\mathcal{H}_1$ must be nearly contained in $\pi_{k_1}(V_{-\infty})$ as claimed. The considerations so far are equally valid if k_1 is replaced by any $k < k_1$ and k_2 by $k + K$, leading to the definition of spaces $\mathcal{H}_{1k}$ and $\mathcal{H}_{2k}$ which are such that the support of functions in the first is essentially on the interval $[k, k+K)$ and of the second on $[k_3,\infty)$.

- Finally, the statement that for $k < k_1$, $\mathcal{H}_{1k}$ is dependent on k only in the sense that all $\mathcal{H}_{1k}Z^{-k}$ are ε-close to each other follows from the fact that if $f \in \ell_2([k,\infty))$ is such that $\|f\| = 1$, $f \perp \text{ran}(\cdot \mathbf{P}_k T)$ and $\text{support}(f) \stackrel{\varepsilon}{\subset} [k, k+K)$, then there exists an $f_1 \in \pi_k\mathcal{H}_o(V_{-\infty})$ such that $\|f - f_1\| < \varepsilon$, which in turn entails that for all integers $\ell \geq 1$, $f_1 Z^{-\ell}$ will be nearly orthogonal on $\text{ran}(\cdot \mathbf{P}_{k-\ell}T)$ so that $\mathcal{H}_{1(k-\ell)}$ will be nearly equal to $\mathcal{H}_{1k}Z^{-\ell}$, by an argument akin to the one used in the proof of theorem 7.9. The stability statement for $\mathcal{H}_{2k}$ follows directly from the fact that $\mathcal{H}_{2k} \stackrel{\varepsilon}{\approx} \pi_k\mathcal{K}_o''$ for all $k \leq k_1$. □

[8] We drop the index k_1 in the definition of the spaces $\mathcal{H}_1$ and $\mathcal{H}_2$ for temporary convenience.

From theorem 7.10 it follows (at least under the hypotheses (A)–(C) stated) that the operator Y which plays a central role in the recursion (7.24) is such that for $k \to -\infty$, Y_k can be forced to have a limit, and that this limit has a form which characterizes the two important spaces $\mathcal{H}_{1k}$ and $\mathcal{H}_{2k}$. The recursion in (7.24) determines the Y_k only up to left unitary equivalence. On the other hand, we know from the definition of Y in equation (7.23) that $Y = \mathbf{P}_0(\mathbf{G}\mathbf{F}_{oT}^*)$, where $\mathbf{G}$ is the orthonormal sliced basis of $\mathcal{H}_o(V)$ and $\mathbf{F}_{oT}$ is the sliced basis for the output state space of T. This specializes to $Y_k = (\pi_k\mathbf{G})(\pi_k\mathbf{F}_{oT})^*$. Hence, if the recursion is arranged in such a way that for very small k,

1. $\pi_k\mathbf{F}_{oT}$ is essentially LTI, while
2. the basis in $\pi_k\mathbf{G}$ decomposes into $\pi_k\mathbf{G}_1$ which is essentially LTI and generates $\mathcal{H}_{1k}$, and $\pi_k\mathbf{G}_2$ which is essentially constant and generates $\pi_k\mathcal{K}_o''$ — see figure 7.4,

then we see that Y_k decomposes in Y_{1k} and Y_{2k} so that

1. $Y_{1k} = (\pi_k\mathbf{G}_1)(\pi_k\mathbf{F}_{oT})^*$ which becomes essentially LTI and converges to a constant matrix with zero (left) kernel, and
2. $Y_{2k} = (\pi_k\mathbf{G}_2)(\pi_k\mathbf{F}_{oT})^*$ which converges to zero, since the support of $\pi_k\mathbf{F}_{oT}$, which is LTI, shifts out of the support of $\pi_k\mathbf{G}_2$, which is constant.

The phenomenom is easily observed in our standard examples. For T as in (7.4), (7.33) we had for $k \geq 1$, $Y_k = \sqrt{3}$ while for $k \to -\infty$, $Y_k \to 0$. Hence, for small k, $\mathcal{H}_{1k} = \{0\}$ and $\mathcal{H}_{2k}$ is one-dimensional and generated by $[\cdots\ \tfrac{1}{4}\ \tfrac{1}{2}\ \boxed{1}\ \tfrac{1}{2}\ \tfrac{1}{4}\ \cdots]$. Dually, suppose that we had tried to find an outer-inner factorization of

$$T = \left[\begin{array}{ccc|ccccc} \ddots & \ddots & & & & & & \\ & 1 & -2 & & & \mathbf{0} & & \\ & & 1 & -2 & & & & \\ \hline & & & \boxed{1} & -1/2 & & & \\ & & & & 1 & -1/2 & & \\ & \mathbf{0} & & & & 1 & -1/2 & \\ & & & & & & 1 & \ddots \\ & & & & & & & \ddots \end{array}\right]. \qquad (7.39)$$

We would have found $Y_k = [\cdot]$ for all k, even for $k \to -\infty$. In this case, T itself is left-outer, it has a bounded left-inverse. Although $\mathcal{H}_o(V_{-\infty})$ is non trivial, we see that $\mathcal{H}_{1k}$ actually is, showing by example that although $\mathcal{H}_o(V_{-\infty})$ is ε-orthogonal on $\mathrm{span}(\mathbf{P}_k T)$, the latter just generates the whole space $\ell_2([k,\infty))$ for all k.

To conclude this section, we make somewhat more general statements on the existence of non-trivial defect spaces such as $\mathcal{K}_o''$. Let us look at the case where the state dimension of the system is the same for all k, all D_k in the state representation for T are square and invertible, and the system is LTI for both $k \to \infty$ and $k \to -\infty$. Let us call these two LTI systems $T_{-\infty}$ and T_∞ respectively. Generalizing what was said in the preceeding paragraph, we can state (and prove easily from the recursion for Y) that if T_∞ is minimal phase, then for all k, $Y_k = [\cdot]$, so that the system is left-outer independently of $T_{-\infty}$. If T_∞ is not minimal phase, then the system is certainly not left-outer and Y_k will be non-trivial, and its behaviour for $k \to \infty$ will be more interesting. If, in that case, $T_{-\infty}$ is minimal phase, then there will be a defect space $\mathcal{K}_o''$ of dimension equal to the degree of non-minimality of T_∞. Let the degrees of non-minimality of $T_{\pm\infty}$ be $\delta_{\pm\infty}$ respectively, then, if $\delta_{-\infty} \leq \delta_\infty$, there will be a defect space of dimension at least $\delta_\infty - \delta_{-\infty}$ and at most δ_∞. We believe that there are examples for any intermediate case. Further classifications and the relation between the right and the left inner-outer factorization of a given operator T merits further study!

7.6 NOTES

The inversion of an operator is of course a central problem in functional analysis, and much theory has been developed for it. Particularly relevant to our case is the formulation in terms of a nest algebra, for which factorization and inversion results have been derived a.o. by Arveson [Arv75], The key ingredient is the inner-outer factorization, also treated in Arveson's paper. However, a more elementary and concrete treatment is based on the classical Beurling-Lax theory, as seen in chapters 6 and 7. In the LTI case, the connection of the inner-outer factorization to the Riccati equation is well known. A parallel treatment of the LTV case can be found in chapter 3 of the book of Halanay and Ionescu [HI94], where the inner-outer factorization is treated as an application of Kalman-Szegö-Popov-Yakubovich systems.

The time-varying inner-outer factorization also provides for a splitting into causal (upper) and anti-causal (lower) parts: a dichotomy. This point has been investigated by Gohberg and co-workers [GKvS84, BAGK94].

II INTERPOLATION AND APPROXIMATION

8 J-UNITARY OPERATORS

J-unitary operators and their siblings, symplectic operators, play an important role in physics and mathematics. Aside from the fact that they describe a physically interesting situation, they are instrumental in interpolation and approximation theory as well. The physical motivation is found in *lossless scattering theory*, which gives an operator description of wave propagation and reflection. An introduction to this is given in section 8.1. We saw in the previous chapters that reachability and observability spaces are instrumental in the realization theory of operators in general. In the case of J-unitary operators these spaces turn out to be rather special, with interesting geometrical properties (sections 8.2 and 8.4).

An important special case of J-unitary or J-isometric operators are those which give a chain description of a lossless scattering operator. We give characterizations of such operators, also for the case where they are of mixed causality. Such operators have been extensively utilized in the H_∞ control literature.

The chapter forms an introduction to chapter 9 in which a number of classical interpolation problems are brought into the general context of time varying systems, and to chapter 10 on optimal approximation of transfer operators and model reduction.

8.1 SCATTERING OPERATORS

Passive media

Let us consider a set of waves impinging onto a physical medium. In general, the medium scatters the waves and even reflects part of the energy back towards the source

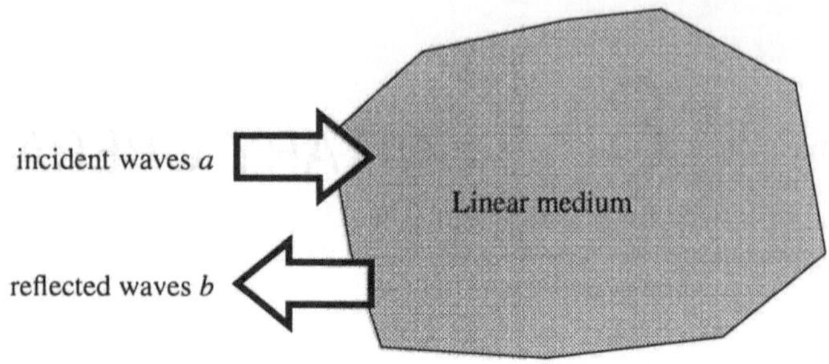

Figure 8.1. Scattering at a linear passive (physical) medium.

as well. We shall restrict our interest to media that are linear, and connected to the outside world with just a finite set of ports (figure 8.1).

Linearity implies, among others, that there is no "harmonic distortion" in the scattering process (no transfer of energy between frequencies), while the assumption of a finite set of ports means that the interaction of the medium with the outside world happens through a finite set of input signals, which are scattered by the medium and transferred to a finite set of output signals. The first set we call *incident waves* while the scattered set consists of the *reflected waves.* We are interested in the energy that the input signals have introduced into the medium at some point in time, and the energy of the signals that flow out of the medium.

With reference to figure 8.1, let the incident wave consist of a sequence $a = [a_k]$. We define the energy brought into the medium from $k = -\infty$ up to and including time t as

$$\mathcal{E}(a,t) = \sum_{k=-\infty}^{t} a_k a_k^* = \sum_{k=-\infty}^{t} \|a_k\|^2 .$$

Similarly, the energy taken out from the medium from $k = -\infty$ up to and including t is

$$\mathcal{E}(b,t) = \sum_{k=-\infty}^{t} b_k b_k^* = \sum_{k=-\infty}^{t} \|b_k\|^2 .$$

The net balance of energy the medium has absorbed from the outside world from creation to time t then becomes

$$\mathcal{E}(t) = \sum_{k=-\infty}^{t} (a_k a_k^* - b_k b_k^*),$$

i.e., the difference between incident and reflected energy.

The incident vector a_k belongs, at each point k, to a finite vector space $\mathcal{M}_k$ while the reflected vector belongs to a (possibly different) space $\mathcal{N}_k$. It is standard practice, usually not limiting, to restrict incident and reflected waves to spaces $\ell_2^{\mathcal{M}}$ and $\ell_2^{\mathcal{N}}$. We shall do so, unless indicated otherwise.

A medium is called passive if it does not contain a source of energy. As stated, the notion is imprecise (for just the physical existence of the medium makes it contain energy). We replace the definition by a more precise, instrumental one which only holds for the linear case. We say that a linear medium is *passive* if for all t and all incident waves with finite energy, the overall reflected energy up to t is smaller than or at most equal to the overall incident energy up to that point in time: $\mathcal{E}(b,t) \le \mathcal{E}(a,t)$. The medium is said to be *lossless* if it is passive and, in addition, for all $a \in \ell_2^{\mathcal{M}}$, $\mathcal{E}(b,\infty) = \mathcal{E}(a,\infty)$. In that case, all incident energy eventually gets reflected (presuming it is finite).

We can always characterize a passive medium by a *scattering operator* which maps the incident wave a to the reflected wave b, and which has the additional property of *causality*. With the basic assumptions of the previous paragraph, this operator must always exist, because

1. if $a_k = 0$ for $-\infty < k \le t$, then $b_k = 0$ for $-\infty < k \le t$ as well, since at any point in time the net absorbed energy $\mathcal{E}(t)$ must be non-negative. This defines causality, *viz.* definition 3.1.

2. The relation (a,b) between incident and reflected waves must be *univocal*: suppose (a,b) and (a,b') are two compatible input-output wave pairs, then by linearity also $(0,b-b')$ must be compatible, and by the passivity assumptions, necessarily for all k, $b_k = b'_k$, *i.e.*, $b \equiv b'$. Hence the relation (a,b) is an operator well defined on an acceptable input space of incident waves.

An additional assumption allows the whole of $\ell_2^{\mathcal{M}}$ as space of incident waves: we say that any $a \in \ell_2^{\mathcal{M}}$ is allowable as incident wave *(solvability assumption)*. From the preceding discussion, we derive the scattering operator $\mathcal{S}$ as

$$\mathcal{S}: \quad \ell_2^{\mathcal{M}} \to \ell_2^{\mathcal{N}}: \quad b = \mathcal{S}(a).$$

The solvability assumption is merely technical: it makes the mathematics work and is reasonably harmless since it could have been obtained by closure on finite input sequences—a tedious procedure which brings no new insights.

The matrix calculus which we described in the earlier chapters allows us now to write

$$b = aS$$

where passivity implies that S is a bounded operator: $S \in \mathcal{X}$, and causality even gives $S \in \mathcal{U}$. The passivity assumption also provides the inequality

$$I - SS^* \ge 0, \tag{8.1}$$

whereas if S is lossless,

$$I - SS^* = 0, \tag{8.2}$$

i.e., S is *isometric*. If, in addition, S^* (which is an anticausal operator) is isometric as well, we have

$$SS^* = I, \qquad S^*S = I \tag{8.3}$$

and S is unitary. We have seen in section 6.4 that a locally finite isometric operator S can often be completed to a unitary one by the addition of a well chosen set of outputs.

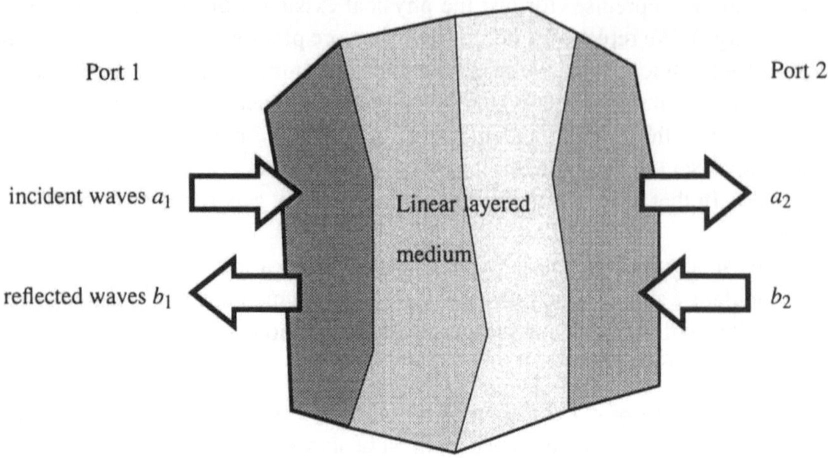

Figure 8.2. Scattering at a layered physical medium.

Layered physical media

Next, consider a layered physical medium for which the inputs and outputs are each partitioned into two sets. The first set we call, for convenience, *port 1*, and the second set *port 2* — see figure 8.2. We split the incident waves accordingly into two sets: $a_1 \in \ell_2^{\mathcal{M}_+}$ to port 1 and $b_2 \in \ell_2^{\mathcal{N}_-}$ to port 2, and the reflected waves $b_1 \in \ell_2^{\mathcal{M}_-}$ and $a_2 \in \ell_2^{\mathcal{N}_+}$. (The "+" subscript goes with the energy transport from left to right, while "−" goes with energy from right to left.) The total energy absorbed up to and including time t by the medium at port 1 is now given by

$$\mathcal{E}_1(t) = \sum_{k=-\infty}^{t} (a_{1,k}a_{1,k}^* - b_{1,k}b_{1,k}^*)$$

and at $t = \infty$ by

$$\mathcal{E}_1(\infty) = [a_1 \quad b_1] \begin{bmatrix} I_{\mathcal{M}_+} & \\ & -I_{\mathcal{M}_-} \end{bmatrix} \begin{bmatrix} a_1^* \\ b_1^* \end{bmatrix}.$$

At port 2, we have

$$\begin{aligned} \mathcal{E}_2(t) &= \sum_{k=-\infty}^{t} (b_{2,k}b_{2,k}^* - a_{2,k}a_{2,k}^*) \\ \mathcal{E}_2(\infty) &= [b_2 \quad a_2] \begin{bmatrix} I_{\mathcal{N}_-} & \\ & -I_{\mathcal{N}_+} \end{bmatrix} \begin{bmatrix} b_2^* \\ a_2^* \end{bmatrix}. \end{aligned}$$

$\mathcal{E}_1(\infty)$ and $\mathcal{E}_2(\infty)$ are expressed in terms of a non-definite inner product characterized by the *signature matrices*

$$J_1 = \begin{bmatrix} I_{\mathcal{M}_+} & \\ & -I_{\mathcal{M}_-} \end{bmatrix}, \quad J_2 = \begin{bmatrix} I_{\mathcal{N}_+} & \\ & -I_{\mathcal{N}_-} \end{bmatrix}.$$

We saw in the previous section that a passive layered medium possesses a causal and contractive scattering operator $\Sigma \in \mathcal{U}$,

$$\begin{array}{l} \Sigma : \ell_2^{\mathcal{M}_+} \times \ell_2^{\mathcal{N}_-} \to \ell_2^{\mathcal{N}_+} \times \ell_2^{\mathcal{M}_-} : \\ [a_2 \quad b_1] = [a_1 \quad b_2] \begin{bmatrix} \Sigma_{11} & \Sigma_{12} \\ \Sigma_{21} & \Sigma_{22} \end{bmatrix} . \end{array} \tag{8.4}$$

It may also happen that the map $[a_1 \quad b_1] \mapsto [a_2 \quad b_2]$ from waves at port 1 to waves at port 2 exists. In that case we say that the medium possesses a *chain scattering operator*. It is commonly denoted with the symbol Θ:

$$\begin{array}{l} \Theta : \ell_2^{\mathcal{M}_+} \times \ell_2^{\mathcal{M}_-} \to \ell_2^{\mathcal{N}_+} \times \ell_2^{\mathcal{N}_-} : \\ [a_2 \quad b_2] = [a_1 \quad b_1] \begin{bmatrix} \Theta_{11} & \Theta_{12} \\ \Theta_{21} & \Theta_{22} \end{bmatrix} . \end{array} \tag{8.5}$$

Since (8.4) and (8.5) describe the same linear relations between a_1, a_2, b_1, b_2, they are connected. In particular, if Σ_{22} is boundedly invertible, then Θ exists as a bounded operator and is given by

$$\Theta = \begin{bmatrix} \Sigma_{11} - \Sigma_{12}\Sigma_{22}^{-1}\Sigma_{21} & -\Sigma_{12}\Sigma_{22}^{-1} \\ \Sigma_{22}^{-1}\Sigma_{21} & \Sigma_{22}^{-1} \end{bmatrix} .$$

Conversely, if Θ is known and Θ_{22}^{-1} is a meaningful operator, then Σ is given by

$$\Sigma = \begin{bmatrix} \Theta_{11} - \Theta_{12}\Theta_{22}^{-1}\Theta_{21} & -\Theta_{12}\Theta_{22}^{-1} \\ \Theta_{22}^{-1}\Theta_{21} & \Theta_{22}^{-1} \end{bmatrix} .$$

Definition 8.1 *A bounded operator $\Theta \in \mathcal{X}$ is a*

- $(J_{\mathcal{M}}, J_{\mathcal{N}})$-isometry *if $\Theta J_{\mathcal{N}} \Theta^* = J_{\mathcal{M}}$,*
- $(J_{\mathcal{N}}, J_{\mathcal{M}})$-coisometry *if $\Theta^* J_{\mathcal{M}} \Theta = J_{\mathcal{N}}$,*
- $(J_{\mathcal{M}}, J_{\mathcal{N}})$-unitary *if both $\Theta J_{\mathcal{N}} \Theta^* = J_{\mathcal{M}}$ and $\Theta^* J_{\mathcal{M}} \Theta = J_{\mathcal{N}}$.*

If Θ is bounded and $(J_{\mathcal{M}}, J_{\mathcal{N}})$-unitary, then Θ^{-1} is bounded as well and given by

$$\Theta^{-1} = J_{\mathcal{N}} \Theta^* J_{\mathcal{M}} .$$

- *Σ is called* inner *if it is causal and unitary,*[1]
- *Σ is* lossless *if it is causal and isometric, $\Sigma\Sigma^* = I$.*

[1] See the extensive discussion on inner operators in the previous chapter. The definition is in line with the notion of "inner" in the theory of matrix analytic functions as given by [Hel64], and does not follow *e.g.*, [FM96a] which does not require unitarity, only isometry. The two definitions cannot be reconciled, because a causal and isometric operator does not necessarily have a causal unitary extension, as shown by the counterexamples in the previous chapter.

Suppose that the scattering operator Σ of a layered medium is inner, and that the corresponding chain scattering operator Θ exists. This will be the case if Σ_{22}^{-1} exists, let us say, as a bounded operator.

- Θ *is* J-inner *[J-lossless] if it is J-unitary [J-isometric], and the corresponding* Σ *exists and is* causal.

A J-inner Θ need not be causal, but it will be J-unitary. It is causal only if Σ_{22}^{-1} is causal.

Similarly, it may be that Θ is J-unitary and that the corresponding Σ exists (which is the case when Θ_{22}^{-1} exists), but such that the corresponding Σ is not causal. Θ is then merely J-unitary, but not J-inner, and it does not correspond to a "physical" scattering system. We will see later that the various cases on Σ and Θ all are of interest.

The following theorem gives a number of properties of the connections between Θ and Σ, where it is only assumed that Θ is J-unitary (not necessarily J-inner).

Theorem 8.2 *Let* $\Theta \in \mathcal{X}(\mathcal{M},\mathcal{N})$ *be a* $(J_{\mathcal{M}},J_{\mathcal{N}})$*-unitary operator with partitioning (8.5). Then*

1. Θ_{22}^{-1} *exists and is bounded,*
2. $\|\Theta_{22}^{-1}\| \le 1$, $\|\Theta_{22}^{-1}\Theta_{21}\| < 1$, $\|\Theta_{12}\Theta_{22}^{-1}\| < 1$.
3. *The corresponding scattering operator* Σ *exists, is unitary, and given by*

$$\begin{aligned}\Sigma &= \begin{bmatrix} \Theta_{11}-\Theta_{12}\Theta_{22}^{-1}\Theta_{21} & -\Theta_{12}\Theta_{22}^{-1} \\ \Theta_{22}^{-1}\Theta_{21} & \Theta_{22}^{-1}\end{bmatrix} \\ &= \begin{bmatrix} \Theta_{11}^{-*} & -\Theta_{12}\Theta_{22}^{-1} \\ \Theta_{12}^{*}\Theta_{11}^{-*} & \Theta_{22}^{-1}\end{bmatrix} \\ &= \begin{bmatrix} \Theta_{11}^{-*} & -\Theta_{11}^{-*}\Theta_{21}^{*} \\ \Theta_{22}^{-1}\Theta_{21} & \Theta_{22}^{-1}\end{bmatrix}.\end{aligned} \tag{8.6}$$

PROOF The proofs are elementary and well known; see *e.g.*, [ADD90, lemma 5.2], [BGK92a].

1. $\Theta J\Theta^* = J$ and $\Theta^* J\Theta = J$ give the relations

$$\Theta_{22}\Theta_{22}^* = I+\Theta_{21}\Theta_{21}^*, \qquad \Theta_{22}^*\Theta_{22} = I+\Theta_{12}^*\Theta_{12}.$$

Hence Θ_{22} and Θ_{22}^* both have closed range and empty kernel. By the closed graph theorem, Θ_{22} is boundedly invertible and both $\Theta_{22}^{-1}\Theta_{22}^{-*} \gg 0$ and $\Theta_{22}^{-*}\Theta_{22}^{-1} \gg 0$.

2. Applying Θ_{22}^{-1} and Θ_{22}^{-*} to the left and right, respectively, of above two expressions yields

$$I = \Theta_{22}^{-1}\Theta_{22}^{-*} + (\Theta_{22}^{-1}\Theta_{21})(\Theta_{22}^{-1}\Theta_{21})^*, \qquad I = \Theta_{22}^{-*}\Theta_{22}^{-1} + (\Theta_{12}\Theta_{22}^{-1})^*(\Theta_{12}\Theta_{22}^{-1}).$$

Hence $\Theta_{22}^{-1}\Theta_{22}^{-*} \le I$ and also $\Theta_{22}^{-*}\Theta_{22}^{-1} \le I$, *i.e.*, $\|\Theta_{22}^{-1}\| \le 1$. Because $\Theta_{22}^{-1}\Theta_{22}^{-*} \gg 0$ and $\Theta_{22}^{-*}\Theta_{22}^{-1} \gg 0$ it follows that $\|\Theta_{22}^{-1}\Theta_{21}\| < 1$ and $\|\Theta_{12}\Theta_{22}^{-1}\| < 1$.

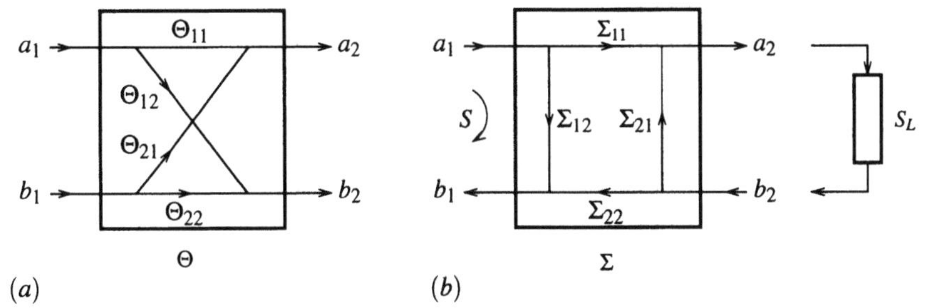

Figure 8.3. The connection between Θ and the corresponding scattering operator Σ.

3. Writing out the expression $[a_1 \quad b_1]\Theta = [a_2 \quad b_2]$ in full gives

$$\begin{cases} a_1\Theta_{11} + b_1\Theta_{21} = a_2 \\ a_1\Theta_{12} + b_1\Theta_{22} = b_2 \end{cases} \Leftrightarrow \begin{cases} a_1(\Theta_{11} - \Theta_{12}\Theta_{22}^{-1}\Theta_{21}) + b_2\Theta_{22}^{-1}\Theta_{21} = a_2 \\ -a_1\Theta_{12}\Theta_{22}^{-1} \qquad + b_2\Theta_{22}^{-1} \qquad = b_1 \end{cases}$$

as Θ_{22} is invertible. The second set of equations is $[a_1 \; b_2]\Sigma = [a_2 \; b_1]$. The fact that Σ is unitary can be verified by computing $\Sigma^*\Sigma$ and $\Sigma\Sigma^*$ in terms of its block entries.

4. Finally, the equalities in (8.6) follow directly from the J-unitarity, in particular the partial equations

$$\Theta_{11}\Theta_{11}^* - \Theta_{12}\Theta_{12}^* = I, \quad \Theta_{21}\Theta_{11}^* = \Theta_{22}\Theta_{12}^*$$

and

$$\Theta_{11}^*\Theta_{11} - \Theta_{21}^*\Theta_{21} = I, \quad \Theta_{11}^*\Theta_{12} = \Theta_{21}\Theta_{22}^*,$$

from which it follows *e.g.*, that Θ_{11}^* is boundedly invertible. □

A signal flow interpretation for the connection between Θ and Σ is given in figure 8.3. We see that the "bottom arrow" is reversed. Because Θ_{22} is invertible, all signals b_2 are admissible (*i.e.*, Θ_{22} has full range), and b_2 can act as an independent input. If we try to reverse the argument and construct Θ from Σ, we must exercise some care, as Σ_{22} is guaranteed to be contractive when Σ is unitary, but Σ_{22}^{-1} need not be upper nor bounded even when Σ is.

Example

Take

$$\begin{aligned} \mathcal{M}_+ &= [\boxed{\mathbb{C}^4}, \emptyset, \emptyset, \emptyset] & \mathcal{M}_- &= [\boxed{\mathbb{C}}, \mathbb{C}, \mathbb{C}, \mathbb{C}] \\ \mathcal{N}_+ &= [\boxed{\mathbb{C}^2}, \mathbb{C}^2, \emptyset, \emptyset] & \mathcal{N}_- &= [\boxed{\emptyset}, \emptyset, \mathbb{C}^2, \mathbb{C}^2]. \end{aligned}$$

A signal in the $\mathcal{M}_+$ space will have the form

$$a_1 = [a_{10}, \cdot, \cdot, \cdot] \in \mathcal{M}_+$$

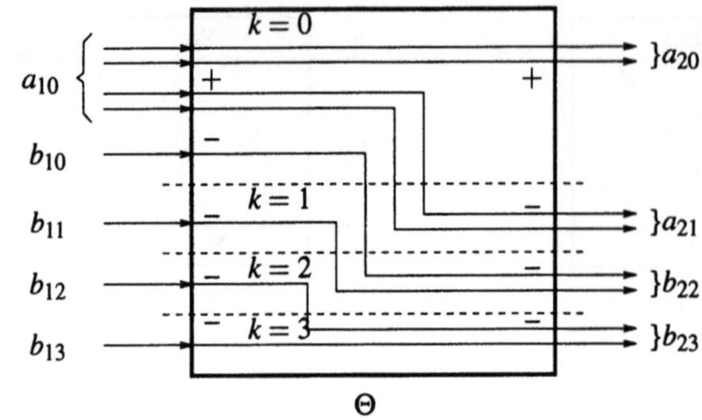

Figure 8.4. Example of non-uniform input and output spaces of a J-unitary operator Θ.

where a_{10} has dimension 4, while

$$\begin{array}{rcll} b_1 & = & [b_{10}, b_{11}, b_{12}, b_{13}] & \in \mathcal{M}_- \\ a_2 & = & [a_{20}, a_{21}, \;\cdot\;, \;\cdot\;] & \in \mathcal{N}_+ \\ b_2 & = & [\;\cdot\;, \;\cdot\;, b_{22}, b_{23}] & \in \mathcal{N}_- \end{array}$$

in which the entries a_{20}, a_{21}, b_{22}, b_{23} have dimension 2, and b_{10}, b_{11}, b_{12}, b_{13} have dimension 1. The other entries are "empty" or non-existent. One possible form Θ could take (see figure 8.4) is

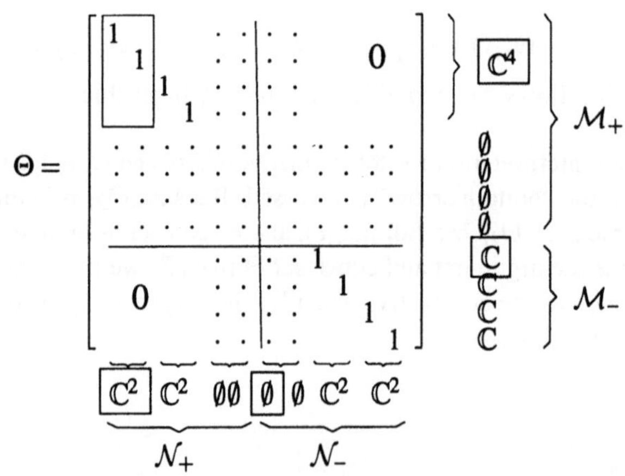

The diagonals in the subblocks of Θ have a somewhat erratic behavior, and some diagonal entries vanish. Such a Θ may occur in practical approximation situations, as is seen from examples in chapter 10. A signal flow diagram for Θ is drawn in figure 8.4. Since Θ is upper, it will be causal. The signal flow for Θ is from left-to-right. It is customary to also indicate the signature of Θ, or the energy flow that goes with a signal (*i.e.*, the signal flow of Σ). This is shown in the figure by '+' and '−'-signs.

Fractional transformations

Given a layered medium characterized by either a scattering operator Σ of a chain scattering operator Θ, we can produce a variety of scattering maps between the waves a_1 and b_1 at port 1, by loading port 2 with a scatterer S_L which realizes the map $b_2 = a_2 S_L$ (as in figure 8.3(*b*)). In that case, and provided the inverse of $(I - S_L\Sigma_{21})$ exists, we have $b_1 = a_1 S$ with

$$\begin{array}{rcl} S &=& \Sigma_{12} + \Sigma_{11}(I - S_L\Sigma_{21})^{-1} S_L \Sigma_{22} \\ S_L &=& (\Theta_{11} + S\Theta_{21})^{-1}(\Theta_{12} + S\Theta_{22}) \\ S &=& (\Theta_{12} - \Theta_{11}S_L)(\Theta_{21}S_L - \Theta_{22})^{-1} \;=:\; T_\Theta[S_L]\,. \end{array} \tag{8.7}$$

The third equation above follows from the second after chasing the denominator and solving for S. The relations between the port quantities can conveniently be expressed, with $A = \Theta_{11} + S\Theta_{21}$, as

$$[I \quad S]\Theta = A[I \quad S_L]\,. \tag{8.8}$$

We shall see that (8.8) is closely related to the interpolation properties of a scattering medium. If $u \in \ell_2^{\mathcal{M}+}$ is an input to the layered system, loaded by S_L, and $y = uS$ is the corresponding output, then we can view the operator $[I \;\; S]$ which maps $u \mapsto [u, \; y]$ as a so-called angle operator. In passive scattering situations, S_L is contractive and Θ will be J-unitary. As a result, S is automatically contractive as well, since

$$\begin{array}{rcl} I - SS^* &=& [I \quad S]J_1 \begin{bmatrix} I \\ S^* \end{bmatrix} \\ &=& [I \quad S]\Theta J_2 \Theta^* \begin{bmatrix} I \\ S^* \end{bmatrix} \\ &=& A[I \quad S_L]J_2 \begin{bmatrix} I \\ S_L^* \end{bmatrix} A^* \\ &=& A(I - S_L S_L^*)A^* \quad \geq 0\,. \end{array}$$

The existence of S can sometimes be asserted even though $I - S_L\Sigma_{21}$ is not invertible. It suffices that Σ corresponds to a lossless system and S_L to a lossy load (*i.e.*, causal and contractive). Under these two hypotheses it will be true that $a_1 a_1^* - b_1 b_1^* \geq 0$, and the map $a_1 \mapsto b_1$ is well defined. There may be a "defect" at the output port of Θ in the sense that all the possible physical b_2's need not span the whole space. Hence it may happen that the domain for $(I - S_L\Sigma_{21})^{-1}$ is restricted (it is the range of Σ_{11}). On the other hand, if Θ exists, then the third expression for S in (8.7) is always well defined for any $\|S_L\| \leq 1$: $(\Theta_{21}S_L - \Theta_{22})^{-1} = (\Theta_{22}^{-1}\Theta_{21}S_L - I)^{-1}\Theta_{22}^{-1}$. From theorem 8.2 we know that Θ_{22}^{-1} is bounded, as well as $(\Theta_{22}^{-1}\Theta_{21}S_L - I)^{-1}$, because $\|\Theta_{22}^{-1}\Theta_{21}\| < 1$.

In a physical context, S describes the reflectivity of the layered medium at port 1. The "thicker" the medium is, the less influence a load S_L will have upon the input scattering function, and the more it will be determined by the intermediate layers. For example, if an incident wave at port 1 needs n sample time slots to travel to port 2 and back, then the first n samples of the reflected wave will not be dependent on the load. Then, writing

$$S = T_\Theta[S_L]$$

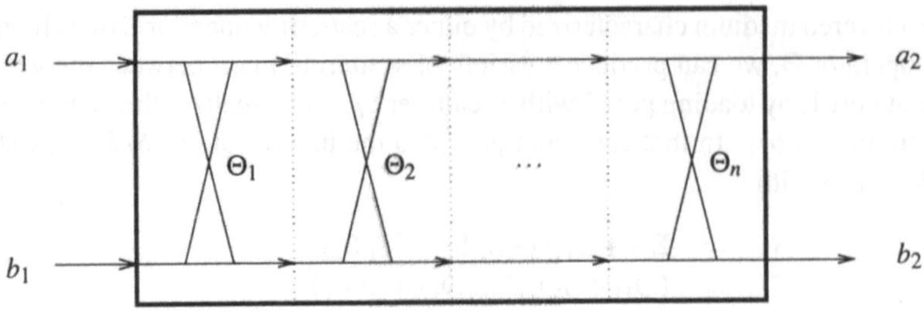

Figure 8.5. Cascade of Θ-sections.

for the input scattering operator at port 1 when port 2 is loaded in S_L, if

$$S_1 = T_\Theta[S_{L_1}] \quad \text{and} \quad S_2 = T_\Theta[S_{L_2}]$$

we shall have that S_1 has the same first n diagonals as S_2 or, alternatively, $S_1 - S_2$ is causally divisible by Z^n — S_1 and S_2 interpolate the same n diagonal "values". In the next chapter we shall see that chain scattering operators can be used to describe many more general instances of interpolation.

One of the main reasons for introducing the chain scattering matrix is the simplicity by which the overall scattering situation of a layerered medium can be expressed in terms of the individual layers. With reference to figure 8.5, we find that the overall chain scattering matrix of the cascade is

$$\Theta = \Theta_1 \Theta_2 \cdots \Theta_n .$$

The corresponding operation on the Σ matrices is complicated and known as the "Redheffer product", after the author who introduced it in the physics literature [Red62].

8.2 GEOMETRY OF DIAGONAL J-INNER PRODUCT SPACES

J-inner products

In the sequel, the "geometry" of the reachability and observability spaces of a J-unitary transformation will play a major role. The important metric, often imposed or induced by the situation, turns out to be indefinite. In this section we collect the main properties of such spaces as we need them. Keeping in line with the policy of working on diagonals as if they were scalars, we define *diagonal* J-inner products. Let $x = [x_1 \;\; x_2]$ and $y = [y_1 \;\; y_2]$ belong to a space of type

$$\mathcal{X}_2^{\mathcal{M}} \times \mathcal{X}_2^{\mathcal{N}}$$

and let $\mathbf{P}_0$ denote, as usual, the projection onto the 0-th diagonal, then the diagonal J-inner product of x with y is given by

$$\{x,y\}_J = \mathbf{P}_0(x_1 x_1^*) - \mathbf{P}_0(x_2 x_2^*) .$$

The scalar J-inner product $\langle x,y\rangle_J$ is the trace of $\{x,y\}_J$. It is easy to verify that it is an ordinary, although indefinite inner product.

Let $\mathcal{H}$ be some subspace of $\mathcal{X}_2^{\mathcal{M}} \times \mathcal{X}_2^{\mathcal{N}}$. We say that

- $\mathcal{H}$ *is J-positive* if, for all $x \in \mathcal{H}$, $\{x,x\}_J \geq 0$ (*i.e.*, entry-wise positive),
- $\mathcal{H}$ *is uniformly J-positive* if there exists an $\varepsilon > 0$ such that for all $x \in \mathcal{H}$, $\{x,x\}_J \geq \varepsilon\{x,x\}$,
- $\mathcal{H}$ *is J-neutral* if, for all $x \in \mathcal{H}$, $\{x,x\}_J = 0$,
- $\mathcal{H}$ *is J-negative* if, for all $x \in \mathcal{H}$, $\{x,x\}_J \leq 0$,
- $\mathcal{H}$ *is uniformly J-negative* if there exists an $\varepsilon > 0$ such that for all $x \in \mathcal{H}$, $-\{x,x\}_J \geq \varepsilon\{x,x\}$,

In many cases $\mathcal{H}$ will have no definite sign, and we call it indefinite. In particular, if $\mathcal{H} = \mathcal{X}_2^{\mathcal{M}}[I\ S]$, S will be contractive, isometric or expansive (*i.e.*, $SS^* - I \geq 0$) if and only if $\mathcal{H}$ is J-positive, J-neutral, or J-negative respectively. Notice that if $\mathcal{H}$ is J-neutral, then for all $x,y \in \mathcal{H}$ it is true that $\{x,y\}_J = 0$. This follows from the trapezium identity which is generally valid for inner product spaces (whether definite or not):

$$\{x,y\} = \frac{1}{4}(\{x+y,x+y\} - \{x-y,x-y\} + i\{x+iy,x+iy\} - i\{x-iy,x-iy\}).$$

Indefinite spaces

Now we move to some basic properties of spaces on which an indefinite inner product is defined, mostly for use in subsequent chapters. Extensive treatments can be found in [Bog74, Kre70, AY89]. Consider the indefinite diagonal inner product $\{\cdot,\cdot\}_J$ defined on $\mathcal{X}_2^{\mathcal{M}} \times \mathcal{X}_2^{\mathcal{N}}$. Note that it is always true that $|\{x,x\}_J| \leq \{x,x\}$, so that all $x \in \mathcal{X}_2^{\mathcal{M}} \times \mathcal{X}_2^{\mathcal{N}}$ have finite $\{x,x\}_J$. We say that a vector x is *isotropic* if $\{x,x\}_J = 0$. The span of a set of isotropic vectors is not necessarily J-neutral as can be seen from $\text{span}\{[1\ \ 1],[1\ \ -1]\}$ in $\mathbb{C}^2$, endowed with the inner product $\{x,y\}_J = x_1y_1 - x_2y_2$.

Let $\mathcal{H}$ be a subspace of $\mathcal{X}_2^{\mathcal{M}} \times \mathcal{X}_2^{\mathcal{N}}$ as before. We denote its orthogonal complement with respect to the indefinite inner product by $\mathcal{H}^{[\perp]}$. It is defined by the rule

$$\mathcal{H}^{[\perp]} = \{x \in \mathcal{X}_2^{\mathcal{M}} \times \mathcal{X}_2^{\mathcal{N}} : \forall y \in \mathcal{H},\ \{x,y\}_J = 0\}.$$

It follows immediately that $\mathcal{H}^{[\perp]} = \mathcal{H}^{\perp}J$ for the orthogonal complement $\mathcal{H}^{\perp}$ in the usual, definite inner product. Hence, $\mathcal{H}^{[\perp]}$ is a closed subspace (with respect to the natural inner product), and $(\mathcal{H}^{[\perp]})^{[\perp]} = \mathcal{H}$. If $\mathcal{H}$ is D-invariant, then so is $\mathcal{H}^{[\perp]}$.

On uniformly J-positive (or J-negative) definite subspaces, the J-inner product is equivalent to the usual inner product: $\varepsilon\{x,x\} \leq |\{x,x\}_J| \leq \{x,x\}$, which ensures that important properties such as completeness and closedness carry over: a uniformly J-definite subspace is a Hilbert space. We are, however, interested in more general cases than just uniformly definite subspaces, namely in cases where subspaces $\mathcal{H}$ can be split into $\mathcal{H} = \mathcal{H}_+ \boxplus \mathcal{H}_-$, where $\mathcal{H}_+$ and $\mathcal{H}_-$ are uniformly J-positive and J-negative subspaces respectively, and "$\boxplus$" denotes the J-orthogonal direct sum:

$$\mathcal{H} = \mathcal{A} \boxplus \mathcal{B} \quad \Leftrightarrow \quad \mathcal{H} = \mathcal{A} \dotplus \mathcal{B}, \quad \mathcal{A}\,[\perp]\,\mathcal{B}.$$

Such spaces are called *Krein spaces*. The indefinite direct sum is the analog of $\oplus$, but in using $\boxplus$, a number of properties that are a matter of course in Hilbert spaces no longer hold. For example, for orthogonal complementation in the usual (definite) inner product, we always have that $\mathcal{H}\cap\mathcal{H}^\perp = 0$ and $\mathcal{H}\oplus\mathcal{H}^\perp = \mathcal{X}_2$. With an indefinite metric, the analogous equations do not hold in general. The intersection of $\mathcal{H}$ and $\mathcal{H}^{[\perp]}$ is not necessarily empty: for example, if $\mathcal{H}$ is a neutral subspace, then $\mathcal{H}\subset\mathcal{H}^{[\perp]}$. One can also show that a subspace and its J-complement do not necessarily span the whole space. E.g., the J-orthogonal complement of span([1 1]) in $\mathbb{C}^2$ is span([1 1]) as well.

The algebraic sum $\mathcal{H}\dot{+}\mathcal{H}^{[\perp]}$ needs no longer be a direct sum: if $x\in\mathcal{H}\dot{+}\mathcal{H}^{[\perp]}$ then the decomposition $x = x_1 + x_2$ with $x_1\in\mathcal{H}$ and $x_2\in\mathcal{H}^{[\perp]}$ need not be unique.

A subspace $\mathcal{H}$ of $\mathcal{X}_2$ is said to be *projectively complete* if $\mathcal{H}\dot{+}\mathcal{H}^{[\perp]} = \mathcal{X}_2$. In this case, each $x\in\mathcal{X}_2$ has at least one decomposition into $x = x_1 + x_2$ with $x_1\in\mathcal{H}$ and $x_2\in\mathcal{H}^{[\perp]}$. A vector $x\in\mathcal{H}$ is called a *J-orthogonal projection* of a vector $y\in\mathcal{X}_2$ if (*i*) $x\in\mathcal{H}$ and (*ii*) $(y-x)\,[\perp]\,\mathcal{H}$.

Let $\mathcal{H}_0 = \mathcal{H}\cap\mathcal{H}^{[\perp]}$. Then $\mathcal{H}_0$ is automatically neutral. $\mathcal{H}$ is called a *non-degenerate* subspace if $\mathcal{H}_0 = \{0\}$. It is straightforward to show that

$$[\mathcal{H}\dot{+}\mathcal{H}^{[\perp]}]^{[\perp]} = \mathcal{H}^{[\perp]}\cap\mathcal{H}^{[\perp][\perp]} = \mathcal{H}^{[\perp]}\cap\mathcal{H} = \mathcal{H}_0$$

so that

$$\mathcal{X}_2 = (\mathcal{H}\dot{+}\mathcal{H}^{[\perp]})\oplus\mathcal{H}_0 J. \tag{8.9}$$

It follows that $\mathcal{H}$ can be projectively complete only if it is non-degenerate: $\mathcal{H}\cap\mathcal{H}^{[\perp]} = \{0\}$. In that case, decompositions are unique, so that *if $\mathcal{H}$ is projectively complete, then $\mathcal{X}_2 = \mathcal{H}\boxplus\mathcal{H}^{[\perp]}$.*

Indefinite Gramians

The situation in which we are interested is as follows: let $\mathcal{H}$ be a locally finite D-invariant subspace in some given "base space" $\mathcal{X}_2^{\mathcal{M}}$, and let $\mathcal{B}$ be the non-uniform space sequence whose dimension $\#\mathcal{B}$ is the sequence of dimensions of the subspace $\mathcal{H}$, as defined in section 4.3. $\mathcal{H}$ has some strong basis representation $\mathbf{F}$ such that $\mathcal{H} = \mathcal{D}_2^{\mathcal{B}}\mathbf{F}$ (*cf.* proposition 4.6). Let $J\in\mathcal{D}(\mathcal{M},\mathcal{M})$ be some given signature operator on the base space $\mathcal{M}$ — it is a diagonal of signature matrices, one for each entry in $\mathcal{M}$:

$$J_k = \begin{bmatrix} (I_{\mathcal{M}+})_k & \\ & (-I_{\mathcal{M}-})_k \end{bmatrix}$$

(the exact form of J is not really important here, the decomposition of $\mathcal{M}$ usually corresponds to a division of ports according to incoming and reflected waves). In analogy to the definition of the Gram operator $\Lambda_\mathbf{F} = \{\mathbf{F},\mathbf{F}\}$ in chapter 4, we define the J-Gram operator of this basis as the diagonal operator

$$\Lambda_\mathbf{F}^J = [\mathbf{F},\mathbf{F}] = \mathbf{P}_0(\mathbf{F}J\mathbf{F}^*) \in \mathcal{D}(\mathcal{B},\mathcal{B}). \tag{8.10}$$

$\mathbf{F}$ is called a *$J_\mathcal{B}$-orthonormal basis representation* when $\Lambda_\mathbf{F}^J = J_\mathcal{B}$, where $J_\mathcal{B}$ is some signature operator on $\mathcal{B}$. The dimensions of $\mathbf{F}$ are $\mathcal{B}\times\mathcal{M}$. We call $\mathcal{H}$ *regular* if the J-Gram operator of some strong basis in the normal metric is boundedly invertible. Since strong bases are related by invertible diagonal transformations R: $\mathbf{F}' = R\mathbf{F}$, the invertibility properties of the Gram operators of all these bases are the same, so that regularity is a property of the subspace. Note that $\Lambda_{\mathbf{F}} \gg 0$ does not imply that $\Lambda_{\mathbf{F}}^J$ is boundedly invertible. The reverse implication is true with some caution: $\Lambda_{\mathbf{F}}^J$ boundedly invertible does not necessarily imply that $\Lambda_{\mathbf{F}}$ is bounded, but if it is, then $\Lambda_{\mathbf{F}} \gg 0$.[2]

If $\Lambda_{\mathbf{F}}^J$ is boundedly invertible, then it has a factorization into $\Lambda_{\mathbf{F}}^J = RJ_{\mathcal{B}}R^*$, where R and $J_{\mathcal{B}}$ are diagonals in $\mathcal{D}(\mathcal{B},\mathcal{B})$, R is invertible and $J_{\mathcal{B}}$ is the signature matrix of $\Lambda_{\mathbf{F}}^J$: it is a diagonal of matrices

$$(J_{\mathcal{B}})_k = \begin{bmatrix} (I_+)_k & \\ & (-I_-)_k \end{bmatrix}$$

and defines a partitioning of $\mathcal{B}$ into $\mathcal{B} = \mathcal{B}_+ \times \mathcal{B}_-$. $J_{\mathcal{B}}$ is again independent of the choice of basis in $\mathcal{H}$. We call $J_{\mathcal{B}}$ the *inertia signature matrix* of the subspace $\mathcal{H}$, and the sequences $\#(\mathcal{B}_+)$ and $\#(\mathcal{B}_-)$ corresponding to the number of positive and negative entries of $J_{\mathcal{B}}$ at each point is called the inertia of $\mathcal{H}$. More general (non regular) subspaces can also have a zero-inertia, corresponding to singularities of $\Lambda_{\mathbf{F}}^J$, but if $\mathcal{H}$ is regular, then it has no zero-inertia. (The zero-inertia is only well defined if the range of $\Lambda_{\mathbf{F}}^J$ is closed, or equivalently, if any of its eigenvalues is either equal to zero or uniformly bounded away from zero.)

Canonical subspace decomposition

The following theorem is proved in [AY89, thm. 1.7.16] for classical Krein spaces, and holds in the present context as well. It is a fairly straightforward consequence of the closed graph theorem of functional analysis [DS63]. We refer the reader to the original paper and standard textbooks if he wishes to explore the matter further.

Theorem 8.3 *Let $\mathcal{H}$ be a locally finite left D-invariant subspace in $\mathcal{X}_2$, and let J be a signature matrix associated to it. The following are equivalent:*

1. *$\mathcal{H}$ is projectively complete; $\mathcal{H} \boxplus \mathcal{H}^{[\perp]} = \mathcal{X}_2$,*
2. *$\mathcal{H}$ is regular,*
3. *$\mathcal{H}$ is a Krein space: $\mathcal{H} = \mathcal{H}_+ \boxplus \mathcal{H}_-$, where $\mathcal{H}_+$ and $\mathcal{H}_-$ are uniformly J-positive (resp. J-negative) subspaces,*
4. *Any element in $\mathcal{X}_2$ has at least one J-orthogonal projection onto $\mathcal{H}$.*

[2] *Counterexample:* Take $J = \left[\begin{smallmatrix} 1 & 0 \\ 0 & -1 \end{smallmatrix}\right]$, ε_n a series of positive numbers for $n = 0, 1, \cdots$ such that $\varepsilon_n \to 0$, and

$$\mathbf{F} = \operatorname{diag}\begin{bmatrix} \sqrt{1+1/\varepsilon_n} & \sqrt{1/\varepsilon_n} \\ \sqrt{1/\varepsilon_n} & \sqrt{1+1/\varepsilon_n} \end{bmatrix},$$

We see that $\Lambda_{\mathbf{F}}^J = J$, so that $\Lambda_{\mathbf{F}}^J$ is boundedly invertible, but $\mathbf{F}_n$ does not have finite regular norm itself, and

$$[\Lambda_{\mathbf{F}}]_n = \begin{bmatrix} 1+2/\varepsilon_n & 2\sqrt{1/\varepsilon_n^2 + 1/\varepsilon_n} \\ 2\sqrt{1/\varepsilon_n^2+1/\varepsilon_n} & 1+2/\varepsilon_n \end{bmatrix} \approx \begin{bmatrix} 2/\varepsilon_n & 2/\varepsilon_n \\ 2/\varepsilon_n & 2/\varepsilon_n \end{bmatrix}.$$

Hence it is not true that $\Lambda_{\mathbf{F}} \gg 0$, not even on its domain.

Corollary 8.4 *Let $\mathcal{H}$ be a locally finite regular left D-invariant subspace in $\mathcal{X}_2^{\mathcal{B}}$ with dimension sequence $\mathcal{B}$, and let $J_{\mathcal{B}}$ be the inertia signature matrix of $\mathcal{H}$. Then $\mathcal{H}$ has a canonical decomposition $\mathcal{H} = \mathcal{H}_+ \boxplus \mathcal{H}_-$ into uniformly definite subspaces, where* sdim $\mathcal{H}_+ = \#\mathcal{B}_+ = \#_+(J_{\mathcal{B}})$ *and* sdim $\mathcal{H}_- = \#\mathcal{B}_- = \#_-(J_{\mathcal{B}})$.

8.3 STATE SPACE PROPERTIES OF J-UNITARY OPERATORS

Block-upper, locally finite, bounded J-unitary operators have remarkable state space properties. If Θ is such an operator, then its canonical input and output state spaces $\mathcal{H}(\Theta)$ and $\mathcal{H}_o(\Theta)$ are closed, regular subspaces. From the theory of the previous section it then follows that the decomposition $\mathcal{H}(\Theta) = \mathcal{H}_+ \boxplus \mathcal{H}_-$ exists, with $\mathcal{H}_+$ and $\mathcal{H}_-$ uniformly definite. We explore this and other state space properties in the present section.

Proposition 8.5 *Let $\Theta \in \mathcal{X}(\mathcal{M},\mathcal{N})$ be a locally finite $(J_{\mathcal{M}},J_{\mathcal{N}})$-unitary operator. Then $\mathcal{H}(\Theta)$ and $\mathcal{H}_o(\Theta)$ are closed subspaces,*

$$\begin{array}{rclcrcl} \mathcal{K}(\Theta) & = & \mathcal{L}_2 Z^{-1}\Theta^* J_{\mathcal{M}}\,, & \qquad & \mathcal{H}(\Theta) & = & \mathcal{L}_2 Z^{-1} \ominus \mathcal{K}(\Theta) \\ \mathcal{K}_o(\Theta) & = & \mathcal{U}_2 \Theta J_{\mathcal{N}}\,, & & \mathcal{H}_o(\Theta) & = & \mathcal{U}_2 \ominus \mathcal{K}_o(\Theta)\,. \end{array} \tag{8.11}$$

Furthermore,

$$\begin{array}{rcl} \mathcal{H}_o(\Theta) & = & \mathcal{H}(\Theta) J_{\mathcal{M}} \Theta \\ \mathcal{H}(\Theta) & = & \mathcal{H}_o(\Theta) J_{\mathcal{N}} \Theta^*\,. \end{array} \tag{8.12}$$

PROOF $\mathcal{L}_2 Z^{-1}\Theta^* J_{\mathcal{M}}$ is contained in the null space of the Hankel operator H_Θ since $\mathcal{L}_2 Z^{-1}\Theta^* J_{\mathcal{M}}\Theta = \mathcal{L}_2 Z^{-1} J_{\mathcal{N}} = \mathcal{L}_2 Z^{-1}$. Define $\mathcal{H} = \mathcal{L}_2 Z^{-1} \ominus \mathcal{L}_2 Z^{-1}\Theta^* J_{\mathcal{M}}$ and let $H'_{\Theta^*} = \mathbf{P}_{\mathcal{L}_2 Z^{-1}}(\,\cdot\,\Theta^*)|_{\mathcal{U}_2}$. For any $x \in \mathcal{H}$, $y = xJ_{\mathcal{M}}\Theta J_{\mathcal{N}}$ is such that $x = y\Theta^*$, and for all $z \in \mathcal{L}_2 Z^{-1}$ it is true that $\{y,z\} = \{x, J_{\mathcal{N}} z \Theta^* J_{\mathcal{M}}\} = 0$, by the definition of x. Hence $y \in \mathcal{U}_2$ and $\mathcal{H}$ is in the range of H'_{Θ^*}, *i.e.,* $\mathcal{H} = \mathcal{H}(\Theta)$. In addition, $\mathcal{H}$ is automatically closed because it is the complement of a subspace, and $\mathcal{K}(\Theta) = \mathcal{L}_2 Z^{-1}\Theta^* J_{\mathcal{M}}$. A dual reasoning produces $\mathcal{K}_o(\Theta)$ and $\mathcal{H}_o(\Theta)$. Equations (8.12) follow in the same vein: let $x \in \mathcal{H}(\Theta)$, then

$$\{xJ\Theta, \mathcal{L}_2 Z^{-1}\} = \{x, \mathcal{L}_2 Z^{-1}\Theta^* J\} = \{x, \mathcal{K}(\Theta)\} = 0\,.$$

□

Proposition 8.5 has great consequences for interpolation theory, as we shall see in chapter 9. Combination with the theory of the previous section produces the following proposition.

Proposition 8.6 *$\mathcal{H}(\Theta)$ and $\mathcal{H}_o(\Theta)$ as defined in proposition 8.5 are regular;*

$$\begin{array}{rcl} \mathcal{L}_2 Z^{-1} & = & \mathcal{H} \boxplus \mathcal{L}_2 Z^{-1}\Theta^* \\ \mathcal{U}_2 & = & \mathcal{H}_o \boxplus \mathcal{U}_2 \Theta\,. \end{array}$$

PROOF $\mathcal{H}_o^{[\perp]} = \mathcal{H}_o^{\perp} J = \mathcal{K}_o J = \mathcal{U}_2 \Theta$ by proposition 8.5. To prove that $\mathcal{U}_2 = \mathcal{H}_o \boxplus \mathcal{H}_o^{[\perp]}$, we show that every $y \in \mathcal{U}_2$ has a J-orthogonal projection onto $\mathcal{H}_o$. Let $y \in \mathcal{U}_2$, and define $y\Theta^{-1} = u_1 + y_1$, with $u_1 \in \mathcal{L}_2 Z^{-1}$ and $y_1 \in \mathcal{U}_2$. Furthermore, define $u_2 = u_1 J \in \mathcal{L}_2 Z^{-1}$. Then $u_2 \in \mathcal{H}$ because $u_2 = \mathbf{P}_{\mathcal{L}_2 Z^{-1}}(y\Theta^{-1})J = \mathbf{P}_{\mathcal{L}_2 Z^{-1}}((yJ)\Theta^*)$. It follows by proposition 8.5 that $y = u_2 J\Theta + y_1\Theta$, where $u_2 J\Theta \in \mathcal{H}_o$ and $y_1\Theta \in \mathcal{H}_o^{[\perp]} = \mathcal{U}_2\Theta$. Hence every $y \in \mathcal{U}_2$ has a J-projection onto $\mathcal{H}_o$, so that according to theorem 8.3 $\mathcal{H}_o$ is regular. A dual proof holds for $\mathcal{H}$. □

Corollary 8.7 *Let $\Theta \in \mathcal{U}(\mathcal{M},\mathcal{N})$ be a locally finite bounded J-unitary operator. If $\mathbf{F}$ is a $J_\mathcal{B}$-orthonormal basis representation of $\mathcal{H}(\Theta)$, then $\mathbf{F}_o = J_\mathcal{B}\mathbf{F}J_\mathcal{M}\Theta$ is a J-orthonormal basis representation of $\mathcal{H}_o(\Theta)$, and in this case the canonical controller realization based on $\mathbf{F}$ (theorem 5.15) and canonical observer realization based on $\mathbf{F}_o$ (theorem 5.18) are equal.*

PROOF Because $\mathcal{H}(\Theta)$ is regular (proposition 8.6), there is a J-orthonormal basis representation $\mathbf{F}$ of $\mathcal{H}$: $\Lambda_\mathbf{F}^J = \mathbf{P}_0(\mathbf{F}J\mathbf{F}^*) = J_\mathcal{B}$. This $\mathbf{F}$ defines a factorization of the Hankel operator of Θ as $H_\Theta = \mathbf{P}_0(\cdot\,\mathbf{F}^*)\,\mathbf{F}_o$ where $\mathbf{F}_o = \Lambda_\mathbf{F}^{-1}\,\mathbf{P}(\mathbf{F}\Theta)$ is a basis of the output state space $\mathcal{H}_o$ of Θ (theorem 5.20). On the other hand, the relation $\mathcal{H}_o = \mathcal{H}J\Theta$ ensures that $\mathbf{F}_a$, defined as $\mathbf{F}_a = \mathbf{F}J_\mathcal{M}\Theta$, is upper and also a J-orthonormal basis representation of $\mathcal{H}_o$. The connection between $\mathbf{F}_a$ and $\mathbf{F}_o$ is $\mathbf{F}_a = \mathbf{F}J_\mathcal{M}\Theta = \mathbf{F}J_\mathcal{M}H_\Theta = \mathbf{P}_0(\mathbf{F}J_\mathcal{M}\mathbf{F}^*)\,\mathbf{F}_o = J_\mathcal{B}\mathbf{F}_o$, so that $\mathbf{F}_o = J_\mathcal{B}\mathbf{F}_a = J_\mathcal{B}\mathbf{F}J_\mathcal{M}\Theta$. It is readily verified that $\mathbf{F}_o$ is also J-orthonormal. Theorem 5.20 claims that the canonical observer realization based on this $\mathbf{F}_o$ is equal to the canonical controller realization of $\mathbf{F}$. □

J-Unitary realizations

A realization matrix $\mathbf{\Theta} \in \mathcal{D}(\mathcal{B}\times\mathcal{M}, \mathcal{B}^{(-1)}\times\mathcal{N})$ with signature matrices $\mathbf{J}_1$, $\mathbf{J}_2$,

$$\mathbf{\Theta} = \begin{bmatrix} A & C \\ B & D \end{bmatrix}, \qquad \mathbf{J}_1 = \begin{bmatrix} J_\mathcal{B} & \\ & J_\mathcal{M} \end{bmatrix}, \quad \mathbf{J}_2 = \begin{bmatrix} J_\mathcal{B}^{(-1)} & \\ & J_\mathcal{N} \end{bmatrix} \tag{8.13}$$

is said to be J-unitary if

$$\mathbf{\Theta}^*\mathbf{J}_1\mathbf{\Theta} = \mathbf{J}_2, \qquad \mathbf{\Theta}\mathbf{J}_2\mathbf{\Theta}^* = \mathbf{J}_1.$$

We call $J_\mathcal{B}$ the state signature matrix of $\mathbf{\Theta}$. With "#" indicating the sequence of dimensions of a space sequence, we have that the total number of positive entries of the signatures at the left-hand side of each equation, for each time instant k, is equal to the total positive signature at the right-hand side, and similarly for the total negative signature (the *inertia theorem*):

$$\begin{aligned} \#\mathcal{B}_+ + \#\mathcal{M}_+ &= \#\mathcal{B}_+^{(-1)} + \#\mathcal{N}_+ \\ \#\mathcal{B}_- + \#\mathcal{M}_- &= \#\mathcal{B}_-^{(-1)} + \#\mathcal{N}_-. \end{aligned} \tag{8.14}$$

As for inner systems, J-unitary systems and J-unitary realizations go together. Proofs of this are similar to the unitary case (theorems 6.3 and 6.4). In particular, the following theorem claims that if Θ is a locally finite bounded J-unitary upper operator, then one can find a minimal realization for Θ which is J-unitary, for an appropriate J-metric defined on the input (or output) canonical state space.

Theorem 8.8 *Let $\Theta \in \mathcal{U}(\mathcal{M},\mathcal{N})$ be a bounded locally finite $(J_\mathcal{M},J_\mathcal{N})$-unitary operator. Let $J_\mathcal{B}$ be the inertia signature matrix of $\mathcal{H}(\Theta)$, and let $\mathbf{F}$ be a J-orthonormal basis representation for $\mathcal{H}(\Theta)$.*

Then the canonical controller realization $\boldsymbol{\Theta}$ based on $\mathbf{F}$ is J-unitary, and identical to the canonical observer realization based on $\mathbf{F}_o = J_\mathcal{B}\mathbf{F}J_\mathcal{M}\Theta$.

PROOF Let $\boldsymbol{\Theta}$ be given by the canonical controller realization (theorem 5.15). This realization satisfies the properties (5.19)–(5.21):

$$\left\{\begin{array}{rcl} Z\mathbf{F} &=& A^*\mathbf{F}+B^*, \\ \mathbf{P}_0(Z^{-1}\cdot\mathbf{F}^*)^{(-1)} &=& \mathbf{P}_0(\cdot[\mathbf{F}^*A+B]) \end{array}\right. \qquad \left\{\begin{array}{rcl} \mathbf{P}_0(\cdot\Theta) &=& \mathbf{P}_0(\cdot[D+\mathbf{F}^*C]), \\ \Theta^* &=& D^*+C^*\mathbf{F}. \end{array}\right. \tag{8.15}$$

To verify that $\boldsymbol{\Theta}^*\mathbf{J}_1\boldsymbol{\Theta} = \mathbf{J}_2$, we have to show that

$$\left\{\begin{array}{rcl} \Theta^*J_\mathcal{M}\Theta &=& J_\mathcal{N} \\ \mathbf{P}_0(\mathbf{F}J_\mathcal{M}\mathbf{F}^*) &=& J_\mathcal{B} \end{array}\right. \quad\Rightarrow\quad \left\{\begin{array}{rcl} A^*J_\mathcal{B}A+B^*J_\mathcal{M}B &=& J_\mathcal{B}^{(-1)} \\ C^*J_\mathcal{B}C+D^*J_\mathcal{M}D &=& J_\mathcal{N} \\ A^*J_\mathcal{B}C+B^*J_\mathcal{M}D &=& 0. \end{array}\right.$$

Indeed,

$$\begin{array}{rcl} \mathbf{P}_0(\mathbf{F}J_\mathcal{M}\mathbf{F}^*) = J_\mathcal{B} \quad\Rightarrow\quad J_\mathcal{B}^{(-1)} &=& \mathbf{P}_0(Z^{-1}(Z\mathbf{F})J_\mathcal{M}\mathbf{F}^*)^{(-1)} \\ &=& \mathbf{P}_0(Z^{-1}[(A^*\mathbf{F}+B^*)J_\mathcal{M}]\mathbf{F}^*)^{(-1)} \\ &=& \mathbf{P}_0([A^*\mathbf{F}+B^*]J_\mathcal{M}[\mathbf{F}^*A+B]) \\ &=& A^*J_\mathcal{B}A+B^*J_\mathcal{M}B, \end{array}$$

$$\begin{array}{rcl} \mathbf{P}_0(\Theta^*J_\mathcal{M}\Theta) = J_\mathcal{N} \quad\Rightarrow && \mathbf{P}_0([D^*+C^*\mathbf{F}]J_\mathcal{M}[D+\mathbf{F}^*C]) \\ && = D^*J_\mathcal{M}D+C^*\mathbf{P}_0(\mathbf{F}J_\mathcal{M}\mathbf{F}^*)C \\ && = D^*J_\mathcal{M}D+C^*J_\mathcal{B}C \quad = J_\mathcal{N}. \end{array}$$

Note further that $\mathbf{P}_0(Z\mathbf{F}J\Theta) = \mathbf{P}_0(ZJ\mathbf{F}_o) = 0$, hence

$$\begin{array}{rcl} \mathbf{P}_0(Z\mathbf{F}J\Theta) = 0 \quad\Rightarrow && \mathbf{P}_0([B^*+A^*\mathbf{F}]J_\mathcal{M}[D+\mathbf{F}^*C]) \\ && = B^*J_\mathcal{M}D+A^*\mathbf{P}_0(\mathbf{F}J_\mathcal{M}\mathbf{F}^*)C \\ && = B^*J_\mathcal{M}D+A^*J_\mathcal{B}C \quad = 0. \end{array}$$

Hence $\boldsymbol{\Theta}^*\mathbf{J}_1\boldsymbol{\Theta} = \mathbf{J}_2$. The relation $\boldsymbol{\Theta}\mathbf{J}_2\boldsymbol{\Theta}^* = \mathbf{J}_1$ can be derived in the same (dual) way as above. The equality of both realizations has been proven in corollary 8.7. □

As was the case with inner operators (*viz.* theorem 6.12), the converse of this theorem is in general true only if, in addition, $\ell_A < 1$. If $\ell_A = 1$, then additional assumptions on the reachability and observability of the realization must be made.

Theorem 8.9 *Let $\boldsymbol{\Theta} = \left[\begin{smallmatrix} A & C \\ B & D \end{smallmatrix}\right]$ be a state realization of a locally finite bounded transfer operator $\Theta \in \mathcal{U}$, with $\ell_A < 1$, and denote by $\Lambda^J_\mathbf{F}$ and $\Lambda^J_{\mathbf{F}_o}$ the reachability and the observability J-Gramians of the given realization respectively. Then*

$$\begin{array}{rcl} \boldsymbol{\Theta}^*\mathbf{J}_1\boldsymbol{\Theta} = \mathbf{J}_2 &\Rightarrow& \Theta^*J_\mathcal{M}\Theta = J_\mathcal{N}, \quad \Lambda^J_\mathbf{F} = J_\mathcal{B}, \\ \boldsymbol{\Theta}\mathbf{J}_2\boldsymbol{\Theta}^* = \mathbf{J}_1 &\Rightarrow& \Theta J_\mathcal{N}\Theta^* = \mathcal{M}, \quad \Lambda^J_{\mathbf{F}_o} = J_\mathcal{B}. \end{array} \tag{8.16}$$

PROOF Since $\ell_A < 1$, a closed form expression for Θ is

$$\Theta = D + BZ(I-AZ)^{-1}C$$

and direct computations give

$$\begin{aligned} J_{\mathcal{N}} - \Theta^* J_{\mathcal{M}} \Theta &= J_{\mathcal{N}} - [D + BZ(I-AZ)^{-1}C]^* J_{\mathcal{M}} [D + BZ(I-AZ)^{-1}C] \\ &= C^* J_{\mathcal{B}} C + C^*(I-Z^*A^*)^{-1} Z^* A^* J_{\mathcal{B}} C + C^* J_{\mathcal{B}} AZ(I-AZ)^{-1} C \\ &\quad - C^*(I-Z^*A^*)^{-1} Z^* \{ J_{\mathcal{B}}^{(-1)} - A^* J_{\mathcal{B}} A \} Z(I-AZ)^{-1} C \end{aligned}$$

since $B^* J_{\mathcal{M}} D = -A^* J_{\mathcal{B}} C$, $B^* J_{\mathcal{M}} B = J_{\mathcal{B}}^{(-1)} - A^* J_{\mathcal{B}} A$ and $J_{\mathcal{N}} - D^* J_{\mathcal{M}} D = C^* J_{\mathcal{B}} C$, and hence

$$\begin{aligned} J_{\mathcal{N}} - \Theta^* J_{\mathcal{M}} \Theta &= C^*(I-Z^*A^*)^{-1} \{ (I-Z^*A^*) J_{\mathcal{B}} (I-AZ) + Z^* A^* J_{\mathcal{B}} (I-AZ) \\ &\quad + (I-Z^*A^*) J_{\mathcal{B}} AZ - J_{\mathcal{B}} + Z^* A^* J_{\mathcal{B}} AZ \} (I-AZ)^{-1} C \\ &= 0. \end{aligned}$$

The second equality follows by an analogous procedure. □

More general versions of these theorems for J-isometric operators are easily deduced and given at the end of the section.

Unitary state representation for Σ in terms of $\boldsymbol{\Theta}$

Let $\boldsymbol{\Theta}$ be a J-unitary realization of a bounded J-unitary operator $\Theta \in \mathcal{U}$, with state signature matrix $J_{\mathcal{B}}$. We have seen (in proposition 8.6) that the input and output state spaces $\mathcal{H}(\Theta)$ and $\mathcal{H}_o(\Theta)$ are regular: there exist definite spaces $\mathcal{H}_+$ and $\mathcal{H}_-$ such that $\mathcal{H} = \mathcal{H}_+ \boxplus \mathcal{H}_-$. This partitioning induces a partitioning of the state space sequence $\mathcal{B}$ into $\mathcal{B} = \mathcal{B}_+ \times \mathcal{B}_-$ conformably to $J_{\mathcal{B}}$. Because the bases chosen for the state spaces are J-orthonormal ($\boldsymbol{\Theta}$ is J-unitary), the basis representation $\mathbf{F}$ can be partitioned into two J-orthonormal bases $\mathbf{F}_+$ and $\mathbf{F}_-$, such that $\mathcal{H}_+ = \mathcal{D}_2^{\mathcal{B}_+} \mathbf{F}_+$ and $\mathcal{H}_- = \mathcal{D}_2^{\mathcal{B}_-} \mathbf{F}_-$. Hence a state $x \in \mathcal{X}_2^{\mathcal{B}}$ is partitioned into $x = [x_+ \;\; x_-] \in \mathcal{X}_2^{\mathcal{B}_+} \times \mathcal{X}_2^{\mathcal{B}_-}$, where x_+ and x_- are the parts of the state that correspond to the positive and negative subspaces in the state space $\mathcal{H}$: $x_+ \mathbf{F}_+ \in \mathcal{H}_+$ and $x_- \mathbf{F}_- \in \mathcal{H}_-$. The decomposition of the state defines a partitioning of $\boldsymbol{\Theta}$ according to the equation

$$[x_+ \;\; x_- \;\; a_1 \;\; b_1] \boldsymbol{\Theta} = [x_+ Z^{-1} \;\; x_- Z^{-1} \;\; a_2 \;\; b_2] \tag{8.17}$$

into

$$\boldsymbol{\Theta} = \begin{array}{c} \\ x_+ \\ x_- \\ a_1 \\ b_1 \end{array} \!\!\begin{array}{c} \begin{array}{cccc} x_+Z^{-1} & x_-Z^{-1} & a_2 & b_2 \end{array} \\ \left[\begin{array}{cc|cc} A_{11} & A_{12} & C_{11} & C_{12} \\ A_{21} & A_{22} & C_{21} & C_{22} \\ \hline B_{11} & B_{12} & D_{11} & D_{12} \\ B_{21} & B_{22} & D_{21} & D_{22} \end{array}\right] \end{array} . \tag{8.18}$$

We have shown, in theorem 8.2, that associated to Θ is a unitary operator Σ such that

$$[a_1 \;\; b_2] \Sigma = [a_2 \;\; b_1] \quad \Leftrightarrow \quad [a_1 \;\; b_1] \Theta = [a_2 \;\; b_2] .$$

The question we adress now is how a given realization $\mathbf{\Theta}$ of Θ gives rise to a realization $\mathbf{\Sigma}$ of Σ.

A reordering of rows and columns in (8.18) with respect to their signatures converts $\mathbf{\Theta}$ into a genuine square-block J-unitary operator, *i.e.*, each matrix

$$\left[\begin{array}{cc|cc} A_{11} & C_{11} & A_{12} & C_{12} \\ B_{11} & D_{11} & B_{12} & D_{12} \\ \hline A_{21} & C_{21} & A_{22} & C_{22} \\ B_{21} & D_{21} & B_{22} & D_{22} \end{array}\right]_k$$

is a square and J-unitary matrix with respect to the signature

$$\begin{bmatrix} I_{(\mathcal{B}_+)_k \times (\mathcal{M}_+)_k} & \\ & -I_{(\mathcal{B}_-)_k \times (\mathcal{M}_-)_k} \end{bmatrix} = \begin{bmatrix} I_{(\mathcal{B}_+)_{k+1} \times (\mathcal{N}_+)_k} & \\ & -I_{(\mathcal{B}_-)_{k+1} \times (\mathcal{N}_-)_k} \end{bmatrix}.$$

In particular, each submatrix

$$\begin{bmatrix} A_{22} & C_{22} \\ B_{22} & D_{22} \end{bmatrix}_k$$

of $\mathbf{\Theta}_k$ is square and invertible, and because $\mathbf{\Theta}$ is J-unitary, the block-diagonal operator constructed from these submatrices is boundedly invertible as well. It follows that the following block-diagonal operators are well defined (*cf.* equation (8.6)):

$$\begin{aligned}
\begin{bmatrix} F_{11} & H_{11} \\ G_{11} & K_{11} \end{bmatrix} &= \begin{bmatrix} A_{11} & C_{11} \\ B_{11} & D_{11} \end{bmatrix} - \begin{bmatrix} A_{12} & C_{12} \\ B_{12} & D_{12} \end{bmatrix} \begin{bmatrix} A_{22} & C_{22} \\ B_{22} & D_{22} \end{bmatrix}^{-1} \begin{bmatrix} A_{21} & C_{21} \\ B_{21} & D_{21} \end{bmatrix} \\
\begin{bmatrix} F_{12} & H_{12} \\ G_{12} & K_{12} \end{bmatrix} &= -\begin{bmatrix} A_{12} & C_{12} \\ B_{12} & D_{12} \end{bmatrix} \begin{bmatrix} A_{22} & C_{22} \\ B_{22} & D_{22} \end{bmatrix}^{-1} \\
\begin{bmatrix} F_{21} & H_{21} \\ G_{21} & K_{21} \end{bmatrix} &= \begin{bmatrix} A_{22} & C_{22} \\ B_{22} & D_{22} \end{bmatrix}^{-1} \begin{bmatrix} A_{21} & C_{21} \\ B_{21} & D_{21} \end{bmatrix} \\
\begin{bmatrix} F_{22} & H_{22} \\ G_{22} & K_{22} \end{bmatrix} &= \begin{bmatrix} A_{22} & C_{22} \\ B_{22} & D_{22} \end{bmatrix}^{-1}
\end{aligned} \tag{8.19}$$

and we obtain the relation

$$[x_+ \quad x_- Z^{-1} \quad a_1 \quad b_2]\mathbf{\Sigma} = [x_+ Z^{-1} \quad x_- \quad a_2 \quad b_1] \tag{8.20}$$

where

$$\mathbf{\Sigma} = \begin{array}{c} \\ x_+ \\ x_- Z^{-1} \\ a_1 \\ b_2 \end{array} \begin{array}{c} \begin{array}{cccc} x_+ Z^{-1} & x_- & a_2 & b_1 \end{array} \\ \left[\begin{array}{cc|cc} F_{11} & F_{12} & H_{11} & H_{12} \\ F_{21} & F_{22} & H_{21} & H_{22} \\ \hline G_{11} & G_{12} & K_{11} & K_{12} \\ G_{21} & G_{22} & K_{21} & K_{22} \end{array}\right] \end{array}. \tag{8.21}$$

See figure 8.6. An important point which can be readily derived from the J-unitarity of $\mathbf{\Theta}$ is the fact that $\mathbf{\Sigma}$ is unitary:

$$\mathbf{\Sigma}\mathbf{\Sigma}^* = I; \quad \mathbf{\Sigma}^*\mathbf{\Sigma} = I.$$

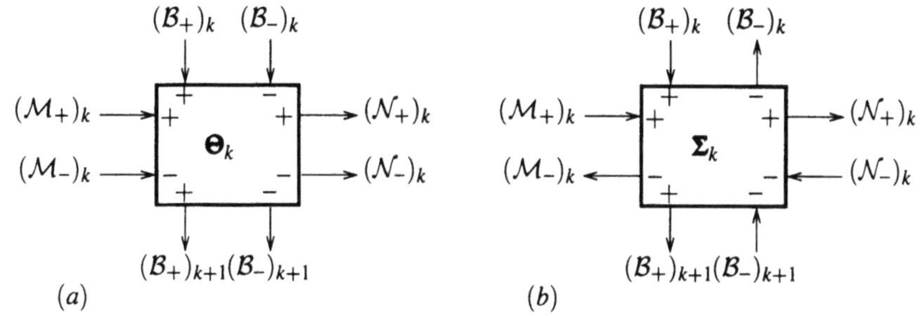

Figure 8.6. (a) The spaces connected with a realization for a J-unitary block-upper operator Θ which transfers $\ell_2^{\mathcal{M}_+} \times \ell_2^{\mathcal{M}_-}$ to $\ell_2^{\mathcal{N}_+} \times \ell_2^{\mathcal{N}_-}$. The state transition operator is marked as $\mathbf{\Theta}$. (b) The corresponding scattering situation.

Because in (8.20) state quantities with and without the factor Z^{-1} appear in the same argument at the left- and right-hand sides, $\mathbf{\Sigma}$ is possibly a kind of generalized or implicit realization for a transfer operator Σ, but is not computable in this form. Σ is guaranteed to exist (because Θ_{22}^{-1} exists—see theorem 8.2) and can be obtained from $\mathbf{\Sigma}$ by elimination of x_- and x_+. $\mathbf{\Sigma}$ can be interpreted as a realization having an upward-going state x_- and a downward state x_+, as depicted in figure 8.6. Recall that although Σ is unitary, it is not necessarily upper. The upward-going state x_- is instrumental in generating the lower triangular (anti-causal) part of Σ. The precise details will be investigated later (proposition 8.14), but a preliminary result is straightforward to derive.

Proposition 8.10 *Let $\mathbf{\Theta}$ be a $(\mathbf{J}_1, \mathbf{J}_2)$-unitary realization for a J-unitary operator Θ. If $J_{\mathcal{B}} = I$, then $\Theta_{22}^{-1} \in \mathcal{U}$, that is, Θ_{22} is outer, Θ is J-inner, and the corresponding unitary operator Σ is upper and hence inner.*

PROOF If $J_{\mathcal{B}} = I$, then the dimension of x_- is zero, so that the implicit state relations $\mathbf{\Sigma}$ for Σ in (8.20) are reduced to ordinary state equations $[x_+ Z^{-1} \;\; a_2 \;\; b_1] = [x_+ \;\; a_1 \;\; b_2]\mathbf{\Sigma}$, which define an upper (causal) operator Σ. □

Θ-matrices with a realization $\mathbf{\Theta}$ which is J-unitary with state signature $J_{\mathcal{B}} = I$ correspond to inner scattering operators Σ. They will play a central role in the interpolation theory of the next chapter.

J-isometric operators

If we only know that Θ is J-isometric ($\Theta^* J \Theta = J$) then a number of properties change. Most importantly, we cannot deduce that the input and output state spaces of Θ are regular. Consequently, we might not be able to find a strong basis that has a boundedly invertible J-Gramian Λ^J, thus precluding a J-isometric realization. Nonetheless, we can find unnormalized realizations that show us whether the corresponding operator is J-isometric. The fact that the J-Gramian is not invertible also implies that the signature (inertia) $J_{\mathcal{B}}$ of Λ^J cannot be very well determined: components might be ε-close to zero. An important implication will be that it is not always possible to extend a given J-isometric operator to a J-unitary operator.

The following theorem is a more general version of theorems 8.8 and 8.9.

Theorem 8.11 *Let $\Theta \in \mathcal{U}$ be a locally finite operator with a u.e. stable realization $\mathbf{\Theta} = \left[\begin{smallmatrix} A & C \\ B & D \end{smallmatrix}\right]$. Then*

$$\Theta^* J \Theta = J \quad \Leftrightarrow \quad \exists M \in \mathcal{D}: \begin{cases} A^*MA + B^*JB &= M^{(-1)} \\ A^*MC + B^*JD &= 0 \\ C^*MC + D^*JD &= J \end{cases}$$

$$\Theta J \Theta^* = J \quad \Leftrightarrow \quad \exists M \in \mathcal{D}: \begin{cases} AM^{(-1)}A^* + CJC^* &= M \\ AM^{(-1)}B^* + CJD^* &= 0 \\ BM^{(-1)}B^* + DJD^* &= J \end{cases}$$

PROOF ($\Theta^* J \Theta = J \Rightarrow \cdots$) Define $\mathbf{F} = (BZ(I-AZ)^{-1})^*$. Since $\ell_A < 1$, properties (8.15) hold. Define $M = \mathbf{P}_0(\mathbf{F}J\mathbf{F}^*)$. The rest of the proof is similar to that of theorem (8.8), replacing $J_{\mathcal{B}}$ by M.

($\Theta^* J \Theta = J \Leftarrow \cdots$) Direct computation as in theorem 8.9.

The other relations follow in a dual way. □

8.4 PAST AND FUTURE SCATTERING OPERATORS

This section dives deeper into the properties of the state space of a general causal J-unitary matrix: one that does not necessarily correspond to a causal (and hence inner) scattering matrix. These properties will appear to be of crucial importance to model reduction theory as treated in chapter 10. We give them here because they form a nice application of Krein space theory and have independent interest.

In section 5.1, we introduced the decomposition $u = u_p + u_f$ for a signal $u \in \mathcal{X}_2$, where $u_p = \mathbf{P}_{\mathcal{L}_2 Z^{-1}}(u) \in \mathcal{L}_2 Z^{-1}$ is the "past" part of the signal (with reference to its 0-th diagonal), and $u_f = \mathbf{P}(u) \in \mathcal{U}_2$ is its "future" part. We also showed how a causal operator T with state realization $\mathbf{T}$ could be split into a past operator T_p which maps u_p to $[x_{[0]} \;\; y_p]$ and a future operator T_f which maps $[x_{[0]} \;\; u_f]$ to y_f. In the present context, let the signals a_1, b_1, a_2, b_2 and the state sequences x_+, x_- be in $\mathcal{X}_2$ and be related by $\mathbf{\Theta}$ as in (8.17). With the partitioning of the signals a_1, etc., into a past and a future part, Θ can be split into operators $(\cdot)\Theta_p : Z^{-1}\mathcal{L}_2^{\mathcal{M}} \to [\mathcal{D}_2^{\mathcal{B}} \;\; Z^{-1}\mathcal{L}_2^{\mathcal{N}}]$ and $(\cdot)\Theta_f : [\mathcal{D}_2^{\mathcal{B}} \;\; \mathcal{U}_2^{\mathcal{M}}] \to \mathcal{U}_2^{\mathcal{N}}$ via

$$\begin{aligned} [a_{1p} \;\; b_{1p}]\Theta_p &= [x_{+[0]} \;\; x_{-[0]} \;\; a_{2p} \;\; b_{2p}] \\ [x_{+[0]} \;\; x_{-[0]} \;\; a_{1f} \;\; b_{1f}]\Theta_f &= [a_{2f} \;\; b_{2f}]. \end{aligned} \tag{8.22}$$

Θ_p and Θ_f are determined once basis representations for the input and output state spaces of Θ have been chosen. The following procedure is as in section 5.1. The splitting of signals into past and future parts associates to Θ an "expanded" version $\hat{\Theta}$, defined such that $(u_p + u_f)\Theta = (y_p + y_f) \Leftrightarrow [u_p \;\; u_f]\hat{\Theta} = [y_p \;\; y_f]$:

$$\hat{\Theta} = \begin{bmatrix} K_\Theta & H_\Theta \\ 0 & E_\Theta \end{bmatrix} \qquad \text{where} \qquad \begin{cases} K_\Theta &=& \mathbf{P}_{\mathcal{L}_2 Z^{-1}}(\cdot\Theta)\big|_{\mathcal{L}_2 Z^{-1}} \\ H_\Theta &=& \mathbf{P}(\cdot\Theta)\big|_{\mathcal{L}_2 Z^{-1}} \\ E_\Theta &=& \mathbf{P}(\cdot\Theta)\big|_{\mathcal{U}_2} \end{cases} \tag{8.23}$$

Let $\mathbf{F}$ be a J-orthonormal basis for $\mathcal{H}(\Theta)$, and let $\mathbf{F}_o = J\mathbf{F}J\Theta$ be the corresponding left DZ-invariant J-orthonormal basis for $\mathcal{H}_o(\Theta)$. Then Θ_p and Θ_f are given by (*cf.* equation (5.18))

$$\Theta_p = [\mathbf{P}_0(\cdot\mathbf{F}^*) \quad K_\Theta], \qquad \Theta_f = \begin{bmatrix} \mathbf{F}_o \\ E_\Theta \end{bmatrix}. \tag{8.24}$$

We first show that Θ_p and Θ_f are J-unitary operators. Then, as a consequence of the theory in the previous section (in particular proposition 8.10) there exist operators $(\cdot)\Sigma_p$, $(\cdot)\Sigma_f$:

$$\begin{array}{rcl} [x_{-[0]} \quad a_{1p} \quad b_{2p}]\Sigma_p &=& [x_{+[0]} \quad a_{2p} \quad b_{1p}] \\ {[x_{+[0]} \quad a_{1f} \quad b_{2f}]\Sigma_f} &=& [x_{-[0]} \quad a_{2f} \quad b_{1f}] \end{array} \tag{8.25}$$

which are scattering operators corresponding to Θ_p and Θ_f, respectively (see figure 8.7(*b*)). The J-unitarity of Θ_p and Θ_f, and hence the existence and unitarity of Σ_p and Σ_f, is asserted in the next proposition.

Proposition 8.12 *Let $\Theta \in \mathcal{U}(\mathcal{M},\mathcal{N})$ be a locally finite J-unitary operator, and let $\boldsymbol{\Theta}$ be a J-unitary realization for Θ. Then Θ_p and Θ_f are J-unitary operators, and Σ_p, Σ_f are well-defined unitary operators.*

PROOF Let $\hat{\Theta}$ be given as in equation (8.23). $\hat{\Theta}$ is, except for ordering of elements, the same operator as Θ. Hence, for appropriate J's it is J-unitary as well, so that

$$\begin{cases} E_\Theta J E_\Theta^* &=& J, \\ H_\Theta J E_\Theta^* &=& 0, \\ H_\Theta J H_\Theta^* + K_\Theta J K_\Theta^* &=& J, \end{cases} \qquad \begin{cases} K_\Theta^* J K_\Theta &=& J, \\ H_\Theta^* J K_\Theta &=& 0, \\ H_\Theta^* J H_\Theta + E_\Theta^* J E_\Theta &=& J. \end{cases} \tag{8.26}$$

Let $\mathbf{F}$ be a J-orthonormal basis for $\mathcal{H}(\Theta)$, and let $\mathbf{F}_o = J\mathbf{F}J\Theta$ be the corresponding J-orthonormal basis for $\mathcal{H}_o(\Theta)$. Note that $\mathbf{F}_o$ is also given by $\mathbf{F}_o = J\mathbf{F}JH_\Theta$, so that $\mathbf{F}_o J E_\Theta^* = J\mathbf{F}JH_\Theta J E_\Theta^* = 0$. With the chosen basis, the Hankel operator has a factorization as $H_\Theta = \mathbf{P}_0(\cdot\mathbf{F}^*)\mathbf{F}_o$ and $H_\Theta^* = \mathbf{P}_0(\cdot\mathbf{F}_o^*)\mathbf{F}$, so that

$$H_\Theta^* J H_\Theta \;=\; \mathbf{P}_0(\cdot\mathbf{F}_o^*)\mathbf{P}_0(\mathbf{F}J\mathbf{F}^*)\mathbf{F}_o \;=\; \mathbf{P}_0(\cdot\mathbf{F}_o^*)J\mathbf{F}_o. \tag{8.27}$$

Θ_f of equation (8.24) has the adjoint $\Theta_f^* = [\mathbf{P}_0(\cdot\mathbf{F}_o^*) \quad E_\Theta^*]$, so that (with (8.26))

$$\begin{array}{rcl} \Theta_f J \Theta_f^* &=& \begin{bmatrix} \mathbf{F}_o \\ E_\Theta \end{bmatrix} J\,[\mathbf{P}_0(\cdot\mathbf{F}_o^*) \quad E_\Theta^*] \\ &=& \begin{bmatrix} \mathbf{P}_0(\mathbf{F}_o J\mathbf{F}_o^*) & \mathbf{F}_o J E_\Theta^* \\ \mathbf{P}_0(E_\Theta J\mathbf{F}_o) & E_\Theta J E_\Theta^* \end{bmatrix} \\ &=& \begin{bmatrix} J_\mathcal{B} & 0 \\ 0 & J_\mathcal{M} \end{bmatrix} \end{array}$$

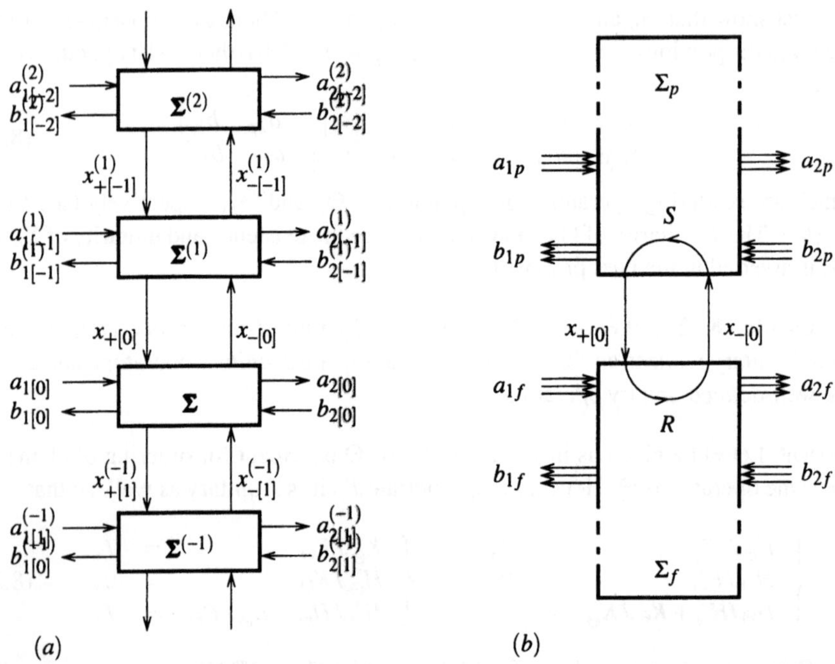

Figure 8.7. (a) The state transition scheme for Σ, (b) The decomposition of Σ into a past operator Σ_p and a future operator Σ_f linked by the state $[x_{+[0]}\ x_{-[0]}]$. This summarizes the figure on the left for all time.

and with (8.27), also

$$\Theta_f^* J \Theta_f = [\mathbf{P}_0(\cdot \mathbf{F}_o^*) \quad E_\Theta^*] J \begin{bmatrix} \mathbf{F}_o \\ E_\Theta \end{bmatrix} = \mathbf{P}_0(\cdot \mathbf{F}_o^*) J \mathbf{F}_o + E_\Theta^* J E_\Theta = J.$$

Hence Θ_f is J-unitary. The J-unitarity of Θ_p follows in a dual way. Θ_f and Θ_p may be considered "normal" J-unitary operators with inputs as given in equations (8.22). Proposition 8.10 then applies and from it follows the existence of unitary scattering matrices Σ_f and Σ_p given by the I/O relations (8.25). □

State space structure of $\Sigma_{22} = \Theta_{22}^{-1}$

Proposition 8.10 shows that if the state signature sequence $J_\mathcal{B} = I$, then $\Sigma_{22} = \Theta_{22}^{-1}$ is upper and Θ is J-inner. In chapter 10 on optimal approximations, an important role is played by J-unitary operators with non-trivial state signature, in which case Θ_{22}^{-1} is generally not upper. In particular, we will be interested in the dimension of the state space sequence $\mathcal{H}(\Theta_{22}^{-*})$ of Θ_{22}^{-1}, determined by the lower (anti-causal) part of Θ_{22}^{-1}: we shall see that this quantity will be characteristic of the complexity of the approximation. To this end, we use in this section a "conjugate-Hankel" operator, defined as

$$H' := H'_{\Theta_{22}^{-1}} = \mathbf{P}'(\cdot \Theta_{22}^{-1})\big|_{\mathcal{U}_2}. \tag{8.28}$$

The definition is such that $\mathcal{H}(\Theta_{22}^{-*}) = \operatorname{ran}(H')$.

Because $\Theta_{22}^{-1} = \Sigma_{22}$, the conjugate-Hankel operator H' defined in (8.28) is a restriction of the partial map $\Sigma_{22} : b_2 \mapsto b_1$. Indeed, $H' : b_{2f} \mapsto b_{1p}$ is such that b_{2f} and b_{1p} satisfy the input-output relations defined by Σ under the conditions $a_1 = 0$ and $b_{2p} = 0$ (see also figure 8.7(b)). H', as a Hankel operator, can be factored as $H' = \sigma\tau$, where the operators

$$\begin{array}{rlll} \sigma: & b_{2f} & \mapsto & x_{-[0]} \\ \tau: & x_{-[0]} & \mapsto & b_{1p} \end{array}$$

can be derived from Σ_f and Σ_p by elimination of $x_{+[0]}$, taking $a_1 = 0$ and $b_{2p} = 0$. We show, in proposition 8.13, that the operator σ is "onto" while τ is "one-to-one", so that the factorization of H' into these operators is minimal. It is even *uniformly* minimal: the state $x_{-[0]}$ is uniformly reachable by b_{2f} (*i.e.*, the range of σ spans $\mathcal{D}_2$), and $x_{-[0]}$ as input of τ is uniformly observable. It follows, in proposition 8.14, that the dimension of $x_{-[0]}$ at each point in time determines the local dimension of the subspace $\mathcal{H}(\Theta_{22}^{-*})$ at that point.

Proposition 8.13 *Let $\Theta \in \mathcal{U}$ be a locally finite J-unitary operator, with J-unitary realization $\mathbf{\Theta}$ such that $\ell_A < 1$. Let $x_+, x_-, a_1, b_1, a_2, b_2$ satisfy (8.22) and (8.25).*

1. *If $a_{1p} = 0$ and $b_{2p} = 0$, then the map $\tau : x_{-[0]} \mapsto b_{1p}$ is one-to-one and boundedly invertible on its range,* i.e.,

$$\exists \varepsilon > 0 : \quad \|b_{1p}\| \geq \varepsilon \|x_{-[0]}\|. \tag{8.29}$$

2. *The map $\sigma: b_{2f} \mapsto x_{-[0]}$ is onto, and moreover, there exists $M < \infty$ such that for any $x_{-[0]}$ there is a b_{2f} in its pre-image such that*

$$\|b_{2f}\| \leq M\|x_{-[0]}\|.$$

PROOF

1. *The map $\tau: x_{-[0]} \mapsto b_{1p}$ is one-to-one.* Since Σ_p is a well-defined bounded operator on the whole space $\mathcal{X}_2^{\mathcal{M}_{1p}} \times \mathcal{X}_2^{\mathcal{N}_{2p}}$, we can put $a_{1p} = 0$ and $b_{2p} = 0$, and specialize equation (8.25) to $[x_{-[0]}\ \ 0\ \ 0]\Sigma_p = [x_{+[0]}\ \ a_{2p}\ \ b_{1p}]$, that is, we have for some $x_{+[0]}$ and a_{2p}
$$[0\ \ b_{1p}]\Theta_p = [x_{+[0]}\ \ x_{-[0]}\ \ a_{2p}\ \ 0]. \tag{8.30}$$
Since Θ_p is bounded, there is an M such that $\|b_{1p}\| < 1 \Rightarrow \|x_{-[0]}\| < M$ and hence, with $\varepsilon = 1/M$: $\|x_{-[0]}\| \geq 1 \Rightarrow \|b_{1p}\| \geq \varepsilon$. It follows that $x_{-[0]} \mapsto b_{1p}$ is one-to-one as claimed, and that (8.29) holds.

2. *The map $\sigma: b_{2f} \mapsto x_{-[0]}$ is onto.* Let be given any $x_{-[0]}$. We have to show that there is a b_{2f} which generates this state via Σ_f. First, with $a_{1p} = 0$ and $b_{2p} = 0$, Σ_p associates a unique b_{1p} and $x_{+[0]}$ to $x_{-[0]}$. Put also $a_{1f} = b_{1f} = 0$, then Θ generates a corresponding b_{2f} as $b_{2f} = b_1\Theta_{22}$. Because Σ_f is well defined, application of Σ_f to $[x_{+[0]}\ 0\ b_{2f}]$ gives again a state $x'_{-[0]}$; but this must be equal to $x_{-[0]}$ because they both generate the same b_{1p} and the map $x_{-[0]} \mapsto b_{1p}$ is one-to-one. Hence this b_{2f} generates the given state $x_{-[0]}$. In addition, we have from $\|b_{1p}\| \leq \|x_{-[0]}\|$ and $\|\Theta\| \leq M < \infty$ that
$$\begin{aligned} \|b_{2f}\| &\leq \|\Theta_{22}\|\,\|b_{1p}\| \\ &\leq M\|x_{-[0]}\|. \end{aligned}$$
This means that the state $x_{-[0]}$ is uniformly reachable by b_{2f} as well. □

Proposition 8.13 is instrumental in proving that the sequence of the number of states x_- of the anti-causal part of Θ_{22}^{-1} is equal to the sequence of ranks of the Hankel operator H'.

Proposition 8.14 *Let $\Theta \in \mathcal{U}$ be a locally finite J-unitary operator, with state signature operator $J_\mathcal{B}$. The s-dimension of $\mathcal{H}(\Theta_{22}^{-*})$ is equal to $\#_-(J_\mathcal{B}) = \#(\mathcal{B}_-)$,* i.e., *the sequence of the number of negative entries in $J_\mathcal{B}$.*

PROOF
$$\begin{aligned} \mathcal{H}(\Theta_{22}^{-*}) &= \mathbf{P}_{\mathcal{L}_2Z^{-1}}(\mathcal{U}_2\Theta_{22}^{-1}) \\ &= \{\mathbf{P}_{\mathcal{L}_2Z^{-1}}(b_{2f}\Theta_{22}^{-1}) : b_{2f} \in \mathcal{U}_2\}. \end{aligned}$$

Put $a_1 = 0$ and $b_{2p} = 0$ so that $b_{1p} = \mathbf{P}_{\mathcal{L}_2Z^{-1}}(b_{2f}\Theta_{22}^{-1})$. The space $\mathcal{H}(\Theta_{22}^{-*}) = \{b_{1p} : b_2 \in \mathcal{U}_2\}$ is generated by the map $H' : b_{2f} \mapsto b_{1p}$. But this map can be split into $\sigma: b_{2f} \mapsto x_{-[0]}$ and $\tau: x_{-[0]} \mapsto b_{1p}$. Because $[x_{-[0]}\ \ 0\ \ 0]\Sigma_p = [x_{+[0]}\ \ a_{2p}\ \ b_{1p}]$, the signal $x_{-[0]}$ determines b_{1p} completely. In proposition 8.13 we showed that $x_{-[0]} \mapsto b_{1p}$ is one-to-one and that $b_{2f} \mapsto x_{-[0]}$ is onto. Hence, the state $x_{-[0]}$ is both uniformly observable in b_{1p} and uniformly reachable by b_{2f}, *i.e.*, its state dimension sequence for the map

$b_{2f} \mapsto b_{1p}$ is minimal at each point in time. Since the number of state variables in $x_{-[0]}$ is given by $\#_-(J_{\mathcal{B}}) = \#(\mathcal{B}_-)$, it follows that

$$\text{sdim } \mathcal{H}(\Theta_{22}^{-*}) = \#(\mathcal{B}_-).$$

□

8.5 J-UNITARY EXTERNAL FACTORIZATION

In this section we investigate external (or J-inner-coprime) factorizations, similar as in chapter 6 for inner factors, but now so that, given $T \in \mathcal{U}(\mathcal{M},\mathcal{N})$ and a signature $J_{\mathcal{M}}$,

$$T^*\Theta = \Delta, \tag{8.31}$$

where $\Theta \in \mathcal{U}$ is J-unitary and Δ is upper. Notice that (8.31) is in a dual form to that used in section 6.2, where we were working with $T = \Delta^* V$, or $VT^* = \Delta$. This is of course not essential, and a dual form of proposition 6.6 holds.

Proposition 8.15 *Let be given operators $T \in \mathcal{U}$ and $\Theta \in \mathcal{U}$. Then $\Delta = T^*\Theta$ is upper if and only if $\mathcal{L}_2 Z^{-1}\Theta^* \subset \mathcal{K}(T)$. If Θ is, in addition, J-unitary, then $\mathcal{L}_2 Z^{-1}\Theta^* J = \mathcal{K}(\Theta)$, and Δ is upper if and only if Θ satisfies*

$$\mathcal{H}(JT) \subset \mathcal{H}(\Theta).$$

The construction of such a Θ is comparable to the construction for inner operators. Assume that T is locally finite, that $\{A,B,C,D\}$ is a realization for T which is uniformly reachable and that $\ell_A < 1$, then $\mathbf{F} = \left(BZ(I-AZ)^{-1}\right)^*$ is a strong basis representation such that $\overline{\mathcal{H}}(T) \subset \mathcal{D}_2\mathbf{F}$ (the latter being necessarily closed). An operator Θ such that $\Delta \in \mathcal{U}$ is obtained by taking $\mathcal{H}(\Theta) = \mathcal{D}_2\mathbf{F}J$, and a J-orthonormal realization of Θ is obtained by making $\mathbf{F}J$ J-orthonormal, which is possible if and only if $\Lambda_{\mathbf{F}}^J = \mathbf{P}_0(\mathbf{F}J\mathbf{F}^*)$ is boundedly invertible, *i.e.*, if $\mathcal{D}_2\mathbf{F}$ is a regular (Krein) space. Let $J_{\mathcal{B}}$ be the signature of $\Lambda_{\mathbf{F}}^J$, then $\Lambda_{\mathbf{F}}^J = R^* J_{\mathcal{B}} R$ for some invertible state transformation R, and hence A_Θ and B_Θ of a J-unitary realization are given by

$$\begin{bmatrix} A_\Theta \\ B_\Theta \end{bmatrix} = \begin{bmatrix} R & \\ & I \end{bmatrix} \begin{bmatrix} A \\ JB \end{bmatrix} R^{-(-1)}. \tag{8.32}$$

It remains to complete this realization such that

$$\mathbf{\Theta} = \begin{bmatrix} A_\Theta & C_\Theta \\ B_\Theta & D_\Theta \end{bmatrix}$$

is $(\mathbf{J}_1,\mathbf{J}_2)$-unitary. This step is less obvious than for inner systems, so we first prove an additional lemma before stating the main theorem.

Lemma 8.16 *Let be given finite matrices α, β, and signature matrices j_1, j_2, j_3 such that*

$$\alpha^* j_1 \alpha + \beta^* j_2 \beta = j_3.$$

Then there exist matrices γ, δ *and a signature matrix* j_4 *such that* $\Theta = \left[\begin{smallmatrix}\alpha & \gamma \\ \beta & \delta\end{smallmatrix}\right]$ *is a* J*-unitary matrix, in the sense*

$$\Theta^* \begin{bmatrix} j_1 & \\ & j_2 \end{bmatrix} \Theta = \begin{bmatrix} j_3 & \\ & j_4 \end{bmatrix}, \qquad \Theta \begin{bmatrix} j_3 & \\ & j_4 \end{bmatrix} \Theta^* = \begin{bmatrix} j_1 & \\ & j_2 \end{bmatrix}.$$

PROOF Suppose that α is an $(m_\alpha \times n_\alpha)$-dimensional matrix, and $\beta : (m_\beta \times n_\alpha)$. It is clear that if an extension exists, then j_4 is specified by the inertia relations:

$$\begin{aligned} \#_+(j_4) &= \#_+(j_1) + \#_+(j_2) - \#_+(j_3) \\ \#_-(j_4) &= \#_-(j_1) + \#_-(j_2) - \#_-(j_3). \end{aligned}$$

Since the first block column of Θ is already J-isometric,

$$[\alpha^* \;\; \beta^*] \begin{bmatrix} j_1 & \\ & j_2 \end{bmatrix} \begin{bmatrix} \alpha \\ \beta \end{bmatrix} = j_3\,,$$

it remains to show that it can be completed to a J-unitary matrix. Because j_3 is non-singular, the n_α columns of $\left[\begin{smallmatrix}\alpha \\ \beta\end{smallmatrix}\right]$ are linearly independent. Choose a matrix $\left[\begin{smallmatrix}c \\ d\end{smallmatrix}\right]$ with $m_\alpha + m_\beta - n_\alpha$ independent columns such that

$$[\alpha^* \;\; \beta^*] \begin{bmatrix} j_1 & \\ & j_2 \end{bmatrix} \begin{bmatrix} c \\ d \end{bmatrix} = 0 \tag{8.33}$$

and such that the columns of $\left[\begin{smallmatrix}c \\ d\end{smallmatrix}\right]$ form a basis for the orthogonal complement of the column span of $\left[\begin{smallmatrix}j_1\alpha \\ j_2\beta\end{smallmatrix}\right]$. We claim that the square matrix $\left[\begin{smallmatrix}\alpha & c \\ \beta & d\end{smallmatrix}\right]$ is invertible. To prove this, it is enough to show that its null space is zero. Suppose that

$$\begin{bmatrix} \alpha & c \\ \beta & d \end{bmatrix} \begin{bmatrix} x_1 \\ x_2 \end{bmatrix} = \begin{bmatrix} 0 \\ 0 \end{bmatrix}$$

then

$$[\alpha^* \;\; \beta^*] \begin{bmatrix} j_1 & \\ & j_2 \end{bmatrix} \begin{bmatrix} \alpha & c \\ \beta & d \end{bmatrix} \begin{bmatrix} x_1 \\ x_2 \end{bmatrix} = \begin{bmatrix} j_3 x_1 \\ 0 \end{bmatrix} = \begin{bmatrix} 0 \\ 0 \end{bmatrix}.$$

Hence $x_1 = 0$ and $\left[\begin{smallmatrix}c \\ d\end{smallmatrix}\right] x_2 = 0$. But the columns of $\left[\begin{smallmatrix}c \\ d\end{smallmatrix}\right]$ are linearly independent, so that $x_2 = 0$. Thus

$$\begin{bmatrix} \alpha^* & \beta^* \\ c^* & d^* \end{bmatrix} \begin{bmatrix} j_1 & \\ & j_2 \end{bmatrix} \begin{bmatrix} \alpha & c \\ \beta & d \end{bmatrix} = \begin{bmatrix} j_3 & \\ & N \end{bmatrix}$$

where N is a square invertible matrix. By the usual inertia argument, the signature of N is equal to j_4, and hence N has a factorization $N = R^* j_4 R$, where R is invertible. Thus putting

$$\begin{bmatrix} \gamma \\ \delta \end{bmatrix} = \begin{bmatrix} c \\ d \end{bmatrix} R^{-1}, \qquad \Theta = \begin{bmatrix} \alpha & \gamma \\ \beta & \delta \end{bmatrix}$$

ensures that Θ is j-unitary as required (we are indepted to H. Dym for this elegant argument).

□

Theorem 8.17 *Let be given a subspace $\mathcal{H} = \mathcal{D}_2^{\mathcal{B}} \mathbf{F} J$ in $\mathcal{Z}^{-1}\mathcal{L}_2^{\mathcal{M}}$, specified by a bounded basis representation $\mathbf{F} = \left(BZ(I-AZ)^{-1}\right)^*$, ($\ell_A < 1$), which is such that $\Lambda_{\mathbf{F}}^J$ is boundedly invertible. Then there exists a bounded J-unitary operator $\Theta \in \mathcal{U}(\mathcal{M},\mathcal{N}_\Theta)$ such that $\mathcal{H} = \mathcal{H}(\Theta)$. Θ is unique up to a right diagonal J-unitary factor. Its output space sequence, $\mathcal{N}_\Theta$ has dimension sequences given by*

$$\begin{aligned} \#(\mathcal{N}_\Theta)_+ &= \#\mathcal{B}_+ - \#\mathcal{B}_+^{(-1)} + \#\mathcal{M}_+ \geq 0 \\ \#(\mathcal{N}_\Theta)_- &= \#\mathcal{B}_- - \#\mathcal{B}_-^{(-1)} + \#\mathcal{M}_- \geq 0\,. \end{aligned} \tag{8.34}$$

PROOF Since $\Lambda_{\mathbf{F}}^J$ is boundedly invertible, there is a signature operator $J_{\mathcal{B}}$ and a boundedly invertible operator $R \in \mathcal{D}$ such that $\Lambda_{\mathbf{F}}^J = R^* J_{\mathcal{B}} R$. The signature $J_{\mathcal{B}}$ implies a space sequence decomposition $\mathcal{B} = \mathcal{B}_+ \times \mathcal{B}_-$, and since $\Lambda_{\mathbf{F}}^J$ satisfies the Lyapunov equation

$$A^*\Lambda_{\mathbf{F}}^J A + B^* J_{\mathcal{M}} B = (\Lambda_{\mathbf{F}}^J)^{(-1)}.$$

A_Θ, B_Θ, given by

$$\begin{bmatrix} A_\Theta \\ B_\Theta \end{bmatrix} = \begin{bmatrix} R & \\ & I \end{bmatrix} \begin{bmatrix} A \\ JB \end{bmatrix} R^{-(-1)}$$

form a J-isometric block column with diagonal entries. We proceed with the construction of a realization $\mathbf{\Theta}$ of the form

$$\mathbf{\Theta} = \begin{bmatrix} A_\Theta & C_\Theta \\ B_\Theta & D_\Theta \end{bmatrix} = \begin{bmatrix} R & \\ & I \end{bmatrix} \begin{bmatrix} A & C' \\ JB & D_\Theta \end{bmatrix} \begin{bmatrix} R^{-(-1)} & \\ & I \end{bmatrix} \tag{8.35}$$

which is a square matrix at each point k, and where C_Θ (or C') and D_Θ are yet to be determined. $\mathbf{\Theta}$ is to satisfy $\mathbf{\Theta}^*\mathbf{J}_1\mathbf{\Theta} = \mathbf{J}_2$, $\mathbf{\Theta}\mathbf{J}_2\mathbf{\Theta}^* = \mathbf{J}_1$, for

$$\mathbf{J}_1 = \begin{bmatrix} J_{\mathcal{B}} & \\ & J_{\mathcal{M}} \end{bmatrix}, \qquad \mathbf{J}_2 := \begin{bmatrix} J_{\mathcal{B}}^{(-1)} & \\ & J_{\mathcal{N}_\Theta} \end{bmatrix}$$

where $J_{\mathcal{N}_\Theta}$ is still to be determined, and with it the dimensionality of the output space sequence $\mathcal{N}_\Theta$. However, since all other signatures are known at this point, these follow from the inertia property (equation (8.14)) as the space sequence with dimensions given by (8.34). To obtain $\mathbf{\Theta}$, it remains to show that $\left[\begin{smallmatrix} A_\Theta \\ B_\Theta \end{smallmatrix}\right]$ can be completed to form $\mathbf{\Theta}$ in (8.35), in such a way that the whole operator is J-unitary. This completion can be achieved for each point k individually with local computations, and exists as was shown in lemma 8.16. Since $\mathbf{\Theta}$ is J-unitary and $\ell_A < 1$, theorem 8.9 implies that the corresponding operator Θ is J-unitary. Finally, $\mathcal{H}(\Theta) = \mathcal{H}$ by construction.

The uniqueness of Θ, up to a left diagonal J-unitary factor, is proven in the same way as for inner operators in the Beurling-Lax like theorem 6.13. Indeed, let Θ_1 be another J-unitary operator such that $\mathcal{H} = \mathcal{H}(\Theta_1)$, then $\mathcal{K} = \mathcal{L}_2 Z^{-1} \ominus \mathcal{H} = \mathcal{L}_2 Z^{-1}\Theta^* = \mathcal{L}_2 Z^{-1}\Theta_1^*$, so that

$$\begin{aligned} \mathcal{L}_2 Z^{-1}\Theta^* J\Theta_1 &= \mathcal{L}_2 Z^{-1} \\ \mathcal{L}_2 Z^{-1}\Theta_1^* J\Theta &= \mathcal{L}_2 Z^{-1} \end{aligned}$$

which implies $\Theta^* J\Theta_1 \in \mathcal{D}$, say $\Theta^* J\Theta_1 = JD$, where $D \in \mathcal{D}$. Then $\Theta_1 = \Theta D$, and D must be J-unitary. □

Corollary 8.18 *Let $T \in \mathcal{U}(\mathcal{M},\mathcal{N})$ be a locally finite operator with uniformly reachable realization $\{A,B,C,D\}$ such that $\ell_A < 1$, and let be given a signature matrix $J_{\mathcal{M}}$. If the solution Λ of the J-Lyapunov equation*

$$A^*\Lambda A + B^* J_{\mathcal{M}} B = \Lambda^{(-1)} \tag{8.36}$$

is such that Λ is boundedly invertible, then there exists a bounded J-unitary operator $\Theta \in \mathcal{U}(\mathcal{M},\mathcal{N}_\Theta)$ such that

$$T^*\Theta = \Delta \in \mathcal{U}.$$

The state signature $J_{\mathcal{B}}$ equal to the inertia of Λ. $\mathcal{N}_\Theta$ and its signature are determined by equation (8.34). In particular, if $\Lambda \gg 0$ then Θ is J-inner.

PROOF The conditions imply that the subspace $\mathcal{H} = \mathcal{D}_2\mathbf{F}J = \mathcal{D}_2\left(BZ(I-AZ)^{-1}\right)^* J$ has $\Lambda_{\mathbf{F}}^J = \Lambda$ boundedly invertible. Hence theorem 8.17 asserts that there is a J-unitary operator Θ such that $\mathcal{H}(\Theta) = \mathcal{H}$. Note that a necessary condition for Λ to be invertible is that the given realization be uniformly reachable, so that

$$\overline{\mathcal{H}}(JT) = \overline{\mathcal{H}}(T)J \subset \mathcal{D}_2\mathbf{F}J = \mathcal{H} = \mathcal{H}(\Theta).$$

This in turn implies that $\Delta = T^*\Theta$ is upper. □

For later use, we evaluate $\Delta = T^*\Theta$. Instead of C_Θ, we use $C' = R^{-1}C_\Theta$ (see equation (8.35)), as A_Δ will become equal to the original A in this case. We also apply the relation $J_{\mathcal{B}}(\mathbf{F}J)J_{\mathcal{M}}\Theta = \mathbf{F}_o$, which in case $\ell_A < 1$ reads

$$(I-Z^*A_\Theta^*)^{-1}Z^*B_\Theta^* J\Theta = J_{\mathcal{B}}(I-A_\Theta Z)^{-1}C_\Theta.$$

Thus

$$\begin{aligned}
\Delta = T^*\Theta &= [D^* + C^*(I-Z^*A^*)^{-1}Z^*B^*]\Theta \\
&= D^*[D_\Theta + B_\Theta Z(I-A_\Theta Z)^{-1}C_\Theta] + C^*R^*(I-Z^*A_\Theta^*)^{-1}Z^*B_\Theta^* J\Theta \\
&= D^*D_\Theta + D^*B_\Theta Z(I-A_\Theta Z)^{-1}C_\Theta + C^*R^*J_{\mathcal{B}}(I-A_\Theta Z)^{-1}C_\Theta \\
&= D^*D_\Theta + D^*JBZ(I-AZ)^{-1}C' + C^*\Lambda^J(I-AZ)^{-1}C'.
\end{aligned}$$

Consequently,

$$\Delta = T^*\Theta = \left\{D^*D_\Theta + C^*\Lambda^J C'\right\} + \left\{D^*JB + C^*\Lambda^J A\right\}Z(I-AZ)^{-1}C', \tag{8.37}$$

where $\Lambda^J = \Lambda$ is given by (8.36) and C' by (8.35).

8.6 J-LOSSLESS AND J-INNER CHAIN SCATTERING OPERATORS

As defined before, a unitary scattering operator Σ is called *inner* if it is also causal. Similarly, an isometric scattering operator which can be embedded into a unitary operator is called *lossless* if it is also causal.[3]. The corresponding chain scattering operator Θ

[3]Just "isometric" is not good enough: we can easily realize a resistor of 1 ohm with an infinite, seemingly lossless transmission line!

is called J-inner respectively J-lossless. Even though we have only considered causal Θ's so far, it is important to note that in general it may be causal, anticausal or of mixed causality, even when the corresponding Σ is inner. The topic of this section and the next is to study conditions under which a general $\Theta \in \mathcal{X}$ corresponds to a lossless system. As done so far, we assume that the chain scattering operators under consideration are bounded and have u.e. stable realizations.

Causal J-lossless and J-inner operators

Proposition 8.19 *Suppose that $\Theta \in \mathcal{U}$ is a locally finite causal J-isometric operator with a canonical realization which is u.e. stable ($\ell_A < 1$). If the output state space $\mathcal{H}_o$ of Θ is uniformly J-positive, then there exists an extension Θ' of Θ which is J-inner.*

PROOF Since it is assumed uniformly J-positive, the output state space $\mathcal{H}_o$ has a J-orthonormal basis $\mathbf{F}_o$ whose J-Gramian is $\Lambda^J_{\mathbf{F}_o} = I_{\mathcal{B}}$. By lemma 8.16 and the actual construction therein, the corresponding J-isometric realization $\mathbf{\Theta} = \left[\begin{smallmatrix} A & C \\ B & D \end{smallmatrix}\right]$ can be completed to a J-unitary realization $\mathbf{\Theta}'$ by the adjunction of an appropriate number of rows:

$$\mathbf{\Theta}' = \begin{bmatrix} A & C \\ B & D \\ B' & D' \end{bmatrix}. \tag{8.38}$$

Since $\mathbf{\Theta}'$ is J-unitary, it is a realization of a J-unitary chain scattering operator Θ', and by proposition 8.10 the corresponding Σ' is inner. □

Anticausal J-inner chain scattering operators

We now consider the case where Θ is J-unitary and *anticausal.* When does Θ correspond to an inner scattering operator Σ, in other words, when is Θ J-inner? Clearly, Θ^* is causal and the previous theory applies to it in a dual sense.

Proposition 8.20 *Suppose that $\Theta \in \mathcal{L}$ is an anticausal, J-unitary operator, and that it has a minimal realization*

$$\Theta = \begin{bmatrix} B_1 \\ B_2 \end{bmatrix} Z^*(I - AZ^*)^{-1} [C_1 \quad C_2] + D$$

with $\ell_A < 1$. The corresponding scattering operator Σ is causal (hence inner) if and only if the Lyapunov-Stein equation

$$-A^* P^{(-1)} A + B_1^* B_1 - B_2^* B_2 = -P \tag{8.39}$$

has a strictly positive definite solution P.

PROOF The realization for Θ can be written as

$$[x_+ \quad x_- \quad a_2 \quad b_2] = [x_+^{(-1)} \quad x_-^{(-1)} \quad a_1 \quad b_1]\mathbf{\Theta}, \qquad \mathbf{\Theta} = \begin{bmatrix} A & C_1 & C_2 \\ B_1 & D_{11} & D_{12} \\ B_2 & D_{21} & D_{22} \end{bmatrix}.$$

The corresponding scattering operator Σ is defined by rearranging state, input and output variables to

$$[x_+ \quad x_-^{(-1)} \quad a_2 \quad b_1] = [x_+^{(-1)} \quad x_- \quad a_1 \quad b_2]\mathbf{\Sigma}.$$

This realization is unitary, and defines a causal operator if and only if the anticausal state dimension is empty: $\#\mathcal{B}_+ = 0$ — dual to proposition 8.14. This is equivalent to requiring that the J-Gramians of the chosen basis representations are uniformly negative, which leads to (8.39) if we do not insist on the realization to be J-unitary.

□

Conjugation

A causal, J-inner operator can be transformed into an anticausal one under a broad general condition of "regularity". This operation, which we introduce in this section, is a form of duality, called *conjugation* here. It has nice applications in signal processing and in interpolation theory, considered in the next chapter. The standard external factorization introduced in chapter 6 does not pull the trick: suppose that $\Theta = U\Delta^*$ with U inner, then the conjugate factor Δ will usually not be J-unitary. However, if we do the external factorization in a block-row fashion, then interesting results appear. Let us assume that external factorizations of the block rows of Θ exist:

$$\begin{aligned} \Theta_+ = [\Theta_{11} \quad \Theta_{12}] &= U[\Theta_{11}^c \quad \Theta_{12}^c] \\ \Theta_- = [\Theta_{21} \quad \Theta_{22}] &= W[\Theta_{21}^c \quad \Theta_{22}^c] \end{aligned} \tag{8.40}$$

where U, W are inner and $\Theta_{ij}^c \in \mathcal{L}$, so that

$$\begin{bmatrix} \Theta_{11}^c & \Theta_{12}^c \\ \Theta_{21}^c & \Theta_{22}^c \end{bmatrix} = \begin{bmatrix} U^* & \\ & W^* \end{bmatrix} \begin{bmatrix} \Theta_{11} & \Theta_{12} \\ \Theta_{21} & \Theta_{22} \end{bmatrix} \tag{8.41}$$

is in $\mathcal{L}$. Now, much more is true: in the next proposition we show that Θ^c is in fact J-inner. To ensure the existence of the factorizations (8.40), we will need the following technical condition

(TC) the part of the reachability Gramian of $\mathbf{\Theta}$ corresponding to the subsystem Θ_- has a closed range.

Proposition 8.21 *Let $\Theta \in \mathcal{U}$ be a causal J-inner operator with a J-unitary realization for which the transition operator $\alpha$for has $\ell_\alpha < 1$ and for which condition (TC) above is satisfied. Then the conjugate operator given by (8.41) is anticausal and J-inner, and has a state transition operator which is a suboperator of α^*.*

PROOF Let the J-unitary realization for the J-inner operator Θ be given by

$$\mathbf{\Theta} = \begin{bmatrix} \alpha & \gamma_1 & \gamma_2 \\ \beta_1 & \delta_{11} & \delta_{12} \\ \beta_2 & \delta_{21} & \delta_{22} \end{bmatrix}$$

such that $\ell_\alpha < 1$. (We use greek symbols here to distinguish the canonical J-unitary realizations.) Since Θ is J-inner, we have, by corollary 8.18, that

$$\alpha^*\alpha + \beta_1^*\beta_1 - \beta_2^*\beta_2 = I. \tag{8.42}$$

Let the positive diagonal operators M_U and M_W be defined by the Lyapunov-Stein equations:

$$\begin{aligned} \alpha^* M_U \alpha + \beta_1^*\beta_1 &= M_U^{(-1)} && (8.43)\\ \alpha^* M_W \alpha + \beta_2^*\beta_2 &= M_W^{(-1)}. && (8.44) \end{aligned}$$

Subtracting (8.43) from (8.42) and adding (8.44) produces

$$\alpha^*(I - M_U + M_W)\alpha = (I - M_U + M_W)^{(-1)}$$

and hence

$$I - M_U + M_W = 0. \tag{8.45}$$

Clearly $M_U \gg 0$, since $M_U = I + M_W$ and $M_W \geq 0$. However, M_W may be singular, which happens if not all states are reachable from the "negative" inputs. Let's first prove the proposition for the case where $M_W \gg 0$. Subsequently we will demonstrate how (under conditions of regularity) the more general situation can be reduced to this case.

- Thus suppose that $M_W \gg 0$. We proceed by computing external factorizations of $[\Theta_{11}\ \Theta_{12}]$ and $[\Theta_{21}\ \Theta_{22}]$. Since $\ell_\alpha < 1$ and the realizations are uniformly reachable ($M_U \gg 0$, $M_W \gg 0$), application of theorem 6.8 produces (unnormalized) realizations for the respective inner factors of the form

$$\mathbf{U} = \begin{bmatrix} \alpha & C_U \\ \beta_1 & D_U \end{bmatrix}, \quad \mathbf{W} = \begin{bmatrix} \alpha & C_W \\ \beta_2 & D_W \end{bmatrix}, \tag{8.46}$$

where, in particular

$$\left\{ \begin{aligned} \alpha M_U^{-(-1)} \alpha^* + C_U C_U^* &= M_U^{-1} \\ \alpha M_W^{-(-1)} \alpha^* + C_W C_W^* &= M_W^{-1}. \end{aligned} \right. \tag{8.47}$$

Realizations for the corresponding external factors are obtained from the result in equation (6.8). For $[\Theta_{11}^c\ \Theta_{12}^c] = U^*[\Theta_{11}\ \Theta_{12}]$ we find that it is anticausal with anticausal realization

$$\begin{bmatrix} \alpha^* & \alpha^* M_U \gamma_1 + \beta_1^* \delta_{11} & \alpha^* M_U \gamma_2 + \beta_1^* \delta_{12} \\ C_U^* & C_U^* M_U \gamma_1 + D_U^* \delta_{11} & C_U^* M_U \gamma_2 + D_U^* \delta_{12} \end{bmatrix}. \tag{8.48}$$

Likewise, $[\Theta_{21}^c\ \Theta_{22}^c] = W^*[\Theta_{21}\ \Theta_{22}]$ is anticausal with realization

$$\begin{bmatrix} \alpha^* & \alpha^* M_W \gamma_1 + \beta_2^* \delta_{21} & \alpha^* M_W \gamma_2 + \beta_2^* \delta_{22} \\ C_W^* & C_W^* M_W \gamma_1 + D_W^* \delta_{21} & C_W^* M_W \gamma_2 + D_W^* \delta_{22} \end{bmatrix}. \tag{8.49}$$

The key is now to show that the two anticausal realizations have equal (1,1), (1,2) and (1,3) entries, so that they can be combined into a single realization with the state dimensions of α^*. This follows directly from $M_U = I + M_W$ and the J-unitarity of the original realization:

$$\alpha^*\gamma_1 + \beta_1^*\delta_{11} - \beta_2^*\delta_{21} = 0, \qquad \alpha^*\gamma_2 + \beta_1^*\delta_{12} - \beta_2^*\delta_{22} = 0.$$

Hence we find as anticausal and unnormalized realization for Θ^c

$$\mathbf{\Theta}^c = \begin{bmatrix} \alpha^* & \alpha^* M_W\gamma_1 + \beta_2^*\delta_{21} & \alpha^* M_W\gamma_2 + \beta_2^*\delta_{22} \\ C_U^* & C_U^*(I+M_W)\gamma_1 + D_U^*\delta_{11} & C_U^*(I+M_W)\gamma_2 + D_U^*\delta_{12} \\ C_W^* & C_W^* M_W\gamma_1 + D_U^*\delta_{21} & C_W^* M_W\gamma_2 + D_W^*\delta_{22} \end{bmatrix}. \tag{8.50}$$

We know already that Θ^c is J-unitary (by construction). It remains to show that it is actually J-inner. By proposition 8.20, this will be the case if there exists a strictly positive solution to the Lyapunov-Stein equation

$$-\alpha P^{(-1)}\alpha^* + C_U C_U^* - C_W C_W^* = -P. \tag{8.51}$$

Using (8.47), it follows that the solution of this equation is

$$P = M_W^{-1}(I+M_W)^{-1}$$

which is indeed strictly positive definite. This proves the proposition for $M_W \gg 0$.

- We now investigate the general case where M_W in (8.45) may be singular. Let $R \in \mathcal{D}$ be a unitary operator such that

$$M_W = R^* \begin{bmatrix} M_W' & 0 \\ 0 & 0 \end{bmatrix} R$$

where $M_W' \in \mathcal{D}$ is of minimal dimensions. Under condition (TC), $M_W' \gg 0$, because M_W is the reachability Gramian of the realization of Θ_-. (See also section 5.3 for this.) If we apply the state transformation R, we obtain a new, equivalent state realization for Θ given by

$$\mathbf{\Theta} = \left[\begin{array}{cc|cc} \alpha_{11} & 0 & \gamma_{11} & \gamma_{12} \\ \alpha_{21} & \alpha_{22} & \gamma_{21} & \gamma_{22} \\ \hline \beta_{11} & \beta_{12} & \delta_{11} & \delta_{12} \\ \beta_{21} & 0 & \delta_{21} & \delta_{22} \end{array}\right],$$

which is still J-unitary, but now exhibits a part of the state space connected to the pair $[\alpha, \beta_2]$ which is unreachable by the "negative" inputs. Hence, this part must be purely inner, and can be factored out as follows. Clearly, $\left[\begin{smallmatrix} \alpha_{22} \\ \beta_{12} \end{smallmatrix}\right]$ is isometric and can be completed to the realization of an inner operator U_1 with unitary realization

$$\mathbf{U}_1 = \begin{bmatrix} \alpha_{22} & C_{U_1} \\ \beta_{12} & D_{U_1} \end{bmatrix}.$$

Since the second block column of $\boldsymbol{\Theta}$ is orthogonal to the others, it follows that

$$\left[\begin{array}{cc|c|c} I & & & \\ & \alpha_{22}^* & \beta_{12}^* & \\ \hline & C_{U_1}^* & D_{U_1}^* & \\ & & & I \end{array}\right]\left[\begin{array}{cc|cc} \alpha_{11} & 0 & \gamma_{11} & \gamma_{12} \\ \alpha_{21} & \alpha_{22} & \gamma_{21} & \gamma_{22} \\ \hline \beta_{11} & \beta_{12} & \delta_{11} & \delta_{12} \\ \beta_{21} & 0 & \delta_{21} & \delta_{22} \end{array}\right] = \left[\begin{array}{cc|cc} \alpha_{11} & 0 & \gamma_{11} & \gamma_{12} \\ 0 & I & 0 & 0 \\ \hline \beta'_{11} & 0 & \delta'_{11} & \delta'_{12} \\ \beta_{21} & 0 & \delta_{21} & \delta_{22} \end{array}\right],$$

for certain β'_{11}, δ'_{11}, δ'_{12} defined by the equation. Thus we have constructed the factorization (*cf.* equation (3.17))

$$\Theta = \begin{bmatrix} U_1 & \\ & I \end{bmatrix} \Theta' \tag{8.52}$$

where $\Theta' \in \mathcal{U}$ has the J-unitary realization

$$\boldsymbol{\Theta}' = \left[\begin{array}{c|cc} \alpha_{11} & \gamma_{11} & \gamma_{12} \\ \hline \beta'_{11} & \delta'_{11} & \delta'_{12} \\ \beta_{21} & \delta_{21} & \delta_{22} \end{array}\right]. \tag{8.53}$$

It is easy to verify that Θ' is J-inner, and it has $M'_U \gg 0$, $M'_W \gg 0$ by construction. This brings us back to the case considered before. □

Θ^c may be called a *conjugate* of Θ'. The inner operators U and W which enter in its construction provide external factorizations for each of its block entries.

8.7 THE MIXED CAUSALITY CASE

We now consider J-isometric and J-unitary chain scattering operators Θ of mixed type, having both a causal and an anticausal part, and give special attention to the J-lossless and J-inner cases. We restrict ourselves to bounded, locally finite operators which have a bounded partitioning into upper and lower triangular parts and have u.e. stable realizations:

$$\Theta = B_1 Z(I - A_1 Z)^{-1} C_1 + D + B_2 Z^* (I - A_2 Z^*)^{-1} C_2 \tag{8.54}$$

in which $\ell_{A_1} < 1$ and $\ell_{A_2} < 1$. Clearly, $B_1 Z(I - A_1 Z)^{-1} C_1$ and $B_2 Z^* (I - A_2 Z^*)^{-1} C_2$ are the strictly causal and anticausal parts of Θ respectively.

A state space description with transfer operator given by (8.54) and corresponding to figure 8.8(a) is given by

$$[x_1^{(-1)} \quad x_2 \mid y] = [x_1 \quad x_2^{(-1)} \mid u] \left[\begin{array}{cc|c} A_1 & & C_1 \\ & A_2 & C_2 \\ \hline B_1 & B_2 & D \end{array}\right]. \tag{8.55}$$

Because J-isometric properties are hard to test on a sum of two realizations, we will be interested in expressing the given realization as the product of a causal and an anticausal factor. Once we have a factorization, it is immediately clear that Θ is J-isometric if its factors are J-isometric, *i.e.*, if the realizations of each of the factors are J-isometric.

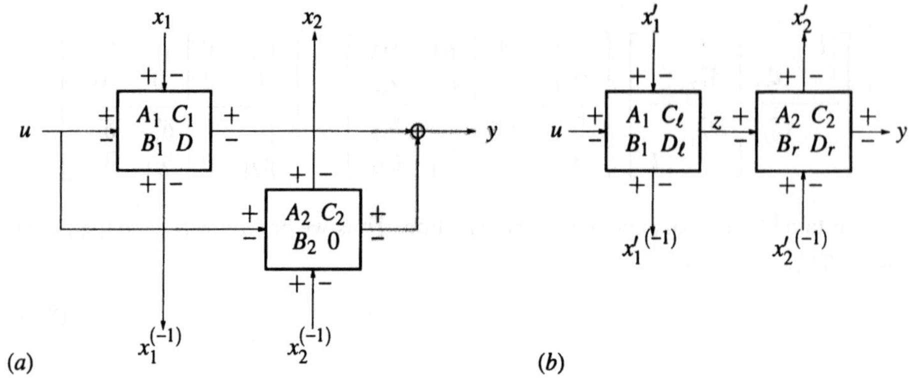

Figure 8.8. (a) A mixed causal/anticausal computational scheme for Θ, (b) an equivalent factored realization, obtained after a state transformation.

Minimal causal-anticausal factorizations

Theorem 8.22 *Let $\Theta \in \mathcal{X}$ be J-isometric ($\Theta J_2 \Theta^* = J_1$), with a locally finite, minimal, u.e. stable realization (8.55) for which (A_2, C_2) is uniformly observable. Define the spaces*

$$\begin{aligned} \mathcal{H}_1 &:= \mathbf{P}'(\mathcal{U}_2\Theta^*) \\ \mathcal{H}_2 &:= \mathbf{P}'(\mathcal{U}_2\Theta) \end{aligned}$$

and suppose that $\mathcal{H}_2$ is a regular space (a Krein space). Then Θ has a factorization as

$$\Theta = \Theta_\ell \Theta_r, \qquad \Theta_\ell \in \mathcal{U},\ \Theta_r \in \mathcal{L},$$

in which Θ_ℓ is J-isometric, Θ_r is J-unitary, and

$$\begin{aligned} \mathcal{H}_1 &= \mathbf{P}'(\mathcal{U}_2\Theta_\ell^*) && \text{(the input state space of } \Theta_\ell) \\ \mathcal{H}_2 &= \mathbf{P}'(\mathcal{U}_2\Theta_r) && \text{(the anti-causal output state space of } \Theta_r). \end{aligned}$$

PROOF $\mathcal{H}_2$ is a (locally finite) regular space with a strong basis generated by (A_2, C_2). Hence, this basis has a non-singular J-Gramian, so that the J-unitary external factorization in corollary 8.18, applied to the lower triangular part of Θ (call it T^*) produces a J-unitary $\Theta_r \in \mathcal{L}$ such that $\Delta = T^*\Theta_r^* \in \mathcal{U}$. Since the upper triangular part of Θ is kept upper, this implies that there exists a factorization $\Theta = \Theta_\ell\Theta_r$ where $\Theta_\ell \in \mathcal{U}$ and $\Theta_r \in \mathcal{L}$ is J-unitary. Moreover, by construction, $\mathcal{H}_2 = \mathbf{P}'(\mathcal{U}_2\Theta_r)$. Since it immediately follows that Θ_ℓ is J-isometric, we only have to show that $\mathcal{H}_1$ is equal to

$$\mathcal{H}' := \mathbf{P}'(\mathcal{U}_2\Theta_\ell^*) = \mathbf{P}'(\mathcal{U}_2\Theta_r J \Theta^*)\,. \tag{8.56}$$

Since Θ_r is J-unitary, we have from proposition 8.5 applied to Θ_r^* that $\mathcal{U}_2 = \mathcal{H}_2 J \Theta_r^* \boxplus \mathcal{U}_2\Theta_r^*$, which implies that $\mathcal{U}_2$ is formed by the span of these two components: $\mathcal{U}_2 =$

$\mathcal{H}_2 J\Theta_r^* \dot{+} \mathcal{U}_2\Theta_r^*$. Using $\Theta_r^* = J\Theta^{-1}J$, it follows that $\mathcal{U}_2\Theta_r = \mathcal{H}_2 \dot{+} \mathcal{U}_2$. (These spaces are actually orthogonal: $\mathcal{U}_2\Theta_r = \mathcal{H}_2 \oplus \mathcal{U}_2$.) Substitution into (8.56) gives

$$\mathcal{H}' = \mathbf{P}'(\mathcal{H}_2 J\Theta^*) \dot{+} \mathbf{P}'(\mathcal{U}_2 J\Theta^*)\,. \tag{8.57}$$

The definition of $\mathcal{H}_2$ and the J-isometry of Θ ensure that $\mathbf{P}'(\mathcal{H}_2 J\Theta^*) \subset \mathbf{P}'(\mathcal{U}_2 J\Theta^*)$:

$$\mathbf{P}'(\mathcal{H}_2 J\Theta^*) = \mathbf{P}'\left(\mathbf{P}'(\mathcal{U}_2\Theta) J\Theta^*\right) \subset \mathbf{P}'\left((\mathcal{U}_2\Theta \dot{+} \mathcal{U}_2) J\Theta^*\right) = \mathbf{P}'(\mathcal{U}_2 J\Theta^*).$$

It follows that $\mathcal{H}' = \mathcal{H}_1$. □

Since Θ_ℓ can be extended to a J-unitary operator if its input state space is regular, we also have the following result.

Corollary 8.23 *Let $\Theta \in \mathcal{X}$ be J-isometric. Under the hypotheses of theorem 8.22, and in addition assuming that $\mathcal{H}_1$ is regular, Θ has a J-unitary extension of the same state complexity.*

The factorization of Θ into $\Theta_\ell\Theta_r$ can be called minimal (in a state complexity sense) since the state complexity of each of the factors add up to the state complexity of Θ itself. Θ_ℓ has a realization given by $\Theta_\ell = D_\ell + B_1 Z(I - A_1 Z)^{-1} C_\ell$ for certain D_ℓ, C_ℓ, and Θ_r has a realization $\Theta_r = D_r + B_r Z^*(I - A_2 Z^*)^{-1} C_2$, for certain D_r, B_r. Since Θ_r is J-unitary, the extension of $[A_2\ C_2]$ to a J-unitary realization is more or less unique, so $[B_r\ D_r]$ can directly be computed. In contrast, B_ℓ and D_ℓ cannot be found by extension, but have to be computed such that the factorization holds. This can be done as follows.

Let

$$\begin{aligned} \Theta_\ell &= D_\ell + B_1 Z(I - A_1 Z)^{-1} C_\ell \\ \Theta_r &= D_r + B_r Z^*(I - A_2 Z^*)^{-1} C_2\,. \end{aligned}$$

In general, the product of two operators of this form (one upper, one lower) is given by

$$\begin{aligned} \Theta_\ell\Theta_r = \ & (D_\ell D_r + B_1 Y^{(-1)} C_2) \ + \ B_1 Z(I - A_1 Z)^{-1}\{C_\ell D_r + A_1 Y^{(-1)} C_2\} \\ & + \{D_\ell B_r + B_1 Y^{(-1)} A_2\} Z^*(I - A_2 Z^*)^{-1} C_2 \end{aligned}$$

where Y is the solution of $Y = A_1 Y^{(-1)} A_2 + C_\ell B_r$. Thus we have

$$\begin{aligned} \Theta = \Theta_\ell\Theta_r \quad & \Leftrightarrow \quad \left\{ \begin{array}{rcl} \underline{Y} &=& A_1 \underline{Y}^{(-1)} A_2 + \underline{C_\ell}\,\underline{B_r} \\ D &=& B_1 \underline{Y}^{(-1)} C_2 + \underline{D_\ell}\,\underline{D_r} \\ C_1 &=& A_1 \underline{Y}^{(-1)} C_2 + \underline{C_\ell}\,\underline{D_r} \\ B_2 &=& B_1 \underline{Y}^{(-1)} A_2 + \underline{D_\ell}\,\underline{B_r} \end{array} \right. \\ & \Leftrightarrow \quad \begin{bmatrix} \underline{Y} & C_1 \\ B_2 & D \end{bmatrix} = \underbrace{\begin{bmatrix} A_1 & \underline{C_\ell} \\ B_1 & \underline{D_\ell} \end{bmatrix}}_{\boldsymbol{\Theta}_\ell} \begin{bmatrix} \underline{Y}^{(-1)} & \\ & I \end{bmatrix} \underbrace{\begin{bmatrix} A_2 & C_2 \\ \underline{B_r} & \underline{D_r} \end{bmatrix}}_{\boldsymbol{\Theta}_r}\,. \end{aligned} \tag{8.58}$$

The underlined variables are unknown and to be computed. Let's assume for simplicity that $[A_2\ C_2]$ has been chosen a J-orthonormal basis: $A_2 J A_2^* + C_2 J C_2^* = J$; the existence of such a basis follows from the regularity assumption on $\mathcal{H}_2$. By theorem 8.17, there is an extension $[B_r\ D_r]$ such that $\mathbf{\Theta}_r$ is a J-unitary realization. Choose any such extension. Upon inverting $\mathbf{\Theta}_r$ (using $\mathbf{\Theta}_r^{-1} = \mathbf{J}\mathbf{\Theta}^*\mathbf{J}$), it follows that

$$\begin{aligned} \begin{bmatrix} A_1 \\ B_1 \end{bmatrix} Y^{(-1)}J &= \begin{bmatrix} Y & C_1 \\ B_2 & D \end{bmatrix} \begin{bmatrix} J & \\ & J \end{bmatrix} \begin{bmatrix} A_2^* \\ C_2^* \end{bmatrix} \\ \begin{bmatrix} C_\ell \\ D_\ell \end{bmatrix} J &= \begin{bmatrix} Y & C_1 \\ B_2 & D \end{bmatrix} \begin{bmatrix} J & \\ & J \end{bmatrix} \begin{bmatrix} B_r^* \\ D_r^* \end{bmatrix}. \end{aligned} \tag{8.59}$$

If we now also assume, for simplicity, that (A_1, B_1) was chosen to be an orthonormal basis, $A_1^*A_1 + B_1^*B_1 = I$, then

$$\begin{aligned} Y^{(-1)}J &= [A_1^*\ \ B_1^*] \begin{bmatrix} Y & C_1 \\ B_2 & D \end{bmatrix} \begin{bmatrix} J & \\ & J \end{bmatrix} \begin{bmatrix} A_2^* \\ C_2^* \end{bmatrix} \\ &= A_1^*(YJ)A_2^* + (A_1^*C_1JC_2^* + B_1^*B_2JA_2^* + B_1^*DJC_2^*). \end{aligned} \tag{8.60}$$

This defines a recursive Lyapunov-type equation for Y, with a unique bounded solution (since $\ell_{A_1} < 1$ and $\ell_{A_2} < 1$). With Y known, C_ℓ and D_ℓ follow from the second equation in (8.59). More in general, a bounded Y can be computed similarly whenever (A_1, B_1) is uniformly reachable and (A_2, C_2) is uniformly observable with a nonsingular J-Gramian. Minimality of the anticausal component of the given realization is important, otherwise the state dimension of Θ_r might become too large, requiring additional states in Θ_ℓ to compensate so that the given structure of the realization of Θ_ℓ is no longer valid.

The connection between the realization of Θ in summation form (fig. 8.8(a)) and the factored realization (fig. 8.8(b)) is via a state transformation in terms of Y. Indeed, let

$$[x_1'\ \ x_2'] = [x_1\ \ x_2] \begin{bmatrix} I & Y \\ 0 & I \end{bmatrix} \quad \Leftrightarrow \quad [x_1\ \ x_2] = [x_1'\ \ x_2'] \begin{bmatrix} I & -Y \\ 0 & I \end{bmatrix} \tag{8.61}$$

be a state transformation. Upon rearranging (8.55), we obtain

$$[x_1\ \ x_2\ \ u] \begin{bmatrix} A_1 & 0 & C_1 \\ 0 & -I & 0 \\ B_1 & B_2 & D \end{bmatrix} = [x_1^{(-1)}\ \ x_1^{(-1)}\ \ y] \begin{bmatrix} I & 0 & 0 \\ 0 & -A_2 & -C_2 \\ 0 & 0 & I \end{bmatrix}.$$

The state transformation then produces

$$[x_1'\ \ x_2'\ \ u] \begin{bmatrix} A_1 & Y & C_1 \\ 0 & -I & 0 \\ B_1 & B_2 & D \end{bmatrix} = [x_1'^{(-1)}\ \ x_2'^{(-1)}\ \ y] \begin{bmatrix} I & Y^{(-1)}A_2 & Y^{(-1)}C_2 \\ 0 & -A_2 & -C_2 \\ 0 & 0 & I \end{bmatrix}.$$

Rearranging back, we obtain

$$\begin{aligned} [x_1'^{(-1)}\ \ x_2'\ \ y] &= [x_1'\ \ x_2'^{(-1)}\ \ u] \begin{bmatrix} A_1 & Y & C_1 \\ 0 & A_2 & C_2 \\ B_1 & B_2 & D \end{bmatrix} \begin{bmatrix} I & Y^{(-1)}A_2 & Y^{(-1)}C_2 \\ 0 & I & 0 \\ 0 & 0 & I \end{bmatrix}^{-1} \\ &= [x_1'\ \ x_2'^{(-1)}\ \ u] \begin{bmatrix} A_1 & Y - A_1Y^{(-1)}A_2 & C_1 - A_1Y^{(-1)}C_2 \\ & A_2 & C_2 \\ B_1 & B_2 - B_1Y^{(-1)}A_2 & D - B_1Y^{(-1)}C_2 \end{bmatrix}. \end{aligned}$$

Suppose we have found $Y, D_\ell, D_r, B_\ell, C_r$ satisfying (8.58), then the realization factors into

$$\begin{bmatrix} A_1 & Y-A_1Y^{(-1)}A_2 & C_1-A_1Y^{(-1)}C_2 \\ & A_2 & C_2 \\ B_1 & B_2-B_1Y^{(-1)}A_2 & D-B_1Y^{(-1)}C_2 \end{bmatrix} = \begin{bmatrix} A_1 & C_\ell B_r & C_\ell D_r \\ & A_2 & C_2 \\ B_1 & D_\ell B_r & D_\ell D_r \end{bmatrix}$$
$$= \begin{bmatrix} A_1 & & C_\ell \\ & I & \\ B_1 & & D_\ell \end{bmatrix} \begin{bmatrix} I & & \\ & A_2 & C_2 \\ & B_r & D_r \end{bmatrix}$$

so that

$$\begin{aligned} [x_1'^{(-1)} \quad z] &= [x_1 \quad u]\boldsymbol{\Theta}_\ell \\ [x_2' \quad y] &= [x_2'^{(-1)} \quad z]\boldsymbol{\Theta}_r \end{aligned}$$

$$\boldsymbol{\Theta}_\ell = \begin{bmatrix} A_1 & C_\ell \\ B_1 & D_\ell \end{bmatrix}, \qquad \boldsymbol{\Theta}_r = \begin{bmatrix} A_2 & C_2 \\ B_r & D_r \end{bmatrix}.$$

Thus, the realization resulting after state transformation factors into the product of two realizations (one causal, the other anticausal), corresponding to a factorization of Θ into $\Theta = \Theta_\ell \Theta_r$.

Condition for J-isometry

We wish to answer the following questions:

1. Under which conditions does the realization (8.55) correspond to a J-isometric or J-unitary transfer operator Θ?

2. Under which conditions is Θ also J-lossless or J-inner?

Because of the expression of the realization as a sum of a causal and an anticausal part, the conditions turn out to be more involved than in the previous cases, but a simple physical reasoning quickly provides sufficient insight which can then be verified under appropriate conditions. Hence, we start the discussion informally and then give a set of properties with proofs.

The guiding physical principle is that (8.55) will correspond to a J-isometric realization when there is a signed hermitian energy metric Q on the states $[x_1 \ x_2]$ which, together with the energy metrics J_1 and J_2 on the inputs and outputs, gets preserved for all compatible input, output and state sequences. Counting the energy carried by $[x_1 \ x_2]$ as algebraically positive (*i.e.*, a positive sign is considered as energy flowing into the circuit shown in figure 8.8(a)), then we obtain the energy balance

$$[x_1^{(-1)} \ x_2^{(-1)} \,|\, y] \left[\begin{array}{cc|c} Q_{11}^{(-1)} & Q_{12}^{(-1)} & 0 \\ Q_{21}^{(-1)} & Q_{22}^{(-1)} & 0 \\ \hline 0 & 0 & J_2 \end{array}\right] \left[\begin{array}{c} x_1^{(-1)*} \\ x_2^{(-1)*} \\ \hline y^* \end{array}\right] = [x_1 \ x_2 \,|\, u] \left[\begin{array}{cc|c} Q_{11} & Q_{12} & 0 \\ Q_{21} & Q_{22} & 0 \\ \hline 0 & 0 & J_1 \end{array}\right] \left[\begin{array}{c} x_1^* \\ x_2^* \\ \hline u^* \end{array}\right].$$

In this equation, the variables x_1, $x_2^{(-1)}$ and u are independent (since Θ was assumed bounded). Substituting the dependent variables $x_1^{(-1)}$, x_2 and y using (8.55), we obtain

$$[x_1\ x_2^{(-1)} \mid u]\left[\begin{array}{cc|c} A_1 & 0 & C_1 \\ 0 & I & C_2 \\ \hline B_1 & 0 & D \end{array}\right]\left[\begin{array}{cc|c} Q_{11}^{(-1)} & Q_{12}^{(-1)} & 0 \\ Q_{21}^{(-1)} & Q_{22}^{(-1)} & 0 \\ \hline 0 & 0 & J_2 \end{array}\right]\left[\begin{array}{cc|c} A_1^* & 0 & B_1^* \\ 0 & I & 0 \\ \hline C_1^* & C_2^* & D^* \end{array}\right]\left[\begin{array}{c} x_1^* \\ x_2^{(-1)*} \\ \hline u^* \end{array}\right]$$
$$= [x_1\ x_2^{(-1)} \mid u]\left[\begin{array}{cc|c} I & 0 & 0 \\ 0 & A_2 & 0 \\ \hline 0 & B_2 & I \end{array}\right]\left[\begin{array}{cc|c} Q_{11} & Q_{12} & 0 \\ Q_{21} & Q_{22} & 0 \\ \hline 0 & 0 & J_1 \end{array}\right]\left[\begin{array}{cc|c} I & 0 & 0 \\ 0 & A_2^* & B_2^* \\ \hline 0 & 0 & I \end{array}\right]\left[\begin{array}{c} x_1^* \\ x_2^{(-1)*} \\ \hline u^* \end{array}\right].$$

Since this must hold for any combination of independent variables, we obtain that the system preserves (Q,J_1,J_2)-energy, if and only if

$$\begin{array}{l}\left[\begin{array}{cc|c} A_1 & 0 & C_1 \\ 0 & I & C_2 \\ \hline B_1 & 0 & D \end{array}\right]\left[\begin{array}{cc|c} Q_{11}^{(-1)} & Q_{12}^{(-1)} & 0 \\ Q_{21}^{(-1)} & Q_{22}^{(-1)} & 0 \\ \hline 0 & 0 & J_2 \end{array}\right]\left[\begin{array}{cc|c} A_1^* & 0 & B_1^* \\ 0 & I & 0 \\ \hline C_1^* & C_2^* & D^* \end{array}\right] = \\ \quad = \left[\begin{array}{cc|c} I & 0 & 0 \\ 0 & A_2 & 0 \\ \hline 0 & B_2 & I \end{array}\right]\left[\begin{array}{cc|c} Q_{11} & Q_{12} & 0 \\ Q_{21} & Q_{22} & 0 \\ \hline 0 & 0 & J_1 \end{array}\right]\left[\begin{array}{cc|c} I & 0 & 0 \\ 0 & A_2^* & B_2^* \\ \hline 0 & 0 & I \end{array}\right].\end{array} \tag{8.62}$$

If the operator Θ is J-isometric, we may expect that there is a (unique) diagonal hermitian operator Q which satisfies (8.62). In case Q is (strictly) positive definite, we expect that Θ will correspond to a *J-lossless* system, since in that case it has a corresponding lossless and causal scattering system. There are, however, additional difficulties. For example, if we look at the (2,2) entries in (8.62), it follows that

$$Q_{22}^{(-1)} + C_2 J_2 C_2^* = A_2 Q_{22} A_2^*. \tag{8.63}$$

When we require Q and hence Q_{22} to be strictly positive definite, it follows that the space $\mathbf{P}'(\mathcal{U}_2\Theta) = \mathcal{D}_2 Z^*(I-A_2Z^*)^{-1}C_2$ has to be a regular space (a Krein space). But it is quite conceivable that there are operators Θ that are J-lossless without this condition being satisfied: the partitioning of the operator into upper and lower triangular parts and forcing a regular state space structure on each part might be too restrictive. We take exemption from such anomalous cases for the sake of simplicity. In the case where Q is not positive definite, we require Q_{22} to be invertible, by posing regularity conditions on certain state spaces related to Θ.

Theorem 8.24 *Let $\Theta \in \mathcal{X}$ have a locally finite, minimal, u.e. stable realization (8.55), for which (A_1,B_1) is uniformly reachable and (A_2,C_2) is uniformly observable. Suppose that $\mathcal{H}_2 := \mathbf{P}'(\mathcal{U}_2\Theta) = \mathcal{D}_2 Z^*(I-A_2Z^*)^{-1}C_2$ is a regular space in the J metric. Then Θ is J-isometric if and only if there exists an operator $Q \in \mathcal{D}$ which satisfies (8.62).*

PROOF
Sufficiency

Suppose that Q satisfies (8.62), then we have to show that Θ is J-isometric. A direct, brute force calculation to verify that $\Theta J_2 \Theta^*$ is equal to J_1 is possible [Yu96, pp. 74-75]. In particular, equation (8.62) specifies

1. $A_1 Q_{11}^{(-1)} A_1^* + C_1 J_2 C_1^* = Q_{11} A$
2. $A_1 Q_{12}^{(-1)} + C_1 J_2 C_2^* = Q_{12} A_2^*$
3. $Q_{22}^{(-1)} = A_2 Q_{22} A_2^* - C_2 J_2 C_2^*$
4. $B_1 Q_{11}^{(-1)} A_1^* - B_2 Q_{21} + D J_2 C_1^* = 0$
5. $B_1 Q_{12}^{(-1)} + D J_2 C_2^* - B_2 Q_{22} A_2^* = 0$
6. $B_1 Q_{11}^{(-1)} B_1^* + D J_2 D^* - B_2 Q_{22} B_2^* = J_1$

By writing out $\Theta J_2 \Theta^*$ and using the above relations, a straightforward but tedious derivation shows that

$$\begin{aligned} \Theta J_2 \Theta^* &= [D + B_1 Z(I - A_1 Z)^{-1} C_1 + B_2 Z^*(I - A_2 Z^*)^{-1} C_2] \cdot J_2 \cdot \\ &\quad \cdot [D^* + C_1^*(I - Z^* A_1^*)^{-1} Z^* B_1^* + C_2^*(I - Z A_2^*)^{-1} B_2^*] \\ &= \cdots = J_1 \end{aligned}$$

and thus that Θ is J-isometric.

Necessity

Now we suppose that Θ is J-isometric ($\Theta J_2 \Theta^* = J_1$) and show that (8.62) holds for a suitable Q. By theorem 8.22, Θ has a minimal factorization into $\Theta = \Theta_\ell \Theta_r$, where $\Theta_\ell \in \mathcal{U}$ is J-isometric, and $\Theta_r \in \mathcal{L}$ is J-unitary, with realizations

$$\begin{aligned} [x_1'^{(-1)} \;\; z] &= [x_1' \;\; u]\boldsymbol{\Theta}_\ell \\ [x_2' \;\; y] &= [x_2'^{(-1)} \;\; z]\boldsymbol{\Theta}_r \end{aligned} \qquad \boldsymbol{\Theta}_\ell = \begin{bmatrix} A_1 & C_\ell \\ B_1 & D_\ell \end{bmatrix}, \qquad \boldsymbol{\Theta}_r = \begin{bmatrix} A_2 & C_2 \\ B_r & D_r \end{bmatrix},$$

see also figure 8.8. The J-isometric properties of these factors translate to the existence of $M \in \mathcal{D}$ and $P \in \mathcal{D}$ such that

$$\begin{aligned} \begin{bmatrix} A_1 & C_\ell \\ B_1 & D_\ell \end{bmatrix} \begin{bmatrix} M^{(-1)} & \\ & J \end{bmatrix} \begin{bmatrix} A_1 & C_\ell \\ B_1 & D_\ell \end{bmatrix}^* &= \begin{bmatrix} M & \\ & J \end{bmatrix} \\ \begin{bmatrix} A_2 & C_2 \\ B_r & D_r \end{bmatrix} \begin{bmatrix} -P & \\ & J \end{bmatrix} \begin{bmatrix} A_2 & C_2 \\ B_r & D_r \end{bmatrix}^* &= \begin{bmatrix} -P^{(-1)} & \\ & J \end{bmatrix} \end{aligned} \qquad (8.64)$$

and the connection of these state space operators to the given realization of Θ is provided by (8.58), *viz.*

$$\begin{cases} Y &= A_1 Y^{(-1)} A_2 + C_\ell B_r \\ D &= B_1 Y^{(-1)} C_2 + D_\ell D_r \\ C_1 &= A_1 Y^{(-1)} C_2 + C_\ell D_r \\ B_2 &= B_1 Y^{(-1)} A_2 + D_\ell B_r . \end{cases}$$

These equations are sufficient to derive (8.62). Indeed, since $\Theta = \Theta_\ell \Theta_r$, an alternative realization for Θ is given by the product realization $[x_1'^{(-1)} \;\; x_2' \;\; y] = [x_1' \;\; x_2'^{(-1)} \;\; u]\boldsymbol{\Theta}'$ where

$$\begin{aligned} \boldsymbol{\Theta}' &= \begin{bmatrix} A_1 & & C_\ell \\ & I & \\ B_1 & & D_\ell \end{bmatrix} \begin{bmatrix} I & & \\ & A_2 & C_r \\ & B_2 & D_r \end{bmatrix} = \begin{bmatrix} A_1 & C_\ell B_r & C_\ell D_r \\ & A_2 & C_2 \\ B_1 & D_\ell B_2 & D_\ell D_r \end{bmatrix} \\ &= \begin{bmatrix} A_1 & Y - A_1 Y^{(-1)} A_2 & C_1 - A_1 Y^{(-1)} C_2 \\ 0 & A_2 & C_2 \\ B_1 & B_2 - B_1 Y^{(-1)} A_2 & D - B_1 Y^{(-1)} C_2 \end{bmatrix}. \end{aligned}$$

The J-isometric properties (8.64) result in a similar property for $\boldsymbol{\Theta}'$:

$$\boldsymbol{\Theta}' \begin{bmatrix} M^{(-1)} & & \\ & -P & \\ & & J \end{bmatrix} \boldsymbol{\Theta}'^* = \begin{bmatrix} M & & \\ & -P^{(-1)} & \\ & & J \end{bmatrix}.$$

Rearranging the center terms gives the equality

$$\begin{aligned} &\begin{bmatrix} A_1 & & C_1 - A_1 Y^{(-1)} C_2 \\ & I & C_2 \\ B_1 & & D - B_1 Y^{(-1)} C_2 \end{bmatrix} \begin{bmatrix} M^{(-1)} & & \\ & P^{(-1)} & \\ & & J \end{bmatrix} [*]^* \\ &\quad = \begin{bmatrix} I & Y - A_1 Y^{(-1)} A_2 & \\ & A_2 & \\ & B_2 - B_1 Y^{(-1)} A_2 & I \end{bmatrix} \begin{bmatrix} M & & \\ & P & \\ & & J \end{bmatrix} [*]^*. \end{aligned} \tag{8.65}$$

Note that

$$\begin{bmatrix} I & A_1 Y^{(-1)} & \\ & I & \\ & B_1 Y^{(-1)} & I \end{bmatrix} \begin{bmatrix} A_1 & & C_1 - A_1 Y^{(-1)} C_2 \\ & I & C_2 \\ B_1 & & D - B_1 Y^{(-1)} C_2 \end{bmatrix} = \begin{bmatrix} A_1 & & C_1 \\ & I & C_2 \\ D_1 & & D \end{bmatrix} \begin{bmatrix} I & Y^{(-1)} & \\ & I & \\ & & I \end{bmatrix},$$

$$\begin{bmatrix} I & A_1 Y^{(-1)} & \\ & I & \\ & B_1 Y^{(-1)} & I \end{bmatrix} \begin{bmatrix} I & Y - A_1 Y^{(-1)} A_2 & \\ & A_2 & \\ & B_2 - B_1 Y^{(-1)} A_2 & I \end{bmatrix} = \begin{bmatrix} I & & \\ & A_2 & \\ & B_2 & I \end{bmatrix} \begin{bmatrix} I & Y & \\ & I & \\ & & I \end{bmatrix}.$$

Thus, premultiplying (8.65) by the first factor and postmultiplying by its conjugate produces

$$\begin{aligned} &\begin{bmatrix} A_1 & & C_1 \\ & I & C_2 \\ B_1 & & D \end{bmatrix} \begin{bmatrix} I & Y^{(-1)} & \\ & I & \\ & & I \end{bmatrix} \begin{bmatrix} M^{(-1)} & & \\ & P^{(-1)} & \\ & & J \end{bmatrix} \begin{bmatrix} I & & \\ Y^{(-1)*} & I & \\ & & I \end{bmatrix} \begin{bmatrix} A_1 & & C_1 \\ & I & C_2 \\ B_1 & & D \end{bmatrix}^* \\ &\quad = \begin{bmatrix} I & & \\ & A_2 & \\ & B_2 & I \end{bmatrix} \begin{bmatrix} I & Y & \\ & I & \\ & & I \end{bmatrix} \begin{bmatrix} M & & \\ & P & \\ & & J \end{bmatrix} \begin{bmatrix} I & & \\ Y^* & I & \\ & & I \end{bmatrix} \begin{bmatrix} I & & \\ & A_2 & \\ & B_2 & I \end{bmatrix}^*. \end{aligned}$$

Hence, we showed that (8.62) holds with

$$Q = \begin{bmatrix} I & Y \\ & I \end{bmatrix} \begin{bmatrix} M & \\ & P \end{bmatrix} \begin{bmatrix} I & \\ Y^* & I \end{bmatrix}. \tag{8.66}$$

□

Clearly, the above factorization of Q induces the same state transformation as used in (8.61) to transform the given realization into a factored realization. This connects the condition on Q for J-isometry to conditions on M and P for J-isometry of each of the factors.

It is straightforward to verify that Θ is J-lossless if both its factors Θ_ℓ and Θ_r are J-lossless, *i.e.*, if both M and P are strictly positive definite. With equation (8.66) in mind, it immediately follows that Θ is J-lossless if Q is strictly positive definite.

Proposition 8.25 *Under the hypotheses of theorem 8.24, the transfer operator $\Theta \in \mathcal{X}$ is J-lossless if and only if the Q satisfying (8.62) is strictly positive definite.*

The various properties of J-lossless scattering operators form the major ingredients of an approach to H_∞ control based on J-external and J-inner-outer factorizations. This approach was pioneered by Kimura in the time-invariant case, see his recent book on the topic [Kim97], and extended to the LTV case in [Yu96]. Since a detailed account of this topic would lead us too far astray, we induce the interested reader to consult the cited literature.

9 ALGEBRAIC INTERPOLATION

In this chapter, we use our knowledge of Hankel operators and chain scattering matrices to solve a set of constrained interpolation problems. These are problems in which one looks for an operator that meets a collection of specifications of the following type: (1) the operator takes specific "values" at specific "points" (we shall make the notion more precise) (2) it is constrained in norm, and (3) it is causal and has minimal state dimensions. We have to limit ourselves to specifications that satisfy a precise structure, but the class is large enough for interesting applications, namely time-varying equivalents of the celebrated "H_∞ optimal control" problem or control for minimal sensitivity. *Algebraic interpolation* is an extension of the notion of interpolation in complex function theory, and we derive algebraic equivalents for very classical interpolation problems such as the Nevanlinna-Pick, Schur, Hermite-Fejer and Nudel'man problems.

The simplest possible formulation of an algebraic interpolation problem in the diagonal taste would be: find an operator for which certain linear combinations of its diagonals have specific values, and which meets additional causality and norm constraints. Even in the case where only a set of diagonals are specified, we do not solve the general constrained interpolation problem in closed form: the specified diagonals are not chosen randomly, they must form a band. This situation resembles the complex function case, and we shall recognize the limitations of the classical theory.

Lossless J-unitary matrices play a central role in the solution of interpolation problems. This can be motivated as follows. Consider the input scattering operator of a time-invariant lossless system with transfer operator $\Sigma(\omega)$ whose output is loaded by the passive scattering operator S_L (figure 9.1). The relation between S_L and S is given

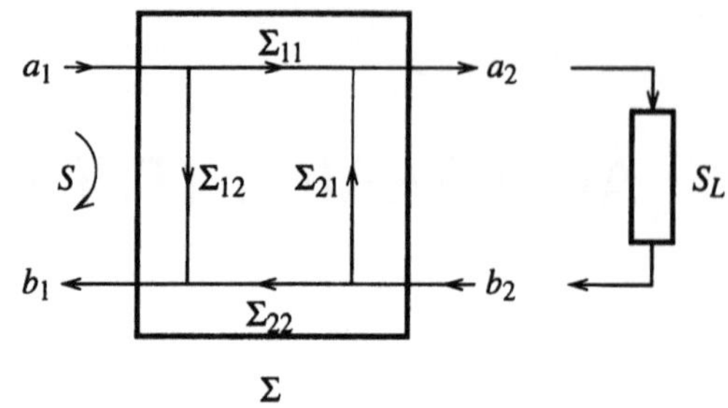

Figure 9.1. Lossless scattering operator loaded by a passive scattering S_L.

by

$$S = \Sigma_{12} + \Sigma_{11}(I - S_L\Sigma_{21})^{-1} S_L \Sigma_{22}.$$

Σ_{12} is the input reflection operator, Σ_{11} is the input transmission operator. Suppose now that for some vector ξ and some complex frequency ω we have that the transmission scattering function satisfies $\xi\Sigma_{11}(\omega) = 0$, then

$$\xi S(\omega) = \xi\Sigma_{12}(\omega) =: \eta,$$

independently of S_L. If η is specified (it is a characteristic of the lossless system and not of the load), then we see that $\xi S(z)$ interpolates η at the frequency ω. At the same time, S is a causal, contractive operator, for physical reasons. In this chapter we shall see how this situation generalizes to the time-varying situation. The frequency ω is known as a transmission zero of the lossless medium, and will be replaced by a diagonal operator in the new theory. Just as in the time-invariant theory (which is subsumed by the more general theory), it will be advantageous to work with chain scattering matrices rather than scattering matrices, because cascading the former gives much simpler expression.

Connections between circuit and system theory problems and the mathematical techniques around interpolation, reproducing kernels and the lifting of a contractive operator had been obtained a decade earlier by Helton [Hel78] in the pursuit of a solution to the broadband matching problem (see also [e.a87]). The connection with the global and recursive solution to the Lossless Inverse Scattering problem was studied in [DVK78, DD81b, DD81a, DD84], and collected in the monograph [Dym89] by Dym. The recursive solution of the Schur-Takagi problem by Limebeer and Green [LG90] can be viewed as an extension of such results to meromorphic (indefinite) interpolation problems. In a parallel development, the state space theory for the interpolation problem was extensively studied in the book [BGR90] by Ball, Gohberg and Rodman. The wide interest in this type of problems was kindled by one of its many applications: the robust (or H_∞-) control problem formulated by Zames in [Zam81] and brought into the context of scattering and interpolation theory by Helton [Hel82].

The general strategy of the interpolation problems studied in this chapter is to connect each interpolation problem to a partially specified realization of an appropriate J-lossless operator (*i.e.*, the chain scattering matrix of a lossless —inner and causal—

system, see definition 8.1). The approach is not unlike the one followed by Ball-Gohberg-Rodman for the classical case [BGR90], but we take a system theoretic tack throughout, which in our view is both more general and numerically more appealing. It is reminiscent of the method adopted by Dym [Dym89], but we do not use reproducing kernel theory since the case of locally finite dimensions can be handled in more elementary ways.

In its simplest, complex-analytic form, the *Nevanlinna-Pick* interpolation problem can be stated as follows:

> Let $\{v_i\}_{i=1,\cdots,n}$ be an indexed set of n points in the open unit disc $\mathbf{D}$ of the complex plane $\mathbb{C}$ and let $\{s_i\}_{i=1,\cdots,n}$ be a set of n values in $\mathbb{C}$,
>
> find a function $S(z)$ which is analytic and contractive in $\mathbf{D}$ (*i.e.*, $\forall z \in \mathbf{D} : |S(z)| \leq 1$), such that $S(v_i) = s_i$.

To translate the classical problem in an algebraic setting, there is one hurdle we must take at the start, since we lack the notion of "point evaluation" in the algebraic setting. What does it mean for a matrix or operator T to "take a value" at a "point", in analogy to the evaluation $S(v_i)$? Keeping in line with our diagonal based methodology, the analogs of v_i and s_i should be diagonals of matrices or operators. Evaluation of an operator on a diagonal was first introduced in [AD90], and studied extensively in [ADD90]. The diagonal version of the Nevanlinna-Pick problem was first solved in [Dew91] in a somewhat restricted setting, and further generalized in [DD92] and a slew of subsequent publications, see *e.g.*, [BGK92a].

9.1 DIAGONAL EVALUATIONS OR THE W-TRANSFORM

Suppose that $T \in \mathcal{U}(\mathcal{M},\mathcal{N})$ is a bounded and upper operator, and that $V \in \mathcal{D}(\mathcal{M},\mathcal{M}^{(1)})$ is a diagonal operator for which the spectral radius $\ell_V = \rho(VZ^*) < 1$. We search for a diagonal $T^\wedge(V) \in \mathcal{D}(\mathcal{M},\mathcal{N})$ which is such that

$$T = T^\wedge(V) + (Z - V)T'$$

for some $T' \in \mathcal{U}$. It turns out that under the conditions stated, $T^\wedge(V)$ exists and is given by a nice series expression which is the equivalent of a Maclaurin series expansion at a given point $z : |z| < 1$ in the complex plane.

Theorem 9.1 *Let $T \in \mathcal{U}(\mathcal{M},\mathcal{N})$ be a bounded, upper operator with diagonal expansion*

$$T = \sum_{i=0}^{\infty} Z^{[i]} T_{[i]}, \tag{9.1}$$

let $V \in \mathcal{D}(\mathcal{M},\mathcal{M}^{(1)})$ be a diagonal operator for which $\ell_V < 1$, and define for $n \geq 0$

$$V^{[n]} = VV^{(1)} \cdots V^{(n-1)} \text{ with } V^{[0]} = I, \tag{9.2}$$

then the sum

$$T^\wedge(V) = \sum_{i=0}^{\infty} V^{[i]} T_{[i]} \tag{9.3}$$

converges in the operator norm, and

$$T = T^{\wedge}(V) + (Z - V)T' \tag{9.4}$$

for a $T' \in \mathcal{U}$. *Moreover, operators* $T^{\wedge}(V) \in \mathcal{D}$ *and* $T' \in \mathcal{U}$ *satisfying (9.3) are unique.*

PROOF A complete analytic proof for the theorem is given in [ADD90]; as we do not need the property directly in the sequel, we suffice with a sketch of the main ingredients. The convergence of the sum (9.3) in operator norm follows from the fact that the diagonals $T_{[i]}$ are uniformly bounded by $\| T \|$, and $\ell_V = \lim_{n\to\infty} \| V^{[n]} \|^{1/n} < 1$ so that the series is majorized in norm by a convergent geometric series. If we now calculate (using again convergent series arguments)

$$T' = Z^*(I - VZ^*)^{-1}T - Z^*(I - VZ^*)^{-1}T^{\wedge}(V),$$

we find that the diagonal coefficients of $Z^{[i]}$ vanish for negative i's, showing that T' is indeed upper. □

In analogy to the complex function case, we say that $T^{\wedge}(V)$ is the "diagonal value" which T takes at the "diagonal point" V.

Definition 9.2 *Given an upper operator* $T \in \mathcal{U}(\mathcal{M},\mathcal{N})$, *then its W-transform* $\mathcal{W}(T)$ *is the map*

$$\mathcal{W}: \quad \{V \in \mathcal{D}(\mathcal{M},\mathcal{M}^{(1)}) : \ell_V < 1\} \to \mathcal{D}(\mathcal{M},\mathcal{N}) : \quad \mathcal{W}(T;V) = \sum_{i=0}^{\infty} V^{[i]} T_{[i]} . \tag{9.5}$$

$\mathcal{W}$ assigns to each diagonal operator V of the proper type the diagonal operator $T^{\wedge}(V)$. If T is a Toeplitz operator, then its z-transform converges in the open unit disc of the complex plane, and its evaluation at the point $z : |z| < 1$ is given by $T(z) = t_0 + zt_1 + z^2 t_2 + \cdots$: exactly the same as the W-transform $T^{\wedge}(zI)$.

The W-transform has interesting properties (again see [ADD90]), in particular:

1. *Chain rule:* $(T_1 T_2)^{\wedge}(V) = (T_1^{\wedge}(V) T_2)^{\wedge}(V)$.

   Remark that the chain rule is not as strong as in the Toeplitz case, where it holds that $(T_1 T_2)(z) = T_1(z) T_2(z)$.

2. $T^{\wedge}(V) = \mathbf{P}_0\left((I - VZ^*)^{-1} T\right)$.

   This useful formula provides a good link with interpolation theory. A direct proof is as follows. For $A, B \in \mathcal{U}$, we have

$$\begin{aligned} \mathbf{P}_0(A^*B) &= \mathbf{P}_0\left(\sum (A_{[n]})^* Z^{-n} B\right) \\ &= \sum (A_{[n]})^* \mathbf{P}_0(Z^{-n} B) \\ &= \sum (A_{[n]})^* B_{[n]} . \end{aligned}$$

   Taking $A = (I - ZV^*)^{-1} = I + ZV^* + ZV^*ZV^* + \cdots$, we have that the n-th diagonal of A is given by $A_{[n]} = (V^{(n-1)})^* \cdots (V^{(1)})^* V^*$, hence putting $(A_{[n]})^* = V^{[n]}$ and $B_{[n]} = T_{[n]}$ and comparing with (9.5) we obtain the result.

3. $T^\wedge(V) = 0 \Leftrightarrow T' := (Z-V)^{-1}T \in \mathcal{U}$.

   This property follows directly from 2.

The interpolation property in 3. is, more generally, characterized by the following proposition which is, in fact, the same as theorem 9.1.

Proposition 9.3 *For $S \in \mathcal{U}$, V, $\eta \in \mathcal{D}$ and $\ell_V < 1$,*

$$\begin{array}{rcl} S^\wedge(V) = \eta & \Leftrightarrow & (Z-V)^{-1}(S-\eta) \in \mathcal{U} \\ & \Leftrightarrow & \mathbf{P}'(\mathcal{D}_2(Z-V)^{-1}(S-\eta)) = 0. \end{array}$$

PROOF

$$\begin{array}{rcl} S^\wedge(V) = \eta & \Leftrightarrow & S^\wedge(V) - \eta = 0 \\ & \Leftrightarrow & S^\wedge(V) - \eta^\wedge(V) = 0 \\ & \Leftrightarrow & (S-\eta)^\wedge(V) = 0 \\ & \Leftrightarrow & (Z-V)^{-1}(S-\eta) \in \mathcal{U}. \end{array}$$

□

9.2 THE ALGEBRAIC NEVANLINNA-PICK PROBLEM

At this point, we assume that we are given a set of diagonals $\{v_i\}_{i=1,\cdots,n} \in \mathcal{D}(\mathcal{M},\mathcal{M}^{(1)})$ and a set of diagonal values $\{s_i\}_{i=1,\cdots,n} \in \mathcal{D}(\mathcal{M},\mathcal{N})$, and are asked for a contractive, upper transfer function $S \in \mathcal{U}(\mathcal{M},\mathcal{N})$ such that

$$S^\wedge(v_i) = s_i. \tag{9.6}$$

This is a straightforward generalization of the classical Nevanlinna-Pick problem to our algebraic context, since (9.6) reduces to the classical case when all the operators are Toeplitz.

It is useful to collect the n data points $\{v_i\}_1^n$ into a single diagonal operator. Let $V \in \mathcal{D}(\mathcal{M}^n,(\mathcal{M}^n)^{(1)})$ be a diagonal operator whose k-th block entry along the diagonal is given by the k-th entry along the diagonal of every v_i, *i.e.*,

$$V_k = \begin{bmatrix} (v_1)_k & & & \\ & (v_2)_k & & \\ & & \ddots & \\ & & & (v_n)_k \end{bmatrix}, \quad (k = -\infty,\cdots,\infty). \tag{9.7}$$

Similarly, define diagonal operators $\xi \in \mathcal{D}(\mathcal{M}^n,\mathcal{M})$ and $\eta \in \mathcal{D}(\mathcal{M}^n,\mathcal{N})$ whose k-th entries along the diagonal are given by

$$\xi_k = \begin{bmatrix} I \\ \vdots \\ I \end{bmatrix}, \qquad \eta_k = \begin{bmatrix} (s_1)_k \\ \vdots \\ (s_n)_k \end{bmatrix}. \tag{9.8}$$

Then the set of n interpolation conditions (9.6) becomes a single condition and the time-varying Nevanlinna-Pick interpolation problem can be stated compactly as follows [AD90].

Basic interpolation problem #1: given operators ξ, η, $V \in \mathcal{D}$, with $\ell_V < 1$, find a strictly contractive operator $S \in \mathcal{U}$ such that

$$(Z-V)^{-1}(\xi S-\eta) \in \mathcal{U}. \tag{9.9}$$

This way of writing the Nevanlinna-Pick problem suggests many generalizations, because V, ξ and η may be replaced by more general structures than (9.7)-(9.8). Some generalizations are considered further on in this chapter. Here we proceed with the solution using the rather general formalism of (9.9).

Let $A := V^*$, $B := \left[\begin{smallmatrix}\xi^* \\ -\eta^*\end{smallmatrix}\right]$, $J = \left[\begin{smallmatrix} I & 0 \\ 0 & -I\end{smallmatrix}\right]$, and define $\mathbf{F}$ by

$$\begin{aligned} \mathbf{F}_1 &= (Z-V)^{-1}\xi \\ \mathbf{F}_2 &= (Z-V)^{-1}\eta \\ \mathbf{F} &= [\mathbf{F}_1 \quad \mathbf{F}_2] = (Z-V)^{-1}[\xi \quad \eta] = (I-AZ)^{-*}Z^*B^*J. \end{aligned} \tag{9.10}$$

The space $\mathcal{H} = \mathcal{D}_2^{\mathcal{M}^n}\mathbf{F}J \subset \mathcal{L}_2\mathcal{Z}^{-1}$ generated by $\mathbf{F}J$ is left D-invariant as well as left invariant for the restricted shift $\mathbf{P}'(Z\cdot)$: it is the input state space of a dynamical system partially described by A and B. The following proposition shows that there is, indeed, an intimate connection between the interpolation problem and the input state space of dynamical systems.

Proposition 9.4 *Let* $\mathbf{F}$ *be defined by (9.10), then* $S \in \mathcal{U}$ *is a solution to the basic interpolation problem #1 if it is strictly contractive and if*

$$\mathbf{P}'(\mathcal{D}_2\mathbf{F}\begin{bmatrix} S \\ -I \end{bmatrix}) = 0. \tag{9.11}$$

PROOF In view of the definition of $\mathbf{F}$, equation (9.11) is nothing but a rewrite of (9.9). □

Let $\Lambda^J_{\mathbf{F}}$ be the J-Gramian associated to $\mathbf{F}$, *i.e.*,

$$\Lambda^J_{\mathbf{F}} = \mathbf{P}_0(\mathbf{F}J\mathbf{F}^*).$$

To proceed with comfort, we impose one more condition on the Nevanlinna-Pick data. Let $\Lambda_{\mathbf{F}_1} = \mathbf{P}_0(\mathbf{F}_1\mathbf{F}_1^*)$ be the Gramian of $\mathbf{F}_1$. We will assume from now on that $\Lambda_{\mathbf{F}_1}$ is bounded and non-singular, *i.e.*, strictly positive: $\Lambda_{\mathbf{F}_1} \gg 0$. This condition enforces a "well posedness" of the problem. It also precludes that data points $\{v_k\}$ coincide. The more general case of interpolation points with multiplicity larger than one is considered in section 9.4.

Proposition 9.5 *Suppose that the given interpolation data is such that* $\Lambda_{\mathbf{F}_1} \gg 0$, *and that the interpolation problem #1 has a strictly contractive solution, then* $\Lambda^J_{\mathbf{F}} \gg 0$.

PROOF Let $S \in \mathcal{U}$ be the strictly contractive solution. By (9.11), we have

$$\mathbf{P}'(\mathbf{F}_1 S - \mathbf{F}_2) = 0.$$

On the other hand, since $\mathbf{F}_2 \in Z^*\mathcal{L}_2$,

$$\mathbf{P}(\mathbf{F}_1 S - \mathbf{F}_2) = \mathbf{P}(\mathbf{F}_1 S).$$

Summing up the two equations, we find $\mathbf{F}_1 S - \mathbf{F}_2 = \mathbf{P}(\mathbf{F}_1 S)$, or

$$\mathbf{F}_2 = \mathbf{P}'(\mathbf{F}_1 S).$$

This implies, in particular, that $\mathbf{P}_0(\mathbf{F}_1 SS^*\mathbf{F}_1^*) \geq \mathbf{P}_0(\mathbf{F}_2\mathbf{F}_2^*)$, so that

$$\Lambda_{\mathbf{F}}^J = \mathbf{P}_0(\mathbf{F}_1\mathbf{F}_1^* - \mathbf{F}_2\mathbf{F}_2^*) \geq \mathbf{P}_0\left(\mathbf{F}_1(I-SS^*)\mathbf{F}_1^*\right).$$

Since S is assumed to be strictly contractive, there will be an $\varepsilon > 0$ such that $I - SS^* \geq \varepsilon I$, and $\Lambda_{\mathbf{F}}^J \geq \varepsilon \Lambda_{\mathbf{F}_1} \gg 0$. □

It should be clear that the converse property

$$\Lambda_{\mathbf{F}}^J \gg 0 \quad \Rightarrow \quad \Lambda_{\mathbf{F}_1} \gg 0$$

holds as well so that the condition of proposition 9.5 is necessary. Now assume that $\Lambda_{\mathbf{F}}^J$ is boundedly invertible. Then $\mathcal{H} := \mathcal{D}_2\mathbf{F}J$ is a closed (regular) subspace, and $S \in \mathcal{U}$ is an interpolant if it is contractive and if

$$\mathbf{P}'(\mathcal{H}\begin{bmatrix} S \\ I \end{bmatrix}) = 0. \tag{9.12}$$

Since $\Lambda_{\mathbf{F}}^J$ is boundedly invertible, there is, by theorem 8.17, a bounded J-unitary operator Θ such that $\mathcal{H} = \mathcal{H}_\Theta$, the input state space of Θ. The following theorem shows that the solution of the interpolation problem reduces to the construction of Θ. This establishes a link between interpolation problems and J-unitary operators, just as in the classical case [Dym89] for interpolation by complex functions. The Gramian $\Lambda_{\mathbf{F}}^J$ plays a central role in interpolation theory and has been dignified with the name *Pick matrix*; in our case it is a Pick operator of a rather general kind.

Theorem 9.6 *Let be given the interpolation data (9.7)-(9.8), define* $\mathbf{F}$ *as in (9.9). Assume furthermore that*

$$\Lambda_{\mathbf{F}_1} = \mathbf{P}_0\left((Z-V)^{-1}\xi\xi^*(Z^*-V^*)^{-1}\right) \gg 0.$$

Then the basic interpolation problem #1 has a strictly contractive solution $S \in \mathcal{U}$ *if and only if*

$$\Lambda_{\mathbf{F}}^J = \mathbf{P}_0((Z-V)^{-1}[\xi\xi^* - \eta\eta^*](Z^*-V^*)^{-1}) \gg 0.$$

In this case, there is a J*-inner operator* Θ *with* $\mathcal{H}_\Theta = \mathcal{D}_2\mathbf{F}J$. *The complete collection of solutions is parametrized by*

$$S = T_\Theta[S_L], \qquad (S_L \in \mathcal{U},\ \|S_L\| < 1),$$

where $T_\Theta[\cdot]$ is defined in (8.7).

PROOF The "only if" part of the proof is the subject of proposition 9.5. Sufficiency goes as follows. If $\Lambda_F^J \gg 0$, then it is boundedly invertible and we can construct a J-unitary operator Θ such that $\mathcal{H} = \mathcal{H}_\Theta$ (theorem 8.17). Because $\Lambda_F^J \gg 0$, we have that the corresponding $\Theta_{22}^{-1} \in \mathcal{U}$ (proposition 8.10). To make the proof complete, we show that (1) if $\Lambda_F^J \gg 0$ and $S_L \in \mathcal{U}$ is some strictly contractive operator, then $S = T_\Theta[S_L]$ is a solution of the interpolation problem, and (2) if S is a solution, it must have the form $T_\Theta[S_L]$ for some strictly contractive $S_L \in \mathcal{U}$.

1. $\Lambda_F^J \gg 0,\ S_L \in \mathcal{U},\ \|S_L\| < 1 \Rightarrow S = T_\Theta[S_L]$ is a solution

   The connection between S_L and S is given by equation (8.7):

$$\begin{array}{ll} & S = (\Theta_{11} - \Theta_{12}S_L)(\Theta_{21}S_L - \Theta_{22})^{-1} \\ \Leftrightarrow & \begin{bmatrix} S \\ -I \end{bmatrix} = \Theta \begin{bmatrix} S_L \\ -I \end{bmatrix} \Phi_o^{-1}, \qquad \Phi_o = \Theta_{22} - \Theta_{21}S_L\,. \end{array} \tag{9.13}$$

   Recall from theorem 8.2 that Θ_{22}^{-1} is upper, and $\|\Theta_{22}^{-1}\Theta_{21}\| < 1$. If $S_L \in \mathcal{U}$ is such that $\|S_L\| \le 1$, then $\Phi_o = \Theta_{22}^{-1}(I - \Theta_{22}^{-1}\Theta_{21})$ is invertible in $\mathcal{U}$: $\Phi_o^{-1} = (I - \Theta_{22}^{-1}\Theta_{21}S_L)^{-1}\Theta_{22}^{-1} \in \mathcal{U}$. Also recall the relation between the input state space $\mathcal{H}(\Theta)$ and output state space $\mathcal{H}_o(\Theta)$ (proposition 8.5): $\mathcal{H}_o = \mathcal{H}J\Theta$. We obtain that $S = T_\Theta[S_L]$ implies

$$\mathcal{H}(\Theta)J \begin{bmatrix} S \\ -I \end{bmatrix} = \mathcal{H}(\Theta)J\Theta \begin{bmatrix} S_L \\ -I \end{bmatrix} \Phi_o^{-1} = \mathcal{H}_o(\Theta) \begin{bmatrix} S_L \\ -I \end{bmatrix} \Phi_o^{-1} \in \mathcal{U}\,,$$

   so that

$$\mathbf{P}'\left(\mathcal{H}(\Theta)J \begin{bmatrix} S \\ -I \end{bmatrix}\right) = 0\,.$$

   By proposition 9.4, S is an interpolant.

2. If S is a solution, then $S = T_\Theta[S_L]$, where S_L is a contraction in $\mathcal{U}$.

   If S is an interpolant, then $\mathbf{P}'(\mathcal{H}(\Theta)J[{S \atop -I}]) = 0$, and we have to show that there is some contractive operator $S_L \in \mathcal{U}$ such that $S = T_\Theta[S_L]$. The proof consists of four steps.

Step 1: $G := \Theta^{-1}\begin{bmatrix} S \\ -I \end{bmatrix}$ *is upper.*

$$\begin{aligned} \mathbf{P}'(\mathcal{U}_2 G) &= \mathbf{P}'(\mathcal{U}_2\Theta^{-1}\begin{bmatrix} S \\ -I \end{bmatrix}) \\ &= \mathbf{P}'(\mathbf{P}'[\mathcal{U}_2\Theta^*]J\begin{bmatrix} S \\ -I \end{bmatrix}) \qquad [\text{since } S \in \mathcal{U}] \\ &= \mathbf{P}'(\mathcal{H}(\Theta)J\begin{bmatrix} S \\ -I \end{bmatrix}) \\ &= 0. \end{aligned}$$

Step 2: *Let G be decomposed in two operators G_1 and G_2 such that*

$$\begin{bmatrix} S \\ -I \end{bmatrix} = \Theta \begin{bmatrix} G_1 \\ G_2 \end{bmatrix} \qquad (G_1, G_2 \in \mathcal{U}), \tag{9.14}$$

then G_2 is boundedly invertible, and $S_L := G_1 G_2^{-1}$ is well defined and contractive. In addition, $S = (\Theta_{11} S_L - \Theta_{12})(\Theta_{22} - \Theta_{21} S_L)^{-1} = T_\Theta[S_L]$, as required.

Θ is boundedly invertible because $\Theta^{-1} = J\Theta^* J$ so that $\|\Theta^{-1}\| = \|\Theta\|$. Hence $\Theta\Theta^* \geq \varepsilon I$ for some $\varepsilon > 0$ and

$$\begin{aligned} G_1^* G_1 + G_2^* G_2 &= [S^* \; I]\Theta\Theta^* \begin{bmatrix} S \\ I \end{bmatrix} \\ &\geq \varepsilon(S^* S + I) \\ &\geq \varepsilon I . \end{aligned} \tag{9.15}$$

From the J-unitarity of Θ, and the contractivity of S we also have that

$$G_1^* G_1 \leq G_2^* G_2.$$

Together, this shows that $G_2^* G_2 \geq \frac{1}{2}\varepsilon I$, and hence G_2 is boundedly invertible (but we have not shown yet that G_2^{-1} is in $\mathcal{U}$). Postmultiplying equation (9.14) with G_2^{-1} gives

$$\begin{aligned} G_2^{-1} &= \Theta_{22} - \Theta_{21} S_L \\ S G_2^{-1} &= \Theta_{11} S_L - \Theta_{12} \end{aligned}$$

and hence $S = (\Theta_{11} S_L - \Theta_{12})(\Theta_{22} - \Theta_{21} S_L)^{-1}$.

Step 3: *Let $X \in \mathcal{X}$ be a strictly contractive operator. Then $(I-X)^{-1} \in \mathcal{U} \Leftrightarrow X \in \mathcal{U}$.*

$\Leftarrow$ is clear. $\Rightarrow$: let $Y = 2(I-X)^{-1} - I = (I+X)(I-X)^{-1}$. Then Y is strictly positive real, *i.e.*, $Y + Y^* \gg 0$, since

$$Y + Y^* = 2(I-X^*)^{-1}(I-X^*X)(I-X)^{-1},$$

and $Y \in \mathcal{U}$ by hypothesis. It follows that $(I+Y)^{-1} \in \mathcal{U}$, for the map $I+Y$ is one to one and onto $\mathcal{U}_2$[1], and the open mapping theorem applies [Rud66]. Since $X = 2(I+G)^{-1} - I$, we have in turn that $X \in \mathcal{U}$.

Step 4: *S_L is upper*

From equation (9.15), we have that $G_2\Theta_{22} = (I - \Theta_{22}^{-1}\Theta_{21} S_L)^{-1}$, and it is known that the left hand side is upper. Hence, by step 3, $\Theta_{22}^{-1}\Theta_{21} S_L$ is upper and since $\Theta_{22} \in \mathcal{U}$, $\Theta_{22} S_L$ is upper, and $G_2^{-1} = \Theta_{22} - \Theta_{21} S_L$ is upper too, so that $S_L = G_1 G_2^{-1} \in \mathcal{U}$. □

[1]The classical argument runs as follows. $I+Y$ is one-to-one since

$$\forall u \in \mathcal{U}_2 : \mathbf{P}_0(u(I+Y)(I+Y^*)u^*) = \mathbf{P}_0(uu^*) + \mathbf{P}_0(u(Y+Y^*)u^*) + \mathbf{P}_0(uYY^*u^*) \geq \mathbf{P}_0(uu^*)$$

and hence $u(I+Y) = 0 \Rightarrow u = 0$. $I+Y$ is onto since (1) by a similar argument, $(I+Y^*)$ is one-to-one, so that the range of $(I+Y)$ is dense in $\mathcal{U}$; and (2) that range must also be closed, because if the sequence $\{v_n \in \mathcal{R}(I+Y)\}$ converges to v, then $v \in \mathcal{R}(I+Y)$, since the corresponding $u_n : v_n = u_n(I+Y)$ forms a Cauchy series.

9.3 THE TANGENTIAL NEVANLINNA-PICK PROBLEM

An immediate extension of the standard Nevanlinna-Pick problem occurs when interpolation is only requested in certain directions. We replace the identity operators in the ξ_k composites in equation (9.8) by more general blocks, or possibly simple vectors, and the values that have to be matched are conformal block-rectangular quantities:

$$\xi_k = \begin{bmatrix} (\xi_1)_k \\ \vdots \\ (\xi_n)_k \end{bmatrix}, \quad \eta_k = \begin{bmatrix} (\eta_1)_k \\ \vdots \\ (\eta_n)_k \end{bmatrix} \qquad (k = -\infty, \cdots, \infty). \tag{9.16}$$

The corresponding tangential interpolation problem has the same formulation as basic interpolation problem #1 in the previous section:

Basic interpolation problem #2: given operators $\xi, \eta, V \in \mathcal{D}$ with $\ell_V < 1$, find a strictly contractive S such that

$$(Z-V)^{-1}[\xi S - \eta] \in \mathcal{U}.$$

Since the formulation is the same, propositions 9.4, 9.5 are valid also for the more general interpolation data (9.16) replacing (9.8). Also theorem 9.6 is valid as stated: the directional interpolation problem has a strictly contractive solution if and only if $\Lambda_{\mathbf{F}}^{J}$ is strictly positive definite.

9.4 THE HERMITE-FEJER INTERPOLATION PROBLEM

The Hermite-Fejer interpolation problem deals with interpolation points of higher multiplicity. We can easily work this problem into the framework of the previous sections provided that certain non-singularity conditions are satisfied. There is a small hurdle that we must take at the start, namely how to define higher order multiplicity in the present context. We find a hint by looking at Z^k. Because $Z : \mathcal{M} \to \mathcal{M}^{(1)}$, a more correct reading of this expression is $Z^{[k]} = ZZ^{(1)} \cdots Z^{(k-1)}$: dimensions change in the product. Extending this observation to $(Z-V)$, where V is a diagonal operator of dimensions $\mathcal{M} \times \mathcal{M}^{(1)}$ conformal to Z, we see that we must consider products of the type

$$(Z-V)^{[k]} = (Z-V) \cdot (Z-V)^{(1)} \cdots (Z-V)^{(k-1)}.$$

A special role will be played by its inverse, which we shall denote by[2]

$$(Z-V)^{-[k]} := (Z-V)^{-(k-1)} \cdots (Z-V)^{-1}.$$

Similarly as before, in a *tangential* Hermite-Fejer problem a directional operator $\xi \in \mathcal{D}(\mathcal{M}^n, \mathcal{N})$ is defined, and we have to consider the operator $(Z-V)^{-[k]}\xi$ as a generalization of $(Z-V)^{-1}\xi$ which occurs in the tangential Nevanlinna-Pick problem.

An interpolation property which puts multiple conditions on the same point V can be formulated, in analogy to the classical case, as

$$\xi S - \left\{ \eta_0 + (Z-V)\eta_1 + \cdots + (Z-V)^{[k-1]}\eta_{k-1} \right\} = (Z-V)^{[k]} S'. \tag{9.17}$$

[2] Recall the shorthand notation $X^{-(k)} := (X^{(k)})^{-1} = (X^{-1})^{(k)}$.

where S' is some upper operator. By rearranging terms, it is directly seen that this equation can be valid only if a number of lower-order interpolation conditions are satisfied as well, *viz.*

$$\begin{aligned} \xi S - \left\{ \eta_0 + (Z-V)\eta_1 + \cdots + (Z-V)^{[k-2]}\eta_{k-2} \right\} &= (Z-V)^{[k-1]} S'' \\ &\vdots \\ \xi S - \eta_0 &= (Z-V) S''^{\cdots\prime} \end{aligned} \tag{9.18}$$

in which the S'', $\cdots$ $S''^{\cdots\prime}$ are upper triangular remainders. At this point, the objective is to make the old strategy work in the present context again, *i.e.*, to construct a basis representation $\mathbf{F}$ from the interpolation data such that S is a solution if and only if $\mathbf{P}'(\mathcal{D}_2\mathbf{F}[{}^{\;S}_{-I}]) = 0$, or $\mathbf{F}[{}^{\;S}_{-I}] \in \mathcal{U}$, and such that $\mathbf{F}$ generates the input state space of some J-lossless system Θ.

To this end, define A, B_1, B_2 as

$$A^* = \begin{bmatrix} V & & & 0 \\ I & V^{(1)} & & \\ & \ddots & \ddots & \\ 0 & & I & V^{(k-1)} \end{bmatrix}.$$

$$B_1 = [\xi^* \quad 0 \;\cdots\; 0], \qquad B_2 = -[\eta_0^* \;\cdots\; \eta_{k-1}^*].$$

Note that, for convenience, we have written A as a matrix of diagonals, whereas we used to have A a diagonal of matrices. The two representations are of course isomorphic and have the same meaning.[3]

Also let

$$\begin{aligned} \mathbf{F}_1 &= Z^*(I-A^*Z^*)^{-1}B_1^* \\ \mathbf{F}_2 &= -Z^*(I-A^*Z^*)^{-1}B_2^* \\ \mathbf{F} &= [\mathbf{F}_1 \quad \mathbf{F}_2] = (I-A^*Z^*)^{-1}[B_1^* \quad B_2^*]J. \end{aligned}$$

To verify that $\mathbf{F}$ does indeed satisfy the interpolation condition $\mathbf{F}[{}^{\;S}_{-I}]$, it has to be evaluated in terms of V, ξ and $\{\eta_i\}$. Thus

$$Z - A^* = \begin{bmatrix} Z-V & & & 0 \\ -I & (Z-V)^{(1)} & & \\ & -I & \ddots & \\ 0 & & -I & (Z-V)^{(k-1)} \end{bmatrix}$$

[3]The construction of A, B_1 and B_2 may appear artificial as given here. However, if follows in a logical way from a study of the 'restricted shift' operator $\mathbf{P}'(Z\cdot)$ applied to $(Z-V)^{-[k]}\xi$. When applying the restricted ship repeatedly, one generates new elements of a subspace, for which a basis consists of all elements of the type $(Z-V)^{-[\ell]}\xi$, $1 \le \ell \le k$. A matrix representation of the restricted shift in that basis is given by the matrix A^*.

$$(Z-A^*)^{-1} = \begin{bmatrix} (Z-V)^{-1} & & & 0 \\ (Z-V)^{-[2]} & (Z-V)^{-(1)} & & \\ \vdots & & \ddots & \\ (Z-V)^{-[k]} & (Z-V)^{-[k-1](1)} & \cdots & (Z-V)^{-(k-1)} \end{bmatrix}$$

and hence

$$\mathbf{F}_1 = (Z-A^*)^{-1}B_1^* = \begin{bmatrix} (Z-V)^{-1}\xi \\ (Z-V)^{-[2]}\xi \\ \vdots \\ (Z-V)^{-[k]}\xi \end{bmatrix} \tag{9.19}$$

$$\mathbf{F}\begin{bmatrix} S \\ -I \end{bmatrix} = \begin{bmatrix} (Z-V)^{-1}(\xi S-\eta_0) \\ (Z-V)^{-[2]}(\xi S-\eta_0)-(Z-V)^{-(1)}\eta_1 \\ \vdots \\ (Z-V)^{-[k]}(\xi S-\eta_0)-(Z-V)^{-[k-1](1)}\eta_1-\cdots-(Z-V)^{-(k-1)}\eta_{k-1} \end{bmatrix}. \tag{9.20}$$

Let us call the last entry of the vector S', then

$$\xi S-\left\{\eta_0+(Z-V)\eta_1+\cdots+(Z-V)^{[k-1]}\eta_{k-1}\right\} = (Z-V)^{[k]}S'.$$

Comparing with (9.17), we obtain that the interpolation condition is satisfied if and only if $S' \in \mathcal{U}$. It is not hard to see that the derived additional interpolation conditions in (9.18) are satisfied if the other entries of the vector in (9.20) are upper. Hence, the interpolation conditions are equivalent to $\mathbf{P}'(\mathcal{D}_2\mathbf{F}[{}^{\,S}_{-I}]) = 0$.

At this point, we are back on familiar grounds. Again, we can find a (strictly) contractive solution S if

$$\mathcal{H} := \mathcal{D}_2\mathbf{F}J = \mathcal{D}_2(Z-A^*)^{-1}[B_1^* \quad B_2^*]$$

is an input state space of a J-lossless operator Θ. That will be the case if and only if the Pick matrix

$$\Lambda_{\mathbf{F}}^J = \mathbf{P}_0\left((Z-A^*)^{-1}[B_1^*B_1-B_2^*B_2](Z^*-A)^{-1}\right) \gg 0$$

because the same chain of arguments which led to theorem 9.6 again applies.

The interpolation problems which we have considered so far can be bootstrapped to an even more general statement, containing interpolation problems #1 and #2 as special cases:

Basic Interpolation Problem #3: Given n diagonal operators $\{V_i\}_{i=1,\cdots,n}$ with all $\ell_{V_i} < 1$, and for each V_i an index $k_i \in \mathbb{N}$ and interpolation data (diagonals) $(\xi_i, \eta_{i0}, \cdots, \eta_{i,k_i-1})$. Find a (strictly) contractive S such that, for $i = 1, \cdots, n$,

$$\exists S_i' \in \mathcal{U}: \quad \xi_i S-\{\eta_{i0}+(Z-V_i)\eta_{i1}+\cdots+(Z-V_i)^{[k_i-1]}\eta_{i,k_i-1}\} = (Z-V_i)^{[k_i]}S_i'.$$

With this data, the above derivations easily leads to the theorem:

Theorem 9.7 *Define*

$$A^* = \text{diag}\left\{ \begin{bmatrix} V_1 & & & 0 \\ I & V_1^{(1)} & & \\ & \ddots & \ddots & \\ 0 & & I & V_1^{(k_1-1)} \end{bmatrix}, \ldots, \begin{bmatrix} V_n & & & 0 \\ I & V_n^{(1)} & & \\ & \ddots & \ddots & \\ 0 & & I & V_n^{(k_n-1)} \end{bmatrix} \right\}$$

$$\begin{array}{rcl} B_1 & = & [(\xi_1^* \;\; 0 \;\; \cdots \;\; 0) \;\; \cdots \;\; (\xi_n^* \;\; 0 \;\; \cdots \;\; 0)] \\ B_2 & = & -[(\eta_{10}^* \;\; \eta_{11}^* \;\; \cdots \;\; \eta_{1,k_1-1}^*) \;\; \cdots \;\; (\eta_{n0}^* \;\; \eta_{n1}^* \;\; \cdots \;\; \eta_{n,k_n-1}^*)], \end{array}$$

and assume that

$$\Lambda_{\mathbf{F}_1} = \mathbf{P}_0\left((Z-A^*)^{-1}B_1^*B_1(Z^*-A)^{-1}\right) \gg 0,$$

then there exists a strictly contractive $S \in \mathcal{U}$ satisfying the interpolation conditions #3, if and only if

$$\Lambda_{\mathbf{F}}^J := \mathbf{P}_0\left((Z-A^*)^{-1}(B_1^*B_1 - B_2^*B_2)(Z^*-A)^{-1}\right) \gg 0.$$

If this is the case, then there is a lossless chain scattering matrix Θ which has $(A, \left[\begin{smallmatrix} B_1 \\ B_2 \end{smallmatrix}\right])$ as reachability pair. All solutions are given as

$$S = T_\Theta[S_L], \qquad S_L \in \mathcal{U}, \; \|S_L\| < 1.$$

PROOF The proof is the same as that of theorem 9.6. □

In chapter 8 we have studied how Θ can actually be computed. A final observation is that in our general notation, even the large, multiple point Hermite-Fejer problem can be formulated as a simple one-point "tangential" Nevanlinna-Pick problem, be it with a very complex single diagonal point given by $A = V^*$, and "tangential" interpolation data given by $B_1 = \xi^*$ and $B_2 = -\eta^*$. It is this fact that reduces all our one sided interpolation problems to a single, simple, general formalism.

9.5 CONJUGATION OF A LEFT INTERPOLATION PROBLEM

For given ξ, η and V with $\ell_V < 1$, let us call a *left interpolation problem*, LIP(V,ξ,η), the problem to find S such that

$$(Z-V)^{-1}(\xi S - \eta) \in \mathcal{U}, \quad S \in \mathcal{U}, \quad \|S\| < 1. \tag{9.21}$$

This covers all the basic interpolation problems considered before, and more. Similarly, a *right interpolation problem*, RIP(V,ζ,ι), is to find S such that

$$(S\zeta - \iota)(Z-V)^{-1} \in \mathcal{U}, \quad S \in \mathcal{U}, \quad \|S\| < 1. \tag{9.22}$$

The right interpolation problem is a dual to the left interpolation problem, and all the 'right' results can be obtained from the 'left' results in a straightforward fashion. However, more is possible. Under certain conditions, a left problem can be converted into a right problem with the same basic operator V, in such a way that a solution for one will exist if and only if a solution for the other exists. The construction is closely related to the conjugation theory of the previous chapter. It turns out that the original problem must satisfy a condition of *non-degeneracy* to be convertible. Degeneracy of an interpolation problem is in itself an interesting property since it leads to (partially) unique solutions given by an inner factor, and a (possibly substantial) reduction of the interpolation problem. We study it in this section together with its connection to conjugation. In a further section we shall use the knowledge obtained to convert a double-sided problem to a single-sided one, after reduction of an eventual degeneracy. To avoid technicalities, we work under certain regularity conditions, which can almost always be assumed in practical problems.

To this end, define

$$\begin{array}{ll} (1) & \Lambda_1 := \mathbf{P}_0[(Z-V)^{-1}\xi\xi^*(Z^*-V^*)^{-1}] \\ (2) & \Lambda_2 := \mathbf{P}_0[(Z-V)^{-1}\eta\eta^*(Z^*-V^*)^{-1}]. \end{array} \tag{9.23}$$

We shall say that the LIP is *non-degenerate* if $\Lambda_1 \gg 0$ and $\Lambda_2 > 0$, *i.e.*, Λ_2 has empty kernel. We say that it is *regular* if

$$\begin{array}{ll} (1) & \Lambda_1 \gg 0 \\ (2) & \Lambda_2 \text{ has closed range.} \end{array} \tag{9.24}$$

Hence a non-degenerate regular LIP has both $\Lambda_1 \gg 0$ and $\Lambda_2 \gg 0$. A degenerate LIP with $\Lambda_1 \gg 0$ can be converted to a non-degenerate one, as shown in the next proposition. Handling the non-regular case is much harder, since the conjugation theory for J-unitary operators of the previous chapter then breaks down. We do not have good results for that case which as far as we know is still open.

Proposition 9.8 *Consider the LIP(V,ξ,η), and let the corresponding $\Lambda_1 \gg 0$. Then all solutions of the LIP are of the form $S = US'$ where $U \in \mathcal{U}$ is inner, and $S' \in \mathcal{U}$ is the solution of a non-degenerate LIP (V',ξ',η'), in which V' is a suboperator of V, and ξ', η' are of comparatively smaller dimensions than ξ, η, respectively. If the original LIP is regular, then so is the deflated LIP.*

PROOF The property is a direct consequence of the conjugation theory for J-inner matrices of section 8.6. Suppose that the interpolation problem has solutions (otherwise there is nothing to prove). Let Θ be the causal J-inner matrix which defines the solutions, *i.e.*, all solutions S are given by $S = T_\Theta[S_L]$ where S_L is causal and strictly contractive. The reachability pair (A,B) for Θ is given by

$$\left(V^*, \begin{bmatrix} \xi^* \\ -\eta^* \end{bmatrix}\right).$$

Let Λ_1 and Λ_2 be as defined above. Concentrating on Λ_2, let R be a unitary transformation such that

$$\Lambda_2 = R^* \begin{bmatrix} \Lambda_2' & 0 \\ 0 & 0 \end{bmatrix} R$$

in which Λ_2' has trivial kernel ($\Lambda_2' > 0$), and let us define a further partitioning of the data, after state transformation by R, as

$$\left[\begin{array}{cc} V_{11}^* & 0 \\ V_{12}^* & V_{22}^* \\ \hline \xi_1^* & \xi_2^* \\ -\eta_1^* & 0 \end{array}\right] := \left[\begin{array}{c} RV^*R^{-(-1)} \\ \hline \xi^* R^{-(-1)} \\ -\eta^* R^{-(-1)} \end{array}\right].$$

Just as in the proof of proposition 8.21 of section 8.6, we see that Θ factors as

$$\Theta = \begin{bmatrix} U & \\ & I \end{bmatrix} \Theta' \tag{9.25}$$

in which U is inner and has a realization of the form

$$\mathbf{U} = \begin{bmatrix} V_{22}^* & C_U^* \\ \xi_2^* & D_U^* \end{bmatrix}$$

for appropriate C_U and D_U, and whereby $\Theta' \in \mathcal{U}$ is J-inner and has a realization with reachability pair

$$\left[\begin{array}{c} V_{11}^* \\ \hline C_U \Lambda_{12} V_{12}^* + D_U \xi_1^* \\ -\eta_1^* \end{array}\right]$$

where Λ_{12} is the unique bounded solution of the Lyapunov equation $V_{22}\Lambda_{12}V_{22}^* + \xi_2\xi_2^* = \Lambda_{12}^{(-1)}$. Hence, with $V' = V_{11}$, $\xi' = V_{12}\Lambda_{12}C_U^* + \xi_1 D_U^*$, $\eta' = \eta_1$, the interpolation problem is reduced to: find S' such that

$$(Z - V')^{-1}(\xi' S' - \eta') \in \mathcal{U}, \tag{9.26}$$

$S' \in \mathcal{U}$, $\|S'\| < 1$. This interpolation problem has $\Lambda_2' > 0$: it is non-degenerate. All solutions to the original interpolation problem are described by $S = T_\Theta[S_L]$. The factorization of Θ in (9.25) forces $S = US'$, where $S' = T_{\Theta'}[S_L]$. (See also figure 9.2.)

□

It should be clear that if the original LIP is regular, then so is the derived problem — the Gramians involved are equivalent under unitary similarity. Proposition 9.8 allows a simple reduction of a partially degenerate LIP (or dually RIP) to a nondegenerate one. We restrict the conjugation theory to purely non-degenerate problems since conjugation does not work on the purely degenerate part: in general, there is no (fixed) diagonal η such that for each S satisfying $\mathbf{P}'(Z-V)^{-1}\xi S = 0$ there exists a conjugate S^c such that $\mathbf{P}' S^c \eta (Z-V)^{-1} = 0$.

Proposition 9.9 *Let (V, ξ, η) describe a non-degenerate, regular LIP with Gramians Λ_1, Λ_2 as defined in (9.23). Then there exist diagonal operators $C_{U_1}, D_{U_1}, C_{U_2}, D_{U_2}$ such that*

$$\mathbf{U}_1 = \begin{bmatrix} V^* & C_{U_1}^* \\ \xi^* & D_{U_1}^* \end{bmatrix}, \quad \mathbf{U}_2 = \begin{bmatrix} V^* & C_{U_2}^* \\ \eta^* & D_{U_2}^* \end{bmatrix} \tag{9.27}$$

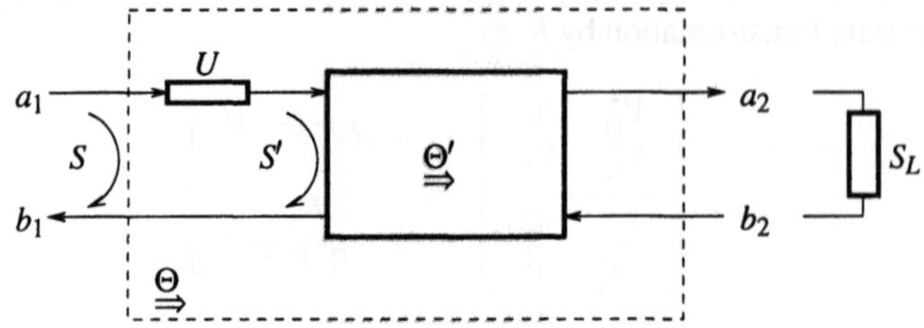

Figure 9.2. The extraction of the degeneracy from an interpolation problem yields an inner factor U and a remainder S'

are realizations of inner operators U_1 and U_2, and the LIP is equivalent to a right interpolation problem given by

$$(C_{U_1} - S'C_{U_2})(Z-V)^{-1} \in \mathcal{U}, \quad S' \in \mathcal{U}, \quad \|S'\| < 1, \tag{9.28}$$

in the sense that if S' is a solution to the RIP, then $S = U_1 S' U_2^*$ is a solution to the LIP, and vice-versa.

PROOF Let Θ be the causal J-inner matrix that solves the LIP problem in equation (9.21) according to theorem 9.6. Θ has reachability pair given by $A := V^*$, $B := [\begin{smallmatrix} \xi^* \\ -\eta^* \end{smallmatrix}]$. By assumption, the Lyapunov-Krein equations $V\Lambda_1 V^* + \xi\xi^* = \Lambda_1^{(-1)}$ and $V\Lambda_2 V^* + \eta\eta^* = \Lambda_2^{(-1)}$ have boundedly invertible solutions, so that by theorem 6.3 the reachability pairs (V^*, ξ^*), (V^*, η^*) can be completed to realizations of inner operators with realizations of the form (9.27). By the conjugation proposition 8.21 of chapter 8 there exists an anticausal, J-inner operator Θ',

$$\Theta' = \begin{bmatrix} U_1^* & \\ & U_2^* \end{bmatrix} \Theta \in \mathcal{L},$$

where Θ' has reachability pair given by

$$(V, \begin{bmatrix} C_{U_1} \\ -C_{U_2} \end{bmatrix}).$$

Let us complete the realization of Θ' in a minimal way with operators B', D' (they will not play a role), so that

$$\Theta' = D' + \begin{bmatrix} C_{U_1} \\ -C_{U_2} \end{bmatrix} (Z-V)^{-1} B'.$$

We show now that if Θ' is loaded in a causal, contractive load S_L, then the resulting $S' = U_1^* S U_2$ satisfies the interpolation problem given by (9.28) — see figure 9.3. The relation between S' and S_L is summarized by the equation

$$[I \quad S']\Theta' = \Phi_i'[I \quad S_L]$$

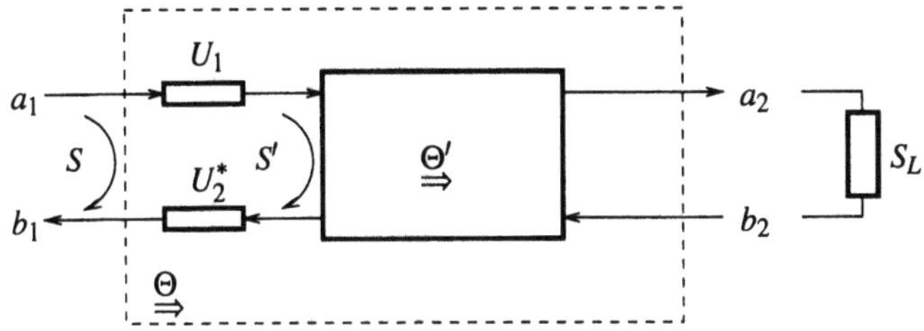

Figure 9.3. The transformation of Θ to Θ' and the resulting transformation of S to S'. The picture is "unphysical" as a signal flow diagram, but both Θ and Θ' are J-inner.

in which

$$\Phi_i{}' = \Theta_{11}'^{-*}(I - S_L \Theta_{12}'{}^{*} \Theta_{11}'^{-*})^{-1} = \Sigma_{11}'(I - S_L \Sigma_{21}')^{-1}.$$

Since Θ' is J-inner, the corresponding Σ' is inner, $\Sigma_{11}' = \Theta_{11}'^{-*}$ is strictly contractive as well as Σ_{21}', and Φ_i' is bounded and upper (its physical meaning should be clear from figure 9.3). Hence we find after substituting Θ'

$$[C_{U_1} - S' C_{U_2}](Z-V)^{-1} B' \;=\; \Phi_i'[I \quad S_L] - [I \quad S'] D'$$

and

$$\mathbf{P}'(C_{U_1} - S' C_{U_2})(Z-V)^{-1} B' \;=\; 0.$$

This is almost the desired interpolation expression: we still have to cancel out B'. That is the subject of the following proposition, which we state as a separate lemma because of its independent interest.

Lemma 9.10 *Suppose that $\ell_A < 1$, (A,B) form a reachable pair, $X \in \mathcal{U}$ and*

$$B(Z-A)^{-1} X \in \mathcal{U}, \qquad (9.29)$$

then $(Z-A)^{-1} X \in \mathcal{U}$.

PROOF of the lemma. By definition of evaluation at a diagonal A we have that $X = X^{\wedge}(A) + (Z-A)X_1$ with $X_1 \in \mathcal{U}$, hence by equation (9.29), $B(Z-A)^{-1}X^{\wedge}(A) \in \mathcal{U}$ so that $B(Z-A)^{-1}X^{\wedge}(A) = 0$. Evaluating this equality term by term we find $B^{(-1)}X^{\wedge}(A) = 0$, $B^{(-2)}A^{(-1)}X^{\wedge}(A) = 0$, etc., or in matrix notation:

$$\begin{bmatrix} B^{(-1)} \\ B^{(-2)}A^{(-1)} \\ B^{(-3)}A^{(-2)}A^{(-1)} \\ \vdots \end{bmatrix} X^{\wedge}(A) = 0. \qquad (9.30)$$

In (9.30) we recognize the reachability operator, which is assumed one-to-one. Hence $X^{\wedge}(A) = 0$, and $(Z-A)^{-1}X = X_1 \in \mathcal{U}$. □

The lemma has to be applied here in its dual (observability) form and yields (9.28). The property is symmetric: if S' solves the RIP interpolation problem (9.28), then, by a theory dual to that given by theorems 9.5 and 9.6, there will be a corresponding lower J-inner Θ'-matrix and a load S_L such that $S' = T_{\Theta'}[S_L]$. This Θ'-matrix will be non-degenerate and regular (due to the symmetrical structure of $\mathbf{U}_1$ and $\mathbf{U}_2$), and will yield, after conjugation, a Θ which solves the original LIP interpolation problem. Hence the RIP and LIP are equivalent in the sense that a solution for one yields a solution for the other and vice versa. □

9.6 TWO SIDED INTERPOLATION

An interesting (and practical) case occurs when doubled sided interpolation data are given and a constrained solution is asked. We shall see that this more general problem has some unique characteristics which make a further generalization of the theory necessary. In the literature it is sometimes referred to as the Nudel'man interpolation problem [Dym89].

Let be given two sets of diagonal operators,

$$(V,\xi,\eta) \quad \text{and} \quad (W,\zeta,\iota), \tag{9.31}$$

with $\ell_V < 1$, $\ell_W < 1$, asked is a strictly contractive $S \in \mathcal{U}$ which satisfies

1. *a left interpolation property,*

$$(Z-V)^{-1}[\xi S-\eta] \in \mathcal{U} \tag{9.32}$$

2. *and a right interpolation property,*

$$[S\zeta-\iota](Z-W)^{-1} \in \mathcal{U}. \tag{9.33}$$

In other words, we wish to solve a left and a right interpolation problem *jointly*. As before, the above description of the problem is such that it fits several types of interpolation problems, such as the tangential Nevanlinna-Pick and Hermite-Fejer problems of the previous sections. Again equations (9.32) and (9.33) can be manipulated in attractive alternative forms.[4] Define

$$\begin{aligned} H_1 &= (Z-V)^{-1}\xi, & G_1 &= (Z-V)^{-1}\eta \\ G_2 &= (I-W^*Z)^{-1}\zeta^*, & H_2 &= (I-W^*Z)^{-1}\iota^* \end{aligned}$$

then $H_1, G_1 \in Z^{-1}\mathcal{L}$ and $G_2, H_2 \in \mathcal{U}$. The interpolation conditions (9.32)–(9.33) can now be written as

$$\mathbf{P}'(\mathcal{D}_2[H_1S-G_1]) = 0 \tag{9.34}$$

$$\mathbf{P}(\mathcal{D}_2[G_2S^*-H_2]) = 0. \tag{9.35}$$

[4] We are indebted to H. Dym for informal information on this matter and providing us with a nice survey of ideas [DF97].

Translation to conditions on Θ

Solutions of the above two-sided interpolation problem turn out to be given in terms of a J-inner chain scattering operator Θ as before, although this time it will be of mixed causality. There are additional complications: it may happen that the complete solution set is not defined in terms of a single operator Θ. We shall explore a more restrictive condition on the interpolation data where the Θ obtained is indeed uniquely determined. A description of the complete set of solutions for the general case can be found in the recent literature, see [DF97]. This is a generalization of what already happens in the linear time invariant case [KKY87].

Proposition 9.11 *Let* $\Theta \in \mathcal{X}$ *be a* J*-inner operator such that*

$$\Theta = \begin{bmatrix} \iota \\ -\zeta \end{bmatrix} (Z-W)^{-1}[C_1 \quad C_2] + \begin{bmatrix} R_{11} \\ R_{12} \end{bmatrix} \qquad (9.36)$$

$$\Theta^{-1} = \begin{bmatrix} C_3 \\ C_4 \end{bmatrix} (Z-V)^{-1}[\xi \quad \eta] + \begin{bmatrix} R_{21} \\ R_{22} \end{bmatrix} \qquad (9.37)$$

in which $R_{ij} \in \mathcal{U}$, *for* $i,j = 1,2$, *and* $\left\{W, \begin{bmatrix} \iota \\ -\zeta \end{bmatrix}, [C_1 \ C_2]\right\}$, $\left\{V, \begin{bmatrix} C_3 \\ C_4 \end{bmatrix}, [\xi \ \eta]\right\}$ *are minimal realizations of the respective anticausal parts, Let* $S = T_\Theta[S_L]$ *for a strictly contractive* $S_L \in \mathcal{U}$. *Then* $(Z-V)^{-1}(\xi S - \eta) \in \mathcal{U}$ *and* $(S\zeta - \iota)(Z-W)^{-1} \in \mathcal{U}$.

PROOF The proposition is derived by using standard properties of a J-lossless Θ and the corresponding lossless scattering operator Σ, *viz.* theorem 8.2. Note that, by definition of losslessness, Σ is causal. Suppose $S = T_\Theta[S_L]$, *i.e.*, (by its definition in equation (8.7)),

$$\begin{aligned} S &= \Sigma_{12} + \Sigma_{11} S_L (I - S_L \Sigma_{21})^{-1} \Sigma_{22} \\ &= (\Theta_{12} - \Theta_{11} S_L)(\Theta_{21} S_L - \Theta_{22})^{-1} \end{aligned}$$

then

$$\begin{bmatrix} -S \\ I \end{bmatrix} \Phi_o = \Theta \begin{bmatrix} -S_L \\ I \end{bmatrix} \qquad (9.38)$$

where $\Phi_o = \Theta_{22} - \Theta_{21} S_L$. From equation (8.6) we have

$$\begin{aligned} \Phi_o &= \Theta_{22}(I - \Theta_{22}^{-1} \Theta_{21} S_L) \\ &= \Sigma_{22}^{-1}(I - \Sigma_{21} S_L). \end{aligned}$$

Since $\| \Sigma_{21} \| < 1$, it follows that Φ_o is invertible once S_L is contractive, with

$$\Phi_o^{-1} = (I - \Sigma_{21} S_L)^{-1} \Sigma_{22}.$$

Since Σ is a *causal* operator, and also $S_L \in \mathcal{U}$, it follows that both $\Phi_o^{-1} \in \mathcal{U}$ and $S = \Sigma_{12} + \Sigma_{11} S_L \Phi_o^{-1} \in \mathcal{U}$. Subsequently postmultiplying (9.37) with (9.38) produces

$$\begin{bmatrix} -S_L \\ I \end{bmatrix} \Phi_o^{-1} = \begin{bmatrix} C_3 \\ C_4 \end{bmatrix} (Z-V)^{-1}(\eta - \xi S) + R_{22} - R_{21} S$$

with R_{22} and R_{21} in $\mathcal{U}$. Because the left hand side is also upper, an invocation of lemma 9.10 together with minimality proves the first interpolation property,

$$(Z-V)^{-1}(\xi S-\eta) \in \mathcal{U}.$$

To show the second interpolation property, recall that $S = T_\Theta[S_L]$ also implies (by equation (8.7))

$$S_L = (\Theta_{11}+S\Theta_{21})^{-1}(\Theta_{12}+S\Theta_{22})$$

or, using $\Theta^{-1} = J\Theta^* J$,

$$\begin{array}{rrcl} & [I \quad S]\Theta & = & \Phi_i[I \quad S_L] \\ \Leftrightarrow & [I \quad -S] & = & \Phi_i[I \quad -S_L]\Theta^* \end{array}$$

where

$$\begin{array}{rcl} \Phi_i & = & \Theta_{11}^{-*}(I-S_L\Theta_{12}^*\Theta_{11}^{-*})^{-1} \\ & = & \Sigma_{11}(I-S_L\Sigma_{21})^{-1} \end{array}$$

using again the connections of Θ with Σ in equation (8.6). For similar reasons as before, it follows from the fact that Σ is upper and S_L is upper and contractive that $\Phi_i \in \mathcal{U}$ and $S \in \mathcal{U}$. Premultiplying (9.36) with $[I \quad S] = \Phi_i[I \quad S_L]\Theta^{-1}$ gives

$$\Phi_i[I \quad S_L] = (\iota - S\zeta)(Z-W)^{-1}[C_1 \quad C_2] + R_{11} + SR_{12},$$

with the R_{ij} in $\mathcal{U}$, and because $\Phi_i[I \quad S_L] \in \mathcal{U}$, also $(\iota - S\zeta)(Z-W)^{-1}[C_1 \quad C_2] \in \mathcal{U}$. An invocation of the dual form of lemma 9.10, using minimality of the realization, produces

$$(\iota - S\zeta)(Z-W)^{-1} \in \mathcal{U}.$$

□

Equations (9.36) and (9.37) completely specify the dynamics of Θ. Since $\Theta^{-1} = J\Theta^* J$, we can write

$$\Theta = \begin{bmatrix} \iota \\ -\zeta \end{bmatrix}(Z-W)^{-1}[C_1 \quad C_2] + D + \begin{bmatrix} \xi^* \\ -\eta^* \end{bmatrix}(Z^*-V^*)^{-1}[C_3^* \quad -C_4^*] \qquad (9.39)$$

in which D and the C_i's are diagonal operators. We explore this form in more detail now.

Lemma 9.12 *$\Theta \in \mathcal{X}$ in equations (9.36) and (9.37)* (i.e., *Θ in (9.39)) has a mixed causality type realization of the form*

$$[x_+^{(-1)} \quad x_- \quad a_2 \quad b_2] = [x_+ \quad x_-^{(-1)} \quad a_1 \quad b_1]\left[\begin{array}{cc|cc} V^* & & C_3^* & -C_4^* \\ & W & C_1 & C_2 \\ \hline \xi^* & \iota & D_{11} & D_{12} \\ -\eta^* & -\zeta & D_{21} & D_{22} \end{array}\right]. \qquad (9.40)$$

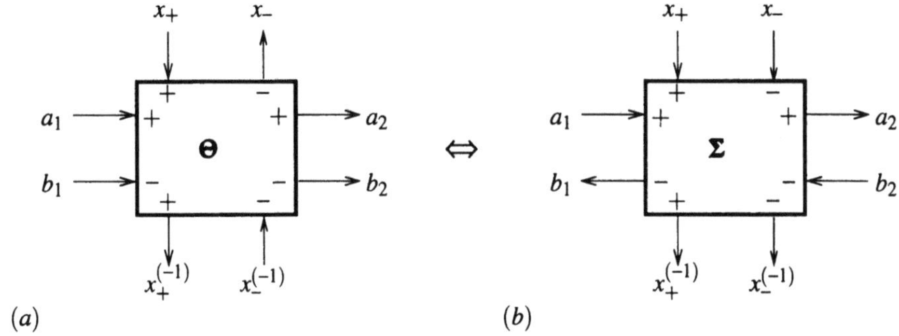

Figure 9.4. (a) The realization for Θ has mixed causality; (b) the realization of the corresponding Σ is causal.

PROOF The first term in (9.39) is generated by the anticausal state equations

$$\begin{cases} x_- &= x_-^{(-1)}W + [a_1\ b_1][\begin{smallmatrix}\iota \\ -\zeta\end{smallmatrix}] \\ y_- &= x_-^{(-1)}[C_1\ C_2] \end{cases}$$

whereas the last term is generated by

$$\begin{cases} x_+^{(-1)} &= x_+V^* + [a_1\ b_1][\begin{smallmatrix}\xi^* \\ -\eta^*\end{smallmatrix}] \\ y_+ &= x_+[C_3^*\ -C_4^*] \end{cases}$$

□

Construction of a J-inner Θ

The central question that has to be resolved now is how and under which conditions suitable C_i's and D_{ij}'s can be found such that the candidate realization $\mathbf{\Theta}$ in (9.40) indeed corresponds to a J-unitary, even J-inner scattering operator. The latter means that the corresponding lossless scattering operator Σ is causal, hence has a causal realization $\mathbf{\Sigma}$. The situation is drawn in figure 9.4.

Our strategy is as follows and uses the knowledge we have of J-inner operators of mixed type. Suppose that we have found matching C_i's and D_{ij}'s to make Θ J-lossless, and that we have computed the corresponding realization $\mathbf{\Sigma}$. We aim at constructing an inner Σ with a state realization $\mathbf{\Sigma}'$ that is *unitary*. It need not be equal to $\mathbf{\Sigma}$, but at least there must be an invertible state transformation R connecting $\mathbf{\Sigma}$ to $\mathbf{\Sigma}'$. We will work out how R transforms $\mathbf{\Theta}$ into $\mathbf{\Theta}'$ (this is not obvious because $\mathbf{\Theta}$ is not a causal realization). A second observation is that if $\mathbf{\Sigma}'$ is unitary, its corresponding $\mathbf{\Theta}'$ is a J-unitary map. That is to say, if we denote $\mathbf{\Theta}' = [\begin{smallmatrix}A' & C' \\ B' & D'\end{smallmatrix}]$, then we must have, among others,

$$A'^*J_BA' + B'^*J_1B' = J_B^{(-1)}, \tag{9.41}$$

for signature matrices J whose partitioning $J = [\begin{smallmatrix} I & 0 \\ 0 & -I \end{smallmatrix}]$ follows the partitioning of $\mathbf{\Theta}$. We will show that A' and B' are determined only by the known data V, W, ι, η, ξ, ζ, and the unknown R (*i.e.*, the unknown C_i's and D_{ij}'s do not enter into A' and B'). It follows that (9.41) gives sufficient conditions to compute a diagonal operator $P := R^*R$, which specifies R as its Cholesky factor. Once R is known, we know A' and B', and it suffices to complete $[\begin{smallmatrix} A' \\ B' \end{smallmatrix}]$ to a square J-unitary operator to find $\mathbf{\Theta}'$. This last step is the same as in section 8.5.

We now work out the details. First, we consider how R transforms $\mathbf{\Theta}$ into $\mathbf{\Theta}'$. Denote the transformed state vector by $[x'_+ \;\; x'_-]$, and define R by

$$[x_+ \quad x_-] = [x'_+ \quad x'_-] \begin{bmatrix} R_{11} & R_{12} \\ 0 & R_{22} \end{bmatrix}, \tag{9.42}$$

where R_{11}, R_{22} are square. We have set $R_{21} = 0$ because in the end R will only be defined as a factor of R^*R. Substituting (9.42) into the state equations (9.40), we obtain

$$\begin{aligned} &[x'^{(-1)}_+ R^{(-1)}_{11} \quad x'_+ R_{12} + x'_- R_{22}] = \\ &= [x'_+ R_{11} \quad x'^{(-1)}_+ R^{(-1)}_{12} + x'^{(-1)}_- R^{(-1)}_{22} \quad a_1 \quad b_1] \left[\begin{array}{cc} V^* & \\ & W \\ \hline \xi^* & \iota \\ -\eta^* & -\zeta \end{array}\right] \end{aligned}$$

and a similar expression in terms of the C_i's and D_{ij}'s which is not of interest. Rearranging the terms of this equation to recover the mapping $[x'^{(-1)}_+ \;\; x'_-] \mapsto [x'_+ \;\; x'^{(-1)}_- \;\; a_1 \;\; b_1]$, we obtain

$$[x'^{(-1)}_+ \quad x'_-] \begin{bmatrix} R^{(-1)}_{11} & -R^{(-1)}_{12} W \\ 0 & R_{22} \end{bmatrix} = [x'_+ \quad x'^{(-1)}_-] \begin{bmatrix} R_{11} V^* & -R_{12} \\ 0 & R^{(-1)}_{22} W \end{bmatrix} + [a_1 \quad b_1] \begin{bmatrix} \xi^* & \iota \\ -\eta^* & -\zeta \end{bmatrix},$$

that is, the leading block column of $\mathbf{\Theta}'$ is

$$\left[\begin{array}{c} A' \\ \hline B' \end{array}\right] = \left[\begin{array}{cc} R_{11} V^* & -R_{12} \\ 0 & R^{(-1)}_{22} W \\ \hline \xi^* & \iota \\ -\eta^* & -\zeta \end{array}\right] \begin{bmatrix} R^{(-1)}_{11} & -R^{(-1)}_{12} W \\ 0 & R_{22} \end{bmatrix}^{-1}.$$

The condition (9.41) that this block column is J-isometric now reads

$$\begin{aligned} &\begin{bmatrix} V R^*_{11} & 0 \\ -R^*_{12} & W^* R^{(-1)}_{22} \end{bmatrix} J \begin{bmatrix} R_{11} V^* & -R_{12} \\ 0 & R^{(-1)}_{22} W \end{bmatrix} + \begin{bmatrix} \xi & -\eta \\ \iota^* & -\zeta^* \end{bmatrix} J \begin{bmatrix} \xi^* & \iota \\ -\eta^* & -\zeta \end{bmatrix} \\ &= \begin{bmatrix} R^{(-1)*}_{11} & 0 \\ -W^* R^{(-1)*}_{12} & R^*_{22} \end{bmatrix} J \begin{bmatrix} R^{(-1)}_{11} & -R^{(-1)}_{12} W \\ 0 & R_{22} \end{bmatrix} \end{aligned}$$

which after some rearrangements becomes

$$\begin{aligned} &\begin{bmatrix} R^{(-1)*}_{11} & 0 \\ -W^* R^{(-1)*}_{12} & W^* R^{(-1)*}_{22} \end{bmatrix} \begin{bmatrix} R^{(-1)}_{11} & -R^{(-1)}_{12} W \\ 0 & R^{(-1)}_{22} W \end{bmatrix} \\ &= \begin{bmatrix} V R^*_{11} & 0 \\ -R^*_{12} & R^*_{22} \end{bmatrix} \begin{bmatrix} R_{11} V^* & -R_{12} \\ 0 & R_{22} \end{bmatrix} + \begin{bmatrix} \xi & -\eta \\ \iota^* & -\zeta^* \end{bmatrix} J \begin{bmatrix} \xi^* & \iota \\ -\eta^* & -\zeta \end{bmatrix}. \end{aligned}$$

With

$$P := R^*R = \begin{bmatrix} R_{11}^* & 0 \\ R_{12}^* & R_{22}^* \end{bmatrix} \begin{bmatrix} R_{11} & R_{12} \\ 0 & R_{22} \end{bmatrix}$$

we find that P satisfies

$$\begin{bmatrix} I & \\ & -W^* \end{bmatrix} P^{(-1)} \begin{bmatrix} I & \\ & -W \end{bmatrix} = \begin{bmatrix} V & \\ & -I \end{bmatrix} P \begin{bmatrix} V^* & \\ & -I \end{bmatrix} + \begin{bmatrix} \xi & \eta \\ \iota^* & \zeta^* \end{bmatrix} J \begin{bmatrix} \xi^* & \iota \\ \eta^* & \zeta \end{bmatrix}. \quad (9.43)$$

P plays the role of "Pick matrix" in the mixed interpolation problem.

Theorem 9.13 *The mixed (Nudel'man) interpolation problem in (9.32)–(9.33) has a strictly contractive solution S if and only if there exists a boundedly invertible, strictly positive, diagonal operator P satisfying (9.43).*

If this is the case, then there exists a J-lossless operator $\Theta \in \mathcal{X}$ with mixed-causality realization of the form (9.40). All S of the form

$$S = T_\Theta[S_L], \qquad S_L \in \mathcal{U}, \; \|S_L\| < 1$$

are solutions to the problem.

Equation (9.43) may have more than one solution which satisfies the positivity condition. In that case there is also more than one Θ which provides solutions to the interpolation problem.
PROOF

Step 1: *If P exists and $P \gg 0$ then there are solutions S. These are given in terms of a lossless chain scattering matrix Θ as $T_\Theta[S_L]$, in which S_L ranges over causal, strictly contractive operators.*

Once P is known, a factorization $P = R^*R$ with R upper triangular gives the state transformation that makes $\left[\begin{smallmatrix} A' \\ B' \end{smallmatrix}\right]$ J-isometric. By theorem 8.17 there exist a completion by matrices C' and D' so that

$$\mathbf{\Theta} = \begin{bmatrix} A' & C' \\ B' & D' \end{bmatrix} \quad (9.44)$$

is a J-unitary state transition matrix mapping $[x'_+, x'^{(-1)}_-, a_1, b_1]$ to $[x'^{(-1)}_+, x'_-, a_2, b_2]$. As a consequence the map

$$\mathbf{\Sigma}' : \quad [x'_+, x'_-, a_1, b_2] \mapsto [x'^{(-1)}_+, x'^{(-1)}_-, a_2, b_1]$$

is not only well-defined but unitary as well, and hence corresponds to a causal lossless system (theorem 6.4). Under the present conditions on the interpolation data, the realizations for $\mathbf{\Sigma}'$, $\mathbf{\Theta}'$ and $\mathbf{\Theta}$ are all minimal. By proposition 9.11, each $S = T_\Theta[S_L]$ with S_L upper and strictly contractive is an interpolant satisfying the interpolation condition (9.32)-(9.33).

Step 2: *If there is a strictly contractive solution S, then equation (9.43) has a strictly positive definite solution P.*

Suppose that we have a strictly contractive S satisfying equations (9.32)-(9.33), or alternatively (9.34)-(9.35), whose notation we use further on. We claim, following the Dym-theory as in [DF97], that the matrix

$$P = \mathbf{P}_0\left(\begin{bmatrix} H_1 & G_1 \\ H_2 & G_2 \end{bmatrix}\begin{bmatrix} I & -S \\ -S^* & I \end{bmatrix}\begin{bmatrix} H_1^* & H_2^* \\ G_1^* & G_2^* \end{bmatrix}\right) \tag{9.45}$$

satisfies (9.43). The proof for the entries P_{11} and P_{22} parallels the earlier developments and follows from the observation that

$$P_{11} = \mathbf{P}_0(H_1H_1^* - G_1S^*H_1^* - H_1SG_1^* + G_1G_1^*) = \mathbf{P}_0(G_1^*G_1 - H_1^*H_1)$$

because $H_1S = G_1 + R_1$ for some $R_1 \in \mathcal{U}$ so that $\mathbf{P}_0(R_1G_1^*) = 0$. Likewise,

$$P_{22} = -\mathbf{P}_0(G_2G_2^* - H_2H_2^*).$$

The interesting calculation is on P_{12} (or its adjoint P_{21}):

$$P_{12} = \mathbf{P}_0(H_1H_2^* - G_1S^*H_2^* - H_1SG_2^* + G_1G_2^*) = -\mathbf{P}_0(H_1SG_2^*) \tag{9.46}$$

since all the other entries are either in $Z\mathcal{U}$ or in $Z^{-1}\mathcal{L}$. Elaborating on (9.46) produces

$$P_{12} = -\mathbf{P}_0(Z^*(I-VZ^*)^{-1}\xi S\zeta(I-Z^*W)^{-1})$$

and

$$P_{12}^{(-1)} = ZP_{12}Z^* = -\mathbf{P}_0((I-VZ^*)^{-1}\xi S\zeta(I-Z^*W)^{-1}Z^*).$$

It follows that

$$\begin{aligned} VP_{12} - P_{12}^{(-1)}W &= -\mathbf{P}_0\left((I-VZ^*)^{-1}\{VZ^*\xi S\zeta - \xi S\zeta Z^*W\}(I-Z^*W)^{-1}\right) \\ &= -\mathbf{P}_0(\xi S\zeta(I-Z^*W)^{-1} - (I-VZ^*)^{-1}\xi S\zeta). \end{aligned}$$

But because of (9.34)-(9.35),

$$\mathbf{P}_0(S\zeta(I-Z^*W)^{-1}) = \mathbf{P}_0((I-W^*Z)^{-1}\zeta^*S^*)^* = \iota$$

and

$$\mathbf{P}_0((I-VZ^*)^{-1}\xi S) = \mathbf{P}_0((I-VZ^*)^{-1}\eta) = \eta$$

the expression simplifies to

$$VP_{12} - P_{12}^{(-1)}W = -\xi\iota + \eta\zeta \tag{9.47}$$

which is precisely the "12" term in (9.43). □

We proceed by showing that the collection of solutions $T_\Theta[S_L]$ to the mixed interpolation problem is complete if the linear map $X \mapsto Y$ on $\mathcal{D}$ defined by

$$VX - X^{(-1)}W = Y \tag{9.48}$$

is boundedly invertible. In the LTI case the latter would happen (only) when the spectra of V and W are disjoint. In the time varying case, V and W may be non-square, so there is more chance for irregular behavior. Conditions on V and W for bounded invertibility of the fundamental equation (9.48) in the LTV case have not been investigated to our knowledge.

An interesting way of proving uniqueness is by converting the mixed interpolation problem to an equivalent one-sided problem, using the conjugation ideas of the previous chapter.

Conjugation of the mixed interpolation problem

In this subsection we shall convert the two sided interpolation problem to a one-sided problem of the same total size, under the additional conditions

(1) (V,ξ,η) defines a regular LIP,

(2) the linear map (9.48), *i.e.*, $X \mapsto Y : VX - X^{(-1)}W = Y$, has a bounded inverse.

These conditions are also sufficient to assure uniqueness of the solution provided by theorem 9.13.

If the original LIP problem with (V,ξ,η) is degenerate, then by proposition 9.8, any solution must be of the form $S = US'$ where U is an inner function defined by the unreachable part in $[\begin{smallmatrix} V^* \\ \eta^* \end{smallmatrix}]$, while S' is the solution of the deflated problem given by (9.26). If (V,ξ,η) is regular, then so is the deflated interpolation problem with data (V',ξ',η'). Moreover, the derived map

$$V'X - X^{(-1)}W = Y$$

is also boundedly invertible if the original map is (the verification is straightforward). Hence we may just as well assume that the original interpolation problem was non-degenerate to start with. Now, let U_1 and U_2 be inner operators with realizations

$$\mathbf{U}_1 = \begin{bmatrix} V^* & C_{U_1}^* \\ \xi^* & D_{U_1}^* \end{bmatrix}, \qquad \mathbf{U}_2 = \begin{bmatrix} V^* & C_{U_2}^* \\ \eta^* & D_{U_2}^* \end{bmatrix},$$

which define as before the conjugate of the LIP(V,ξ,η) as a right interpolation problem

$$(C_{U_1} - S'C_{U_2})(Z-V)^{-1} \in \mathcal{U}. \tag{9.49}$$

By proposition 9.9, we know that S is a solution to LIP(V,ξ,η) if and only if

$$S' = U_1^* S U_2 \tag{9.50}$$

is a solution to (9.49). Propagation of S' to the second interpolation condition $(\iota - S\zeta)(Z-W)^{-1} \in \mathcal{U}$ produces

$$(\iota - U_1 S' U_2^* \zeta)(Z-W)^{-1} \in \mathcal{U}. \tag{9.51}$$

This problem does not quite look like an interpolation problem, but it turns out that (9.51) can be converted to a normal (right-) interpolation condition, if the bounded invertibility of (9.48) holds. The remainder of this section is devoted to proving that there exists interpolation data (W, ι', ζ') so that (9.51) is equivalent to the interpolation condition

$$(\iota' - S'\zeta')(Z-W)^{-1} \in \mathcal{U} \tag{9.52}$$

on the transformed scattering function S'.

Lemma 9.14 *If the linear map (9.48) is boundedly invertible, then the equations*

$$(U_1\iota' - \iota)(Z-W)^{-1} \in \mathcal{U} \tag{9.53}$$
$$(U_2\zeta' - \zeta)(Z-W)^{-1} \in \mathcal{U} \tag{9.54}$$

have unique diagonal solutions ι' and ζ'.

PROOF Let's concentrate on (9.53) — the proof for the second one will be similar. Premultiplying it with U_1^* we obtain the condition for ι':

$$\iota'(Z-W)^{-1} - U_1^*\iota(Z-W)^{-1} = U_1^* Y \tag{9.55}$$

for some $Y \in \mathcal{U}$. Since the left hand side of this equation is in $Z^* \mathcal{L}$, we have that the right hand side is actually of the form $C_U(Z-V)^{-1}D$ for some diagonal D (it is obtained by projecting it on $Z^*\mathcal{L}$.) Now,

$$\begin{aligned} U_1^*\iota(Z-W)^{-1} &= D_{U_1}\iota(Z-W)^{-1} + C_{U_1}(Z-V)^{-1}\xi\iota(Z-W)^{-1} \\ &= [D_{U_1}\iota - C_{U_1}X](Z-W)^{-1} + C_{U_1}(Z-V)^{-1}X^{(-1)} \end{aligned}$$

in which X is the unique solution of

$$VX - X^{(-1)}W = \xi\iota.$$

Hence, (9.53) will be satisfied if and only if

$$(\iota' - D_{U_1}\iota + C_{U_1}X)(Z-W)^{-1} + C_{U_1}(Z-V)^{-1}(X^{(-1)} - D) = 0.$$

We find a solution if we choose $D = X^{(-1)}$ and

$$\iota' = D_{U_1}\iota - C_{U_1}X.$$

We now have to show that the solution found is unique. This we do by an invocation of lemma 9.10. Equation (9.4) has the form

$$\alpha(Z-W)^{-1} + C_{U_1}(Z-V)^{-1}\beta = 0$$

for some diagonals α and β which we must show to be necessarily zero. Postmultiplying with $(Z-W)$ yields

$$\alpha + C_{U_1}(Z-V)^{-1}\beta(Z-W) = 0$$

which is of the form required by the lemma (the pair (V, C_{U_1}) is reachable by construction). We conclude that

$$(Z-V)^{-1}\beta(Z-W) \in \mathcal{U}$$

but it cannot be anything else than a diagonal, say α_1, due to the form of the left member of the inclusion. Hence

$$\beta(Z-W) = (Z-V)\alpha_1$$

from which it follows immediately that $\alpha_1 = \beta^{(1)}$, and

$$V\beta^{(1)} - \beta W = 0.$$

This equation is now of the form (9.48) which we assumed to have a unique solution, $\beta^{(1)} = 0$ in this case. Hence also $\alpha_1 = 0$ and following also $\alpha = C_{U_1}\alpha = 0$. This shows uniqueness.

Likewise, the second equation will be solved by $\zeta' = D_{U_2}\zeta - C_{U_2}Y$ in which $VY - Y^{(-1)}W = \eta\zeta$, which, again, has a unique solution by assumption. □

Based on the lemma we can now show that (9.51) is equivalent to the interpolation condition (9.52). The easiest way to see this is via an adaptation of the properties of the W-transform to its dual. Mutatis mutandis, we write the right version of the W-transform as $\cdot^{\vee}$ so that we can write (9.53) as

$$\begin{cases} (U_1\iota')^{\vee}(W) & = & \iota \\ (U_2\zeta')^{\vee}(W) & = & \zeta \end{cases}$$

while the interpolation problem to be shown is expressed by

$$(S'\zeta')^{\vee}(W) = \iota'.$$

Using the chain rule, which in this case reads $(AB)^{\vee}(W) = (A(B)^{\vee}(W))^{\vee}(W)$, the equation $U_1S'\zeta' = SU_2\zeta'$ and the lemma, we obtain the following chain of equivalences:

$$\begin{array}{rcl} (S'\zeta')^{\vee}(W) = \iota' & \Leftrightarrow & (U_1(S'\zeta')^{\vee}(W))^{\vee}(W) = \iota \\ & \Leftrightarrow & (U_1S'\zeta')^{\vee}(W) = \iota \\ & \Leftrightarrow & (SU_2\zeta')^{\vee}(W) = \iota \\ & \Leftrightarrow & (S(U_2\zeta')^{\vee}(W))^{\vee}(W) = \iota \\ & \Leftrightarrow & (S\zeta)^{\vee}(W) = \iota \end{array}$$

which is the equivalence to be proven. We have established the following theorem:

Theorem 9.15 *Consider the two-sided interpolation problem: find $S \in \mathcal{U}$, $\|S\| \le 1$, such that*

$$\begin{array}{rcl} (Z-V)^{-1}(\xi S - \eta) & \in & \mathcal{U} \\ (S\zeta - \iota)(Z-W)^{-1} & \in & \mathcal{U}. \end{array}$$

Suppose that (V,ξ,η) defines a non-degenerate and regular left interpolation problem, and that the linear map given by (9.48), i.e., $X \mapsto Y : VX - X^{(-1)}W = Y$, is boundedly invertible.

Then the interpolation problem is equivalent to a one-sided interpolation problem given by the interpolation conditions (9.49)–(9.52), viz.

$$\begin{array}{rcl} (C_{U_1} - S'C_{U_2})(Z-V)^{-1} & \in & \mathcal{U} \\ (\iota' - S'\zeta')(Z-W)^{-1} & \in & \mathcal{U} \end{array}$$

in the sense that the solution of one produces a solution of the other and vice-versa, the relation being given by (9.50). In particular, the Θ-matrix defined in theorem 9.13 is unique except for a diagonal J-unitary right factor, and provides a complete set of solutions.

Theorem 9.15 settles the uniqueness question for the case that the double-sided interpolation problem has a non-degenerate LIP part, and the linear map (9.48) is non-singular. It is easy to relax the condition of non-degeneracy. Indeed, if the LIP under consideration is regular but degenerate, then we know, by proposition 9.9, that we can handle the degenerate part of the interpolation problem separately, whose solution is essentially unique and described by a single inner function U. The overall solution will then be characterized by the cascade

$$\Theta = \begin{bmatrix} U & \\ & I \end{bmatrix} \Theta'$$

in which Θ' solves a reduced, but now non-degenerate (mixed-mode) interpolation problem. If the other condition, namely the non-singularity of equation (9.48), is satisfied, then uniqueness is again assured. In practice, it will always pay to take the degenerate part of an interpolation problem out, since its solution is so much simpler than the general solution.

9.7 THE FOUR BLOCK PROBLEM

An important application and illustration of interpolation theory is the solution of control problems for optimal sensitivity. Let us assume that we are given a block-partitioned, strictly anticausal transfer operator

$$T = \begin{bmatrix} T_{11} & T_{12} \\ T_{21} & T_{22} \end{bmatrix} \in \mathcal{Z}^*\mathcal{L}. \tag{9.56}$$

The question is to find $R \in \mathcal{U}$ such that

$$\left\| \begin{bmatrix} T_{11}+R & T_{12} \\ T_{21} & T_{22} \end{bmatrix} \right\| < 1, \tag{9.57}$$

and to describe all possible solutions.

This problem is known as the "four block problem", and it is a prototype problem for a variety of questions in optimal control and game theory. It has received quite some attention in the literature, see *e.g.*, [GL95, CSK94, IO96]. In our present formalism it has a simple and straightforward solution. We follow roughly the treatment of [CSK94] adapted to the powerful interpolation results described earlier in this chapter. The problem is amenable to various interesting extensions, but we limit ourselves to the standard, basic case. It is an extension of the Nehari problem in section 10.6 to block-partitioned matrices for which only the (1,1) block is allowed to be modified. Our treatment here uses an ancillary result from the orthogonal embedding theory of chapter 12, but we give it here nonetheless because of its strong connection to classical interpolation theory. There is also a connection to the Schur-Takagi type interpolation theory, and we give a discussion of that connection and a resulting algorithm for the one-block case in section 10.6.

We will assume that T is locally finite, and that it has a left inner coprime factorization (*cf.* theorem 6.8):

$$T := B(Z-V)^{-1}C = U^*\Delta \tag{9.58}$$

with $\Delta \in \mathcal{U}$ and U inner, $U^* = D_U + B(Z-V)^{-1}C_U$ for some D_U, C_U which make U inner. Inserting (9.57) and (9.58) in (9.56) after premultiplication with U, we find that (9.57) is equivalent to

$$\left\| \begin{bmatrix} \Delta_{11} + U_{11}R & \Delta_{12} \\ \Delta_{21} + U_{21}R & \Delta_{22} \end{bmatrix} \right\| < 1. \tag{9.59}$$

A necessary condition is certainly $\|[\begin{smallmatrix}\Delta_{12}\\ \Delta_{22}\end{smallmatrix}]\| < 1$. Let us write

$$X := \begin{bmatrix} \Delta_{11} + U_{11}R \\ \Delta_{21} + U_{21}R \end{bmatrix}, \qquad H := \begin{bmatrix} \Delta_{12} \\ \Delta_{22} \end{bmatrix}$$

and define G to be the left outer factor in $\mathcal{U}$ satisfying the "embedding" relation

$$GG^* = I - HH^*.$$

(Orthogonal embedding is treated in detail in chapter 12, *viz.* theorem 12.14.) Equation (9.59) transforms to $I - HH^* - XX^* = GG^* - XX^* \gg 0$ and hence we must have

$$XX^* \ll GG^*.$$

This inequality implies that there exists $S \in \mathcal{U}$, $\|S\| < 1$, such that

$$X = GS \tag{9.60}$$

(see *e.g.*, theorem 12.6 due to Douglas [Dou66]). In particular, since G is left outer, it will have a left pseudoinverse $G^\dagger$ on a dense subset of $\mathcal{U}$, and we can take $S := G^\dagger X$. Premultiplying (9.60) with U^* we find

$$\begin{bmatrix} T_{11} \\ T_{21} \end{bmatrix} + \begin{bmatrix} R \\ 0 \end{bmatrix} = U^*GS \tag{9.61}$$

in which $R \in \mathcal{U}$ is unknown. A necessary condition for (9.61) to be satisfied is

$$B(Z-V)^{-1}C_1 = \mathbf{P}'(U^*GS), \tag{9.62}$$

in which C is decomposed as $C = [C_1\ C_2]$ conformal to the block structure of T. This condition is also sufficient. Indeed, if it is satisfied and $X := GS$, then R follows from $R = [I\ \ 0]U^*GS - T_{11} \in \mathcal{U}$, and $[0\ \ I]U^*X = T_{21}$ automatically. It turns out that (9.62) actually defines a left interpolation problem. To see this we evaluate the right hand side:

$$\begin{aligned} \mathbf{P}'(U^*GS) &= \mathbf{P}'[B(Z-V)^{-1}C_U GS] \\ &= B\mathbf{P}'[(Z-V)^{-1}C_U GS] \\ &= B(Z-V)^{-1}(C_U GS)^\wedge(V). \end{aligned} \tag{9.63}$$

From the chain rule for the W-transform of section 9.1, we have

$$(C_U GS)^\wedge(V) = ((C_U G)^\wedge(V)S)^\wedge(V),$$

hence writing $\eta := C_1$ and $\xi = (C_U G)^\wedge(V)$, we find that

$$\mathbf{P}'(U^* GS) = (Z-V)^{-1}[\eta - \xi S] \in \mathcal{U}. \tag{9.64}$$

Hence a necessary and sufficient condition for the solution of the (suboptimal) four block problem is that $S \in \mathcal{U}$, $\|S\| < 1$ and S satisfies the left interpolation condition (9.64), in which the "data" (V, ξ, η) comes from the original problem. A necessary and sufficient condition for the existence of a solution is then that the Gramian

$$\mathbf{P}_0((Z-V)^{-1}[\xi\xi^* - \eta\eta^*](Z^* - V^*)^{-1}) \gg 0.$$

The reachability pair $(V^*, [\begin{smallmatrix} \xi^* \\ -\eta^* \end{smallmatrix}])$ defines a J-inner Θ-matrix, and all possible solutions for S are given by $S = T_\Theta[S_L]$ with $S_L \in \mathcal{U}$ and strictly contractive. With a little more effort one shows that all solutions of the non-strict problem are given by the same homographic transformation, but now with the condition $\|S_L\| \leq 1$.

10 HANKEL-NORM MODEL REDUCTION

In the previous chapters, we assumed that a given upper operator or matrix T has a computational model of a sufficiently low order to warrant the (possibly expensive) step of deriving its state realization. Once a state model is known, we showed how multiplication by T or its inverse can be done efficiently, using the model rather than the entries of T. We also derived some useful factorizations, such as the external and inner-outer (~ QR) factorization. A spectral factorization/Cholesky factorization result is given in chapter 13.

However, if the ranks of the sequence of Hankel matrices of T are not sufficiently low, then the system order of the computational model will be large. This can already happen if T is modified only slightly, *e.g.*, caused by numerical imprecisions, as the rank of a matrix is a very sensitive (ill-conditioned) parameter. Hence one wonders whether, for a given $T \in \mathcal{U}$, there is an approximating system T_a close to it such that T_a has a low system order. Such an approximation is useful also when T is known exactly, but if for analysis purposes one would like to work with a low complexity, yet accurate approximating model.

In this chapter, we derive a suitable model approximation theory, using a norm which generalizes the Hankel norm of classical LTI systems. We obtain a parametrization of all solutions of the model order reduction problem in terms of a fractional representation based on a non-stationary J-unitary operator constructed from the data. In the stationary case, the problem was solved by Adamyan, Arov and Krein in their paper on Schur-Takagi interpolation [AAK71]. Our approach extends that theory to cover general, non-Toeplitz upper operators or matrices.

10.1 INTRODUCTION

One standard way to find an approximant of a matrix (A, say) goes via the singular value decomposition (SVD). This decomposition yields a diagonal matrix of singular values. Setting those singular values that are smaller than some tolerance level ε equal to zero produces an approximant $\hat{A}$ such that $\| A-\hat{A} \| < \varepsilon$ and rank(A) is equal to the remaining number of non-zero singular values. One can show that the approximant thus obtained is optimal in the operator norm (matrix 2-norm), and also in the Hilbert-Schmidt norm (matrix Frobenius norm). Since the state complexity of the operator/matrix T is given by the rank sequence of H_T rather than the rank of T itself (corollary 5.7), it seems logical to approximate each H_k by some $\hat{H}_k$ of lower rank. However, the Hankel matrices have many entries in common, and approximating one of them by a matrix of low rank might make it impossible for all other $\hat{H}_k$ to acquire a low rank: a local optimum might prevent a global one. In this respect, the approximation error norm used is also of importance: the Hilbert-Schmidt (Frobenius) norm is rather strong:

$$\min_{\text{rank}\hat{A}\leq d} \| A-\hat{A} \|_{HS}$$

has only one (unique) solution $\hat{A}$, obtained by setting all but the first d singular values equal to zero, and keeping the first d untouched. The operator norm approximation problem

$$\min_{\text{rank}\hat{A}\leq d} \| A-\hat{A} \|$$

has many solutions, since only the largest singular value of the difference $E = A-\hat{A}$ is minimized, and $d-1$ others are free, as long as they remain smaller. For sequences of Hankel matrices, the extra freedom in each of the $\hat{H}_k$ can be used to reduce the rank of the other H_k. The problem can be described in two ways: by

$$\min_{\text{rank}\hat{H}_k\leq d_k} \| H_k-\hat{H}_k \|, \qquad \text{(for all } k\text{)},$$

which is the *model error reduction problem* for given target ranks d_k, and by

$$\min\{\text{rank}\hat{H}_k : \| H_k-\hat{H}_k \| \leq \varepsilon_k\}, \qquad \text{(for all } k\text{)}, \tag{10.1}$$

the *model order reduction problem* for given tolerance levels ε_k. The latter problem description is the one which we take up in this chapter. The error criterion (10.1) leads to the definition of the *Hankel norm*, which is a generalization of the Hankel norm for time-invariant systems:

$$\| T \|_H = \| H_T \|. \tag{10.2}$$

$\| T \|_H$ is the supremum over the operator norm of each individual Hankel matrix H_k. It is a reasonably strong norm: if T is a strictly upper triangular matrix and $\| T \|_H \leq 1$, then each row and column of T has vector norm smaller than 1. The main approximation theorem that we derive can be stated as follows.

Theorem 10.1 *Let $T \in \mathcal{U}$, and let $\Gamma = \text{diag}(\gamma_i) \in \mathcal{D}$ be a Hermitian operator. Let H_k be the Hankel operator of $\Gamma^{-1}T$ at stage k, and suppose that an $\varepsilon > 0$ exists such that,* *for each k, none of the singular values of H_k are in the interval $[1-\varepsilon, 1+\varepsilon]$. Then there exists a strictly upper triangular operator T_a with system order at stage k at most equal to the number of singular values of H_k that are larger than 1, such that*

$$\| \Gamma^{-1}(T-T_a) \|_H \leq 1. \qquad (10.3)$$

The error tolerance diagonal Γ parametrizes the problem. As ε in (10.1), it can be used to influence the local approximation error: if $\Gamma = \gamma I$, then $\|T-T_a\|_H \leq \gamma$ and the approximation error is uniformly distributed over T. If one of the γ_i is made larger than γ, then the error at the i-th row of T can become larger also, which might result in an approximant T_a of lower system order. Hence Γ can be chosen to yield an approximant that is accurate at certain points but less tight at others, and whose complexity is minimal.

Although we have seen that, given the same tolerance level, the operator norm allows more freedom than the Hilbert-Schmidt norm, the computational task still seems formidable: there is an infinity of minimization problems, all coupled to each other. It is remarkable that the problem allows a clean and straightforward solution (as we show in this chapter), which can even be obtained in a non-iterative way. The clue is in the fact that the condition (10.3) translates to the computation of contractive operators E, which, as we saw in chapter 8, are linked to the computation of a J-unitary operator Θ, "loaded" by a contractive operator S_L. This is the way that J-unitary systems enter into the picture. The general solution using this approach was originally published in [DvdV93], and specializations to finite matrices were made in [vdVD94b].

Hankel norm approximation theory originates as a special case of the solution to the Schur-Takagi interpolation problem in the context of complex function theory. Suppose that a number of complex values are given at a set of points in the interior of the unit disc of the complex plane, then this problem consists in finding a complex function (a) which interpolates these values at the given points (multiplicities counted), (b) which is meromorphic with at most k poles inside the unit disc, and (c) whose restriction to the unit circle (if necessary via a limiting procedure from inside the unit disc) belongs to L_∞ with minimal norm. The Schur-Takagi problem can be seen as an extension problem whereby the "conjugate-analytic" or anti-causal part of a function is given, and it is desired to extend it to a function which is meromorphic with at most k poles inside the unit disc, and belongs to L_∞ with minimal norm. (Translated into our context, the objective would be to determine an extension of an operator $T \in \mathcal{L}Z^{-1}$ to an operator $T' \in \mathcal{X}$, such that T' is contractive and has an upper part with state dimension sequence smaller than a given sequence.) The L_∞ problem was studied by Adamjan, Arov and Krein (AAK)[AAK71], based on properties of the SVD of infinite dimensional Hankel matrices with Hankel structure, and associated approximation problems of bounded analytical functions by rational functions. (See *e.g.*, [CC92] for a more recent introduction to this class of problems.)

It was remarked by Bultheel and Dewilde [BD80] and subsequently worked out by a number of authors (Kung-Lin [KL81], Genin-Kung [GK81a], Ball-Helton [BH83], Glover [Glo84]) that the procedure of AAK could be utilized to solve the problem of optimal model-order reduction of a dynamical time-invariant system. The computational problem with the general theory is that it involves an operator which maps a

Hilbert space of input sequences to a Hilbert space of output sequences, and which is thus intrinsically non-finite. In [BD80] it was shown that the computations are finite if one assumes the context of a system of finite (but possibly large) degree, *i.e.*, an approximant to the original system of high order. The resulting computations involve only the realization matrices $\{A,B,C,D\}$ of the approximating system and can be done using classical matrix calculus. They can also be done in a recursive fashion, see Limebeer-Green [LG90] as a pioneering paper in this respect. The recursive method is based on the interpolation theory of the Schur-Takagi type.

For time-invariant systems, the Hankel-norm model reduction method may be compared with another popular method for model reduction, known as the balanced model reduction method. In this method, a reduced-order model is obtained by setting all small singular values of the Hankel matrix equal to zero, and using the resulting truncated column space and row space in the construction of a state model [KL81]. Alternatively, one may start from a high-order balanced model (one for which the reachability and observability Gramians are diagonal and equal to each other), and delete all states variables that correspond to small entries in the Gramians [PS82, Moo81]. These methods also give good approximation results, although no tight upper bounds on the modeling error have been derived. An extensive study on error bounds was made by Glover [Glo84], and by Glover-Curtain-Partington [GCP88] for the infinite-dimensional time-invariant case.

Numerical example

As an example of the use of theorem 10.1, we consider a matrix T and determine an approximant T_a. Let the matrix to be approximated be

$$T = \left[\begin{array}{ccc|ccc} 0 & .800 & .200 & .050 & .013 & .003 \\ 0 & 0 & .600 & .240 & .096 & .038 \\ 0 & 0 & 0 & .500 & .250 & .125 \\ \cline{4-6} 0 & 0 & 0 & 0 & .400 & .240 \\ 0 & 0 & 0 & 0 & 0 & .300 \\ 0 & 0 & 0 & 0 & 0 & 0 \end{array}\right].$$

The position of the Hankel matrix H_4 is indicated. Taking $\Gamma = 0.1I$, the non-zero singular values of the Hankel operators of $\Gamma^{-1}T$ are

H_1	H_2	H_3	H_4	H_5	H_6
	8.26	6.85	6.31	5.53	4.06
		0.33	0.29	0.23	
			0.01		

Hence T has a state-space realization which grows from zero states ($i = 1$) to a maximum of 3 states ($i = 4$), and then shrinks back to 0 states ($i > 6$). The number of Hankel singular values of $\Gamma^{-1}T$ that are larger than one is 1 ($i = 2, \cdots, 6$). This is to correspond to the number of states of the approximant at each point. Using the technique detailed in this chapter, we obtain

$$T_a = \begin{bmatrix} 0 & .790 & .183 & .066 & .030 & .016 \\ 0 & 0 & .594 & .215 & .098 & .052 \\ 0 & 0 & 0 & .499 & .227 & .121 \\ 0 & 0 & 0 & 0 & .402 & .214 \\ 0 & 0 & 0 & 0 & 0 & .287 \\ 0 & 0 & 0 & 0 & 0 & 0 \end{bmatrix}$$

with non-zero Hankel singular values (scaled by Γ)

H_1	H_2	H_3	H_4	H_5	H_6
	8.15	6.71	6.16	5.36	3.82

The number of non-zero singular values indeed corresponds to the number of Hankel singular values of $\Gamma^{-1}T$ that are larger than 1. The modeling error is

$$T - T_a = \begin{bmatrix} 0 & .010 & .017 & -.016 & -.017 & -.013 \\ 0 & 0 & .006 & .025 & -.002 & -.014 \\ 0 & 0 & 0 & .001 & .023 & .004 \\ 0 & 0 & 0 & 0 & -.002 & .026 \\ 0 & 0 & 0 & 0 & 0 & .013 \\ 0 & 0 & 0 & 0 & 0 & 0 \end{bmatrix}$$

and indeed, the Hankel norm of $\Gamma^{-1}(T - T_a)$ is less than 1:

$$\|\Gamma^{-1}(T - T_a)\|_H = \sup\{0.334,\ 0.328,\ 0.338,\ 0.351,\ 0.347\} = 0.351$$

The realization algorithm (algorithm 3.9) yields as realization for T

$$\mathbf{T}_1 = \left[\begin{array}{c|c} \cdot & \cdot \\ \hline -.826 & 0 \end{array}\right] \qquad \mathbf{T}_2 = \left[\begin{array}{cc|c} .246 & -.041 & -.968 \\ \hline -.654 & -.00 & 0 \end{array}\right]$$

$$\mathbf{T}_3 = \left[\begin{array}{ccc|c} .397 & -.044 & .000 & -.917 \\ .910 & .140 & .040 & .388 \\ \hline -.573 & .00 & .00 & 0 \end{array}\right] \qquad \mathbf{T}_4 = \left[\begin{array}{cc|c} .487 & .037 & -.873 \\ .853 & -.237 & .465 \\ .189 & .971 & .147 \\ \hline -.466 & .00 & 0 \end{array}\right]$$

$$\mathbf{T}_5 = \left[\begin{array}{c|c} -.515 & -.858 \\ .858 & -.515 \\ \hline .300 & 0 \end{array}\right] \qquad \mathbf{T}_6 = \left[\begin{array}{c|c} \cdot & 1 \\ \hline \cdot & 0 \end{array}\right].$$

A realization of the approximant is determined via algorithm 10.5 in section 10.3 as

$$\mathbf{T}_{a,1} = \left[\begin{array}{c|c} \cdot & \cdot \\ \hline -.993 & 0 \end{array}\right] \qquad \mathbf{T}_{a,2} = \left[\begin{array}{c|c} .293 & -.795 \\ \hline -.946 & 0 \end{array}\right]$$

$$\mathbf{T}_{a,3} = \left[\begin{array}{c|c} .410 & -.629 \\ \hline -.901 & 0 \end{array}\right] \qquad \mathbf{T}_{a,4} = \left[\begin{array}{c|c} .525 & -.554 \\ \hline -.837 & 0 \end{array}\right]$$

$$\mathbf{T}_{a,5} = \left[\begin{array}{c|c} -.651 & -.480 \\ \hline .729 & 0 \end{array}\right] \qquad \mathbf{T}_{a,6} = \left[\begin{array}{c|c} \cdot & .393 \\ \hline \cdot & 0 \end{array}\right].$$

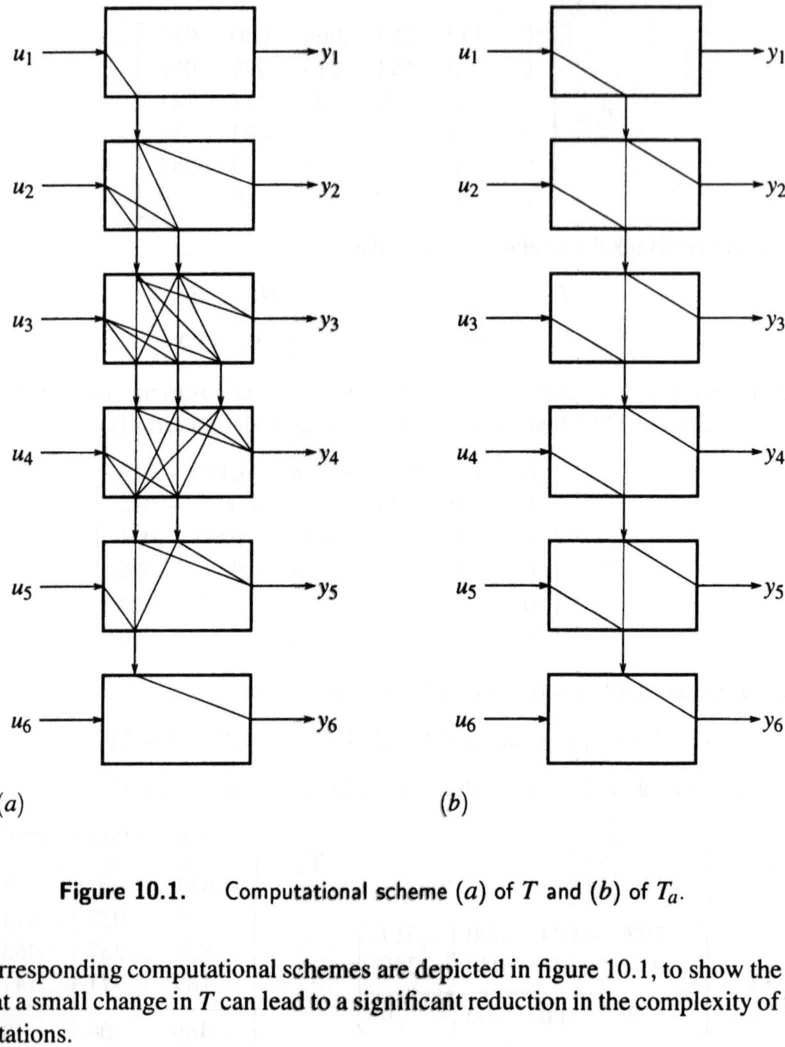

Figure 10.1. Computational scheme (a) of T and (b) of T_a.

The corresponding computational schemes are depicted in figure 10.1, to show the effect that a small change in T can lead to a significant reduction in the complexity of the computations.

Hankel norm

As mentioned in the introduction, we compute approximants which are optimal in the Hankel norm, defined as

$$\| T \|_H = \| H_T \| .$$

It is a norm on $\mathcal{U}$, a semi-norm on $\mathcal{X}$. Since this is not such a familiar norm as, for example, the operator norm of T, we first determine its relation to the latter. The Hankel norm can also be compared to another norm, which we call for simplicity the *diagonal 2-norm*. Let T_i be the i-th row of a block matrix representation of $T \in \mathcal{X}$, then

$$\begin{aligned} D \in \mathcal{D}: &\ \| D \|_{\mathcal{D}2} = \| D \| = \sup_i \| D_i \|, \\ T \in \mathcal{X}: &\ \| T \|_{\mathcal{D}2}^2 = \| \mathbf{P}_0(TT^*) \|_{\mathcal{D}2} = \sup_i \| T_i T_i^* \| . \end{aligned}$$

For diagonals, it is equal to the operator norm, but for more general operators, it is the supremum over the ℓ_2-norms of each row of T.

Proposition 10.2 *The Hankel norm satisfies the following ordering:*

$$T \in \mathcal{X}: \qquad \|T\|_H \le \|T\| \qquad (10.4)$$
$$T \in Z\mathcal{U}: \qquad \|T\|_{\mathcal{D}2} \le \|T\|_H. \qquad (10.5)$$

PROOF The first norm inequality is proven by

$$\begin{aligned} \|T\|_H &= \sup_{u\in \mathcal{L}_2Z^{-1}, \|u\|_{HS}\le 1} \|\mathbf{P}(uT)\|_{HS} \\ &\le \sup_{u\in \mathcal{L}_2Z^{-1}, \|u\|_{HS}\le 1} \|uT\|_{HS} \\ &\le \sup_{u\in \mathcal{X}_2, \|u\|_{HS}\le 1} \|uT\|_{HS} = \|T\|. \end{aligned}$$

For the second norm inequality, we first prove $\|T\|_{\mathcal{D}2}^2 \le \sup_{D\in\mathcal{D}_2, \|D\|_{HS}\le 1} \|DTT^*D^*\|_{HS}$. Indeed,

$$\begin{aligned} \|T\|_{\mathcal{D}2}^2 &= \|\mathbf{P}_0(TT^*)\|_{\mathcal{D}2}^2 \\ &= \sup_{D\in\mathcal{D}_2, \|D\|_{\mathcal{D}2}\le 1} \|D\mathbf{P}_0(TT^*)D^*\|_{\mathcal{D}2} \\ &= \sup_{D\in\mathcal{D}_2, \|D\|_{HS}\le 1} \|D\mathbf{P}_0(TT^*)D^*\|_{HS} \\ &\le \sup_{D\in\mathcal{D}_2, \|D\|_{HS}\le 1} \|DTT^*D^*\|_{HS}. \end{aligned}$$

Then (10.5) follows, with use of the fact that $T \in Z\mathcal{U}$, by

$$\begin{aligned} \|T\|_{\mathcal{D}2}^2 &\le \sup_{D\in\mathcal{D}_2, \|D\|_{HS}\le 1} \|DTT^*D^*\|_{HS} \\ &= \sup_{D\in\mathcal{D}_2, \|D\|_{HS}\le 1} \|DZ^*TT^*ZD^*\|_{HS} \\ &= \sup_{D\in\mathcal{D}_2, \|D\|_{HS}\le 1} \|\mathbf{P}(DZ^*T)\,[\mathbf{P}(DZ^*T)]^*\|_{HS} \\ &\le \sup_{u\in \mathcal{L}_2Z^{-1}, \|u\|_{HS}\le 1} \|\mathbf{P}(uT)\,[\mathbf{P}(uT)]^*\|_{HS} \\ &= \|T\|_H^2. \end{aligned}$$

□

We see that the Hankel norm is not as strong as the operator norm, but is stronger than the row-wise uniform least square norm.

10.2 APPROXIMATION VIA INDEFINITE INTERPOLATION

Approximation recipe

In the present section we outline a procedure to obtain a reduced-order approximant, and put the various relevant facts in perspective. Details are proven in subsequent sections.

Let $T \in \mathcal{U}$ be a given bounded, locally finite, strictly upper operator. The decision to assume that T is strictly upper is made for convenience and is without serious consequences: $D = \mathbf{P}_0(T)$ has no influence on the Hankel (semi-)norm, so that there are no conditions on the D operator of the approximant. Let $\Gamma \in \mathcal{D}$ be a diagonal and Hermitian operator. As discussed in the introduction, the objective is to determine an operator $T_a \in \mathcal{U}$ such that $\|\Gamma^{-1}(T - T_a)\|_H \le 1$. Instead of working with T_a directly, we look for a bounded operator $T' \in \mathcal{X}$ such that

$$\|\Gamma^{-1}(T - T')\| \le 1, \qquad (10.6)$$

and such that the strictly upper part of T' has state-space dimensions of low order — as low as possible for a given Γ. Let T_a be the strictly causal part of T'. Proposition 10.2 showed that

$$\begin{aligned} \|\Gamma^{-1}(T-T_a)\|_H &= \|\Gamma^{-1}(T-T')\|_H \\ &\leq \|\Gamma^{-1}(T-T')\| \leq 1, \end{aligned} \tag{10.7}$$

so that T_a is a Hankel-norm approximant of T (parametrized by Γ) whenever T' is an operator-norm approximant. T' can be viewed as an extension of T_a which is such that $\|\Gamma^{-1}(T-T_a)\|_H \leq \|\Gamma^{-1}(T-T')\|$. A generalization of Nehari's theorem to the present setting would state that $\inf\|E\|$ over all possible extensions $E \in \mathcal{X}$ of a given part $E_a \in \mathcal{U}$ actually equals $\|E_a\|_H$ (see section 10.6).

The construction of an operator T' satisfying (10.6) consists of three steps, specified in the following lemma. (The definitions and notation in this lemma will be kept throughout the rest of the section.)

Lemma 10.3 (recipe for a hankel-norm approximant) *Let $T \in \mathcal{U}(\mathcal{M},\mathcal{N})$ be strictly upper, and let $\Gamma \in \mathcal{D}(\mathcal{M},\mathcal{M})$ be a given diagonal Hermitian operator. Then, provided the indicated factorizations exist, an operator $T' \in \mathcal{X}$ such that $\|\Gamma^{-1}(T-T')\| \leq 1$ is obtained by performing the following steps:*

1. *an external factorization (inner-coprime factorization; theorem 6.8):*

$$T = \Delta^* U \qquad (U,\Delta \in \mathcal{U}, U \text{ unitary}), \tag{10.8}$$

2. *a J-inner coprime factorization (corollary 8.18):*

$$[U^* \quad -T^*\Gamma^{-1}]\Theta = [A' \quad -B'] \in [\mathcal{U} \quad \mathcal{U}] \qquad (\Theta \in \mathcal{U},\ J\text{-unitary}), \tag{10.9}$$

3. *with a block-decomposition of Θ as in (8.5),*

$$T'^* = B'\Theta_{22}^{-1}\Gamma. \tag{10.10}$$

PROOF If the factorizations exist, then Θ_{22} is boundedly invertible so that $\Sigma_{12} = -\Theta_{12}\Theta_{22}^{-1}$ exists and is contractive (theorem 8.2). From (10.9) we have $B' = -U^*\Theta_{12} + T^*\Gamma^{-1}\Theta_{22}$. Substitution of (10.10) leads to

$$\begin{aligned} T'^*\Gamma^{-1} &= T^*\Gamma^{-1} - U^*\Theta_{12}\Theta_{22}^{-1} \\ &= T^*\Gamma^{-1} - U^*\Sigma_{12} \end{aligned}$$

and it follows that $(T^* - T'^*)\Gamma^{-1} = -U^*\Sigma_{12}$. Because Σ_{12} is contractive and U unitary,

$$\begin{aligned} \|(T^* - T'^*)\Gamma^{-1}\| &= \|-U^*\Sigma_{12}\| \\ &= \|\Sigma_{12}\| \leq 1, \end{aligned}$$

so that $T' = \left(B'\Theta_{22}^{-1}\Gamma\right)^*$ is indeed an approximant with an admissible modeling error. □

In anticipation of a proof of theorem 10.1, it remains to show that the strictly upper part T_a of T' has at most the specified number of states, and to verify the relation with the Hankel singular values of $\Gamma^{-1}T$. This is done in the remaining part of this section. The definition of T' in (10.10) can be generalized by the introduction of a contractive operator S_L that parametrizes the possible approximants, which is the subject of section 10.4. The crucial step in the procedure is step 2. The computation of Θ can be viewed as the solution of an interpolation problem

$$U^*[I \quad S]\Theta \in [\mathcal{U} \quad \mathcal{U}], \qquad S = -UT^*\Gamma^{-1} = -\Delta\Gamma^{-1}, \tag{10.11}$$

where the interpolation subspace is determined by U. If $\Theta_{22}^{-1} \in \mathcal{U}$, then an exact representation of S in Θ is obtained as $S = -\Theta_{12}\Theta_{22}^{-1}$. In this case, the interpolation problem is definite: the relevant J-Gramian is positive definite, which happens if $\Gamma^{-1}T$ is strictly contractive. In addition, $T'^* = B'\Theta_{22}^{-1}\Gamma$ is upper, and the approximant T_a is zero, which matches one's expectation in view of $\|\Gamma^{-1}T\| < 1$. If $\Gamma^{-1}T$ is not contractive then Θ_{22}^{-1} is not upper, and this is the situation which leads to approximations and which is considered in this chapter.

Construction of Θ

We now determine sufficient conditions on a state-space realization $\{A,B,C,0\}$ of T for the existence of the two factorizations in the above lemma. Assuming $\ell_A < 1$, the external factorization in the first step can be computed from the given realization if it is uniformly observable (theorem 6.8). Without loss of generality, we can (and do) assume that such a realization has been normalized, so that $AA^* + CC^* = I$. Then, a realization for the inner factor U of the external factorization is given by

$$\mathbf{U} = \begin{bmatrix} A & C \\ B_U & D_U \end{bmatrix}$$

where B_U and D_U are obtained by locally completing $[A_k \;\; C_k]$ to a square and unitary matrix.

The second step is to derive expressions for Θ to satisfy the interpolation condition (10.9). $[U^* \;\; -T^*\Gamma^{-1}]^*$ has a realization

$$\begin{bmatrix} U \\ -\Gamma^{-1}T \end{bmatrix} = \begin{bmatrix} D_U \\ 0 \end{bmatrix} + \begin{bmatrix} B_U \\ -\Gamma^{-1}B \end{bmatrix} Z(I-AZ)^{-1}C,$$

so that, according to corollary 8.18, there is a J-unitary operator Θ mapping $[U^* \;\; -T^*\Gamma^{-1}]$ to upper if the relevant J-Gramian $\Lambda := \Lambda^J$ (as defined in (8.10)) is boundedly invertible. With the above realization of $[U^* \;\; -T^*\Gamma^{-1}]^*$, Λ satisfies the J-Lyapunov equation (*cf.* equation (8.36))

$$\Lambda^{(-1)} = A^*\Lambda A + B_U^*B_U - B^*\Gamma^{-2}B.$$

Substituting the relation $A^*A + B_U^*B_U = I$ yields $I - \Lambda^{(-1)} = A^*(I-\Lambda)A + B^*\Gamma^{-2}B$. With the additional definition of $M = I - \Lambda$, it is seen that M satisfies

$$M^{(-1)} = A^*MA + B^*\Gamma^{-2}B$$

so that M is the reachability Gramian of the given realization of $\Gamma^{-1}T$. It follows that the J-inner coprime factorization exists if $I-M$ is boundedly invertible, that is, if 1 is a regular point for the operator M [AG81]. With M known (and hence Λ), Θ is determined along the lines of the proof of theorem 8.17. In particular, the input state space of Θ is defined by

$$\mathcal{H}(\Theta) = \mathcal{D}_2^{\mathcal{B}}(I-Z^*A^*)^{-1}Z^*\begin{bmatrix} B_U^* & B^*\Gamma^{-1}\end{bmatrix}. \tag{10.12}$$

Let $\Lambda = R^*J_BR$ be a factorization of Λ, then

$$\begin{bmatrix} A_\Theta \\ B_\Theta \end{bmatrix} = \begin{bmatrix} R & & \\ & I & \\ & & I \end{bmatrix}\begin{bmatrix} A \\ B_U \\ \Gamma^{-1}B \end{bmatrix} R^{-(-1)}$$

is J-isometric, and a J-unitary realization for Θ is of the form

$$\mathbf{\Theta} = \begin{bmatrix} A_\Theta & C_\Theta \\ B_\Theta & D_\Theta \end{bmatrix} = \begin{bmatrix} R & & \\ & I & \\ & & I \end{bmatrix}\begin{bmatrix} A & C_1 & C_2 \\ B_U & D_{11} & D_{12} \\ \Gamma^{-1}B & D_{21} & D_{22} \end{bmatrix}\begin{bmatrix} R^{-(-1)} & & \\ & I & \\ & & I \end{bmatrix} \tag{10.13}$$

and is obtained by completing A_Θ and B_Θ with certain diagonal operators C_Θ and D_Θ to a square J-unitary matrix. Corollary 8.18 claims that this is always possible under the present conditions (Λ boundedly invertible), and the procedure to do so is given in lemma 8.16. Since the realization $\mathbf{\Theta}$ is J-unitary, the corresponding transfer operator Θ is also J-unitary and has the specified input state space. The third step in lemma 10.3 is always possible (*cf.* theorem 8.2).

We have proven the following lemma:

Lemma 10.4 *Let $T \in \mathcal{U}(\mathcal{M},\mathcal{N})$ be a strictly upper locally finite operator, with output normal realization $\{A,B,C,0\}$ such that $\ell_A < 1$, and let Γ be a Hermitian diagonal operator. If the solution M of the Lyapunov equation*

$$M^{(-1)} = A^*MA + B^*\Gamma^{-2}B \tag{10.14}$$

is such that $\Lambda = I - M$ is boundedly invertible, then the conditions mentioned in lemma 10.3 are satisfied: there exists an external factorization $T = \Delta^*U$, a J-unitary block upper operator Θ such that

$$[U^* \quad -T^*\Gamma^{-1}]\Theta \in [\mathcal{U} \quad \mathcal{U}],$$

and an operator $T' \in \mathcal{X}$ such that $\| \Gamma^{-1}(T-T') \| \le 1$, according to the recipe in lemma 10.3.

Let $\mathcal{M}$, $\mathcal{N}$ and $\mathcal{B}$ be the input, output and state space sequences of T and its realization, and let $\mathcal{M}_U$ be the input space sequence for U: its index sequence is specified by

$$\#\mathcal{M}_U = \#\mathcal{B}^{(-1)} - \#\mathcal{B} + \#\mathcal{N}.$$

In:	**T**	(model in output normal form for a strictly upper matrix T)
	Γ	(approximation parameters)
Out:	$\boldsymbol{\Theta}$	(realization for Θ satisfying (10.9))

$$\begin{aligned} M_1 &= [\cdot] \\ R_1 &= [\cdot] \\ J_{\mathcal{B}_1} &= [\cdot] \end{aligned}$$

for $k = 1, \cdots, n$

$$\left[\begin{array}{l} M_{k+1} = A_k^* M_k A_k + B_k^* \Gamma_k^{-2} B_k \\ R_{k+1}^* J_{\mathcal{B}_{k+1}} R_{k+1} := I - M_{k+1} \\ [B_{U,k}\; D_{U,k}] = [A_k\; C_k]^{\perp} \\ \begin{bmatrix} \alpha \\ \beta \end{bmatrix} = \begin{bmatrix} R_k A_k \\ \hline B_{U,k} \\ \Gamma_k^{-1} B_k \end{bmatrix} R_{k+1}^{-1} \\ \begin{bmatrix} c \\ d \end{bmatrix} = \begin{bmatrix} J_{\mathcal{B}_k} \alpha \\ J_1 \beta \end{bmatrix}^{\perp} \\ r^* J_2 r := [c \quad d]^* \begin{bmatrix} J_{\mathcal{B}_k} & \\ & J_1 \end{bmatrix} \begin{bmatrix} c \\ d \end{bmatrix} \\ \begin{bmatrix} \gamma \\ \delta \end{bmatrix} = \begin{bmatrix} c \\ d \end{bmatrix} r^{-1} \\ \boldsymbol{\Theta}_k = \begin{bmatrix} \alpha & \gamma \\ \beta & \delta \end{bmatrix} \end{array}\right.$$

end

Figure 10.2. Indefinite interpolation: step 1 and 2 of lemma 10.3.

The signature $J_{\mathcal{B}}$ of Λ determines a decomposition of $\mathcal{B}$ into $\mathcal{B} = \mathcal{B}_+ \times \mathcal{B}_-$. Let $\Theta^* J_1 \Theta = J_2$, $\Theta J_2 \Theta^* = J_1$, where J_1 and J_2 are shorthand for $J_1 = J_{\mathcal{M}_\Theta}$ and $J_2 = J_{\mathcal{N}_\Theta}$. The space sequence $\mathcal{M}_\Theta$ is equal to $\mathcal{M}_\Theta = \mathcal{M}_U \times \mathcal{M}$, and the corresponding signature operator J_1 follows this partitioning. The dimensions of the positive and negative parts of the output sequence space of Θ, and hence the signature J_2, are then given by inertia rules as (*cf.* corollary 8.18)

$$\begin{aligned} \#(\mathcal{N}_\Theta)_+ &= \#\mathcal{B}_+ - \#\mathcal{B}_+^{(-1)} + \#\mathcal{M}_U \\ \#(\mathcal{N}_\Theta)_- &= \#\mathcal{B}_- - \#\mathcal{B}_-^{(-1)} + \#\mathcal{M} \end{aligned}$$

Algorithm 10.2 summarizes the construction in lemma 10.4 and can be used to compute Θ satisfying equation (10.9). The inner factor U of T is computed *en passant*.

Connection to the Hankel operator

We continue by establishing the link between the Lyapunov equation (10.14) and the Hankel operator of $\Gamma^{-1}T$.

Lemma 10.5 *Let $T \in \mathcal{U}$ be a locally finite strictly upper operator, with u.e. stable realization $\{A,B,C,0\}$ in output normal form. Let H_k be the Hankel operator of $\Gamma^{-1}T$ at stage k, and suppose that an $\varepsilon > 0$ exists such that, for each k, none of the singular values of H_k are in the interval $[1-\varepsilon, 1+\varepsilon]$. Let N_k be equal to the number of singular values of H_k that are larger than 1. Then the solution M of the Lyapunov equation*

$$M^{(-1)} = A^*MA + B^*\Gamma^{-2}B \tag{10.15}$$

is such that $\Lambda = I - M$ is boundedly invertible and has a signature operator $J_\mathcal{B}$ with N_k negative entries at point k.

PROOF The solutions of the two Lyapunov equations associated to the realization of $\Gamma^{-1}T$ (corresponding to the reachability and observability Gramians),

$$\begin{aligned} M^{(-1)} &= A^*MA + B^*\Gamma^{-2}B \\ Q &= AQ^{(-1)}A^* + CC^* \end{aligned}$$

may be expressed in terms of the reachability and observability operators of $\Gamma^{-1}T$,

$$\mathcal{C} := \begin{bmatrix} (\Gamma^{-1}B)^{(+1)} \\ (\Gamma^{-1}B)^{(+2)}A^{(+1)} \\ (\Gamma^{-1}B)^{(+3)}A^{(+2)}A^{(+1)} \\ \vdots \end{bmatrix} \qquad \mathcal{O} := \begin{bmatrix} C & AC^{(-1)} & AA^{(-1)}C^{(-2)} & \cdots \end{bmatrix}$$

as $M = \mathcal{C}^*\mathcal{C}$, $Q = \mathcal{O}\mathcal{O}^*$. The Hankel operator H_k of $\Gamma^{-1}T$ at time instant k satisfies the decomposition $H_k = \mathcal{C}_k\mathcal{O}_k$. Hence

$$H_kH_k^* = \mathcal{C}_k\mathcal{O}_k\mathcal{O}_k^*\mathcal{C}_k^*.$$

The state realization of T is assumed to be in output normal form, so that $Q_k = \mathcal{O}_k\mathcal{O}_k^* = I$. With the current locally finiteness assumption, the non-zero eigenvalues of $H_kH_k^* = \mathcal{C}_k\mathcal{C}_k^*$ are the same as those of $\mathcal{C}_k^*\mathcal{C}_k = M_k$. In particular, the number of singular values of H_k that are larger than 1 is equal to the number of eigenvalues of M_k that are larger than 1. Writing $\Lambda_k = I - M_k$, this is in turn equal to the number of negative eigenvalues of Λ_k. □

Figure 10.3 shows a simple instance of the application of the theory developed in this section, emphasizing the dimensions of the input, output and state space sequences related to the Θ operator. We assume in the figure that one singular value of the Hankel operator of $\Gamma^{-1}T$ at time 1 is larger than 1, so that the state signature $J_\mathcal{B}$ of Θ has one negative entry in total. We know from equation (8.20) that the negative entries of $J_\mathcal{B}$ determine the number of upward arrows in the diagram of the unitary scattering operator $\mathbf{\Sigma}$. We show, in the following subsection, that this number also determines the number of states of the Hankel-norm approximant T_a of T.

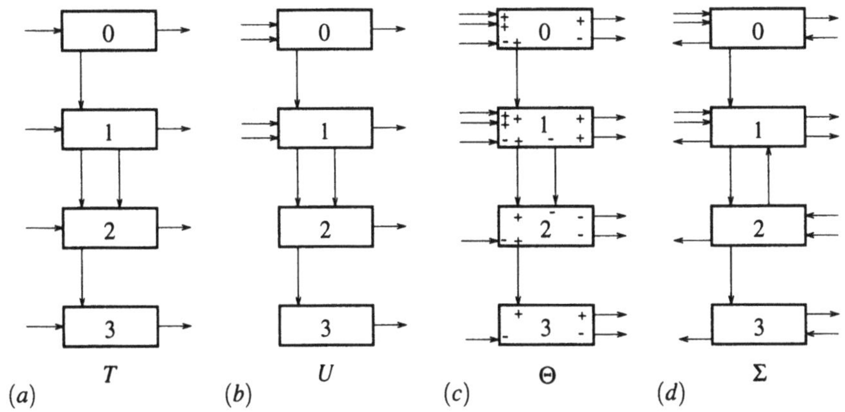

Figure 10.3. (a) State-space realization scheme for T and (b) for U. (c) State-space realization scheme for a possible Θ, where it is assumed that one singular value of the Hankel operator of $\Gamma^{-1}T$ at time 1 is larger than 1, and (d) for the corresponding scattering operator Σ.

Complexity of the approximant

At this point we have proven the first part of theorem 10.1: we have constructed a J-unitary operator Θ and from it an operator T' with strictly upper part T_a which is a Hankel-norm approximant of T. It remains to verify the complexity assertion, which stated that the sequence of dimensions of the state space of T_a is at most equal to the sequence N: the number of Hankel singular values of $\Gamma^{-1}T$ that are larger than 1. In view of lemmas 10.4 and 10.5, N is equal to the number of negative entries in the state signature $J_{\mathcal{B}}$ of Θ. We now show that the state dimension sequence of T_a is smaller than or equal to N. (Later, we will show that equality holds.) The proof is, again, based on the determination of the natural input state space for T_a, which can be derived in terms of the realization of the scattering operator Σ that is connected to Θ.

Suppose that the conditions of lemma 10.3 are fulfilled so that Θ satisfies

$$[U^* \quad -T^*\Gamma^{-1}]\Theta = [A' \quad -B']$$

with $A', B' \in \mathcal{U}$. Let $T'^*\Gamma^{-1} = B'\Theta_{22}^{-1}$ as in lemma 10.3. The approximating transfer function T_a is, in principle, given by the strictly upper part of T' (see lemma 10.3 for the summary of the procedure). It might not be a bounded operator, since operators in $\mathcal{X}$ do not necessarily have a decomposition into an upper and lower part in $\mathcal{X}$. However, its extension T' is bounded, and hence its Hankel operator $H_{T_a} = H_{T'}$ is well defined and bounded. We have the following lemma.

Lemma 10.6 *Under the conditions of lemma 10.4, the natural input state space of $\Gamma^{-1}T_a$ satisfies*

$$\mathcal{H}(\Gamma^{-1}T_a) \subset \mathcal{H}(\Theta_{22}^{-*})\,. \tag{10.16}$$

PROOF From the definition of $\mathcal{H}$ in equation (5.3) and the operators we have

$$\begin{aligned} \mathcal{H}(\Gamma^{-1}T_a) &= \mathbf{P}'(\mathcal{U}_2 T_a^* \Gamma^{-1}) \\ &= \mathbf{P}'(\mathcal{U}_2 T'^* \Gamma^{-1}) \\ &= \mathbf{P}'(\mathcal{U}_2 B' \Theta_{22}^{-1}) \\ &\subset \mathbf{P}'(\mathcal{U}_2 \Theta_{22}^{-1}) \qquad \text{[since } B' \in \mathcal{U}\text{]} \\ &= \mathcal{H}(\Theta_{22}^{-*}) \,. \end{aligned}$$

□

Hence the sequence of dimensions of the subspace $\mathcal{H}(\Theta_{22}^{-*})$ is of interest. According to proposition 8.14, this dimension sequence is equal to $N = \#(\mathcal{B}_-)$, *i.e.*, the number of negative entries in the state signature sequence $J_{\mathcal{B}}$ of Θ. Combining this result with the lemmas in this section proves the model reduction theorem, theorem 10.1, repeated below:

Theorem 10.7 *Let $T \in \mathcal{U}$ be a locally finite strictly upper operator with a uniformly observable u.e. stable realization, and let $\Gamma = \operatorname{diag}(\gamma_i) \in \mathcal{D}$ be a Hermitian operator. Let H_k be the Hankel operator of $\Gamma^{-1}T$ at stage k, and suppose that an $\varepsilon > 0$ exists such that, for each k, none of the singular values of H_k are in the interval $[1-\varepsilon, 1+\varepsilon]$. Then there exists a strictly upper triangular operator T_a with system order at stage k at most equal to the number of singular values of H_k that are larger than 1, such that*

$$\| \Gamma^{-1}(T - T_a) \|_H \leq 1 \,.$$

PROOF Under the present conditions on T, lemma 10.3 can be applied. Indeed, lemma 10.5 claims that the reachability Gramian M of the realization (normalized to output normal form) is such that $\Lambda = I - M$ is boundedly invertible, where Λ satisfies the same J-Lyapunov equation as in lemma 10.4. This lemma showed that the necessary conditions to apply the procedure in lemma 10.3 are satisfied. Thus construct T' and T_a using lemma 10.3, so that $\| \Gamma^{-1}(T - T_a) \|_H \leq 1$. According to lemma 10.6, the state dimension sequence of T_a is less than or equal to the state dimension sequence of the causal part of Θ_{22}^{-*}, which is equal to the number of negative entries of the state signature sequence $J_{\mathcal{B}}$ (proposition 8.14), in turn equal to N (lemma 10.5). Hence T_a has the claimed state complexity, so that it is a Hankel norm approximant of T for the given Γ. □

10.3 STATE REALIZATION OF THE APPROXIMANT

Theorem 10.7 shows the existence of a Hankel norm approximant T_a under certain conditions. The proof uses a construction of this approximant (lemma 10.3), but this construction is at the operator level. However, it is also possible to obtain a *state realization* for T_a directly. We will derive this result in the present section.

Throughout this section, we take signals a_1, a_2, b_1, b_2 to be elements of $\mathcal{X}_2$, generically related by

$$[a_1 \quad b_1]\Theta = [a_2 \quad b_2]$$

where Θ is as constructed in the previous section. In particular, Θ is a bounded operator, and Θ_{22}^{-1} exists and is bounded. In section 10.2 we constructed Θ via a J-unitary realization $\boldsymbol{\Theta}$, with state signature matrix $J_{\mathcal{B}}$. $\boldsymbol{\Theta}$ is bounded by construction (because of the assumption that none of the Hankel singular values of $\Gamma^{-1}T$ are equal or "asymptotically close" to 1), and is u.e. stable because T is assumed to be so. As before, the part of an operator $u \in \mathcal{X}_2$ that is in $\mathcal{L}_2 Z^{-1}$ is denoted by $u_p = \mathbf{P}'(u)$, and the part in $\mathcal{U}_2$ is $u_f = \mathbf{P}(u)$. Associated to the transfer operator Θ is the scattering operator Σ which relates

$$[a_1 \quad b_1]\Theta = [a_2 \quad b_2] \quad \Leftrightarrow \quad [a_1 \quad b_2]\Sigma = [a_2 \quad b_1].$$

We have derived in theorem 8.2 a representation $\boldsymbol{\Sigma} = \{F, G, H, K\}$ in terms of entries $\{A_\Theta, B_\Theta, C_\Theta, D_\Theta\}$ in $\boldsymbol{\Theta}$, according to the relation

$$\begin{aligned} [x_+ \quad x_- \quad a_1 \quad b_1]\boldsymbol{\Theta} &= [x_+ Z^{-1} \quad x_- Z^{-1} \quad a_2 \quad b_2] \\ [x_+ \quad x_- Z^{-1} \quad a_1 \quad b_2]\boldsymbol{\Sigma} &= [x_+ Z^{-1} \quad x_- \quad a_2 \quad b_1]. \end{aligned}$$

The above realizations act on operators in $\mathcal{X}_2$. Taking the k-th diagonal of each operator yields the following state recursions on diagonals, which we use throughout the section:

$$\begin{aligned} [x_{+[k]} \quad x_{-[k]} \quad a_{1[k]} \quad b_{1[k]}]\boldsymbol{\Theta} &= [x^{(-1)}_{+[k+1]} \quad x^{(-1)}_{-[k+1]} \quad a_{2[k]} \quad b_{2[k]}] \\ [x_{+[k]} \quad x^{(-1)}_{-[k+1]} \quad a_{1[k]} \quad b_{2[k]}]\boldsymbol{\Sigma} &= [x^{(-1)}_{+[k+1]} \quad x_{-[k]} \quad a_{2[k]} \quad b_{1[k]}]. \end{aligned}$$

In order to compute a realization of T_a, we first determine a model for the strictly upper part of Θ_{22}^{-*} from the model $\boldsymbol{\Sigma}$. It is given in terms of operators S and R defined as[1]

$$\begin{aligned} x_{-[0]}S &= x_{+[0]} \qquad \text{when } a_{1p} = 0, b_{2p} = 0 \\ x_{+[0]}R &= x_{-[0]} \qquad \text{when } a_{1f} = 0, b_{2f} = 0, \end{aligned} \tag{10.17}$$

which can be obtained from $\boldsymbol{\Sigma}$ in terms of two recursive equations. S is, for example, obtained as the input scattering matrix of a ladder network consisting of a semi-infinite chain of contractive (*i.e.*, lossy) scattering matrices F_{ij}.

Lemma 10.8 *The relations*

$$\begin{aligned} x_{-[0]}S &= x_{+[0]} \qquad \text{when } a_{1p} = 0, b_{2p} = 0 \\ x_{+[0]}R &= x_{-[0]} \qquad \text{when } a_{1f} = 0, b_{2f} = 0, \end{aligned} \tag{10.18}$$

define bounded maps which are strictly contractive: $\|S\| < 1$, $\|R\| < 1$.

PROOF S exists as a partial map of Σ_p, taking $a_{1p} = 0$, $b_{2p} = 0$. In this situation,

$$[0 \quad b_{1p}]\Theta_p = [x_{+[0]} \quad x_{-[0]} \quad a_{2p} \quad 0],$$

and we have

$$\|x_{-[0]}\|^2 = \|x_{+[0]}\|^2 + \|b_{1p}\|^2 + \|a_{2p}\|^2.$$

According to proposition 8.13, there is an ε, $0 < \varepsilon \le 1$, such that $\|b_{1p}\|^2 \ge \varepsilon^2 \|x_{-[0]}\|^2$, and hence

$$\|x_{-[0]}\|^2 \ge \|x_{+[0]}\|^2 + \varepsilon^2 \|x_{-[0]}\|^2.$$

[1]Here, S is not the same as S in (10.11), and no connection is intended.

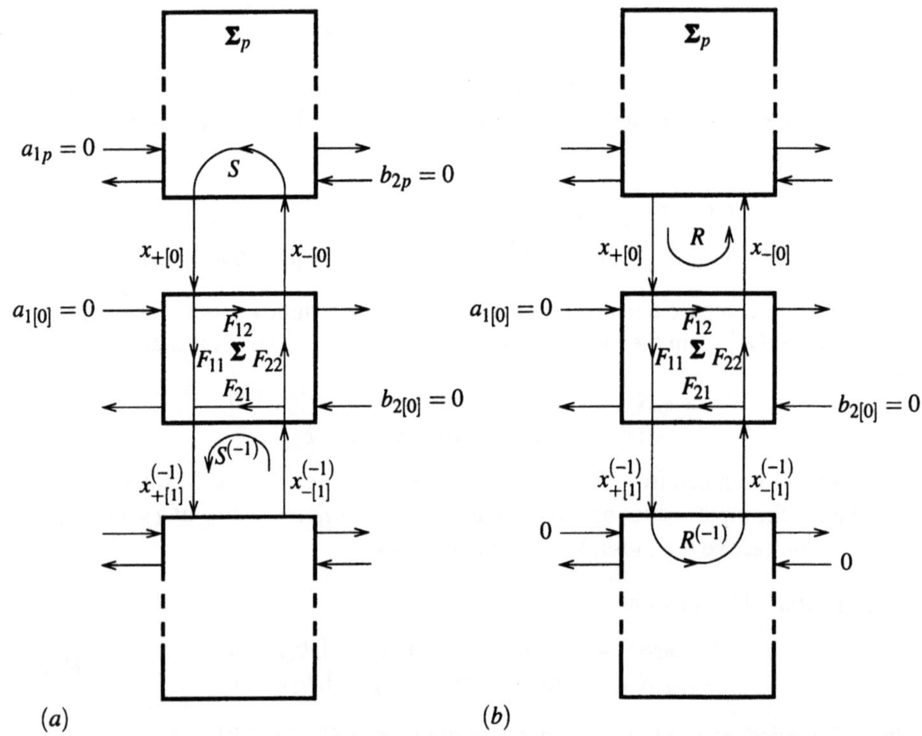

Figure 10.4. (*a*) The propagation of S, (*b*) the propagation of R.

Consequently, there is a constant μ $(0 \le \mu < 1)$ such that $\| x_{+[0]} \|^2 \le \mu^2 \| x_{-[0]} \|^2$ (take $\mu = \sqrt{1-\varepsilon^2}$). This shows that $\| S \| < 1$. A similar argument holds for R. $\square$

Proposition 10.9 *The operators S and R defined in (10.18) are determined in terms of $\boldsymbol{\Sigma}$ (with block entries as in (8.21)) by the following recursions:*

$$\begin{aligned} S &= \left(F_{21} + F_{22}(I - SF_{12})^{-1} SF_{11}\right)^{(+1)} \\ R &= F_{12} + F_{11}(I - R^{(-1)} F_{21})^{-1} R^{(-1)} F_{22} . \end{aligned} \tag{10.19}$$

A state-space model $\{A_a, B_a, C_r\}$ of the strictly upper part of Θ_{22}^{-*} is given in terms of S, R by

$$\begin{aligned} A_a &= \left(F_{22}(I - SF_{12})^{-1}\right)^* \\ B_a &= \left(H_{22} + F_{22}(I - SF_{12})^{-1} SH_{12}\right)^* \\ C_r &= (I - SR)^{-*} \left[G_{22} + G_{21}(I - R^{(-1)} F_{21})^{-1} R^{(-1)} F_{22}\right]^* . \end{aligned} \tag{10.20}$$

This model is uniformly minimal, with contractive reachability and observability Gramians.

PROOF The existence and contractivity of $S \in \mathcal{D}$ and $R \in \mathcal{D}$ has already been determined (lemma 10.8). First observe that although S satisfies by definition $x_{-[0]} S = x_{+[0]}$ $(a_{1p} = b_{2p} = 0)$, it also satisfies $x_{-[1]} S = x_{+[1]}$ $(a_{1p} = b_{2p} = 0$ and $a_{1[0]} = b_{2[0]} = 0)$, etc. This is readily obtained by applying inputs $Z^{-1} a_1$, etc., so that we get states $Z^{-1} x_+$ and $Z^{-1} x_-$. If $(Z^{-1} a_1)_p = Z^{-1} a_{1p} + Z^{-1} a_{1[0]} = 0$, then $(Z^{-1} x_-)_{[0]} S = (Z^{-1} x_+)_{[0]}$. But $(Z^{-1} x_-)_{[0]} = x_{-[1]}$, and likewise $(Z^{-1} x_+)_{[0]} = x_{+[1]}$. Hence $x_{-[1]} S = x_{+[1]}$.

To determine a state realization for the strictly upper part of $\Sigma_{22}^* = \Theta_{22}^{-*}$, we start from the definition of $\boldsymbol{\Sigma}$ (8.20), and specialize to the 0-th diagonal to obtain

$$[x_{+[0]} \quad x_{-[1]}^{(-1)} \quad a_{1[0]} \quad b_{2[0]}] \boldsymbol{\Sigma} = [x_{+[1]}^{(-1)} \quad x_{-[0]} \quad a_{2[0]} \quad b_{1[0]}] .$$

Taking $a_1 = 0$ throughout this proof, inserting the partitioning of $\boldsymbol{\Sigma}$ in (8.21) gives

$$\left\{ \begin{aligned} x_{+[1]}^{(-1)} &= x_{+[0]} F_{11} + x_{-[1]}^{(-1)} F_{21} + b_{2[0]} G_{21} \\ x_{-[0]} &= x_{+[0]} F_{12} + x_{-[1]}^{(-1)} F_{22} + b_{2[0]} G_{22} \\ b_{1[0]} &= x_{+[0]} H_{12} + x_{-[1]}^{(-1)} H_{22} + b_{2[0]} K_{21} \end{aligned} \right. \tag{10.21}$$

With $b_{2p} = 0$ and $b_{2[0]} = 0$, these equations yield an expression for $S^{(-1)}$:

$$\left\{ \begin{aligned} x_{+[1]}^{(-1)} &= x_{-[1]}^{(-1)} S^{(-1)} = x_{-[0]} S F_{11} + x_{-[1]}^{(-1)} F_{21} \\ x_{-[0]} &= x_{-[0]} S F_{12} + x_{-[1]}^{(-1)} F_{22} \end{aligned} \right.$$

$$\Leftrightarrow$$

$$\left\{ \begin{aligned} x_{-[0]} &= x_{-[1]}^{(-1)} F_{22}(I - SF_{12})^{-1} \\ x_{-[1]}^{(-1)} S^{(-1)} &= x_{-[1]}^{(-1)} \{F_{22}(I - SF_{12})^{-1} SF_{11} + F_{21}\} \end{aligned} \right. \tag{10.22}$$

(note that $(I-SF_{12})^{-1}$ is bounded because $\|S\|<1$ and $\|F_{12}\|\leq 1$), and hence S satisfies the indicated recursive relations (see also figure 10.4). The recursion for R is determined likewise.

In view of proposition 8.13, we can take x_- as the state of a minimal realization of the strictly upper part of Θ_{22}^{-*}. Let $\{A_a, B_a, C_r\}$ be a corresponding state realization, so that the strictly lower part of Θ_{22}^{-1} has an anti-causal state realization

$$\begin{cases} x_{-[0]} &= x_{-[1]}^{(-1)} A_a^* + b_{2[0]} C_r^* \\ b_{1[0]} &= x_{-[1]}^{(-1)} B_a^* . \end{cases}$$

The unknowns A_a, B_a and C_r can be expressed in terms of F, G, H by substitution in equations (10.21), and using S and R as intermediate quantities. Doing so with $b_2 = 0$, the first equation in (10.22) yields the expression for A_a in (10.20) and B_a can be determined in terms of S from the last equation in (10.21). C_r^* is obtained as the transfer $b_{2[0]} \mapsto x_{-[0]}$ for $a_1 = 0$ and $b_2 = b_{2[0]} \in \mathcal{D}_2$, so that $x_{-[0]}S = x_{+[0]}$ and $x_{-[1]}^{(-1)} = x_{+[1]}^{(-1)} R^{(-1)}$. Inserting the latter expression into the first equation of (10.21) twice yields

$$x_{-[1]}^{(-1)} = x_{+[0]} F_{11} (I - R^{(-1)} F_{21})^{-1} R^{(-1)} + b_{2[0]} G_{21} (I - R^{(-1)} F_{21})^{-1} R^{(-1)} .$$

Inserting this in the second equation of (10.21), and using $x_{+[0]} = x_{-[0]}S$ results in

$$\begin{aligned} x_{-[0]} &= x_{-[0]} S F_{12} + x_{-[0]} S F_{11} (I - R^{(-1)} F_{21})^{-1} R^{(-1)} F_{22} \\ &\quad + b_{2[0]} G_{21} (I - R^{(-1)} F_{21})^{-1} R^{(-1)} F_{22} + b_{2[0]} G_{22} \\ \Rightarrow \quad & \\ x_{-[0]} (I - SR) &= b_{2[0]} (G_{22} + G_{21} (I - R^{(-1)} F_{21})^{-1} R^{(-1)} F_{22}) \end{aligned}$$

which gives the expression for C_r.

We have defined, in equation (8.28), the conjugate-Hankel operator $H' = \mathbf{P}'(\cdot\,\Theta_{22}^{-1})|_{\mathcal{U}_2}$. In proposition 8.13 we showed that H' has a factorization $H' = \sigma\tau$, where the maps $\sigma: b_{2f} \mapsto x_{-[0]}$ and $\tau: x_{-[0]} \mapsto b_{1p}$ are onto and one-to-one, respectively, and both contractive. In particular, we can write $H' = \mathbf{P}_0(\cdot\,\mathbf{F}_r^*)\,\mathbf{F}_a$, where $\tau = \mathbf{F}_a = [B_a Z (I - A_a Z)^{-1}]^*$ (if $\ell_{A_a} < 1$) and $\sigma = \mathbf{P}_0(\cdot\,\mathbf{F}_r^*)$ with $\mathbf{F}_r = (I - A_a Z)^{-1} C_r$ (if $\ell_{A_a} < 1$). The properties of σ and τ imply that the derived model $\{A_a, B_a, C_r\}$ is uniformly minimal, with contractive reachability/observability Gramians. □

The second step in the construction of a realization for T_a is to determine a state realization for B'. This is done by evaluating $[U^* \;\; -T^*\Gamma^{-1}]\Theta = [A' \;\; -B']$. This has already been done in equation (8.37), which gives, with the state model for Θ written as

$$\mathbf{\Theta} = \begin{bmatrix} A_\Theta & C_\Theta \\ B_\Theta & D_\Theta \end{bmatrix} = \left[\begin{array}{c|c} R & \\ \hline & I \end{array}\right] \left[\begin{array}{c|cc} A & C_1 & C_2 \\ \hline B_U & D_{11} & D_{12} \\ \Gamma^{-1}B & D_{21} & D_{22} \end{array}\right] \left[\begin{array}{c|c} R^{-(-1)} & \\ \hline & I \end{array}\right],$$

$$
\begin{aligned}
[U^* \quad -T^*\Gamma^{-1}]\Theta &= \{[D_U^* \quad 0]D_\Theta + C^*\Lambda[C_1 \quad C_2]\} + \\
&\quad + \left\{[D_U^* \quad 0]\begin{bmatrix} B_U \\ \Gamma^{-1}B \end{bmatrix} + C^*\Lambda A\right\} Z(I-AZ)^{-1}[C_1 \quad C_2] \\
&= \{[D_U^* \quad 0]D_\Theta + C^*\Lambda[C_1 \quad C_2]\} + \\
&\quad + C^*(\Lambda - I)AZ(I-AZ)^{-1}[C_1 \quad C_2]
\end{aligned}
$$

(in which we used $C^*A + D_U^*B_U = 0$). Since this expression is equal to $[A' \quad -B']$ and $M = I - \Lambda$, we obtain a state-space model for B' as

$$B' = \left\{-D_U^*D_{12} - C^*(I-M)C_2\right\} + C^*MAZ(I-AZ)^{-1}C_2. \qquad (10.23)$$

We are now in a position to determine a state realization for T_a.

Theorem 10.10 *Let T, Γ, U and Θ be as in lemma 10.3, so that $[U^* \quad -T^*\Gamma^{-1}]\Theta = [A' \quad -B']$. Let $\{A,B,C,0\}$ be an output normal u.e. stable state realization for T, let M satisfy the Lyapunov equation (10.14), and let $\{A,B_U,C,D_U\}$ be a realization for U. Denote the block entries of $\mathbf{\Theta}$ as in (10.13), and let $\mathbf{\Sigma}$ corresponding to $\mathbf{\Theta}$ have a partitioning (8.21).*

Then the approximant T_a, defined as the strictly upper part of $T' = \Gamma\Theta_{22}^{-*}B'^*$, has a state realization $\{A_a, \Gamma B_a, C_a, 0\}$, where A_a, $B_a \in \mathcal{D}$ are defined by (10.20), and C_a is given by

$$C_a = C_r\left[-D_{12}^*D_U - C_2^*(I-M)C\right] + A_aY^{(-1)}A^*MC, \qquad (10.24)$$

where C_r is defined in (10.20), and $Y \in \mathcal{D}$ satisfies the recursion $Y = A_aY^{(-1)}A^* + C_rC_2^*$

PROOF The state realization for T_a is obtained by multiplying the model for B' in (10.23) by the model $\{A_a, B_a, C_r\}$ of the strictly upper part of Θ_{22}^{-*} as obtained in proposition 10.9. From this proposition, we have a state model of Θ_{22}^{-1} as

$$\Theta_{22}^{-1} = [\text{upper}] + C_r^*\mathbf{F}_a.$$

$\mathbf{F}_a$ is the selected basis representation of $\mathcal{H}(\Theta_{22}^{-*})$, satisfying $\mathbf{F}_a = (I - A_aZ)^{-*}Z^*B_a^* \in \mathcal{L}Z^{-1}$ when $\ell_{A_a} < 1$, and more in general the recursive equation

$$\mathbf{F}_a = Z^*B_a^* + Z^*A_a^*\mathbf{F}_a.$$

The model of B' is given in (10.23) as $B' = D' + C^*MAZ\mathbf{F}_o$, where

$$
\begin{aligned}
D' &= -D_{12}^*D_U - C_2^*(I-M)C, \\
\mathbf{F}_o &= (I-AZ)^{-1}C_2, \qquad \mathbf{F}_o = C_2 + AZ\mathbf{F}_o.
\end{aligned}
$$

Hence T_a is given by

$$
\begin{aligned}
T_a^*\Gamma^{-1} &= \mathbf{P}'(B'\Theta_{22}^{-1}) \\
&= D'C_r^*\mathbf{F}_a + C^*MA\mathbf{P}'(Z\mathbf{F}_o\Theta_{22}^{-1}).
\end{aligned}
$$

It remains to evaluate $\mathbf{P}'(Z\mathbf{F}_o\Theta_{22}^{-1})$. Because $\mathbf{P}'(\mathcal{D}_2\mathbf{F}_o\Theta_{22}^{-1}) \in \mathcal{H}(\Theta_{22}^{-*})$, we can write $\mathbf{P}'(\mathbf{F}_o\Theta_{22}^{-1}) = Y^*\mathbf{F}_a$, for some $Y \in \mathcal{D}$. Consequently,

$$\begin{aligned}\mathbf{P}'(ZY^*\mathbf{F}_a) &= Y^{*(-1)}\mathbf{P}'(Z\mathbf{F}_a)\\ &= Y^{*(-1)}A_a^*\mathbf{F}_a.\end{aligned}$$

Because also $\mathbf{P}'(ZY^*\mathbf{F}_a) = \mathbf{P}'(Z\mathbf{P}'(Y^*\mathbf{F}_a)) = \mathbf{P}'(Z\mathbf{F}_o\Theta_{22}^{-1})$, we obtain

$$T_a^*\Gamma^{-1} = \left\{D'C_r^* + C^*MAY^{*(-1)}A_a^*\right\}\mathbf{F}_a,$$

which gives the expression for C_a in (10.24). Finally, the indicated recursion for Y follows via

$$\begin{aligned}A\mathbf{P}'(Z\mathbf{F}_o\Theta_{22}^{-1}) &= \mathbf{P}'(AZ\mathbf{F}_o\Theta_{22}^{-1})\\ &= \mathbf{P}'(\mathbf{F}_o\Theta_{22}^{-1}) - \mathbf{P}'(C_2\Theta_{22}^{-1})\\ \Leftrightarrow \quad AY^{*(-1)}A_a^*\mathbf{F}_a &= Y^*\mathbf{F}_a - C_2C_r^*\mathbf{F}_a\\ \Leftrightarrow \quad AY^{*(-1)}A_a^* &= Y^* - C_2C_r^*,\end{aligned}$$

where in the last step we used that $\mathbf{F}_a$ is a strong basis representation (proposition 10.9). □

A check on the dimensions of A_a reveals that this state realization for T_a has indeed a state dimension sequence given by $N = \#(\mathcal{B}_-)$: at each point in time it is equal to the number of singular values larger than 1 of the Hankel operator of T at that point. The realization is given in terms of four recursions: two for M and S that run forward in time, the other two for R and Y that run backward in time and depend on S. One implication of this is that it is not possible to compute part of an optimal approximant of T if the model of T is known only partly, say up to time instant k. Based on theorem 10.10, the algorithm in figure 10.5 computes a model $\{A_a, B_a, C_a, 0\}$ for the Hankel norm approximant T_a in terms of Γ and a model $\{A, B, C, 0\}$ for T.

There are a few remaining issues. T_a, as the strictly upper part of some operator in $\mathcal{X}$, is possibly unbounded. This occurs if the strictly upper part of Θ_{22}^{-*} is unbounded. We do not know whether this can actually occur. The realization of T_a is well defined, because Θ_{22}^{-1} is bounded, as well as projections of the kind $\mathbf{P}'(\cdot\,\Theta_{22}^{-1})$, so that in particular the Hankel operator H' which defines that realization is bounded. (In fact, one could probably set up a realization theory for unbounded operators with bounded Hankel operators.) A related second issue is that possibly $\ell_{A_a} = 1$. Although this seems unlikely in view of the assumptions on ℓ_A and the singular values of H_T that we have made, we have no proof yet that this cannot occur. Note that the proof of theorem 10.10 is not dependent on ℓ_{A_a} being strictly smaller than 1. Finally, an alternative derivation of a model for T_a is possible via an inner-outer factorization of Θ_{22}. This gives rise to different expressions but still produces a two-directional recursive algorithm.

10.4 PARAMETRIZATION OF ALL APPROXIMANTS

At this point, we can study the description of all possible solutions to the Hankel norm approximation problem that have order at most equal to N, where $N = \text{sdim } \mathcal{H}(\Theta_{22}^{-*})$ is the sequence of dimensions of the input state space of Θ_{22}^{-*}. We determine all possible

In:	$\mathbf{T}$	(model in output normal form for a strictly upper matrix T)
	Γ	(approximation parameters)
Out:	$\mathbf{T}_a$	(model for Hankel norm approximant T_a)

do algorithm 10.2: gives $M_k, \boldsymbol{\Theta}_k, J_{\mathcal{B}_k}, C_{2,k}, D_{12,k}, D_{U,k}$ $(k = 1,\cdots,n)$
$S_1 = [\cdot]$

for $k = 1,\cdots,n$

$$\left[\begin{array}{l} \text{Compute } \boldsymbol{\Sigma}_k \text{ from } \boldsymbol{\Theta}_k \text{ using (8.19): gives } F_{ij}, G_{ij}, H_{ij} \\ S_{k+1} = F_{21,k} + F_{22,k}(I - S_k F_{12,k})^{-1} S_k F_{11,k} \end{array}\right.$$

end

$R_{n+1} = [\cdot]$
$Y_{n+1} = [\cdot]$

for $k = n,\cdots,1$

$$\left[\begin{array}{lcl} R_k & = & F_{12,k} + F_{11,k}(I - R_{k+1}F_{21,k})^{-1} R_{k+1} F_{22,k} \\ C_{r,k}^* & = & \left\{G_{22,k} + G_{21,k}(I - R_{k+1}F_{21,k})^{-1} R_{k+1} F_{22,k}\right\}(I - S_k R_k)^{-1} \\ A_{a,k} & = & \left\{F_{22,k}(I - S_k F_{12,k})^{-1}\right\}^* \\ B_{a,k} & = & \left\{H_{22,k} + F_{22,k}(I - S_k F_{12,k})^{-1} S_k H_{12,k}\right\}^* \\ Y_k & = & A_{a,k} Y_{k+1} A_k^* + C_{r,k} C_{2,k}^* \\ C_{a,k} & = & C_{r,k}\left\{-D_{12,k}^* D_{U,k} - C_{2,k}^*(I - M_k) C_k\right\} + A_{a,k} Y_{k+1} A_k^* M_k C_k \end{array}\right.$$

end

Figure 10.5. The approximation algorithm.

bounded operators of mixed causality type T' for which it is true that

$$\begin{array}{ll} (1) & \|\Gamma^{-1}(T-T')\| = \|S^*U\| \le 1, \\ \text{and } (2) & \text{the state dimension sequence of } T_a = (\text{upper part of } T') \\ & \text{is at most equal to } N. \end{array}$$

(Note that we do not assume boundedness of T_a.) As we show in theorem 10.17 below, there are no Hankel norm approximants satisfying (1) and (2) with state dimension lower than N. The result is that all solutions are obtained by a linear fractional transform (chain scattering transformation) of Θ with an upper and contractive parameter S_L. That this procedure effectively generates all approximants of locally finite type of s-degree at most equal to the sequence N can be seen from the fact that if $\|\Gamma^{-1}(T-T_a)\|_H \le 1$, then an extension T' of T_a must exist such that $\|\Gamma^{-1}(T-T')\| \le 1$. This is a consequence of a theorem on the Nehari problem (see section 10.6).

The notation is as in the previous sections. We started out with an operator $T \in \mathcal{ZU}$, and we assumed it to be locally finite, with a state realization in output normal form and related factorization $T = \Delta^* U$, where $\Delta \in \mathcal{U}$ and $U \in \mathcal{U}$, unitary and locally finite. Then we proceeded to solve the interpolation problem $[U^* \;\; -T^*\Gamma^{-1}]\Theta = [A' \;\; -B'] \in [\mathcal{U} \;\; \mathcal{U}]$, and we saw that the problem was solvable provided a related Lyapunov-Stein equation had a boundedly invertible solution. The solution was given in terms of an operator $T' = \Gamma^{-1}\Theta_{22}^{-*}B'^*$ in $\mathcal{X}$ of mixed causality type, and the approximant T_a of low order was given by the strictly upper part of T'. In the present section we shall first show that a large class of Hankel-norm approximants can be given in terms of the same J-unitary operator Θ and an arbitrary upper, contractive parameter S_L. Our previous result is the special case $S_L = 0$. Then we move on to show that all approximants of s-degree at most N are obtained in this way.

We first derive a number of preliminary facts which allow us to determine the state dimension sequence of a product of certain matrices.

Preliminary facts

Proposition 10.11 *Let $B = I - X$, where $X \in \mathcal{X}$ and $\|X\| < 1$. Then $\mathbf{P}(\cdot B)|_{\mathcal{U}_2}$ and $\mathbf{P}(\cdot B^{-1})|_{\mathcal{U}_2}$ are Hilbert space isomorphisms on $\mathcal{U}_2$. Likewise, $\mathbf{P}'(\cdot B)|_{\mathcal{L}_2Z^{-1}}$ and $\mathbf{P}'(\cdot B^{-1})|_{\mathcal{L}_2Z^{-1}}$ are isomorphisms on $\mathcal{L}_2Z^{-1}$.*

PROOF B is invertible because $\|X\| < 1$. Since also

$$X_p := \mathbf{P}'(\cdot X)|_{\mathcal{L}_2Z^{-1}}, \qquad X_f := \mathbf{P}(\cdot X)|_{\mathcal{U}_2}$$

are strictly contractive: $\|X_p\| < 1$, $\|X_f\| < 1$, it follows that $B_p = I - X_p = \mathbf{P}'(\cdot B)|_{\mathcal{L}_2Z^{-1}}$ is invertible in $\mathcal{L}$, and $B_f = I - X_f$ is invertible in $\mathcal{U}$. In particular, for $u \in \mathcal{L}_2Z^{-1}$, the decomposition $uB =: y_1 + u_1$ (with $y_1 \in \mathcal{U}_2$, $u_1 = uB_p \in \mathcal{L}_2Z^{-1}$) satisfies

$$\|u_1\| \ge \varepsilon \|u\|, \qquad \text{some } \varepsilon > 0. \tag{10.25}$$

Take $y \in \mathcal{U}_2$, $y \ne 0$. To show that $\mathbf{P}(\cdot B^{-1})|_{\mathcal{U}_2}$ is one-to-one, we will show that the norm of the upper part of yB^{-1} is uniformly bounded from below: $y_2 := \mathbf{P}(yB^{-1})$ has $\|y_2\| \ge \varepsilon_1 \|y\|$ (with $\varepsilon_1 > 0$).

Indeed, put $yB^{-1} =: y_2 + u_2$ ($y_2 \in \mathcal{U}_2$, $u_2 \in \mathcal{L}_2 \mathcal{Z}^{-1}$). Since $u_2 B = y - y_2 B$, and B_p is invertible, we can apply the relation (10.25) proven above, in the form $\mathbf{P}'(u_2 B) \geq \varepsilon_2 \| u_2 \|$, to obtain

$$\| \mathbf{P}'(y_2 B) \| = \| \mathbf{P}'(u_2 B) \| \geq \varepsilon_2 \| u_2 \| \qquad (\varepsilon_2 > 0).$$

Because B is bounded: $\|B\| < 2$, it follows that $\| y_2 \| > 1/2 \| y_2 B \| > 1/2 \varepsilon_2 \| u_2 \|$, or

$$\| y_2 \| > \varepsilon_3 \| u_2 \|, \qquad \varepsilon_3 = 1/2\varepsilon_2 > 0.$$

Hence, at this point we have $yB^{-1} = y_2 + u_2$ with $\| y_2 \| > \varepsilon_3 \| u_2 \|$ ($\varepsilon_3 > 0$). Because B^{-1} is boundedly invertible, there exists $\varepsilon_4 > 0$ such that $\| yB^{-1} \| \geq \varepsilon_4 \| y \|$, and we have

$$\| y_2 \| (1 + \frac{1}{\varepsilon_3}) \; > \; \| y_2 \| + \| u_2 \| \; > \; \| y_2 + u_2 \| \; > \; \varepsilon_4 \| y \|.$$

We finally obtain that

$$\| y_2 \| \; > \; \frac{\varepsilon_4}{1 + 1/\varepsilon_3} \| y \| \; =: \; \varepsilon_1 \| y \|$$

so that $\mathbf{P}(\cdot B^{-1})|_{\mathcal{U}_2}$ is one-to-one.

To show that $\mathbf{P}(\cdot B^{-1})|_{\mathcal{U}_2}$ is onto: $\mathbf{P}(\mathcal{U}_2 B^{-1}) = \mathcal{U}_2$, we have to show that for all $y_2 \in \mathcal{U}_2$, there exists an $y \in \mathcal{U}_2$ such that

$$\mathbf{P}(yB^{-1}) = y_2,$$

i.e., given $y_2 \in \mathcal{U}_2$ find $y \in \mathcal{U}_2$ such that $yB^{-1} = y_2 + u_2$ (some $u_2 \in \mathcal{L}_2 \mathcal{Z}^{-1}$), or equivalently, $y_2 B = y + u_2 B$. We will use the fact that $B_p = \mathbf{P}'(\cdot B)|_{\mathcal{L}_2 \mathcal{Z}^{-1}}$ is invertible so that $\mathbf{P}'(u_2 B) = u_2 B_p$ uniquely determines u_2. Indeed, given y_2, u_2 is computed as $u_2 = \mathbf{P}'(y_2 B) B_p^{-1}$, and then $y \in \mathcal{U}_2$ is given by $y = (u_2 + y_2) B$.

The property on $\mathbf{P}'(\cdot B^{-1})|_{\mathcal{L}_2 \mathcal{Z}^{-1}}$ is proven in a similar manner. □

Proposition 10.11 allows us to conclude, in particular, that if $\mathcal{A}$ is a sliced subspace in $\mathcal{U}_2$ and B is as in the proposition, then

$$\text{sdim } \mathbf{P}(\mathcal{A}B^{-1}) \; = \; \text{sdim } \mathcal{A}$$

and if $\mathcal{B}$ is another sliced subspace in $\mathcal{U}_2$, then $\mathcal{B} \subset \mathcal{A} \Leftrightarrow \mathbf{P}(\mathcal{B}B^{-1}) \subset \mathbf{P}(\mathcal{A}B^{-1})$.

Proposition 10.12 *If $B = I - X$, $X \in \mathcal{X}$ and $\| X \| < 1$, and if $\mathcal{B} = \mathbf{P}(\mathcal{L}_2 \mathcal{Z}^{-1} B)$, then*

$$\mathbf{P}(\mathcal{B}B^{-1}) \; = \; \mathbf{P}(\mathcal{L}_2 \mathcal{Z}^{-1} B^{-1}).$$

PROOF We show mutual inclusion.

(1) $\mathbf{P}(\mathcal{B}B^{-1}) \subset \mathbf{P}(\mathcal{L}_2 \mathcal{Z}^{-1} B^{-1})$. Let $y \in \mathbf{P}(\mathcal{B}B^{-1})$. Then there exist $u \in \mathcal{L}_2 \mathcal{Z}^{-1}$ and $u_1 \in \mathcal{L}_2 \mathcal{Z}^{-1}$ such that $y = \mathbf{P}\left((uB + u_1) B^{-1} \right) = \mathbf{P}(u_1 B^{-1})$. Hence $y \in \mathbf{P}(\mathcal{L}_2 \mathcal{Z}^{-1} B^{-1})$.

(2) $\mathbf{P}(\mathcal{L}_2\mathcal{Z}^{-1}B^{-1}) \subset \mathbf{P}(\mathcal{B}B^{-1})$. Assume $y = \mathbf{P}(u_1 B^{-1})$ for some $u_1 \in \mathcal{L}_2\mathcal{Z}^{-1}$. Since $B_p = \mathbf{P}'(\cdot B)|_{\mathcal{L}_2\mathcal{Z}^{-1}}$ is an isomorphism (proposition 10.11), a $u \in \mathcal{L}_2\mathcal{Z}^{-1}$ exists such that $\mathbf{P}'(uB) = -u_1$. It follows that

$$\begin{aligned} y &= \mathbf{P}(u_1 B^{-1}) \\ &= \mathbf{P}\big((uB + u_1)B^{-1}\big) \\ &= \mathbf{P}\big((uB - \mathbf{P}'(uB))\, B^{-1}\big) \\ &= \mathbf{P}\big(\mathbf{P}(uB)B^{-1}\big) \quad \in \mathbf{P}(\mathcal{B}B^{-1}). \end{aligned}$$

□

Proposition 10.13 *If $A \in \mathcal{L}$ and $A^{-1} \in \mathcal{X}$ and if $\mathcal{A} = \mathbf{P}(\mathcal{L}_2\mathcal{Z}^{-1}A^{-1})$, then*

$$\mathcal{L}_2\mathcal{Z}^{-1}A^{-1} = \overline{\mathcal{A}} \oplus \mathcal{L}_2\mathcal{Z}^{-1}.$$

PROOF (Note that $\mathcal{A}$, as the range of a Hankel operator, need not be closed.) The left-to-right inclusion is obvious. To show the right-to-left inclusion, we show first that $\mathcal{L}_2\mathcal{Z}^{-1} \subset \mathcal{L}_2\mathcal{Z}^{-1}A^{-1}$. Assume that $u \in \mathcal{L}_2\mathcal{Z}^{-1}$. Then $u = (uA)A^{-1}$. But since $A \in \mathcal{L}$, we have $uA \in \mathcal{L}_2\mathcal{Z}^{-1}$, and $u \in \mathcal{L}_2\mathcal{Z}^{-1}A^{-1}$. The fact that $\overline{\mathcal{A}}$ is also in the image follows by complementation: $\mathcal{L}_2\mathcal{Z}^{-1}A^{-1} \ominus \mathcal{L}_2\mathcal{Z}^{-1} = \overline{\mathbf{P}(\mathcal{L}_2\mathcal{Z}^{-1}A^{-1})}$. □

Theorem 10.14 *Let $A \in \mathcal{L}$, $A^{-1} \in \mathcal{X}$, and suppose that the space $\mathcal{A} = \mathbf{P}(\mathcal{L}_2\mathcal{Z}^{-1}A^{-1})$ is locally finite of s-dimension N. Let $B = I - X$ with $X \in \mathcal{X}$ and $\|X\| < 1$. Then*

$$\operatorname{sdim} \mathbf{P}(\mathcal{L}_2\mathcal{Z}^{-1}A^{-1}B^{-1}) = N + p \quad \Rightarrow \quad \operatorname{sdim} \mathbf{P}(\mathcal{L}_2\mathcal{Z}^{-1}BA) = p.$$

PROOF

$$\begin{aligned} \mathbf{P}(\mathcal{L}_2\mathcal{Z}^{-1}A^{-1}B^{-1}) &= \mathbf{P}\big((\mathcal{L}_2\mathcal{Z}^{-1} \oplus \overline{\mathcal{A}})\, B^{-1}\big) && \text{[prop. 10.13]} \\ &= \mathbf{P}(\mathcal{L}_2\mathcal{Z}^{-1}B^{-1}) + \mathbf{P}(\overline{\mathcal{A}}B^{-1}) && \text{[linearity]} \\ &= \mathbf{P}(\mathcal{B}B^{-1}) + \mathbf{P}(\overline{\mathcal{A}}B^{-1}) && \text{[prop. 10.12]} \end{aligned}$$

where $\mathcal{B} = \mathbf{P}(\mathcal{L}_2\mathcal{Z}^{-1}B)$.

In the sequel of the proof, we use the following two properties. The closure of a D-invariant locally finite linear manifold $\mathcal{H}$ yields a locally finite D-invariant subspace $\overline{\mathcal{H}}$ with the same sdim. Secondly, let $\mathcal{M}$ be another locally finite D-invariant subspace and let X be a bounded operator on $\mathcal{X}_2$, then $\mathcal{H}X = [\mathbf{P}_{\mathcal{M}}(\mathcal{H})]X$ if $\mathcal{M}^{\perp}X = 0$.

Since $\overline{\mathcal{A}}$ and $\mathcal{B}$ are spaces in $\mathcal{U}_2$, and since according to proposition 10.11, $\mathbf{P}(\cdot B^{-1})|_{\mathcal{U}_2}$ is an isomorphism mapping $\overline{\mathcal{A}}$ and $\mathcal{B}$ to $\mathbf{P}(\overline{\mathcal{A}}B^{-1})$ and $\mathbf{P}(\mathcal{B}B^{-1})$, respectively, we obtain that sdim $(\overline{\mathcal{A}} + \mathcal{B}) = N + p$. With $\mathcal{A}^{\perp} = \mathcal{U}_2 \ominus \overline{\mathcal{A}}$, it follows that $\mathbf{P}_{\mathcal{A}^{\perp}}(\mathcal{B})$ has sdim equal to p, because sdim $\overline{\mathcal{A}} = N$. The proof terminates by showing that

(1) $\mathbf{P}(\mathcal{L}_2\mathcal{Z}^{-1}BA) = \mathbf{P}(\mathbf{P}_{\mathcal{A}^{\perp}}(\mathcal{B})A)$, for

$$\begin{aligned} \mathbf{P}(\mathcal{L}_2\mathcal{Z}^{-1}BA) &= \mathbf{P}(\mathbf{P}(\mathcal{L}_2\mathcal{Z}^{-1}B)A) \\ &= \mathbf{P}(\mathcal{B}A) \\ &= \mathbf{P}(\mathbf{P}_{\mathcal{A}^{\perp}}(\mathcal{B})A)\,, \end{aligned}$$

because $\overline{\mathcal{A}}A \subset \mathcal{L}_2 Z^{-1}$.

(2) $\mathbf{P}(\mathbf{P}_{\mathcal{A}^\perp}(\mathcal{B})A)$ is isomorphic to $\mathbf{P}_{\mathcal{A}^\perp}(\mathcal{B})$, which follows from the fact that the map $\mathbf{P}(\cdot A)\big|_{\mathcal{A}^\perp}$ is one-to-one, for $\mathbf{P}(xA) = 0 \Rightarrow x \in \overline{\mathcal{A}} \oplus \mathcal{L}_2 Z^{-1}$, and the kernel of $\mathbf{P}(\cdot A)\big|_{\mathcal{A}^\perp}$ is thus just $\{0\}$.

Consequently, sdim $\mathbf{P}(\mathcal{L}_2 Z^{-1} BA) =$ sdim $\mathbf{P}(\mathbf{P}_{\mathcal{A}^\perp}(\mathcal{B})A) =$ sdim $\mathbf{P}_{\mathcal{A}^\perp}(\mathcal{B}) = p$.

□

In the above theorem, we had $A \in \mathcal{L}$. A comparable result for $A \in \mathcal{U}$ follows directly by considering a duality property, and yields the corollary below.

Corollary 10.15 *Let $A \in \mathcal{U}$, $X \in \mathcal{X}$, $B = I - X$ and $\|X\| < 1$, and let A be invertible in $\mathcal{X}$. Suppose that $\mathcal{A} = \mathbf{P}'(\mathcal{U}_2 A^{-1})$ has s-dimension N. Then*

$$\text{sdim } \mathbf{P}'(\mathcal{U}_2 B^{-1} A^{-1}) = N + p \qquad \Rightarrow \qquad \text{sdim } \mathbf{P}'(\mathcal{U}_2 AB) = p\,.$$

PROOF For any bounded operator, the dimension of its range is equal to the dimension of its co-range. Hence for $T \in \mathcal{X}$, we have that sdim $\text{ran}(H_T) =$ sdim $\text{ran}(H_T^*)$, or

$$\text{sdim } \mathbf{P}(\mathcal{L}_2 Z^{-1} T) \;=\; \text{sdim } \mathbf{P}'(\mathcal{U}_2 T^*)\,.$$

□

Generating new solutions of the interpolation problem

Throughout the remainder of the section we use the notion of *causal state dimension sequence* of an operator $T \in \mathcal{X}$ as the s-dimension N of the space $\mathcal{H}(T) = \mathbf{P}'(\mathcal{U}_2 T^*)$. N is thus a sequence of numbers $\{N_i : i \in \mathbb{Z}\}$ where all N_i in our case are finite. Dually, we call the s-dimension of $\mathbf{P}'(\mathcal{U}_2 T)$ the *anti-causal state dimension sequence.* We use the following lemma, in which we must assume that Θ is constructed according to the recipe given in corollary 8.18, so that its input state space $\mathcal{H}(\Theta)$ is generated by (*viz.* equation (10.12))

$$\mathcal{H}(\Theta) \;=\; \mathcal{D}_2^{\mathcal{B}}(I - Z^* A^*)^{-1} Z^* \begin{bmatrix} B_U^* & B^* \Gamma^{-1} \end{bmatrix}.$$

Lemma 10.16 *Let T, Γ and Θ be as in lemma 10.4, such that $T = \Delta^* U$ is a factorization of T with $\Delta \in \mathcal{U}$ and $U \in \mathcal{U}$ is inner, and Θ is the J-unitary operator with input state space given by (10.12) and defined by the realization (10.13). Then*

$$\begin{array}{lcl} [U^* \;\; 0]\Theta & \in & [\mathcal{L} \;\; \mathcal{L}] \\ [-\Delta^* \;\; \Gamma]\Theta & \in & [\mathcal{L} \;\; \mathcal{L}]. \end{array}$$

PROOF We prove this by brute-force calculations on the realizations of U and Θ, as in (10.13):

$$\begin{aligned} [U^* \;\; 0]\Theta = & \left\{D_U^* + C^*(I - Z^* A^*)^{-1} Z^* B_U^*\right\}\left\{[D_{11} \;\; D_{12}] + B_U Z (I - AZ)^{-1} [C_1 \;\; C_2]\right\} \\ = & \; D_U^* [D_{11} \;\; D_{12}] + D_U^* B_U Z (I - AZ)^{-1} [C_1 \;\; C_2] + \\ & + C^* (I - Z^* A^*)^{-1} Z^* B_U^* [D_{11} \;\; D_{12}] + \\ & + C^* \underline{(I - Z^* A^*)^{-1} Z^* B_U^* B_U Z (I - AZ)^{-1}} [C_1 \;\; C_2]. \end{aligned}$$

Upon using the identities $D_U^* B_U + C^* A = 0$, $B_U^* B_U + A^* A = I$, and

$$(I - Z^* A^*)^{-1} Z^* (I - A^* A) Z (I - AZ)^{-1} = AZ(I - AZ)^{-1} + (I - Z^* A^*)^{-1},$$

it is seen that the terms with $(I - AZ)^{-1}$ cancel each other, so that

$$\begin{aligned}[U^* \;\; 0]\Theta &= D_U^* [D_{11} \;\; D_{12}] + C^* [C_1 \;\; C_2] + \\ &\quad + C^* (I - Z^* A^*)^{-1} Z^* \{A^* [C_1 \;\; C_2] + B_U^* [D_{11} \;\; D_{12}]\} \\ &\in [\mathcal{L} \;\; \mathcal{L}].\end{aligned}$$

In much the same way,

$$\begin{aligned}[-\Delta^* \;\; \Gamma]\Theta &= [\{-DD_U^* - BB_U^* - (DC^* + BA^*) Z^* (I - A^* Z^*)^{-1} B_U^*\} \;\; \Gamma] \times \\ &\quad \times \left\{ \begin{bmatrix} D_{11} & D_{12} \\ D_{21} & D_{22} \end{bmatrix} + \begin{bmatrix} B_U \\ \Gamma^{-1} B \end{bmatrix} Z (I - AZ)^{-1} [C_1 \;\; C_2] \right\} \\ &= \text{(lower)} + \{(-DD_U^* - BB_U^*) B_U + B\} Z (I - AZ)^{-1} [C_1 \;\; C_2] + \\ &\quad + (-DC^* - BA^*) \underline{Z^* (I - A^* Z^*)^{-1} B_U^* B_U Z (I - AZ)^{-1}} [C_1 \;\; C_2] \\ &= \text{(lower)} + \{-DD_U^* B_U - BB_U^* B_U + B - DC^* A - BA^* A\} Z (I - AZ)^{-1} [C_1 \;\; C_2] \\ &= \text{(lower)} + \{DC^* A - B + BA^* A + B - DC^* A - BA^* A\} Z (I - AZ)^{-1} [C_1 \;\; C_2] \\ &= \text{(lower)} + 0.\end{aligned}$$

□

Theorem 10.17 *Let $T \in \mathcal{ZU}$ be a locally finite operator with u.e. stable output normal realization $\{A, B, C, 0\}$, let Γ be an invertible Hermitian diagonal operator. Let H_k be the Hankel operator of $\Gamma^{-1} T$ at time point k, and suppose that an $\varepsilon > 0$ exists such that, for each k, none of the singular values of H_k are in the interval $[1 - \varepsilon, 1 + \varepsilon]$. Let N be the sequence of the numbers N_k of singular values of H_k that are larger than 1.*

Define U to be the inner factor of an external factorization (theorem 6.8), with unitary realization $\{A, B_U, C, D_U\}$, and let Θ be a J-unitary block-upper operator such that its input state space $\mathcal{H}(\Theta)$ is given by (10.12).

(1) If $S_L \in \mathcal{U}$ is contractive, then $\Theta_{22} - \Theta_{21} S_L$ is boundedly invertible, and

$$S = (\Theta_{11} S_L - \Theta_{12})(\Theta_{22} - \Theta_{21} S_L)^{-1} \qquad (10.26)$$

is contractive.

(2) Let, furthermore, $T' = T + \Gamma S^* U$. Then

(a) $\|\Gamma^{-1}(T - T')\| = \|S^* U\| \leq 1$,

(b) *the causal state dimension sequence of T_a = (upper part of T') is precisely equal to N.*

That is, T_a is a Hankel norm approximant of T.

PROOF (1) By the J-unitarity of Θ, Θ_{22} is boundedly invertible and $\|\Theta_{22}^{-1}\Theta_{21}\| < 1$, whence $\Theta_{22} - \Theta_{21} S_L = \Theta_{22}(I - \Theta_{22}^{-1}\Theta_{21} S_L)$ is boundedly invertible. Hence S exists as a bounded operator. Its contractivity follows by the usual direct calculations on scattering operators (see *e.g.*, [DD92]).

(2*a*) follows immediately since $\Gamma^{-1}(T-T') = S^*U$ and U is unitary.
(2*b*) The proof uses the following equality:

$$\begin{aligned} T'^*\Gamma^{-1} &= [U^* \quad -T^*\Gamma^{-1}]\begin{bmatrix} S \\ -I \end{bmatrix} \\ &= [U^* \quad -T^*\Gamma^{-1}]\begin{bmatrix} \Theta_{11} & \Theta_{12} \\ \Theta_{21} & \Theta_{22} \end{bmatrix}\begin{bmatrix} S_L \\ -I \end{bmatrix}(\Theta_{22}-\Theta_{21}S_L)^{-1} \\ &= [A' \quad -B']\begin{bmatrix} S_L \\ -I \end{bmatrix}(\Theta_{22}-\Theta_{21}S_L)^{-1} \\ &= (A'S_L+B')\,(\Theta_{22}-\Theta_{21}S_L)^{-1}\,. \end{aligned}$$

Since $(A'S_L+B') \in \mathcal{U}$, the anti-causal state dimension sequence of T'^* is at each point in time *at most* equal to the anti-causal state dimensions of $(\Theta_{22}-\Theta_{21}S_L)^{-1}$ at that point. Because the latter expression is equal to $(I-\Theta_{22}^{-1}\Theta_{21}S_L)^{-1}\Theta_{22}^{-1}$, and $\|\Theta_{22}^{-1}\Theta_{21}S_L\| < 1$, application of corollary 10.15 with $A = \Theta_{22}$ and $B = I-\Theta_{22}^{-1}\Theta_{21}S_L$ shows that this sequence is equal to the anti-causal state dimension sequence of Θ_{22}^{-1}, *i.e.*, equal to N. Hence sdim $\mathcal{H}(T') \le N$ (pointwise).

The proof terminates by showing that also sdim $\mathcal{H}(T') \ge N$, so that in fact sdim $\mathcal{H}(T') = N$. Define

$$\begin{cases} G_2 &= (\Theta_{22}-\Theta_{21}S_L)^{-1} \\ G_1 &= S_L G_2 \end{cases}$$

so that

$$\begin{bmatrix} S \\ -I \end{bmatrix} = \Theta \begin{bmatrix} G_1 \\ -G_2 \end{bmatrix}.$$

Because Θ is J-inner: $\Theta^* J\Theta = J$, this equality is equivalent to $[G_1^* \quad G_2^*] := [S^* \quad I]\Theta$, and using $S = -\Delta\Gamma^{-1} + UT'^*\Gamma^{-1}$ we obtain

$$\Gamma[G_1^* \quad G_2^*] = T'[U^* \quad 0]\Theta + [-\Delta^* \quad \Gamma]\Theta. \qquad (10.27)$$

However, according to lemma 10.16,

$$\begin{aligned} [U^* \quad 0]\Theta &\in [\mathcal{L} \quad \mathcal{L}] \\ [-\Delta^* \quad \Gamma]\Theta &\in [\mathcal{L} \quad \mathcal{L}]. \end{aligned}$$

This implies $\mathcal{H}(G_2^*) \subset \mathcal{H}(T')$ (same proof as in lemma 10.6). Hence sdim $\mathcal{H}(T') \ge$ sdim $\mathcal{H}(G_2^*) = N$. □

Thus, all S of the form $S = (\Theta_{11}S_L - \Theta_{12})(\Theta_{22}-\Theta_{21}S_L)^{-1}$ with $S_L \in \mathcal{U}$, $\|S_L\| \le 1$ give rise to Hankel norm approximants of T. We encountered this expression earlier in chapter 8: it is a chain-scattering transformation of S_L by Θ. Consequently, S is the transfer of port a_1 to b_1 if $b_2 = a_2 S_L$, as in figure 10.6.

The reverse question is: are all Hankel norm approximants obtained this way? That is, given some T' whose strictly upper part is a Hankel norm approximant of T, is there a corresponding upper and contractive S_L such that T' is given by $T' = T + \Gamma S^* U$, with S as in equation (10.26) above. This problem is addressed in the following theorem.

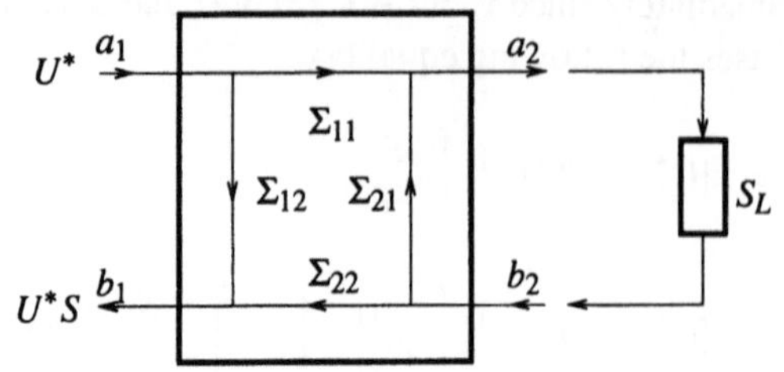

Figure 10.6. Θ (or Σ) generates Hankel norm approximants via S and parametrized by S_L.

The main issue is to prove that S_L as defined by the equations is upper; the proof is an extension of the proof that S_L generated all interpolants in the definite interpolation problem in section 9.2 (theorem 9.6), although some of the items are now more complicated.

Generating all approximants

Theorem 10.18 *Let T, Γ, U and Θ be as in theorem 10.17, and let N be the number of Hankel singular values of $\Gamma^{-1}T$ that are larger than 1. Let be given a bounded operator $T' \in \mathcal{X}$ such that*

(1) $\quad \|\Gamma^{-1}(T-T')\| \leq 1,$

(2) *the state dimension sequence of T_a = (upper part of T') is at most equal to N.*

Define $S = U(T'^* - T^*)\Gamma^{-1}$. Then there is an operator S_L with ($S_L \in \mathcal{U}, \|S_L\| \leq 1$) such that

$$S = (\Theta_{11}S_L - \Theta_{12})(\Theta_{22} - \Theta_{21}S_L)^{-1}$$

(i.e., Θ generates all Hankel-norm approximants). The state dimension of T_a is precisely equal to N.

PROOF The proof parallels in a certain sense the time-invariant proof as given e.g. in [BGR90], but differs in detail. In particular, the "winding number" argument to determine state dimensions must be replaced by theorem 10.14 and its corollary 10.15. The proof consists of five steps.

1. From the definition of S, and using the factorization $T = \Delta^* U$, we know that

$$\|S\| = \|U(T'^* - T^*)\Gamma^{-1}\| = \|\Gamma^{-1}(T' - T)\| \leq 1$$

so S is contractive. Since $S = -\Delta\Gamma^{-1} + UT'^*\Gamma^{-1}$, where Δ and U are upper, the anti-causal state dimension sequence of S is at most equal to N, since it depends exclusively on T'^*, for which N is the anti-causal state dimension sequence.

2. Define

$$[G_1^* \quad G_2^*] := [S^* \quad I]\Theta. \tag{10.28}$$

Then $\mathcal{H}(G_1^*) \subset \mathcal{H}(T')$ and $\mathcal{H}(G_2^*) \subset \mathcal{H}(T')$.

PROOF Using $S = -\Delta\Gamma^{-1} + UT'^*\Gamma^{-1}$, equation (10.28) can be rewritten as

$$\Gamma[G_1^* \quad G_2^*] = T'[U^* \quad 0]\Theta + [-\Delta^* \quad \Gamma]\Theta.$$

According to lemma 10.16,

$$\begin{array}{lcl} [U^* \quad 0]\Theta & \in & [\mathcal{L} \quad \mathcal{L}] \\ [-\Delta^* \quad \Gamma]\Theta & \in & [\mathcal{L} \quad \mathcal{L}]. \end{array}$$

As in the proof of theorem 10.17, this implies $\mathcal{H}(G_1^*) \subset \mathcal{H}(T')$ and $\mathcal{H}(G_2^*) \subset \mathcal{H}(T')$.

3. Equation (10.28) can be rewritten using $\Theta^{-1} = J\Theta^*J$ as

$$\begin{bmatrix} S \\ -I \end{bmatrix} = \Theta \begin{bmatrix} G_1 \\ -G_2 \end{bmatrix}. \tag{10.29}$$

G_2 is boundedly invertible, and S_L defined by $S_L = G_1 G_2^{-1}$ is well defined and contractive: $\| S_L \| \le 1$. In addition, S satisfies $S = (\Theta_{11}S_L - \Theta_{12})(\Theta_{22} - \Theta_{21}S_L)^{-1}$ as required.

PROOF As in the proof of theorem 9.6, step 2, we have, for some $\varepsilon > 0$,

$$G_1^*G_1 + G_2^*G_2 \ge \varepsilon I, \qquad G_1^*G_1 \le G_2^*G_2. \tag{10.30}$$

Together, this shows that $G_2^*G_2 \ge 1/2\varepsilon I$, and hence G_2 is boundedly invertible (but not necessarily in $\mathcal{U}$). With $S_L = G_1G_2^{-1}$, equation (10.30) shows that $S_L^*S_L \le 1$, and hence $\| S_L \| \le 1$. Evaluating equation (10.29) gives

$$\begin{array}{rcl} G_2^{-1} & = & \Theta_{22} - \Theta_{21}S_L \\ SG_2^{-1} & = & \Theta_{11}S_L - \Theta_{12} \end{array} \tag{10.31}$$

and hence $S = (\Theta_{11}S_L - \Theta_{12})(\Theta_{22} - \Theta_{21}S_L)^{-1}$.

4. $G_2^{-1} \in \mathcal{U}$, the space $\mathcal{H}(T')$ has the same dimension as $\mathcal{H}(\Theta_{22}^{-*})$, and $\mathcal{H}(G_1^*) \subset \mathcal{H}(G_2^*)$.

PROOF According to equation (10.31), G_2^{-1} satisfies

$$\begin{array}{rcl} G_2^{-1} & = & \Theta_{22}(I - \Theta_{22}^{-1}\Theta_{21}S_L) \\ G_2 & = & (I - \Theta_{22}^{-1}\Theta_{21}S_L)^{-1}\Theta_{22}^{-1}. \end{array}$$

Let p be the dimension sequence of anti-causal states of G_2^{-1}, and $N_2 \le N$ be the number of anti-causal states of G_2, with N the number of anti-causal states of Θ_{22}^{-1}. Application of corollary 10.15 with $A = \Theta_{22}$ and $B = (I - \Theta_{22}^{-1}\Theta_{21}S_L)$ shows that $N_2 = N + p$, and hence $N_2 = N$ and $p = 0$: $G_2^{-1} \in \mathcal{U}$, and $\mathcal{H}(G_2^*)$ has dimension N. Step 2 claimed $\mathcal{H}(G_2^*) \subset \mathcal{H}(T')$, and because T' has at most N anti-causal states, we must have that in fact $\mathcal{H}(G_2^*) = \mathcal{H}(T')$, and hence $\mathcal{H}(G_1^*) \subset \mathcal{H}(G_2^*)$, by step 2.

5. $S_L \in \mathcal{U}$.

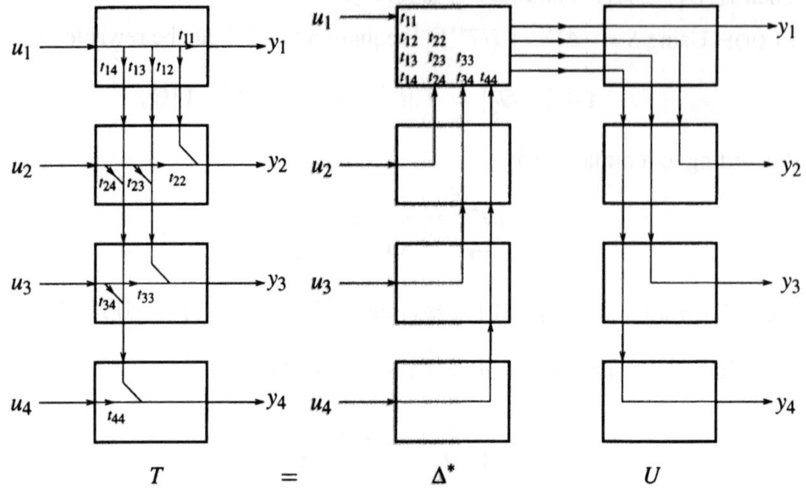

Figure 10.7. Trivial external factorization of T.

PROOF This can be inferred from $G_2^{-1} \in \mathcal{U}$, and $\mathcal{H}(G_1^*) \subset \mathcal{H}(G_2^*)$, as follows. $S_L \in \mathcal{U}$ is equivalent to $\mathbf{P}'(\mathcal{U}_2 S_L) = 0$, and

$$\begin{aligned} \mathbf{P}'(\mathcal{U}_2 S_L) &= \mathbf{P}'(\mathcal{U}_2 G_1 G_2^{-1}) \\ &= \mathbf{P}'(\mathbf{P}'(\mathcal{U}_2 G_1) G_2^{-1}) \end{aligned}$$

since $G_2^{-1} \in \mathcal{U}$. Using $\mathcal{H}(G_1^*) \subset \mathcal{H}(G_2^*)$, or $\mathbf{P}'(\mathcal{U}_2 G_1) \subset \mathbf{P}'(\mathcal{U}_2 G_2)$ we obtain that

$$\begin{aligned} \mathbf{P}'(\mathcal{U}_2 S_L) &\subset \mathbf{P}'(\mathbf{P}'(\mathcal{U}_2 G_2) G_2^{-1}) \\ &= \mathbf{P}'(\mathcal{U}_2 G_2 G_2^{-1}) \qquad (\text{since } G_2^{-1} \in \mathcal{U}) \\ &= 0. \end{aligned}$$

□

10.5 SCHUR-TYPE RECURSIVE INTERPOLATION

The global state-space procedure of the previous sections constructs, for a given $T \in \mathcal{U}$, an inner factor U and an interpolating operator Θ. The procedure can be specialized and applied to the case where T is a general upper triangular matrix without an a priori known state structure. The specialization produces a generalized Schur recursion, which we derive for an example T.

Consider a 4×4 strictly upper triangular matrix T,

$$T = \begin{bmatrix} \boxed{0} & t_{12} & t_{13} & t_{14} \\ & \underline{0} & t_{23} & t_{24} \\ & & \underline{0} & t_{34} \\ & & & \underline{0} \end{bmatrix},$$

where the $(1,1)$ entry is indicated by a square and the main diagonal by underscores. For convenience of notation, and without loss of generality, we may take $\Gamma = I$, and thus seek for T_a (a 4×4 matrix) such that $\| T - T_a \| \le 1$. A trivial (but non-minimal) state realization for T that has $AA^* + CC^* = I$ is obtained by selecting $\{[0\ 0\ 1], [0\ 1\ 0], [1\ 0\ 0]\}$ as a basis for the row space of the second Hankel matrix $H_2 = [t_{12}\ t_{13}\ t_{14}]$, and likewise we select trivial bases for H_3 and H_4. Omitting the details, the realizations for T and an inner factor U that result from this choice turn out to be

$$\mathbf{T}_1 = \left[\begin{array}{ccc|c} \cdot & \cdot & \cdot & \cdot \\ \hline t_{14} & t_{13} & t_{12} & 0 \end{array}\right] \qquad \mathbf{U}_1 = \left[\begin{array}{ccc|c} \cdot & \cdot & \cdot & \cdot \\ \hline 1 & & & \\ & 1 & & \\ & & 1 & \\ & & & 1 \end{array}\right]$$

$$\mathbf{T}_2 = \left[\begin{array}{cc|c} 1 & & \\ & 1 & \\ & & 1 \\ \hline t_{24} & t_{23} & 0 \end{array}\right] \qquad \mathbf{U}_2 = \left[\begin{array}{cc|c} 1 & & \\ & 1 & \\ & & 1 \\ \hline \cdot & \cdot & \cdot \end{array}\right]$$

$$\mathbf{T}_3 = \left[\begin{array}{c|c} 1 & \\ & 1 \\ \hline t_{34} & 0 \end{array}\right] \qquad \mathbf{U}_3 = \left[\begin{array}{c|c} 1 & \\ & 1 \\ \hline \cdot & \cdot \end{array}\right]$$

$$\mathbf{T}_4 = \left[\begin{array}{c|c} \cdot & 1 \\ \hline \cdot & 0 \end{array}\right] \qquad \mathbf{U}_4 = \left[\begin{array}{c|c} \cdot & 1 \\ \hline \cdot & \cdot \end{array}\right]$$

('$\cdot$' stands for an entry with zero dimensions). The corresponding matrices U and $\Delta = UT^*$ are

$$U = \left[\begin{array}{cccc} \boxed{1} & & & \\ & 1 & & \\ & & 1 & \\ & & & 1 \end{array}\right], \qquad \Delta = \left[\begin{array}{cccc} \boxed{0} & & & \\ t_{12}^* & 0 & & \\ t_{13}^* & t_{23}^* & 0 & \\ t_{14}^* & t_{24}^* & t_{34}^* & 0 \end{array}\right]$$

with input space sequence $\mathbb{C}^4 \times \mathbb{C}^0 \times \mathbb{C}^0 \times \mathbb{C}^0$, and output space sequence $\mathbb{C}^1 \times \mathbb{C}^1 \times \mathbb{C}^1 \times \mathbb{C}^1$. All inputs of U and Δ are concentrated at point 1, and hence the causality requirement is always satisfied: $U \in \mathcal{U}$ and $\Delta \in \mathcal{U}$. The structure of Δ and U is clarified by figure 10.7.

The global realization procedure would continue by computing a sequence M

$$M_{k+1} = A_k^* M_k A + B_k^* B_k, \qquad M_1 = [\cdot]$$

and using this to derive Θ as demonstrated in section 10.2. Note that it is not necessary to have a *minimal* realization for T (or U). The extra states correspond to eigenvalues of M that are zero, and hence are of no influence on the negative signature of $\Lambda = I - M$ (independently of Γ). Hence our non-minimal choice of the realization for T does not influence the complexity of the resulting approximant T_a. For a recursive derivation of an interpolating matrix Θ, however, we proceed as follows. The (trivial) state realizations $\mathbf{T}$ and $\mathbf{U}$ are not needed, but the resulting U is used. The interpolation problem is to determine a J-unitary and causal Θ (whose signature will be determined by the construction) such that

$$[U^* \quad -T^*]\Theta \in [\mathcal{U} \quad \mathcal{U}].$$

Assume that $\Theta \in \mathcal{U}(\mathcal{M}_\Theta, \mathcal{N}_\Theta)$. The signature matrix $J_1 := J_{\mathcal{M}_\Theta}$ is known from the outset and is according to the decomposition $[U^* \; -T^*]$. Although the signature $J_2 := J_{\mathcal{N}_\Theta}$ is not yet known at this point, the number of outputs of Θ (*i.e.*, the space sequence $\mathcal{N}_\Theta$) is already determined by the condition that each $\mathbf{\Theta}_k$ is a square matrix. With the above (trivial) realizations of T and U, it follows that Θ has a constant number of two outputs at each point in time. The signature of each output ($+1$ or -1) is determined in the process of constructing Θ, which is done in two steps: $\Theta = \tilde{\Theta}\Pi$. Here, $\tilde{\Theta}$ is such that $[U^* \; -T^*]\tilde{\Theta} \in [\mathcal{U} \; \mathcal{U}]$, where the dimension sequences of each $\mathcal{U}$ are constant and equal to 1 at each point; for example

$$\begin{array}{c} \begin{array}{cccccccc} + & + & + & + & - & - & - & - \end{array} \\ \left[\begin{array}{cccc|cccc} 1 & & & & -t_{11}^* & & & \\ & 1 & & & -t_{12}^* & -t_{22}^* & & \\ & & 1 & & -t_{12}^* & -t_{23}^* & -t_{33}^* & \\ & & & 1 & -t_{14}^* & -t_{24}^* & -t_{34}^* & -t_{44}^* \end{array}\right] \end{array} \tilde{\Theta} = \begin{array}{c} \begin{array}{cccccccc} + & + & - & - & + & + & - & - \end{array} \\ \left[\begin{array}{cccc|cccc} * & * & * & * & * & * & * & * \\ & \underline{*} & * & * & & \underline{*} & * & * \\ & & \underline{*} & * & & & \underline{*} & * \\ & & & \underline{*} & & & & \underline{*} \end{array}\right] \end{array}$$

where the first upper triangular matrix at the right-hand side corresponds to the first output of each section of $\tilde{\Theta}$, and the second to the second output. At this point, the signature of each column at the right-hand side can be positive of negative: the output signature matrix of $\tilde{\Theta}$ is $\tilde{J}_2$, which is an *unsorted* signature matrix such that $\tilde{\Theta}\tilde{J}_2\tilde{\Theta}^* = J_1$ (the signature of the right-hand side in the equation above is just an example). See also figure 10.8. The second step is to sort the columns according to their signature, by introducing a permutation matrix $\Pi \in \mathcal{D}$, such that $J_2 = \Pi^*\tilde{J}_2\Pi$ is a conventional (sorted) signature matrix. The permutation does not change the fact that $[U^* \; -T^*]\Theta \in [\mathcal{U} \; \mathcal{U}]$, but the output dimension sequences of each $\mathcal{U}$ are different now, and are in general no longer constant. For the above example signature, $[A' \; -B']$ has the form

$$\begin{array}{c} \begin{array}{cccccccc} + & + & + & + & - & - & - & - \end{array} \\ \left[\begin{array}{cccc|cccc} 1 & & & & -t_{11}^* & & & \\ & 1 & & & -t_{12}^* & -t_{22}^* & & \\ & & 1 & & -t_{12}^* & -t_{23}^* & -t_{33}^* & \\ & & & 1 & -t_{14}^* & -t_{24}^* & -t_{34}^* & -t_{44}^* \end{array}\right] \end{array} \Theta = \begin{array}{c} \begin{array}{cccccccc} + & + & + & + & - & - & - & - \end{array} \\ \left[\begin{array}{cccccc|cccccc} * & * & * & * & \cdot & \cdot & \cdot & \cdot & * & * & * & * \\ & & \underline{*} & \underline{*} & \cdot & \cdot & & \underline{\cdot} & * & * & * & * \\ & & & & \underline{\cdot} & \cdot & & & \underline{*} & \underline{*} & * & * \\ & & & & & \underline{\cdot} & & & & & \underline{*} & \underline{*} \end{array}\right] \end{array}$$

$$= [A' \quad -B']$$

where A' has as output sequence $\mathbb{C}^2 \times \mathbb{C}^2 \times \mathbb{C}^0 \times \mathbb{C}^0$, and B' has as output sequence $\mathbb{C}^0 \times \mathbb{C}^0 \times \mathbb{C}^2 \times \mathbb{C}^2$. We now consider these operations in more detail.

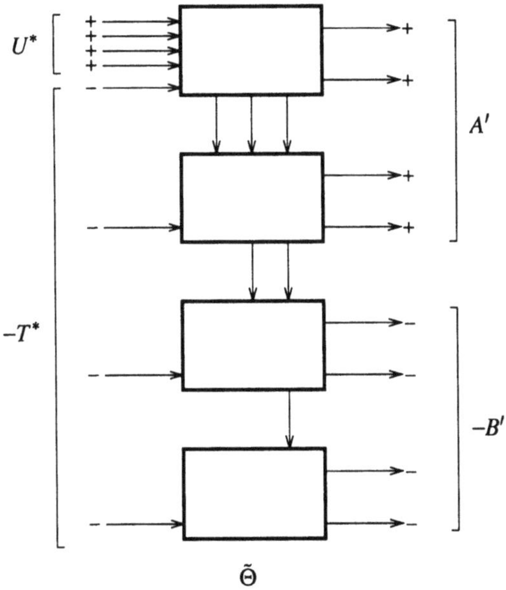

Figure 10.8. Computational structure of $\tilde{\Theta}$, with example signature at the outputs.

Computational structure

$\tilde{\Theta}$ can be determined recursively in n steps: $\tilde{\Theta} = \tilde{\Theta}_{(1)}\tilde{\Theta}_{(2)}\cdots\tilde{\Theta}_{(n)}$, in the following way. The columns of $\tilde{\Theta}$ act on the columns of U^* and $-T^*$. Its operations on U^* are always causal because all columns of U^* correspond to the first point of the recursion ($k = 1$). However, for Θ to be causal, the k-th column of Θ can act only on the first k columns of T^*. Taking this into consideration, we are led to a recursive algorithm of the form

$$[A_{(k)}\;\; B_{(k)}]\tilde{\Theta}_{(k)} = [A_{(k+1)}\;\; B_{(k+1)}]$$

initialized by $A_{(1)} = U^*$, $B_{(1)} = -T^*$, and where $\tilde{\Theta}_{(k)}$ involves

- using columns $n, n-1, \cdots, k+1$ of $A_{(k)}$ in turn, make the last $(n-k)$ entries of the k-th column of $A_{(k)}$ equal to 0. In particular, the $(k+i)$-th column of $A_{(k)}$ is used to make the $(k+i)$-th entry of the k-th column of $A_{(k)}$ equal to zero.

The operations required to carry out each of these steps are elementary J-unitary rotations that act on two columns at a time and make a selected entry of the second column equal to zero. The precise nature of a rotation depends on the corresponding signature and is in turn dependent on the data — this will be detailed later. We first verify that this recursion leads to a solution of the interpolation problem.

$k=1$: using 3 elementary rotations, the entries $t_{14}^*, t_{13}^*, t_{12}^*$ are zeroed in turn. This produces

$$\Rightarrow \left[\begin{array}{cccc|cccc} 1 & * & * & * & \boxed{*} & & & \\ 0 & * & * & * & 0 & -t_{22}^* & & \\ 0 & 0 & * & * & 0 & -t_{23}^* & -t_{33}^* & \\ 0 & 0 & 0 & * & 0 & -t_{24}^* & -t_{34}^* & -t_{44}^* \end{array}\right]$$

$k=2$:

$$\Rightarrow \left[\begin{array}{cccc|cccc} 1 & * & * & * & \boxed{*} & * & & \\ 0 & * & * & * & 0 & * & & \\ 0 & 0 & * & * & 0 & 0 & -t_{33}^* & \\ 0 & 0 & 0 & * & 0 & 0 & -t_{34}^* & -t_{44}^* \end{array}\right]$$

$k=3$:

$$\Rightarrow \left[\begin{array}{cccc|cccc} 1 & * & * & * & \boxed{*} & * & * & \\ 0 & * & * & * & 0 & * & * & \\ 0 & 0 & * & * & 0 & 0 & * & \\ 0 & 0 & 0 & * & 0 & 0 & 0 & -t_{44}^* \end{array}\right]$$

$k=4$: no rotations are required.

The resulting matrices are upper triangular. The signal flow corresponding to this computational scheme is outlined in figure 10.9(a). Note that the computations have introduced an implicit notion of state, formed by the arrows that cross a dotted line between two stages, so that a (non-minimal) realization of Θ can be inferred from the elementary operations.

Elementary rotations: keeping track of signatures

We now consider the elementary operations in the above recursions. An elementary rotation θ such that $\theta^* j_1 \theta = j_2$ (j_1 and j_2 are 2×2 signature matrices) is defined by

$$[u \quad t]\theta = [* \quad 0],$$

where u,t are scalars, and where '$*$' stands for some resulting scalar. Initially, one would consider θ of a traditional J-unitary form:

$$\theta = \begin{bmatrix} 1 & -s \\ -s^* & 1 \end{bmatrix} \frac{1}{c^*}, \qquad cc^* + ss^* = 1\,,\ c \neq 0$$

which satisfies

$$\theta^* \begin{bmatrix} 1 & \\ & -1 \end{bmatrix} \theta = \begin{bmatrix} 1 & \\ & -1 \end{bmatrix}.$$

However, since $|s| < 1$, a rotation of this form is appropriate only if $|u| > |t|$. In the recursive algorithm, this is the case only if $TT^* < I$ which corresponds to a 'definite'

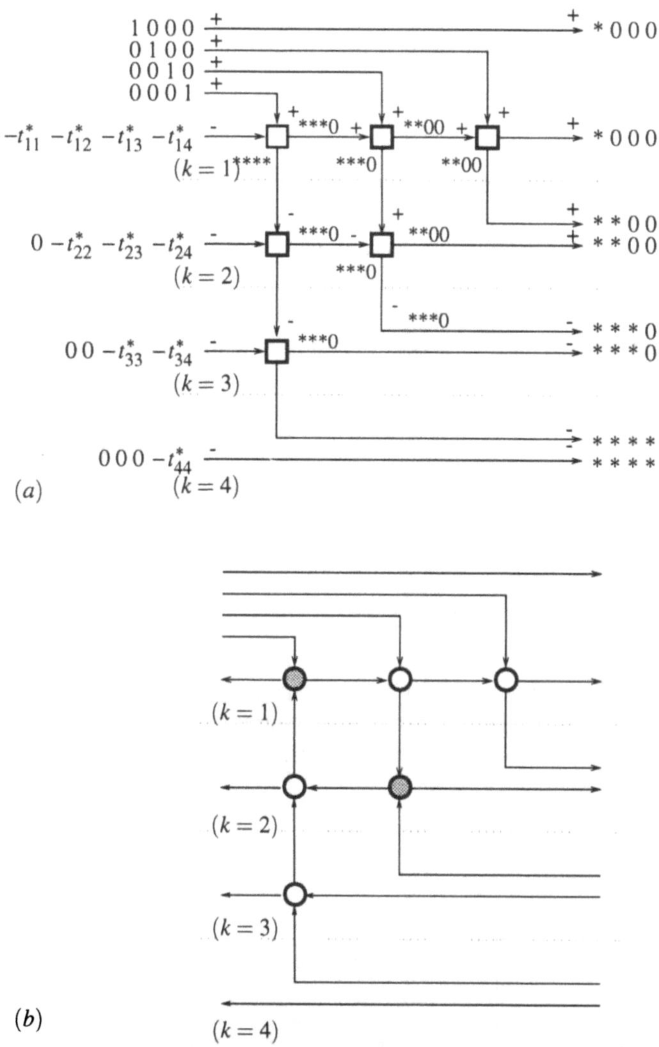

Figure 10.9. Computational structure of a recursive solution to the interpolating problem. (a) $\tilde{\Theta}$, with elementary rotations of mixed type (both circular and hyperbolic); (b) one possible corresponding Σ, with circular elementary rotations. The type of sections in (a) and the signal flow in (b) depend on the data of the interpolation problem. The rotations which cause an upward arrow (ultimately: a state for T_a) are shaded.

interpolation problem and leads to an approximant $T_a = 0$. Our situation is more general. If $|u| < |t|$, we require a rotational section of the form

$$\tilde{\theta} = \begin{bmatrix} -s & 1 \\ 1 & -s^* \end{bmatrix} \frac{1}{c^*},$$

resulting in $[u \;\; t]\tilde{\theta} = [* \;\; 0]$. $\tilde{\theta}$ has signature pairs determined by

$$\tilde{\theta}^* \begin{bmatrix} 1 & \\ & -1 \end{bmatrix} \tilde{\theta} = \begin{bmatrix} -1 & \\ & 1 \end{bmatrix}.$$

This shows that the signature of the 'energy' of the output vector of such a section is reversed: if $[a_1 \;\; b_1]\theta_2 = [a_2 \;\; b_2]$, then $a_1a_1^* - b_1b_1^* = -a_2a_2^* + b_2b_2^*$. Instead of ordinary (j_1, j_2)-unitary elementary rotations, we thus have to work with J-unitary rotations $\tilde{\theta}$ with respect to *unsorted* signature matrices $(\tilde{j}_1, \tilde{j}_2)$.

Because the signature can be reversed at each elementary step, we have to keep track of it to ensure that the resulting global Θ-matrix is J-unitary with respect to a certain signature. Thus assign to each column in $[U^* \;\; -T^*]$ a signature (+1 or −1), which is updated after each elementary operation, in accordance to the type of rotation. Initially, the signature of the columns of U^* is chosen +1, and those of $-T^*$ are chosen −1. Because $\tilde{\Theta} = \tilde{\Theta}_{(1)}\tilde{\Theta}_{(2)}\cdots\tilde{\Theta}_{(n)}$, where $\tilde{\Theta}_{(i)}$ is an embedding of the i-th elementary rotation $\tilde{\theta}_{(i)}$ into one of full size, it is seen that keeping track of the signature at each intermediate step ensures that

$$\tilde{\Theta}^* \begin{bmatrix} I & \\ & -I \end{bmatrix} \tilde{\Theta} = \tilde{J}_2,$$

where $\tilde{J}_2$ is the unsorted signature matrix given by the signatures of the columns of the final resulting upper triangular matrices. The types of signatures that can occur, and the appropriate elementary rotations $\tilde{\theta}$ to use, are listed below. These form the processors in figure 10.9(a).

1. $\begin{bmatrix} \overset{+}{u} & \overset{-}{t} \end{bmatrix} \begin{bmatrix} 1 & -s \\ -s^* & 1 \end{bmatrix} \frac{1}{c^*} = \begin{bmatrix} \overset{+}{*} & \overset{-}{0} \end{bmatrix},$ if $|u| > |t|$

2. $\begin{bmatrix} \overset{+}{u} & \overset{-}{t} \end{bmatrix} \begin{bmatrix} -s & 1 \\ 1 & -s^* \end{bmatrix} \frac{1}{c^*} = \begin{bmatrix} \overset{-}{*} & \overset{+}{0} \end{bmatrix},$ if $|u| < |t|$

3. $\begin{bmatrix} \overset{-}{u} & \overset{+}{t} \end{bmatrix} \begin{bmatrix} -s & 1 \\ 1 & -s^* \end{bmatrix} \frac{1}{c^*} = \begin{bmatrix} \overset{+}{*} & \overset{-}{0} \end{bmatrix},$ if $|u| < |t|$

4. $\begin{bmatrix} \overset{-}{u} & \overset{+}{t} \end{bmatrix} \begin{bmatrix} 1 & -s \\ -s^* & 1 \end{bmatrix} \frac{1}{c^*} = \begin{bmatrix} \overset{-}{*} & \overset{+}{0} \end{bmatrix},$ if $|u| > |t|$

5. $\begin{bmatrix} \overset{+}{u} & \overset{+}{t} \end{bmatrix} \begin{bmatrix} c & s \\ -s^* & c^* \end{bmatrix} = \begin{bmatrix} \overset{+}{*} & \overset{+}{0} \end{bmatrix}$

6. $\begin{bmatrix} \overset{-}{u} & \overset{-}{t} \end{bmatrix} \begin{bmatrix} c & s \\ -s^* & c^* \end{bmatrix} = \begin{bmatrix} \overset{-}{*} & \overset{-}{0} \end{bmatrix}$

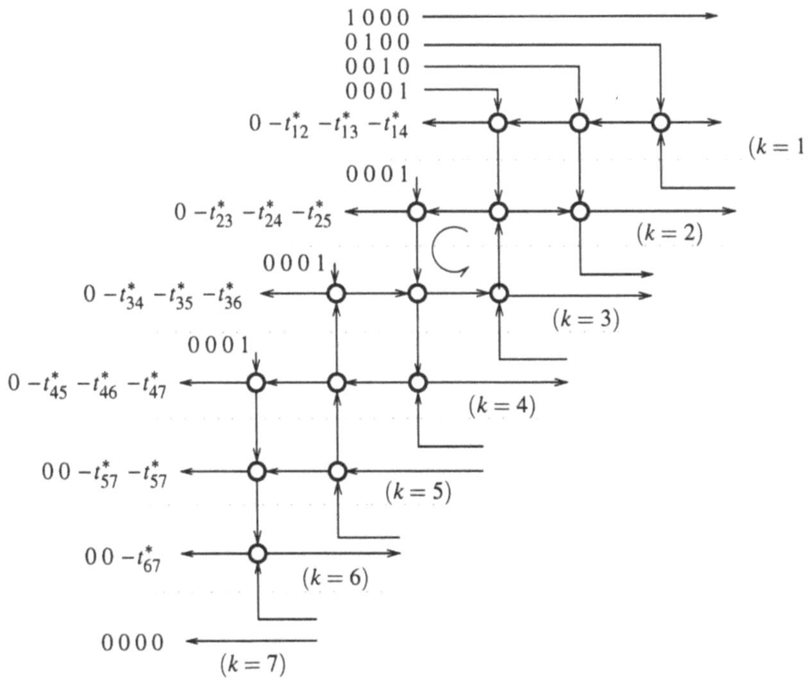

Figure 10.10. Computational network of an interpolating Σ-matrix of a band-matrix (7×7 matrix, band width 3).

(The case $|u| = |t|$ could occur, which leads to an exception.) We can associate, as usual, with each J-unitary rotation a corresponding unitary rotation, which is obtained by rewriting the corresponding equations such that the '+' quantities appear on the left-hand side and the '−' quantities on the right-hand side. The last two sections are already circular rotation matrices. By replacing each of the sections of Θ by the corresponding unitary section, a unitary Σ matrix that corresponds to Θ is obtained. A signal flow scheme of a possible Σ in our 4×4 example is depicted in figure 10.9(b). The matching of signatures at each elementary rotation in the algorithm effects in figure 10.9(b) that the signal flow is well defined: an arrow leaving some section will not bounce into a signal flow arrow that leaves a neighboring section.

Finally, a solution to the interpolation problem $[U^* \; -T^*]\Theta = [A' \; -B']$ is obtained by *sorting* the columns of the resulting upper triangular matrices obtained by the above procedure according to their signature, such that all positive signs correspond to A' and all negative signs to B'. The columns of Θ are sorted likewise. The solution that is obtained this way is reminiscent of the state-space solution in the previous section, and in fact can be derived from it by factoring Θ into elementary operations as above. Again, the network of Σ is not computable since it contains loops.

To give an example of the foregoing, suppose that T is a band matrix. It may be verified that computations on entries off the band reduce to identity operations and can therefore be omitted. The corresponding computational scheme is, for a typical example, depicted in figure 10.10. A number of '0' entries that are needed to match the sequences in the correct way have been suppressed in the figure: as many trailing '0's as needed must be postpended to make each sequence have length 7. The recursive procedure can be specialized even further to handle staircase matrices as well, for which even more of the elementary computations are rendered trivial and can be omitted. The structure of the diagram will reflect the structure of the staircase.

The recursion and the resulting computational network is a further generalization (to include indefinite interpolation) of the generalized Schur algorithm introduced in [DD88]. However, the formalism by which the matrices are set up to initiate the algorithm is new.

Computation of the approximant

With Θ and B' available, there are various ways to obtain the Hankel norm approximant T_a. The basic relations are given in terms of T' (the upper triangular part of which is equal to T_a) and the operator Σ associated to Θ:

$$\begin{aligned} T'^* &= T^* + U^*\Sigma_{12} \\ T'^* &= B'\Theta_{22}^{-1}, \qquad \Theta_{22}^{-1} = \Sigma_{22}\,. \end{aligned}$$

Ideally, one would want to use the computational network of Σ to derive either $U^*\Sigma_{12}$ or $B'\Theta_{22}^{-1}$. However, the network that has been constructed in the previous step of the algorithm is not *computable*: it contains delay-free loops, and hence it cannot be used directly. A straightforward alternative is to extract Θ_{22} from the network of Θ (by applying an input of the form $[0 \;\; I]$), and subsequently use any technique to invert this matrix and apply it to B'. A second alternative is to work with the (non-causal) state realization for Σ which is available at this point. From this one can derive a realization for the upper triangular part of Θ_{22}^{-*}, by using the recursions given in section 10.3.

The first solution can be made more or less 'in style' with the way Θ has been constructed, to the level that only elementary, unitary operations are used. However, the overall solution is a bit crude: after extracting the matrix Θ_{22}, the computational network of Θ is discarded, although it reveals the structure of Θ_{22} and Θ_{22}^{-1}, and the algorithm continues with a matrix inversion technique that is not very specific to its current application. The state-space technique, on the other hand, uses half of the computational network structure of Θ (the 'vertical' segmentation into stages), but does not use the structure within a stage. The algorithm operates on (state-space) matrices, rather than at the elementary level, and is in this respect 'out of style' with the recursive computation of Θ. It is as yet unclear whether an algorithm can be devised that acts directly on the computational network of Θ using elementary operations.

10.6 THE NEHARI PROBLEM

The classical Nehari problem is to determine the distance — in the infinity norm — of a given scalar function in L_∞ to the space of bounded analytical functions H_∞ [Neh57,

AAK71]. Put in another way, it asks to extend a given analytic function to a function in L_∞ such that the norm of the result is as small as possible. Usually, a sub-optimal version of the problem is defined: the norm of the result should be smaller than a given bound.

For time-invariant systems, the solutions are well-known and derived using interpolation or Beurling-Lax representation theory. For time-varying systems, an early statement and proof appears in the work of Arveson [Arv75, thm. 1.1] on operators in a nest algebra. A comparable result has been obtained by Gohberg, Kaashoek and Woerdeman [Woe89, GKW89, GMW91] in the context of block matrix and operator matrix extensions. Their solutions are recursive on the entries of the block matrix: it is possible to work from top to bottom, adding rows to the extension found so far, in such a way that the resulting matrices remain contractive. The time-varying Nehari problem was perhaps first solved in [DvdV93]. An independent solution appears in [HI94], which however assumes the invertibility of A.

Placed in our context, the Nehari problem is to find, if it exists, an extension to a given operator $T \in \mathcal{U}$ to $T' \in \mathcal{X}$ such that the norm of T' is as small as possible, or smaller than a given bound. The theorems given in section 10.2 contain an implicit solution of such a problem, for operators T which have a u.e. stable, uniformly observable realization. If Γ in (10.7) is chosen such that all local Hankel singular values are uniformly smaller than 1, then $T' = (B'\Theta_{22}^{-1}\Gamma)^*$ obtained through lemma 10.3 is a lower ($\in \mathcal{L}$) operator and the state sequence x_- is of dimension zero: $\#(\mathcal{B}_-) = 0$ and $J_{\mathcal{B}} = I$. Such a T' is known as the Nehari extension of T: it is such that $\|\Gamma^{-1}(T-T')\| \le 1$ so that, when $\|\Gamma^{-1}T\|_H < 1$, there exists an extension $E \in \mathcal{X}$ such that the upper part of E is equal to $\Gamma^{-1}T$ and E is contractive. The Nehari problem is to find E or, equivalently, T'. This problem can also be viewed as a distance problem: given $T \in \mathcal{ZU}$, find an operator $T' \in \mathcal{L}$ that is closest to it, in the sense that $\|T-T'\|$ is minimized.

Theorem 10.19 *If T is a bounded upper operator which has a locally finite u.e. stable and uniformly observable realization, then*

$$\|T\|_H = \inf_{T' \in \mathcal{L}} \|T-T'\|. \tag{10.32}$$

PROOF Let $d = \|T\|_H$ and consider the operator $(d+\varepsilon)^{-1}T$ for some $\varepsilon > 0$. Then, with $\Gamma = d+\varepsilon$, $r := \|(d+\varepsilon)^{-1}\Gamma^{-1}T\|_H < 1$ and lemma 10.4 applies. Since the largest singular value of any local Hankel operator of $(d+\varepsilon)^{-1}T$ is majorized by r, we have that the sequence of singular values larger than one is zero, so that $\Theta_{22}^{-1} \in \mathcal{U}$ and $T' = (B'\Theta_{22}^{-1}(d+\varepsilon))^*$ is a lower operator. Lemma 10.4 ensures that

$$\|(d+\varepsilon)^{-1}(T-T')\| \le 1$$

by construction, and hence

$$\|T-T'\| \le d+\varepsilon.$$

Letting $\varepsilon \downarrow 0$ achieves (10.32). The reverse inequality is obvious from proposition 10.2.

□

All possible Nehari extensions are parameterized by the set of contractive upper operators S_L, as a special case of theorem 10.18.

A state-space realization of the "maximum entropy" or "central" Nehari extension T' for which $S_L = 0$ can be obtained as a special instance of the method presented in section 10.3, and does not need the upward recursions because the dimension of x_- is zero. The result is a closed-form solution: it is specified solely in terms of the given state realization operators of T.

Theorem 10.20 *Let $T \in \mathcal{U}$ be a strictly upper locally finite operator with realization $\{A,B,C,0\}$ in output normal form. If $\|T\|_H < 1$ then T has a Nehari extension $E = T - T' \in \mathcal{X}$ such that E is contractive and the strictly upper part of E is equal to T (i.e., $T'^* \in \mathcal{U}$). A realization of T'^*, i.e., the upper part of $-E^*$, is given by*

$$\begin{aligned} A_e &= A(I-(I-A^*MA)^{-1}B^*B) \\ B_e &= C^*MA(I-(I-A^*MA)^{-1}B^*B) \\ C_e &= A(I-A^*MA)^{-1}B^* \\ D_e &= C^*MA(I-A^*MA)^{-1}B^* \end{aligned} \tag{10.33}$$

where M satisfies $M^{(-1)} = A^*MA + B^*B$.

PROOF The existence of the Nehari extension has already been proven: with $\Gamma = I$, it suffices to take $T'^* = B'\Theta_{22}^{-1}$, where B' and Θ are as in lemma 10.3 and 10.4. Let B_U and D_U be such that

$$\mathbf{U} = \begin{bmatrix} A & C \\ B_U & D_U \end{bmatrix}$$

is a unitary realization of the inner factor U of the external factorization of T. The realization $\boldsymbol{\Theta}$ has the general form of equation (10.13) (with $\Gamma = I$), but since $J_{\mathcal{B}} = I$, all negative signature is associated with D_{22}, which implies that D_{22}^{-1} exists and is bounded, and also that D_{21} can be chosen equal to zero (as in [DD92, thm. 3.1]). Hence we consider a realization of Θ of the form

$$\boldsymbol{\Theta} = \begin{bmatrix} R & & \\ & I & \\ & & I \end{bmatrix} \left[\begin{array}{c|cc} A & C_1 & C_2 \\ \hline B_U & D_{11} & D_{12} \\ B & 0 & D_{22} \end{array}\right] \begin{bmatrix} R^{-(-1)} & & \\ & I & \\ & & I \end{bmatrix}$$

where the first column of the operator matrix in the middle is specified, and an extension by a second and third column is to be determined, as well as a state transformation R, such that $\boldsymbol{\Theta}$ is J-unitary. We use the fact that $\mathbf{U}$ is unitary to derive expressions for entries in $\boldsymbol{\Theta}$. Let, as before, Λ be the J-Gram operator, which is here equal to $\Lambda = R^*R$ (recall that $J_{\mathcal{B}} = I$). The remainder of the proof consists of 6 steps.

1. $$\begin{aligned} C_1 &= \Lambda^{-1}C\alpha, \\ D_{11} &= D_U\alpha, \qquad \text{where } \alpha = (C^*\Lambda^{-1}C + D_U^*D_U)^{-1/2}. \end{aligned}$$

   PROOF The J-unitarity relations between the first and second block column of $\boldsymbol{\Theta}$ lead to

   $$\begin{aligned} A^*\Lambda C_1 + B_U^*D_{11} &= 0 \\ C_1^*\Lambda C_1 + D_{11}^*D_{11} &= I. \end{aligned}$$

The first equation shows that, for some scaling α,

$$\begin{bmatrix} \Lambda C_1 \\ D_{11} \end{bmatrix} = \begin{bmatrix} A \\ B_U \end{bmatrix}^{\perp} \alpha = \begin{bmatrix} C \\ D_U \end{bmatrix} \alpha .$$

The scaling α follows from the second equation.

2. $C_2^* C + D_{12}^* D_U = 0$.

PROOF The J-unitarity conditions between the second and third column lead to

$$\begin{array}{rcl} C_1^* \Lambda C_2 + D_{11}^* D_{12} & = & 0 \\ \Rightarrow \quad \alpha^* C^* C_2 + \alpha^* D_U^* D_{12} & = & 0 . \end{array}$$

3. $B' = C^* M (I - AZ)^{-1} C_2$.

PROOF A state-space model of B' was given in equation (10.23) as

$$B' = \left\{ -D_U^* D_{12} - C^* (I - M) C_2 \right\} + C^* MAZ (I - AZ)^{-1} C_2 .$$

Using the result of step 2 gives the intended simplification.

4. $T'^* = B' \Theta_{22}^{-1} = C^* M (I - [A - C_2 D_{22}^{-1} B] Z)^{-1} C_2 D_{22}^{-1}$.

PROOF Let $A_e = A - C_2 D_{22}^{-1} B$. Then, because $\Theta_{22}^{-1} \in \mathcal{U}$,

$$\begin{array}{rcl} T'^* = B' \Theta_{22}^{-1} & = & [C^* M (I - AZ)^{-1} C_2] \, [D_{22}^{-1} - D_{22}^{-1} BZ (I - A_e Z)^{-1} C_2 D_{22}^{-1}] \\ & = & C^* M (I - AZ)^{-1} \left[I - C_2 D_{22}^{-1} BZ (I - A_e Z)^{-1} \right] C_2 D_{22}^{-1} \\ & = & C^* M (I - AZ)^{-1} \left[(I - A_e Z) - C_2 D_{22}^{-1} BZ \right] (I - A_e Z)^{-1} C_2 D_{22}^{-1} \\ & = & C^* M (I - AZ)^{-1} (I - AZ)(I - A_e Z)^{-1} C_2 D_{22}^{-1} . \end{array}$$

5. $C_2 D_{22}^{-1} = A (I - A^* MA)^{-1} B^*$.

PROOF The J-unitarity conditions imply

$$\begin{bmatrix} A & C_1 \\ B_U & D_{11} \\ B & 0 \end{bmatrix}^* \begin{bmatrix} \Lambda & & \\ & I & \\ & & -I \end{bmatrix} \begin{bmatrix} C_2 \\ D_{12} \\ D_{22} \end{bmatrix} = 0$$

$$\Rightarrow \quad \begin{bmatrix} A & C_1 \\ B_U & D_{11} \end{bmatrix}^* \begin{bmatrix} \Lambda & \\ & I \end{bmatrix} \begin{bmatrix} C_2 \\ D_{12} \end{bmatrix} = \begin{bmatrix} B^* \\ 0 \end{bmatrix} D_{22}$$

$$\begin{array}{rcl} \Rightarrow \quad \begin{bmatrix} C_2 D_{22}^{-1} \\ D_{12} D_{22}^{-1} \end{bmatrix} & = & \begin{bmatrix} \Lambda^{-1} & \\ & I \end{bmatrix} \begin{bmatrix} A & C_1 \\ B_U & D_{11} \end{bmatrix}^{-*} \begin{bmatrix} B^* \\ 0 \end{bmatrix} \\ & = & \begin{bmatrix} \Lambda^{-1} & \\ & I \end{bmatrix} \begin{bmatrix} \Lambda A & C\alpha \\ B_U & D_U \alpha \end{bmatrix} \begin{bmatrix} (\Lambda^{(-1)} + B^* B)^{-1} & \\ & I \end{bmatrix} \begin{bmatrix} B^* \\ 0 \end{bmatrix} \\ & = & \begin{bmatrix} A (\Lambda^{(-1)} + B^* B)^{-1} B^* \\ B_U (\Lambda^{(-1)} + B^* B)^{-1} B^* \end{bmatrix} \end{array}$$

where we have used the fact that

$$\begin{bmatrix} A^* & B_U^* \\ C_1^* & D_{11}^* \end{bmatrix} \begin{bmatrix} \Lambda A & C\alpha \\ B_U & D_U\alpha \end{bmatrix} = \begin{bmatrix} \Lambda^{(-1)} + B^*B & \\ & I \end{bmatrix}$$

Finally, using $M = I - \Lambda$, where M satisfies $M^{(-1)} = A^*MA + B^*B$ gives $\Lambda^{(-1)} + B^*B = I - A^*MA$.

6. $T'^* = D_e + B_eZ(I - A_eZ)^{-1}C_e$, where $\{A_e, B_e, C_e, D_e\}$ are as in equation (10.33).

   PROOF From step 4,

$$\begin{aligned} T'^* &= C^*M(I-A_eZ)^{-1}C_2D_{22}^{-1} \\ &= C^*MC_2D_{22}^{-1} + C^*MA_eZ(I-A_eZ)^{-1}C_2D_{22}^{-1} \end{aligned}$$

   where $A_e = A - C_2D_{22}^{-1}B$. It remains to substitute $C_2D_{22}^{-1} = A(I - A^*MA)^{-1}B^*$. □

Numerical example

We illustrate theorem 10.20 with a numerical example. Let T be given by the strictly upper matrix

$$T = \begin{bmatrix} 0 & .326 & .566 & .334 & .078 & -.008 & -.012 & -.003 \\ 0 & 0 & .326 & .566 & .334 & .078 & -.008 & -.012 \\ 0 & 0 & 0 & .326 & .566 & .334 & .078 & -.008 \\ 0 & 0 & 0 & 0 & .326 & .566 & .334 & .078 \\ 0 & 0 & 0 & 0 & 0 & .326 & .566 & .334 \\ 0 & 0 & 0 & 0 & 0 & 0 & .326 & .566 \\ 0 & 0 & 0 & 0 & 0 & 0 & 0 & .326 \\ 0 & 0 & 0 & 0 & 0 & 0 & 0 & 0 \end{bmatrix}.$$

The norm of T is computed as $\| T \| = 1.215$, and T has Hankel singular values equal to

H_1	H_2	H_3	H_4	H_5	H_6	H_7	H_8
	.7385	.9463	.9856	.9866	.9856	.9463	.7385
		.2980	.3605	.3661	.3605	.2980	
			.0256	.0284	.0256		

so that $\| T \|_H = .9866 < 1$. The objective is to extend T with a lower triangular part such that the operator norm of the result is less than 1.

A realization for T is obtained via algorithm 3.9 as

$$\mathbf{T}_1 = \left[\begin{array}{c|c} \cdot & \cdot \\ \hline -.739 & .000 \end{array}\right] \qquad \mathbf{T}_2 = \left[\begin{array}{cc|c} .733 & -.517 & -.442 \\ \hline -.738 & .000 & .000 \end{array}\right]$$

$$\mathbf{T}_3 = \left[\begin{array}{ccc|c} .733 & -.517 & .000 & -.442 \\ .508 & -.012 & -.084 & .857 \\ \hline -.738 & -.000 & .000 & .000 \end{array}\right] \qquad \mathbf{T}_4 = \left[\begin{array}{ccc|c} .733 & -.517 & -.000 & -.442 \\ .508 & -.012 & -.084 & .857 \\ .430 & .836 & -.212 & -.265 \\ \hline -.738 & -.000 & .000 & .000 \end{array}\right]$$

$$\mathbf{T}_5 = \left[\begin{array}{ccc|c} .738 & -.509 & .000 & -.442 \\ .509 & -.005 & .076 & .857 \\ .424 & .845 & .192 & -.264 \\ \hline -.734 & .000 & .000 & .000 \end{array}\right] \qquad \mathbf{T}_6 = \left[\begin{array}{cc|c} .780 & .441 & -.444 \\ .506 & -.026 & .862 \\ -.369 & .897 & .244 \\ \hline -.654 & -.000 & .000 \end{array}\right]$$

$$\mathbf{T}_7 = \left[\begin{array}{c|c} -.867 & -.499 \\ .499 & -.867 \\ \hline .326 & .000 \end{array}\right] \qquad \mathbf{T}_8 = \left[\begin{array}{c|c} \cdot & 1.000 \\ \hline \cdot & .000 \end{array}\right]$$

Theorem 10.20 gives a realization of T'^* as

$$\mathbf{T}'_1 = \left[\begin{array}{c|c} \cdot & \cdot \\ \hline .000 & .000 \end{array}\right] \qquad \mathbf{T}'_2 = \left[\begin{array}{cc|c} .021 & -.517 & -.965 \\ \hline -.005 & .125 & .233 \end{array}\right]$$

$$\mathbf{T}'_3 = \left[\begin{array}{ccc|c} -.124 & -.517 & .000 & -1.161 \\ .025 & -.012 & -.084 & -.654 \\ \hline .074 & .281 & -.018 & .494 \end{array}\right] \qquad \mathbf{T}'_4 = \left[\begin{array}{ccc|c} -.130 & -.517 & -.000 & -1.168 \\ .023 & -.012 & -.084 & -.657 \\ .296 & .836 & -.212 & -.183 \\ \hline .084 & .306 & -.024 & .519 \end{array}\right]$$

$$\mathbf{T}'_5 = \left[\begin{array}{ccc|c} -.121 & -.509 & -.000 & -1.171 \\ .027 & -.005 & .076 & -.656 \\ .295 & .845 & .192 & -.176 \\ \hline .080 & .303 & .021 & .519 \end{array}\right] \qquad \mathbf{T}'_6 = \left[\begin{array}{cc|c} .062 & .441 & -1.098 \\ .114 & -.026 & -.601 \\ -.289 & .897 & .122 \\ \hline -.006 & -.272 & .494 \end{array}\right]$$

$$\mathbf{T}'_7 = \left[\begin{array}{c|c} -.702 & -.504 \\ .404 & .290 \\ \hline .324 & .233 \end{array}\right] \qquad \mathbf{T}'_8 = \left[\begin{array}{c|c} \cdot & .000 \\ \hline \cdot & .000 \end{array}\right]$$

and the resulting Nehari extension follows as

$$E = T - T' = \begin{bmatrix} 0 & .326 & .566 & .334 & .078 & -.008 & -.012 & -.003 \\ 0 & -.233 & .326 & .566 & .334 & .078 & -.008 & -.012 \\ 0 & .076 & -.494 & .326 & .566 & .334 & .078 & -.008 \\ 0 & .003 & .267 & -.519 & .326 & .566 & .334 & .078 \\ 0 & -.011 & -.050 & .295 & -.519 & .326 & .566 & .334 \\ 0 & .003 & -.013 & -.050 & .267 & -.494 & .326 & .566 \\ 0 & .000 & .003 & -.011 & .003 & .076 & -.233 & .326 \\ 0 & 0 & 0 & 0 & 0 & 0 & 0 & 0 \end{bmatrix}$$

E is indeed contractive: $\|E\| = .9932$.

10.7 CONCLUDING REMARKS

In this chapter, we have presented an approximation scheme to derive, for a given upper triangular matrix T, a Hankel-norm approximant T_a of lower complexity. A model of T_a can be computed starting from a high-order model of T (obtained *e.g.*, by algorithm 3.9) by applying algorithm 10.5. However, the derivation of a model for T can be computationally intensive: it involves a sequence of SVDs to compute the relevant subspaces. An alternative approach is via the algorithm discussed in section 10.5, which acts directly on the entries of T. Only local computations are needed to obtain Θ and B'. Further research is required to efficiently compute T_a as the upper part of $(B'\Theta_{22}^{-1})^*$: a direct computation is not really satisfactory in view of the fact that Θ is obtained in a factored form.

A second open problem is the selection of a suitable error tolerance matrix Γ. At present, one has to choose some Γ, which then results in an approximant with a certain complexity. It is, as yet, unclear how to obtain the reverse, *i.e.*, how to derive, for a given desired complexity of the approximant, the tolerance Γ that will achieve this ·omplexity.

11 LOW-RANK MATRIX APPROXIMATION AND SUBSPACE TRACKING

The usual way to compute a low-rank approximant of a matrix H is to take its singular value decomposition (SVD) and truncate it by setting the small singular values equal to 0. However, the SVD is computationally expensive. Using the Hankel-norm model reduction techniques in chapter 10, we can devise a much simpler generalized Schur-type algorithm to compute similar low-rank approximants. Since rank approximation plays an important role in many linear algebra applications, we devote an independent chapter to this topic, even though this leads to some overlap with previous chapters.

For a given matrix H which has d singular values larger than γ, we find all rank d approximants $\hat{H}$ such that $H - \hat{H}$ has operator norm (matrix 2-norm) less than γ. The set of approximants includes the truncated SVD approximation. The advantages of the Schur algorithm are that it has a much lower computational complexity (similar to a QR factorization), and directly produces a description of the column space of the approximants. This column space can be updated and downdated in an on-line scheme, amenable to implementation on a parallel array of processors.

11.1 INTRODUCTION

Fast adaptive subspace estimation plays an increasingly important role in modern signal processing. It forms the key ingredient in many sensor array signal processing algorithms, system identification, and several recently derived blind signal separation and equalization algorithms (*e.g.*, [MDCM95, Slo94, vdVP96, vdVTP97]).

The generic subspace estimation problem in these applications might be stated as follows. Suppose that we are given a matrix $H : m \times n$, consisting of measurement data which becomes available column-by-column. Furtermore, suppose that it satisfies the model $H = \tilde{H} + \tilde{N}$, where $\tilde{H}$ is a low rank matrix and $\tilde{N}$ is a disturbance. Knowing only H, we can try to estimate $\tilde{H}$ by solving

$$\min_{\hat{H}} \|H - \hat{H}\| \quad \text{s.t. } \operatorname{rank}(\hat{H}) = d \tag{11.1}$$

where $\|\cdot\|$ denotes the matrix 2-norm (largest singular value). The value of the rank d is either given or is estimated from the singular values of H. The usual truncated SVD (TSVD) solution is to set all but the largest d singular values of H equal to zero. In subspace estimation, we are primarily interested in the column span of $\tilde{H}$. For the TSVD solution, this space is estimated by the span of the first d left singular vectors of H, the so-called principal subspace.

Continuing efforts on SVD algorithms have reduced its computational complexity to be mainly that of reducing a matrix to a bidiagonal form: not much more than the complexity of a QR factorization. However, a remaining disadvantage of the SVD in demanding applications is that it is difficult to update the decomposition for a growing number of columns of H. Indeed, there are important applications in signal processing (e.g. adaptive beamforming, model identification, adaptive least squares filters) that require on-line estimation of the principal subspace, for growing values of n. A number of other methods have been developed that alleviate the computational requirements, yet retain important information such as numerical rank and principal subspaces. Some of these techniques are the URV decomposition [Ste92], which is a rank revealing form of a complete orthogonal decomposition [GV89], and the rank revealing QR decomposition (RRQR), [Fos86, Cha87, CH90, CH92, BS92, CI94], see [CI94] for a review. Both the RRQR and the URV algorithms require estimates of the conditioning of certain submatrices at every step of the iteration. This is a global and data-dependent operation: not a very attractive feature. The SVD and URV decomposition can be updated [BN78, Ste92], which is still an iterative process, although it has been shown recently that a simpler scheme is feasible if the updating vectors satisfy certain stationarity assumptions [MVV92, MDV93]. An initial computation of the RRQR consists of an ordinary QR, followed by an iteration that makes the decomposition rank revealing. As a one-sided decomposition, the RRQR is easier to update than an SVD, but also requires (incremental) condition estimations at each updating step. Alternatively, there are efficient subspace tracking algorithms which under stationary conditions gradually converge towards the principal subspace, *e.g.*, [Yan95].

As an alternative, we consider a technique based on the Hankel-norm approximation theory of chapter 10. It is based on the knowledge of an upper bound to the noise, $\|\tilde{N}\| \le \gamma$, and gives a parametrization for all $\hat{H}$ that satisfy

$$\min_{\hat{H}} \operatorname{rank}(\hat{H}) \quad \text{s.t.} \quad \|H - \hat{H}\| \le \gamma. \tag{11.2}$$

It is readily shown that the resulting approximants $\hat{H}$ have rank d, where d is equal to the number of singular values of H that are larger than γ. The TSVD is within the class, but it is not explicitly identified. The prime advantage of the resulting technique is that it gives subspace estimates that have the correct dimension and a known performance (projecting H onto the estimated subspace gives an $\hat{H}$ such that $\| H-\hat{H} \| \le \gamma$), but are substantially easier to compute and update than the TSVD.

The connection to the theory in chapter 10 is obtained by looking at a special case of our usual operator $T \in \mathcal{U}(\mathcal{M},\mathcal{N})$, in which

$$\begin{array}{rclcccccccccc} \mathcal{M} & = & \cdots & \oplus & \emptyset & \oplus & \mathcal{M}_1 & \oplus & \emptyset & \oplus & \emptyset & \oplus \cdots \\ \mathcal{N} & = & \cdots & \oplus & \emptyset & \oplus & \emptyset & \oplus & \mathcal{N}_2 & \oplus & \emptyset & \oplus \cdots . \end{array}$$

A matrix $T \in \mathcal{U}$ has the form

$$T = \begin{bmatrix} \cdot & \cdot & \cdot & \cdot \\ & \cdot & T_{12} & \vdots \\ & & \cdot & \vdots \\ & & & \cdot \end{bmatrix} \equiv [T_{12}] \, ,$$

that is, $T = T_{12}$ is just any matrix of any size. Its only nonzero Hankel matrix is $H = T_{12}$. In this chapter, we work out the implications of this specialization.

The computation of the "Schur subspace estimators" (SSE) that result from this technique is based on an implicit signed Cholesky factorization

$$HH^* - \gamma^2 I \;=:\; BB^* - AA^*$$

where A, B have minimal dimensions. Thus, the spectrum of HH^* is shifted such that the small eigenvalues become negative, which enables their separation from the large eigenvalues. It is readily shown from inertia considerations that, even though A and B are not unique, if H has d singular values larger than γ and $m-d$ less than γ, then B has d columns and A has $m-d$ columns. The main result in this chapter is that, *for any such pair* (A,B), *all principal subspace estimates leading to approximants* $\hat{H}$ *satisfying (11.2) are given by the column span of* $B-AM$, *where* M *is any matrix of compatible size with* $\|M\| \le 1$. The factorization can be computed via a hyperbolic factorization

$$[\gamma I \quad H]\Theta \;=\; [(A \;\; 0) \quad (B \;\; 0)]$$

where Θ is a J-unitary matrix.

Straightforward generalizations are possible. Suppose that instead of $\|\tilde{N}\| < \gamma$, we know $\tilde{N}\tilde{N}^* \le \gamma^2 R_N$, where R_N could be an estimate of the noise covariance matrix. An implicit factorization of $HH^* - \gamma^2 R_N$ leads to minimal rank approximants $\hat{H}$ such that $\| R_N^{-1/2}(H-\hat{H}) \| \le \gamma$. The subspace estimates are computed from $[N \;\; H]\Theta = [(A \;\; 0) \;\; (B \;\; 0)]$ where N is any matrix such that $NN^* = \gamma^2 R_N$, and are still given by the range of $B-AM$, for any $\|M\| \le 1$. Hence, without extra effort, we can take knowledge of the noise covariance matrix into account. Note that, asymptotically, a suitable N simply consists of scaled sample vectors of the noise process. If we have access to this process (or can estimate noise vectors via subtraction of the estimated $\tilde{H}$), then it is interesting to consider updating schemes for N as well as for H.

11.2 J-UNITARY MATRICES

At this point, we review and specialize some material on J-unitary matrices from earlier chapters. A square matrix Θ is J-unitary if it satisfies $\Theta^* J \Theta = J$, $\;\Theta J \Theta^* = J$, where J is

In: $[r \; x]$ with signature $\tilde{j}_1$; **out**: $\tilde{\theta}$, $\tilde{j}_2$ such that $[r \; x]\tilde{\theta} = [r' \; 0]$, $\tilde{\theta}\tilde{j}_2\tilde{\theta}^* = \tilde{j}_1$:

case 1. $\tilde{j}_1 = \begin{bmatrix} 1 & \\ & -1 \end{bmatrix}, |r| > |x| \;\Rightarrow\; \tilde{j}_2 = \begin{bmatrix} 1 & \\ & -1 \end{bmatrix}, \quad s = x/r,\ \tilde{\theta} = \begin{bmatrix} 1 & -s \\ -s^* & 1 \end{bmatrix} \frac{1}{c}$

case 2. $\tilde{j}_1 = \begin{bmatrix} 1 & \\ & -1 \end{bmatrix}, |r| < |x| \;\Rightarrow\; \tilde{j}_2 = \begin{bmatrix} -1 & \\ & 1 \end{bmatrix}, \quad s = r/x,\ \tilde{\theta} = \begin{bmatrix} -s^* & 1 \\ 1 & -s \end{bmatrix} \frac{1}{c}$

case 3. $\tilde{j}_1 = \begin{bmatrix} -1 & \\ & 1 \end{bmatrix}, |r| < |x| \;\Rightarrow\; \tilde{j}_2 = \begin{bmatrix} 1 & \\ & -1 \end{bmatrix}, \quad s = r/x,\ \tilde{\theta} = \begin{bmatrix} -s^* & 1 \\ 1 & -s \end{bmatrix} \frac{1}{c}$

case 4. $\tilde{j}_1 = \begin{bmatrix} -1 & \\ & 1 \end{bmatrix}, |r| > |x| \;\Rightarrow\; \tilde{j}_2 = \begin{bmatrix} -1 & \\ & 1 \end{bmatrix}, \quad s = x/r,\ \tilde{\theta} = \begin{bmatrix} 1 & -s \\ -s^* & 1 \end{bmatrix} \frac{1}{c}$

case 5. $\tilde{j}_1 = \begin{bmatrix} 1 & \\ & 1 \end{bmatrix} \;\Rightarrow\; \tilde{j}_2 = \begin{bmatrix} 1 & \\ & 1 \end{bmatrix}, \quad s = \frac{x}{\sqrt{|r|^2+|x|^2}},\ \tilde{\theta} = \begin{bmatrix} c^* & -s \\ s^* & c \end{bmatrix}$

case 6. $\tilde{j}_1 = \begin{bmatrix} -1 & \\ & -1 \end{bmatrix} \;\Rightarrow\; \tilde{j}_2 = \begin{bmatrix} -1 & \\ & -1 \end{bmatrix}, \quad s = \frac{x}{\sqrt{|r|^2+|x|^2}},\ \tilde{\theta} = \begin{bmatrix} c^* & -s \\ s^* & c \end{bmatrix}$

where $c = \sqrt{1-|s|^2}$

Figure 11.1. Elementary J-unitary zeroing rotations

a signature matrix which follows some prescribed $(p+q)\times(p+q)$ block-partitioning of Θ:

$$\Theta = \begin{matrix} & p & q \\ p & \Theta_{11} & \Theta_{12} \\ q & \Theta_{21} & \Theta_{22} \end{matrix}, \qquad J = \begin{bmatrix} I_p & \\ & -I_q \end{bmatrix}. \tag{11.3}$$

If Θ is applied to a block-partitioned matrix $[A \;\; B]$, then $[A \;\; B]\Theta = [C \;\; D] \Rightarrow AA^* - BB^* = CC^* - DD^*$. Hence, J associates a positive signature to the columns of A, C, and a negative signature to those of B, D.

For updating purposes, it is necessary to work with column permutations of $[A \;\; B]$ and $[C \;\; D]$, which induces row and column permutations of Θ. Thus we introduce matrices $\tilde{\Theta}$ that are J-unitary with respect to *unsorted* signature matrices $\tilde{J}$ (the tilde reminds of the absence of sorting), satisfying $\tilde{\Theta}^*\tilde{J}_1\tilde{\Theta} = \tilde{J}_2$, $\tilde{\Theta}\tilde{J}_2\tilde{\Theta}^* = \tilde{J}_1$, where $\tilde{J}_1$ and $\tilde{J}_2$ are diagonal matrices with diagonal entries equal to ± 1. If $M\tilde{\Theta} = N$, then $M\tilde{J}_1M^* = N\tilde{J}_2N^*$, so that $\tilde{J}_1$ associates its signature to the columns of M, and $\tilde{J}_2$ associates its signature to the columns of N. By inertia, the total number of positive entries in $\tilde{J}_1$ has to be equal to that in $\tilde{J}_2$, and likewise for the negative entries.

A 2×2 matrix $\tilde{\theta}$ is an elementary J-unitary rotation if it satisfies $\tilde{\theta}^*\tilde{j}_1\tilde{\theta} = \tilde{j}_2$, $\tilde{\theta}\tilde{j}_2\tilde{\theta}^* = \tilde{j}_1$, for unsorted signature matrices $\tilde{j}_1$, $\tilde{j}_2$. Similar to Givens rotations, it can be used to zero specific entries of vectors: for a given vector $[r \;\; x]$ and signature $\tilde{j}_1$, we can find $\tilde{\theta}$, r', and $\tilde{j}_2$ such that $[r \;\; x]\tilde{\theta} = [r' \;\; 0]$. The precise form that $\tilde{\theta}$ assumes depends on $\tilde{j}_1$ and whether $|r| > |x|$ or $|r| < |x|$, as listed in figure 11.1. Cases 5 and 6 in the table occur when $\tilde{j}_1$ is definite and lead to ordinary circular (unitary) rotations. Situations where $|r| = |x|$ with an indefinite signature $\tilde{j}_1$ are degenerate ($c = 0$): the result $[0 \;\; 0]$ is well defined but θ must be considered unbounded.

A matrix A is said to be $\tilde{J}$-*nonsingular*, with respect to a certain signature matrix $\tilde{J}$, if $A\tilde{J}A^*$ is nonsingular. It is immediate that if A is $\tilde{J}_1$-nonsingular and $\tilde{\Theta}$ is a $(\tilde{J}_1,\tilde{J}_2)$-unitary matrix, then $A\tilde{\Theta}$ is $\tilde{J}_2$-nonsingular. The following basic result claims that J-nonsingular matrices can be factored (*cf.* corollary 8.18):

Theorem 11.1 *A matrix $A : m\times(m+n)$ is $\tilde{J}_1$-nonsingular if and only if there exists a signature matrix $\tilde{J}_2$ and a $(\tilde{J}_1,\tilde{J}_2)$-unitary matrix $\tilde{\Theta}$ such that*

$$A\tilde{\Theta} = [X \quad 0_{m\times n}], \qquad X : m\times m, \text{ invertible.} \tag{11.4}$$

PROOF Sufficiency is obvious. As to necessity, assume that A is $\tilde{J}$-nonsingular. Then we can factor $A\tilde{J}_1A^*$ as

$$A\tilde{J}_1A^* = X\tilde{J}'X^*, \qquad X : m\times m, \text{ invertible,}$$

for some $m\times m$ signature matrix $\tilde{J}'$. This factorization exists and can in principle be computed from an LDU factorization with pivoting, or from an eigenvalue decomposition of $A\tilde{J}_1A^*$. Since A is $\tilde{J}_1$-nonsingular, it is also nonsingular in the ordinary sense, so that there exists a matrix $T : (m+n)\times m$, such that $AT = X$. T is not unique. Because X is invertible, we can take

$$T = \tilde{J}_1A^*(A\tilde{J}_1A^*)^{-1}X\,.$$

Using $(A\tilde{J}_1A^*)^{-1} = X^{-*}\tilde{J}'X^{-1}$, it is directly verified that this T satisfies $T^*\tilde{J}_1T = \tilde{J}'$. The remainder of the proof is technical: we have to show that T can be extended to a square, J-unitary matrix. For this, see the proof of lemma 8.16. □

Corollary 11.2 *Let $A : m\times(m+n)$ be $\tilde{J}_1$-nonsingular. Denote by $A_{1..i,\cdot}$ the submatrix of A, consisting of its first i rows. Then there exists a signature matrix $\tilde{J}_2$, and a $(\tilde{J}_1,\tilde{J}_2)$-unitary matrix $\tilde{\Theta}$ such that*

$$A\tilde{\Theta} = [R \quad 0_{m\times n}], \qquad R : m\times m, \text{ lower triangular, invertible}$$

if and only if $A_{1..i,\cdot}$ is $\tilde{J}_1$-nonsingular, for $i = 1,\ldots,m$. If the diagonal entries of R are chosen to be positive, then R is unique.

Such a factorization was proven in [BG81] for square matrices A and upper triangular R, but this result extends directly to the rectangular case. In [BG81], it was called the HR-decomposition, and it is also known as the hyperbolic QR factorization [OSB91].

11.3 APPROXIMATION THEORY

Central approximant

For a given $m\times n$ data matrix H and threshold γ, denote the SVD of H as

$$H = U\Sigma V^* = [U_1 \quad U_2]\begin{bmatrix} \Sigma_1 & \\ & \Sigma_2 \end{bmatrix}\begin{bmatrix} V_1^* \\ V_2^* \end{bmatrix} \qquad (\Sigma_1)_{ii} > \gamma, \quad (\Sigma_2)_{ii} \le \gamma. \tag{11.5}$$

Here, U and V are unitary matrices, and Σ is a diagonal matrix which contains the singular values σ_k of H. The matrices are partitioned such that Σ_1 contains the singular values that are strictly larger than γ, and Σ_2 contains those that are equal to or smaller than γ.

Suppose that d singular values of H are larger than γ, and that none of them are equal to γ. Our approximation theory is based on an implicit factorization of

$$HH^* - \gamma^2 I = BB^* - AA^* . \tag{11.6}$$

This is a Cholesky factorization of an indefinite Hermitian matrix. A and B are chosen to have full column rank. They are not unique, but by Sylvester's inertia law, their dimensions are well-defined. Using the SVD of H, we obtain one possible decomposition as

$$HH^* - \gamma^2 I = U_1(\Sigma_1^2 - \gamma^2 I)U_1^* + U_2(\Sigma_2^2 - \gamma^2 I)U_2^* ,$$

where the first term is positive semidefinite and has rank d, and the second term is negative semidefinite and has rank $m-d$. Hence, B has d columns, and A has $m-d$ columns.

To obtain an implicit factorization which avoids computing HH^*, we make use of theorem 11.1.

Theorem 11.3 *Let $H : m \times n$ have d singular values larger than γ, and none equal to γ. Then there exists a J-unitary matrix Θ such that*

$$[\gamma I_m \quad H]\Theta = [A' \quad B'] \tag{11.7}$$

where $A' = [A \quad 0_{m\times d}]$, $B' = [B \quad 0_{m\times n-d}]$, $A: m \times (m-d)$, $B: m \times d$, and $[A \quad B]$ is of full rank.

PROOF The matrix $[\gamma I_m \quad H]$ is J-nonsingular: by assumption, $\gamma^2 I - HH^*$ has d negative, $m-d$ positive, and no zero eigenvalues. Hence theorem 11.1 implies that there exists $\tilde{\Theta} : [\gamma I_m \quad H]\tilde{\Theta} = [X \quad 0_{m\times n}]$. The columns of X are the columns of $[A, B]$, in some permuted order, where A, B correspond to columns of X that have a positive or negative signature, respectively. After sorting the columns of $[X \quad 0]$ according to their signature, equation (11.7) results. □

Note that, by the preservation of J-inner products, equation (11.7) implies (11.6). From the factorization (11.7), we can immediately derive a 2-norm approximant satisfying the conditions in (11.2). To this end, partition Θ according to its signature J into 2×2 blocks, like in (11.3).

Theorem 11.4 *Let $H : m \times n$ have d singular values larger than γ, and none equal to γ. Define the factorization $[\gamma I_m \quad H]\Theta = [A' \quad B']$ as in theorem 11.3. Then*

$$\hat{H} = B'\Theta_{22}^{-1} \tag{11.8}$$

is a rank d approximant such that $\| H - \hat{H} \| < \gamma$.

PROOF $\hat{H}$ is well-defined because Θ_{22} is invertible (*cf.* theorem 8.2). It has rank d because $B' = [B \quad 0]$ has rank d. By equation (11.7), $B' = \gamma I\Theta_{12} + H\Theta_{22}$, hence $H - \hat{H} =$

$-\gamma\Theta_{12}\Theta_{22}^{-1}$. The proof follows from the fact that $\Theta_{12}\Theta_{22}^{-1}$ is contractive (theorem 8.2). □

We mentioned in the introduction that the column span (range) of the approximant is important in signal processing applications. From theorem 11.4, it is seen that this column span is equal to that of B: it is directly produced by the factorization. However, remark that $[A \;\; B]$ in (11.7) is not unique: for any J-unitary matrix Θ_1, $[A_1 \;\; B_1] = [A \;\; B]\Theta_1$ also satisfies $\gamma^2 I - HH^* = A_1 A_1^* - B_1 B_1^*$, and could also have been produced by the factorization. E.g., for some choices of Θ_1, we will have $\text{ran}(B) = \text{ran}(U_1)$, and $\text{ran}(A) = \text{ran}(U_2)$. Using Θ_1, we can find more approximants.

Parametrization of all approximants

We will now give a formula of all possible 2-norm approximants $\hat{H}$ of H of rank equal to d; there are no approximants of rank less than d. As usual, the set of all minimal-rank 2-norm approximants will be parametrized by matrices $S_L : m \times n$, with 2×2 block partitioning as

$$S_L = \begin{array}{c} \scriptstyle m-d \\ \scriptstyle d \end{array} \overset{\begin{array}{cc} \scriptstyle d & \scriptstyle n-d \end{array}}{\begin{bmatrix} (S_L)_{11} & (S_L)_{12} \\ (S_L)_{21} & (S_L)_{22} \end{bmatrix}}, \tag{11.9}$$

and satisfying the requirements

$$\begin{array}{ll} (i) & \text{contractive: } \|S_L\| \le 1, \\ (ii) & \text{block lower: } (S_L)_{12} = 0. \end{array} \tag{11.10}$$

The first condition on S_L will ensure that $\|H - \hat{H}\| \le \gamma$, whereas the second condition is required to have $\hat{H}$ of rank d.

Theorem 11.5 *With the notation and conditions of theorem 11.4, all rank d 2-norm approximants $\hat{H}$ of H are given by*

$$\hat{H} = (B' - A'S_L)(\Theta_{22} - \Theta_{21}S_L)^{-1},$$

where S_L satisfies (i): $\|S_L\| \le 1$, and (ii): $(S_L)_{12} = 0$. The approximation error is

$$S := H - \hat{H} = \gamma(\Theta_{11}S_L - \Theta_{12})(\Theta_{22} - \Theta_{21}S_L)^{-1}. \tag{11.11}$$

PROOF The proof is a special case of the proof of theorem 10.18. See also [vdV96]. □

By this theorem, an estimate of the principal subspace of H is given by $\mathcal{R}(\hat{H}) = \mathcal{R}(B' - A'S_L) = \mathcal{R}(B - A(S_L)_{11})$, for any valid choice of S_L. Note that $(S_L)_{11}$ ranges over the set of all contractive $(m-d) \times d$ matrices, so that all suitable principal subspace estimates are given by

$$\text{ran}(B - AM), \qquad \|M\| \le 1.$$

The distance of a subspace estimate with the actual principal subspace, $\text{ran}(U_1)$, is measured only implicitly, in the sense that there exists an approximant $\hat{H}$ with this column span that is γ-close to H. Actually, for each subspace estimate there are many such approximants, since the subspace estimate only depends on $(S_L)_{11}$, whereas the approximant also depends on $(S_L)_{21}$ and $(S_L)_{22}$.

The choice of a particular approximant $\hat{H}$, or subspace estimate ran($\hat{H}$), boils down to a suitable choice of the parameter S_L. Various choices are interesting:

1. The approximant $\hat{H}$ in theorem 11.4 is obtained by taking $S_L = 0$. This approximant is the simplest to compute; the principal subspace estimate is equal to the range of B. The approximation error is given by $\gamma\|\Theta_{12}\Theta_{22}^{-1}\|$. Note that, even if all nonzero singular values of H are larger than γ so that it is possible to have $\hat{H} = H$, the choice $S_L = 0$ typically does not give zero error. Hence, this simple choice of S_L could lead to 'biased' estimates.

2. As the truncated SVD solution satisfies the requirements, there is an S_L which yields this particular solution and minimizes the approximation error. However, computing this S_L requires an SVD, or a hyperbolic SVD [OSB91].

3. It is sometimes possible to obtain a uniform approximation error. First write equation (11.11) in a more implicit form,

$$\begin{bmatrix} \gamma^{-1}SG \\ -G \end{bmatrix} = \begin{bmatrix} \Theta_{11} & \Theta_{12} \\ \Theta_{21} & \Theta_{22} \end{bmatrix} \begin{bmatrix} S_L \\ -I_n \end{bmatrix},$$

   where G is an invertible $n \times n$ matrix. This equation implies

$$G^*(\gamma^{-2}S^*S - I_n)G = S_L^*S_L - I_n.$$

   Suppose $m \le n$. If we can take S_L to be an isometry, $S_L S_L^* = I_m$, then rank$(S_L^*S_L - I_n) = n-m$. It follows that $\gamma^{-1}S$ must also be an isometry, so that all singular values of $S = H - \hat{H}$ are equal to γ: the approximation error is uniform. S_L can be an isometry and have $(S_L)_{12} = 0$ only if $d \ge m-d$, *i.e.*, $d \ge m/2$. In that case, we can take for example $S_L = [I_m \;\; 0]$. This approximant might have relevance in signal processing applications where a singular data matrix is distorted by additive uncorrelated noise with a covariance matrix $\sigma^2 I_m$.

4. If we take $S_L = \Theta_{11}^{-1}\Theta_{12}$, then we obtain $\hat{H} = H$ and the approximation error is zero. Although this S_L is contractive, it does not satisfy the condition $(S_L)_{12} = 0$, unless $d = m$ or $d = n$. Simply putting $(S_L)_{12} = 0$ might make the resulting S_L noncontractive. To satisfy both conditions on S_L, a straightforward modification is by setting

$$S_L = \Theta_{11}^{-1}\Theta_{12} \begin{bmatrix} I_d & \\ & 0_{n-d} \end{bmatrix} = \begin{bmatrix} (\Theta_{11}^{-1}\Theta_{12})_{11} & 0 \\ (\Theta_{11}^{-1}\Theta_{12})_{21} & 0 \end{bmatrix}. \tag{11.12}$$

   The corresponding approximant is

$$\hat{H}^{(1)} := (B' - A'\Theta_{11}^{-1}\Theta_{12}[\begin{smallmatrix} I & 0 \\ 0 & 0 \end{smallmatrix}])(\Theta_{22} - \Theta_{21}\Theta_{11}^{-1}\Theta_{12}[\begin{smallmatrix} I & 0 \\ 0 & 0 \end{smallmatrix}])^{-1}, \tag{11.13}$$

   and the corresponding principal subspace estimate is given by the range of

$$B^{(1)} := B - A(\Theta_{11}^{-1}\Theta_{12})_{11}. \tag{11.14}$$

The subspace estimate is "unbiased" in a sense discussed below, and is usually quite accurate when σ_d is not very close to γ. Its efficient computation is discussed in section 11.5.

The approximation error is determined by

$$S = H - \hat{H}^{(1)} = \gamma\Theta_{12}\begin{bmatrix} 0_d & \\ & -I_{n-d} \end{bmatrix}(\Theta_{22} - \Theta_{21}\Theta_{11}^{-1}\Theta_{12}[\begin{smallmatrix} I & 0 \\ 0 & 0 \end{smallmatrix}])^{-1}. \qquad (11.15)$$

This shows that the rank of S is at most equal to $\min(m, n-d)$. If $m = n$, then the rank of S is $m-d$, *i.e.*, the error has the same rank as a truncated SVD solution would give.

5. To improve on the approximation error, we propose to take $(S_L)_{11} = (\Theta_{11}^{-1}\Theta_{12})_{11}$, as in the previous item, and use the freedom provided by $(S_L)_{21}$ and $(S_L)_{22}$ to minimize the norm of the error. The subspace estimate is only determined by $(S_L)_{11}$ and is the same as before. Instead of minimizing in terms of S_L, which involves a non-linear function and a contractivity constraint, we make use of the fact that we know already the column span of the approximant: we are looking for $\hat{H} = B^{(1)}N$, with $B^{(1)}$ given by (11.14) and $N : d \times n$ a minimizer of

$$\min_N \| H - B^{(1)}N \|.$$

A solution is given by $N = B^{(1)\dagger}H$, and the resulting approximant is

$$\begin{aligned} \hat{H} &= B^{(1)}B^{(1)\dagger}H \\ &=: \hat{H}^{(2)}, \end{aligned} \qquad (11.16)$$

the projection of H onto $\mathrm{ran}(B^{(1)})$. Although we do not compute the S_L to which this approximant corresponds, the residual error is guaranteed to be less than or equal to γ, because it is at most equal to the norm of S in (11.15). Hence, there will be some S_L that satisfies the constraints, although we never compute it explicitly. For this S_L, the rank of the residual error is always at most equal to $m-d$, the rank of $I_m - B^{(1)}B^{(1)\dagger}$.

One other important feature of the subspace estimate $B^{(1)}$ in (11.14) is that it is *unbiased*, in the following sense.

Lemma 11.6 $\mathrm{ran}(B^{(1)}) \subset \mathrm{ran}(H)$.

PROOF From $[(A\ \ 0)\ \ (B\ \ 0)] = [A'\ \ B'] = [\gamma I\ \ H]\Theta$, we have

$$\begin{cases} [A\ \ 0] &= \gamma\Theta_{11} + H\Theta_{21} \\ [B\ \ 0] &= \gamma\Theta_{12} + H\Theta_{22} \end{cases}$$

Hence

$$\begin{aligned} [B^{(1)}\ \ 0] &= [B\ \ 0] - [A\ \ 0]\Theta_{11}^{-1}\Theta_{12}\begin{bmatrix} I & \\ & 0 \end{bmatrix} \\ &= (\gamma\Theta_{12} + H\Theta_{22}) - (\gamma\Theta_{11} + H\Theta_{21})\Theta_{11}^{-1}\Theta_{12}\begin{bmatrix} I & \\ & 0 \end{bmatrix} \\ &= H(\Theta_{22} - \Theta_{21}\Theta_{11}^{-1}\Theta_{12})\begin{bmatrix} I & \\ & 0 \end{bmatrix} + H\Theta_{22}\begin{bmatrix} 0 & \\ & I \end{bmatrix} + \gamma\Theta_{12}\begin{bmatrix} 0 & \\ & I \end{bmatrix} \end{aligned}$$

so that

$$B^{(1)} = H(\Theta_{22} - \Theta_{21}\Theta_{11}^{-1}\Theta_{12})\begin{bmatrix} I \\ 0 \end{bmatrix}.$$

□

We also have

$$\| B^{(1)} \| \leq \| H \|. \tag{11.17}$$

This shows that, although norms of J-unitary matrices may be large, this particular subspace estimate is bounded in norm by the matrix it was derived from.

Because they will be used throughout the chapter, we will give names to the two "Schur subspace estimates" B and $B^{(1)}$:

$$\text{SSE-1}: \qquad U_{SSE1} = B \tag{11.18}$$

$$\text{SSE-2}: \qquad U_{SSE2} = B - AM_\Theta, \quad M_\Theta = [I_{m-d} \;\; 0]\Theta_{11}^{-1}\Theta_{12}\begin{bmatrix} I_d \\ 0 \end{bmatrix}. \tag{11.19}$$

11.4 HYPERBOLIC QR FACTORIZATION

In this section, we consider the computation of the SSE-1 subspace estimate, *i.e.*, the actual construction of a J-unitary matrix Θ such that

$$[\gamma I \;\; H]\Theta = [A' \;\; B'], \qquad J = \begin{bmatrix} I_m & \\ & -I_n \end{bmatrix}.$$

We are looking for algorithms that do not square the data and that allow easy updating of the factorization as more and more columns of H are included (growing n). Θ will be computed in two steps: $\Theta = \tilde{\Theta}\Pi$, where $\tilde{\Theta}$ is a $(J, \tilde{J}_2)$-unitary matrix with respect to J and an unsorted signature $\tilde{J}_2$ and is such that

$$\begin{bmatrix} \overset{+}{\gamma I_m} & \overset{-}{H} \end{bmatrix}\tilde{\Theta} = \begin{bmatrix} \overset{\pm}{R} & \overset{\pm}{0_{m\times n}} \end{bmatrix}, \qquad R: m\times m. \tag{11.20}$$

Π is any permutation matrix such that $\Pi\tilde{J}_2\Pi^* = J$ is a sorted signature matrix. The latter factorization can be viewed as a hyperbolic QR factorization, in case R has a triangular form, and can be computed in a number of ways. Hyperbolic Householder transformations have been employed for this purpose [BG81, OSB91], zeroing full rows at each step, but the most elementary way is to use elementary rotations to create one zero entry at a time, like Givens rotations for QR factorizations. Such techniques are known as (generalized) Schur algorithms, because of their similarity to the Schur method for Toeplitz matrices.

Indefinite Schur algorithm

To compute the factorization (11.20), elementary rotations $\tilde{\theta}$ as in figure 11.1 are embedded in plane rotations $\tilde{\Theta}_{(i,k)}$ which are applied to the columns of $[\gamma I \;\; H]$ in the same way as Givens rotations are used for computing a QR factorization. Each plane rotation produces a zero entry in H; specifically, $\tilde{\Theta}_{(i,k)}$ annihilates entry (i,k). A difference with QR is that we have to keep track of the signatures associated to the columns of the

$[X \quad Y] := [\gamma I_m \quad H]$

$\tilde{J} := \begin{bmatrix} I_m & \\ & -I_n \end{bmatrix}$

$\tilde{\Theta} = I_{m+n}$

for $k = 1$ to n and $i = 1$ to m,

  $[a \quad b] := [X(i,i) \quad Y(i,k)]$

  $\tilde{j}_1 := \begin{bmatrix} \tilde{J}(i,i) & 0 \\ 0 & \tilde{J}(m+k, m+k) \end{bmatrix}$

  Compute $\tilde{\theta}, \tilde{j}_2$ from $a, b, \tilde{j}_1$ s.t. $[a \quad b]\tilde{\theta} = [* \quad 0]$

  Embed $\tilde{\theta}$ into $\tilde{\Theta}_{(i,k)}$

  $[X \quad Y] := [X \quad Y]\tilde{\Theta}_{(i,k)}$

  $\tilde{\Theta} := \tilde{\Theta}\tilde{\Theta}_{(i,k)}$

  $\tilde{J}(i,i) := (\tilde{j}_2)_{1,1}$

  $\tilde{J}(m+k, m+k) := (\tilde{j}_2)_{2,2}$

end

$\tilde{J}_2 := \tilde{J}$

Figure 11.2. Schur algorithm to compute the factorization $[\gamma I \quad H]\tilde{\Theta} = [X \quad 0]$ from H.

matrix to determine which type of rotations to use. The general scheme, however, goes as follows:

$$
\begin{aligned}
[\gamma I \quad H] \quad = \quad & \begin{matrix} + & + & + & & - & - & - & - \end{matrix} \\
& \left[\begin{array}{ccc|cccc} \underline{\gamma} & & & \underline{\times} & \times & \times & \times \\ & \gamma & & \times & \times & \times & \times \\ & & \gamma & \times & \times & \times & \times \end{array}\right] \overset{\tilde{\Theta}_{(1,1)}}{\longrightarrow} \\
& \begin{matrix} - & + & + & & + & - & - & - \end{matrix} \\
& \left[\begin{array}{ccc|cccc} \times & & & 0 & \times & \times & \times \\ \times & \underline{\gamma} & & \underline{\times} & \times & \times & \times \\ \times & & \gamma & \times & \times & \times & \times \end{array}\right] \overset{\tilde{\Theta}_{(2,1)}}{\longrightarrow} \\
& \begin{matrix} - & + & + & & + & - & - & - \end{matrix} \\
& \left[\begin{array}{ccc|cccc} \times & & & 0 & \times & \times & \times \\ \times & \times & & 0 & \times & \times & \times \\ \times & \times & \underline{\gamma} & \underline{\times} & \times & \times & \times \end{array}\right] \longrightarrow \\
& \begin{matrix} - & + & - & & + & + & - & - \end{matrix} \\
\cdots \overset{\tilde{\Theta}_{(m,n)}}{\longrightarrow} \quad & \left[\begin{array}{ccc|cccc} \times & & & 0 & 0 & 0 & 0 \\ \times & \times & & 0 & 0 & 0 & 0 \\ \times & \times & \times & 0 & 0 & 0 & 0 \end{array}\right] = [R \quad 0],
\end{aligned}
$$

$$\tilde{\Theta} = \tilde{\Theta}_{(1,1)}\tilde{\Theta}_{(2,1)} \cdots \tilde{\Theta}_{(m,1)} \cdot \tilde{\Theta}_{(1,2)} \cdots \tilde{\Theta}_{(2,2)} \cdots \tilde{\Theta}_{(m,n)} .$$

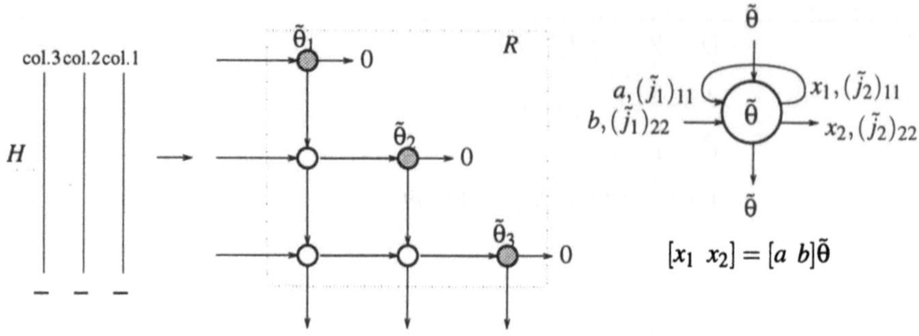

Figure 11.3. Signal flow graph of the Schur algorithm. Associated to every matrix entry is also its signature (+ or −). R contains a permutation of $[A \;\; B]$ and is initialized by γI. The shaded processors compute rotation parameters as in figure 11.1.

(Except for the first matrix, the signatures of the columns in the above matrices are examples, as they are data dependent.) The pivot elements at each step are underlined; these entries, along with the signatures of the two columns in which they appear, determine the elementary rotation $\tilde{\theta}$ that will be used at that step, as well as the resulting signature $\tilde{j}_2$. This signature is the new signature of these two columns, after application of the rotation. The algorithm is summarized in figure 11.2.

The nulling scheme ensures that $[\gamma I \;\; H]\tilde{\Theta} = [R \;\; 0]$, where R is a resulting lower triangular invertible matrix; it contains the columns of A and B in some permuted order. The columns of R with a positive signature are the columns of A, the columns with a negative signature are those of B. Hence, the final step (not listed figure 11.2) is to sort these columns, such that $[R \;\; 0]\Pi = [A \;\; 0 \;\; B \;\; 0] = [A' \;\; B']$. Then $\Theta = \tilde{\Theta}\Pi$ is J-unitary with respect to J, and $[\gamma I \;\; H]\Theta = [A' \;\; B']$.

The complexity of the algorithm is similar to that of the QR factorization: about $1/2m^2n$ rotations, or $2m^2n$ flops. The Schur algorithm has a direct implementation on a systolic array of processors. This array is entirely similar to the classical Gentleman-Kung triangular Givens array [GK81b], except that, now, all data entries have a signature associated to them, and the processors have to perform different types of rotations, depending on these signatures. The corresponding array is shown in figure 11.3.

Updating and downdating

The Schur method is straightforward to update as more and more columns of H become known. If $[\gamma I \;\; H_n]\tilde{\Theta}_{(n)} = [R_n \;\; 0]$ is the factorization at point n and $H_{n+1} = [H_n \;\; \mathbf{h}_{n+1}]$, then, because the algorithm works column-wise,

$$[\gamma I \quad H_{n+1}]\tilde{\Theta}_{(n+1)} = [R_{n+1} \quad 0] \quad \Rightarrow \quad \begin{array}{l} [R_n \quad 0 \quad \mathbf{h}_{n+1}]\tilde{\theta}^{(n+1)} = [R_{n+1} \quad 0 \quad 0] \\ \tilde{\Theta}_{(n+1)} = \tilde{\Theta}_{(n)}\tilde{\theta}^{(n+1)}, \end{array}$$

for some J-unitary matrix $\tilde{\theta}^{(n+1)}$ acting on the columns of R_n and on $\mathbf{h}_{n+1}$. Hence, we can continue with the result of the factorization that was obtained at the previous step. Each update requires about $1/2m^2$ rotations.

The downdating problem is to compute the factorization for H_n with its first column $\mathbf{h}_1$ removed, from a factorization of H_n. It can be converted to an updating problem, where the old column $\mathbf{h}_1$ is now introduced with a positive signature,

$$\begin{matrix} \pm & + \\ [R_n & \mathbf{h}_1] \end{matrix}\,\tilde{\theta}^{(n+1)} = [R_{n+1} \;\; 0]\,.$$

This is possible because, implicitly, we factor $\gamma^2 I - H_n H_n^* + \mathbf{h}_1\mathbf{h}_1^* = R_n \tilde{J} R_n^* + \mathbf{h}_1\mathbf{h}_1^*$. The uniqueness of the hyperbolic QR factorization into triangular matrices with positive diagonals ([BG81], *viz.* corollary 11.2) implies that the result R_{n+1} is precisely the same as if $\mathbf{h}_1$ had never been part of H_n at all.

Breakdown

In section 11.4, we had to assume that the data matrix H was such that at no point in the algorithm $[a \;\; b]\,\tilde{j}_1[a \;\; b]^*$ is equal to zero. If the expression is zero, then there is no J-unitary rotation $\tilde{\theta}$ such that $[a \;\; b]\tilde{\theta} = [* \;\; 0]$. Note that the condition in theorem 11.3 that none of the singular values of H are equal to γ does not preclude this case, but merely ascertains that there *exists* a $\tilde{\Theta}$ which will zero H. One simple example is obtained by taking $H = [1 \;\; 1]^T$, $\gamma = 1$. It is straightforward to show that there is no J-unitary $\tilde{\Theta}$ such that

$$\left[\begin{array}{cc|c} 1 & & 1 \\ & 1 & 1 \end{array}\right]\tilde{\Theta} = \left[\begin{array}{cc|c} \times & 0 & 0 \\ \times & \times & 0 \end{array}\right] \tag{11.21}$$

as the J-norms of the first row will not be equal. Hence Θ cannot be obtained by the recursive algorithm. However, a more general $\tilde{\Theta}$ does exist, such that

$$\begin{array}{c} \begin{array}{ccc} + & + & - \end{array} \\ \left[\begin{array}{cc|c} 1 & & 1 \\ & 1 & 1 \end{array}\right] \end{array}\tilde{\Theta} = \frac{1}{\sqrt{2}}\begin{array}{c} \begin{array}{ccc} + & - & + \end{array} \\ \left[\begin{array}{cc|c} 1 & 1 & 0 \\ -1 & 1 & 0 \end{array}\right] \end{array}$$

viz.

$$\tilde{\Theta} = \frac{1}{\sqrt{2}}\begin{bmatrix} 1 & -1 & \sqrt{2} \\ -1 & -1 & \sqrt{2} \\ 0 & 2 & -\sqrt{2} \end{bmatrix}, \qquad \tilde{J}_1 = \begin{bmatrix} 1 & & \\ & 1 & \\ & & -1 \end{bmatrix}, \qquad \tilde{J}_2 = \begin{bmatrix} 1 & & \\ & -1 & \\ & & 1 \end{bmatrix}.$$

The difference is that, in this factorization, the resulting matrix R is no longer lower triangular. Theorem 11.7 gives necessary and sufficient conditions on the singular values of H and a collection of submatrices of H, so that the Schur algorithm does not break down.

Theorem 11.7 *Let $H : m \times n$ be a given matrix, and $\gamma \geq 0$. Denote by $H_{1..i,1..k}$ the submatrix, consisting of the first to the i-th row and the first k columns of H. The Schur algorithm does not break down if and only if none of the singular values of $H_{1..i,1..k]}$ is equal to γ, for $i = 1, \ldots, m$ and $k = 1, \ldots, n$.*

PROOF *(Necessity)* When processing the k-th column of H by the Schur algorithm, we are in fact computing a triangular factorization of $[\gamma I_m \; H_{1..m,1..k}]$. Corollary 11.2 claims that a suitable J-unitary operator exists if and only if $[\gamma I_i \; H_{[i,k]}]$ is J-nonsingular, for $i = 1,\ldots,m$, *i.e.*, if and only if none of the singular values of $H_{1..i,1..k]}$ is equal to 1. The triangularization is done for $k = 1,2,\ldots,n$ in turn.

(Sufficiency) Sufficiency at stage (i,k) follows recursively from the factorization at the previous stage and the existence and uniqueness of the factorization at the current stage. □

Similar results are known for the case where the factorization is computed via hyperbolic Householder transformations where all zeros in a row are generated at the same time. In this case there are less conditions [BG81], *viz.* theorem 11.2. It should be noted that the conditions in theorem 11.7 are quite elaborate, as only one condition (none of the singular values of H are equal to γ) suffices for the *existence* of Θ. Numerically, we might run into problems also if one of the singular values is close to γ, in which case the corresponding hyperbolic rotation has a large norm. How serious this is depends on a number of factors, and a careful numerical analysis is called for. One example where a large rotation is not fatal is the case where the singularity occurs while processing the last entry of a column ($i = m$). Although the rotation will be very large, the resulting R remains bounded and becomes singular: $R_{m,m} = 0$. Hence, the subspace information is still accurate, and R varies in a continuous way across the γ-boundary; only its signature is necessarily discontinuous. Pivoting schemes could in principle be used to prevent large hyperbolic rotations. A more attractive scheme results in the computation of the SSE-2, as discussed in section 11.5.

Comparison of SSE-1 and SSE-2

We demonstrate some of the properties of the approximation scheme by means of a simple example. We take $H(\sigma_2) = U\Sigma(\sigma_2)V^*$ to be a sequence of 3×4 matrices, with U and V randomly selected constant unitary matrices, and with singular values equal to

$$(20, \sigma_2, 0.5)\,, \qquad \sigma_2 = 0, 0.01, \ldots, 3.99, 4\,.$$

The approximation tolerance is set to $\gamma = 1$. We compare the approximants $\hat{H}^{(0)}$ given by $S_L = 0$, $\hat{H}^{(1)}$ given by equation (11.13), $\hat{H}^{(2)}$ given by (11.16), and $\hat{H}^{(1)}$ when the factorization is computed with pivoting. The pivoting scheme consists of column permutations, except when processing the last column, in which case we switch to row permutations. The pivoting is applied in its extreme form, *i.e.*, whenever this leads to elementary rotation matrices with a smaller norm. The approximants are compared on the following aspects: *(a)* $\|\Theta\|$, with and without pivoting; *(b)* $\|H - \hat{H}\|$, for each of the mentioned approximants; *(c)* the accuracy of the subspace estimates, compared to the principal subspace of H (the column span of the singular vectors with corresponding singular values larger than 1). The distance between two subspaces $\mathcal{A}$ and $\mathcal{B}$ is defined as $\text{dist}(\mathcal{A},\mathcal{B}) = \|\mathbf{P}_{\mathcal{A}} - \mathbf{P}_{\mathcal{B}}\|$, where $\mathbf{P}_{\mathcal{A}}$ is the orthogonal projection onto $\mathcal{A}$ [GV89].

Figure 11.4(a) shows $\|\Theta\|$ as a function of σ_2. Without pivoting, there are a number of peaks, corresponding to the values of σ_2 where one of the submatrices $H_{[i,k]}$ has a singular value equal to 1. In the range $0 \le \sigma_2 \le 4$, this occurred for $(i,k) = (3,4)$,

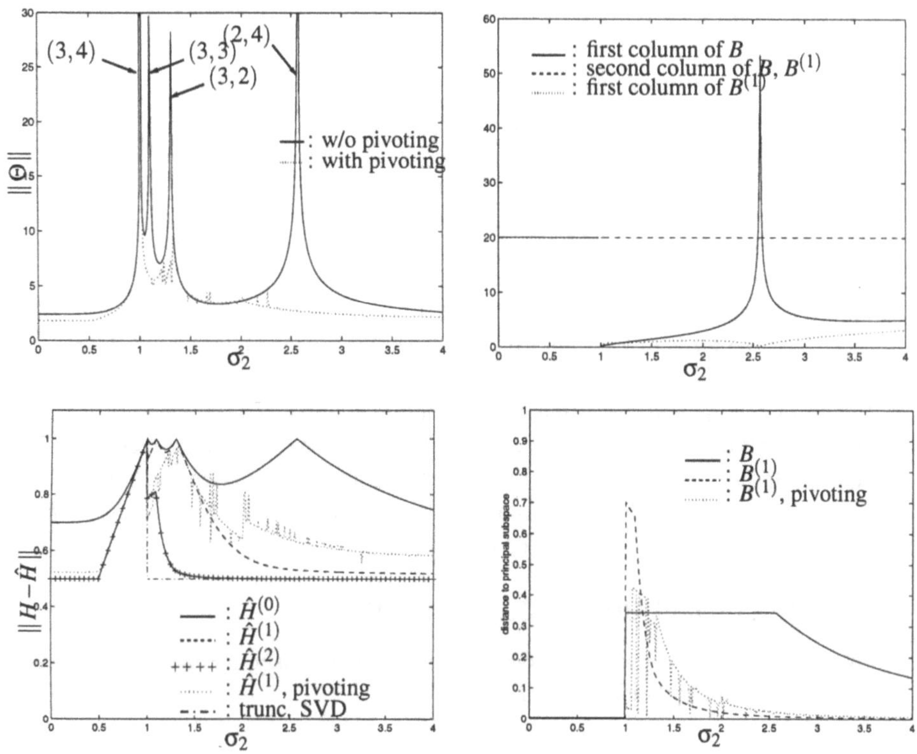

Figure 11.4. (a) Norm of Θ. $\|\Theta\| \to \infty$ for certain values of σ_2 when the indicated entry (i, j) of H is processed. (b) The norm of the first and second column of B and $B^{(1)}$. (c) Norm of the approximation error. (d) Distance between the principal and estimated subspaces.

(3,3), (3,2) and (2,4), respectively. When pivoting is applied, the peak at $\sigma_2 = 1$ is, necessarily, still present, but the other peaks are mostly smoothed out. Figure 11.4(b) shows the norm of the columns of B, in the scheme without pivoting. For $\sigma_2 < 1$, the rank of the approximant is 1. At $\sigma_2 = 1$, the dimension of B increases, although at first, the new column has a very small norm. For larger values of σ_2, the norm grows and the subspace becomes better defined. Figure 11.4 also shows that no peak occurs for the norm of the columns of the SSE-2 subspace estimate $B^{(1)}$ of equation (11.14), on which both $\hat{H}^{(1)}$ and $\hat{H}^{(2)}$ are based. This is as predicted by lemma 11.6: $\| B^{(1)} \| \le \| H \| = 20$. Instead of having a peak, the norm of the first column of $B^{(1)}$ dips to about 0.12.

In figure 11.4(c), the norm of $H - \hat{H}$ is shown, for the various choices of $\hat{H}$ that we discussed in section 10.4. The lowest line corresponds to the truncated SVD solution, which gives the lowest attainable error. It is seen that, for all approximants, the approximation error is always less than $\gamma \equiv 1$. The approximation error for $\hat{H}^{(0)}$ is in this example always higher than the error for $\hat{H}^{(1)}$, $\hat{H}^{(2)}$, and the error for $\hat{H}^{(1)}$ is always higher than the error for $\hat{H}^{(2)}$, since the latter approximant minimizes this error while retaining the same subspace estimate. The approximation error for $\hat{H}^{(2)}$ is almost identically close to the theoretical minimum, except in a small region $1 \le \sigma_2 \le 1.5$. The errors for $\hat{H}^{(0)}$ and $\hat{H}^{(1)}$ touch a number of times on the $(\gamma = 1)$-line. For $\hat{H}^{(0)}$ this can be explained as follows. The error for $S_L = 0$ is given by equation (11.11) as $-\gamma \Theta_{12} \Theta_{22}^{-1}$. Because the J-unitarity of Θ implies $\Theta_{22}^{-*} \Theta_{22}^{-1} + (\Theta_{22}^{-*} \Theta_{12}^{*})(\Theta_{12} \Theta_{22}^{-1}) = I$, it follows that whenever $\| \Theta_{22} \| \to \infty$, necessarily $\| \Theta_{12} \Theta_{22}^{-1} \| \to 1$.

Figure 11.4(d) depicts the distance between the principal and estimated subspaces. For $\sigma_2 < 1$, this distance is very close to zero ($< .0002$) for each of the methods. The distance jumps up when σ_2 crosses 1: the subspace increases in dimension but is at first only weakly defined. For $B^{(1)}$, the distance goes down again quickly, whereas for B, it stays constant for a while before going down.

11.5 HYPERBOLIC URV DECOMPOSITION

Let $N : m \times n_1$ and $H : m \times n_2$ be given matrices. (Previously, we had $N = \gamma I$.) We consider implicit factorizations of $HH^* - NN^*$ as

$$HH^* - NN^* = BB^* - AA^*, \tag{11.22}$$

where A and B together have m columns. A and B follow from the factorization

$$m \begin{bmatrix} \overset{n_1}{\underset{+}{}} & \overset{n_2}{\underset{-}{}} \\ N & H \end{bmatrix} \Theta = m \begin{bmatrix} \overset{n_1}{\underset{+}{}} & \overset{n_2}{\underset{-}{}} \\ A' & B' \end{bmatrix}; \qquad A' = m \begin{bmatrix} \overset{m-d}{} & \overset{n_1-m+d}{} \\ A & 0 \end{bmatrix}, \quad B' = m \begin{bmatrix} \overset{d}{} & \overset{n_2-d}{} \\ B & 0 \end{bmatrix} \tag{11.23}$$

where Θ is a J-unitary matrix partioned conform the equation. According to theorem 11.1, the factorization always exists although Θ will be unbounded when $HH^* - NN^*$ is singular. However, the factorization is not unique.

In section 11.4, we computed the factorization (11.23) by means of a hyperbolic QR factorization

$$[\overset{+}{N} \;\; \overset{-}{H}] \tilde{\Theta} = [\overset{\pm}{R} \;\; \overset{\pm}{0}_{m \times (n1+n2-m)}], \tag{11.24}$$

in which R is a lower or upper triangular $m \times m$ matrix. Although this factorization is simple to update, it has the drawback that it does not always exist: the triangular form of R is too restrictive (corollary 11.2, theorem 11.7). The set of exceptions is finite, but in the neighborhood of an exception it may happen that A and B are unbounded with nearly collinear column spans.

To get around this, introduce a QR factorization of $[A\ \ B]$: $R = [R_A\quad R_B] = Q^*[A\quad B]$, where R is triangular and Q is unitary. This leads to the more general two-sided decomposition

$$Q^*[\overset{+}{N}\quad \overset{-}{H}]\Theta = [\overset{+}{R_A}\quad \overset{+}{0} \mid \overset{-}{R_B}\quad \overset{-}{0}]\,. \tag{11.25}$$

Note that still $[A\ \ 0 \mid B\ \ 0] = [N\ \ H]\Theta$. This two-sided decomposition always exists. We can choose to have R upper triangular or lower triangular, or even permute the columns of $[A\ \ B]$ before introducing the QR factorization. It is convenient to take R lower triangular: if we split $Q = [Q_A\ \ Q_B]$ accordingly, then

$$\operatorname{ran}(B) = \operatorname{ran}(Q_B)\,.$$

Hence, for this choice, Q_B is an *orthonormal* basis of the (central) principal subspace estimate. If our objective is to estimate a null space basis, then we would swap (A,B) or take R upper triangular so that $\operatorname{ran}(A) = \operatorname{ran}(Q_A)$.

We are interested in SSE-2 subspace estimates, as defined in (11.19). This definition involves the inversion of submatrices of Θ, which is not attractive, also because the size of these submatrices is not constant but grows with n_1 and n_2. We will now show how this can be avoided by posing additional structural restrictions on Θ, which is possible because A, B and Θ are not unique. We can use this freedom to transform M_Θ in (11.19) to zero, as shown in the following lemma.

Lemma 11.8 *For given A, B, Θ, consider a transformation by a J-unitary matrix Θ_M:*

$$\begin{aligned} [A\ \ 0 \mid B\ \ 0]\Theta_M &= [A'\ \ 0 \mid B'\ \ 0] && (11.26)\\ \Theta\Theta_M &= \Theta' && (11.27) \end{aligned}$$

where Θ_M only acts on the columns of A, B (and corresponding columns of Θ).

Then $\operatorname{ran}(B - AM_\Theta) = \operatorname{ran}(B' - A'M_{\Theta'})$, i.e., *the SSE-2 subspace is invariant under Θ_M. Furthermore, there exists a Θ_M such that $M_{\Theta'} = 0$,* i.e., *such that* $\operatorname{ran}(B')$ *is the SSE-2 subspace.*

PROOF The proof is rather technical and is given in the appendix. □

Hence, there is a matrix Θ_M which transforms Θ to $\Theta' = \Theta\Theta_M$, such that after the transformation we simply take B' and have the desired SSE-2 subspace basis. Knowing this, there are easier ways to find this transformation. Suppose $[\Theta_{11}\ \ \Theta_{12}]$ is partitioned as

$$[\Theta_{11}\quad \Theta_{12}] = \begin{matrix} \scriptstyle m-d \\ \scriptstyle n_1-(m-d) \end{matrix} \overset{\begin{matrix}\scriptstyle m-d & \scriptstyle n_1-(m-d) & & \scriptstyle d & \scriptstyle n_2-d\end{matrix}}{\left[\begin{array}{cc|cc} (\Theta_{11})_{11} & (\Theta_{11})_{12} & (\Theta_{12})_{11} & (\Theta_{12})_{12} \\ * & * & * & * \end{array}\right]}.$$

From the definition of M_Θ in (11.19), it is seen that to have $M_{\Theta'} = 0$, it suffices to find a transformation on Θ such that $\Theta_{11}'^{-1}\Theta_{12}'$ has a zero (11)-block. This will be the case, for example, if both $(\Theta_{11}')_{12} = 0$ and $(\Theta_{12}')_{11} = 0$. The latter can always be effected by a suitably chosen Θ_M which cancels $(\Theta_{12})_{11}$ against $(\Theta_{11})_{11}$. However, to apply lemma 11.8, Θ_M is not allowed to change the columns of $(\Theta_{11})_{12}$. To zero this block, we may apply any invertible transformation T_e to the *rows* of $[\Theta_{11}\ \ \Theta_{12}]$:

$$[\Theta'_{11}\quad \Theta'_{12}] = T_e[\Theta_{11}\quad \Theta_{12}]$$

because $\Theta'^{-1}_{11}\Theta'_{12} = \Theta^{-1}_{11}\Theta_{12}$ is invariant under T_e. This leads to a new characterization of SSE-2 estimates:

Theorem 11.9 *The following factorization provides an SSE-2 subspace estimate. For given $N : m\times n_1$, $H : m\times n_2$, with $n_1 \geq m$, find the subspace dimension d, Q (unitary), Θ (J-unitary), $R = [R_A\ \ R_B]$ (lower triangular), $T : (m-d)\times n_1$ (full rank) such that*

$$Q^* \begin{array}{c} \\ \left[\begin{array}{cc} \overset{n_1}{\underset{N}{+}} & \overset{n_2}{\underset{H}{-}} \end{array}\right]\end{array}\Theta = \left[\begin{array}{cccc} \overset{m-d}{\underset{R_A}{+}} & \overset{n_1-(m-d)}{\underset{0}{+}} & \overset{d}{\underset{R_B}{-}} & \overset{n_2-d}{\underset{0}{-}} \end{array}\right], \tag{11.28}$$

$$T \left[\begin{array}{cc} \overset{n_1}{\underset{I}{+}} & \overset{n_2}{\underset{0}{-}} \end{array}\right]\Theta = \left[\begin{array}{cccc} \overset{m-d}{\underset{I}{+}} & \overset{n_1-(m-d)}{\underset{0}{+}} & \overset{d}{\underset{0}{-}} & \overset{n_2-d}{\underset{*}{-}} \end{array}\right]. \tag{11.29}$$

With the partitioning $Q = [Q_A\ \ Q_B]$, an orthonormal basis for the SSE-2 subspace estimate is given by Q_B.

PROOF We only have to show that $M_\Theta = 0$. Let T_e be an extension of T to a full rank $n_1\times n_1$ matrix, then

$$T[I_{n_1}\quad 0]\Theta = [(T_e\Theta_{11})_{11}\ \ (T_e\Theta_{11})_{12}\ \ (T_e\Theta_{12})_{11}\ \ (T_e\Theta_{12})_{12}] = [I_{m-d}\ \ 0\ \ \ 0\ \ *].$$

Hence $M_\Theta = [I_{m-d}\ \ 0]\Theta^{-1}_{11}\Theta_{12}\left[\begin{smallmatrix} I_d \\ 0\end{smallmatrix}\right] = [I_{m-d}\ \ 0]\Theta^{-1}_{11}T_e^{-1}T_e\Theta_{12}\left[\begin{smallmatrix} I_d \\ 0\end{smallmatrix}\right] = [*\ \ 0]\left[\begin{smallmatrix} 0 \\ *\end{smallmatrix}\right] = 0.$ □

By virtue of theorem 11.1, the above factorization always exists. If $HH^* - NN^*$ is singular, then certain columns of Θ are unbounded and corresponding columns of R are identically zero. Note that the factorization, and hence the SSE-2 subspace, is still not unique: some freedom is remaining in the generation of the zero entries. With proper choices for Q, Θ and T in terms of the left and right singular vectors of H, one can show that the TSVD (principal) subspace is within the class of SSE-2 subspaces, and has R diagonal as distinctive feature (see the appendix at the end of this chapter).

The factorization in (11.28) is reminiscent of the URV decomposition [Ste92], but with a J-unitary Θ. The following corollary shows that the factorization has certain desirable norm properties as well.

Corollary 11.10 *The factorization (11.28)-(11.29) is such that*

$$\operatorname{ran}(Q_B) \subset \operatorname{ran}(H),\ \ \|R_B\| \leq \|H\|,\ \ \|R_A\| \leq \|N\|.$$

PROOF Using the fact that $M_\Theta = 0$, lemma 11.6 implies $BB^* \leq HH^*$, $AA^* \leq NN^*$. It remains to apply the definition $[A\ \ B] = [Q_A\ \ Q_B][R_A\ \ R_B]$ where Q is unitary and R is lower. □

col:	1. Compute $\tilde{\theta}$ and $\tilde{j}_2$ s.t. $[R_{i,i}\ \ c_i]\tilde{\theta} = [*\ \ 0]$, with $\tilde{j}_1 = \mathrm{diag}[J_i, j_c]$ 2. Apply $\tilde{\theta}$ to the i-th column of R and $\mathbf{c}$; update signatures J_i, j_c
row:	1. Determine q s.t. $q^*\left[\begin{smallmatrix} c_i \\ c_{i+1} \end{smallmatrix}\right] = \left[\begin{smallmatrix} 0 \\ * \end{smallmatrix}\right]$. 2. Apply q^* to rows $(i, i+1)$ of R; apply q to columns $(i, i+1)$ of Q 3. Compute $\tilde{\theta}$ and $\tilde{j}_2$, as in figure 11.1, s.t. $[R_{i,i}\ \ R_{i,i+1}]\tilde{\theta} = [*\ \ 0]$ 4. Apply $\tilde{\theta}$ to columns $(i, i+1)$ of R; update signatures J_i, J_{i+1}

Figure 11.5. Two ways to zero c_i

11.6 UPDATING THE SSE-2

Now that we have identified (11.28)-(11.29) as a factorization which provides an SSE-2 subspace, we investigate how this factorization can be updated when new columns for H and N become available. The update consists of two phases, one to update (11.28), and a second to restore the zero structure of (11.29).

Several updating algorithms are possible, depending on one's objectives. The direction taken here follows from an interest in parallel and pipelined multi-processor architectures for high-throughput signal processing applications. A very desirable aspect then is to have a localized, data-independent and one-directional computational flow, perhaps at the expense of some additional operations. At the same time, we would like to minimize the number of *hyperbolic* rotations, since these are a potential source of numerical instability. This induces a tradeoff.

Updating $Q^*[N\ \ H]\Theta$

Suppose we have already computed the decomposition $Q^*[N\ \ H]\tilde{\Theta} = [R\ \ 0]$, where $R = [R_A\ \ R_B]$ is lower triangular and sorted according to signature. In principle, updating the factorization with new columns of H or N is straightforward. Indeed, let us say that we want to find a new factorization $Q'^*[N'\ \ H']\tilde{\Theta}' = [R'\ \ 0]$, where either $N' = [N\ \ \mathbf{n}]$, $H' = H$ if we want to add a new column to N, or $N' = N$, $H' = [H\ \ \mathbf{h}]$ if we augment H. Making use of the previously computed decomposition, it suffices to find Q_c and $\tilde{\Theta}_c$ such that

$$
\begin{array}{rcl}
Q_c^* \begin{bmatrix} \overset{\substack{m-d \\ +}}{R_A} & \overset{\substack{d \\ -}}{R_B} & \overset{\substack{1 \\ j_c}}{\mathbf{c}} \end{bmatrix} \tilde{\Theta}_c & = & \begin{bmatrix} \overset{\substack{m-d' \\ +}}{R'_A} & \overset{\substack{d' \\ -}}{R'_B} & \overset{\substack{1 \\ j'_c}}{0} \end{bmatrix} \\
Q' & := & QQ_c\,,
\end{array}
\tag{11.30}
$$

where $\mathbf{c} = Q^*\mathbf{n}$ if we add a column $\mathbf{n}$ to N or $\mathbf{c} = Q^*\mathbf{h}$ if we add a column $\mathbf{h}$ to H. (Note that we need to store and update Q to apply this transformation. Storage of $\tilde{\Theta}$ will not be needed.) In the first case, $\mathbf{c}$ has a positive signature $j_c = 1$; in the second case, $j_c = -1$. Denote the signature of R by $J = I_{m-d} \oplus -I_d$, and let J_i denote the i-th diagonal entry of J.

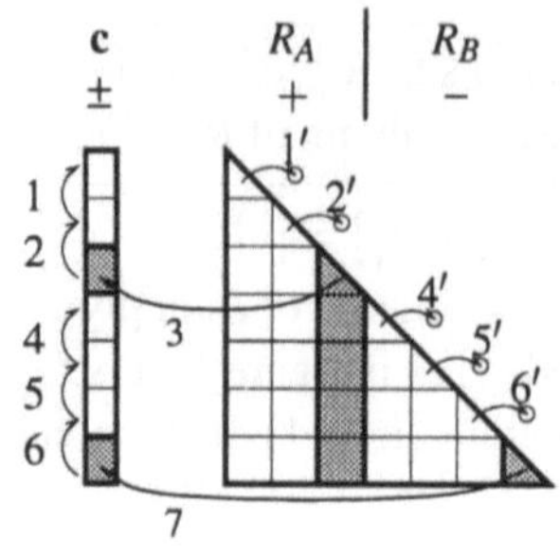

Figure 11.6. Order in which zero entries are created by algorithm *zero-c*. Only column operations (rotations 3 and 7) are possibly hyperbolic and may lead to signature changes

To compute the factorization (11.30), the entries $c_1, c_2, \cdots, c_m$ of $\mathbf{c}$ are zeroed in turn. As listed in figure 11.5, there are two possibilities to do this: by elementary column rotations $\tilde{\theta}$ or by elementary row rotations q. The "*col*" scheme to zero entry c_i is the most natural and efficient, and directly zeros c_i against $R_{i,i}$. The "*row*" scheme first computes an elementary circular (unitary) rotation q to zero c_i against c_{i+1}, and then a $\tilde{\theta}$-rotation to zero the resulting fill-in in $R_{i,i+1}$ against $R_{i,i}$.

For reasons of numerical stability, it is desirable to minimize the number of *hyperbolic* rotations, i.e. rotations $\tilde{\theta}$ that act on columns with unequal signatures. Such hyperbolic rotations also might lead to an interchange of signatures, thus destroying the sorting of the columns of R. Hence, we propose to zero most entries c_i using row operations, in spite of the added complexity, and to use column operations only for the zeroing of c_{m-d} and c_m.

A graphical representation of this scheme is given in figure 11.6. Hyperbolic rotations and signature changes are only possible in steps $m-d$ and m. The θ-rotations in the row stages act on columns of equal signatures, so that they are circular rotations without signature changes. The resulting signature of R depends on the initial and final signature of $\mathbf{c}$, *i.e.*, j_c and j'_c. A list of possibilities is given in figure 11.7.

The second phase is to restore the sorting of the columns of R according to their signature. This is only necessary in cases (b) and (d_2) of figure 11.7, and it suffices to move the last column of R by a series of d swaps with its right neighbors. After each permutation, the resulting fill-in in $R_{i,i+1}$ has to be zeroed by a q-rotation. If desired, this phase can be made data-independent by always performing the permutations, independent of the signatures.

At this point, $\mathbf{c}' = 0$, and R' is lower triangular and sorted into $R' = [\overset{+}{R'_A}\ \overset{-}{R'_B}]$, so that we have obtained the updated factorization (11.30). The number of columns d' of R'_B, *i.e.*, the principal subspace dimension after the update, depends on j_c and j'_c: d stays constant if $j'_c = j_c$, it increases if $(j_c = -1, j'_c = 1)$ and decreases if $(j'_c = -1, j_c = 1)$, *i.e.*, $d' = d + \frac{1}{2}(j'_c - j_c)$. The columns of the matrix block Q_B form an orthonormal basis for the updated subspace estimate B.

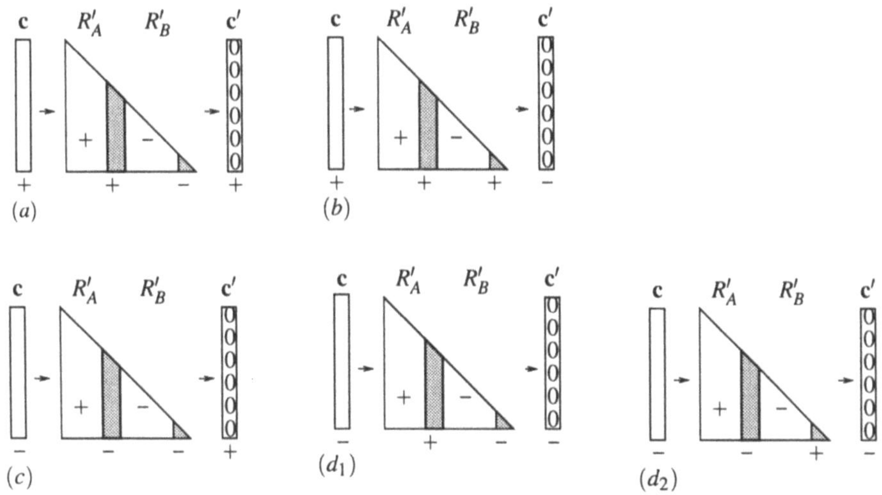

Figure 11.7. The four possible signature changes of $\mathbf{c}$, $\mathbf{c}'$, and the resulting possible signatures J' (after *zero-c*, before sorting). Only columns $m-d$ and d of R may have changed signature.

Updating the structure of Θ

The next step is to modify the candidate Q_c and $\tilde{\Theta}_c$ by some Q_M and $\tilde{\Theta}_M$ in order to satisfy the structural conditions (11.29) on Θ. Equation (11.29) shows that we do not have to keep track of T and Θ at all: we only have to update a matrix $[I_{m-d}\ 0_{m-d \times d}]$. The columns marked '$*$' in (11.29) never change, so we do not have to track them. Obviously, we do not have to store $[I_{m-d}\ 0]$. Hence, updating is possible by only storing matrices Q and $R = [R_A\ R_B]$. In update notation, the structural requirements take the following form:

$$Q_M^* Q_c^*{}_m \left[\begin{array}{ccc} \overset{m-d}{\underset{+}{}} & \overset{d}{\underset{-}{}} & \overset{1}{\underset{j_c}{}} \\ R_A & R_B & \mathbf{c} \end{array} \right] \tilde{\Theta}_c \tilde{\Theta}_M = \left[\begin{array}{ccc} \overset{m-d}{\underset{+}{}} & \overset{d}{\underset{-}{}} & \overset{1}{\underset{j'_c}{}} \\ R'_A & R'_B & 0 \end{array} \right] \qquad (11.31)$$

$$Q' := Q Q_c Q_M \qquad (11.32)$$

$$\text{if } j_c = +1: \quad T_M \begin{array}{c} {\scriptstyle m-d} \\ {\scriptstyle 1} \end{array} \left[\begin{array}{ccc} \overset{m-d}{\underset{+}{}} & \overset{d}{\underset{-}{}} & \overset{1}{\underset{j_c}{}} \\ I & 0 & 0 \\ 0 & 0 & 1 \end{array} \right] \tilde{\Theta}_c \tilde{\Theta}_M = \begin{array}{c} {\scriptstyle m-d'} \\ {\scriptstyle 1+d'-d} \end{array} \left[\begin{array}{ccc} \overset{m-d'}{\underset{+}{}} & \overset{d'}{\underset{-}{}} & \overset{1}{\underset{j'_c}{}} \\ I & 0 & \mathbf{e}'_c \\ * & * & * \end{array} \right] \qquad (11.33)$$

$$\text{if } j_c = -1: \quad T_M \ {\scriptstyle m-d} \left[\begin{array}{ccc} \overset{m-d}{\underset{+}{}} & \overset{d}{\underset{-}{}} & \overset{1}{\underset{j_c}{}} \\ I & 0 & 0 \end{array} \right] \tilde{\Theta}_c \tilde{\Theta}_M = \begin{array}{c} {\scriptstyle m-d'} \\ {\scriptstyle d'-d} \end{array} \left[\begin{array}{ccc} \overset{m-d'}{\underset{+}{}} & \overset{d'}{\underset{-}{}} & \overset{1}{\underset{j'_c}{}} \\ I & 0 & \mathbf{e}'_c \\ * & * & * \end{array} \right] \qquad (11.34)$$

$$\text{where} \qquad \mathbf{e}'_c = \begin{cases} 0, & j'_c = +1 \\ *, & j'_c = -1 \end{cases}$$

The last set of equations (11.33)-(11.34) represent (11.29). Let us summarize (11.33)-(11.34) by $T_M E\tilde{\Theta}_c\tilde{\Theta}_M = E'$, where the structure of E and E' depends on j_c, j'_c. We thus have to investigate four cases ($j_c = \pm 1$, $j'_c = \pm 1$). Depending on the case at hand, we have to ensure that selected parts of E' are zero. We can use T_M and additional rotations $\tilde{\Theta}_M$ for this purpose, *i.e.*, $E' = T_M(E\tilde{\Theta}_c)\tilde{\Theta}_M$, and we try to minimize the number of rotations in $\tilde{\Theta}_M$ since they might be hyperbolic and create fill-ins in R that have to be zeroed by additional rotations Q_M (viz. step 5). Note that $\tilde{\Theta}_M$ is not allowed to act on the last column of E (by definition, and because such an operation would destroy $\mathbf{c}' = 0$). Also note that the fill-in in $E\tilde{\Theta}_c$ is caused only by the two (hyperbolic) rotations that are present in $\tilde{\Theta}_c$ because of algorithm *zero-c*, hence consists only of 6 entries.

The following investigation of each of the cases separately is technical, but the result is simple: only in one case a specific action is required. This covers steps 2-4 from the outline in the previous section. The labeling of the cases follows figure 11.7, but we also assume that the sorting by *sort-R* has been carried out at this point.

(a) $j_c = +1$, $j'_c = +1$ ($d' = d$):

$$E\tilde{\Theta}_c = \begin{array}{c} \\ \\ m-d-1 \\ 1 \\ 1 \end{array} \begin{array}{c} \begin{array}{ccccc} m-d-1 & 1 & d-1 & 1 & 1 \\ + & + & - & - & + \end{array} \\ \left[\begin{array}{cc|cc|c} * & 0 & 0 & 0 & 0 \\ 0 & * & 0 & * & * \\ \hline 0 & * & 0 & * & * \end{array}\right] \end{array}$$

$$\Rightarrow \quad E' = T_M E\tilde{\Theta}_c\tilde{\Theta}_M = \begin{array}{c} \\ \\ m-d-1 \\ 1 \\ 1 \end{array} \begin{array}{c} \begin{array}{ccccc} m-d-1 & 1 & d-1 & 1 & 1 \\ + & + & - & - & + \end{array} \\ \left[\begin{array}{cc|cc|c} I & 0 & 0 & 0 & 0 \\ 0 & 1 & 0 & 0 & 0 \\ \hline 0 & 0 & 0 & * & * \end{array}\right] \end{array}$$

The first *col*-rotation from algorithm *zero-c* is in fact circular, and can be undone by choosing a similar rotation for T_M. Because matrix multiplication is associative, this automatically clears the fill-in in the top part of the last column as well. After the second θ-rotation, no fill-in in top part of the last column is created, and since the last row is unconstrained, we end up with the required structure. No extra rotations $\tilde{\Theta}_c$ result in this case, so that it is not necessary to actually perform the T_M-operations.

(b) $j_c = +1$, $j'_c = -1$ ($d' = d-1$): after sorting, we have, respectively,

$$E\tilde{\Theta}_c = \begin{array}{c} \\ \\ m-d-1 \\ 1 \\ 1 \end{array} \begin{array}{c} \begin{array}{ccccc} m-d-1 & 1 & 1 & d-1 & 1 \\ + & + & + & - & - \end{array} \\ \left[\begin{array}{ccc|c|c} * & 0 & 0 & 0 & 0 \\ 0 & * & * & 0 & * \\ 0 & * & * & 0 & * \\ \hline \end{array}\right] \end{array}$$

$$\Rightarrow \quad E' = \begin{array}{c} \\ \\ m-d-1 \\ 1 \\ 1 \end{array} \begin{array}{c} \begin{array}{ccccc} m-d-1 & 1 & 1 & d-1 & 1 \\ + & + & + & - & - \end{array} \\ \left[\begin{array}{ccc|c|c} I & 0 & 0 & 0 & 0 \\ 0 & 1 & 0 & 0 & 0 \\ 0 & 0 & 1 & 0 & * \\ \hline \end{array}\right] \end{array}$$

In this case, the fill-in by the first rotation in $\tilde{\Theta}_c$ is removed by a single circular rotation for T_M, as in the previous case. The second rotation only calls for a scaling of the last row by T_M; the last column has a negative signature ($j'_c = -1$) so is not constrained. Again, no extra Θ-operations are generated, so that it is not necessary to actually compute T_M.

(c) $j_c = -1$, $j'_c = +1$ $(d' = d+1)$:

$$E\tilde{\Theta}_c = \begin{array}{r} \\ \\ m-d-1 \\ 1 \end{array} \begin{array}{c|ccc|c} m-d-1 & 1 & d-1 & 1 & 1 \\ + & - & - & - & + \\ \hline * & 0 & 0 & 0 & 0 \\ \hline 0 & * & 0 & * & * \end{array}$$

$$\Rightarrow \quad E' = \begin{array}{r} \\ \\ m-d-1 \\ 1 \end{array} \begin{array}{c|ccc|c} m-d-1 & 1 & d-1 & 1 & 1 \\ + & - & - & - & + \\ \hline I & 0 & 0 & 0 & 0 \\ \hline 0 & * & 0 & * & * \end{array}$$

In this case, $d' = d+1$, hence the last row of E' is unconstrained. No operations are required.

(d) $j_c = -1$, $j'_c = -1$ $(d' = d)$. This case covers two possibilities: one in which no sign-changes occurred during the hyperbolic rotations, and one in which there was a double sign-change. After sorting, we have

$$E\tilde{\Theta}_c = \begin{array}{r} \\ \\ m-d-1 \\ 1 \end{array} \begin{array}{cc|cc|c} m-d-1 & 1 & d-1 & 1 & 1 \\ + & + & - & - & - \\ \hline * & 0 & 0 & 0 & 0 \\ 0 & * & 0 & * & * \\ \hline \end{array}$$

$$\text{or } E\tilde{\Theta}_c = \begin{array}{r} \\ \\ m-d-1 \\ 1 \end{array} \begin{array}{cc|cc|c} m-d-1 & 1 & 1 & d-1 & 1 \\ + & + & - & - & - \\ \hline * & 0 & 0 & 0 & 0 \\ 0 & * & * & 0 & * \\ \hline \end{array}$$

$$\Rightarrow \quad E' = \begin{array}{r} \\ \\ m-d-1 \\ 1 \end{array} \begin{array}{cc|cc|c} m-d-1 & 1 & 1 & d-1 & 1 \\ + & + & - & - & - \\ \hline I & 0 & 0 & 0 & 0 \\ 0 & 1 & 0 & 0 & * \\ \hline \end{array}$$

The last column is unconstrained ($j'_c = -1$), but the fill-in in the second block of the last row has to be zeroed, after which the row has to be scaled properly. This creates a situation that cannot be handled using T_M only: we need a (hyperbolic) *column* rotation $\tilde{\theta}$ to zero the selected entry. Before we can do this, the first possibility for $E\tilde{\Theta}_c$ requires us to place the two columns that are involved right next to each other, by column permutations. This will generate extra q-rotations as well, to keep R'_B lower triangular. After sorting, a single hyperbolic rotation $\tilde{\Theta}_M = \tilde{\theta}$ suffices. The resulting signature of the two columns involved in this rotation is sorted as $[+1 \ -1]$ automatically, because the total J-norm of this row is invariant: it is still $+1$, and the only other nonzero entry has a negative signature. After this rotation, the $+$-entry can be scaled by T_M to become 1. Again, this scaling need not be actually carried out. However, $\tilde{\theta}$ has to be applied to R' as well, and the resulting fill-in has to be zeroed using an additional q-rotation.

In: $\mathbf{c}$, j_c; R (lower), $J = \text{diag}[J_1, \cdots, J_m]$ (sorted), Q (unitary); d
Out: updated versions of R, J, Q, d, according to (11.31)-(11.34)

Algorithm SSE2-update:

zero-c: $\mathbf{c} := Q^* \mathbf{c}$
    $e_c = 0,\ e_{m-d} = 1,\ e_m = 0,$
    for $i = 1$ to m
        if $i = m-d$ or $i = m$
            zero c_i using **col** $(\tilde{\theta})$
            $[e_i\ e_c] := [e_i\ e_c]\tilde{\theta}$
        else
            zero c_i using **row**
    end

sort-R: for $i = m-1$ down to $m-d+1$
        permute columns i and $i+1$ of R (and J_i, J_{i+1})
        compute q to zero the fill-in $R_{i,i+1}$ against $R_{i+1,i+1}$
        apply q to rows $(i, i+1)$ of R and columns of Q
    end

M_Θ-trans.: if $J_{m-d} = -1$ and $J_{m-d+1} = +1$, (case (d))
        compute $\tilde{\theta}, \tilde{j}_2$ s.t. $[e_{m-d}\ e_m]\tilde{\theta} = [*\ 0]$, $\tilde{j}_1 = \text{diag}[J_{m-d}, J_{m-d+1}]$
        apply $\tilde{\theta}$ to columns $(m-d, m-d+1)$ of R, update signatures
        compute q to zero fill-in $R_{m-d,m-d+1}$ against $R_{m-d+1,m-d+1}$
        apply q to rows $(m-d, m-d+1)$ of R and columns of Q
    end
    update d : $d := d + \frac{1}{2}(j_c' - j_c)$

Figure 11.8. SSE-2 updating algorithm

Hence, only in case (d) do we have to perform an additional θ-rotation to effect the M_Θ-transformation. Note that we never have to act on the I_{m-d-1} matrix, only three entries of the last row of E are needed.

The resulting algorithm is summarized in figure 11.8, where the entries of E that we need to keep track of are denoted by e_{m-d}, e_m, e_c. The M_Θ-transformation, if needed, consists of a single θ-rotation on the columns of R, followed by a q-rotation on the rows of R to zero the fill-in. The sorting stage is slightly different than before: for simplicity, it now sorts unconditionally, and only up to column $m-d+1$. Possibly, one additional permutation is required (in case (d_2)). This permutation is a side effect of the M_Θ-transformation (*i.e.,* the sorting effect of the hyperbolic rotation of case (d)).

The last column is unconstrained ($j'_c = -1$), but the fill-in in the second block of the last row has to be zeroed, after which the row has to be scaled properly. This creates a situation that cannot be handled using T_M only: we need a (hyperbolic) column rotation $\tilde{\theta}$ to zero the selected entry. Before we can do this, the first possibility for $E\tilde{\Theta}_c$ requires us to place the two columns that are involved right next to each other, by column permutations. This will generate extra q-rotations as well, to keep R'_B lower triangular. After sorting, a single hyperbolic rotation $\tilde{\Theta}_M = \tilde{\theta}$ suffices. It can be shown from inertia considerations that the resulting signature of the two columns involved in this rotation will be sorted as $[+1 \ \ -1]$ automatically. $\tilde{\Theta}_M$ has to be applied to R' as well, and the resulting fill-in has to be zeroed using an additional q-rotation. A final observation is that we never have act on the I_{m-d-1} matrix, only three entries of the last row of E are needed.

The updating algorithm can be initialized by $R = 0$, $d = 0$, $Q = I_m$. The computational complexity is assessed as m^2 multiplications (for the initial transformation of $\mathbf{c}$ by Q), and about $2m^2 + 2md$ elementary rotations. This is four times more than the original HQR scheme for computing the SSE-1.

11.7 NOTES

The updating algorithm which we derived has the following properties. Its main feature is a localized, piecewise regular, data-independent computational flow using plane J-unitary rotations. The algorithm consists of two phases: a forward phase to zero the update vector, and a backward phase to restore the sorting and at the same time satisfy a structural constraint. Each phase is fully pipelineable, but unfortunately the combination is not, unless they can be meshed together (with some effort, this is sometimes possible, *cf.* [MDV93]). Per update vector, there are at most 3 hyperbolic rotations, which is not minimal, but significantly less than the HQR updating algorithm. Updating and downdating uses the same computational structure, since downdating H by a vector $\mathbf{h}$ can be done by updating N by $\mathbf{h}$. Exponential windowing and several interesting updating/downdating schemes are possible.

Two closely related subspace tracking algorithms are RRQR and URV. These are similar to SSE in that they update non-iteratively a rank-revealing factorization with respect to a specified threshold level. The tolerance for RRQR is a (soft) upper bound on the approximation error in matrix 2-norm, as in (11.2). URV on the other hand puts an upper bound on the error in *Frobenius* norm. All three algorithms roughly have the same number of operations, but RRQR and URV use only circular rotations and are numerically stable. The main distinctive feature is that RRQR and URV rely on a condition estimation to detect changes in rank, which can be regarded as their Achilles heel. The condition estimation results in a long critical path and makes the computational flow data-dependent. Also, the condition estimate is not perfect: in critical cases with a singular value close to the threshold, the rank decision is observed to become erratic.

Appendix 11.A: Proof of lemma 11.8

PROOF of lemma 11.8. Equation (11.27) implies $[\Theta_{11} \;\; \Theta_{12}]\Theta_M = [\Theta'_{11} \;\; \Theta'_{12}]$, which implies

$$\Theta_M \begin{bmatrix} -\Theta_{11}'^{-1}\Theta'_{12} \\ I_{n_2} \end{bmatrix} = \begin{bmatrix} -\Theta_{11}^{-1}\Theta_{12} \\ I_{n_2} \end{bmatrix} T$$

where T is some invertible matrix. Moreover, since Θ_M only acts on the columns of A, B, it is seen that T has to be block upper. Using this result and equation (11.26), we obtain

$$B' - A'M_{\Theta'} = [A \;\; 0 \,|\, B \;\; 0]\Theta_M \begin{bmatrix} -\Theta_{11}'^{-1}\Theta'_{12} \\ I \end{bmatrix} \begin{bmatrix} I \\ 0 \end{bmatrix} = [B - AM_\Theta \quad *]T \begin{bmatrix} I \\ 0 \end{bmatrix}.$$

Since T is block upper, $\text{ran}(B - AM_\Theta) = \text{ran}(B' - A'M_{\Theta'})$. To show the second part, *i.e.*, there exists Θ_M such that $M_{\Theta'} = 0$, it suffices to compute a J-unitary matrix Θ_M (compatible in size with Θ) such that

$$\overset{\begin{matrix} + & + & - & - \end{matrix}}{\begin{bmatrix} I_{m-d} & 0 & M_\Theta & 0 \end{bmatrix}}\Theta_M = \overset{\begin{matrix} + & + & - & - \end{matrix}}{\begin{bmatrix} * & 0 & 0 & 0 \end{bmatrix}}.$$

As $\|M_\Theta\| < 1$, such a matrix which does not change signatures exists and is bounded. Premultiplying with A, it follows that $[A \;\; 0 \,|\, AM_\Theta \;\; 0]\Theta_M = [* \;\; 0 \,|\, 0 \;\; 0]$, so that

$$\begin{aligned} [A \;\; 0 \,|\, B \;\; 0]\Theta_M &= [0 \;\; 0 \,|\, B - AM_\Theta \;\; 0]\Theta_M + [A \;\; 0 \,|\, AM_\Theta \;\; 0]\Theta_M \\ &= [* \;\; 0 \,|\, B' \;\; 0] + [* \;\; 0 \,|\, 0 \;\; 0] \quad =: [A' \;\; 0 \,|\, B' \;\; 0]. \end{aligned}$$

It is clear that $\text{ran}(B') = \text{ran}(B - AM_\Theta)$. Since $\text{ran}(A')$ complements $\text{ran}(B')$, the invariance of $\text{ran}(B - AM_\Theta)$ implies $M_{\Theta'} = 0$.

□

Appendix 11.B: The principal subspace is an SSE-2

We show that for a given matrix H, there is a decomposition (11.28)–(11.29) such that SSE-2 subspace Q_B is equal to the left principal subspace of H. Suppose for simplicity of notation that H is square, with SVD $H = U_1\Sigma_1 V_1^* + U_2\Sigma_2 V_2^*$, where $\Sigma_1 > \gamma I$ and $\Sigma_2 < \gamma I$. Define Q, Θ, T as

$$\begin{aligned} Q &= [U_2 \;\; U_1] \\ \Theta &= \left[\begin{array}{cc|cc} U_2\gamma & U_1\Sigma_1 & -U_1\gamma & -U_2\Sigma_2 \\ \hline -V_2\Sigma_2 & -V_1\gamma & V_1\Sigma_1 & V_2\gamma \end{array}\right] \left[\begin{array}{cc|cc} \gamma^2 - \Sigma_2^2 & & & \\ & \Sigma_1^2 - \gamma^2 & & \\ \hline & & \Sigma_1^2 - \gamma^2 & \\ & & & \gamma^2 - \Sigma_2^2 \end{array}\right]^{-1/2} \\ T &= (\gamma^2 - \Sigma_2^2)^{1/2}\gamma^{-1}U_2^* \end{aligned}$$

It is readily verified that Q is unitary, Θ is J-unitary, and that

$$\begin{aligned} Q^*[\gamma I \quad H]\Theta &= \left[\begin{array}{cc|cc} (\gamma^2-\Sigma_2^2)^{1/2} & 0 & 0 & 0 \\ 0 & 0 & (\Sigma_1^2-\gamma^2)^{1/2} & 0 \end{array}\right] \\ T[I \quad 0]\Theta &= [I \; 0 \; 0 \; \gamma^{-1}I]. \end{aligned}$$

so that (11.28)–(11.29) hold. Since $Q_B = U_1$, the SSE-2 subspace is equal to the principal subspace.

III FACTORIZATION

12 ORTHOGONAL EMBEDDING

In chapter 5, we saw how a state realization of a time-varying transfer operator T can be computed. The realizations which we obtained were in principle either in input normal form ($A^*A + B^*B = I$) or in output normal form ($AA^* + CC^* = I$). In chapter 6, we considered unitary systems V with unitary realizations. Such realizations are both in input normal form and in output normal form, and satisfy the additional property that both $\|V\| = 1$ and $\|\mathbf{V}\| = 1$, while for $\mathbf{T}$ in either normal form, we have $\|\mathbf{T}\| \geq 1$, whether $\|T\|$ is small or not. Since $\|\mathbf{T}\|$ tells something about the sensitivity of the realization, *i.e.*, the transfer of errors in either the input or the current state to the output and the next state, it is interesting to know whether it is possible to have a realization of T for which $\|\mathbf{T}\| \leq 1$ when $\|T\| \leq 1$. This issue can directly be phrased in terms of the problem which is the topic in this chapter: the *orthogonal embedding problem.* This problem is, given a transfer operator $T \in \mathcal{U}$, to extend this system by adding more inputs and outputs to it such that the resulting system Σ, a 2×2 block operator with entries in $\mathcal{U}$,

$$\Sigma = \begin{bmatrix} \Sigma_{11} & \Sigma_{12} \\ \Sigma_{21} & \Sigma_{22} \end{bmatrix},$$

is inner and has T as its partial transfer when the extra inputs are forced to zero: $T = \Sigma_{11}$. See figure 12.1. Since the unitarity of Σ implies $T^*T + T_c^*T_c = I$, (where $T_c = \Sigma_{21}$), it will be possible to find solutions to the embedding problem only if T is contractive: $I - T^*T \geq 0$, so that $\|T\| \leq 1$. Since Σ is inner, it has a unitary realization $\mathbf{\Sigma}$, and a possible realization $\mathbf{T}$ of T is at each point k in time a submatrix of $\mathbf{\Sigma}_k$ (with the same A_k, and smaller dimensional B_k, C_k, D_k), and hence $\mathbf{T}$ is a contractive realization.

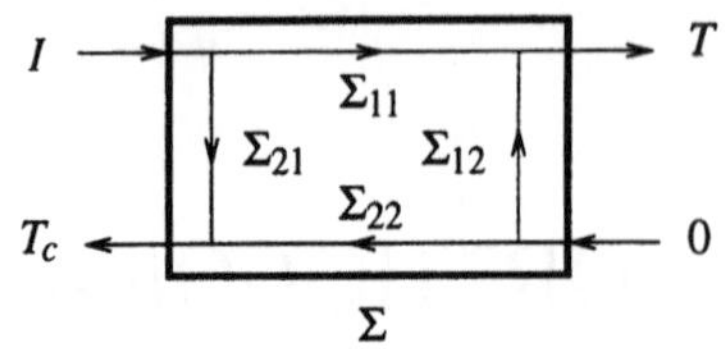

Figure 12.1. Embedding of a contractive time-varying operator T.

The orthogonal embedding problem, and algorithms to solve it, are the central issues in this chapter. The orthogonal embedding problem is known in other fields as well: it is called the unitary extension problem in operator theory, and the equations governing its solution (in a state-space context) are known in control theory as the discrete-time bounded real lemma.

12.1 INTRODUCTION AND CONNECTIONS

In this chapter, we present a constructive solution to the embedding problem, under the assumption that the number of states of T is finite at any point in time (locally finite systems). The construction is done in a state-space context and gives rise to (again) a time-varying Riccati equation. While it is clear that the contractivity of T is a necessary condition for the existence of an embedding, we show in the sequel that this condition is, also in the time-varying context, sufficient to construct a solution when T is locally finite and u.e. stable. (It is known that not all contractive transfer operators have an orthogonal embedding, see chapter 7 where we show this negative result for isometric operators. This generalizes what already happens in the time-invariant case [Dew76].) We first derive such a solution for the case where T is strictly contractive. The extension to the boundary case invokes some mathematical complications but in the end, almost the same algorithm is obtained [vdVD94a].

Besides the above application, the orthogonal embedding problem is typically the first step in digital filter synthesis problems in which filters (contractive operators) are realized as the partial transfer operator of a lossless multi-port filter Σ. Once such a Σ is obtained, it can be factored into various kinds of "ladder" or "lattice" cascade realizations consisting of elementary lossless degree-1 sections. Such a factorization is known in classical (time-invariant) circuit theory as a Darlington synthesis [Dar39, AV73], and provides a structured way to realize a given operator ('filter') in elementary components (in the circuit case, gyrators and a single resistor). In our case, each section is constructed with two elementary (Givens) rotors which have time-varying rotation angles, and the network that is obtained can, for example, be of the form depicted in figure 1.4. In this figure, the transfer function T is from (block) input u_1 to output y_1 if the secondary input u_2 is made equal to zero (the secondary output y_2 is not used). The structural factorization is the topic of chapter 14.

An application of the embedding problem in an operator or linear algebra context is the (Cholesky or spectral) factorization of a positive definite operator Ω into factors

$\Omega = W^*W$, where W is an upper operator. The transition to the embedding problem is obtained by a Cayley transformation, which transforms $\Omega > 0$ to an upper strictly contractive operator T: a scattering operator. From the orthogonal embedding Σ, a factor W can be derived via a few straightforward manipulations. This subsumes the generalized Schur method [DD88] that has also been used for this application, and in which an embedding Σ is obtained in cascaded form. However, the Schur method is order recursive, and can indeed give rise to a fairly large order, whereas the embedding procedure in this chapter can be used to obtain an embedding Σ and a factor W of minimal order. This connection is described in chapter 13.

The time-invariant orthogonal embedding problem in its simplest form acts on transfer functions $T(z)$ and uses a spectral factorization: with

$$T(z) = \frac{h(z)}{f(z)}, \quad T_c(z) = \frac{g(z)}{f(z)} \tag{12.1}$$

where f, g, h are polynomials of finite degree, it is derived that $g(z)$ (and hence $T_c(z)$) can be determined from a spectral factorization of

$$g(z)g_*(z) = f(z)f_*(z) - h(z)h_*(z)$$

where $f_*(z) = \overline{f(\bar{z}^{-1})}$ [Bel68]. The solution of the spectral factorization problem involves finding the zeros of $g(z)g_*(z)$. Note that in equation (12.1) we use the knowledge that T_c can have the same poles as T.

Polynomial spectral factorization for multi-input/multi-output systems is rather complicated, see *e.g.*, [Dew76]. A solution strategy that is easier to handle (and that carries over to the time-varying case too) is obtained when the problem is cast into a state space context. Such an approach is discussed in [AV73] for continuous-time systems, and implies what is called the bounded real lemma. This lemma states that $T(s)$ is contractive if and only if certain conditions on the state-space matrices are fulfilled. If this is the case, the conditions are such that they imply a realization for $T_c(s)$ such that $[T(s) \;\; T_c(s)]$ is lossless and has the same A and C matrices as the realization of T. To determine this solution, a Riccati equation has to be solved. The bounded real lemma can without much effort be stated in the discrete-time context by means of a bilinear transformation [AHD74]. A derivation based on the conservation of energy appears in [GS84], and a proof independent of a continuous-time equivalent is given in [Vai85a]. A Riccati equation which describes the problem is stated in [Des91], which forms the basis of a cascade factorization. Control applications of the bounded real lemma include H_∞-optimal state regulation and state estimation [YS91].

In the present chapter, the aim is to extend the above classical time-invariant theory to the time-varying context. To introduce the strategy for solving the time-varying embedding problem in a state-space context, consider the following simplified problem. Just for the moment, let T be a single-input, single-output system, with state-space realization $\mathbf{T}$ of constant dimensions. The objective is to determine a lossless embedding system Σ, having two inputs and two outputs, and with state-space realization $\mathbf{\Sigma}$ of the form

$$\mathbf{\Sigma} = \begin{bmatrix} R & & \\ & I & \\ & & I \end{bmatrix} \left[\begin{array}{cc|c} A & C & C_2 \\ B & D & D_{12} \\ \hline B_2 & D_{21} & D_{22} \end{array}\right] \begin{bmatrix} \left[R^{(-1)}\right]^{-1} & & \\ & I & \\ & & I \end{bmatrix},$$

(all entries in this expression are diagonals). $\mathbf{\Sigma}$ contains the given realization $\mathbf{T}$, suitably state-space transformed by some boundedly invertible $R = \mathrm{diag}(R_i)$, which does not alter the input-output characteristics, hence Σ_{11} is equal to the given T. $\mathbf{\Sigma}$ is extended by matrix operators B_2, C_2, D_{21}, D_{12}, D_{22} corresponding to the second input and output. If Σ is to be inner, it must have a unitary realization $\mathbf{\Sigma}$ (theorem 6.3). Conversely, if $\mathbf{\Sigma}$ is unitary and $\ell_A < 1$, then the corresponding transfer operator Σ is inner; see theorem 6.4), and hence a way to solve the embedding problem using state-space methods is to require $\mathbf{\Sigma}$ to be unitary.

The embedding problem is thus reduced to finding the state transformation R, and the embedding matrices B_2 etc., such that $\mathbf{\Sigma}$ is unitary. The problem can be split into two parts:

1. Determine R, B_2, D_{21} to make the columns of $\mathbf{\Sigma}_a$ isometric and orthogonal to each other, with

$$\mathbf{\Sigma}_a = \begin{bmatrix} R & & \\ & I & \\ & & I \end{bmatrix} \left[\begin{array}{cc} A & C \\ B & D \\ \hline B_2 & D_{21} \end{array}\right] \begin{bmatrix} \left[R^{(-1)}\right]^{-1} & \\ & I \end{bmatrix}.$$

   That is, $(\mathbf{\Sigma}_a)^* \mathbf{\Sigma}_a = I$.

2. Add one orthonormal column $\mathbf{\Sigma}_b$ to $\mathbf{\Sigma}_a$ to make $\mathbf{\Sigma} = [\mathbf{\Sigma}_a \;\; \mathbf{\Sigma}_b]$ unitary. The realization $\mathbf{\Sigma}$ that is obtained consists of a diagonal sequence of square finite-dimensional matrices, hence this can always be done.

The key step in the above construction is step 1. With proper attention to the dimensions of the embedding, it is always possible to find solutions to step 2 since in general, $\mathbf{\Sigma}_b$ is just the orthogonal complement of the columns of $\mathbf{\Sigma}_a$.

The orthonormality conditions of step 1 translate to a set of equations whose solution depends at each time instant i on the (strict) positivity of a matrix $M_i = R_i^* R_i$, which, as we will show, can be computed recursively from the given state-space realization as

$$\begin{aligned} M_{i+1} &= A_i^* M_i A_i + B_i^* B_i + \\ &\quad [A_i^* M_i C_i + B_i^* D_i]\,(I - D_i^* D_i - C_i^* M_i C_i)^{-1}\,[D_i^* B_i + C_i^* M_i A_i]\,. \end{aligned} \tag{12.2}$$

This recursion is again a Riccati-type recursion. The problem with such recursions is the term $(I - D_i^* D_i - C_i^* M_i C_i)$, which can potentially become negative and cause M_{i+1} to become negative (or indefinite) too. The main contribution of the theory given in the rest of the chapter is to show that the recursion does not break down (*i.e.*, all M_i are uniformly positive, hence we can find a sequence of invertible state-space transformations R_i), under the condition that T is strictly contractive and the given realization

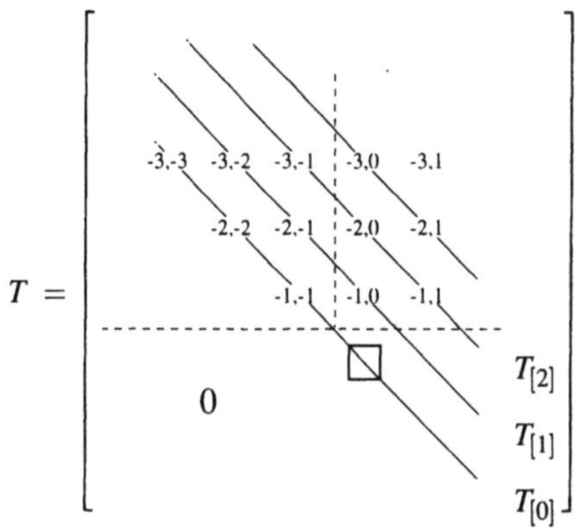

$$K_0 = \begin{bmatrix} T_{-1,-1} & & & \mathbf{0} \\ T_{-2,-1} & T_{-2,-2} & & \\ T_{-3,-1} & T_{-3,-2} & T_{-3,-3} & \\ \cdots & & & \ddots \end{bmatrix}$$

$$H_0 = \begin{bmatrix} T_{-1,0} & T_{-1,1} & T_{-1,2} & \cdots \\ T_{-2,0} & T_{-2,1} & T_{-2,2} & \\ T_{-3,0} & T_{-3,1} & T_{-3,2} & \\ \cdots & & & \ddots \end{bmatrix} \qquad V_0 = \begin{bmatrix} T_{-1,0} \\ T_{-2,0} \\ T_{-3,0} \\ \cdots \end{bmatrix}$$

Figure 12.2. K_i, H_i and V_i are submatrices of T.

for T is uniformly reachable. Subsequently, we show in section 12.3 that a slightly altered recursion also does not break down if T is contractive (but not necessarily in the strict sense), but then we have to impose more requirements on $\mathbf{T}$, for example that it be uniformly observable. These requirements are sufficient but possibly too restrictive.

Preliminary relations

We recall some notations and definitions from chapters 2, 4 and 5, and define some additional ones as well. Let $T \in \mathcal{U}$. We will use the following partial transfer operators on a restricted domain and range (*cf.* equation (5.2)):

$$\begin{array}{lll} H_T: & \mathcal{L}_2Z^{-1} \to \mathcal{U}_2, & uH_T = \mathbf{P}(uT) \\ K_T: & \mathcal{L}_2Z^{-1} \to \mathcal{L}_2Z^{-1}, & uK_T = \mathbf{P}'(uT) \\ V_T: & \mathcal{L}_2Z^{-1} \to \mathcal{D}_2, & uV_T = \mathbf{P}_0(uT)\,. \end{array}$$

For $u \in \mathcal{L}_2Z^{-1}$ we have that $uT = uK_T + uH_T$. V_T is a further restriction of H_T.

We have already used the fact that H_T is a left D-invariant operator, and hence has "snapshots" H_i (definition 4.1), which can be viewed as a sequence of *time-varying* matrices that would have a Hankel structure in the time-invariant case. In the same way, matrix representations are obtained for K_i and vector representations for V_i:

$$H_i = \begin{bmatrix} T_{i-1,i} & T_{i-1,i+1} & T_{i-1,i+2} & \cdots \\ T_{i-2,i} & T_{i-2,i+1} & & \ddots \\ T_{i-3,i} & & & \\ \vdots & & & \end{bmatrix}$$

$$V_i = \begin{bmatrix} T_{i-1,i} \\ T_{i-2,i} \\ T_{i-3,i} \\ \vdots \end{bmatrix} \qquad K_i = \begin{bmatrix} T_{i-1,i-1} & & & \mathbf{0} \\ T_{i-2,i-1} & T_{i-2,i-2} & & \\ T_{i-3,i-1} & T_{i-3,i-2} & T_{i-3,i-3} & \\ \vdots & & \vdots & \ddots \end{bmatrix}.$$

Again because H_T, K_T and V_T are D invariant, they also have diagonal expansions $\tilde{H}_T$, $\tilde{K}_T$ and $\tilde{V}_T$, as follows. Define the diagonal expansions of signals u in $\mathcal{L}_2Z^{-1}$ and

y in $\mathcal{U}_2$ as

$$\begin{aligned} u &= Z^{-1}u_{[-1]} + Z^{-2}u_{[-2]} + \cdots = u_{[-1]}^{(+1)}Z^{-1} + u_{[-2]}^{(+2)}Z^{-2} + \cdots \\ \tilde{u} &= \begin{bmatrix} u_{[-1]}^{(+1)} & u_{[-2]}^{(+2)} & \cdots \end{bmatrix} \in \ell_2^-(\mathcal{D}). \end{aligned}$$

$$\begin{aligned} y &= y_{[0]} + Zy_{[1]} + Z^2y_{[2]} + \cdots = y_{[0]} + y_{[1]}^{(-1)}Z + y_{[2]}^{(-2)}Z^2 + \cdots \\ \tilde{y} &= \begin{bmatrix} y_{[0]} & y_{[1]}^{(-1)} & y_{[2]}^{(-2)} & \cdots \end{bmatrix} \in \ell_2^+(\mathcal{D}). \end{aligned}$$

Induced by this isomorphy, the definitions

$$\begin{array}{lllcrcl} y_f = uH_T & \in & \mathcal{U}_2 & \Leftrightarrow & \tilde{y}_f & = & \tilde{u}\tilde{H}_T \\ y_p = uK_T & \in & \mathcal{L}_2Z^{-1} & \Leftrightarrow & \tilde{y}_p & = & \tilde{u}\tilde{K}_T \\ D = uV_T & \in & \mathcal{D}_2 & \Leftrightarrow & D & = & \tilde{u}\tilde{V}_T. \end{array}$$

lead to

$$\tilde{H}_T = \begin{bmatrix} T_{[1]} & T_{[2]}^{(-1)} & T_{[3]}^{(-2)} & \cdots \\ T_{[2]} & T_{[3]}^{(-1)} & & \\ T_{[3]} & & \ddots & \\ \vdots & & & \end{bmatrix}$$

$$\tilde{V}_T = \begin{bmatrix} T_{[1]} \\ T_{[2]} \\ T_{[3]} \\ \vdots \end{bmatrix} \qquad \tilde{K}_T = \begin{bmatrix} T_{[0]}^{(+1)} & & \mathbf{0} & \\ T_{[1]}^{(+1)} & T_{[0]}^{(+2)} & & \\ T_{[2]}^{(+1)} & T_{[1]}^{(+2)} & T_{[0]}^{(+3)} & \\ \vdots & & \vdots & \ddots \end{bmatrix}. \tag{12.3}$$

The connection of $\tilde{H}_T$ with H_i is obtained by selecting the i-th entry of each diagonal in $\tilde{H}_T$ and constructing a matrix from it. Similarly, the sequence K_i forms a matrix representation of the operator K_T and likewise V_i is the vector representation of the operator V_T, obtained by selecting the i-th entry of each diagonal in the representation of $\tilde{V}_T$.

Recall from chapter 5 that H_i has a factorization $H_i = \mathcal{C}_i\mathcal{O}_i$, where $\mathcal{C}_i$ and $\mathcal{O}_i$ are the reachability and observability matrices as defined in (3.23). In terms of diagonal expansions, it is straightforward to show that $\tilde{H}_T$ has a decomposition $\tilde{H}_T = \mathcal{C}\mathcal{O}$, where $\mathcal{C}$ and $\mathcal{O}$ are defined as

$$\mathcal{C} := \begin{bmatrix} B^{(1)} \\ B^{(2)}A^{(1)} \\ B^{(3)}A^{(2)}A^{(1)} \\ \vdots \end{bmatrix} \qquad \mathcal{O} := \begin{bmatrix} C & AC^{(-1)} & AA^{(-1)}C^{(-2)} & \cdots \end{bmatrix}.$$

Note that $\mathcal{C}$ and $\mathcal{O}$ are diagonal expansions of the reachability and observability operators $\mathbf{P}_0(\cdot\mathbf{F}^*) = \mathbf{P}_0(\cdot BZ(I-AZ)^{-1}$ and $\cdot\mathbf{F}_o = \cdot(I-AZ)^{-1}C$. In turn, $\mathcal{C}_i$ and $\mathcal{O}_i$ are snapshots of these operators.

Since $\tilde{V}_T$ is the first column of $\tilde{H}_T$, we have that $\tilde{V}_T$ has a decomposition

$$\tilde{V}_T = \mathcal{C} \cdot C. \tag{12.4}$$

Finally, it is clear from equation (12.3) that $\tilde{K}_T$ satisfies the relation

$$\tilde{K}_T^{(-1)} = \begin{bmatrix} T_{[0]} & 0 \\ \tilde{V}_T & \tilde{K}_T \end{bmatrix}. \tag{12.5}$$

This relation is seen to correspond to a recursive relation: it specifies that

$$K_{i+1} = \begin{bmatrix} T_{ii} & 0 0 \cdots \\ V_i & K_i \end{bmatrix}$$

for all time instants i. K_i 'grows' when i increases as the history of the system grows — in particular, K_∞ is just a mirrored version of T.

12.2 STRICTLY CONTRACTIVE SYSTEMS

As indicated in the introduction, an orthogonal embedding of a transfer operator $T \in \mathcal{U}$ is possible only if T is at least contractive. In this section, we explore the consequences of assuming the strict contractivity of T, which leads to sufficient conditions for an embedding to exist if T is strictly contractive. This is done in two steps. Lemma 12.3 derives a general relation in terms of $\tilde{V}_T$ and $\tilde{K}_T$ which is a direct consequence of the strict contractivity of T. Theorem 12.4 uses this relation to show that some quantity $M \in \mathcal{D}$, defined by $M = \mathcal{C}^*(I - \tilde{K}_T\tilde{K}_T^*)^{-1}\mathcal{C}$, is strictly positive definite, and gives a recursion for this M in terms of state-space quantities of T. The point is that this recursion is precisely the same as the recursion for M in the embedding problem (*viz.* equation (12.2)). This proves the essential step in the embedding problem for strictly contractive operators (section 12.4). The case where T is contractive, but not necessarily strictly contractive, is deferred to section 12.3.

Contractivity of a transfer operator

Recall proposition 4.3 on the positivity, respectively the strict positivity of a Hermitian operator $A \in \mathcal{X}$:

$$\begin{array}{lcl} A \geq 0 & \Leftrightarrow & \{uA, u\} \geq 0, \quad (\text{all } u \in \mathcal{X}_2) \\ A \gg 0 & \Leftrightarrow & \exists \varepsilon > 0: \{uA, u\} \geq \varepsilon \{u, u\}, \quad (\text{all } u \in \mathcal{X}_2). \end{array}$$

Let T be a transfer operator in $\mathcal{U}$. We have defined, in section 4.2, to call T contractive, respectively strictly contractive, if

$$I - TT^* \geq 0, \qquad \text{resp.} \qquad I - TT^* \gg 0.$$

In the latter case, $I - TT^*$ is boundedly invertible. In this section, our focus is on the case that T is strictly contractive. The more general case is treated in section 12.3. $I - TT^* \gg 0$ implies that $I - T^*T \gg 0$, because of the identity $I + T^*(I - TT^*)^{-1}T = (I - T^*T)^{-1}$.

Lemma 12.1 *If T is strictly contractive, then K_T and $\tilde{K}_T$ are strictly contractive.*

PROOF Let $u \in \mathcal{L}_2 Z^{-1}$, and $y = uK_T$. Since T is strictly contractive, we have from the above definition that

$$\begin{aligned} \mathbf{P}_0(uu^*) - \mathbf{P}_0(yy^*) &= \mathbf{P}_0[u(I - K_T K_T^*)u^*] \\ &\geq \mathbf{P}_0[u(I - TT^*)u^*] \\ &\geq \varepsilon \mathbf{P}_0(uu^*) \qquad (\text{some } \varepsilon > 0). \end{aligned}$$

Since, by definition of the diagonal expansion, $\mathbf{P}_0(uu^*) = \tilde{u}\tilde{u}^*$ and $\mathbf{P}_0(yy^*) = \tilde{y}\tilde{y}^*$, and by definition of $\tilde{K}_T$, $\tilde{y} = \tilde{u}\tilde{K}_T$, we obtain that

$$\begin{aligned} \tilde{u}(I - \tilde{K}_T \tilde{K}_T^*)\tilde{u}^* &= \tilde{u}\tilde{u}^* - \tilde{y}\tilde{y}^* \\ &= \mathbf{P}_0(uu^*) - \mathbf{P}_0(yy^*) \\ &\geq \varepsilon \mathbf{P}_0(uu^*) = \varepsilon \tilde{u}\tilde{u}^* \qquad (\text{some } \varepsilon > 0), \end{aligned}$$

which shows that we also have that $\tilde{K}_T$ is strictly contractive: $I - \tilde{K}_T \tilde{K}_T^* \gg 0$, $I - \tilde{K}_T^* \tilde{K}_T \gg 0$. □

The fact that K_T is strictly contractive implies in turn that all K_i are strictly contractive.

Strict contractivity in terms of a state-space realization

The purpose of this section is to find conditions in state-space quantities on the contractivity of a transfer operator T. To this end, we use K_T rather than T, and in particular the fact that $I - K_T K_T^*$ is boundedly invertible and strictly positive when T is contractive. Since $\tilde{K}_T^{(-1)}$ can be specified in terms of $\tilde{K}_T$ and an extra column of diagonals (equation (12.5)), it is possible to derive a (recursive) formula for $(I - \tilde{K}_T \tilde{K}_T^*)^{(-1)}$ in terms of $\tilde{K}_T$ and the newly introduced column. The following lemma is standard and will be instrumental.

Lemma 12.2 (Schur complements/inversion formula) *Let X be a block-partitioned operator,*

$$X = \begin{bmatrix} A & B^* \\ B & C \end{bmatrix},$$

where A, B and C are bounded operators on Hilbert spaces, and let A and C be self-adjoint. Then

$$X \gg 0 \quad \Leftrightarrow \quad \begin{cases} (1) & C \gg 0 \\ (2) & A - B^* C^{-1} B \gg 0. \end{cases}$$

If $X \gg 0$, then

$$\begin{aligned} \begin{bmatrix} A & B^* \\ B & C \end{bmatrix}^{-1} &= \begin{bmatrix} I & 0 \\ -C^{-1}B & I \end{bmatrix} \begin{bmatrix} (A - B^*C^{-1}B)^{-1} & 0 \\ 0 & C^{-1} \end{bmatrix} \begin{bmatrix} I & -B^*C^{-1} \\ 0 & I \end{bmatrix} \\ &= \begin{bmatrix} 0 & 0 \\ 0 & C^{-1} \end{bmatrix} + \begin{bmatrix} I \\ -C^{-1}B \end{bmatrix} (A - B^*C^{-1}B)^{-1} \begin{bmatrix} I & -B^*C^{-1} \end{bmatrix}. \end{aligned}$$

PROOF $X \gg 0$ implies that $C \gg 0$. If $C \gg 0$, then C^{-1} exists, and

$$\begin{bmatrix} A & B^* \\ B & C \end{bmatrix} = \begin{bmatrix} I & B^*C^{-1} \\ & I \end{bmatrix} \begin{bmatrix} A - B^*C^{-1}B & \\ & C \end{bmatrix} \begin{bmatrix} I & \\ C^{-1}B & I \end{bmatrix}$$

Because the first and third factors in this decomposition are invertible,

$$\begin{bmatrix} A & B^* \\ B & C \end{bmatrix} \gg 0 \quad \Leftrightarrow \quad \begin{bmatrix} A - B^*C^{-1}B & \\ & C \end{bmatrix} \gg 0$$

$$\Leftrightarrow \quad \begin{cases} (1) & C \gg 0 \\ (2) & A - B^*C^{-1}B \gg 0. \end{cases}$$

This proves the first part of the lemma. The second part is immediate from the above factorization of X. □

Lemma 12.3 *Let be given a transfer operator $T \in \mathcal{U}$. If T is strictly contractive, then*

$$I - T_{[0]}^* T_{[0]} - \tilde{V}_T^* (I - \tilde{K}_T \tilde{K}_T^*)^{-1} \tilde{V}_T \gg 0.$$

PROOF Since T is strictly contractive, lemma 12.1 ensures that $\tilde{K}_T$ and $\tilde{K}_T^{(-1)}$ are also strictly contractive. Using equation (12.5), we have that

$$I - \tilde{K}_T^{(-1)*} \tilde{K}_T^{(-1)} = \begin{bmatrix} I - T_{[0]}^* T_{[0]} - \tilde{V}_T^* \tilde{V}_T & -\tilde{V}_T^* \tilde{K}_T \\ -\tilde{K}_T^* \tilde{V}_T & I - \tilde{K}_T^* \tilde{K}_T \end{bmatrix}. \tag{12.6}$$

Now apply lemma 12.2. It is seen that this expression is strictly positive definite if and only if

$$\begin{cases} (1) & I - \tilde{K}_T^* \tilde{K}_T \gg 0 \\ (2) & I - T_{[0]}^* T_{[0]} - \tilde{V}_T^* \tilde{V}_T - \tilde{V}_T^* \tilde{K}_T (I - \tilde{K}_T^* \tilde{K}_T)^{-1} \tilde{K}_T^* \tilde{V}_T \gg 0. \end{cases}$$

The first condition is satisfied because T is strictly contractive. The second condition is equal to the result, because of the equality $I + \tilde{K}_T (I - \tilde{K}_T^* \tilde{K}_T)^{-1} \tilde{K}_T^* = (I - \tilde{K}_T \tilde{K}_T^*)^{-1}$.

□

Theorem 12.4 *Let $T \in \mathcal{U}$ be a locally finite transfer operator with state realization $\{A, B, C, D\}$, where $A \in \mathcal{D}(\mathcal{B}, \mathcal{B}^{(-1)})$ is u.e. stable ($\ell_A < 1$). If T is strictly contractive, then $M \in \mathcal{D}(\mathcal{B}, \mathcal{B})$, defined by*

$$M = \mathcal{C}^* (I - \tilde{K}_T \tilde{K}_T^*)^{-1} \mathcal{C}, \tag{12.7}$$

satisfies the relations $I - D^*D - C^*MC \gg 0$, and

$$M^{(-1)} = A^*MA + B^*B + \left[A^*MC + B^*D\right] (I - D^*D - C^*MC)^{-1} \left[D^*B + C^*MA\right].$$

If in addition the state-space realization is uniformly reachable, then $M \gg 0$.

PROOF M is well defined if T is strictly contractive, which also implies that $M \geq 0$. If in addition the state-space realization is uniformly reachable, $\mathcal{C}^*\mathcal{C} \gg 0$, then $M \gg 0$ and hence M is invertible.

With the definition of M and using the fact that $D = T_{[0]}$ and $\tilde{V}_T = \mathcal{C} \cdot C$ (equation (12.4)), the positivity of $I - D^*D - C^*MC$ follows directly from lemma 12.3.

The recursive relation for M is obtained by an application of Schur's inversion formula (lemma 12.2) to equation (12.6), which gives

$$\begin{aligned}\left[I - \tilde{K}_T^{(-1)*}\tilde{K}_T^{(-1)}\right]^{-1} &= \begin{bmatrix} 0 & \\ & (I - \tilde{K}_T^*\tilde{K}_T)^{-1} \end{bmatrix} + \\ &\quad + \begin{bmatrix} I \\ (I - \tilde{K}_T^*\tilde{K}_T)^{-1}\tilde{K}_T^*\tilde{V}_T \end{bmatrix} \Phi^{-2} \begin{bmatrix} I & \tilde{V}_T^*\tilde{K}_T(I - \tilde{K}_T^*\tilde{K}_T)^{-1} \end{bmatrix}\end{aligned} \tag{12.8}$$

with

$$\begin{aligned}\Phi^2 &= I - T_{[0]}^*T_{[0]} - \tilde{V}_T^*\tilde{V}_T - \tilde{V}_T^*\tilde{K}_T(I - \tilde{K}_T^*\tilde{K}_T)^{-1}\tilde{K}_T^*\tilde{V}_T \\ &= I - D^*D - C^*MC.\end{aligned}$$

The invertibility of Φ^2 was already shown. Inserting this expression into the definition of $M^{(-1)}$, and using the relations that have been summarized above, $M^{(-1)}$ is obtained as

$$\begin{aligned}M^{(-1)} &= \mathcal{C}^{(-1)*}\left[I - \tilde{K}_T^{(-1)}\tilde{K}_T^{(-1)*}\right]^{-1}\mathcal{C}^{(-1)} \\ &= \mathcal{C}^{(-1)*}\left[I + \tilde{K}_T^{(-1)}\left(I - \tilde{K}_T^{(-1)*}\tilde{K}_T^{(-1)}\right)^{-1}\tilde{K}_T^{(-1)*}\right]\mathcal{C}^{(-1)} \\ &= \begin{bmatrix} B^* & A^*\mathcal{C}^* \end{bmatrix}\begin{bmatrix} B \\ \mathcal{C}A \end{bmatrix} + \begin{bmatrix} B^* & A^*\mathcal{C}^* \end{bmatrix}\begin{bmatrix} T_{[0]} & \\ \tilde{V}_T & \tilde{K}_T \end{bmatrix} \cdot \\ &\quad \cdot\left(\begin{bmatrix} 0 & \\ & (I - \tilde{K}_T^*\tilde{K}_T)^{-1} \end{bmatrix} + \begin{bmatrix} I \\ (I - \tilde{K}_T^*\tilde{K}_T)^{-1}\tilde{K}_T^*\tilde{V}_T \end{bmatrix}\Phi^{-2}\begin{bmatrix} I & \tilde{V}_T^*\tilde{K}_T(I - \tilde{K}_T^*\tilde{K}_T)^{-1} \end{bmatrix}\right) \cdot \\ &\quad \cdot\begin{bmatrix} T_{[0]} & \\ \tilde{V}_T & \tilde{K}_T \end{bmatrix}^*\begin{bmatrix} B \\ \mathcal{C}A \end{bmatrix} \\ &= B^*B + A^*\mathcal{C}^*\mathcal{C}A + A^*\mathcal{C}^*\tilde{K}_T(I - \tilde{K}_T^*\tilde{K}_T)^{-1}\tilde{K}_T^*\mathcal{C}A + \\ &\quad + \left(B^*D + A^*\mathcal{C}^*\left[I + \tilde{K}_T(I - \tilde{K}_T^*\tilde{K}_T)^{-1}\tilde{K}_T^*\right]\mathcal{C}C\right) \cdot \Phi^{-2} \cdot \\ &\quad \cdot\left(D^*B + C^*\mathcal{C}^*\left[I + \tilde{K}_T^*(I - \tilde{K}_T^*\tilde{K}_T)^{-1}\tilde{K}_T\right]\mathcal{C}A\right) \\ &= B^*B + A^*MA + (A^*MC + B^*D)\Phi^{-2}(D^*B + C^*MA).\end{aligned}$$

□

The equation (12.20) for M is actually a recursive equation, which becomes apparent if we write $M = \text{diag}[M_i]$ and take the i-th entry of every diagonal in the equation: this produces the Riccati recursion (12.2). Theorem 12.4 claims that for a strictly contractive system, the Riccati recursion has a positive solution M, which is given in explicit form. In section 12.4 this M plays a crucial role in the construction of such an embedding. It also furnishes part of the proof of the bounded real lemma.

12.3 CONTRACTIVE SYSTEMS: THE BOUNDARY CASE[1]

We will now derive an equivalent of theorem 12.4 for the case where T is contractive but not necessarily strictly contractive: $I-TT^* \geq 0$. While the mathematical derivation is more complicated now, the resulting theorem is only slightly altered. It will turn out that K_T is not strictly contractive, and that, instead of $(I-\tilde{K}_T\tilde{K}_T^*)^{-1}$, we will have to use the pseudo-inverse of $(I-\tilde{K}_T^*\tilde{K}_T)$. Mathematical complications arise because the range of $(I-\tilde{K}_T^*\tilde{K}_T)$ is not necessarily closed, so that its pseudo-inverse can be unbounded.

Schur inversion formulas for positive semi-definite operators

Let be given some operator A on a Hilbert space $\mathcal{H}$. For better correspondence with results from other papers, as well as for historical reasons, we work in this section with operators written from the right to the left, and thus denote the 'left' range of A as $\mathcal{R}(A) = \{Ax : x \in \mathcal{H}\}$, and its nullspace is as $\mathcal{N}(A) = \{x : Ax = 0\}$, which is a closed subspace. An orthogonal complement is denoted by $\perp$. The operator pseudo-inverse of A is defined as follows (following Beutler and Root [BR76]).

Definition 12.5 *Let $\mathcal{H}$ be a Hilbert space, and A be a bounded linear operator defined on $\mathcal{H}$. The linear operator $A^\dagger : \mathcal{H} \to \mathcal{H}$ is a pseudo-inverse of A if and only if it is defined on $\mathcal{R}(A) \oplus \mathcal{R}(A)^\perp$ (which is dense in $\mathcal{H}$) and satisfies the following conditions:*

$$\begin{array}{lrcl} (1) & \mathcal{N}(A^\dagger) & = & \mathcal{R}(A)^\perp \\ (2) & \overline{\mathcal{R}(A^\dagger)} & = & \mathcal{N}(A)^\perp \quad (= \overline{\mathcal{R}(A^*)}) \\ (3) & AA^\dagger x & = & x \quad \text{for all } x \in \mathcal{R}(A). \end{array}$$

It is proven in [BR76] that $(A^\dagger)^\dagger = A$, $(A^\dagger)^* = (A^*)^\dagger$, $(A^*A)^\dagger = A^\dagger A^{*\dagger}$, and that $A^\dagger$ is bounded if and only if $\mathcal{R}(A)$ is closed. We will also apply a result of Douglas [Dou66] on majorization of operators on Hilbert spaces:

Theorem 12.6 *Let A and B be bounded operators on a Hilbert space $\mathcal{H}$. The following are equivalent:*

$$\begin{array}{lrcl} (1) & AA^* & \leq & \lambda^2 BB^* \quad (\text{some } \lambda > 0), \\ (2) & \mathcal{R}(A) & \subset & \mathcal{R}(B), \\ (3) & A & = & BC \quad \text{for some bounded operator } C \text{ on } \mathcal{H}. \end{array}$$

If (1)-(3) are valid, then a unique operator C exists such that

$$\begin{array}{lrcl} (a) & \|C\| & = & \inf\{\mu : AA^* \leq \mu BB^*\}, \\ (b) & \mathcal{N}(A) & = & \mathcal{N}(C), \\ (c) & \mathcal{R}(C) & \subset & \overline{\mathcal{R}(B^*)}. \end{array}$$

The 'unique operator C' in this theorem is in fact $C = B^\dagger A$, since also $B^\dagger$ is uniquely defined and $B^\dagger A$ qualifies for C. Consequently, if $AA^* \leq BB^*$, then this C satisfies $\|C\| <$ 1.

[1] This section may be skipped without loss of continuity.

Using pseudo-inverses, the Schur inversion formula (lemma 12.2) can be extended to the case where X is not uniformly positive.

Lemma 12.7 *With $\mathcal{H}_1$ and $\mathcal{H}_2$ Hilbert spaces, let $A:\mathcal{H}_1\to\mathcal{H}_2$, $B:\mathcal{H}_1\to\mathcal{H}_2$, $C:\mathcal{H}_2\to\mathcal{H}_2$ be bounded operators, and let A and C be self-adjoint. Then*

$$X:=\begin{bmatrix} A & B^* \\ B & C\end{bmatrix}\geq 0 \Leftrightarrow \begin{cases} (1) & C\geq 0, \\ (2) & \mathcal{R}(B)\subset\mathcal{R}(C^{1/2});\ \text{i.e. } B_1=C^{\dagger/2}B \text{ is bounded,} \\ (3) & A-B_1^*B_1\geq 0. \end{cases}$$

Lemma 12.8 *Let A,B,C,X be as in lemma 12.7. Let $X\geq 0$ and write $B_1=C^{\dagger/2}B$. Define the operator $W^\ddagger$:*

$$W^\ddagger=\begin{bmatrix} (A-B_1^*B_1)^{\dagger/2} & \\ & I\end{bmatrix}\begin{bmatrix} I & -B_1^* \\ & I\end{bmatrix}\begin{bmatrix} I & \\ & C^{\dagger/2}\end{bmatrix}.$$

Then $W^\ddagger$ is well-defined and bounded on $\mathcal{R}(X^{1/2})$. If v is some bounded operator with range in $\mathcal{R}(X^{1/2})$, and if

$$v_1 = X^{\dagger/2}v, \qquad v_2 = W^\ddagger v$$

then v_1 and v_2 are bounded, and $v_1^*v_1=v_2^*v_2$.

The proof of both lemmas appears as an appendix at the end of the chapter. Note that $W^\ddagger\neq X^{\dagger/2}$, but rather $W^\ddagger=UX^{\dagger/2}$ on $\mathcal{R}(X^{1/2})$, where U is some Hilbert space isometry such that $U^*U=I$. The point is that $W^\ddagger$ is specified in terms of A,B,C, whereas it is hard to do so for $X^{\dagger/2}$.

Contractivity in terms of a state space realization

We are now ready to derive a solution to the embedding problem along the lines of section 12.2 for the case where T is contractive, but not necessarily strictly contractive. Recall the definition of H_T and K_T of section 12.1.

Lemma 12.9 *Let T be an input-output operator in $\mathcal{U}$. If T is contractive, then*

$$I-K_TK_T^*\geq H_TH_T^*\geq 0, \tag{12.9}$$

and hence K_T and $\tilde{K}_T$ are contractive.

PROOF Let $u\in\mathcal{L}_2Z^{-1}$, and put $y=uT=uK_T+uH_T$. The contractivity of T implies

$$\begin{array}{rl} & \mathbf{P}_0(uu^*)-\mathbf{P}_0(yy^*)\geq 0 \\ \Leftrightarrow & \mathbf{P}_0(u[I-TT^*]u^*)\geq 0 \\ \Leftrightarrow & \mathbf{P}_0(u[I-K_TK_T^*-H_TH_T^*]u^*)\geq 0 \\ \Leftrightarrow & \mathbf{P}_0(u[I-K_TK_T^*]u^*)\geq \mathbf{P}_0(uH_TH_T^*u^*)\geq 0. \end{array}$$

Hence $I-K_TK_T^*\geq 0$ on $\mathcal{L}_2Z^{-1}$. $\tilde{K}_T$ is isometrically isomorphic to K_T and is also contractive.

□

Corollary 12.10 *If* **T** *is a uniformly observable realization of* T*, then* $\mathcal{R}(\tilde{K}_T^* \mathcal{C}) \subset \mathcal{R}(I - \tilde{K}_T^* \tilde{K}_T)^{1/2}$ *and hence* $\mathcal{C}_1$ *defined by*

$$\mathcal{C}_1 = (I - \tilde{K}_T^* \tilde{K}_T)^{\dagger/2} \tilde{K}_T^* \mathcal{C} \qquad (12.10)$$

is bounded.

PROOF Apply theorem 12.6 to (12.9). From $I - K_T K_T^* \geq H_T H_T^*$ it follows that $H_T = (I - K_T K_T^*)^{1/2} N$, for some operator N with $\|N\| \leq 1$. Taking diagonal expansions, we have that $\tilde{H}_T = (I - \tilde{K}_T \tilde{K}_T^*)^{1/2} \tilde{N}$, and with $\tilde{H}_T = \mathcal{C}\mathcal{O}$ such that $\mathcal{O}\mathcal{O}^* \gg 0$, we obtain

$$\begin{aligned} \tilde{K}_T^* \mathcal{C} &= \tilde{K}_T^* \mathcal{C}\mathcal{O}\mathcal{O}^* (\mathcal{O}\mathcal{O}^*)^{-1} \\ &= \tilde{K}_T^* \tilde{H}_T \mathcal{O}^* (\mathcal{O}\mathcal{O}^*)^{-1} \\ &= \tilde{K}_T^* (I - \tilde{K}_T \tilde{K}_T^*)^{1/2} \tilde{N} \mathcal{O}^* (\mathcal{O}\mathcal{O}^*)^{-1} \\ &= (I - \tilde{K}_T^* \tilde{K}_T)^{1/2} \mathcal{C}_1 \end{aligned}$$

where $\mathcal{C}_1 = \tilde{K}_T^* \tilde{N} \cdot \mathcal{O}^* (\mathcal{O}\mathcal{O}^*)^{-1}$ is bounded. □

For $\mathcal{C}_1$ defined in (12.10), define the operator $M \in \mathcal{D}$ by

$$M = \mathcal{C}^* \mathcal{C} + \mathcal{C}_1^* \mathcal{C}_1 . \qquad (12.11)$$

M is bounded, and $M \gg 0$ if $\mathcal{C}^* \mathcal{C} \gg 0$, *i.e.*, if the realization is uniformly reachable. This definition of M is compatible with the definition of M in (12.7) if T is strictly contractive, viz. $M = \mathcal{C}^* (I - \tilde{K}_T \tilde{K}_T^*)^{-1} \mathcal{C}$, because then $\mathcal{C}_1^* \mathcal{C}_1 = \mathcal{C}^* \tilde{K}_T (I - \tilde{K}_T^* \tilde{K}_T)^{-1} \tilde{K}_T^* \mathcal{C}$, and $I + \tilde{K}_T (I - \tilde{K}_T^* \tilde{K}_T)^{-1} \tilde{K}_T^* = (I - \tilde{K}_T \tilde{K}_T^*)^{-1}$. The latter relation is however not necessarily valid if a pseudo-inverse is used.

The following theorem subsumes theorem 12.4.

Theorem 12.11 *Let* $T \in \mathcal{U}$ *be an input-output operator with a u.e. stable state space realization* $\{A, B, C, D\}$*. If* T *is contractive and the realization is uniformly observable, then* M *defined by (12.10) and (12.11) is bounded,* $M \geq 0$*, and*

$$M^{(-1)} = A^* M A + B^* B \; + \; \left([A^* M C + B^* D] \Phi^\dagger\right) \cdot \left(\Phi^\dagger [D^* B + C^* M A]\right) \qquad (12.12)$$

with $\Phi = (I - D^* D - C^* M C)^{1/2}$ *and* $I - D^* D - C^* M C \geq 0$*. If, in addition, the state space realization is [uniformly] reachable then* $M > 0$ *[*$M \gg 0$*].*

PROOF The proof uses the expressions for $\tilde{V}_T$, $\tilde{K}_T$ and $\mathcal{C}$ as given by equations (12.4) and (12.5). To find an expression for $M^{(-1)}$, put

$$X \;=\; (I - \tilde{K}_T^* \tilde{K}_T)^{(-1)} \;=\; \begin{bmatrix} I - T_{[0]}^* T_{[0]} - \tilde{V}_T^* \tilde{V}_T & -\tilde{V}_T^* \tilde{K}_T \\ -\tilde{K}_T^* \tilde{V}_T & I - \tilde{K}_T^* \tilde{K}_T \end{bmatrix} .$$

According to lemma 12.9, $X \geq 0$. Lemma 12.7 then implies that $\mathcal{R}(\tilde{K}_T^* \tilde{V}_T) \subset \mathcal{R}(I - \tilde{K}_T^* \tilde{K}_T)^{1/2}$ so that $(I - \tilde{K}_T^* \tilde{K}_T)^{\dagger/2} \tilde{K}_T^* \tilde{V}_T = \mathcal{C}_1 C$ is bounded. (This result would also follow from corollary 12.10 because $\mathcal{R}(\tilde{K}_T^* \tilde{V}_T) = \mathcal{R}(\tilde{K}_T^* \mathcal{C} C) \subset \mathcal{R}(\tilde{K}_T^* \mathcal{C})$.) Let

$$\begin{aligned} \Phi &= \left[I - T_{[0]}^* T_{[0]} - \tilde{V}_T^* \tilde{V}_T - C^* \mathcal{C}_1^* \mathcal{C}_1 C\right]^{1/2} \\ &= [I - D^* D - C^* (\mathcal{C}^* \mathcal{C} + \mathcal{C}_1^* \mathcal{C}_1) C]^{1/2} \\ &= (I - D^* D - C^* M C)^{1/2} . \end{aligned}$$

The third item of lemma 12.7 implies that $I-D^*D-C^*MC \geq 0$. Put

$$\begin{aligned} W^{\ddagger} &= \begin{bmatrix} \Phi^{\dagger} & \\ & I \end{bmatrix} \begin{bmatrix} I & C^*C_1^* \\ & I \end{bmatrix} \begin{bmatrix} I & \\ & (I-\tilde{K}_T^*\tilde{K}_T)^{\dagger/2} \end{bmatrix} \\ v &= [\tilde{K}_T^*C]^{(-1)} = \tilde{K}_T^{*(-1)} \begin{bmatrix} B \\ CA \end{bmatrix} = \begin{bmatrix} D^*B+C^*C^*CA \\ \tilde{K}_T^*CA \end{bmatrix}. \end{aligned}$$

Then lemma 12.8 yields that the operator $v_1 = X^{\dagger/2}v = C_1^{(-1)}$ is bounded, and $v_2 = W^{\ddagger}v$ is such that $v_1^*v_1 = v_2^*v_2$. Evaluation of v_2 gives

$$\begin{aligned} v_2 = W^{\ddagger}v &= \begin{bmatrix} \Phi^{\dagger} & \\ & I \end{bmatrix} \begin{bmatrix} I & C^*C_1^* \\ & I \end{bmatrix} \begin{bmatrix} I & \\ & (I-\tilde{K}_T^*\tilde{K}_T)^{\dagger/2} \end{bmatrix} \begin{bmatrix} D^*B+C^*C^*CA \\ \tilde{K}_T^*CA \end{bmatrix} \\ &= \begin{bmatrix} \Phi^{\dagger} & \\ & I \end{bmatrix} \begin{bmatrix} I & C^*C_1^* \\ & I \end{bmatrix} \begin{bmatrix} D^*B+C^*C^*CA \\ C_1A \end{bmatrix} \\ &= \begin{bmatrix} \Phi^{\dagger}(D^*B+C^*MA) \\ C_1A \end{bmatrix}. \end{aligned}$$

Hence

$$\begin{aligned} [C_1^*C_1]^{(-1)} &= v_1^*v_1 = v_2^*v_2 \\ &= A^*C_1^*C_1A + ([B^*D+A^*MC]\Phi^{\dagger}) \cdot (\Phi^{\dagger}[D^*B+C^*MA]) \end{aligned}$$

and with $C^{(-1)} = \begin{bmatrix} B \\ CA \end{bmatrix}$ we finally obtain

$$\begin{aligned} M^{(-1)} &= [C^*C]^{(-1)} + [C_1^*C_1]^{(-1)} \\ &= B^*B + A^*C^*CA + A^*C_1^*C_1A + ([B^*D+A^*MC]\Phi^{\dagger}) \cdot (\Phi^{\dagger}[D^*B+C^*MA]) \\ &= A^*MA + B^*B + ([B^*D+A^*MC]\Phi^{\dagger}) \cdot (\Phi^{\dagger}[D^*B+C^*MA]) \,. \end{aligned}$$

□

The result of this section is thus a relatively simple extension of theorem 12.4: in the case that T is not strictly contractive, we can use the recursion

$$\begin{aligned} \Phi &= (I-D^*D-C^*MC)^{1/2} \\ M^{(-1)} &= A^*MA + B^*B \, + \, [A^*MC + B^*D]\,\Phi^{\dagger} \cdot \Phi^{\dagger}\,[D^*B + C^*MA] \end{aligned}$$

although we need the given realization to be uniformly observable. This condition is sufficient, but too strong: we only need "observability at the boundary", but this is hard to express (for time-invariant systems, the usual condition is that the realization should be 'stabilizable'). The recursion for M is very close to (and encompasses) the expression that we have obtained before in the strictly contractive case. Note that we know only that $\Phi^{\dagger}(D^*B+C^*MA)$ is bounded, but not necessarily $\Phi^{\dagger}\Phi^{\dagger}(D^*B+C^*MA)$: we have to evaluate $\Phi^{\dagger}(D^*B+C^*MA)$, and then square this expression in order to get a correct answer.

The above theorem will allow the embedding theorems in the next section to include contractive systems that need not be strictly contractive. It also gives part of the proof of the Bounded Real Lemma.

12.4 LOSSLESS EMBEDDING

We are now ready to solve the embedding problem as defined in the introduction: given a bounded causal transfer operator of a locally finite system T, determine a lossless system Σ such that $\Sigma_{11} = T$. The strategy is as outlined in the introduction: the prime quantity to be determined is a state transformation operator R such that the transformed realization of T is part of the realization of Σ.

We start with an intermediate result.

Isometric embedding

Theorem 12.12 (Isometric embedding) *Let $T \in \mathcal{U}(\mathcal{M},\mathcal{N})$ be a locally finite input-output operator with u.e. stable state realization $\mathbf{T} = \{A,B,C,D\}$. If $I - T^*T \gg 0$, or $I - T^*T \geq 0$ and T is uniformly observable, then T has an extension $\Sigma_a \in \mathcal{U}(\mathcal{M}\times\mathcal{N},\mathcal{N})$,*

$$\Sigma_a = \begin{bmatrix} T \\ \Sigma_{21} \end{bmatrix}$$

such that $\Sigma_a^*\Sigma_a = I$ and $A_{\Sigma_a} = A$. A realization for Σ_{21} is

$$\mathbf{\Sigma}_{21} = \begin{bmatrix} A & C \\ B_2 & D_{21} \end{bmatrix} = \begin{bmatrix} A & C \\ -\Phi^\dagger(D^*B + C^*MA) & \Phi \end{bmatrix} \qquad (12.13)$$

where $\Phi = (I - D^*D - C^*MC)^{1/2}$ and M is as defined in (12.11).

PROOF Let $\mathbf{\Sigma}_a$ be of the form

$$\mathbf{\Sigma}_a = \left[\begin{array}{cc} A & C \\ B & D \\ \hline B_2 & D_{21} \end{array}\right] \qquad (12.14)$$

in which B_2 and D_{21} are to be determined such that $\Sigma_a^*\Sigma_a = I$. Using corollary 6.5 in section 6.1, this is the case if there is an $M \geq 0$ such that

$$\begin{cases} A^*MA + B^*B + B_2^*B_2 = M^{(-1)} \\ A^*MC + B^*D + B_2^*D_{21} = 0 \\ C^*MC + D^*D + D_{21}^*D_{21} = I. \end{cases} \qquad (12.15)$$

We will show that M given by equation (12.11) is a positive semidefinite solution to these equations. Indeed, under the conditions imposed on T, theorem 12.4 [theorem 12.11] ensures that this M satisfies $M \geq 0$, $I - D^*D - C^*MC \gg 0$ [$I - D^*D - C^*MC \geq 0$], and

$$M^{(-1)} = A^*MA + B^*B + \left([A^*MC + B^*D]\Phi^\dagger\right) \cdot \left(\Phi^\dagger[D^*B + C^*MA]\right), \qquad (12.16)$$

where $\Phi = (I - D^*D - C^*MC)^{1/2}$. With B_2 and D_{21} as in (12.13), it immediately follows that equations (12.15) are satisfied.

□

In the above theorem, M can be interpreted as the reachability Gramian of Σ, and since $M = \mathcal{C}^*\mathcal{C} + \mathcal{C}_1^*\mathcal{C}_1$ with $\mathcal{C}_1$ as in (12.10), it is seen that $\mathcal{C}_1^*\mathcal{C}_1$ is the reachability Gramian of Σ_{21}. (A more detailed analysis shows that $-\mathcal{C}_1$ is its reachability operator.)

Suppose that $\|T\| < 1$ so that $I - T^*T$ is invertible. A result of Arveson [Arv75], which is applicable in the present context, claims that there is a factor Σ_{21} of $I - T^*T$ which is *outer, i.e.*, such that $\Sigma_{21}^{-1} \in \mathcal{U}$. We will show that our choice for Σ_{21}, as defined by the realization $\mathbf{\Sigma}_{21}$ in (12.13), is in fact outer. To this end, we will look at a possible realization for Σ_{21}^{-1}, viz. proposition 7.7 in section 7.3,

$$\mathbf{\Sigma}_{21}^{\times} = \begin{bmatrix} A^{\times} & C^{\times} \\ B^{\times} & D^{\times} \end{bmatrix} = \begin{bmatrix} A - CD_{21}^{-1}B_2 & -CD_{21}^{-1} \\ D_{21}^{-1}B_2 & D_{21}^{-1} \end{bmatrix} \tag{12.17}$$

and will show that this realization is u.e. stable: $\ell_{A^{\times}} < 1$. In that case, we can conclude that $\Sigma_{21}^{-1} \in \mathcal{U}$.

Proposition 12.13 *Suppose* $\|T\| < 1$. *Define* $\mathbf{\Sigma}_{21}^{\times}$ *as in (12.17) and theorem 12.12. Then* $\ell_{A^{\times}} < 1$, *and* Σ_{21} *is outer.*

PROOF We first assert that the reachability operator of $\mathbf{\Sigma}_{21}^{\times}$ is given by $\mathcal{C}^{\times} = -(I - \tilde{K}_T^*\tilde{K}_T)^{-1}\tilde{K}_T^*\mathcal{C}$. It is sufficient to show that the given formula of $\mathcal{C}^{\times}$ satisfies the recursion $\mathcal{C}^{\times(-1)} = \left[\begin{smallmatrix} B^{\times} \\ \mathcal{C}^{\times}A^{\times} \end{smallmatrix}\right]$. Indeed, with equations (12.4), (12.5), (12.8),

$$\begin{aligned}
\mathcal{C}^{\times(-1)} &= -(I - \tilde{K}_T^*\tilde{K}_T)^{-(-1)}\tilde{K}_T^{*(-1)}\mathcal{C}^{(-1)} = \\
&= -\left(\begin{bmatrix} 0 & \\ & (I-\tilde{K}_T^*\tilde{K}_T)^{-1} \end{bmatrix} + \begin{bmatrix} I \\ (I-\tilde{K}_T^*\tilde{K}_T)^{-1}\tilde{K}_T^*\tilde{V}_T \end{bmatrix} \cdot \right. \\
&\qquad \left. \cdot \Phi^{-2}\begin{bmatrix} I & \tilde{V}_T^*\tilde{K}_T(I-\tilde{K}_T^*\tilde{K}_T)^{-1} \end{bmatrix}\right) \begin{bmatrix} D^* & \tilde{V}_T^* \\ 0 & \tilde{K}_T^* \end{bmatrix} \begin{bmatrix} B \\ \mathcal{C}A \end{bmatrix} \\
&= -\begin{bmatrix} 0 \\ (I-\tilde{K}_T^*\tilde{K}_T)^{-1}\tilde{K}_T^*\mathcal{C}A \end{bmatrix} - \begin{bmatrix} I \\ (I-\tilde{K}_T^*\tilde{K}_T)^{-1}\tilde{K}_T^*\mathcal{C}C \end{bmatrix} \Phi^{-2}(D^*B + C^*MA) \\
&= \begin{bmatrix} -\Phi^{-2}(D^*B + C^*MA) \\ -(I-\tilde{K}_T^*\tilde{K}_T)^{-1}\tilde{K}_T^*\mathcal{C}[A + C\Phi^{-2}(D^*B + C^*MA)] \end{bmatrix} \\
&= \begin{bmatrix} D_{21}^{-1}B_2 \\ \mathcal{C}^{\times}(A - CD_{21}^{-1}B_2) \end{bmatrix}.
\end{aligned}$$

The reachability Gramian of $\mathbf{\Sigma}_{21}^{\times}$ is $\Lambda^{\times} = \mathcal{C}^*\tilde{K}_T(I - \tilde{K}_T^*\tilde{K}_T)^{-2}\tilde{K}_T^*\mathcal{C}$, which is bounded because the inverse is bounded and $\mathcal{C}^*\mathcal{C}$ is bounded. According to a result of Anderson and Moore [AM81, thm. 4.3] (see also [Nic92]), if $\Lambda^{\times}$ is bounded and $\ell_A < 1$, then $\ell_{A^{\times}} < 1$.[2] It follows that $\Sigma_{21}^{-1} \in \mathcal{U}$, so that Σ_{21} is outer. □

[2] The actual condition in [AM81] is that $(A - CD_{21}^{-1}B_2, D_{21}^{-1}B_2)$ is *uniformly stabilizable*, but it is also shown that this is the case if and only if $(A, D_{21}^{-1}B_2)$ is uniformly stabilizable. For this, it is sufficient that $\ell_A < 1$.

Orthogonal embedding

Using theorem 12.12, it is straightforward to solve the lossless embedding problem.

Theorem 12.14 (Orthogonal embedding) *Let $T \in \mathcal{U}(\mathcal{M}_1,\mathcal{N}_1)$ be a locally finite input-output operator with u.e. stable state realization $\mathbf{T} = \{A,B,C,D\}$. If $I - T^*T \gg 0$, or $I - T^*T \geq 0$ and $\mathbf{T}$ is uniformly observable, and if the realization $\mathbf{T}$ is uniformly reachable, then the lossless embedding problem has a solution $\Sigma \in \mathcal{U}(\mathcal{M}_1 \times \mathcal{N}_1, \mathcal{N}_1 \times \mathcal{N}_2)$ such that Σ is inner, $\Sigma_{11} = T$, Σ_{21} is outer, and Σ has a unitary realization $\mathbf{\Sigma}$ where A_Σ is state equivalent to A. If $A \in \mathcal{D}(\mathcal{B},\mathcal{B}^{(-1)})$, then $\mathcal{N}_2$ is specified by $\#(\mathcal{N}_2) = \#(\mathcal{B}) - \#(\mathcal{B}^{(-1)}) + \#(\mathcal{M}_1)$.*

PROOF The proof is by construction. Let $\mathbf{\Sigma}$ be of the form

$$
\begin{aligned}
[\mathbf{\Sigma}_a \ \mathbf{\Sigma}_b] &= \left[\begin{array}{cc|c} A & C & C_2 \\ B & D & D_{12} \\ B_2 & D_{21} & D_{22} \end{array}\right] \\
\mathbf{\Sigma} &= \begin{bmatrix} R & & \\ & I & \\ & & I \end{bmatrix} [\mathbf{\Sigma}_a \ \mathbf{\Sigma}_b] \begin{bmatrix} [R^{(-1)}]^{-1} & & \\ & I & \\ & & I \end{bmatrix} = [\mathbf{\Sigma}'_a \ \ \mathbf{\Sigma}'_b],
\end{aligned}
\tag{12.18}
$$

in which $R \in \mathcal{D}(\mathcal{B},\mathcal{B})$ is a boundedly invertible state transformation. R, B_2, D_{12}, D_{21}, D_{22} are to be determined such that $\mathbf{\Sigma}$ is unitary, in which case Σ is inner (theorem 6.4 in section 6.1).

First, determine M, B_2, D_{12} and hence $\mathbf{\Sigma}_a$ as in theorem 12.12. Because $\mathbf{T}$ is uniformly reachable, $M \gg 0$. If we define the state transformation R by $M = R^*R$, then R is invertible, and $\mathbf{\Sigma}'_a$ is an isometry ($\mathbf{\Sigma}'^*_a \mathbf{\Sigma}'_a = I$). The extension of a rectangular isometric matrix to a square unitary matrix by adding columns is a standard linear algebra procedure that always has a solution. The same holds for diagonals of matrices. Hence, we can extend $\mathbf{\Sigma}'_a$ to a unitary matrix $\mathbf{\Sigma}$, which is the realization of an inner system Σ. The resulting dimension sequence of $\mathbf{\Sigma}$ is given by $[\#(\mathcal{B}) + \#(\mathcal{M}_1) + \#(\mathcal{N}_1)]$, and the number of columns of each diagonal entry of $\mathbf{\Sigma}'_a$ is the sequence $[\#(\mathcal{B}^{(-1)}) + \#(\mathcal{N}_1)]$, hence the number of columns to be added is equal to $\#(\mathcal{N}_2) = \#(\mathcal{B}) - \#(\mathcal{B}^{(-1)}) + \#(\mathcal{M}_1)$. This number is non-negative because the columns of $\mathbf{\Sigma}'_a$ are linearly independent. □

As was the case with the inner-outer factorization in chapter 7, one difference with the time-invariant situation is that the solution of the embedding problem gives rise to a time-varying number of added extra outputs if the number of states of T is time-varying ($\mathcal{B} \neq \mathcal{B}^{(-1)}$), even if the number of inputs and outputs of T is fixed. Another difference is that, for the boundary case, we need both uniform reachability and uniform observability in order to construct an embedding. From chapter 5 we know that not every time-varying system admits such a realization, not even if it has a finite state dimension; the condition is that the range of H_T must be closed.

Bounded real lemma

A reformulation of theorem 12.12 and proposition 12.13 leads to the bounded real lemma which appears in system and control theory.

Theorem 12.15 (Time-varying bounded real lemma, I)
Let $T \in \mathcal{U}(\mathcal{M},\mathcal{N})$ be a bounded causal locally finite input-output operator, with u.e. stable state realization $\mathbf{T} = \{A,B,C,D\}$, and $A \in \mathcal{D}(\mathcal{B},\mathcal{B}^{(-1)})$.

- $\|T\| < 1$ *if and only if there exists* $M \in \mathcal{D}(\mathcal{B},\mathcal{B})$, $B_2 \in \mathcal{D}(\mathcal{N},\mathcal{B}^{(-1)})$, $D_{21} \in \mathcal{D}(\mathcal{N},\mathcal{N})$ *solving*

$$\left\{ \begin{array}{lllllll} A^*MA & + & B^*B & + & B_2^*B_2 & = & M^{(-1)} \\ C^*MC & + & D^*D & + & D_{21}^*D_{21} & = & I \\ A^*MC & + & B^*D & + & B_2^*D_{21} & = & 0 \end{array} \right. \tag{12.19}$$

with $M \ge 0$, $I - D^*D - C^*MC \gg 0$ *and* $\ell_{A-CD_{21}^{-1}B_2} < 1$.

- *If* $\mathbf{T}$ *is uniformly observable, then* $\|T\| \le 1$ *if and only if (12.19) has a solution* M, B_2, D_{21} *such that* $M \ge 0$.

PROOF The 'only if' part is directly derived from theorem 12.12 and proposition 12.13. The 'if' part is a corollary of theorem 12.12: given such M, it follows that there exists an isometric embedding Σ_a such that $\Sigma_a^*\Sigma_a = T^*T + \Sigma_{21}^*\Sigma_{21} = I$, so that $\Sigma_{21}^*\Sigma_{21} = I - T^*T \ge 0$. If in addition D_{21} is invertible and $\ell_{A-CD_{21}^{-1}B_2} < 1$, then by proposition 12.13 we can conclude that Σ_{21} is invertible, so that $I - T^*T \gg 0$, *i.e.*, $\|T\| < 1$. □

An alternative version of this theorem is given in 13.5, where the connection between unitary embedding and spectral factorization is explored.

12.5 NUMERICAL ISSUES

Initial point for the recursion

Suppose that we are given a realization of a system T that meets the requirements of the embedding theorem. How do we go about determining a realization of Σ? The embedding theorem is constructive, and $\mathbf{\Sigma}_i$ (the realization of Σ at time instant i) can be determined from knowledge of $\mathbf{T}_i$ and both M_i and M_{i+1}. In addition, equation (12.16) can be used to determine M_{i+1} from M_i:

$$M_{i+1} = A_i^*M_iA_i + B_i^*B_i + \left[A_i^*M_iC_i + B_i^*D_i\right](I - D_i^*D_i - C_i^*M_iC_i)^{-1}\left[D_i^*B_i + C_i^*M_iA_i\right], \tag{12.20}$$

and this is the only recursive aspect of the procedure. The single missing item is the initial point of this recursion: the value of $M_{-\infty}$, or rather M_{k_0}, where k_0 is the point in time at which the solution of the embedding problem starts to be of interest.

As was the case before in the solution of the Lyapunov equation and the inner-outer factorization (section 7.2), it is possible to find an initial value for the recursion for certain specific time-varying systems. In fact, the Riccati equation (12.20) is very similar to the one which occurred in the inner-outer factorization (*cf.* equation (7.19)), so that also the solutions are obtained in similar ways.

The first (and simplest) class is the case where the state dimension of T is zero at a certain point in time k_0. Consider, for example, a finite $n \times n$ upper triangular (block)-matrix T, then the input space sequence is

$$\mathcal{M}_1 = \cdots \times \emptyset \times \emptyset \times \underbrace{\boxed{\mathbb{C}} \times \mathbb{C} \times \cdots \times \mathbb{C}}_{n} \times \emptyset \times \emptyset \times \cdots$$

and output space sequence $\mathcal{N}_1 = \mathcal{M}_1$. A reachable realization of T obviously has a state-space sequence $\mathcal{B}$ also with $\mathcal{B}_i = \emptyset$ for $(i < 0, i \geq n)$, and hence an initial value of the recursion for M is $M_0 = [\cdot]$.

A second example is the case where T is time invariant before a certain point in time ($i = 0$ say). T has a time-invariant realization $\{a, b, c, d\}$ for $i < 0$, and there is a time-invariant solution for M also: $M_{i+1} = M_i =: m \ (i < 0)$. The recursion (12.20) becomes an eigenvalue (Riccati) equation

$$m = a^*ma + b^*b + \left[a^*mc + b^*d\right](I - d^*d - c^*mc)^{-1}\left[d^*b + c^*ma\right]. \qquad (12.21)$$

This equation has exact solutions m which can be obtained analytically as in section 7.2, or numerically by using a Newton-Raphson iteration. An overview of these and other methods can be found in the collection [BLW91]. It is well known that the analytical (eigenvalue) methods usually provide more than one solution that satisfies the Riccati equation; the solution $M = \mathcal{C}^*(I - \tilde{K}_T\tilde{K}_T^*)^{-1}\mathcal{C}$ corresponds to the "stable" solution, for which $M \geq 0$. The stable solution is also the only solution of the Riccati equation that is stable to a small perturbation when it is plugged in the Riccati recursion (12.20). In fact, one way to solve (12.21) is to use the *recursion* (12.20) for an initial value of $M_{-\infty} = 0$, and to iterate till convergence. It is known that this occurs if the eigenvalues of a are strictly smaller than 1, and that the recursion will monotonically converge to the stable solution of the Riccati equation.

We can do the same for time-varying systems, which will then apply to other specific situations as well, such as periodic systems. The claim is that if $M'_0 = 0$ is taken as the initial value of the recursion (12.20) which gives a sequence M'_i, then $M'_i \to M_i$ as $i \to \infty$. An elegant proof is possible, not based on numerical properties of the Riccati equation but rather on the knowledge that $M = \mathcal{C}^*(I - \tilde{K}_T\tilde{K}_T^*)\mathcal{C}$ is the solution of the recursion that we are looking. Details of this proof are however cumbersome because many time indices will appear, but we give an outline of it below. (A formal proof of convergence of a related Riccati equation appears in section 13.4.

Proposition 12.16 *Let $\{A, B, C, D\}$ be a u.e. stable realization ($\ell_A < 1$) of a locally finite strictly contractive transfer operator $T \in \mathcal{U}$. Let $M_i = \mathcal{C}_i^*(I - K_iK_i^*)\mathcal{C}_i$ be the exact solution of the Riccati equation (12.20), and let M'_i be the solution, obtained by starting the recursion with $M'_0 = 0$. Then $M'_i \to M_i$ for $i \to \infty$ (strong convergence).*

PROOF (outline). The initial value $M'_0 = 0$ is the *exact* initial point of a recursion for M' of a system T' which is related to T: $T'_{ij} = 0$ for $i < 0$, and $T'_{ij} = T_{ij}$ for $i \geq 0$. The sequence M_i corresponds to T and is at each point i in time given by $M_i = \mathcal{C}_i^*(I - K_iK_i^*)^{-1}\mathcal{C}_i$. For $i \geq 0$, we can define a partitioning of K_i and $\mathcal{C}_i$ as

$$K_i = \begin{bmatrix} K'_i & 0 \\ H_0^r & K_0 \end{bmatrix} \qquad \mathcal{C}_i = \begin{bmatrix} \mathcal{C}'_i \\ \mathcal{C}_0 A^{[0..i-1]} \end{bmatrix}$$

where K'_i is an $(i \times i)$ matrix, $\mathcal{C}'_i$ is equal to the first i rows of $\mathcal{C}_i$,

$$A^{[0..n-1]} := A_0 A_1 \cdots A_{n-1},$$

and H_0^r is related to the Hankel operator H_0, but has a finite number (i) of columns, which are in *reversed* order in comparison with H_0. In terms of these quantities, M' is given at time $i \geq 0$ by $M'_i = C_i'^*(I - K'_i K'^*_i)^{-1} C'_i$. Using this decomposition of K_i, and a variant of Schur's inversion lemma (lemma 12.2), one can derive that, for $i \geq 0$,

$$\begin{aligned}(I - K_i K_i^*)^{-1} &= \begin{bmatrix} (I - K'_i K'^*_i)^{-1} & \\ & 0 \end{bmatrix} + \\ &+ \begin{bmatrix} (I - K'_i K'^*_i)^{-1} K'_i (H_0^r)^* \\ I \end{bmatrix} \Phi^{-2} \begin{bmatrix} H_0^r K'^*_i (I - K'_i K'^*_i)^{-1} & I \end{bmatrix}\end{aligned}$$

where

$$\Phi^2 = I - K_0 K_0^* - H_0^r (I - K'^*_i K'_i)^{-1} (H_0^r)^* \gg 0$$

and hence its inverse is bounded. Inserting the expression for C_i and defining $H_0^r = C_0 \mathcal{O}_0^r$ yields

$$\begin{aligned}M_i &= M'_i + \left[C_i'^*(I - K'_i K'^*_i)^{-1} K'_i (\mathcal{O}_0^r)^* + (A^{[0..i-1]})^*\right] C_0^* \Phi^{-2} C_0 \cdot \\ &\quad \cdot \left[\mathcal{O}_0^r K'^*_i (I - K'_i K'^*_i)^{-1} C'_i + A^{[0..i-1]}\right].\end{aligned}$$

An examination of the term $\mathcal{O}_0^r K'^*_i (I - K'_i K'^*_i)^{-1} C'_i$ that is more detailed than we wish to include at this point reveals that it consists of a summation of i terms, each of which has a factor $A^{[0..k-1]}$ and $A^{[k+1..i-1]}$ (for $0 \leq k \leq i$). The stability condition $\ell_A < 1$ implies that $\varepsilon > 0$ exists such that, in the limit, products of the form $A^{[k..k+n]}$ are bounded in norm by $(1-\varepsilon)^n$ which goes to 0 strongly and uniformly in k as $n \to \infty$. Since Φ^{-2} is bounded, this equation gives $M'_i \to M_i$ as $i \to \infty$. □

"Square-root" solution of the Riccati equation

The embedding algorithm can be implemented along the lines of the proof of the embedding theorem. However, as was the case with the solution of the inner-outer factorization problem in chapter 7, the Riccati recursions on M_i can be replaced by more efficient algorithms that recursively compute the square root of M_i, i.e., R_i, instead of M_i itself. The square-root algorithm is given in figure 12.3. The algorithm acts on data known at the k-th step: the state matrices A_k, B_k, C_k, D_k, and the state transformation R_k obtained at the previous step. This data is collected in a matrix $\mathbf{T}_{e,k}$:

$$\mathbf{T}_e = \begin{bmatrix} R & & \\ & I & \\ & & I \end{bmatrix} \begin{bmatrix} A & C \\ B & D \\ 0 & I \end{bmatrix} \tag{12.22}$$

The key of the algorithm is the construction of a J-unitary operator $\Theta \in \mathcal{D}^{3\times 3}$, satisfying $\Theta^* J \Theta = J$, where

$$\Theta = \begin{bmatrix} \Theta_{11} & \Theta_{12} & \Theta_{13} \\ \Theta_{21} & \Theta_{22} & \Theta_{23} \\ \Theta_{31} & \Theta_{32} & \Theta_{33} \end{bmatrix} \qquad J = \begin{bmatrix} I & & \\ & I & \\ & & -I \end{bmatrix},$$

In: $\{\mathbf{T}_k\}$ (a reachable realization of T, $\|T\| < 1$)
Out: $\{\mathbf{\Sigma}_k\}$ (a unitary realization of embedding Σ)

$R_1 = [\cdot]$

for $k = 1, \cdots, n$

$$\mathbf{T}_{e,k} = \begin{bmatrix} R_k & & \\ & I & \\ & & I \end{bmatrix} \begin{bmatrix} A_k & C_k \\ B_k & D_k \\ 0 & I \end{bmatrix}$$

$$\mathbf{T}'_{e,k} := \Theta_k \mathbf{T}_{e,k}, \quad \Theta_k \text{ such that } \mathbf{T}'_{e,k}(2,2) = \mathbf{T}'_{e,k}(1,2) = \mathbf{T}'_{e,k}(2,1) = 0$$

$$\mathbf{T}'_{e,k} =: \begin{bmatrix} R_{k+1} & 0 \\ 0 & 0 \\ B_{2,k} & D_{21,k} \end{bmatrix}$$

$$\mathbf{\Sigma}_{1,k} = \begin{bmatrix} R_k & & \\ & I & \\ & & I \end{bmatrix} \left[\begin{array}{cc} A_k & C_k \\ B_k & D_k \\ \hline B_{2,k} & D_{21,k} \end{array}\right] \begin{bmatrix} R_{k+1}^{-1} & \\ & I \end{bmatrix}$$

$$\mathbf{\Sigma}_k = \begin{bmatrix} \mathbf{\Sigma}_{1,k} & \mathbf{\Sigma}_{1,k}^{\perp} \end{bmatrix}$$

end

Figure 12.3. The embedding algorithm for finite $n \times n$ matrices.

such that certain entries of $\mathbf{T}'_e := \Theta\mathbf{T}_e$ are zero. (General J-unitary operators are the subject of chapter 8.) It turns out that, because Θ is J-unitarity, we have that $\mathbf{T}'^*_e J\,\mathbf{T}_e = \mathbf{T}^*_e J\,\mathbf{T}_e$; writing these equations out, it follows that the remaining non-zero entries of $\mathbf{T}'_e$ are precisely the unknowns $R^{(-1)}$, B_2 and D_{21}.

Proposition 12.17 *Let $T \in \mathcal{U}$ be a strictly contractive operator, and let $\{A,B,C,D\}$ be a uniformly reachable realization of T. Define $\mathbf{T}_e$ as in equation (12.22).*

Then there is a J-unitary operator $\Theta \in \mathcal{D}^{3\times 3}$ such that $\mathbf{T}'_e := \Theta\mathbf{T}_e$ has zeros at the entries $(2,2)$, $(1,2)$ and $(2,1)$. $\mathbf{T}'_e$ is of the form

$$\mathbf{T}'_e = \Theta T_e = \begin{bmatrix} R^{(-1)} & 0 \\ 0 & 0 \\ B_2 & D_{21} \end{bmatrix}$$

where $M = R^*R$, B_2, D_{21} satisfy the embedding equations (12.15).

PROOF Assume first that such an operator Θ exists. A direct computation reveals that (with $M = R^*R$)

$$\mathbf{T}^*J\mathbf{T} = \begin{bmatrix} A^*MA + B^*B & A^*MC + B^*D \\ (A^*MC + B^*D)^* & -(I - D^*D - C^*MC) \end{bmatrix}$$

$$\mathbf{T}'^*J\mathbf{T}' = \begin{bmatrix} M^{(-1)} - B_2^*B_2 & -D_{21}^*B_2 \\ -B_2^*D_{21} & -D_{21}^*D_{21} \end{bmatrix}$$

Since Θ is J-unitary, we must have $\mathbf{T}^*J\mathbf{T} = \mathbf{T}^{*\prime}J\mathbf{T}'$, which produces the relations (12.15):

$$\begin{cases} A^*MA + B^*B + B_2^*B_2 & = & M^{(-1)} \\ C^*MC + D^*D + D_{21}^*D_{21} & = & I \\ A^*MC + B^*D + B_2^*D_{21} & = & 0 \end{cases}$$

i.e., the equations that constituted the Riccati equations. It remains to verify the existence of a J-unitary Θ such that $\mathbf{T}'_e$ has zeros at the entries $(2,2)$, $(1,2)$ and $(2,1)$. Choose Θ of the form

$$\Theta = \Theta_3\Theta_2\Theta_1 = \begin{bmatrix} \Theta^3_{11} & \Theta^3_{12} & 0 \\ \Theta^3_{21} & \Theta^3_{22} & 0 \\ 0 & 0 & I \end{bmatrix} \begin{bmatrix} \Theta^2_{11} & 0 & \Theta^2_{13} \\ 0 & I & 0 \\ \Theta^2_{31} & 0 & \Theta^2_{33} \end{bmatrix} \begin{bmatrix} I & 0 & 0 \\ 0 & \Theta^1_{22} & \Theta^1_{23} \\ 0 & \Theta^1_{32} & \Theta^1_{33} \end{bmatrix}$$

where the submatrix $\{\Theta^3_{ij}\}^2_{i,j=1}$ is unitary, while the submatrices $\{\Theta^2_{ij}\}$ and $\{\Theta^1_{ij}\}$ are J-unitary with with signature matrix $J_1 = \left[\begin{smallmatrix} I & 0 \\ 0 & -I \end{smallmatrix}\right]$. The submatrices are determined by the requirements

$$\begin{aligned} \begin{bmatrix} \Theta^1_{22} & \Theta^1_{23} \\ \Theta^1_{32} & \Theta^1_{33} \end{bmatrix} \begin{bmatrix} D \\ I \end{bmatrix} &= \begin{bmatrix} 0 \\ (I - D^*D)^{1/2} \end{bmatrix} \\ \begin{bmatrix} \Theta^2_{11} & \Theta^2_{13} \\ \Theta^2_{31} & \Theta^2_{33} \end{bmatrix} \begin{bmatrix} RC \\ (I - D^*D)^{1/2} \end{bmatrix} &= \begin{bmatrix} 0 \\ (I - D^*D - C^*MC)^{1/2} \end{bmatrix} \\ \begin{bmatrix} \Theta^3_{11} & \Theta^3_{12} \\ \Theta^3_{21} & \Theta^3_{22} \end{bmatrix} \begin{bmatrix} \Theta^2_{11} & 0 & \Theta^2_{13} \\ 0 & I & 0 \end{bmatrix} \begin{bmatrix} I & 0 \\ 0 & \Theta^1_{22} \\ 0 & \Theta^1_{32} \end{bmatrix} \begin{bmatrix} RA \\ B \end{bmatrix} &= \begin{bmatrix} * \\ 0 \end{bmatrix}. \end{aligned}$$

Hence necessary requirements are $I-D^*D \geq 0$ and $I-D^*D-C^*MC \geq 0$, respectively. In the present case, because T is strictly contractive, we know that $I-D^*D \gg 0$ and $I-D^*D-C^*MC \gg 0$, and these conditions ensure that the J-unitary submatrices $\{\Theta^1_{ij}\}$ and $\{\Theta^2_{ij}\}$ are well defined, and for example, of the form of a Halmos extension [DD92]

$$H(K) = \begin{bmatrix} (I-KK^*)^{-1/2} & 0 \\ 0 & (I-K^*K)^{-1/2} \end{bmatrix} \begin{bmatrix} I & K \\ K^* & I \end{bmatrix}.$$

The unitary submatrix $\{\Theta^3_{ij}\}$ is always well defined. □

It is also a standard technique to factor Θ even further down into elementary (J)-unitary operations that each act on only two scalar entries of $\mathbf{T}_e$, and zero one of them by applying an elementary J-unitary rotation of the form

$$\theta = \frac{1}{c}\begin{bmatrix} 1 & s \\ s & 1 \end{bmatrix}, \qquad c^*c+s^*s = 1.$$

With B_2 and D_{21} known, it is conjectured that it is not really necessary to apply the state transformation by R and to determine the orthogonal complement of $\mathbf{\Sigma}_1$ if, in the end, only a cascade factorization of T is required, much as in [LK92]. Cascade factorizations are the subject of chapter 14.

12.6 NOTES

Many control applications give rise to the Riccati equation (12.2). Usually, the existence of a stabilizing solution is of importance. In the context of our embedding problem, this would be a solution for which $A-CD_{21}^{-1}B_2$ is u.e. stable, or Σ_{21} is outer. The uniqueness of such a solution is a standard result which is straightforward to prove.

More is known on time-varying Riccati equations, and on its connection to embedding, positivity, and spectral factorization. We mention in particular the papers [Nic92], in which detailed attention is paid to the convergence of the recursion to maximal/minimal solutions, and [HI93], where the solution of a Kalman-Szegö-Popov-Yakubovich (KSPY) system of equations is presented. (See also [HI94] for further details.) The equations (12.19) can be viewed as a particular instance of these equations. Although [HI93] gives solutions to a more general class of problems, the boundary case is not considered. A major difference with [HI93] is in the *proofs* of the results: whereas the latter heavily relies on insights gained in optimal control theory, the approach taken in this chapter is more based on first principles: $I-T^*T = \Sigma_{21}^*\Sigma_{21} \Leftrightarrow I-\tilde{K}_T^*\tilde{K}_T = \tilde{K}_{\Sigma_{21}}^*\tilde{K}_{\Sigma_{21}}$. The analysis of the latter equation directly leads to a recursion in which the given expressions for M, D_{21}, B_2 turn up, along with an explicit expression for the reachability operator of the realization for Σ_{21}. Similar analysis of $\tilde{K}_{\Sigma_{21}^{-1}} = (\tilde{K}_{\Sigma_{21}})^{-1}$ leads to the realization of the inverse, the given expression for the reachability operator, and the fact that our choice for Σ_{21} is outer.

In an operator-theoretic setting, additional research on the existence of isometric extensions was done by Feintuch and Markus [FM96b, FM96a], in the context of a nest algebra.

Appendix 12.A: Derivation of lemmas 12.7 and 12.8

The contents of lemmas 12.7 and 12.8 are well known for finite matrices (see *e.g.*, [CHM74, BCHM74]) for generalized inverse formulas involving Schur complements). The matrix case is readily extended to operators if the operators are assumed to have closed range. Without this condition, complications arise because the pseudo-inverses that are involved are unbounded operators.

We will repeatedly use theorem 12.6 in the following form. Let $X \geq 0$ be a bounded operator on a Hilbert space $\mathcal{H}$. If v is a bounded operator whose range is in $\mathcal{R}(X)$, then $v = Xv_1$, for some bounded $v_1 \in \overline{\mathcal{R}(X^*)}$ for which we can take $v_1 = X^\dagger v$.

A second fact that is used in the proof of lemma 12.8 is that $X^\dagger X = \mathbf{P}_{\mathcal{X}^*}$: the orthogonal projector onto $\overline{\mathcal{R}(X^*)}$, with domain $\mathcal{H}$ [BR76].

Proof of lemma 12.7

Suppose first that $X \geq 0$; we show that (1), (2), (3) hold. It is immediate that $A \geq 0$, $C \geq 0$.

$\mathcal{R}(B) \subset \mathcal{R}(C^{1/2})$ is proven by showing that there exists λ such that $BB^* \leq \lambda C$; Douglas' theorem then implies the result. The proof is by contradiction. Suppose that there is not such a λ. Then there exists a sequence $\{x_n : n \in \mathrm{N}\}$ such that

$$(BB^* x_n, x_n) \;\geq\; n(Cx_n, x_n) \;>\; 0. \tag{12.A.1}$$

where $(\cdot,\cdot)$ denotes the inner product in $\mathcal{H}$. In particular, $\| B^* x_n \| > 0$ (all n). For any u_n, $X \geq 0$ implies

$$\left(\begin{bmatrix} A & B^* \\ B & C \end{bmatrix} \begin{bmatrix} u_n \\ x_n \end{bmatrix}, \begin{bmatrix} u_n \\ x_n \end{bmatrix} \right) \geq 0$$

i.e., $(Au_n, u_n) + (B^* x_n, u_n) + (Bu_n, x_n) + (Cx_n, x_n) \;\geq\; 0$. Choose $u_n = -\frac{1}{\sqrt{n}} B^* x_n$. Using (12.A.1), we obtain

$$(B \left\{ \frac{A}{n} - \frac{2}{\sqrt{n}} + \frac{I}{n} \right\} B^* x_n, x_n) \;\geq\; 0.$$

But if $n > \| I + A \|^2$, the term in braces is smaller than $-1/\sqrt{n}$, which gives a contradiction. Hence $\mathcal{R}(B) \subset \mathcal{R}(C^{1/2})$.

Define $L = C^{1/2}$ (although $L = L^*$, we will not use this), and let $B_1 = L^\dagger B$. Then B_1 is bounded, and $B = LB_1$ with $\mathcal{R}(B_1) \subset \overline{\mathcal{R}(L^*)}$, which implies

$$\mathcal{N}(B_1^*) \supset \mathcal{N}(L). \tag{12.A.2}$$

To prove $A - B_1^* B_1 \geq 0$, we will show that

$$X = \begin{bmatrix} A & B_1^* L^* \\ LB_1 & LL^* \end{bmatrix} \geq 0 \;\;\Rightarrow\;\; \begin{bmatrix} A & B_1^* \\ B_1 & I \end{bmatrix} \geq 0 \tag{12.A.3}$$

from which $A - B_1^* B_1 \geq 0$ follows directly by applying vectors of the form $\left[\begin{smallmatrix} I \\ -B_1 \end{smallmatrix} \right] a$.

Thus for $x \in \mathcal{H}_1 \oplus \mathcal{H}_2$, take x of the form

$$x = \begin{bmatrix} u \\ x_1 + x_2 \end{bmatrix} \in \begin{bmatrix} \mathcal{H}_1 \\ \mathcal{N}(L) \oplus \mathcal{R}(L^*) \end{bmatrix}$$

where $x_1 \in \mathcal{N}(L)$ and $x_2 \in \mathcal{R}(L^*)$. Note that $\mathcal{N}(L) \oplus \mathcal{R}(L^*)$ is dense in $\mathcal{H}_2$. Then $\mathcal{N}(B_1^*) \supset \mathcal{N}(L)$ implies $B_1^* x_1 = 0$, while $x_2 \in \mathcal{R}(L^*)$ implies that $x_2 = L^* x_2'$, for some bounded x_2'. Using these observations, it follows that

$$\begin{aligned}
&\left(\begin{bmatrix} A & B_1^* \\ B_1 & I \end{bmatrix}\begin{bmatrix} u \\ x_1 + x_2 \end{bmatrix}, \begin{bmatrix} u \\ x_1 + x_2 \end{bmatrix}\right) \\
&\quad = (Au,u) + (B_1^* x_1, u) + (B_1 u, x_1) + (x_1, x_1) + (B_2^* x_2, u) + (B_1 u, x_2) + (x_2, x_2) \\
&\quad \geq (Au,u) + (B_1^* x_2, u) + (B_1 u, x_2) + (x_2, x_2) \\
&\quad = (Au,u) + (B^* x_2', u) + (B_1 u, x_2') + (x_2', x_2') \\
&\quad = \left(X \begin{bmatrix} u \\ x_2' \end{bmatrix}, \begin{bmatrix} u \\ x_2' \end{bmatrix}\right) \geq 0.
\end{aligned}$$

Hence relation (12.A.3) holds on a dense subset of $\mathcal{H}_1 \oplus \mathcal{H}_2$. By continuity, it holds everywhere, and consequently $A - B_1^* B_1 \geq 0$.

It remains to prove the reverse implication: $X \geq 0$ if the three conditions are satisfied. Because $C \geq 0$ a decomposition of C as $C = LL^*$ is defined. Using this decomposition and $B = LB_1$,

$$X = \begin{bmatrix} A & B_1^* L^* \\ LB_1 & LL^* \end{bmatrix} = \begin{bmatrix} I & B_1^* \\ & L \end{bmatrix} \begin{bmatrix} A - B_1^* B_1 & \\ & I \end{bmatrix} \begin{bmatrix} I & \\ B_1 & L^* \end{bmatrix}.$$

Under the stated conditions, the operator

$$W = \begin{bmatrix} I & \\ & L \end{bmatrix} \begin{bmatrix} I & B_1^* \\ & I \end{bmatrix} \begin{bmatrix} (A - B_1^* B_1)^{1/2} & \\ & I \end{bmatrix} \qquad (12.A.4)$$

is well defined, and is a factor of X such that $X = WW^*$. Hence $X \geq 0$. □

Proof of lemma 12.8

Let $X \geq 0$ have a factorization $X = WW^*$, then $\mathcal{R}(X^{1/2}) = \mathcal{R}(W)$ (again by theorem 12.6). It can be inferred from Beutler and Root [BR76] that

$$X^\dagger = W^{*\dagger} W^\dagger = X^{\dagger/2} X^{\dagger/2},$$

hence if $\mathcal{R}(v) \subset \mathcal{R}(X^{1/2}) = \mathcal{R}(W)$, then v_1 and v_2 defined by

$$\begin{aligned}
v_1 &= X^{\dagger/2} v, \qquad && \mathcal{R}(v_1) \subset \overline{\mathcal{R}(X^{1/2})} \\
v_2 &= W^\dagger v, && \mathcal{R}(v_2) \subset \overline{\mathcal{R}(W^*)}
\end{aligned}$$

are bounded, and[3] $v_1^* v_1 = v_2^* v_2$.

[3] We are careful here not to write $X^\dagger v$. Although $\overline{\mathcal{R}(X)} = \overline{\mathcal{R}(X^{1/2})}$, we only have that $\mathcal{R}(X) \subset \mathcal{R}(X^{1/2})$, and hence $X^\dagger v$ can be unbounded with $\mathcal{R}(v) \in \mathcal{R}(X^{1/2})$.

Let $L = C^{1/2}$, $B_1 = L^\dagger B$ and put W as in (12.A.4), so that $X = WW^*$. Define the operator $W^\ddagger$ by

$$W^\ddagger = \begin{bmatrix} (A-B_1^*B_1)^{\dagger/2} & \\ & I \end{bmatrix} \begin{bmatrix} I & -B_1^* \\ & I \end{bmatrix} \begin{bmatrix} I & \\ & L^\dagger \end{bmatrix}.$$

We prove that $W^\ddagger = W^\dagger$ on $\mathcal{R}(W)$. The result will be, for a bounded operator v with $\mathcal{R}(v) \subset \mathcal{R}(X^{1/2}) = \mathcal{R}(W)$, that $W^\dagger v = W^\ddagger v$, so that $v_1 := X^{\dagger/2} v$ and $v_2 := W^\ddagger v$ are bounded and satisfy $v_1^* v_1 = v_2^* v_2$.

For any v with range in $\mathcal{R}(W)$ we have that the operator $v_1 = W^\dagger v$ is bounded and such that $v = Wv_1$. Hence $W^\ddagger v = W^\ddagger W v_1 = W^\dagger W v_1 = W^\dagger v$, so that $W^\ddagger = W^\dagger$ on $\mathcal{R}(W)$ if and only if

$$W^\ddagger W = W^\dagger W \quad \text{on } \overline{\mathcal{R}(W^*)}.$$

To analyze $W^\ddagger W$, we first prove that $B_1^* - B_1^* L^\dagger L = 0$. Indeed, if $x \in \mathcal{N}(L)$ then $x \in \mathcal{N}(B_1^*)$ (by equation (12.A.2)), and hence both $B_1^* x = 0$ and $Lx = 0$. If, on the other hand, $x \in \mathcal{N}(L)^\perp$, then $L^\dagger L x = x$ since $L^\dagger L$ is the projector onto $\mathcal{N}(L)^\perp$, and hence $B_1^* L^\dagger L x = B_1^* x$.

With the definition of $W^\ddagger$ and the above result,

$$\begin{aligned} W^\ddagger W &= \begin{bmatrix} (A-B_1^*B_1)^{\dagger/2} & \\ & I \end{bmatrix} \begin{bmatrix} I & -B_1^* \\ & I \end{bmatrix} \begin{bmatrix} I & \\ & L^\dagger \end{bmatrix} \cdot \\ &\quad \cdot \begin{bmatrix} I & \\ & L \end{bmatrix} \begin{bmatrix} I & B_1^* \\ & I \end{bmatrix} \begin{bmatrix} (A-B_1^*B_1)^{1/2} & \\ & I \end{bmatrix} \\ &= \begin{bmatrix} (A-B_1^*B_1)^{\dagger/2} & \\ & I \end{bmatrix} \begin{bmatrix} I & B_1^* - B_1^* L^\dagger L \\ & L^\dagger L \end{bmatrix} \begin{bmatrix} (A-B_1^*B_1)^{1/2} & \\ & I \end{bmatrix} \\ &= \begin{bmatrix} (A-B_1^*B_1)^{\dagger/2}(A-B_1^*B_1)^{1/2} & \\ & L^\dagger L \end{bmatrix} =: \begin{bmatrix} \mathbf{P}_1 & \\ & \mathbf{P}_2 \end{bmatrix} \end{aligned}$$

$\mathbf{P}_1$ and $\mathbf{P}_2$ are projectors onto $\overline{\mathcal{R}(A-B_1^*B_1)^{1/2}}$ and $\overline{\mathcal{R}(L^*)}$, respectively. Now, using

$$W^* = \begin{bmatrix} (A-B_1^*B_1)^{1/2} & \\ & I \end{bmatrix} \cdot \begin{bmatrix} I & \\ B_1 & L^* \end{bmatrix}$$

and $\mathcal{R}(B_1) \subset \overline{\mathcal{R}(L^*)}$, we have that

$$\overline{\mathcal{R}(W^*)} \subset \overline{\mathcal{R}} \begin{bmatrix} (A-B_1^*B_1)^{1/2} & \\ & L^* \end{bmatrix}$$

Since $W^\dagger W$ is the projector onto $\overline{\mathcal{R}(W^*)}$, and $W^\ddagger W$ is the projector onto the range at the right-hand side of the expression, this proves that $W^\ddagger W = W^\dagger W$ on $\overline{\mathcal{R}(W^*)}$, as required. Hence $W^\ddagger = W^\dagger$ on $\mathcal{R}(W)$, which also implies that $W^\ddagger$ is well defined on $\mathcal{R}(W)$. □

13 SPECTRAL FACTORIZATION

In this chapter we give a simple and straightforward treatment of the spectral factorization problem of a positive operator $\Omega \in \mathcal{X}$ into $\Omega = W^*W$, where $Wi \in \mathcal{U}$ is outer. We only consider the case where Ω is a *strictly* positive operator and where its causal part is bounded and has a u.e. stable realization. This leads to a recursive Riccati equation with time-varying coefficients for which the minimal positive definite solution leads to the outer factor. The theory also includes a formulation of a time-varying (strictly-) positive real lemma. In addition, we provide connections with related problems discussed in previous chapters in which Riccati equations appear as well, such as inner-outer factorization and orthogonal embedding. The results can no doubt be formulated in a more general way where strict positivity is not assumed, but we consider these extensions as laying outside the scope of the book.

13.1 INTRODUCTION

The term "spectral factorization" as commonly used refers in its most simple form to the problem of splitting a polynomial $p(s)$ in two factors $p(s) = p_1(s)p_2(s)$ so that the zeros of $p_1(s)$ are within a given open region of the complex plane and the zeros of $p_2(s)$ strictly in the complement of its closure. The problem has a solution if and only if $p(s)$ has no zeros on the boundary. One could extent the problem by allowing zeros on the boundary and counting them with one or the other region (or both). More interesting is the matrix function extension, which necessitates a definition of a "zero" consistent with Smith-McMillan theory. Spectral factorization became a hot topic when it was seen to be an essential step towards the solution of estimation and embedding problems. The standard problem then became: given a matrix function $M(s)$ which is positive definite on the imaginary axis, find a causal and causally invertible $T(s)$ so that $M(j\omega) = T(j\omega)^*T(j\omega)$. This more specific case became identified as the generic one, and a splitting of the necessarily even number of zeros on the imaginary axis became a part of the factorization as well. If $M(s)$ is rational, then $T(s)$ can be found by splitting both the zeros and the poles of $M(s)$ with the imaginary axis as boundary. It is remarkable that this is always possible. The first (complicated) algorithm to do so is due to Oono and Yasuura [OY54], based on the properties of the Smith-McMillan form [CC92]. Much more attractive schemes came later, *e.g.*, based on the state space description of the causal part of $M(s)$, and resulting in an algebraic Riccati equation and a criterion for positivity known as the positive real lemma. (A later extension to a more general metric is known as the Kalman-Yacubovitch-Popov lemma, for a very nice introduction to the topic and its ramifications, see [AV73].

In the very general and abstract context of "nest algebras" of operators on a Hilbert space, Arveson [Arv75] studied spectral factorization as the factorization of a positive operator $\Omega = W^*W$ in which W has a causality property related to the nest algebra. Arveson shows that in this very general set-up, the factorization is always possible provided that the original operator Ω is invertible. In the traditional special case of operators belonging to $L_\infty^{n\times n}$, (the LTI case), it is known that a spectral factorization for $M(s)$ with Hermitian $M(j\omega) \geq 0$ will exist if and only if the so-called Szegö condition

$$\int_{-\infty}^{\infty} \log\det M(j\omega)\frac{d\omega}{1+\omega^2} > -\infty$$

is satisfied (see [Hel64] for a general treatment). For example, the power spectrum of an ideal low pass filter will not qualify because it will have large intervals on which $M(j\omega) = 0$ and hence $\log\det M(j\omega) = -\infty$. An extension of this famous result to the time-varying case is not available, at least not to the knowledge of the authors. Therefore we shall adopt Arveson's result and put ourselves squarely in the situation where Ω is a bounded and invertible operator in $\mathcal{X}$, a case which is subsumed by the Szegö condition, but considerably less general. Still, a further word of caution is needed. The boundedness and invertibility of Ω does not entail the boundedness of the projection $\mathbf{P}(\Omega)$ as we already know from counterexamples in chapter 2. So, and in order to achieve finite computations, we introduce a further assumption, namely that $\Omega = T + T^*$ in which T is a causal and u.e. stable operator.

In our discussion on the inner-outer factorization problem and the embedding problem (see chapters 6, 7, 12), we have obtained solutions governed by Riccati equations. In many other problems in time-invariant system and H_∞-control theory, for example linear quadratic optimal control, optimal filtering and sensitivity minimization, Riccati equations play an important role as well. There is a family of related forms of this equation, and the precise form depends on the application. Underlying these problems is typically a spectral factorization problem. The equation usually has more than one solution, and important issues are the existence and computation of solutions which are Hermitian and maximally positive or negative, as these conditions imply minimal-phase properties of spectral factors, or the stability of closed-loop transfer operators constructed from the solution. Such solutions are, for time-invariant systems, obtained by an analysis of the eigenvalues and invariant subspaces of an associated Hamiltonian matrix.

For general time-varying systems, the Riccati equation becomes a recursion with time-varying coefficients that can also have time-varying dimensions. For such equations, much less is known on the structure of solutions. One reason for this is that the usual eigenvalue analysis to classify stable and unstable systems is no longer applicable: A_k need not even be square. Some results, *e.g.*, on the convergence of solutions starting from an approximate initial point, have already been obtained in the solution of the embedding problem (chapter 12).

In this chapter, we approach the time-varying Riccati equation from a different angle, by starting from the spectral factorization problem. The same approach is followed in [SA73] although, in that paper, the starting point is the existence of the Cholesky factor of a positive definite, finite size matrix. The Riccati recursion in these factorization problems emerges once a state realization for the operator is assumed.

The spectral factorization problem is treated in section 13.2, where also a (related) time-varying version of the positive real lemma is formulated. Some computational issues are discussed in section 13.3. It is argued in section 13.4 that under certain conditions the Riccati recursion converges to the exact solution even if the recursion is started from an approximate initial point. This allows us to compute spectral factors of more general time-varying positive operators, even if they are not constant or periodically varying before some point in time. Finally, in section 13.5, we discuss some connections of the spectral factorization theory with related problems in which a Riccati equation occurs, in particular the orthogonal embedding problem of contractive operators and the inner-outer factorization problem.

13.2 SPECTRAL FACTORIZATION

We recall the definitions of outer operators from section 7.2. An operator $W_\ell \in \mathcal{U}(\mathcal{M},\mathcal{N})$ is defined to be left outer if

$$\overline{\mathcal{U}_2^{\mathcal{M}} W_\ell} = \mathcal{U}_2^{\mathcal{N}} .$$

W_r is right outer if

$$\overline{\mathcal{L}_2 Z^{-1} W_r^*} = \mathcal{L}_2 Z^{-1} .$$

Arveson [Arv75] has shown, in the general context of nest algebras which also applies to our model of time-varying systems, that if $\Omega \in \mathcal{X}$ is a positive invertible operator, i.e. if Ω is a strictly positive operator, then an operator $W \in \mathcal{U}$ exists such that

$$\Omega = W^* W .$$

W can be chosen to be outer, in which case the factorization is called a spectral factorization, in the strict sense described in the introduction. Related to this fact is another theorem by Arveson in the same paper, which claims that operators in a Hilbert space have an inner-outer factorization

$$W = U W_r$$

where U is a co-isometry ($U^*U = I$) and W_r is right outer.[1] Hence, if Ω is uniformly positive definite, then Ω has the factorization $\Omega = W_r^* W_r$ where W_r is both left and right outer and invertible, and hence $\mathcal{L}_2 Z^{-1} W_r^* = \mathcal{L}_2 Z^{-1}$ (no closure is needed) and $W_r^{-1} \in \mathcal{U}$. Any other invertible factor W can be written as $W = UW_r$, where U is now invertible and hence inner.

In this section, we derive an algorithm to compute a time-varying spectral factorization of operators with a state-space realization. The computation amounts to the (recursive) solution of a Riccati equation. Such equations have in general a collection of solutions. We show that in order to obtain an outer spectral factor, one must select a uniformly positive solution of the Riccati equation, and we show that this solution is unique. We need a number of preliminary results.

Realization for T^*T

We first derive a formula to compute a realization of the upper part of the operator T^*T, when a realization of $T \in \mathcal{U}$ is given.

Lemma 13.1 *Let $T \in \mathcal{U}$ be given by the state realization $T = D + BZ(I-AZ)^{-1}C$, where $\ell_A < 1$. Then a state realization of the upper part of T^*T is*

$$\begin{bmatrix} A & C \\ D^*B + C^*\Lambda A & D^*D + C^*\Lambda C \end{bmatrix}$$

where $\Lambda \in \mathcal{D}$ is the (unique) operator satisfying the Lyapunov equation $\Lambda^{(-1)} = A^*\Lambda A + B^*B$.

PROOF Evaluation of T^*T gives

$$\begin{aligned} T^*T &= [D^* + C^*(I-Z^*A^*)^{-1}Z^*B^*]\,[D + BZ(I-AZ)^{-1}C] \\ &= D^*D + C^*(I-Z^*A^*)^{-1}Z^*B^*D + D^*BZ(I-AZ)^{-1}C + \\ &\quad + C^*(I-Z^*A^*)^{-1}Z^*B^*BZ(I-AZ)^{-1}C. \end{aligned}$$

The expression $(I-Z^*A^*)^{-1}Z^*B^*BZ(I-AZ)^{-1}$ evaluates as

$$(I-Z^*A^*)^{-1}Z^*B^*BZ(I-AZ)^{-1} = (I-Z^*A^*)^{-1}Z^*X + \Lambda(I-AZ)^{-1}$$

where $X = A^*\Lambda$, and Λ is given by the Lyapunov equation $\Lambda^{(-1)} = A^*\Lambda A + B^*B$. Λ is unique if $\ell_A < 1$, and

$$\begin{aligned} T^*T &= [D^*D + C^*\Lambda C] + [D^*B + C^*\Lambda A]Z(I-AZ)^{-1}C \\ &\quad + C^*(I-Z^*A^*)^{-1}Z^*[A^*\Lambda C + B^*D]\,. \end{aligned}$$

□

[1] Actually, Arveson uses a slightly different definition of outerness (not requiring $\ker(\cdot W_r)|_{\mathcal{L}_2 Z^{-1}} = 0$), so that U can be chosen inner. The resulting inner-outer factorizations are the same when W is invertible. See chapter 7.)

Properties of outer factors

The input and output state spaces of an outer factor in a spectral factorization of a strictly positive definite operator have certain characteristic properties, which we formulate in proposition 13.4. The recursive version of these properties then produces a Riccati equation, and the existence of the outer factor implies the existence of a (positive) solution to this equation. Other properties of the equation can be derived from the link with outer factors as well.

Proposition 13.2 *Let $T \in \mathcal{U}(\mathcal{M},\mathcal{M})$ be an outer invertible[2] operator, with state realization $\mathbf{T} = \{A,B,C,D\}$. Then $S = T^{-1} \in \mathcal{U}(\mathcal{M},\mathcal{M})$ has a state realization given by*

$$\mathbf{S} = \begin{bmatrix} A-CD^{-1}B & -CD^{-1} \\ D^{-1}B & D^{-1} \end{bmatrix}. \tag{13.1}$$

Moreover, $\mathbf{T}$ is [uniformly] reachable if and only if $\mathbf{S}$ is [uniformly] reachable, $\mathbf{T}$ is [uniformly] observable if and only if $\mathbf{S}$ is [uniformly] observable. Let $A^\times = A - CD^{-1}B$. If $\ell_A < 1$ and $\mathbf{T}$ is reachable or observable, then $\ell_{A^\times} < 1$.

PROOF Since T is outer and invertible, $T^{-1} \in \mathcal{U}$, so $S = T^{-1}$ has a realization which is causal. Let $y = uT$, where $u, y \in \mathcal{X}_2^{\mathcal{M}}$. Then $u = yS$, and

$$\left\{ \begin{array}{lcl} x_{[k+1]}^{(-1)} & = & x_{[k]}A + u_{[k]}B \\ y_{[k]} & = & x_{[k]}C + u_{[k]}D \end{array} \right. \Leftrightarrow \left\{ \begin{array}{lcll} x_{[k+1]}^{(-1)} & = & x_{[k]}(A-CD^{-1}B) & + \; y_{[k]}D^{-1}B \\ u_{[k]} & = & -x_{[k]}CD^{-1} & + \; y_{[k]}D^{-1} \end{array} \right.$$

so that S has a state realization as in (13.1). To prove the remaining properties, let, as in (5.1), T be decomposed as

$$\begin{array}{ll} \cdot T\big|_{\mathcal{L}_2Z^{-1}} = K_T + H_T : & \cdot H_T = \mathbf{P}(\cdot T\big|_{\mathcal{L}_2Z^{-1}}); \quad \cdot K_T = \mathbf{P}'(\cdot T\big|_{\mathcal{L}_2Z^{-1}}) \\ \cdot T\big|_{\mathcal{U}_2} = E_T & \end{array}$$

Since $\cdot T\big|_{\mathcal{X}_2}$ is an invertible operator, the same is true for K_T, because a decomposition of $\cdot T|_{\mathcal{X}_2}$ along $\mathcal{X}_2 = \mathcal{L}_2Z^* \oplus \mathcal{U}_2$ in the input and output spaces gives

$$\cdot T\big|_{\mathcal{X}_2} = \begin{bmatrix} K_T & H_T \\ 0 & E_T \end{bmatrix}.$$

The Hankel operator H_T has a factorization in terms of the reachability and observability operators $\mathbf{F}$ and $\mathbf{F}_o$ defined as $H_T = \mathbf{P}_0(\cdot \mathbf{F}^*)\mathbf{F}_o$. Partition $u \in \mathcal{X}_2^{\mathcal{M}}$ into a past and a future part: $u = u_p + u_f \in \mathcal{L}_2Z^{-1} \oplus \mathcal{U}_2$, and partition y likewise. Then

$$y = uT \quad \Leftrightarrow \quad \left\{ \begin{array}{lcl} y_p & = & u_pK_T \\ x_{[0]} & = & \mathbf{P}_0(u_p\mathbf{F}^*) \\ y_f & = & u_fT + x_{[0]}\mathbf{F}_o. \end{array} \right.$$

[2]Thus, both left and right outer.

Because T is invertible in $\mathcal{U}$, K_T is invertible, and hence, the above set of equations is equivalent to

$$u = yS \quad \Leftrightarrow \quad \begin{cases} u_p &= y_p K_T^{-1} \\ x_{[0]} &= \mathbf{P}_0(y_p K_T^{-1} \mathbf{F}^*) \\ u_f &= y_f T^{-1} - x_{[0]} \mathbf{F}_o T^{-1}. \end{cases}$$

It follows that $\mathbf{S}$ has reachability and observability operators given by

$$\mathbf{F}_S = \mathbf{F} K_T^{-*}, \qquad \mathbf{F}_{o,S} = -\mathbf{F}_o T^{-1}.$$

These operators inherit the one-to-one and onto properties of the reachability and observability operators of T.

Finally, to show that $\ell_{A^\times} < 1$ if $\ell_A < 1$ and $\Lambda_\mathbf{F} > 0$ or $\Lambda_{\mathbf{F}_o} > 0$, we invoke the following extension of proposition 5.14: if $\Lambda_\mathbf{F} > 0$, then

$$\mathbf{F} \text{ is bounded on } \mathcal{X}_2 \quad \Leftrightarrow \quad \ell_A < 1.$$

Applying this result twice yields, if $\Lambda_\mathbf{F} > 0$,

$$\ell_A < 1 \quad \Rightarrow \quad \mathbf{F} \text{ bounded on } \mathcal{X}_2 \quad \Rightarrow \quad \mathbf{F}_S \text{ bounded on } \mathcal{X}_2 \quad \Rightarrow \quad \ell_A^\times < 1.$$

A similar result holds if $\Lambda_{\mathbf{F}_o} > 0$. □

Proposition 13.3 *Let $W \in \mathcal{U}$ be boundedly invertible (in $\mathcal{X}$), with inner-outer factorization $W = UW_r$, and suppose that W and W_r have u.e. stable realizations with the same (A,C): $W = D + BZ(I-AZ)^{-1}C$, $W_r = D_r + B_r Z(I-AZ)^{-1}C$, $\ell_A < 1$. Let Λ and Λ_r be the reachability Gramians of W and W_r, respectively. Then*

$$\Lambda \geq \Lambda_r, \qquad \Lambda = \Lambda_r \;\; \textit{iff} \;\; U \in \mathcal{D}.$$

PROOF Since W_r is outer, a realization of $W_r^{-1} \in \mathcal{U}$ is given by

$$W_r^{-1} = D_r^{-1} - D_r^{-1} BZ(I - A^\times Z)^{-1} C, \qquad A^\times = A - CD^{-1}B,$$

so that $U = WW_r^{-1}$ has main diagonal $\mathbf{P}_0(U) = DD_r^{-1}$. Since $U^*U = I$, this implies that $D_r^{-*} D^* D D_r^{-1} \leq I$.

Using $W^*W = W_r^* W_r$ and evaluating each term by means of lemma 13.1 yields the equalities

$$\begin{aligned} D^*D + C^*\Lambda C &= D_r^* D_r + C^* \Lambda_r C \\ D^*B + C^*\Lambda A &= D_r^* B_r + C^* \Lambda_r A \end{aligned}$$

where the reachability Gramians Λ and Λ_r are specified (uniquely) by

$$\begin{aligned} \Lambda^{(-1)} &= A^*\Lambda A + B^*B \\ \Lambda_r^{(-1)} &= A^*\Lambda_r A + B_r^* B_r. \end{aligned}$$

The first equation is equivalent to

$$D_r^{-*} C^* (\Lambda - \Lambda_r) C D_r^{-1} = I - D_r^{-*} D^* D D_r^{-1}$$

and since $D_r^{-*} D^* D D_r^{-1} \leq I$, it follows that $\Lambda \geq \Lambda_r$.

The proof that $\Lambda = \Lambda_r \Leftrightarrow U \in \mathcal{D}$ is also a straightforward consequence of these equations. □

Positive real lemmas

The following proposition is of crucial importance in proving that there is a solution to the Riccati equation associated to the time-varying spectral factorization problem which gives an outer factor W, and in characterizing this solution. Recall the definitions of input and output state spaces of T as $\mathcal{H}(T) = \mathbf{P}'(\mathcal{U}_2 T)$, $\mathcal{H}_o(T) = \mathbf{P}(\mathcal{L}_2 Z^{-1} T)$, *viz.* equations (5.3) and (5.5).

Proposition 13.4 *Let $T \in \mathcal{U}(\mathcal{M},\mathcal{M})$ be such that $T^* + T \gg 0$. In addition, let $W \in \mathcal{U}(\mathcal{M},\mathcal{M})$ be an invertible factor of $T^* + T = W^*W$. Then $\mathcal{H}_o(T) \subset \mathcal{H}_o(W)$. If W is outer, $\mathcal{H}_o(T) = \mathcal{H}_o(W)$. In particular, there exists a realization of an outer W that has the same (A,C) pair as a realization of T.*

PROOF Arveson's theorem on spectral factorization[Arv75] is applicable in this case, so we may infer the existence of an invertible operator $W \in \mathcal{U}$ such that

$$T^* + T = W^*W .$$

In general, $\mathcal{L}_2 Z^{-1} W^* \subset \mathcal{L}_2 Z^{-1}$, and $\mathcal{L}_2 Z^{-1} W^* = \mathcal{L}_2 Z^{-1}$ if and only if W is outer. Thus

$$\begin{aligned} \mathcal{H}_o(T) &= \mathbf{P}(\mathcal{L}_2 Z^{-1} T) \\ &= \mathbf{P}(\mathcal{L}_2 Z^{-1}[T+T^*]) \qquad \text{[since } T^* \in \mathcal{L}\text{]} \\ &= \mathbf{P}(\mathcal{L}_2 Z^{-1} W^* W) \\ &\subset \mathbf{P}(\mathcal{L}_2 Z^{-1} W) = \mathcal{H}_o(W) . \end{aligned}$$

If W is outer, then $\mathcal{L}_2 Z^{-1} W^* = \mathcal{L}_2 Z^{-1}$ and the inclusion in the above derivation becomes an identity: W outer $\Rightarrow \mathcal{H}_o(T) = \mathcal{H}_o(W)$. If $\{A,B,C,D\}$ is a realization of T with $\ell_A < 1$, then $\mathcal{H}_o(T) = \mathcal{D}_2(I-AZ)^{-1}C$ (if the realization of T is uniformly reachable) or, more generally, $\mathcal{H}_o(T) \subset \mathcal{D}_2(I-AZ)^{-1}C$. Hence, it is clear that a realization of an outer W can have the same (A,C)-pair as a realization of T. □

Note that not necessarily $\mathcal{H}_o(T) = \mathcal{H}_o(W) \Rightarrow W$ outer, as a simple time-invariant example shows. The proposition, along with lemma 13.3, assures that a minimal degree factor W of $T + T^* \gg 0$ is obtained by taking a realization of W to have the (A,C)-pair as a realization of T, and that this factor is outer if the reachability Gramian of this realization is as small as possible. This observation forms the main part of the proof of the following theorems, which can be used to actually compute the realization of the outer factor if a realization of T is given.

Theorem 13.5 *Let $T \in \mathcal{U}(\mathcal{M},\mathcal{M})$ be a locally finite operator with an observable state realization $\{A,B,C,D\}$ such that $\ell_A < 1$. Then $T^* + T \gg 0$ if and only if a solution $\Lambda \in \mathcal{D}$, $\Lambda \geq 0$ exists of*

$$\Lambda^{(-1)} = A^*\Lambda A + \left[B^* - A^*\Lambda C\right](D + D^* - C^*\Lambda C)^{-1}\left[B - C^*\Lambda A\right] \tag{13.2}$$

such that $D + D^* - C^*\Lambda C \gg 0$.

If $T^* + T \gg 0$, let $W \in \mathcal{U}(\mathcal{M},\mathcal{M})$ be an invertible factor of $T^* + T = W^*W$. A realization $\{A, B_W, C, D_W\}$ for W such that W is outer is then given by the smallest solution $\Lambda \geq 0$, and

$$\begin{cases} D_W = (D + D^* - C^*\Lambda C)^{1/2} \\ B_W = D_W^{-*}\left[B - C^*\Lambda A\right]. \end{cases}$$

The realization of W is observable and [uniformly] reachable, if T is so.

PROOF Let the realization of T satisfy the given requirements, and suppose that $T+T^* \gg 0$. Then $T+T^* = W^*W$, where W is outer. According to proposition 13.4, W can have the same (A,C) pair as T. Hence assume that $W = D_W + B_W Z(I-AZ)^{-1}C$, and denote its reachability Gramian by Λ. Then, with help of lemma 13.1, this realization satisfies

$$\begin{array}{rcll} D+D^* &=& D_W^* D_W + C^*\Lambda C, & D_W^* D_W \gg 0 \\ BZ(I-AZ)^{-1}C &=& [D_W^* B_W + C^*\Lambda A]\, Z(I-AZ)^{-1}C & \\ \Lambda^{(-1)} &=& A^*\Lambda A + B_W^* B_W\,, & \Lambda \geq 0\,. \end{array}$$

Because the realization of T is observable, the operator $\cdot\,(I-AZ)^{-1}C$ is one-to-one by definition, and the above set of equations reduce to

$$\begin{array}{rcll} D+D^* &=& D_W^* D_W + C^*\Lambda C, & D_W^* D_W \gg 0 \\ B &=& D_W^* B_W + C^*\Lambda A & \\ \Lambda^{(-1)} &=& A^*\Lambda A + B_W^* B_W\,, & \Lambda \geq 0\,. \end{array}$$

$$\begin{array}{rrcl} \Rightarrow & D_W &=& (D+D^* - C^*\Lambda C)^{1/2} \\ & B_W &=& D_W^{-*}[B - C^*\Lambda A] \\ & \Lambda^{(-1)} &=& A^*\Lambda A + [B^* - A^*\Lambda C]\,(D+D^* - C^*\Lambda C)^{-1}\,[B - C^*\Lambda A]\,, \end{array}$$

(D_W, and hence B_W, are determined up to a left diagonal unitary factor), so that Λ satisfies the given Riccati equation. In fact, we showed that if $T+T^* \gg 0$, the existence of an outer factor implies that there is a solution Λ of the Riccati equation which is positive semi-definite, and such that also $D+D^* - C^*\Lambda C \gg 0$. The converse, to show that $T+T^* \gg 0$ if these quantities are positive semi-definite, resp. uniformly positive, follow almost directly from the construction, since it specifies a realization of an invertible factor W of $T+T^*$. If this solution Λ is the smallest possible solution, then, by lemma 13.3, W is outer. □

The above theorem can be extended to observable realizations without reachability constraint.

Theorem 13.6 *Theorem 13.5 holds also if the realization of T is not observable.*

The proof of this theorem is technical and given in the appendix at the end of the chapter.

Theorems 13.5 and 13.6 can also be specified in two alternate forms, familiar from the time-invariant context [AV73, Den75]:

Corollary 13.7 (The time-varying positive real lemma) *Let $T \in \mathcal{U}$ be a locally finite operator with state realization $\{A,B,C,D\}$ such that $\ell_A < 1$.*

Then $T^* + T \gg 0$ if and only if there exist diagonal operators Λ, Q, B'_W with $\Lambda \geq 0$ and $Q \gg 0$ satisfying

$$\begin{array}{rcl} \Lambda^{(-1)} &=& A^*\Lambda A + B_W'^* Q B'_W \\ B_W'^* Q &=& B^* - A^*\Lambda C \\ Q &=& D + D^* - C^*\Lambda C\,. \end{array}$$

PROOF In view of theorems 13.5 and 13.6, it suffices to make the connection $Q = D_W^* D_W$ and $B_W = D_W B'_W$. □

Corollary 13.8 (Time-varying spectral factorization) *Let $\Omega \in \mathcal{X}$ be a Hermitian operator whose upper part is locally finite with state realization $\{A,B,C,D\}$ satisfying $\ell_A < 1$, i.e.,*

$$\Omega = D + BZ(I-AZ)^{-1}C + C^*(I-Z^*A^*)^{-1}Z^*B^* .$$

Then $\Omega \gg 0$ if and only if there exists a solution $\Lambda \in \mathcal{D}, \Lambda \geq 0$ of

$$\Lambda^{(-1)} = A^*\Lambda A + \left[B^* - A^*\Lambda C\right](D - C^*\Lambda C)^{-1}\left[B - C^*\Lambda A\right] , \tag{13.3}$$

such that $D - C^*\Lambda C \gg 0$.

If $\Omega \gg 0$ and Λ is the smallest positive solution, then a realization $\{A, B_W, C, D_W\}$ for an outer factor W of Ω is given by

$$\begin{aligned} D_W &= (D - C^*\Lambda C)^{1/2} \\ B_W &= D_W^{-*}\left[B - C^*\Lambda A\right] . \end{aligned}$$

If the realization $\{A,B,C,D\}$ is observable and reachable resp. uniformly reachable, then $\Lambda > 0$ resp. $\Lambda \gg 0$: the realization for W is observable and [uniformly] reachable.

13.3 COMPUTATIONAL ISSUES

We now consider some computational issues that play a role in actually computing a spectral factorization of a uniformly positive operator Ω with a locally finite observable realization given as in (13.3). First, note that by taking the k-th entry along each diagonal of (13.3), we obtain the Riccati *recursion*

$$\Lambda_{k+1} = A_k^*\Lambda_k A_k + \left[B_k^* - A_k^*\Lambda_k C_k\right](D_k - C_k^*\Lambda_k C_k)^{-1}\left[B_k - C_k^*\Lambda_k A_k\right] , \tag{13.4}$$

and with Λ_k known, $(B_W)_k$, $(D_W)_k$ also follow locally:

$$\begin{aligned} (D_W)_k^*(D_W)_k &= D_k - C_k^*\Lambda_k C_k \\ (B_W)_k &= (D_W^{-*})_k\left[B_k - C_k^*\Lambda_k A_k\right] . \end{aligned}$$

Hence all that is needed in practical computations is an initial point for the recursion of Λ_k. Special cases where such an initial point can indeed be obtained are familiar from previous chapters.

One general observation is that, since there may be more than one positive solution Λ, there also may be more than one initial point Λ_k. Outer factors are obtained by choosing the smallest positive solution, which implies taking the smallest positive initial point: since $\Lambda \leq \Lambda' \Rightarrow \Lambda_k \leq \Lambda'_k (\forall k)$, a single Λ_k is part of the smallest solution if and only if the corresponding Λ is the smallest.

Finite matrices

Exact initial conditions can be obtained in the case where $\Omega \in \mathcal{X}(\mathcal{M},\mathcal{M})$ is actually a finite matrix, *i.e.*, where

$$\mathcal{M} = \cdots \times \emptyset \times \emptyset \times \mathcal{M}_1 \times \mathcal{M}_2 \times \cdots \times \mathcal{M}_n \times \emptyset \times \cdots .$$

In this case, Ω is a finite $n\times n$ (block) matrix, and a realization for Ω can start off with no states at point 1 in time. Since the dimension of Λ follows that of A, an exact initial point for the recursion is $\Lambda_1 = [\cdot]$ (a 0×0 matrix). The spectral factorization reduces for finite matrices to a Cholesky factorization, and the resulting algorithm is an efficient way to compute Cholesky factorizations for (large) matrices with a sparse state space.

Initial time-invariance

A second class of systems are systems which are time invariant before some point in time, say $k=1$. Then, before point $k=1$, all Λ_k are equal to each other, and in particular $\Lambda_0 = \Lambda_1$. Hence the recursion for Λ reduces to an algebraic equation

$$\Lambda_0 = A_0^*\Lambda_0 A_0 + \left[B_0^* - A_0^*\Lambda C_0\right](D_0 - C_0^*\Lambda_0 C_0)^{-1}\left[B_0 - C_0^*\Lambda_0 A_0\right],$$

which is the classical time-invariant Riccati equation. A solution to this equation can be obtained in one of the classical ways, *e.g.*, as the solution of a Hamiltonian equation. Multiple solutions exist, and in order to obtain an outer spectral factor W, the smallest positive solution of the above equation must be chosen. Because the Λ_k for $k>0$ are determined without freedom by Λ_0 via the recursion (13.4), the resulting Λ will also be the smallest positive solution for all time.

Periodic systems

If Ω is periodically time varying, with period n say, then one can apply the usual time-invariance transformation, by considering a block system consisting of n consecutive state realization sections. Since the block-system is time invariant, one can compute the smallest positive solution Λ_1 from the resulting block-Riccati equation with the classical techniques, and Λ_1 is an exact initial condition to compute the realization of the spectral factor for time points $2,\cdots,n$. As usual, such a technique may not be attractive if the period is large.

Unknown initial conditions

Finally, we consider the more general case where Ω is not completely specified but only, say, its "future" submatrix $[\Omega_{i,j}]_0^\infty$ is known. The unknown "past" of Ω is assumed to be such that $\Omega \gg 0$. In this case, the exact initial point for the recursion of Λ_k is unknown. It is possible to start the recursion (13.4) from an approximate initial point, for which typically $\hat{\Lambda}_0 = 0$ is chosen. The convergence of this choice is investigated in the following section. It is shown in proposition 13.10 that when the realization $\{A,B,C,D\}$ is observable and has $\ell_A < 1$, then $\hat{\Lambda}_k$ (corresponding to the recursion (13.4) with initial point $\hat{\Lambda}_0 = 0$) converges to Λ_k, the exact solution obtained with the correct initial point Λ_0.

13.4 CONVERGENCE OF THE RICCATI RECURSION

We study the convergence of an approximate solution $\hat{\Lambda}_k$ ($k\geq 0$) to the Riccati recursion (13.4), if the recursion is started with $\hat{\Lambda}_0 = 0$ rather than the exact initial point Λ_0. It is shown that $\hat{\Lambda}_k \to \Lambda_k$ for $k\to\infty$, when $\Omega \gg 0$, $\ell_A < 1$ and the given realization is observable. Similar results are well known for the time-invariant case, and for the time-varying case some results are known from the connection of the Riccati recursion with Kalman filtering (*cf.* [AK74, AM79]). However, the derivation given below is more general because state dimensions are allowed to vary, and hence A_k cannot be assumed to be square and invertible, as required in [AK74].

Consider the following block decomposition of the matrix representation of $\Omega = W^*W$, and a related operator $\hat{\Omega} = \hat{W}^*\hat{W}$:

$$\Omega = \left[\begin{array}{c|cc} \underline{\Omega}_{11} & \underline{\Omega}_{12} & \underline{\Omega}_{13} \\ \hline \underline{\Omega}_{12}^* & \underline{\Omega}_{22} & \underline{\Omega}_{23} \\ \underline{\Omega}_{13}^* & \underline{\Omega}_{23}^* & \underline{\Omega}_{33} \end{array}\right], \qquad W = \left[\begin{array}{c|cc} \underline{W}_{11} & \underline{W}_{12} & \underline{W}_{13} \\ \hline & \underline{W}_{22} & \underline{W}_{23} \\ & & \underline{W}_{33} \end{array}\right]$$
$$\hat{\Omega} = \left[\begin{array}{c|cc} \underline{\Omega}_{11} & 0 & 0 \\ \hline 0 & \underline{\Omega}_{22} & \underline{\Omega}_{23} \\ 0 & \underline{\Omega}_{23}^* & \underline{\Omega}_{33} \end{array}\right], \qquad \hat{W} = \left[\begin{array}{c|cc} \hat{\underline{W}}_{11} & 0 & 0 \\ \hline & \hat{\underline{W}}_{22} & \hat{\underline{W}}_{23} \\ & & \hat{\underline{W}}_{33} \end{array}\right]. \tag{13.5}$$

In these decompositions,[3] $\underline{\Omega}_{11}$ corresponds to $[\Omega_{i,j}]_{-\infty}^{-1}$, $\underline{\Omega}_{22} = [\Omega_{i,j}]_0^{n-1}$ is a finite $n \times n$ matrix (where n is some integer to be specified later), and $\underline{\Omega}_{33}$ corresponds to $[\Omega_{i,j}]_n^{\infty}$. The point of introducing the operator $\hat{\Omega}$ is that $\hat{\Lambda}_0$ is the *exact* (and smallest positive) initial point of the Riccati recursion (13.4) for a spectral factorization of the lower right part of $\hat{\Omega}$, and leads to an outer spectral factor $\hat{W}$ such that $\hat{\Omega} = \hat{W}^*\hat{W}$, of which only the lower right part is computed. This is seen by putting $A_{-1} = 0$, $B_{-1} = 0$ in the Riccati recursion for Λ, which leads to $\hat{\Lambda}_0 = 0$. The convergence of $\hat{\Lambda}_k$ to Λ_k is studied from this observation.

As a preliminary step, the following lemma considers a special case of the above Ω.

Lemma 13.9 *Let be given an operator $\Omega \in \mathcal{X}$, $\Omega \gg 0$, with block decomposition*

$$\Omega = \begin{bmatrix} \underline{\Omega}_{11} & \underline{\Omega}_{12} & 0 \\ \underline{\Omega}_{12}^* & \underline{\Omega}_{22} & \underline{\Omega}_{23} \\ 0 & \underline{\Omega}_{23}^* & \underline{\Omega}_{33} \end{bmatrix}$$

where $\underline{\Omega}_{22}$ is an $n \times n$ matrix. Let the upper triangular part of Ω be locally finite and u.e. stable. Then

$$\underline{(\Omega^{-1})}_{33} \to (\underline{\Omega}_{33} - \underline{\Omega}_{23}^* \underline{\Omega}_{22}^{-1} \underline{\Omega}_{23})^{-1} \quad \text{as } n \to \infty$$

(strong convergence). Hence $\underline{(\Omega^{-1})}_{33} \to \underline{(\hat{\Omega}^{-1})}_{33}$, where $\hat{\Omega}$ is equal to Ω, but with $\hat{\underline{\Omega}}_{12} = 0$.

[3] The underscore is used in this section to denote that we take block submatrices rather than entries of Ω.

PROOF Let $\{A,B,C,D\}$ be a realization of the upper triangular part of Ω with $\ell_A < 1$. Let $\underline{\Omega}_{12} = \underline{C}_1\underline{O}_1$, $\underline{\Omega}_{23} = \underline{C}_2\underline{O}_2$, where

$$\underline{C}_1 = \begin{bmatrix} \vdots \\ B_{-3}A_{-2}A_{-1} \\ B_{-2}A_{-1} \\ B_{-1} \end{bmatrix}, \qquad \underline{C}_2 = \begin{bmatrix} B_0A_1\cdots A_{n-1} \\ \vdots \\ B_{n-3}A_{n-2}A_{n-1} \\ B_{n-2}A_{n-1} \\ B_{n-1} \end{bmatrix},$$

$$\begin{aligned} \underline{O}_1 &= [C_0 \quad A_0C_1 \quad A_0A_1C_2 \quad \cdots \quad A_0\cdots A_{n-2}C_{n-1}] \\ \underline{O}_2 &= [C_n \quad A_nC_{n+1} \quad A_nA_{n+1}C_{n+2} \quad \cdots]. \end{aligned}$$

Then $\underline{O}_1\underline{C}_2$ is a summation of n terms, each containing a product $A_0\cdots A_{i-1}$ and a product $A_{i+1}\cdots A_{n-1}$. Because $\ell_A < 1$ implies that products of the form $A_k\cdots A_{k+n} \to 0$ as $n\to\infty$ strongly and uniformly in k, we obtain $\underline{O}_1\underline{C}_2 \to 0$ if $n\to\infty$.

Write $X_3 = (\Omega^{-1})_{33}$. By repeated use of Schur's inversion formula (lemma 12.2), X_3 is given by the recursion

$$X_1 = \underline{\Omega}_{11}^{-1}, \qquad X_{k+1} = (\underline{\Omega}_{k+1,k+1} - \underline{\Omega}_{k,k+1}^* X_k \underline{\Omega}_{k,k+1})^{-1}. \tag{13.6}$$

We first consider a special case, where $\underline{\Omega}_{k,k} = I$ $(k=1,2,3)$. In the derivation below, for ease of discussion it is assumed that also $\underline{O}_k\underline{O}_k^* = I$, *i.e.*, the realization is uniformly observable and in output normal form, although this is not an essential requirement. The recursion (13.6) becomes

$$\begin{aligned} Y_k &= \underline{C}_k^* X_k \underline{C}_k \\ X_{k+1} &= (I - \underline{O}_k^* Y_k \underline{O}_k)^{-1} = I + \underline{O}_k^* \left[Y_k + Y_k^2 + \cdots\right] \underline{O}_k, \end{aligned}$$

so that, in particular,

$$Y_2 = \underline{C}_2^*\underline{C}_2 + \underline{C}_2^*\underline{O}_1^* \left[Y_1(I-Y_1)^{-1}\right] \underline{O}_1\underline{C}_2.$$

For large n, $Y_2 \to \underline{C}_2^*\underline{C}_2$ and becomes independent of Y_1 and $\underline{C}_1$, and

$$X_3 \to (I - \underline{O}_2^*\underline{C}_2^*\underline{C}_2\underline{O}_2)^{-1} = (\underline{\Omega}_{33} - \underline{\Omega}_{23}^*\underline{\Omega}_{22}^{-1}\underline{\Omega}_{23})^{-1}$$

independently of $\underline{C}_1$. The expression on the right-hand side is the same as the value obtained for $\underline{C}_1 = 0$, *i.e.*, $\underline{\Omega}_{12} = 0$.

The general case reduces to the above special case by a pre- and post-multiplication by

$$\begin{bmatrix} \underline{\Omega}_{11}^{-1/2} & & \\ & \underline{\Omega}_{22}^{-1/2} & \\ & & \underline{\Omega}_{33}^{-1/2} \end{bmatrix}.$$

This maps $\underline{\Omega}_{k,k}$ to I, $\underline{C}_k$ to $\underline{\Omega}_{k,k}^{-1/2}\underline{C}_k$, and $\underline{O}_k$ to $\underline{O}_k\underline{\Omega}_{k+1,k+1}^{-1/2}$. The latter two mappings lead to realizations with different B_i and C_i, but the A_i remain the same, and in particular the convergence properties of $\underline{C}_2\underline{O}_1$ remain unchanged. It follows that $\underline{(\Omega^{-1})}_{33} \to (\underline{\Omega}_{33} - \underline{\Omega}_{23}^*\underline{\Omega}_{22}^{-1}\underline{\Omega}_{23})^{-1}$ also in the general case. □

We now return to the spectral factorization problem, with Ω given as in (13.5).

Proposition 13.10 *Let $\Omega \in \mathcal{X}$, $\Omega \gg 0$ have an upper triangular part which is locally finite and given by an observable realization $\{A,B,C,D\}$ where $\ell_A < 1$. Let $\Lambda \in \mathcal{D}$ be the unique solution of (13.3) so that its entries Λ_n satisfy the recursive Riccati equation (13.4). Let $\hat{\Lambda}_n$ $(n \geq 0)$ be the sequence obtained from the same recursion, but starting from $\hat{\Lambda}_0 = 0$.*

Then $\hat{\Lambda}_n \to \Lambda_n$ as $n \to \infty$ (strong convergence).

PROOF Let Ω, $\hat{\Omega}$ have block decompositions as in (13.5), where $\underline{\Omega}_{22}$ is an $n \times n$ matrix. Let $\Omega = W^*W$, $\hat{\Omega} = \hat{W}^*\hat{W}$, where $W, \hat{W}$ are outer spectral factors, then Λ, $\hat{\Lambda}$ are the reachability Gramians of the realization of W, $\hat{W}$ given in corollary 13.8. Denote

$$\begin{aligned} W_{12} &= \underline{\mathcal{C}}_{W,1}\underline{\mathcal{O}}_1 \\ W_{23} &= \underline{\mathcal{C}}_{W,2}\underline{\mathcal{O}}_2 \\ W_{13} &= \underline{\mathcal{C}}_{W,1}A_0A_1\cdots A_{n-1}\underline{\mathcal{O}}_2 \,. \end{aligned}$$

Because $\ell_A < 1$, we have that $W_{13} \to 0$ as $n \to \infty$ (strongly), so that for large enough n, $\Lambda_n \approx \underline{\mathcal{C}}_{W,2}^*\underline{\mathcal{C}}_{W,2}$ and hence

$$\begin{aligned} \underline{\Omega}_{33} &= W_{33}^*W_{33} + W_{23}^*W_{23} + W_{13}^*W_{13} \\ &\approx W_{33}^*W_{33} + \underline{\mathcal{O}}_2^*\Lambda_n\underline{\mathcal{O}}_2 \,. \end{aligned}$$

Consequently, $\underline{\mathcal{O}}_2^*(\Lambda_n - \hat{\Lambda}_n)\underline{\mathcal{O}}_2 \approx \hat{W}_{33}^*\hat{W}_{33} - W_{33}^*W_{33}$.

The next step is to show that $\hat{W}_{33}^*\hat{W}_{33} - W_{33}^*W_{33} \to 0$ for large n, so that, if the realization is observable, $\hat{\Lambda}_n \to \Lambda_n$. Let $X_3 = (W_{33}^*W_{33})^{-1}$, and $\hat{X}_3 = (\hat{W}_{33}^*\hat{W}_{33})^{-1}$. Since $\Omega^{-1} = W^{-1}W^{-*}$, and W is outer so that $W^{-1} \in \mathcal{U}$, it follows that $X_3 = (\Omega^{-1})_{33}$ and $\hat{X}_3 = (\hat{\Omega}^{-1})_{33}$. Lemma 13.9 proves that, if $\ell_A < 1$, then $(\Omega^{-1})_{33} \to (\hat{\Omega}^{-1})_{33}$ as $n \to \infty$, so that $X_3 \to \hat{X}_3$, and hence $\hat{\Lambda}_n \to \Lambda_n$. □

Finally, we remark that always $\hat{\Lambda}_k \leq \Lambda_k$. This is a consequence of the fact that

$$\hat{\Lambda}_k \leq \Lambda_k \quad \Rightarrow \quad \hat{\Lambda}_{k+1} \leq \Lambda_{k+1} \,, \tag{13.7}$$

which is proven directly from the Riccati recursion (13.4) in a way similar to [AM79, ch. 9]. Indeed, let the matrix G_{X,Λ_k} be given by

$$G_{X,\Lambda_k} = \begin{bmatrix} X - A_k^*\Lambda_kA_k & B_k - C_k^*\Lambda_kA_k \\ B_k^* - A_k^*\Lambda_kC_k & D_k - C_k^*\Lambda_kC_k \end{bmatrix} = \begin{bmatrix} X & B_k \\ B_k^* & D_k \end{bmatrix} - \begin{bmatrix} A_k^* \\ C_k^* \end{bmatrix} \Lambda_k [A_k \quad C_k] \,,$$

parameterized by some matrix $X = X^*$. Using Schur's complements, it follows that, if $D_k - C_k^*\Lambda_kC_k > 0$, then

$$G_{X,\Lambda_k} \geq 0 \quad \Rightarrow \quad X - A_k^*\Lambda_kA_k - \left[B_k^* - A_k^*\Lambda C_k\right](D_k - C_k^*\Lambda_kC_k)^{-1}\left[B_k - C_k^*\Lambda_kA_k\right] \geq 0 \,.$$

Hence $\Lambda_{k+1} = \min\{X : G_{X,\Lambda_k} \geq 0\}$. But if $\hat{\Lambda}_k \leq \Lambda_k$, then $G_{\Lambda_{k+1},\hat{\Lambda}_k} \geq G_{\Lambda_{k+1},\Lambda_k} \geq 0$. It follows that $\Lambda_{k+1} \geq \hat{\Lambda}_{k+1}$, since $\hat{\Lambda}_{k+1}$ is the smallest matrix X for which $G_{X,\hat{\Lambda}_k} \geq 0$. This proof also supplements the remark made in section 13.3 that the "smallest solution" is well defined: if Λ_k is the smallest solution at one point, the resulting diagonal operator Λ is the smallest solution at all points.

13.5 CONNECTIONS

Spectral factorization is intimately connected to the various incarnations of the time-varying Riccati equation that we encountered earlier, in the solution of the orthogonal embedding problem (chapter 12) and inner-outer factorizations (chapter 6).

Orthogonal embedding

Recall the orthogonal embedding problem: given a transfer operator T of a bounded causal discrete-time linear system, extend this system by adding more inputs and outputs to it such that the resulting system Σ,

$$\Sigma = \begin{bmatrix} \Sigma_{11} & \Sigma_{12} \\ \Sigma_{21} & \Sigma_{22} \end{bmatrix},$$

is inner and has $T = \Sigma_{11}$ as its partial transfer operator when the extra inputs are forced to zero. One way to solve the embedding problem is to start out from a spectral factorization of $\Sigma_{21}^*\Sigma_{21} = I - T^*T$, which gives Σ_{21}, and next to follow the embedding procedure for isometric operators of chapter 7 (we know already that even in very simple cases the embedding may not lead to an inner operator, but it certainly will under the additional assumption that $\ell_A < 1$ for the transition operator of T. Hence, the solution of the embedding problem can also be obtained starting from the spectral factorization theorems 13.5 and 13.6. This leads to a variant of the embedding theorems (theorem 12.12 and 12.14) and the bounded real lemma (theorem 12.15).

Theorem 13.11 (Time-varying bounded real lemma, II) *Let $T \in \mathcal{U}(\mathcal{M}_1, \mathcal{N}_1)$ be a locally finite operator with a state realization $\{A, B, C, D\}$ such that $\ell_A < 1$. Then $I - T^*T \gg 0$ if and only if there exists a solution $M \in \mathcal{D}(\mathcal{B}, \mathcal{B})$, $M \geq 0$ of*

$$M^{(-1)} = A^*MA + B^*B + \left[A^*MC + B^*D\right](I - D^*D - C^*MC)^{-1}\left[D^*B + C^*MA\right] \quad (13.8)$$

such that $I - D^*D - C^*MC \gg 0$. If in addition the realization of T is observable and [uniformly] reachable, then M is [uniformly] positive.

If $I - T^*T \gg 0$, let $W \in \mathcal{U}(\mathcal{N}_1, \mathcal{N}_1)$ be a factor of $I - T^*T = W^*W$. A realization $\{A, B_W, C, D_W\}$ for W such that W is outer is then given by the smallest solution $M \geq 0$ of the above equation, and

$$\begin{cases} D_W &= (I - D^*D - C^*MC)^{1/2} \\ B_W &= -D_W^{-*}[D^*B + C^*MA]. \end{cases} \quad (13.9)$$

PROOF Since $\ell_A < 1$, the Lyapunov equation

$$\Lambda^{(-1)} = A^*\Lambda A + B^*B$$

has a unique solution $\Lambda \geq 0$. By lemma 13.1, an expression for $I - T^*T$ is

$$\begin{aligned} I - T^*T \;=\; & (I - D^*D - C^*\Lambda C) - [D^*B + C^*\Lambda A]Z(I - AZ)^{-1}C \\ & - C^*(I - Z^*A^*)^{-1}Z^*[B^*D + A^*\Lambda C]. \end{aligned}$$

The implied realization for the upper part of $I-T^*T$ need not be reachable. Theorem 13.6 claims that $I-T^*T \gg 0$ if and only if there exists a solution $P \in \mathcal{D}$ of

$$P^{(-1)} = A^*PA + \left[B^*D + A^*(\Lambda+P)C\right](I-D^*D-C^*(\Lambda+P)C)^{-1}\left[D^*B+C^*(\Lambda+P)A\right]$$

such that $I-D^*D-C^*(\Lambda+P)C \gg 0$ and $P \geq 0$. We can add the equation $\Lambda^{(-1)} = A^*\Lambda A + B^*B$ to obtain

$$\begin{aligned}(\Lambda+P)^{(-1)} &= A^*(\Lambda+P)A + B^*B + \\ &+ \left[B^*D+A^*(\Lambda+P)C\right](I-D^*D-C^*(\Lambda+P)C)^{-1}\left[D^*B+C^*(\Lambda+P)A\right].\end{aligned}$$

As a consequence, the operator $M = \Lambda + P$ is positive semi-definite and satisfies equation (13.8). If the realization of T is observable and [uniformly] reachable, then $\Lambda > 0$ $[\Lambda \gg 0]$, and the same holds for M.

Theorem 13.6 in addition shows that the realization $\{A, B_W, C, D_W\}$, with D_W, B_W as given in (13.9), defines an outer factor W of $I - T^*T = W^*W$ if M is the smallest positive semi-definite solution. □

Inner-outer factorization

A realization of the right outer factor T_r in an inner-outer factorization of T can also be computed via a Riccati equation, as was shown in theorem 7.4. A realization of the outer factor followed from a observable realization $\{A,B,C,D\}$ of T as

$$\mathbf{T}_r = \begin{bmatrix} I & \\ & R^* \end{bmatrix} \begin{bmatrix} A & C \\ C^*MA+D^*B & C^*MC+D^*D \end{bmatrix} \tag{13.10}$$

where $M \geq 0$ is the solution of maximal rank of

$$M^{(-1)} = A^*MA + B^*B - \left[A^*MC+B^*D\right](D^*D+C^*MC)^{\dagger}\left[D^*B+C^*MA\right] \tag{13.11}$$

and R is a minimal (full range) factor of $RR^* = (D^*D+C^*MC)^{\dagger}$. Let T_r be invertible, so that the pseudo-inverse becomes an ordinary and bounded inverse. Using lemma 13.1 and assuming $T^*T \gg 0$, one can verify that, indeed, $T^*T = T_r^*T_r$, by deriving that the realizations of the upper parts are equal. With lemma 13.1, the realization of the upper part of $T_r^*T_r$ is obtained from (13.10) as

$$\begin{bmatrix} A & C \\ (D^*B+C^*MA)+C^*\Lambda'A & (D^*D+C^*MC)+C^*\Lambda'C \end{bmatrix} \tag{13.12}$$

where Λ' is the unique operator satisfying the Lyapunov equation

$$\Lambda' = A^*\Lambda'A + \left[B^*D+A^*MC\right](D^*D+C^*MC)^{-1}\left[D^*B+C^*MA\right].$$

Consequently, $(\Lambda'+M)^{(-1)} = A^*(\Lambda'+M)A + B^*B$, so that $\Lambda = \Lambda' + M$ satisfies the Lyapunov equation $\Lambda^{(-1)} = A^*\Lambda A + B^*B$. With Λ, the realization (13.12) becomes

$$\begin{bmatrix} A & C \\ B^*D+C^*\Lambda A & D^*D+C^*\Lambda C \end{bmatrix},$$

which is the same realization as that of T^*T in lemma 13.1. Conversely, one can try to derive theorem 7.4 from the spectral factorization theorem in this manner, for the special case where T^*T is invertible (theorem 7.4 is more general).

In other words, if Λ is the reachability Gramian of the realization of T, and Λ' is the smallest positive solution of (13.2) so that it is the reachability Gramian of a realization of the right outer factor of T^*T, then $M = \Lambda - \Lambda'$ is the solution of (13.11) to obtain the inner-outer factorization. This gives some interpretation of M in that equation. From lemma 13.3 we know that of all factors of T^*T with the same (A,C), the right outer factor T_r provides the smallest reachability Gramian. Hence it follows that $M \geq 0$.

Cholesky factorization and Schur recursions

As noted before, the spectral factorization of a finite-sized positive matrix reduces to Cholesky factorization. For time-invariant systems (Toeplitz operators), one efficient technique to compute a Cholesky factorization makes use of Schur recursions [Sch17, Kai86]. The Schur algorithm can be generalized in various ways to apply to triangular factorizations of general matrices [ADM82], structured operators which have a displacement structure [KKM79, LK84, LK86, LK91], *cf.* section 3.6, and approximate factorizations on a staircase band [DD88]. See [Chu89] for an overview.

The key step in the traditional and also generalized Schur and Levinson algorithms is the translation of the original context (Ω, with $\Omega > 0$) to a scattering context (contractive operators). A standard transition to the scattering context is obtained by finding upper triangular operators Γ, Δ, such that $\Omega = \Gamma\Gamma^* - \Delta\Delta^*$. Using $\mathbf{P}(\Omega)$, the upper triangular part of Ω, and assuming Ω has been scaled such that $\mathbf{P}_0(\Omega) = I$, a suitable Γ and Δ are defined by

$$\begin{array}{rclcl} \Omega_1 & = & 2\mathbf{P}(\Omega) - I & & \\ \Gamma & = & \frac{1}{2}(\Omega_1 + I) & = & \mathbf{P}(\Omega) \\ \Delta & = & \frac{1}{2}(\Omega_1 - I) & = & \mathbf{P}(\Omega) - I \end{array}$$

It is readily verified that, indeed, $\Omega = \Gamma\Gamma^* - \Delta\Delta^*$. In general there is no guarantee that Γ is bounded (see counterexamples in chapter 2) so it is necessary to put the additional assumption that $\mathbf{P}(\Omega)$ is bounded, but it will certainly be boundedly invertible, since Ω is strictly positive definite. Then $S := \Gamma^{-1}\Delta = (\Omega_1 + I)^{-1}(\Omega_1 - I)$ is a well-defined and contractive operator: $\|S\| < 1$. The definition of S may be recognized as a Cayley transformation of Ω_1. It has a direct relation with Ω:

$$\mathbf{P}(\Omega) = (I - S)^{-1}; \qquad S = I - [\mathbf{P}(\Omega)]^{-1}.$$

Since S is strictly contractive and $\mathbf{P}(\Omega)$ is upper triangular, the first expression ensures that S is upper triangular. S is even strictly upper triangular because Δ is so. Also the state structure is preserved: S has the same number of states as $\mathbf{P}(\Omega)$, and its model can be directly derived from the model of $\mathbf{P}(\Omega)$ using equation (1.3).

The standard way to obtain a Cholesky factorization of Ω continues as follows. Compute any J-unitary matrix Θ such that

$$[\Gamma \quad \Delta]\Theta = [A_1 \quad 0], \tag{13.13}$$

A consequence of the J-unitarity of Θ is that

$$A_1A_1^* = [\Gamma \quad \Delta]\Theta J\Theta\begin{bmatrix}\Gamma^* \\ \Delta^*\end{bmatrix} = \Gamma\Gamma^* - \Delta\Delta^* = \Omega.$$

Hence A_1 is a factor of Ω: $\Omega = A_1A_1^*$. With Θ, a factor of Ω^{-1} is obtained by computing

$$[I \quad I]\Theta = [A_2 \quad B_2], \tag{13.14}$$

as it is readily verified using (13.13) and the J-unitarity of Θ that $\Omega^{-1} = A_2A_2^* = B_2B_2^*$. Hence knowledge of Θ provides both a factorization of Ω and of its inverse. Θ can be computed recursively using a generalized Schur algorithm (as *e.g.*, in [DD88]) which amounts to a repetition of (*i*) shifting the rows of Γ one position to the right to align with Δ (*i.e.*, post-multiplication by Z), and (*ii*) using an elementary Θ "section" to cancel the front diagonal of Δ against the corresponding diagonal of Γ. It is thus an order-recursive algorithm. For finite upper triangular matrices of size $n\times n$, the algorithm can be carried out in a finite number of steps and yields a Θ-matrix having at most $n-1$ states. It is possible to obtain an approximate factor by making Δ zero only on a staircase band. This leads to approximate factors A_2' of Ω^{-1} that are zero outside the staircase band, and whose inverse matches the factor A_1 of Ω on the band [DD88, Nel89].

The above algorithm is just one way to compute a Cholesky factorization of a given positive matrix Ω. Efficient ("fast") algorithms are based on exploiting knowledge on the structure of Ω. For example, if Ω is a Toeplitz matrix, then Θ can be computed using the same algorithm but now acting only on the top row of Γ and the top row of Δ (the "generators" of Γ and Δ). This yields the traditional Schur method. More general displacement structures obeying a relation of the form "$G - F_1^*GF_2$ has rank α" are treated in much the same way [Chu89, LK91].

Using the embedding technique given in chapter 12, one other possibility to compute the Cholesky factor via Θ is the following. Assume that a computational model for $\mathbf{P}(\Omega)$, the upper triangular part of Ω, is known. We have already noted that, since S is also upper triangular, a computational model for the associated scattering operator S follows without special effort. The next step is to do an embedding: using theorem 12.14, construct a lossless embedding matrix Σ for S, which is a unitary (2×2) block matrix computed such that $\Sigma_{12} = S$. The J-unitary Θ-matrix associated to Σ is defined as usual by

$$\Theta = \begin{bmatrix} \Sigma_{11} - \Sigma_{12}\Sigma_{22}^{-1}\Sigma_{21} & -\Sigma_{12}\Sigma_{22}^{-1} \\ \Sigma_{22}^{-1}\Sigma_{21} & \Sigma_{22}^{-1}\end{bmatrix}$$

Σ_{22} is outer, so that Σ_{22}^{-1} and hence Θ are again upper. Σ and Θ satisfy by construction the relations (for some $A_1' \in \mathcal{U}$)

$$[I \quad 0]\Sigma = [A_1' \quad S] \qquad \Leftrightarrow \qquad [I \quad S]\Theta = [A_1' \quad 0]$$

and since $S = \Gamma^{-1}\Delta$, multiplication by Γ shows that Θ indeed satisfies

$$[\Gamma \quad \Delta]\Theta = [A_1 \quad 0].$$

From the model of Θ, factors B_2 and $W = B_2^{-1}$ of Ω^{-1} and Ω, respectively, follow using equation (13.14). The whole algorithm can be put into a single recursion. Not surprisingly, the resulting recursion for W is precisely the Riccati equation in corollary 13.8.

13.6 NOTES

The discrete-time Riccati equation corresponding to the spectral factorization problem was originally studied in [And67, AHD74] in the LTI context. Recent overviews of solution methods, as well as many references to older literature, can be found in the collection [BLW91] and in [LR95]. The time-varying case has only been studied recently. The present chapter is based on [vdV93b], although several of the results are reported in the book by Halanay and Ionescu as well [HI94].

Appendix 13.A: Proof of theorem 13.6

The proof of theorem 13.6, *i.e.*, theorem 13.5 without the observability constraint, uses properties of outer operators given in proposition 13.2.

PROOF of theorem 13.6. We will first transform the given realization into one that is observable. Factor the observability Gramian $\Lambda_{\mathbf{F}_o}$ of the given realization as

$$\Lambda_{\mathbf{F}_o} = X^* \begin{bmatrix} \Lambda_{11} & \\ & 0 \end{bmatrix} X,$$

where X is an invertible state transformation and $\Lambda_{11} > 0$. Applying X^{-1} as state transformation to $\mathbf{T}$ leads to a realization $\mathbf{T}' = \{A', B', C', D\}$ given by

$$\begin{bmatrix} A' & C' \\ B' & D \end{bmatrix} = \begin{bmatrix} X^{-*} & \\ & I \end{bmatrix} \begin{bmatrix} A & C \\ B & D \end{bmatrix} \begin{bmatrix} X^{*(-1)} & \\ & I \end{bmatrix}.$$

Partition A', B', C' conformably to the partitioning of Λ. It follows that

$$A' = \begin{bmatrix} A_{11} & A_{12} \\ 0 & A_{22} \end{bmatrix}, \qquad B' = [B_1 \;\; B_2], \qquad C' = \begin{bmatrix} C_1 \\ 0 \end{bmatrix}.$$

The subsystem $\{A_{11}, B_1, C_1, D\}$ is an observable realization of T, with $\ell_{A_{11}} < 1$.

Suppose P is a Hermitian solution of

$$P^{(-1)} = A'^* P A' + [B'^* - A'^* P C'] (D + D^* - C'^* P C')^{-1} [B' - C'^* P A'] \tag{13.A.1}$$

Partition P conformably to the partitioning of A: $P = \left[\begin{smallmatrix} P_{11} & P_{12} \\ P_{12}^* & P_{22} \end{smallmatrix}\right]$. Then equation (13.A.1) is equivalent to the three equations

$$\begin{aligned}
(a)\quad P_{11}^{(-1)} &= A_{11}^* P_{11} A_{11} + [B_1^* - A_{11}^* P_{11} C_1] (D + D^* - C_1^* P_{11} C_1)^{-1} [B_1 - C_1^* P_{11} A_{11}] \\
(b)\quad P_{12}^{(-1)} &= (A_{11}^{\times})^* P_{12} A_{22} + A_{11}^* P_{11} A_{12} + \\
&\quad + [B_1^* - A_{11}^* P_{11} C_1] (D + D^* - C_1^* P_{11} C_1)^{-1} [B_2 - C_1^* P_{11} A_{12}] \\
(c)\quad P_{22}^{(-1)} &= A_{22}^* P_{22} A_{22} + A_{12}^* P_{11} A_{12} + A_{12}^* P_{12} A_{22} + A_{22}^* P_{12}^* A_{12} + \\
&\quad + [B_2^* - (A_{12}^* P_{11} + A_{22}^* P_{12}^*) C_1] (D + D^* - C_1^* P_{11} C_1)^{-1} [B_2 - C_1^* (P_{11} A_{12} + P_{12} A_{22})]
\end{aligned}$$

where $A_{11}^{\times} := A_{11} - C_1 (D + D^* - C_1^* P_{11} C_1)^{-1} [B_1 - C_1^* P_{11} A_{11}]$.

According to theorem 13.5, the first equation has solutions $P_{11} \geq 0$ such that $D + D^* - C_1^* P_{11} C_1 \gg 0$, if and only if $T + T^* \gg 0$. Take P_{11} to be the smallest positive solution, then W is outer and the has an observable realization $\{A_{11}, B_{W1}, C_1, D_W\}$ with D_W and B_{1W} given by

$$\begin{aligned} D_W^* D_W &= D + D^* - C_1^* P_{11} C_1 \\ B_{1W} &= D_W^{-*} [B_1 - C_1^* P_{11} A_{11}]. \end{aligned}$$

According to proposition 13.2, W^{-1} has a realization with A-operator given by $A_{11}^{\times} = A_{11} - C_1 D_W^{-1} B_{1W} = A_{11} - C_1 (D + D^* - C_1^* P_{11} C_1)^{-1} [B_1 - C_1^* P_{11} A_{11}]$, and satisfying $\ell_{A_{11}^{\times}} < 1$ (since $\ell_{A_{11}} < 1$ and the realization of W is observable). The second equation is a kind of Lyapunov equation in P_{12}, as only the first term of the right-hand side is dependent on P_{12}. Given P_{11}, it has a unique bounded solution since $\ell_{A_{11}^{\times}} < 1$ and $\ell_{A_{22}} < 1$. The last equation is a Lyapunov equation in P_{22}, and also has a unique bounded solution.

Also note that $D + D^* - C_1^* P_{11} C_1 = D + D^* - C'^* P C'$. Hence we showed

$$T + T^* \gg 0 \quad \Leftrightarrow \quad \exists\, P \text{ satisfying (13.A.1), such that } D + D^* - C'^* P C' \gg 0.$$

The latter also implies $P \geq 0$. With $\Lambda = X^{-1} P X^{-*}$, Λ is in fact independent of the chosen state transformation X and satisfies the statements of the theorem.

The realization of W can be extended to a non-minimal one that is specified in terms of P as $\{A', B'_W, C', D_W\}$, where the newly introduced quantity B'_W is given by $B'_W = D_W^{-*} [B' - C'^* P A'] = [B_{1W} \;\; B_{2W}]$, for a certain B_{2W}. Upon state-transforming this realization by X, we obtain a realization of W as $\{A, B_W, C, D_W\}$, where D_W is as before, and B_W is specified in terms of Λ as $B_W = B'_W X^{-*(-1)} = D_W^{-*} (B - C^* \Lambda A)$. □

14 LOSSLESS CASCADE FACTORIZATIONS

In chapter 12, we showed how a contractive u.e. stabletransfer operator T can be embedded into an inner operator Σ. We now derive minimal structural factorizations of locally finite inner transfer operators into elementary inner operators of degree one. The resulting lossless cascade networks provide a canonical realization of T into a network of minimal degree and with a minimal number of coefficients. For a better understanding of the problem, we first review some aspects of cascade factorizations for time-invariant systems.

14.1 TIME-INVARIANT CASCADE FACTORIZATIONS

Overview

An important and recurring subject in network theory concerns the synthesis (implementation, or actual realization) of a desired transfer function using elementary components. For continuous-time systems, these components would be resistors, capacitances, inductors and transformers. In the discrete-time context, the elementary operator is the basic processor which performs the actual calculations on the digital signals: typically a multiplier-adder, but other elementary processors are certainly possible. While one can directly use the given $\{A,B,C,D\}$ realization as the actual realization of the transfer operator, doing so is often unsatisfactory. The number of multiplications in an arbitrary state realization of the given system is not minimal: a single-input single output system with n states would require $(n+1)^2$ multiplications. Typically, such an implementation is also rather sensitive to small changes in the values of the coefficients: a small change (*e.g.*, because of finite word length effects) can sometimes even make the modified system unstable. For digital filters, a third issue is the occurrence of limit cycles and register overflow. The above-mentioned effects are mitigated by a deliberate use of the freedom of state transformations on the given state realization. By selecting certain canonical forms of the A matrix, such as a companion form or a diagonal form (which is not always possible), filters specified by a minimal number of coefficients are obtained [Kai80].

The coefficient sensitivity issue is a more complicated matter. The central idea is that one of the few ways to make the locations of poles and zeros of the resulting system well defined is to factor the given transfer function into a cascade of elementary (degree one) transfer functions:

$$T(z) = T_1(z) \cdot T_2(z) \cdots T_n(z). \tag{14.1}$$

Each elementary transfer function realizes a pole and a zero of $T(z)$. For an n-th order system $T(z)$, the factorization is minimal if it consists of n degree one sections. In this case, the factorization into n elementary factors is canonical and leads to a minimum of coefficients, for SISO systems $2n+1$, *i.e.*, n coefficients for the poles, n for the zeros, and one coefficient for the overall scaling.

The synthesis of passive transfer functions via cascade factorizations has a long history in classical network theory. The first results were concerned with the factorization of a lossless (inner) transfer function of degree n into a product of n degree-1 lossless transfer functions, by recursively extracting a degree-1 subnetwork. This procedure is known as Darlington synthesis of lossless multiports [Dar39], and produces ladder filters with well-known properties [Bel68]. The use of a lossless (unitary) state realization of the inner operator gave the synthesis procedures by Youla and Tissi [YT66], while the synthesis of more general J-unitary operators was considered by Fettweis [Fet70] in connection to wave-digital filters.

The cascade realization of inner operators leads to a realization procedure of any passive (contractive) rational transfer function, via a lossless embedding of the contractive transfer function $T(z)$. Thus, one obtains a realization of $T(z)$ in which either the poles or the zeros of $T(z)$ are localized in the elementary sections. State-space versions of this procedure are discussed in Roberts and Mullis' book [RM87].

Although it is more general, Darlington synthesis is closely connected to the Levinson algorithm used in estimation filter theory of stationary stochastic processes [DVK78]. The estimation filters are prediction (AR) filters with their transmission zeros at infinity, but the filter structure that is obtained is also a ladder filter which can be derived recursively from the covariance matrix of the stochastic process. The synthesis procedure thus constitutes a recursive Cholesky factorization of positive Toeplitz matrices. The Toeplitz matrices can be generalized to include the covariance matrices of more general α-stationary processes [KKM79, FMKL79], and leads to a generalized Schur parametrization of structured (α-stationary) matrices, *i.e.*, matrices with a low displacement rank [LK84]. The paper by Genin *et al.*[GDK+83] explored the relation between lossless state realizations and the characterization of structured matrices via a cascade of elementary lossless sections. Finally, there are many parallel results in operator theory: Potapov [Pot60] obtained a complete description of (not necessarily rational) J-unitary and J-contractive matrix functions in terms of general cascade decompositions, while the lossless embedding and subsequent factorization of contractive functions in the setting of colligations was considered by Livsic and Brodskii [BL58, Liv72]. The Darlington synthesis procedure is also closely connected, via the Lossless Inverse Scattering problem, to classical interpolation problems of the Nevanlinna-Pick type; see [DVK78, DD81b, DD81a, DD84].

Besides a factorization of a lossless embedding of T, it is also possible to determine a direct factorization (14.1) [DBN71, VD77, BGK79, DD81c, Rak92]. Such factorizations realize both a zero and a pole of T in each elementary section, which makes them attractive in some applications, but they are also more complicated to derive. One can act directly on the transfer function $T(z)$, and in this case the complication is that non-square factors can occur [VD77], giving rise to a plethora of possible elementary sections. The situation is easier to describe in state-space terms. Let $T(z)$ be a bounded system, and suppose that it has a factorization $T = T_1T_2$, where T_1, T_2 are again bounded systems, with minimal realizations $\mathbf{T}_1 = \{A_1, B_1, C_1, D_1\}, \mathbf{T}_2 = \{A_2, B_2, C_2, D_2\}$. A realization for T is thus given by

$$\mathbf{T} = \left[\begin{array}{cc|c} A_1 & & C_1 \\ & I & \\ \hline B_1 & & D_1 \end{array}\right] \left[\begin{array}{cc|c} I & & \\ & A_2 & C_2 \\ \hline & B_2 & D_2 \end{array}\right] = \left[\begin{array}{cc|c} A_1 & C_1B_2 & C_1D_2 \\ 0 & A_2 & C_2 \\ \hline B_1 & D_1B_2 & D_1D_2 \end{array}\right]. \tag{14.2}$$

Note that $A = A_T$ is block upper triangular. If D_1 and D_2 are both invertible, then T^{-1} has a realization given by the product of the realizations of T_1^{-1} and T_2^{-1}, which turns out to have

$$A^\times = A_{T^{-1}} = \begin{bmatrix} A_1^\times & 0 \\ -C_2D_2^{-1}D_1^{-1}B_1 & A_2^\times \end{bmatrix},$$

(where $A^\times := A - BD^{-1}C$ is the A-matrix of the inverse system, whose eigenvalues are the zeros of T). This matrix is block lower triangular. It can be shown, e.g. [BGKD80, DD81c], that T can be factorized minimally into factors T_1, T_2 if and only if it has a minimal realization $\mathbf{T}$ in which A_T is block upper triangular and $A_T^\times$ is block lower triangular. The factorization problem is thus reduced to finding a state-space transformation acting on a given realization of T and a partitioning into 2×2 blocks such that A_T and $A_T^\times$ have the required forms. To this end, one has either to determine the solutions of a certain Riccati equation (this replaces the Riccati equation that occurs in the embedding step), or to compute eigenvalue decompositions (Schur decompositions) of both A and $A^\times$, describing the poles and zeros of the given transfer function. However, in the subsequent factorization procedure, the conditioning of certain inverses can be problematic [BGKD80]. Such problems do not occur with the factorization of inner or J-inner functions, as in this case the poles of the system also determine the zeros: for inner functions Σ with unitary realizations, $\mathbf{\Sigma}^*$ is a realization of $\Sigma^{-1} = \Sigma^*$, and hence $A^\times = A^*$. We only consider the cascade realization of inner functions Σ from now on.

Repetition of the above factorization into two systems leads to a factorization of a degree-n system into n systems of degree 1: the elementary sections. A particular realization of the elementary sections produces orthogonal digital filters. Here, the elementary operator is not a multiplication, but a plane rotation, where the rotation angle is the coefficient of the section. The advantage of such filters is that (with ideal rotors) they are inherently lossless and stable, even if the coefficients are imprecise, and that no limit cycles or overflow oscillations can occur. Another advantage is that the filters are typically cascade arrays of identical processors with only nearest neighbor connections, which allows for VLSI implementation. Some other issues to be considered are the pipelinability and computability of the array, which are not always guaranteed. A number of orthogonal filter structures are possible, depending on the precise factorization of the inner transfer operator, and on whether a factorization of Σ, or its associated J-unitary operator Θ is performed. The factorization can also be done directly on the transfer function $T(z)$, if it is specified as a ratio of two polynomials, or on the state-space matrices. In both cases, a preliminary embedding step is necessary. The main reference papers on orthogonal filter realizations are by Deprettere, Dewilde, P. Rao and Nouta [DD80, DDR84, DDN84, Dew85], S.K. Rao and Kailath [RK84], Vaidyanathan [Vai85b], Regalia, Mitra and Vaidyanathan [RMV88], and Roberts and Mullis' book [RM87]. More recent references are [JM91, Des91].

Givens rotations

We say that $\hat{\Sigma}$ is an elementary orthogonal rotation if $\hat{\Sigma}$ is a 2×2 unitary matrix of the form

$$\hat{\Sigma} = \begin{bmatrix} e^{j\phi_1} & \\ & e^{j\phi_2} \end{bmatrix} \begin{bmatrix} c & -s \\ s & c \end{bmatrix}, \tag{14.3}$$

with $c^2 + s^2 = 1$. If we operate in the real domain, then the first factor is of course omitted. An important property of elementary rotations is that they can be used to zero a selected entry of a given matrix: for given a and b, an elementary orthogonal rotation $\hat{\Sigma}$ exists such that

$$\hat{\Sigma}^* \begin{bmatrix} a \\ b \end{bmatrix} = \begin{bmatrix} a' \\ 0 \end{bmatrix}, \tag{14.4}$$

In this case, $\hat{\Sigma}$ is called a Givens rotation, and we write $\hat{\Sigma} = \text{givens}[a;b]$ in algorithms. We do not need ϕ_1 for zeroing entries.

Givens rotations are used to factor a given state realization into elementary rotations, or certain generic groups of rotations called elementary sections. Acting on state realizations, the 2×2 elementary rotation matrix is typically extended by identity matrices, say

$$\mathbf{\Sigma}_i = \left[\begin{array}{ccc|c} I_{i-1} & & & \\ & \times & & \times \\ & & I_{n-i} & \\ \hline & \times & & \times \end{array}\right], \tag{14.5}$$

where the four '×'-s together form the 2×2 unitary matrix. We use a hat symbol to denote this elementary 2×2-matrix, *i.e.*, we write it as $\hat{\mathbf{\Sigma}}_i$.

An elementary J-unitary rotation $\hat{\Theta}$ can be obtained from $\hat{\Sigma}$ in (14.3) if $c \neq 0$ as

$$\hat{\Theta} = \begin{bmatrix} e^{j\phi_1} & \\ & 1 \end{bmatrix} \begin{bmatrix} 1 & s \\ s & 1 \end{bmatrix} \frac{1}{c} \begin{bmatrix} 1 & \\ & e^{-j\phi_2} \end{bmatrix}$$

It can also be used to zero entries of vectors,

$$[a \quad b]\hat{\Theta}^{-1} = [a' \quad 0],$$

but only if $a^*a - b^*b = a'^*a' > 0$.

Orthogonal digital filter synthesis

Assume that Σ is known, along with a unitary realization $\mathbf{\Sigma}$. As was shown in equation (14.2), a necessary condition for factorization of Σ is that A_Σ is upper triangular. From the given realization, this can be ensured via a unitary state-space transformation Q obtained from a Schur decomposition of the given A-matrix:

$$QA_\Sigma Q^* = R,$$

where R is upper triangular. This decomposition always exists (in the complex domain), and amounts to a computation of the poles of the system. With A_Σ upper triangular, the second phase of the factorization procedure is the factoring of $\mathbf{\Sigma}$ into a minimal number of elementary (degree-1) factors. Here, one makes use of the fact that the product of two unitary matrices is again unitary. A consequence of this fact is that, in equation (14.2) (where all matrices are unitary now), any $\mathbf{\Sigma}_1$ such that $\mathbf{\Sigma}_1^*\mathbf{\Sigma}$ has zero block entries $(2,1)$ and $(3,1)$ leads to $\mathbf{\Sigma}_2$ of the required form. Since the $(2,1)$ entry is already equal to zero, it follows that $\mathbf{\Sigma}_1$ can be of the form indicated in (14.2): using $\mathbf{\Sigma}_1^*$, one only has to cancel entry $(3,1)$ using entry $(1,1)$. The unitarity of the product $\mathbf{\Sigma}_1^*\mathbf{\Sigma}$ ensures that also its entries $(1,2)$ and $(1,3)$ are zero. Upon factoring $\mathbf{\Sigma}$ down to the scalar level, it follows that the elementary unitary factors have the form $\mathbf{\Sigma}_i$ in (14.5). If Σ is of degree d, then the factorization consists of d degree-1 factors and is of the form $\mathbf{\Sigma} = \mathbf{\Sigma}_1 \cdots \mathbf{\Sigma}_d$, where

$$\begin{aligned}
\mathbf{\Sigma} &= \left[\begin{array}{ccc|c} a_{11} & \times & \times & \times \\ & a_{22} & \times & \times \\ & & a_{dd} & \times \\ \hline \times & \times & \times & \times \end{array}\right] \\
&= \left[\begin{array}{ccc|c} a_{11} & & & \times \\ & 1 & & \\ & & 1 & \\ \hline \times & & & \times \end{array}\right] \left[\begin{array}{ccc|c} 1 & & & \\ & a_{22} & & \times \\ & & 1 & \\ \hline & \times & & \times \end{array}\right] \cdots \left[\begin{array}{ccc|c} 1 & & & \\ & 1 & & \\ & & a_{dd} & \times \\ \hline & & \times & \times \end{array}\right] .
\end{aligned} \tag{14.6}$$

The a_{ii} are the diagonal entries of A_Σ, which are the poles of the system. Hence, each elementary section realizes a pole of Σ. In (14.6), we assumed that Σ is a SISO system. For multi-input multi-output systems, the procedure is an extension of the above, and gives (for an example of a system with two inputs and two outputs)

$$\begin{aligned}
\mathbf{\Sigma} &= [\mathbf{\Sigma}_{1,1}\mathbf{\Sigma}_{2,1}] \cdot [\mathbf{\Sigma}_{1,2}\mathbf{\Sigma}_{2,2}] \cdots [\mathbf{\Sigma}_{1,d}\mathbf{\Sigma}_{2,d}]\, \mathbf{\Sigma}' \qquad (14.7) \\
&= \left[\begin{array}{ccc|cc} \times & & & \times & \\ & 1 & & & \\ & & 1 & & \\ \hline \times & & & \times & \\ & & & & 1 \end{array}\right] \left[\begin{array}{ccc|cc} \times & & & & \times \\ & 1 & & & \\ & & 1 & & \\ \hline & & & 1 & \\ \times & & & & \times \end{array}\right] \left[\begin{array}{ccc|cc} 1 & & & & \\ & \times & & \times & \\ & & 1 & & \\ \hline & \times & & \times & \\ & & & & 1 \end{array}\right] \left[\begin{array}{ccc|cc} 1 & & & & \\ & \times & & & \times \\ & & 1 & & \\ \hline & & & 1 & \\ & \times & & & \times \end{array}\right] \cdots
\end{aligned}$$

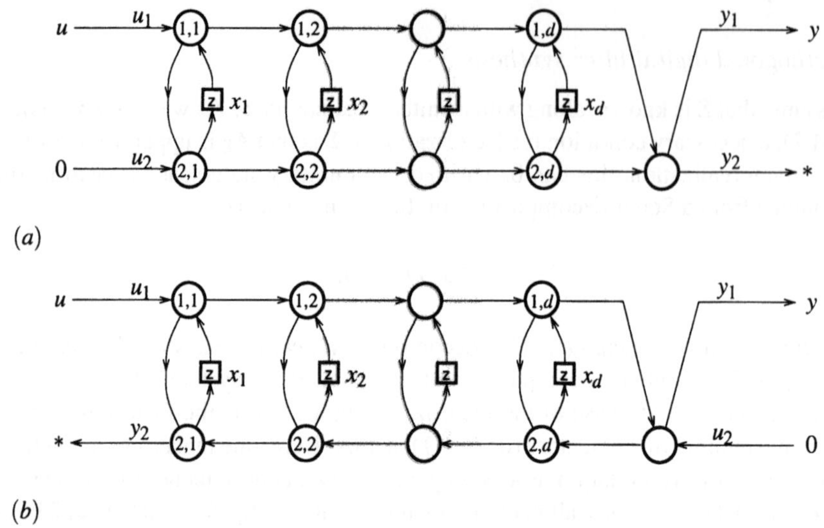

Figure 14.1. (a) Σ-based cascade factorization, based on a Schur decomposition of A_Σ. Σ is a unitary embedding of $T : u \to y$ which is the transfer of u_1 to y_1 if $u_2 = 0$. (b) Θ-based cascade factorization, based on a Schur decomposition of A_Θ, where Θ is the J-unitary chain scattering operator associated to Σ.

$$\cdots \left[\begin{array}{ccc|cc} 1 & & & & \\ & 1 & & & \\ & & \times & \times & \\ \hline & & \times & \times & \\ & & & & 1 \end{array}\right] \left[\begin{array}{ccc|cc} 1 & & & & \\ & 1 & & & \\ & & \times & & \times \\ \hline & & & 1 & \\ & & \times & & \times \end{array}\right] \left[\begin{array}{ccc|cc} 1 & & & & \\ & 1 & & & \\ & & 1 & & \\ \hline & & & \times & \times \\ & & & \times & \times \end{array}\right].$$

Σ' is the terminating section of degree $\bar{0}$. It is in general a unitary matrix itself, which can also be factored into elementary Givens rotations, and finally a unit-norm scaling. The network structure that is obtained is drawn in figure 14.1, which is straightforwardly derived from (14.7) by considering how a vector $[x_1\ x_2\ \cdots x_d\ u_1\ u_2]$ is transformed in elementary steps to $[x_1'\ x_2'\ \cdots x_d'\ y_1\ y_2]$. The network is pipelinable: the signal flow is strictly unidirectional (from the left to the right). It is also computable: given the current values of the inputs and of the states, the outputs and the next states can be computed. The network is specified by a minimal number of $2d+1$ Givens rotations (parametrized by rotation angles and complex phase rotations). Any strictly contractive LTI system T can be realized in this way, by embedding T into an inner system Σ such that $T = \Sigma_{11}$. As a matter of fact, it is not necessary to compute the embedding completely: if Σ has a realization as in (12.18), viz.

$$\boldsymbol{\Sigma} = \begin{bmatrix} R & & \\ & I & \\ & & I \end{bmatrix} \begin{bmatrix} A & C & C_2 \\ B & D & D_{12} \\ B_2 & D_{21} & D_{22} \end{bmatrix} \begin{bmatrix} R^{-1} & & \\ & I & \\ & & I \end{bmatrix},$$

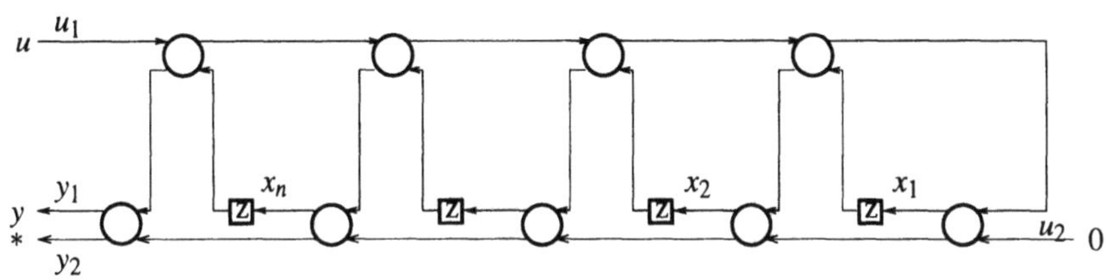

Figure 14.2. Hessenberg lossless filter structure.

where $\{A,B,C,D\}$ is the given realization of T, and R, B_2, D_{21} are computed via a Riccati equation, then only A, B and B_2 determine the factors $\mathbf{\Sigma}_{i,j}$ ($i = 1,\cdots,d,\ j = 1,2$), and C_2, D_{12} and D_{22} are not needed. As far as the cascade factorization is concerned, it is even possible to omit the state transformation by R [LK92], although this is at the expense of a number of other matrix inversions, and we still have to compute R to determine the extension by B_2, D_{21} anyway. As an alternative to the above factorization of Σ, one can convert Σ to a J-unitary Θ operator with realization $\mathbf{\Theta}$ (*cf.* theorem 8.2), factor $\mathbf{\Theta}$ in a comparable way as done for $\mathbf{\Sigma}$, and convert the factors back to the scattering domain. This gives network structures as depicted in figure 14.1(b).

In the above two solutions to the factorization problem, the trick to determine a minimal factorization was to compute a Schur decomposition of A_Σ (or A_Θ), which introduced as many zero entries in $\mathbf{\Sigma}$ as possible. The remaining $2d+1$ non-zero entries below the main diagonal of $\mathbf{\Sigma}$ induced a factorization of $\mathbf{\Sigma}$ into $2d+1$ elementary factors. There are other structures of $\mathbf{\Sigma}$, not requiring an (expensive) Schur decomposition step, which still result in a factorization of $\mathbf{\Sigma}$ into $2d+1$ elementary factors. However, this time we do not obtain a factorization of Σ itself into a product $\Sigma_1 \cdots \Sigma_d$, so that the individual elementary sections do not realize poles and zeros of Σ, and the implementation is not truly a cascade factorization in the sense used before. One possible structure that can be obtained via a unitary state transformation is a Hessenberg structure of A and the first row of B, which can be computed non-recursively:

$$
\mathbf{\Sigma} = \left[\begin{array}{cccc|cc}
\times & \times & \times & \times & \times & \times \\
\times & \times & \times & \times & \times & \times \\
 & \times & \times & \times & \times & \times \\
 & & \times & \times & \times & \times \\
\hline
 & & & \times & \times & \times \\
\times & \times & \times & \times & \times & \times
\end{array}\right].
$$

$\mathbf{\Sigma}$ can be be brought into the same form (14.7) as before, by a simple row permutation operation. This does not induce any mathematical operations, but will change the apparent structure of the filter. After the permutations, the factorization proceeds in the same way. The resulting network structure is as depicted in figure 14.2 (viz. [RK84, Dew85, Des91, vdVV96]). The network is again pipelinable. If the realization is real-valued, then a Hessenberg structure can keep all parameters real. (A Schur structure needs more complicated sections to handle complex pole pairs.)

14.2 PARSIMONIOUS PARAMETRIZATION OF CONTRACTIVE LTI SYSTEMS

The preceding technique can be further refined. As can be shown, the Hessenberg structure has a minimal number of parameters. However, without further constraints on the Givens rotations (14.3), there is a continuum of equivalent parameter values that all give the same transfer operator T. For the purpose of identification, which often rely on nonlinear optimization schemes for parameter fitting, it is interesting to have a *canonical* parametrization, where there is a one-to-one relation between the parameters and the transfer operator. Thus, there has been an active search for *canonical* system representations, *i.e.*, minimal parametrizations by which any dynamical system $T(z)$ of a given class and order may uniquely be represented. For multi-input multi-output systems, a number of canonical forms are known, based *e.g.*, on the observer or controller canonical forms or on balanced realizations [Obe91, Cho94].

For optimization purposes, an important deficiency in most canonical representations for real-valued systems is that they require both continuously varying parameters (in a subset of $\mathbb{R}$), and discrete parameters (in a subset of $\mathbb{N}$). The latter parameters are extra parameters that specify the structure of the system, such as the Kronecker indices or the number of equal Hankel singular values (for balanced parametrizations). Alternatively, one can say that the space of all real-valued rational LTI systems cannot be covered by a single continuous parametrization, but at best by a set of overlapping parametrizations (indexed by the discrete parameters), each of which on itself does not cover the whole set. For model identification, these "structural" parameters are a nuisance, since they have to be selected *a priori*, and modified if the resulting continuous parametrization is not sufficiently accurate. In fact, since they have little physical meaning, the only way to solve the optimization problem is to enumerate over a sufficient range of structural parameter values to cover all systems of a given order, and to perform a non-linear search for each such choice. Obviously, this is not a very attractive solution.

The purpose of this section is to further refine the Hessenberg structure and derive that the class of contractive asymptotically stable rational LTI systems is covered by a minimal representation *without* any structural parameters. The representation is not unique, but for each system $T(z)$ there is only a finite number of equivalent descriptions (unless the system is overparametrized). Because the solutions are isolated, this should not pose a problem for numerical optimization techniques.

Both the realvalued and the complex case can be treated by the same procedure, but with slightly different elementary rotations.

In the real-valued case, define for $-1 \le s \le 1$, $c = \sqrt{1-s^2}$, and integers d, m, n, the plane rotations

$$Q_{ij}(s) = \begin{array}{c} \begin{matrix} \quad\quad & i & \quad & d+j & \quad \end{matrix} \\ \begin{bmatrix} I & & & & \\ & c & & -s & \\ & & I & & \\ & s & & c & \\ & & & & I \end{bmatrix} \end{array} \in \mathbb{R}^{(d+m+n)\times(d+m+n)}, \tag{14.8}$$

$$Z_{ij}(s) = \begin{array}{c} \begin{array}{ccccc} & d+i & & d+n+j & \end{array} \\ \begin{bmatrix} I_d & & & & \\ & c & & -s & \\ & & I & & \\ & s & & c & \\ & & & & I \end{bmatrix} \end{array} \in \mathbb{R}^{(d+n+m)\times(d+n+m)} . \tag{14.9}$$

In the complex-valued case, we take $-1 \le s \le 1$, $c = \sqrt{1-s^2}$, and $-\frac{\pi}{2} \le \phi \le \frac{\pi}{2}$, and define

$$Q_{ij}(s) = \begin{array}{c} \begin{array}{ccccc} & i & & d+j & \end{array} \\ \begin{bmatrix} I & & & & \\ & c & & -s & \\ & & I & & \\ & se^{j\phi} & & ce^{j\phi} & \\ & & & & I \end{bmatrix} \end{array} \in \mathbb{C}^{(d+m+n)\times(d+m+n)} , \tag{14.10}$$

$$Z_{ij}(s) = \begin{array}{c} \begin{array}{ccccc} & d+i & & d+n+j & \end{array} \\ \begin{bmatrix} I_d & & & & \\ & c & & -s & \\ & & I & & \\ & se^{j\phi} & & ce^{j\phi} & \\ & & & & I \end{bmatrix} \end{array} \in \mathbb{C}^{(d+n+m)\times(d+n+m)} . \tag{14.11}$$

Also define permutations $\Pi_{1,d+1}$ and Π_D by

$$\Pi_{1,d+1}^{*} \begin{bmatrix} x_1 \\ \vdots \\ x_{d+1} \\ x_{d+2} \\ \vdots \end{bmatrix} = \begin{bmatrix} x_2 \\ \vdots \\ x_{d+1} \\ x_1 \\ x_{d+2} \\ \vdots \end{bmatrix} , \qquad \Pi_D^{*} = \begin{bmatrix} I_d & & \\ & 0 & I_n \\ & I_m & 0 \end{bmatrix} . \tag{14.12}$$

Theorem 14.1 *There is a minimal continuous parametrization with $d(m+n)+mn$ bounded [real or complex] coefficients which covers the set of all [real or complex]-valued rational stable contractive LTI systems with m inputs, n outputs and d states.*

In particular, every such system may be specified in terms of two matrices $S^{(1)}$: $(m+n)\times d$, $S^{(2)}$: $m\times n$ with entries $|s_{ij}^{(\cdot)}| \le 1$ as $T(z) = D + Bz(I-Az)^{-1}C$ where

$$\begin{array}{c} \begin{array}{cc} d & n \end{array} \\ \begin{array}{c} d \\ m \end{array} \begin{bmatrix} A & C \\ B & D \end{bmatrix} \end{array} = \begin{bmatrix} I_{d+m} & 0_{(d+m)\times n} \end{bmatrix} \cdot \Pi_{1,d+1} Q_{11} Q_{21} \cdots Q_{m+n,d} \cdot \Pi_D \cdot Z_{11} Z_{21} \cdots Z_{m,n} \begin{bmatrix} I_{d+n} \\ 0_{m\times(d+n)} \end{bmatrix} \tag{14.13}$$

for $Q_{ij} := Q_{ij}(s_{ij}^{(1)})$, $Z_{ij} := Z_{ij}(s_{ij}^{(2)})$. The parametrization is not unique, but for strictly contractive systems which are reachable via the first input, only a finite (discrete) set of parameter matrices lead to the same $T(z)$.

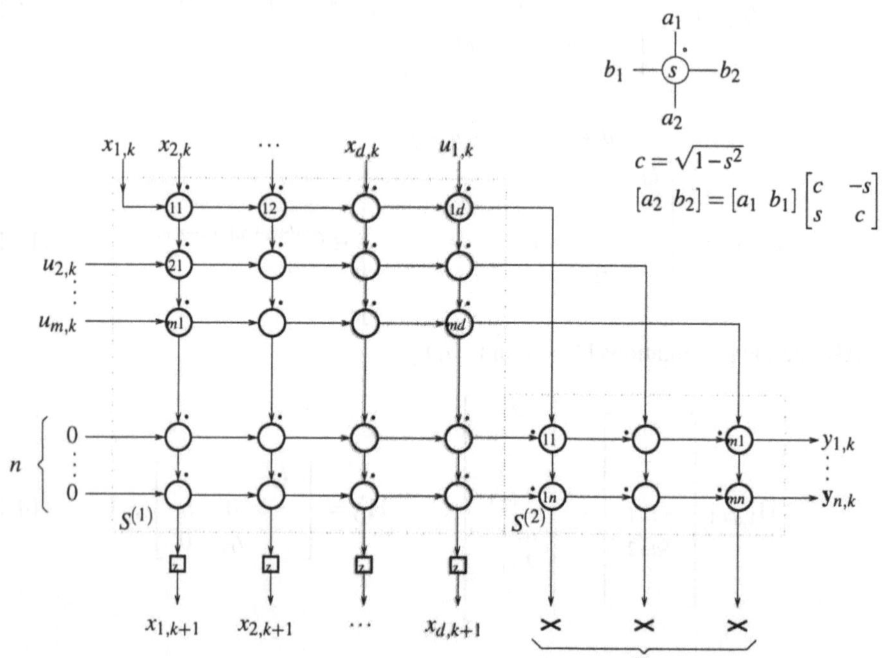

Figure 14.3. Hessenberg structure (m inputs, n outputs, d states)

In the real-valued case, each parameter specifies a rotation as in (14.8) or (14.9). In the complex case, the parameter is $se^{j\phi}$, which specifies both s and ϕ for a complex rotation. This is possible since ϕ is restricted to $-\frac{\pi}{2} \le \phi \le \frac{\pi}{2}$.

The structure of this parametrization is perhaps better understood from figure 14.3, which shows the state space mapping

$$[\mathbf{x}_{k+1} \quad \mathbf{y}_k] = [\mathbf{x}_k \quad \mathbf{u}_k] \begin{bmatrix} A & C \\ B & D \end{bmatrix}$$

in terms of the factorization (14.13). This is a generalization of the structure in figure 14.2 to multiple inputs and outputs.

The proof of theorem 14.1 is by construction, in three steps.

Step 1: Lossless embedding Assume that $T(z)$ is specified in terms of a minimal realization (A,B,C,D). Step 1 is to find an invertible state transformation R, and state matrices B_2, D_{12} such that

$$\mathbf{\Sigma}_1 = \begin{bmatrix} R & & \\ & I & \\ & & I \end{bmatrix} \begin{matrix} d \\ m \\ n \end{matrix} \overset{\begin{matrix} d & \quad n \end{matrix}}{\begin{bmatrix} A & C \\ B & D \\ B_2 & D_{21} \end{bmatrix}} \begin{bmatrix} R^{-1} & \\ & I \end{bmatrix} \tag{14.14}$$

is isometric: $\mathbf{\Sigma}_1^*\mathbf{\Sigma}_1 = I$. Upon defining $M = R^*R$, the condition $\mathbf{\Sigma}_1^*\mathbf{\Sigma}_1 = I$ is equivalent to solving

$$\begin{array}{ccccccc} A^*MA & + & B^*B & + & B_2^*B_2 & = & M \\ C^*MA & + & D^*B & + & D_{21}^*B_2 & = & 0 \\ C^*MC & + & D^*D & + & D_{21}^*D_{21} & = & I. \end{array} \tag{14.15}$$

Under the conditions of theorem 14.1, the embedding theorem (theorem 12.12) claims that solutions $M > 0$ exist, and that for each solution $I - D^*D - C^*MC \ge 0$ (> 0 holds if T is strictly contractive). M is not unique but solutions are isolated.

Take any solution M. Then D_{21} and B_2 follow from

$$\begin{array}{rcl} D_{21}^*D_{21} & = & I - D^*D - C^*MC \\ B_2 & = & -D_{21}^{\dagger}(C^*MA + D^*B). \end{array} \tag{14.16}$$

D_{21} is a square root of a positive semidefinite matrix. We choose D_{21} to be *upper triangular* with $\operatorname{diag}(D_{21}) \ge 0$ (and real-valued). If T is strictly contractive, then this D_{21} is unique and $\operatorname{diag}(D_{21}) > 0$, otherwise $D_{21}^*D_{21}$ might be singular with a continuum of suitable factors.

Step 2: Transformation into Hessenberg form Suppose at this point that we have

$$\mathbf{\Sigma}_1 =: \begin{matrix} d \\ m \\ n \end{matrix} \overset{\begin{matrix} d & \quad n \end{matrix}}{\begin{bmatrix} A & C \\ B & D \\ B_2 & D_{21} \end{bmatrix}}$$

where $\mathbf{\Sigma}_1^*\mathbf{\Sigma}_1 = I$, D_{21} is upper triangular and $\mathrm{diag}(D_{21}) \geq 0$. Step 2 is to find a unitary state transformation Q such that

$$\mathbf{\Sigma}_1' = \begin{bmatrix} Q^* & & \\ & I & \\ & & I \end{bmatrix} \mathbf{\Sigma}_1 \begin{bmatrix} Q & \\ & I \end{bmatrix} =$$

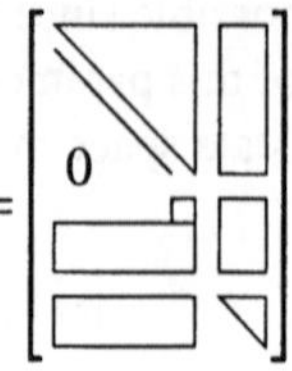

i.e., denoting by $\mathbf{b}_1$ the first row of B,

$$\begin{bmatrix} A' \\ \mathbf{b}_1' \end{bmatrix} = \begin{bmatrix} Q^* & \\ & 1 \end{bmatrix} \begin{bmatrix} A \\ \mathbf{b}_1 \end{bmatrix} Q =$$

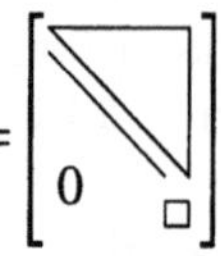

is in upper Hessenberg form. Some freedom is left; we can use it to guarantee that all entries on the sub-diagonal are nonnegative and real (a "positive upper Hessenberg" form). This is always possible by scaling these entries in turn, starting with the lower-right entry.

The entries of the sub-diagonal of $\left[\begin{smallmatrix} A' \\ \mathbf{b}_1' \end{smallmatrix}\right]$ are strictly positive and Q is unique if and only if the system is reachable via its first input. Indeed, consider the (finite and reversed) reachability matrix of $(A', \mathbf{b}_1')$,

$$\begin{bmatrix} \mathbf{b}_1'(A')^{d-1} \\ \vdots \\ \mathbf{b}_1'(A')^2 \\ \mathbf{b}_1'A' \\ \mathbf{b}_1' \end{bmatrix} = \begin{bmatrix} \mathbf{b}_1 A^{d-1} \\ \vdots \\ \mathbf{b}_1 A^2 \\ \mathbf{b}_1 A \\ \mathbf{b}_1 \end{bmatrix} Q =: R.$$

The structure of $\left[\begin{smallmatrix} A' \\ \mathbf{b}_1' \end{smallmatrix}\right]$ ensures that R is upper triangular with nonnegative main diagonal. The system is reachable via the first input if and only if R is nonsingular; also, the QR factorization is unique if and only if R is nonsingular; in that case there can be no other Q that will produce an upper-triangular R.

On the other hand, suppose that the system is not reachable via its first input, *i.e.*, suppose an entry $(k+1, k)$ of the sub-diagonal of $\left[\begin{smallmatrix} A' \\ \mathbf{b}_1' \end{smallmatrix}\right]$ is zero, then Q is not unique: for $k > 1$ any 2×2 rotation acting on columns and rows $k-1$ and k of $\left[\begin{smallmatrix} A' \\ \mathbf{b}_1' \end{smallmatrix}\right]$ will keep the Hessenberg structure invariant, for $k = 1$, the freedom is a ± 1 scaling of the first row and column. Hence if $k > 1$ a continuum of suitable Q is obtained.

Step 3: Factorization of $\mathbf{\Sigma}_1$ Suppose at this point that we have an embedding $\mathbf{\Sigma}_1$, isometric, in the required positive Hessenberg form, and with D_{21} upper triangular with nonnegative real-valued main diagonal. The final step is to factor $\mathbf{\Sigma}_1$ into elementary Givens rotations, producing the actual parameters of the state space model. It suffices for our purposes to consider rotations of the form

$$q(s) = \begin{bmatrix} c & -s \\ s & c \end{bmatrix}, \qquad c = \sqrt{1-s^2}, \qquad -1 \leq s \leq 1,$$

or

$$q(\tilde{s}) = \begin{bmatrix} 1 & \\ & e^{j\phi} \end{bmatrix} \begin{bmatrix} c & -s \\ s & c \end{bmatrix},$$
$$-1 \le s \le 1, \quad -\tfrac{\pi}{2} \le \phi \le \tfrac{\pi}{2}, \qquad c = \sqrt{1-s^2}, \quad \tilde{s} = s e^{j\phi}.$$

For ease of description, we move column 1 of $\mathbf{\Sigma}_1$ behind column $n+1$, giving

$$\Phi = \Pi^*_{1,d+1} \mathbf{\Sigma}_1 =$$

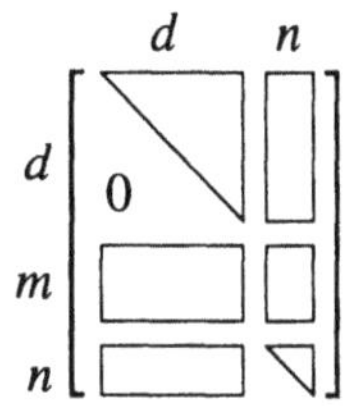

$$=: \begin{matrix} d \\ m \\ n \end{matrix} \overset{\begin{matrix} d & \quad n \end{matrix}}{\begin{bmatrix} A & C \\ B & D \\ B_2 & D_{21} \end{bmatrix}}$$

where the permutation $\Pi_{1,d+1}$ is defined in equation (14.12). (Note that we redefined $A, B, \cdots$ for ease of notation.) Subsequently, we apply a sequence of rotations to the rows of Φ to reduce it to a submatrix of the identity matrix, taking care that A and D_{21} remain upper triangular with nonnegative diagonal entries throughout the transformations.

- Apply a Givens rotation q^*_{11} to rows 1 and $d+1$ of Φ, to cancel b_{11} against a_{11}, *i.e.*, $q^*_{11}\left[\begin{smallmatrix} a_{11} \\ b_{11} \end{smallmatrix}\right] = \left[\begin{smallmatrix} a'_{11} \\ 0 \end{smallmatrix}\right]$. In the real-valued case, the rotation is specified by

$$s = \frac{b_{11}}{(a_{11}^2 + b_{11}^2)^{1/2}}, \qquad c = \sqrt{1-s^2}.$$

  (If both $a_{11} = 0$ and $b_{11} = 0$, then we may select any s in the range $[-1, 1]$.) Because $a_{11} \ge 0$, $c \ge 0$ and $\operatorname{sign}(s) = \operatorname{sign}(b_{11})$, we have $a'_{11} \ge 0$, so that the positivity property of the main diagonal of A is invariant.

  In the complex-valued case, ϕ is used to map b_{11} to the real domain first. a_{11} is already real, and this property is retained by the rotation.

- In the same way, use the transformed a_{11} to zero all entries of the first column of $\left[\begin{smallmatrix} B \\ B_2 \end{smallmatrix}\right]$. This defines a sequence of Givens rotations $q^*_{21}, \cdots, q^*_{m+n,1}$ which are applied in turn to Φ. Because Φ is isometric, the norm of each row is 1. This property is retained by the rotations, so that after the transformations we must have $a_{11} = 1$ (and not -1 since the property $a_{11} \ge 0$ is invariant).

- It is clear that A remains upper triangular during the rotations. We have to show that D_{21} also remains upper triangular, with nonnegative main diagonal, and that the first row of C is zero. This nontrivial fact follows from the orthonormality of the columns of Φ, which is invariant under the transformations. Indeed, after the first column of B has been zeroed, $a_{11} > 0$ because the realization is reachable. After $(B_2)_{11}$ has been zeroed, we have for the transformed Φ,

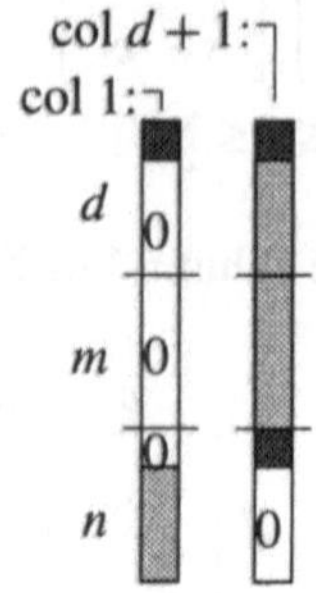

(14.17)

Since the rows are orthonormal, the first entry of the $d+1$-st column, the transformed c_{11}, must be zero at this point. Hence, subsequent rotations of the first row and rows 2 to n of $[B_2 \;\; D_{21}]$ do not destroy the zeros on the $d+1$-st column. The same holds for columns $d+2, \cdots$, so that D_{21} stays upper triangular while B_2 is made zero. The fact that $(D_{21})_{11} \geq 0$ after rotation $q_{m+1,1}$ follows directly from the small lemma below. Thus, the property $\operatorname{diag}(D_{21}) \geq 0$ is invariant under the transformations as well.

Lemma 14.2 *Suppose*

$$\begin{bmatrix} c & s \\ -s & c \end{bmatrix} \begin{bmatrix} a \\ b \end{bmatrix} = \begin{bmatrix} 0 \\ r \end{bmatrix}$$

where $b \geq 0$, $c = \sqrt{1-s^2} \geq 0$. *Then* $r \geq 0$.

PROOF The two solutions to $ca + sb = 0$, $s^2 + c^2 = 1$ are

$$s = \frac{a}{(a^2+b^2)^{1/2}}, \quad c = -\frac{b}{(a^2+b^2)^{1/2}}$$

and

$$s = -\frac{a}{(a^2+b^2)^{1/2}}, \quad c = \frac{b}{(a^2+b^2)^{1/2}}.$$

Since both $b \geq 0$ and $c \geq 0$, the first solution cannot occur. The second solution has $\operatorname{sign}(s) = -\operatorname{sign}(a)$. Hence, $r = -sa + cb \geq 0$. □

■ At this point, we have obtained

$$Q^*_{m+n,1} \cdots Q^*_{11} \Pi_{1,d+1} \Sigma_1 =$$

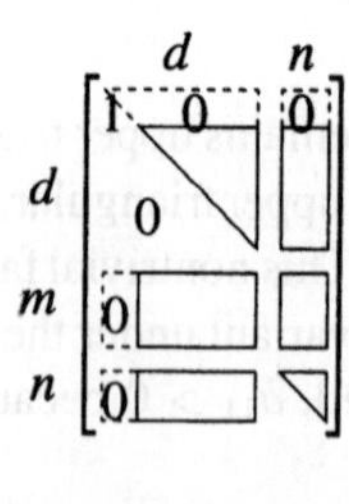

where each Q_{ij} is an embedding of q_{ij}, as defined in (14.8). The zeroing of the second through the d-th column of $\left[\begin{smallmatrix} B \\ B_2 \end{smallmatrix}\right]$ proceeds similarly. This gives

$$\Phi' = Q^*_{m+n,d} \cdots Q^*_{11} \Pi^*_{1,d+1} \mathbf{\Sigma}_1 = \begin{array}{c} \\ d \\ m \\ n \end{array} \begin{array}{c} \begin{array}{cc} d & n \end{array} \\ \left[\begin{array}{cc} I & 0 \\ 0 & D' \\ 0 & D'_{21} \end{array}\right] \end{array} \tag{14.18}$$

where D'_{21} is upper triangular with nonnegative main diagonal. In similar ways, we now use the main diagonal entries of D'_{21} to zero the entries of D'. For notational convenience, first permute $\left[\begin{smallmatrix} D' \\ D'_{21} \end{smallmatrix}\right]$ to $\left[\begin{smallmatrix} D'_{21} \\ D' \end{smallmatrix}\right]$,

$$\Pi^*_D \Phi' = \begin{array}{c} \\ d \\ n \\ m \end{array} \begin{array}{c} \begin{array}{cc} d & n \end{array} \\ \left[\begin{array}{cc} I & 0 \\ 0 & D'_{21} \\ 0 & D' \end{array}\right] \end{array}$$

where Π_D is defined in (14.12). Use $(D'_{21})_{11}$ to zero the top column of D', by a sequence of Givens rotations $z^*_{11}, \cdots, z^*_{1m}$. By orthonormality of the columns, after the transformations we must have $(D'_{21})_{11} = 1$, and the entries at the right of $(D'_{21})_{11}$ have become zero as a side effect. Hence, we can continue with using $(D'_{21})_{22}$ to zero the second column of D', etcetera. In the end, we obtain

$$Z^*_{mn} \cdots Z^*_{21} Z^*_{11} \Pi^*_D \Phi' = \left[\begin{array}{c} I_{d+n} \\ 0 \end{array}\right]$$

where each Z_{ij} is an embedding of z_{ij} as defined in (14.9). Conversely, after substituting (14.18) and inverting all rotations, we have

$$\mathbf{\Sigma}_1 = \left[I_{d+m} \quad 0_{(d+m)\times n}\right] \cdot \Pi_{1,d+1} Q_{11} Q_{21} \cdots Q_{m+n,d} \cdot \Pi_D \cdot Z_{11} Z_{21} \cdots Z_{m,n} \,.$$

Since $T(z)$ is specified by the first $m+n$ columns of $\mathbf{\Sigma}_1$, it follows that equation (14.13) holds. □

14.3 TIME-VARYING Σ-BASED CASCADE FACTORIZATION

The time-invariant cascade factorization results are readily extended to the context of time-varying systems. The procedure is roughly the same three-stage algorithm:

1. Embed a given realization for T into a lossless system Σ.

2. Using unitary state transformations, bring $\mathbf{\Sigma}$ into a form that allows a minimal factorization. We choose a Schur form, in which the A matrix of $\mathbf{\Sigma}$ is upper triangular.

3. Using Givens rotations, factor $\mathbf{\Sigma}$ into a product of such elementary sections. From this factorization, the lossless cascade network follows directly.

For time-invariant systems, we considered a state transformation to Hessenberg form to avoid eigenvalue computations and to lead to a parsimonious parametrization even in case of real systems with complex pole pairs. In the time-varying setting, eigenvalue computations are in a natural way replaced by recursions consisting of QR factorizations of the A_k, so this seems no longer to be an issue. The actual factorization is similar to the time-invariant procedure, and can be carried out locally. The main difference is that for time-varying systems the dimensions of the state-space matrices need not be constant, and a distinction has to be made between shrinking and growing state-space dimensions.

It is shown that it is still possible to obtain a factorization $\Sigma = \Sigma_1 \cdots \Sigma_n \Sigma'$, where $n = \max d_k$ is the maximal local state dimension over all stages, and each Σ_i is a section of local degree at most equal to 1. In a sense, the result is evident: by adding extra inputs and outputs, it is possible to expand the realization of Σ to a non-minimal realization which has d states at each point. However, the theorem is more specific: the local state dimensions of the factors add up to the local degree of Σ, and we obtain a cascade network with a minimal number of coefficients as well.

Time-varying embedding

Let $T \in \mathcal{U}(\mathcal{M}_1, \mathcal{N}_1)$ be a locally finite input-output operator with u.e. stable state realization $\mathbf{T} = \{A, B, C, D\}$. Assume that T is strictly contractive (this can always be obtained by a suitable scaling) and that $\mathbf{T}$ is uniformly reachable. Then the embedding theorem (theorem 12.14) claims that T admits a lossless embedding $\Sigma \in \mathcal{U}(\mathcal{M}_1 \times \mathcal{N}_1, \mathcal{N}_1 \times \mathcal{N}_2)$ such that $\Sigma_{11} = T$, and with unitary realization of the form

$$\mathbf{\Sigma} = \begin{bmatrix} R & & \\ & I & \\ & & I \end{bmatrix} \begin{bmatrix} A & C & C_2 \\ B & D & D_{12} \\ B_2 & D_{21} & D_{22} \end{bmatrix} \begin{bmatrix} [R^{(-1)}]^{-1} & & \\ & I & \\ & & I \end{bmatrix}.$$

D_{21} is a diagonal of square matrices, and we can arrange it such that each of these matrices is upper triangular, with nonnegative main diagonal.

Time-varying "Schur decomposition"

We continue by working on the unitary realization $\mathbf{\Sigma}$. Let $A = A_\Sigma \in \mathcal{D}(\mathcal{B}, \mathcal{B}^{(-1)})$ be the A-operator of $\mathbf{\Sigma}$. The factorization algorithm continues by finding a locally square unitary state transformation $Q \in \mathcal{D}(\mathcal{B}, \mathcal{B})$ such that

$$Q A Q^{(-1)*} = R, \tag{14.19}$$

where $R \in \mathcal{D}(\mathcal{B}, \mathcal{B}^{(-1)})$ has R_k upper triangular. If A_k is not square, say of size $d_k \times d_{k+1}$, then R_k will be of the same size and also be rectangular. In this case, "upper triangular" is to be made more precise: it means $(R_k)_{i,j} = 0$ for $i > j + (d_k - d_{k+1})$ (figure 14.4). In the case where $d_{k+1} > d_k$ (figure 14.4(c)), and if the increase in the number of states is 2 or more, it is possible to introduce extra zero entries in B too, as indicated in the figure. These play a role later in this chapter. Note that, for minimal realizations, the growth in state dimension is at most equal to the number of inputs at that point in time, so that the extra zero entries only appear in B and not in B_2.

In the time-invariant case, expression (14.19) would read $QAQ^* = R$, and the solution is then precisely the Schur decomposition of A. In this context, the main diagonal of A consists of its eigenvalues, which are the (inverses of the) poles of the system. In the present context, relation (14.19) is effectively the (unshifted) QR iteration algorithm that is sometimes used to compute the eigenvalues of $A = A_k$, if all A_k are the same [GV89]. The iteration (or rather recursion) is obtained by expanding the diagonal relation into its entries: $Q_k A_k Q_{k+1}^* = R_k$, or

$$\begin{array}{rcll} & \vdots & & \\ Q_1 A_1 & =: & R_1 Q_2 & \to Q_2, R_1 \\ Q_2 A_2 & =: & R_2 Q_3 & \to Q_3, R_2 \\ Q_3 A_3 & =: & R_3 Q_4 & \\ & \vdots & & \end{array} \tag{14.20}$$

Each step in the computation amounts to a multiplication by the previously computed Q_k, followed by a QR factorization of the result, yielding Q_{k+1} and R_k. Given an initial Q_{k_0}, *e.g.*, $Q_{k_0} = I$, the above recursion can be carried out in two directions, both forward and backward in time. For example, take $k_0 = 1$, then the forward recursion is given by (14.20), while the backward decomposition is

$$\begin{array}{rcll} A_0 Q_1^* & = & Q_0^* R_0 & \to Q_0, R_0 \\ A_{-1} Q_0^* & = & Q_{-1}^* R_{-1} & \to Q_{-1}, R_{-1} \\ A_{-2} Q_{-1}^* & = & Q_{-2}^* R_{-2} & \\ & \vdots & & \end{array}$$

Since we can start at any k_0 with any unitary Q_{k_0}, the decomposition (14.19) is not unique, although it always exists. For later reference, we formulate this result in the following proposition.

Proposition 14.3 *Let $A \in \mathcal{D}(\mathcal{B}, \mathcal{B}^{(-1)})$ be locally finite. Then there is a unitary state transformation $Q \in \mathcal{D}(\mathcal{B}, \mathcal{B})$ such that $QAQ^{(-1)*} = R$ is a diagonal operator with all R_k upper triangular matrices with nonnegative main diagonals: if A_k has size $d_k \times d_{k+1}$, then $(R_k)_{i,j} = 0$ for $i > j + (d_k - d_{k+1})$.*

In the context of finite upper triangular matrices whose state realization starts with 0 states at instant $k = 1$, we can take as initial transformation $Q_1 = [\cdot]$. If the A_k are equal to each other, then the above recursion is precisely the (unshifted) QR iteration for computing the eigenvalues (or Schur decomposition) of A. It is known (see [GV89]) that the unshifted QR iteration will converge if the absolute values of the eigenvalues of A are unequal to each other, and that the rate of convergence is dependent on the smallest ratio between those absolute eigenvalues. For periodically time-varying systems, with period n say, an initial state transformation Q_1 such that $Q_k = Q_{n+k}$ is also periodical can be computed by considering the conjunction of n consecutive stages. Writing $A_p = A_1 A_2 \cdots A_n$, the Schur decomposition of A_p ($Q_1 A_p Q_1^* = R_p$) gives Q_1, while (14.19) gives $Q_2, \cdots, Q_n$ in turn. Recent investigations show that one can compute the Schur decomposition of a product of matrices without ever explicitly evaluating the

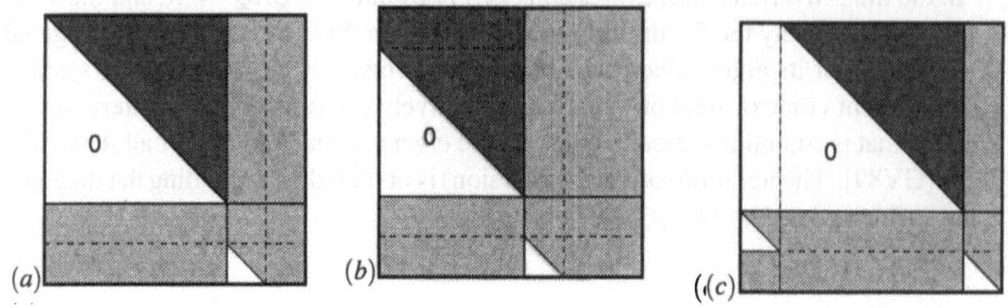

Figure 14.4. Schur forms of Σ_k. (*a*) Constant state dimension, (*b*) shrinking state dimension, (*c*) growing state dimension.

product [BGD92]. The procedure is called the periodic QR algorithm, and consists basically of an implicit shifted QR algorithm acting over a sequence of matrices, rather than just one matrix.

Structure of a factored lossless stage

A single stage $\mathbf{\Sigma}_k$ of $\mathbf{\Sigma}$ has, after transformation to Schur form, one of the three structures displayed in figure 14.4, depending on whether the state dimension of $\mathbf{\Sigma}$ is constant, shrinking or growing at point k. The factorization procedure is to factor each stage into elementary Givens rotations of the general type in (14.3). As before in the time-invariant case, it suffices to consider parsimonious rotations

$$\begin{bmatrix} 1 & \\ & e^{j\phi} \end{bmatrix} \begin{bmatrix} c & -s \\ s & c \end{bmatrix} \qquad -\tfrac{\pi}{2} \le \phi \le \tfrac{\pi}{2}, \quad -1 \le s \le 1 .$$

If the state dimension of Σ_k is constant, then its factorization is precisely the same as in the time-invariant case. E.g., suppose

$$\mathbf{\Sigma}_k = \left[\begin{array}{cccc|cc} a & \times & \times & \times & \times & \times \\ & \times & \times & \times & \times & \times \\ & & \times & \times & \times & \times \\ & & & \times & \times & \times \\ \hline b_1 & \times & \times & \times & \times & \times \\ b_2 & \times & \times & \times & \times & \times \end{array}\right]$$

then two rotations factor $\mathbf{\Sigma}$ into

$$(\mathbf{\Sigma}_1^*)_k \mathbf{\Sigma}_k = \left[\begin{array}{cccc|cc} 1 & 0 & 0 & 0 & 0 & 0 \\ & \times & \times & \times & \times & \times \\ & & \times & \times & \times & \times \\ & & & \times & \times & \times \\ \hline 0 & \times & \times & \times & \times & \times \\ 0 & \times & \times & \times & \times & \times \end{array}\right] .$$

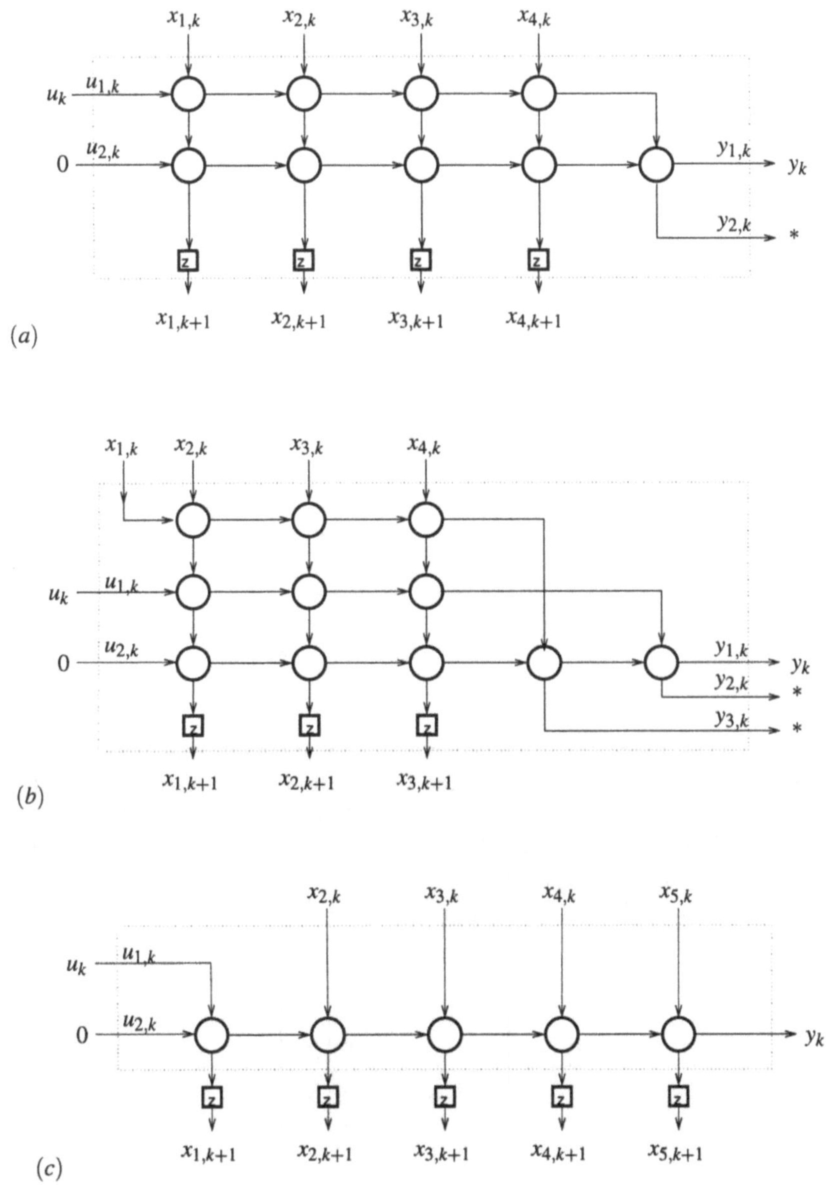

Figure 14.5. Lossless cascade realizations of a contractive system T, stage k. (a) Constant state dimension, (b) shrinking state dimension, (c) growing state dimension.

Continuing recursively, we obtain a factorization as $\mathbf{\Sigma}_k = \mathbf{\Sigma}_{1,k} \cdots \mathbf{\Sigma}_{d_k,k} \mathbf{\Sigma}'_k$. $\mathbf{\Sigma}'_k$ is the residue $\left[\begin{array}{c|c} I & \\ \hline & D'_k \end{array}\right]$ where D'_k is a unitary matrix and can also be factored into elementary operations. The corresponding network structure of a single stage is depicted in figure 14.5(*a*).

In the case of a shrinking state dimension, we have for example

$$\mathbf{\Sigma}_k = \left[\begin{array}{cccc|ccc} \cdot & \times & \times & \times & \times & \times & \times \\ & \times & \times & \times & \times & \times & \times \\ & & \times & \times & \times & \times & \times \\ & & & \times & \times & \times & \times \\ \hline \cdot & \times & \times & \times & \times & \times & \times \\ \cdot & \times & \times & \times & \times & \times & \times \end{array}\right].$$

We first perform a row permutation, which produces

$$\Pi_k^* \mathbf{\Sigma}_k = \left[\begin{array}{ccc|ccc} \times & \times & \times & \times & \times & \times \\ & \times & \times & \times & \times & \times \\ & & \times & \times & \times & \times \\ \hline \times & \times & \times & \times & \times & \times \\ \times & \times & \times & \times & \times & \times \\ \times & \times & \times & \times & \times & \times \end{array}\right],$$

so that, effectively, the first state has become an input of the subsequent factors. At this point, the factorization is equivalent to the factorization of a realization with constant state dimension. The resulting network structure of the lossless stage is shown in figure 14.5(*b*). More in general, if the state dimension of $\mathbf{\Sigma}_k$ shrinks by n states, then a number of n states are permuted to become inputs.

Finally, if the state dimension of Σ_k grows, for example

$$\mathbf{\Sigma}_k = \left[\begin{array}{cccc|c} 0 & \times & \times & \times & \times \\ & & \times & \times & \times \\ & & & \times & \times \\ \hline b_1 & \times & \times & \times & \times \\ b_2 & \times & \times & \times & \times \end{array}\right],$$

then a permutation of the first row of B produces

$$\Pi_k^* \mathbf{\Sigma}_k = \left[\begin{array}{cccc|c} b_1 & \times & \times & \times & \times \\ 0 & \times & \times & \times & \times \\ & & \times & \times & \times \\ & & & \times & \times \\ \hline b_2 & \times & \times & \times & \times \end{array}\right].$$

The first input $u_{1,k}$ has effectively been mapped to a new state. $\Phi_k^*\Sigma_k$ can subsequently be factored as a section with constant state dimensions:

$$(\mathbf{\Sigma}_1^*)_k\mathbf{\Sigma}_k = \left[\begin{array}{cccc|c} 1 & 0 & 0 & 0 & 0 \\ & \times & \times & \times & \times \\ & & \times & \times & \times \\ & & & \times & \times \\ \hline 0 & \times & \times & \times & \times \end{array}\right].$$

The corresponding network is depicted in figure 14.5(c). If, more in general, the first n columns of A would have been zero, then the first n rows of B are permuted to become states. For minimality of the factorization, we must require that the top left $n \times n$ submatrix of B_Σ has been made upper triangular by suitable unitary state transformations, in the process of the transformation to Schur form (as indicated in figure 14.4(c)).

With the three types of stages shown in figure 14.5, we can describe all possible stages that can occur in locally finite unitary realizations that are in Schur form. It has already been mentioned that the stages can be factored independently of each other. The cascade network structure of the complete state realization $\mathbf{\Sigma}$ then follows by piecing together the results of the individual stages. An example network is shown in figure 14.6(a). In the example, we consider a 10×10 strictly contractive upper triangular matrix T, with 1 input and 1 output at each point, and a state dimension sequence $\mathcal{B}$ given by

$$\#\mathcal{B} = [0,1,2,3,3,3,2,3,2,1,0].$$

T has an embedding into an inner operator Σ. Hence T is the partial transfer operator of Σ from the first input to the first output when the secondary input is put to zero.

In the time-invariant case, the Schur form produces a factorization of the realization into a cascade of elementary sections, each of degree 1. The question at this point is whether the time-varying cascaded network obtained in figure 14.6(a) also produced such a factorization. Obviously, with time varying state dimensions, the elementary sections now have to be time-varying. In the remainder of the section, we show that T is realized by a cascade of $d = \max d_k$ elementary time-varying sections, each of local degree 0 or 1. We start by making a few more general observations.

Factorization into two factors

The factorization result (equation (14.2)), which stated that a time-invariant rational transfer operator T has a factorization $T = T_1T_2$ if and only if its realization has a certain structure, admits a straightforward generalization to time-varying inner systems.

Proposition 14.4 *Let* $\mathbf{\Sigma} \in \mathcal{D}(\mathcal{B}\times\mathcal{M}, \mathcal{B}^{(-1)}\times\mathcal{N})$ *be unitary, with locally finite dimensions, and have a block partitioning as*

$$\mathbf{\Sigma} = \left[\begin{array}{cc|c} A_{11} & A_{12} & C_1 \\ 0 & A_{22} & C_2 \\ \hline B_1 & B_2 & D \end{array}\right] \qquad (14.21)$$

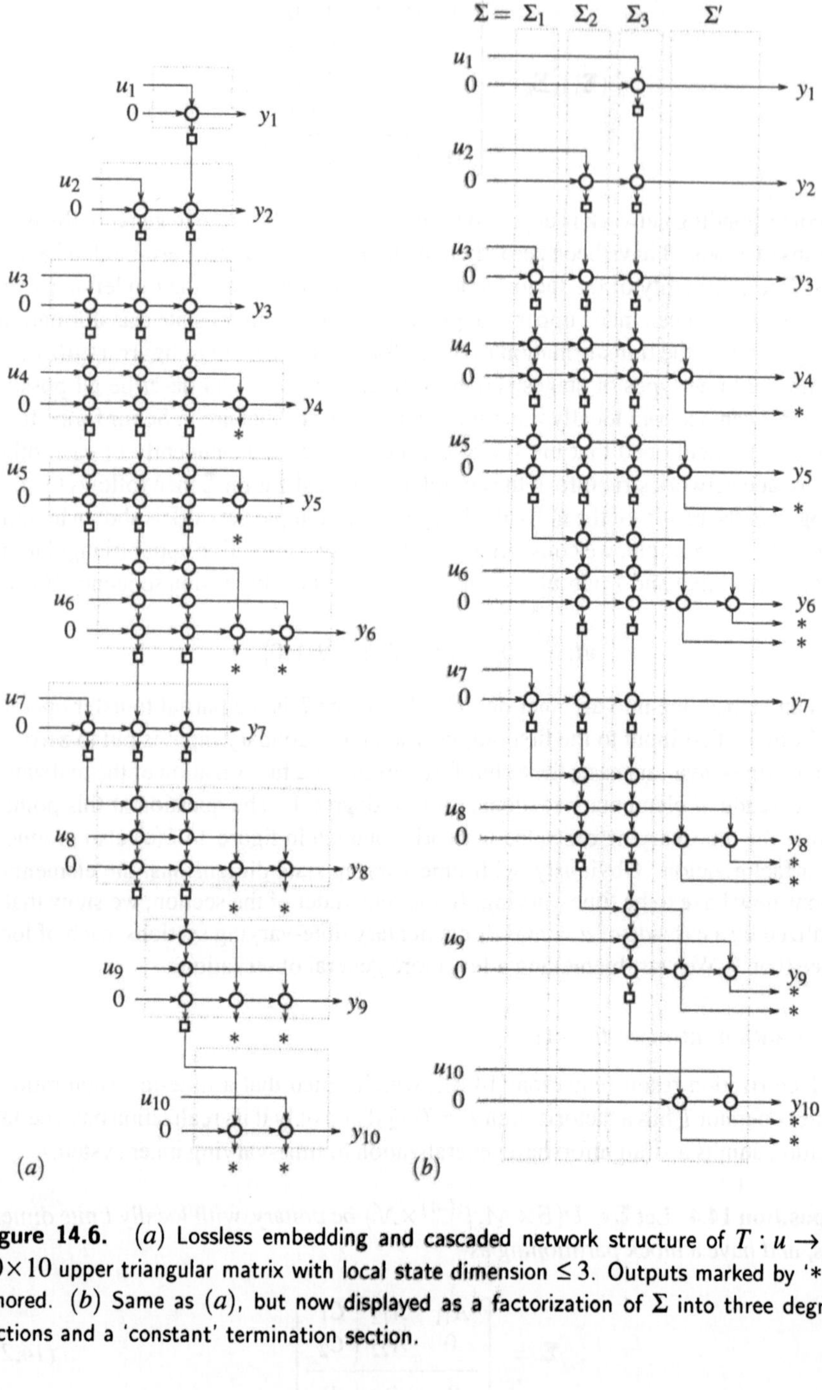

Figure 14.6. (a) Lossless embedding and cascaded network structure of $T : u \to y$, a 10×10 upper triangular matrix with local state dimension ≤ 3. Outputs marked by '$*$' are ignored. (b) Same as (a), but now displayed as a factorization of Σ into three degree-1 sections and a 'constant' termination section.

where $A_{11} \in \mathcal{D}(\mathcal{B}_1, \mathcal{B}_1^{(-1)})$ for some state-space sequence $\mathcal{B}_1 \subset \mathcal{B}$. Define the space sequences $\mathcal{N}_1$ and $\mathcal{B}_2$ by the relations $\mathcal{B}_1 \times \mathcal{M} = \mathcal{B}_1^{(-1)} \times \mathcal{N}_1$, and $\mathcal{B} = \mathcal{B}_1 \times \mathcal{B}_2$.

1. *Then unitary operators $\hat{\mathbf{\Sigma}}_1, \hat{\mathbf{\Sigma}}_2$ exist, with $\hat{\mathbf{\Sigma}}_1 = \{A_{11}, B_1, C_1', D_1\} \in \mathcal{D}(\mathcal{B}_1 \times \mathcal{M}, \mathcal{B}_1^{(-1)} \times \mathcal{N}_1)$, $\hat{\mathbf{\Sigma}}_2 = \{A_{22}, B_2', C_2, D_2\} \in \mathcal{D}(\mathcal{B}_2 \times \mathcal{N}_1, \mathcal{B}_2^{(-1)} \times \mathcal{N})$, such that*

$$\mathbf{\Sigma} = \left[\begin{array}{cc|c} A_{11} & & C_1' \\ & I & \\ \hline B_1 & & D_1 \end{array}\right] \left[\begin{array}{cc|c} I & & \\ & A_{22} & C_2 \\ \hline & B_2' & D_2 \end{array}\right] =: \mathbf{\Sigma}_1 \mathbf{\Sigma}_2 . \qquad (14.22)$$

2. *If Σ is an inner operator with unitary realization $\mathbf{\Sigma}$ of the form (14.21), with $\ell_{A_\Sigma} < 1$, then $\Sigma = \Sigma_1 \Sigma_2$, where Σ_1, Σ_2 are inner operators with unitary realizations given by $\hat{\mathbf{\Sigma}}_1, \hat{\mathbf{\Sigma}}_2$, with $\ell_{A_{11}} < 1$, $\ell_{A_{22}} < 1$. The sequence of state dimensions of Σ_1, Σ_2 add up to the sequence of state dimensions of Σ: the factorization is minimal.*

PROOF

1. Consider $[A_{11}^* \;\; B_1^*]^*$. It is an isometry in $\mathcal{D}$ because $A_{11}^* A_{11} + B_1^* B_1 = I$. Choose $C_1', D_1 \in \mathcal{D}$ such that, for each point k,

$$(\mathbf{\Sigma}_1)_k = \begin{bmatrix} (A_{11})_k & (C_1')_k \\ (B_1)_k & (D_1)_k \end{bmatrix}$$

is a unitary matrix. Then $\mathbf{\Sigma}_1$ is a unitary operator in $\mathcal{D}$ as required, and the number of added outputs is $\#(\mathcal{N}_1) = \#(\mathcal{B}_1) - \#(\mathcal{B}_1^{(-1)}) + \#(\mathcal{M})$. Because $[A_{11}^* \; 0 \; B_1^*]^*$ is also the first column of $\mathbf{\Sigma}$, it directly follows that $\mathbf{\Sigma}_1^* \mathbf{\Sigma} = \mathbf{\Sigma}_2$ has the form specified in (14.21).

2. The fact $\ell_{A_\Sigma} < 1 \;\Rightarrow \ell_{A_{11}} < 1, \ell_{A_{22}} < 1$ is straightforward to show. With $\ell_{A_{11}} < 1, \ell_{A_{22}} < 1$, the unitary realizations $\hat{\mathbf{\Sigma}}_1$, $\hat{\mathbf{\Sigma}}_2$ define inner operators Σ_1, Σ_2 (theorem 6.4). The cascade $\Sigma_1 \Sigma_2$ has a realization $\mathbf{\Sigma}_1 \mathbf{\Sigma}_2 = \mathbf{\Sigma}$ as in (14.21), and hence $\Sigma = \Sigma_1 \Sigma_2$. The factorization is minimal because (with $\ell_A < 1$) $\hat{\mathbf{\Sigma}}_1$, $\hat{\mathbf{\Sigma}}_2$ are minimal realizations, whose degrees add up to the degree of $\mathbf{\Sigma}$.

□

Some remarks are apposite here. First note that if $\ell_{A_\Sigma} = 1$, and $\mathbf{\Sigma}$ is a unitary realization with reachability and observability Gramians equal to the identity, then $\hat{\mathbf{\Sigma}}_1$ inherits the fact that the reachability Gramian is I, but if $\ell_{A_{11}} = 1$, then nothing can be said, at first sight, of its observability Gramian, and hence the fact that Σ_1 is inner is not proven in this case. Second, note that all computations can be carried out locally (separately) for each stage k. The state dimension sequence $\mathcal{B}_1$ determines the degree of the factors, and also the number of outputs (inputs) of Σ_1 (Σ_2). The choice of $\mathcal{B}_1$ is restricted by the required form of (14.21), *i.e.*, the fact that $A_{21} = 0$.

The above proposition can be formulated in a different way that provides some additional (more fundamental) insight.

Proposition 14.5 *Let Σ be a locally finite inner operator. Then*

$$\Sigma = \Sigma_1 \Sigma_2 \quad \Rightarrow \quad \mathcal{H}(\Sigma) = \mathcal{H}(\Sigma_1) \oplus \mathcal{H}(\Sigma_2)\Sigma_1^*,$$

where Σ_1 and Σ_2 are inner operators. Conversely, let Σ_1 be an inner operator, then

$$\mathcal{H}(\Sigma_1) \subset \mathcal{H}(\Sigma) \quad \Rightarrow \quad \Sigma = \Sigma_1 \Sigma_2,$$

where Σ_2 is an inner operator.

PROOF For an inner operator Σ_2, we have that $\mathcal{U}_2\Sigma_2^* = \mathcal{U}_2 \oplus \mathcal{H}(\Sigma_2)$ (proposition 6.1). Consequently, $\mathcal{U}_2\Sigma^* = \mathcal{U}_2\Sigma_1^* \oplus \mathcal{H}(\Sigma_2)\Sigma_1^*$, and because $\Sigma_1^* \in \mathcal{L}$,

$$\begin{aligned} \mathcal{H}(\Sigma) &= \mathbf{P}_{\mathcal{L}_2 Z^{-1}}(\mathcal{U}_2\Sigma^*) \\ &= \mathcal{H}(\Sigma_1) \oplus \mathbf{P}_{\mathcal{L}_2 Z^{-1}}(\mathcal{H}(\Sigma_2)\Sigma_1^*) \\ &= \mathcal{H}(\Sigma_1) \oplus \mathcal{H}(\Sigma_2)\Sigma_1^*. \end{aligned}$$

Conversely, the fact that $\Sigma_2 = \Sigma_1^*\Sigma$ is a unitary operator is clear, and we have to show that it is in fact upper. Indeed, since $\Sigma \in \mathcal{U}$,

$$\begin{aligned} \mathbf{P}_{\mathcal{L}_2 Z^{-1}}(\mathcal{U}_2\Sigma_2) &= \mathbf{P}_{\mathcal{L}_2 Z^{-1}}(\mathcal{U}_2\Sigma_1^*\Sigma) \\ &= \mathbf{P}_{\mathcal{L}_2 Z^{-1}}(\mathcal{H}(\Sigma_1)\Sigma) \\ &\subset \mathbf{P}_{\mathcal{L}_2 Z^{-1}}(\mathcal{H}(\Sigma)\Sigma) \quad = 0 \qquad \text{[prop. 6.1]} \end{aligned}$$

so that the lower triangular part of Σ_2 is zero. □

Hence, in order to obtain a factorization of Σ, we can select any inner Σ_1 such that $\mathcal{H}(\Sigma_1) \subset \mathcal{H}(\Sigma)$. A suitable Σ_1 is again obtained from equation (14.21): a minimal realization based on A_{11} and B_1 has

$$\mathcal{H}(\Sigma_1) = \mathcal{D}_2 \left[B_1 Z(I - A_{11}Z)^{-1}\right]^* = [\mathcal{D}_2 \;\; 0]\left[[B_1 \;\; B_2]Z(I - AZ)^{-1}\right]^*$$

because $A_{21} = 0$, so that indeed $\mathcal{H}(\Sigma_1) \subset \mathcal{H}(\Sigma)$. Σ_1 is obtained, as in the proof of proposition 14.4, by extending $[A_{11}^* \;\; B_1^*]^*$ to a unitary state-space operator. Special cases occur if $(\mathcal{B}_1)_k = 0$ for some k, although the propositions remains valid. The following two situations are typical.

- If $\#(\mathcal{B}_1)_{k+1} = 0$, with $\#(\mathcal{B}_1)_k = n \geq 0$, then $(A_{11})_k$ is a $(n \times 0)$-matrix. In this case, $\mathbf{\Sigma}_k$ has the form

$$\mathbf{\Sigma}_k = \left[\begin{array}{cc|c} \cdot & A_{12} & C_1 \\ \cdot & A_{22} & C_2 \\ \hline \cdot & B_2 & D \end{array}\right]$$

(as before, '$\cdot$' stands for an entry of zero dimensions) so that

$$(\hat{\mathbf{\Sigma}}_1)_k = \left[\begin{array}{c|cc} \cdot & I_n & 0 \\ \hline \cdot & 0 & I \end{array}\right], \qquad (\mathbf{\Sigma}_1)_k = \left[\begin{array}{c|c|c} \cdot & 0 & I_n & 0 \\ \cdot & I & 0 & 0 \\ \hline \cdot & 0 & 0 & I \end{array}\right].$$

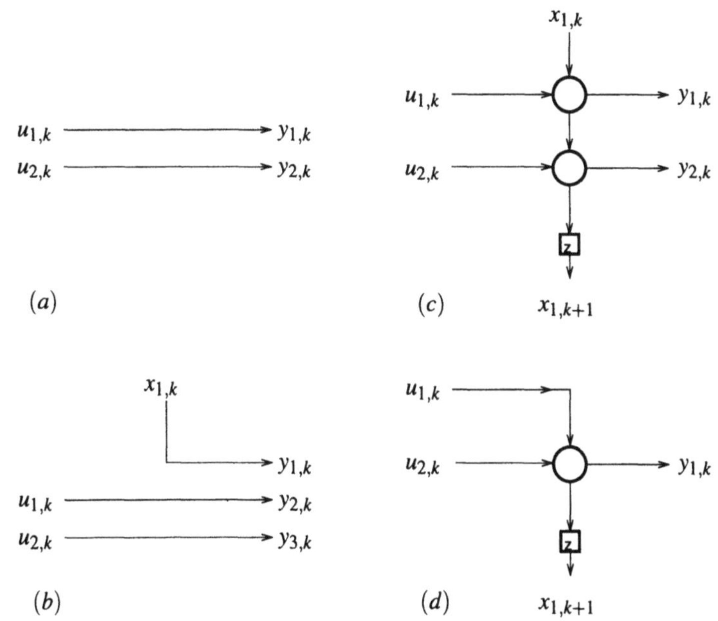

Figure 14.7. Elementary sections in a stage. (a) $C(0)$ constant section with zero states, (b) S section, going from 1 state to 0, (c) $C(1)$ section with a constant number of 1 states, (d) G section, going from 0 to 1 state. The number of inputs/outputs have arbitrarily been set to 2.

$\mathbf{\Sigma}_1$ is a trivial state-space operator mapping its first n states to n outputs. If $n = 0$, then $(\mathbf{\Sigma}_1)_k = I$.

- If $\#(\mathcal{B}_1)_k = 0$, $\#(\mathcal{B}_1)_{k+1} = n \geq 0$, then $(\hat{\mathbf{\Sigma}}_1)_k$ is obtained as the extension of $(B_1)_k$ to a unitary matrix:

$$(\hat{\mathbf{\Sigma}}_1)_k = \left[\begin{array}{c|c} \cdot & \cdot \\ \hline (B_1)_k & (D_1)_k \end{array}\right].$$

Note that this case can only happen if $(A_\Sigma)_k$ has its first n columns equal to zero:

$$(A_\Sigma)_k = \begin{bmatrix} \cdot & \cdot \\ 0 & (A_{22})_k \end{bmatrix},$$

that is, in view of figure 14.4, this can only happen at points where the state dimension of Σ grows with at least n states.

Elementary lossless stage sections

We apply proposition 14.4 to the most elementary type of state dimension sequence $\mathcal{B}_1$: $\mathcal{B}_1$ with entries having dimensions $\#(\mathcal{B}_1)_k \in \{0, 1\}$. In a later section, we discuss the choice of $\mathcal{B}_1$; here, we consider the factorization of a single stage of $\mathbf{\Sigma}$, and pay detailed attention to the fact that input/output and state dimensions can be time varying. With a partitioning of $\mathbf{\Sigma}$ as before in (14.21), a factor $\hat{\mathbf{\Sigma}}_1$ of $\mathbf{\Sigma}$ is determined by finding a unitary extension of the matrices $(A_{11})_k$ and $(B_1)_k$. The purpose of this section is to show how an extension can be obtained in factored form using elementary Givens rotations. With $\#(\mathcal{B}_1)_k \in \{0,1\}$ and $\#(\mathcal{B}_1)_{k+1} \in \{0,1\}$, the submatrix $(A_{11})_k$ can have only the following sizes:

$$\begin{cases} C(0): & 0\times 0, \qquad S: \quad 1\times 0, \\ C(1): & 1\times 1, \qquad G: \quad 0\times 1. \end{cases}$$

The cases $C(0)$ and $C(1)$ describe sections with a constant state dimension, while G,S stand for sections with growing and shrinking state dimensions, respectively. We discuss these sections in turn.

$C(0)$: $(\hat{\mathbf{\Sigma}}_1)_k$ has the form $(\hat{\mathbf{\Sigma}}_1)_k = \left[\begin{array}{c|c} \cdot & \cdot \\ \hline \cdot & I \end{array}\right]$. See figure 14.7($a$). Obviously, a $C(0)$ section can always be extracted, but doing so does not lead to a degree reduction. Nonetheless, it plays a role as padding section in the description of a global factorization of Σ into a constant number of sections, later in this chapter.

S: $(\hat{\mathbf{\Sigma}}_1)_k$ has the form $(\hat{\mathbf{\Sigma}}_1)_k = \left[\begin{array}{c|cc} \cdot & 1 & 0 \\ \hline \cdot & 0 & I \end{array}\right]$. See figure 14.7($b$).

$C(1)$: Let $a = (A_{11})_k$, and suppose that Σ has n inputs at point k, so that $b = (B_1)_k$ is an $n\times 1$ vector. Then $(\hat{\mathbf{\Sigma}}_1)_k$ is a unitary extension of the vector $[a^*\ b_1^*\ \cdots\ b_n^*]^*$. Of the many possible extensions, one that results in a minimal number of coefficients is obtained using Givens rotations, which gives the extension directly in factored form:

$$(\hat{\mathbf{\Sigma}}_1)_k = (\hat{\mathbf{\Sigma}}_1)_{1,k} \cdots (\hat{\mathbf{\Sigma}}_1)_{n,k} \tag{14.23}$$

where $(\hat{\mathbf{\Sigma}}_1)_{i,k}$ is used to zero entry $(i+1)$ of the vector $(\hat{\mathbf{\Sigma}}_1)^*_{i-1,k} \cdots (\hat{\mathbf{\Sigma}}_1)^*_{1,k} \left[\begin{smallmatrix} a \\ b \end{smallmatrix}\right]$ against the first entry. The computational structure (for $n=2$) is shown in figure 14.7(c).

G: In this case, $(A_{11})_k = [\cdot]$, and $(\hat{\mathbf{\Sigma}}_1)_k$ is a unitary extension of the vector $b = (B_1)_k$. Again, the extension can be found in factored form, now requiring $n-1$ Givens rotations. See figure 14.7(d).

The four types of elementary stage sections in figure 14.7 form the building blocks of the cascade network realizations based on the Schur form. General structures are obtained by connecting these sections horizontally (realizing a single stage in factored form) and vertically (realizing an elementary degree-1 factor of Σ). The result of horizontal connections into stages has already been discussed before, see figure 14.6(a). It remains to discuss the connection into vertical elementary degree-1 factors.

Factorization into degree-1 lossless sections

Let be given a locally finite inner operator Σ, with state dimension sequence $\mathcal{B}$. The objective is to compute a factorization $\Sigma = \Sigma_1 \cdots \Sigma_n \Sigma'$ into a minimal number of n degree-

1 sections, and a terminating diagonal unitary operator Σ' (a 'constant' section). A related question is: what is the minimal value of n? It is clear that n is at least equal to the maximal number $\max_k \#(\mathcal{B})_k$ of states of Σ that are present at any stage. We show that n can in fact be equal to this number.

In view of proposition 14.4, it remains to determine a possible state sequence $\mathcal{B}_1$ of the first factor Σ_1. The other factors are then obtained recursively, by factoring $\Sigma_1^*\Sigma$, until the state dimension has been reduced to zero. The remainder $\Sigma_n^* \cdots \Sigma_1^*$ is then equal to the constant section Σ'. The number of states $\#(\mathcal{B}_1)_k$ of the first factor is allowed to be at most equal to 1 at each stage k, in order to obtain a degree-1 section. The other constraint on $\mathcal{B}_1$ is the fact that $(A_{21})_k$ in (14.21) must be equal to zero (or have vanishing dimensions) for each k. The discussions in the previous paragraph have shown that, as a consequence, within a stage it is not possible to extract a $C(1)$ section before an S section or a G section. A trivial $C(0)$ section can always be extracted.

The following choice of $\mathcal{B}_1$ satisfies the constraints. Let $n = \max_k \#(\mathcal{B})_k$. Then $\mathcal{B}_1$ is given by

$$\#(\mathcal{B}_1)_k = \begin{cases} 1, & \text{if } \#(\mathcal{B})_k = n\,, \\ 0, & \text{otherwise}\,. \end{cases} \tag{14.24}$$

Indeed, with this $\mathcal{B}_1$, we extract as many stages with $C(0)$ sections as possible (which do not have constraints), and only extract other sections where factors Σ_2 till Σ_n must have states anyway. At the same time, $\mathcal{B}_1$ is such that it reduces the degree of Σ: $\Sigma_1^*\Sigma$ has a maximal state dimension $n-1$. Acting recursively, we obtain a factorization of Σ into n sections, each of which has local degree at most 1. The results are summarized in the following theorem.

Theorem 14.6 *Let Σ be an inner operator which is locally finite with state dimension sequence $\mathcal{B}$, and u.e. stable. Let $n = \max_k \#(\mathcal{B})_k$. Then Σ has a factorization*

$$\Sigma = \Sigma_1 \cdots \Sigma_n \Sigma'\,,$$

where each Σ_i is a u.e. stable inner section of local degree at most 1 ($\max_k \#(\mathcal{B}_i)_k = 1$), and whose local degrees add up to the local degree of Σ ($\sum_i \#(\mathcal{B}_i)_k = \#(\mathcal{B})_k$). Σ' is a unitary diagonal operator.

PROOF According to theorem 6.3, Σ has a unitary realization $\mathbf{\Sigma}$. The realization can be put into Schur form by unitary state transformations (proposition 14.3). Next, choose $\mathcal{B}_1$ according to equation (14.24). We first show that $\mathcal{B}_1$ generates a partitioning of $A = A_\Sigma$ such that, for all k, $(A_{21})_k = 0$ or has vanishing dimensions. Indeed, as long as $\#(\mathcal{B})_k < n$ and $\#(\mathcal{B})_{k+1} < n$, we have $\#(\mathcal{B}_1)_k = 0$ and $\#(\mathcal{B}_1)_{k+1} = 0$ so that $(A_{21})_k = [\cdot]$. At a certain point k, $\#(\mathcal{B})_k < n$ and $\#(\mathcal{B})_{k+1} = n$, and figure 14.4($c$) shows that in this case we can put $\#(\mathcal{B}_1)_{k+1} = 1$, which makes $(A_{21})_k$ equal to the first column, consisting only of zero entries. While $\#(\mathcal{B})_k = n$ and $\#(\mathcal{B})_{k+1} = n$, A_k is an upper triangular matrix, so that we can put $\#(\mathcal{B}_1)_k = 1$, $\#(\mathcal{B}_1)_{k+1} = 1$ to obtain $(A_{21})_k = 0$. Finally, when $\#(\mathcal{B})_k = n$ and $\#(\mathcal{B})_{k+1} < n$, A_k has the form shown in figure 14.4(b), so that we have to put $\#(\mathcal{B}_1)_{k+1} = 0$, which gives $(A_{21})_k = [\cdot]$. Hence $\mathcal{B}_1$ satisfies the requirements, so that, according to proposition 14.4, we can extract a factor Σ_1. We can continue in the same way with $\Sigma_1^*\Sigma$, which has a maximal state dimension equal to $n-1$. This degree reduction is because we had $\#(\mathcal{B}_1)_k = 1$ whenever $\#(\mathcal{B})_k = n$. Acting recursively, we end with $\Sigma' = \Sigma_n^* \cdots \Sigma_1^* \Sigma$ having 0 states, and hence Σ' is a unitary diagonal constant. □

We can write the 10×10 example in figure 14.6(a) in factored form, as obtained by the above theorem. The resulting cascade factorization is displayed in figure 14.6(b). The actual structure is the same as given in figure 14.6(a), but the elementary stage sections are now grouped vertically into sections, rather than horizontally into stages.

Computational complexity

The computational complexity of the cascade network is, at each stage, linear in the number of elementary operations. This is in contrast to a direct network realization of a given state realization $\{A, B, C, D\}$, which would have quadratical complexity. If the network consists of N stages and if the average number of states in a stage is d, then the number of elementary operations required for a vector-matrix multiplication using the cascade network is of order $O(2dN)$ rotations, rather than $O(\frac{1}{2}N^2)$ multiplications for a direct vector-matrix multiplication. (The complexity of a rotation operation is 4 multiplications for a direct implementation). Hence, if $d \ll N$, a considerable reduction in the number of operations is achieved. In addition, the network is numerically stable. All elementary operations are rotations, which means that the network is lossless and does not amplify numerical errors introduced at any point in the computation.

14.4 TIME-VARYING Θ-BASED CASCADE FACTORIZATION

In the previous section, we embedded the given contractive operator T in a unitary operator Σ, and subsequently factored this operator into elementary sections. The result was a computational network consisting of unitary Givens rotations, with a data flow strictly from the left to the right, and from the top to the bottom. An alternative cascade factorization is obtained by computing the J-unitary operator Θ associated with Σ,[1] factoring Θ into elementary J-unitary sections Θ_i, and converting each of the sections to their unitary equivalent. The result is again a minimal factorization of the unitary realization $\mathbf{\Sigma}$ of Σ into degree-1 realizations, although the factorization is different from the one we obtained earlier. The order of the computations in this factorization is such that the corresponding cascade factorization of Σ can no longer be written as a product of elementary unitary sections.

The reason for studying Θ-based factorizations is at least twofold. Firstly, they lead to different realizations of unitary operators Σ, also specified by a minimal number of parameters. These realizations may have different numerical properties with respect to parameter sensitivity (although we do not go to that level of detail). Secondly, the same type of networks are obtained in the solution of a number of other problems. For example, the solution of certain constrained interpolation problems, such as the Nevanlinna-Pick interpolation problem in chapter 9, or the solution of the Nehari problem and (more in general) the model approximation problem in chapter 10, leads to

[1] In this section, we assume that the reader has knowledge of the contents of section 8.1.

Θ-based cascade networks. This is of course not coincidental: the description of the solution of these interpolation problems also gives rise to J-unitary operators Θ. Upon factorization of Θ, each factor implements a single interpolation constraint of the original problem. Other problems where networks of the same type occur are in the Generalized Schur algorithm for inverse Cholesky factorization [Dep81, DD88], and (time-varying) prediction-error filters and RLS adaptive filters [Hay91].

We will first derive some factorization results for J-unitary upper operators, and then specialize to the case where the state signature sequence equals $J_{\mathcal{B}} = I$. Subsequently, we derive the corresponding factorization of $\mathbf{\Sigma}$, and the computational network that this factorization of $\mathbf{\Sigma}$ represents.

Factorization into J-unitary elementary sections

The J-unitary factorization into elementary sections is again straightforward once a general factorization into two J-unitary factors has been derived. The latter is formulated in the following proposition, comparable to proposition 14.4.

Proposition 14.7 *Let $\mathbf{\Theta} \in \mathcal{D}(\mathcal{B}\times\mathcal{M}, \mathcal{B}^{(-1)}\times\mathcal{N})$ be J-unitary with state signature $\mathbf{J}_{\mathcal{B}} = I$, and with a block partitioning as*

$$\mathbf{\Theta} = \left[\begin{array}{cc|c} A_{11} & A_{12} & C_1 \\ 0 & A_{22} & C_2 \\ \hline B_1 & B_2 & D \end{array}\right] \qquad (14.25)$$

where $A_{11} \in \mathcal{D}(\mathcal{B}_1, \mathcal{B}_1^{(-1)})$ for some state-space sequence $\mathcal{B}_1 \subset \mathcal{B}$. Define the space sequences $\mathcal{N}_1$ and $\mathcal{B}_2$ by the relations $\mathcal{B}_1\times\mathcal{M} = \mathcal{B}_1^{(-1)}\times\mathcal{N}_1$, and $\mathcal{B} = \mathcal{B}_1\times\mathcal{B}_2$.

1. *J-unitary operators $\hat{\mathbf{\Theta}}_1, \hat{\mathbf{\Theta}}_2$ exist, with $\hat{\mathbf{\Theta}}_1 = \{A_{11}, B_1, C_1', D_1\} \in \mathcal{D}(\mathcal{B}_1\times\mathcal{M}, \mathcal{B}_1^{(-1)}\times\mathcal{N}_1)$, $\hat{\mathbf{\Theta}}_2 = \{A_{22}, B_2', C_2, D_2\} \in \mathcal{D}(\mathcal{B}_2\times\mathcal{N}_1, \mathcal{B}_2^{(-1)}\times\mathcal{N})$, such that*

$$\mathbf{\Theta} = \left[\begin{array}{cc|c} A_{11} & & C_1' \\ & I & \\ \hline B_1 & & D_1 \end{array}\right]\left[\begin{array}{cc|c} I & & \\ & A_{22} & C_2 \\ \hline & B_2' & D_2 \end{array}\right] = \mathbf{\Theta}_1\mathbf{\Theta}_2. \qquad (14.26)$$

2. *If $\Theta \in \mathcal{U}$ is a J-unitary operator with a J-unitary realization $\mathbf{\Theta}$ of the form (14.25), and if $\ell_{A_\Theta} < 1$, then $\Theta = \Theta_1\Theta_2$, where Θ_1, Θ_2 are J-unitary operators with J-unitary realizations given by $\hat{\mathbf{\Theta}}_1$, $\hat{\mathbf{\Theta}}_2$, with $\ell_{A_{11}} < 1$, $\ell_{A_{22}} < 1$. The factorization is minimal.*

PROOF The proof is the same as in proposition 14.4, except that now a J-unitary extension of $[A_{11}^* \;\; B_1^*]^*$ must be found. The existence of such an extension was proven in lemma 8.16. The extension yields $\hat{\mathbf{\Theta}}_1$, and $\hat{\mathbf{\Theta}}_2$ then follows from $\mathbf{\Theta}_1^{-1}\mathbf{\Theta} = \mathbf{\Theta}_2$, which has the form specified in (14.25). □

In order to obtain a factorization into elementary sections of local degree ≤ 1, we choose $\mathcal{B}_1$ as in equation (14.24), viz.

$$\#(\mathcal{B}_1)_k = \begin{cases} 1 & \text{if } \#(\mathcal{B})_k = n, \\ 0 & \text{otherwise}. \end{cases}$$

With this choice, theorem 14.6 can be adapted to J-unitary operators:

Theorem 14.8 *Let $\Theta \in \mathcal{U}$ be a J-unitary operator which is locally finite with state dimension sequence $\mathcal{B}$, u.e. stable, and with positive state signature. Let $n = \max_k \#(\mathcal{B})_k$. Then Θ has a factorization*

$$\Theta = \Theta_1 \cdots \Theta_n \Theta',$$

where each Θ_i is a u.e. stable J-unitary section of local degree ≤ 1 ($\max_k \#(\mathcal{B}_i)_k = 1$), and the local degrees of the Θ_i add up to the local degree of Θ ($\sum_i \#(\mathcal{B}_i)_k = \#(\mathcal{B})_k$). Θ' is a J-unitary diagonal operator.

PROOF The proof is the same as that of theorem 14.6, but now refers to proposition 14.7. □

It remains to investigate the structure of an elementary J-unitary section.

Elementary Θ sections

We now describe the factorization of an elementary J-unitary section of local degree at most equal to 1 into (J-unitary) Givens rotations. The resulting *structure* of the factored section is the same as in the unitary case, because the same sequence of operations is used to do the factorization. However, the *type* of each elementary operation is now either a unitary or a J-unitary Givens rotation. To keep the discussion manageable, we assume from now on that all state signatures are positive, as this will be the case in our future application.

As in the unitary case, we assume that a J-isometric column $[A_{11}^* \;\; B_1^*]^* \in \mathcal{D}$ is given, where each matrix $(A_{11})_k$ of the diagonal has dimensions at most equal to 1. This column is extended to a J-unitary realization $\hat{\mathbf{\Theta}}_1$, to be obtained in factored form. It is sufficient at this point to look only at the factorization of a single stage of the degree-1 section. With $\#(\mathcal{B}_1)_k \in \{0,1\}$ and $\#(\mathcal{B}_1)_{k+1} \in \{0,1\}$, the four possible sections in a stage are again described by the dimension of $(A_{11})_k$ as

$$\left\{ \begin{array}{llll} C(0): & 0\times 0, & S: & 1\times 0, \\ C(1): & 1\times 1, & G: & 0\times 1. \end{array} \right.$$

The cases $C(0)$ and S result in the same (trivial) sections as before:

$$(\hat{\mathbf{\Theta}}_1)_k = \left[\begin{array}{c|c} \cdot & \cdot \\ \hline \cdot & I \end{array}\right] \qquad \text{resp.} \qquad (\hat{\mathbf{\Theta}}_1)_k = \left[\begin{array}{c|cc} \cdot & 1 & 0 \\ \hline \cdot & 0 & I \end{array}\right]$$

(see figure 14.8(a),(b)). The case $C(1)$ is more interesting and follows from a factorization with Givens rotations of vectors of the form

$$\left[\begin{array}{c} a \\ \hline b_+ \\ b_- \end{array}\right] = \left[\begin{array}{c} (A_{11})_k \\ (B_1)_k \end{array}\right],$$

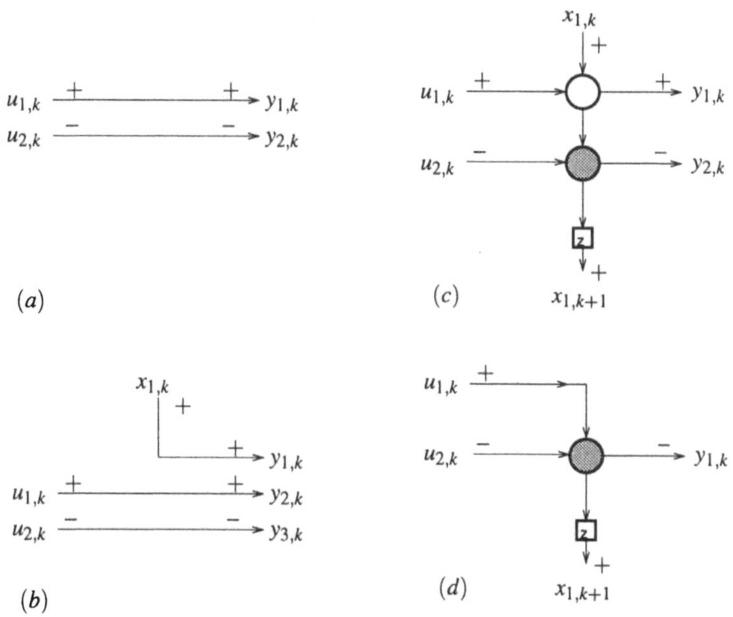

Figure 14.8. Elementary J-unitary sections in a stage. (a) $C(0)$ constant section with zero states, (b) S section, going from 1 state to 0, (c) $C(1)$ section with a constant number of 1 states, (d) G section, going from 0 to 1 state. The number of inputs/outputs have arbitrarily been set to 2: one with positive signature, the other with negative signature. The shaded circles represent J-unitary Givens rotations.

where a is a scalar and $b = \left[\begin{smallmatrix} b_+ \\ b_- \end{smallmatrix}\right]$ is partitioned according to the signature of the inputs at that point. The factorization is obtained in two steps,

$$(\hat{\mathbf{\Theta}}_{1,1}^{-1})_k \begin{bmatrix} a \\ \hline b_+ \\ b_- \end{bmatrix} = \begin{bmatrix} a' \\ \hline 0 \\ b_- \end{bmatrix}, \qquad (\hat{\mathbf{\Theta}}_{1,2}^{-1})_k \begin{bmatrix} a' \\ \hline 0 \\ b_- \end{bmatrix} = \begin{bmatrix} a'' \\ \hline 0 \\ 0 \end{bmatrix}.$$

Here, $(\hat{\mathbf{\Theta}}_{1,1})_k$ consists solely of unitary Givens rotations, used to cancel the entries of b_+ against a, while $(\hat{\mathbf{\Theta}}_{1,2})_k$ consists only of J-unitary Givens rotations. See figure 14.8(c). Note that the unitary scattering operator $(\hat{\mathbf{\Sigma}}_{1,1})_k$ corresponding to $(\hat{\mathbf{\Theta}}_{1,1})_k$ is the same because it is already unitary: $(\hat{\mathbf{\Sigma}}_{1,1})_k = (\hat{\mathbf{\Theta}}_{1,1})_k$. The factorization of a G section obviously results in a comparable structure, and can also be described as $(\hat{\mathbf{\Theta}}_1)_k = (\hat{\mathbf{\Theta}}_{1,1})_k(\hat{\mathbf{\Theta}}_{1,2})_k = (\hat{\mathbf{\Sigma}}_{1,1})_k(\hat{\mathbf{\Theta}}_{1,2})_k$. As the same can obviously be done for the $C(0)$- and the S sections, the overall result is as follows.

Lemma 14.9 *Let* $[A_{11}^* \;\; B_1^*]^* \in \mathcal{D}(\mathcal{B}_1 \times \mathcal{M}, \mathcal{B}_1^{(-1)})$ *be* $\{I, J_{\mathcal{M}}\}$*-isometric:*

$$[A_{11}^* \;\; B_1^*] \begin{bmatrix} I & \\ & J_{\mathcal{M}} \end{bmatrix} \begin{bmatrix} A_{11} \\ B_1 \end{bmatrix} = I,$$

and assume that its state dimension sequence $\mathcal{B}_1$ has dimension at most equal to 1 *at each point. Then this column has a J-unitary extension to* $\hat{\mathbf{\Theta}}_1 \in \mathcal{D}(\mathcal{B}_1 \times \mathcal{M}, \mathcal{B}_1^{(-1)} \times \mathcal{N})$ *such that*

$$\begin{aligned} \hat{\mathbf{\Theta}}_1 &= \hat{\mathbf{\Theta}}_{1,1}\hat{\mathbf{\Theta}}_{1,2} = \hat{\mathbf{\Sigma}}_{1,1}\hat{\mathbf{\Theta}}_{1,2} \\ &= \left[\begin{array}{c|cc} \times & \times & \\ \hline \times & \times & \\ & & I \end{array}\right]\left[\begin{array}{c|cc} \times & & \times \\ \hline & I & \\ \times & & \times \end{array}\right] \end{aligned}$$

(where partitionings are according to $J_{\mathcal{M}}$).

With theorem 14.8, the result is that if Θ is a J-unitary operator which has a J-unitary realization $\mathbf{\Theta}$ with state signature sequence $J_{\mathcal{B}} = I$, then $\mathbf{\Theta}$ has a factorization into unitary and J-unitary factors as

$$\mathbf{\Theta} = [\mathbf{\Sigma}_{1,1}\mathbf{\Theta}_{1,2}] \cdot [\mathbf{\Sigma}_{2,1}\mathbf{\Theta}_{2,2}] \cdots [\mathbf{\Sigma}_{n,1}\mathbf{\Theta}_{n,2}] \cdot \mathbf{\Theta}' . \tag{14.27}$$

Lemma 14.10 *If $\mathbf{\Theta}$ has factorization (14.27), then the corresponding $\mathbf{\Sigma}$ has factorization*

$$\mathbf{\Sigma} = [\mathbf{\Sigma}_{1,1}\mathbf{\Sigma}_{2,1}\cdots\mathbf{\Sigma}_{n,1}]\, \mathbf{\Sigma}'\, [\mathbf{\Sigma}_{n,2}\cdots\mathbf{\Sigma}_{2,2}\mathbf{\Sigma}_{1,2}] \tag{14.28}$$

in which $\mathbf{\Theta}_{i,2} \leftrightarrow \mathbf{\Sigma}_{i,2}$, $\mathbf{\Theta}' \leftrightarrow \mathbf{\Sigma}'$.

PROOF We first argue that $\mathbf{\Theta}$ in (14.27) can be written as

$$\mathbf{\Theta} = [\mathbf{\Sigma}_{1,1}\mathbf{\Sigma}_{2,1}\cdots\mathbf{\Sigma}_{n,1}] \cdot [\mathbf{\Theta}_{1,2}\mathbf{\Theta}_{2,2}\cdots\mathbf{\Theta}_{n,2}] \cdot \mathbf{\Theta}' \tag{14.29}$$

Indeed, because $\mathbf{\Sigma}_{i,1}$ and $\mathbf{\Theta}_{j,2}$, for $i \neq j$, act on different state variables and on different inputs, their order of application may be reversed: $\mathbf{\Theta}_{j,2}\mathbf{\Sigma}_{i,1} = \mathbf{\Sigma}_{i,1}\mathbf{\Theta}_{j,2}$. This allows to transform (14.27) into (14.29). Omitting the details, we note that the transition from a Θ-representation to a Σ-representation is obtained by reversing the computational direction of the secondary inputs and outputs. This does not affect $[\mathbf{\Sigma}_{1,1}\mathbf{\Sigma}_{2,1}\cdots\mathbf{\Sigma}_{n,1}]$ as only the primary inputs and outputs are involved, while $[\mathbf{\Theta}_{1,2}\mathbf{\Theta}_{2,2}\cdots\mathbf{\Theta}_{n,2}] \cdot \mathbf{\Theta}' \leftrightarrow \mathbf{\Sigma}' \cdot [\mathbf{\Sigma}_{n,2}\cdots\mathbf{\Sigma}_{2,2}\mathbf{\Sigma}_{1,2}]$. This leads to equation (14.28). □

The structure of Θ according to the above factorization of $\mathbf{\Theta}$ is depicted in figure 14.9(*a*). It is the same as the structure of the network of Σ given in figure 14.6(*b*), but contains both unitary and J-unitary rotations (represented by shaded circles). The structure of Σ corresponding to this factorization of $\mathbf{\Theta}$ (figure 14.9(*b*)) is again the same, but the order in which computations are done is not only from left to right, but partially also from right to left. Within a single stage, suppose that the inputs and the current state variables are known. In order to compute the next states and the outputs, first all rotations going from left to right have to be performed, and only then the next state variables and the output at the left can be computed. The network is said to be *non-pipelinable*, and the computational dependency, going from the left to the right and back to the left again, is said to be the computational bottleneck. This bottleneck is not present in the network in figure 14.6, and hence, from a computational point of view, a direct factorization of $\mathbf{\Sigma}$ yields a more attractive network.

Note that this network of Σ is a special case of the type of networks that has been obtained in the model reduction problem (*cf.* figure 10.10). In chapter 10, more general networks were obtained because the state signature of Θ was allowed to contain negative entries too.

Θ-*based cascade factorization of* T

Let $T \in \mathcal{U}$ be a given strictly contractive locally finite transfer operator. The process of realizing T via a Θ-based cascade starts with the orthogonal embedding of T in a unitary operator Σ, such that

$$\Sigma = \begin{bmatrix} \Sigma_{11} & T \\ \Sigma_{21} & \Sigma_{22} \end{bmatrix} \tag{14.30}$$

where we have set $\Sigma_{12} = T$. The next step is to convert Σ to Θ, which requires the invertibility of Σ_{22}:

$$\Theta = \begin{bmatrix} \Sigma_{11} - \Sigma_{12}\Sigma_{22}^{-1}\Sigma_{21} & -\Sigma_{12}\Sigma_{22}^{-1} \\ \Sigma_{22}^{-1}\Sigma_{21} & \Sigma_{22}^{-1} \end{bmatrix}$$

Θ is an upper operator only if Σ_{22}^{-1} is upper. As the factorization of Θ in the previous subsection required Θ to be upper (so that it has a causal realization), we see that Σ_{22} should be outer and invertible in order to obtain a Θ-based cascade factorization of Σ. If this requirement is satisfied, then a J-unitary realization $\mathbf{\Theta}$ of Θ is obtained in terms of a unitary realization $\mathbf{\Sigma}$ of Σ as

$$\mathbf{\Sigma} = \left[\begin{array}{c|cc} A & C_1 & C_2 \\ \hline B_1 & D_{11} & D_{12} \\ B_2 & D_{21} & D_{22} \end{array}\right] \;\Rightarrow\; \mathbf{\Theta} = \left[\begin{array}{c|cc} A - C_2D_{22}^{-1}B_2 & C_1 - C_2D_{22}^{-1}D_{21} & -C_2D_{22}^{-1} \\ \hline B_1 - D_{12}D_{22}^{-1}B_2 & D_{11} - D_{12}D_{22}^{-1}D_{21} & -D_{12}D_{22}^{-1} \\ D_{22}^{-1}B_2 & D_{22}^{-1}D_{21} & D_{22}^{-1} \end{array}\right]. \tag{14.31}$$

Note that if Σ_{22}^{-1} would not be upper, then we would by necessity obtain $\ell_{A_\Theta} > 1$ at this point. The factorization proceeds with a state transformation to make A_Θ upper triangular at each stage, which requires the time-varying Schur decomposition discussed in section 14.3. $\mathbf{\Theta}$ is subsequently factored into elementary sections, and conversion to scattering operators finally produces a factorization of $\mathbf{\Sigma}$ as in equation (14.28), and in a computational network as in figure 14.9(b). In this figure, T is the transfer operator $u \to y$ if the inputs at the right are put to zero.

However, the above is only possible when Σ_{22}^{-1} is outer and invertible. With Σ given as (14.30), when is this the case? A necessary condition for invertibility is that $\Sigma_{22}^*\Sigma_{22} \gg 0$, and since $\Sigma_{22}^*\Sigma_{22} = I - T^*T$, it follows that T must be strictly contractive. In this case, proposition 12.13 has shown that the embedding algorithm yields Σ_{22} as an outer spectral factor of $I - T^*T$. Hence, if T is strictly contractive, Σ_{22} is outer and invertible automatically, and T has a Θ-based cascade realization. This is the reason why we have put $\Sigma_{12} = T$ in equation (14.30).

The Θ-based cascade network of Σ represents a filter structure which is well known in its time-invariant incarnation. In this context, one typically chooses $\Sigma_{11}(z) = T(z)$, because then the transmission zeros of $\Sigma(z)$, the zeros of $\Sigma_{11}(z)$, are equal to those of $T(z)$. Simultaneously, the zeros of $\Sigma_{22}(z)$ are directly related to those of $\Sigma_{11}(z)$ (they are

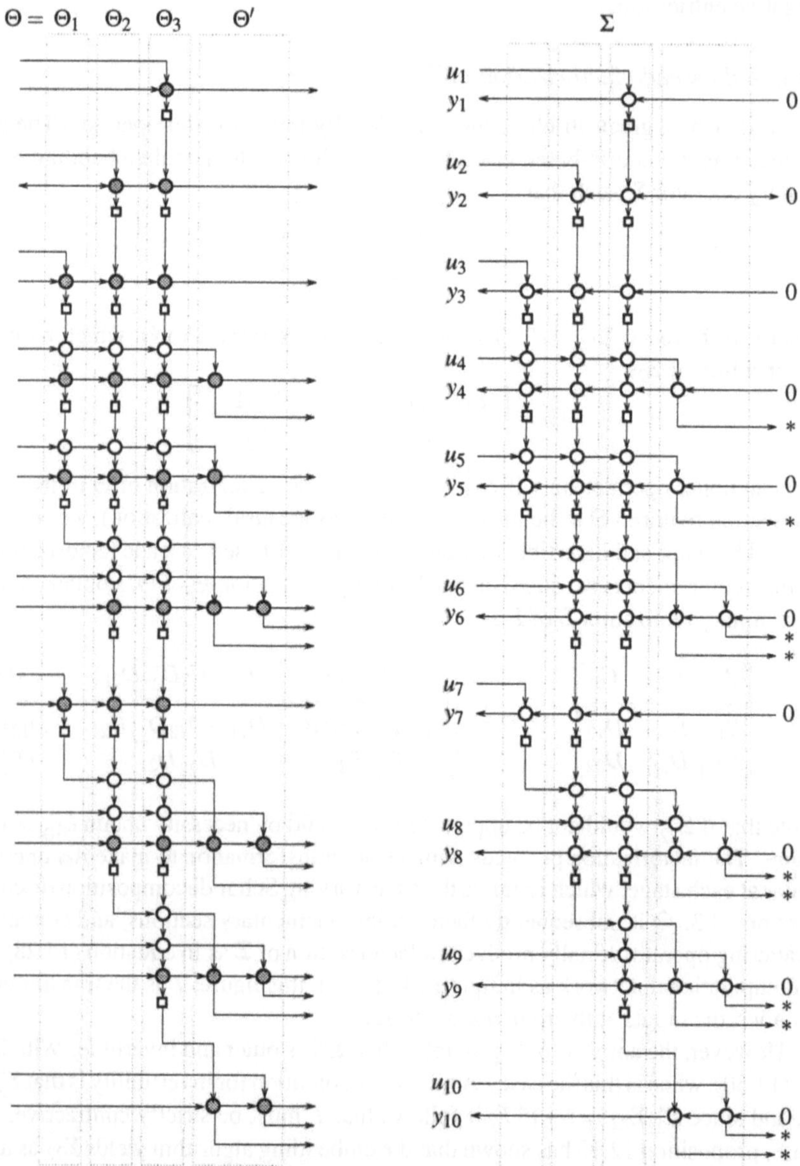

Figure 14.9. *(a)* a J-unitary cascade factorization has the same structure as a unitary cascade factorization, but contains J-unitary rotations (shaded circles), *(b)* Lossless embedding and Θ-cascade factorization of a strictly contractive upper operator $T : u \to y$.

'reflected' in the unit circle). The point of using this filter structure is that these zeros appear as the zeros of the individual sections of the cascade, and hence they are individually determined directly by the parameters of the corresponding section, rather than by the combined effect of all parameters. It follows that the zeros of $T(z)$ are highly insensitive to parameter changes of the cascade, which makes the construction of filters with a well-defined stopband possible, even if approximate parameters (finite word-length implementations) are used.

However, note that in the time-varying case, using the above-described procedure, it is not possible to choose $\Sigma_{11} = T$, because Σ_{22} will in general not be outer and in this case A_Θ in (14.31) is not stable: $\ell_{A_\Theta} > 1$. In the time-invariant case, this does not pose real problems: even with the eigenvalues of A_Θ larger than 1, it is possible to factor Θ in the same way as before, which ultimately results in a stable cascade filter back in the scattering domain. There is no apparent reason why the same would not work in the time-varying domain: currently, the limitation seems to lie in the fact that we always require our realizations to be stable, in order to associate a transfer operator to it via $(I-AZ)^{-1}$. The foregoing factors provide reason to investigate (in other research) cases where the A-matrix contains both a stable and an anti-stable part. Because of state transformations, these parts can become mixed, and one of the first issues to address would be, given an A operator, to decouple it into stable and anti-stable parts.

15 CONCLUSION

As a concluding chapter of this book we offer some thoughts on the likeness and the differences between linear time invariant and linear time varying systems, and a short summary of possible applications beyond the realm of the computational theory that we have presented.

On the likeness and difference between LTI and LTV

It is sometimes said that "linear time-varying system theory is but a slight extension of the time-invariant case". Such a sweeping proposition has only a very partial claim to truth! While it is true that a number of methods carry over from an LTI to an LTV context, it is also true that central (and fairly deep) properties of LTI systems do not hold in the LTV case—in other words: the LTV case is considerably richer. Having got to the end of this book we can appreciate the similarities and the differences and give a reasoned account of them.

Let us start with *realization theory*. In both LTI and LTV theory, the Hankel operator plays a central role. Its minimal factorization into a reachability and an observability operator allows to derive the state realization operators $\{A,B,C\}$. In the LTI case, the Hankel operator has a classical Hankel structure: its matrix representation has the form

$$\begin{bmatrix} h_1 & h_2 & h_3 & \cdots \\ h_2 & h_3 & & \\ h_3 & & & \\ \vdots & & & \ddots \end{bmatrix}.$$

In the LTV case, the Hankel operator is actually a tensor with three indices, the third index being necessary because of its time-varying nature. The typical Hankel structure is not any more a simple algebraic condition of equality between specific entries in a matrix but among entries of the tensor. At the same time, a factorization of the "snapshots" of the operator, obtained by keeping one of the indices of the tensor fixed, gives rise to range conditions from which a realization can be derived. Range properties are capable of characterizing a Hankel operator also in the LTI case, a fact that has been exploited to great advantage in modern system identification theory under the name 4SID: state space subspace system identification (see *e.g.*, [Vib95]). In the LTV case, however, it is this property that really counts and whose exploitation yields the desired realization, as was shown in chapters 3 and 5. An interesting corollary is the fact that if an u.e. stable LTI system has a minimal realization of a certain degree, then there will not exist an LTV realization for it of lower degree (not even locally). In this respect, LTV does not provide much more freedom as far as realization theory is concerned!

Moving to the *realization theory for inner and J-inner operators* (chapters 6, 7 and 8), significant differences between LTI and LTV appear. In the LTV case it is conceivable, even quite common, to find a transfer operator with finite state space and a unitary realization, but which is not inner. This situation cannot occur in the LTI case where we can show that systems with a finite dimensional orthogonal state space realization are necessarily inner. Defective cases show up much more easily in the LTV case. A case in point is an example where the finite-degree LTV system starts out as an LTI system (for large, negative values of t), changes around $t = 0$, and finally stabilizes again to LTI for large positive values of t. If the number of zeros inside the unit circle of the two extreme LTI systems are different, then the isometric operator in an inner-outer factorization will not be inner. The reason is the presence of a doubly shift invariant subspace (see chapter 7). The situation is not at all exotic and manifests itself already in simple examples. The corresponding LTI case involves the notion of "full range" shift invariant spaces, with great system theoretic and analytic importance, but of a very different nature: the defective case necessarily entails systems with infinite dimensional state spaces which in the discrete time context will be rather exotic. Be that as it may, it turns out that the lack of doubly invariant defect spaces plays an important role in inner embedding theory in both cases equally.

Most of the *approximation theory* for systems is based on constrained interpolation theory—see chapters 9 and 10. It turns out that here LTI and LTV parallel each other. In fact, LTV throws a new light on the LTI theory which was previously based on analytic properties of transfer functions (as in the classical paper of Adamyan, Arov and Krein). Much to our surprise when we developed the theory originally, the time-varying version of the Schur-Takagi theory (and of course all the other versions of much simpler classical interpolation problems) appears to be completely and exclusively algebraic. No analytic property has to be used to derive all the pertinent results. This makes interpolation a smooth and transparent piece of theory of great import for many problems in system theory such as optimal control and model reduction theory.

Spectral factorization theory offers an interesting piece of comparison between the two cases. Although the W-transform provides for a kind of surrogate spectral theory in the LTV case, it is a rather weak tool, mainly of use in interpolation theory. LTV theory thus misses a strong notion of spectrum on which a splitting of "time-varying poles" and zeros with regard to stability can be founded. On the other hand, if calculations are based on inner-outer and spectral factorizations, and expressed in state space terms, then the two have an obvious parallel, as seen from the resulting Riccati equations. In chapters 12 and 13 we have given a closed form solution to the Riccati equation which arises in embedding, directly in terms of the transfer operator to be embedded. In the LTI case, solving the algebraic Riccati equation leads directly to an eigenvalue problem for the related Hamiltonian. No such luxury exists in the LTV case, where the Riccati equation is in name recursive but can be partially recursive and partially algebraic, or even completely algebraic as in the LTI case, which is anyway a special case. The existence of a closed form solution is of great help not only to show existence of a solution, but also to prove convergence of the recursion to the true solution when started from an approximate initial point.

Finally, *parametrization of state space representations* works equally well for the LTI case as for the LTV case, and according to the same principles. Since the LTI theory is the most contentious, we have worked it out in detail in chapter 14, but the technique applies equally well to the LTV case, and has been inspired by it.

We have presented the development in such a way that the LTI case appears as a special case of the LTV theory—as it should be. Likewise, classical matrix algebra can be viewed as another special case of the LTV theory (disjoint from the LTI case, of course). It is remarkable that a single theory is capable to cover all cases. Specializing LTV to LTI gives sharper results in some key instances, especially when external or inner-outer factorizations are considered, but in many other cases, LTV works just as well and yields much more general properties. Yet, there are other cases where a specialization to LTI from LTV does not give all results, *e.g.*, in Hankel-norm model reduction for LTI Hankel matrices, the LTV theory applies but one would still have to show that the resulting approximant is LTI.

Applications

A relatively weak point of LTV theory has been the presumed lack of major applications. Two major reasons for this are (*i*) the impossibility of identification from a single input-output pair, thus precluding adaptive (tracking) applications unless further assumptions are made, and (*ii*) the absence of a spectral theory (no convenient z-transform). Major results such as a generalized interpolation theory and the corresponding model reduction techniques give new directions but are still very new. Thus, LTV theory has been slow in coming of age, and quite a few related problems were considered intractable by people working in control, signal processing or numerical algebra. Gradually, major applications are now appearing, and we expect many more to come. A short summary:

- model reduction for finite element models of large scale integrated circuits using "Schur type interpolation" [ND91, DN90];

- precalculated control for minimal sensitivity of switched networks, *e.g.*, power distribution systems [Yu96, SV96];

- new preconditioners for calculating eigenvalues of large sparse matrices using Krylov subspace methods (*e.g.*, [Saa96]);
- subspace estimation and tracking; stable large matrix inversion; low complexity inversion of matrices with (multiple) thin bands or other forms of sparseness;
- the design of time-varying filter banks and appropriate inverses for image coding, especially in high quality applications such as medical images [Heu96].

However, many applications are still to be developed, even for the cases just mentioned. Given the present high level of understanding of LTV theory, we believe that many new applications will arise in the coming years, and that they will be based on the sound system theoretical principles that we have tried to develop in this book.

Appendix A
Hilbert space definitions and properties

This appendix contains a brief review of those Hilbert space definitions and results that are relevant to this book. The material in this chapter is basic and can be found in textbooks such as Akhiezer-Glazman [AG81] (which we follow here), Halmos [Hal51], and Fuhrmann [Fuh81, chap. 2]. The main focus is on the properties of subspaces of a Hilbert space.

Linear manifolds

In this section, we consider *complex vector spaces* whose elements ('vectors') are not further specified (they could, for example, be vectors in the usual n-dimensional Euclidean space $\mathbb{C}^n$, or more in general, be infinite-dimensional vectors). In a complex vector space $\mathcal{H}$ two operations are defined: the addition of two elements of $\mathcal{H}$ and the multiplication of an element of $\mathcal{H}$ by a complex number, and $\mathcal{H}$ should contain a unique null element for addition. Elements $f_1, f_2, \cdots, f_n$ in $\mathcal{H}$ are called *linearly independent* if (for complex numbers α_i)

$$\alpha_1 f_1 + \alpha_2 f_2 + \cdots \alpha_n f_n = 0 \quad \Leftrightarrow \quad \alpha_1, \cdots, \alpha_n = 0 .$$

$\mathcal{H}$ is finite dimensional (say n-dimensional) if at most a finite number of n elements are linearly independent. Such spaces are studied in linear algebra and yield a specialization of Hilbert space theory. A set $\mathcal{M}$ of elements of a complex vector space $\mathcal{H}$ is called a *linear manifold* if for all complex scalars α, β,

$$f \in \mathcal{M},\, g \in \mathcal{M} \quad \Rightarrow \quad \alpha f + \beta g \in \mathcal{M}$$

A set $\mathcal{M}$ is called the *direct sum* of a finite number of linear manifolds $\mathcal{M}_k \subset \mathcal{H}$,

$$\mathcal{M} = \mathcal{M}_1 \dotplus \cdots \dotplus \mathcal{M}_n , \tag{A.1}$$

if for every $g \in \mathcal{M}$ there is one and only one expression in the form of a sum

$$g = g_1 + g_2 + \cdots + g_n$$

where $g_k \in \mathcal{M}_k$, and if any sum of this form is in $\mathcal{M}$. $\mathcal{M}$ is a linear manifold itself. A set of n linear manifolds $\{\mathcal{M}_k\}_1^n$ is called linearly independent if

$$f_1 + f_2 + \cdots + f_n = 0 \quad (f_i \in \mathcal{M}_i) \quad \Rightarrow \quad f_1, \cdots, f_n = 0.$$

Linear independence is both a necessary and a sufficient condition for the construction of the direct sum in (A.1).

Metric space

A *metric space* is a set $\mathcal{H}$ for which a *distance* $d(f,g)$ is defined, which satisfies

$$\begin{array}{llll} (i) & d(f,g) & = & d(g,f) > 0 \quad \text{when } f \neq g \\ (ii) & d(f,f) & = & 0 \\ (iii) & d(f,g) & \leq & d(f,h) + d(g,h) \quad \text{(triangle ineq.)} \end{array}$$

A sequence of elements f_n in $\mathcal{H}$ has a strong limit the point $f \in \mathcal{H}$ if

$$\lim_{n\to\infty} d(f_n, f) = 0. \tag{A.2}$$

We write $f_n \to f$, and say that $\{f_n\}$ converges to f in norm. This is called *strong* or *norm convergence*. From (*iii*) it follows that (A.2) implies

$$\lim_{m,n\to\infty} d(f_n, f_m) = 0. \tag{A.3}$$

A sequence $\{f_n\}$ that satisfies (A.3) is called a *Cauchy sequence*. There are metric spaces $\mathcal{H}$ in which a Cauchy sequence $\{f_n\}$ does not necessarily converge to an element of the set: (A.3) does not imply (A.2). If it does, then $\mathcal{H}$ is called *complete*.

A limit point of a set $\mathcal{M} \subset \mathcal{H}$ is any point $f \in \mathcal{H}$ such that any ε-neighborhood $\{g : d(f,g) < \varepsilon\}$ $(\varepsilon > 0)$ of f contains infinitely many points of $\mathcal{M}$. A set that contains all its limit points is said to be *closed*. The process of adding to $\mathcal{M}$ all its limit points is called *closure*, the set yielded is denoted by $\overline{\mathcal{M}}$: the closure of $\mathcal{M}$. A set is *dense* in another set if the closure of the first set yields the second set. As an example, the set of rational numbers is dense in $\mathbb{R}$, for the usual notion of distance.

If in a metric space there is a countable set whose closure coincides with the whole space, then the space is said to be *separable*. In this case, the countable set is *everywhere dense*.

Inner product

A complex vector space $\mathcal{H}$ is an *inner product space* if a functional $(\cdot,\cdot) : \mathcal{H}\times\mathcal{H} \to \mathbb{C}$ is defined such that, for every $f, g \in \mathcal{H}$ and $\alpha_1, \alpha_2 \in \mathbb{C}$,

$$\begin{array}{llll} (i) & (g,f) & = & \overline{(f,g)} \\ (ii) & (\alpha_1 f_1 + \alpha_2 f_2, g) & = & \alpha_1 (f_1,g) + \alpha_2 (f_2,g) \\ (iii) & (f,f) \geq 0; & & (f,f) = 0 \quad \Leftrightarrow \quad f = 0. \end{array}$$

The overbar denotes complex conjugation. The *norm* of $f \in \mathcal{H}$, induced by the inner product, is defined by

$$\|f\|_2 = (f,f)^{1/2}.$$

Some properties that follow from the definitions (*i*)–(*iii*) are

$$\begin{array}{lcll} \|\alpha f\|_2 & = & |\alpha|\cdot\|f\|_2 & (\alpha \in \mathbb{C}) \\ |(f,g)| & \leq & \|f\|_2\cdot\|g\|_2 & \text{(Schwarz's inequality)} \\ \|f+g\|_2 & \leq & \|f\|_2+\|g\|_2 & \text{(triangle inequality).} \end{array}$$

Orthogonality

Two vectors f, g are said to be *orthogonal*, $f \perp g$, if $(f,g) = 0$. Given a set $\mathcal{M}$, we write $f \perp \mathcal{M}$ if for all $m \in \mathcal{M}$, $f \perp m$. A set of vectors $\{f_i\}$ is an *orthogonal set* if for $i \neq j$, $(f_i, f_j) = 0$. A vector f is *normalized* if $\|f\|_2 = 1$. An *orthonormal set* is an orthogonal set of normalized vectors.

Hilbert space

A *Hilbert space* is an inner product space that is complete, relative to the metric defined by the inner product. The prime example of a Hilbert space is the space ℓ_2 of sequences $f = [\cdots\ f_0\ f_1\ f_2\ \cdots] = [f_i]_{-\infty}^{\infty}$ of complex numbers f_i such that $\|f\|_2 < \infty$. The inner product in this space is defined by[1]

$$(f,g) = \sum_{-\infty}^{\infty} f_i \overline{g_i}\,.$$

This space is separable: a countable set whose closure is equal to ℓ_2 is for example the set of all vectors with a finite number of non-zero rational components f_i. The space ℓ_2 is complete, and it is infinite dimensional since the unit vectors

$$\begin{array}{lcl} & \vdots & \\ e_0 & = & [\cdots\ 0\ 1\ 0\ 0\ \cdots] \\ e_1 & = & [\cdots\ 0\ 0\ 1\ 0\ \cdots] \\ e_2 & = & [\cdots\ 0\ 0\ 0\ 1\ \cdots] \\ & \vdots & \end{array} \tag{A.4}$$

are linearly independent.

A *closed* linear manifold in a Hilbert space $\mathcal{H}$ is called a *subspace*. A subspace is itself a Hilbert space. An example of a subspace is, given some vector $y \in \mathcal{H}$, the set $\{x \in \mathcal{H} : (x,y) = 0\}$. (The main issue in proving that this set is a subspace is the proof that it is closed; this goes via the fact that $x_n \to x \Rightarrow (x_n, y) \to (x,y)$. See [AG81].) Given a set $\mathcal{M} \subset \mathcal{H}$, we define

$$\mathcal{M}^\perp = \{x \in \mathcal{H} : (x,y) = 0, \forall y \in \mathcal{M}\}\,.$$

Again, $\mathcal{M}^\perp$ is a subspace. If $\mathcal{M}$ is a subspace, then $\mathcal{M}^\perp$ is called the *orthogonal complement* of $\mathcal{M}$. Given a subspace $\mathcal{M}$ and a vector $f \in \mathcal{H}$, there exists a *unique* vector

[1] The meaning of the infinite sum is defined via a limit process of sums over finite sets, in case these sums converge. See Halmos [Hal51, §7].

$f_1 \in \mathcal{M}$ such that $\| f - f_1 \|_2 < \| f - g \|_2$ for all $g \in \mathcal{M}$ $(g \neq f_1)$. This vector f_1 is called the *component* of f in $\mathcal{M}$, or the *orthogonal projection* of f onto the subspace $\mathcal{M}$. The vector $f_2 = f - f_1$ is readily shown to be orthogonal to $\mathcal{M}$, *i.e.*, $f_2 \in \mathcal{M}^\perp$. With respect to $\mathcal{H}$, we have obtained the decomposition

$$\mathcal{H} = \mathcal{M} \oplus \mathcal{M}^\perp, \tag{A.5}$$

where '$\oplus$' denotes the direct sum ($\dot{+}$) of orthogonal spaces. The orthogonal complement $\mathcal{M}^\perp$ is likewise written as

$$\mathcal{M}^\perp = \mathcal{H} \ominus \mathcal{M}.$$

Projection onto a finite-dimensional subspace

Let $\{e_i\}_1^n$ be a set of n orthonormal vectors in a Hilbert space $\mathcal{H}$, and let $\mathcal{M}$ be the finite-dimensional subspace spanned by linear combinations of the $\{e_i\}$:

$$\mathcal{M} = \{m : m = \alpha_1 e_1 + \alpha_2 e_2 + \cdots + \alpha_n e_n, \ \text{all } \alpha_i \in \mathbb{C}\}.$$

Because the $\{e_i\}$ are linearly independent, any $m \in \mathcal{M}$ can be written as a unique linear combination of the $\{e_i\}$. It immediately follows that $(m, e_i) = \alpha_i$, so that

$$m = \sum_1^n (m, e_i) e_i$$

where $(m, e_i) e_i$ can be regarded as the projection of m onto e_i. Let $f \in \mathcal{H}$, then there is a unique decomposition $f = f_1 + f_2$, with $f_1 \in \mathcal{M}$, $f_2 \in \mathcal{M}^\perp$. Since $(f_2, e_i) = 0$, we have $(f, e_i) = (f_1, e_i)$ and hence

$$f = \sum_1^n (f, e_i) e_i + f_2 \qquad (f_2 \in \mathcal{M}^\perp).$$

Hence the projection of f onto $\mathcal{M}$ is obtained explicitly as $\sum_1^n (f, e_i) e_i$. The projection formula can be extended to infinite dimensional subspaces which are spanned by a countable sequence of orthonormal elements $\{e_i\}_1^\infty$.

Basis

For a given separable Hilbert space $\mathcal{H}$ and sequence of vectors $\{\phi_i\}_1^\infty$ in $\mathcal{H}$, if every subset of $\{\phi_i\}$ is linearly independent and the span of the $\{\phi_i\}$ is dense in $\mathcal{H}$, then $\{\phi_i\}$ is called a *basis*. This means that every vector $f \in \mathcal{H}$ can be expanded in a unique way in a series

$$f = \sum_1^\infty \alpha_i \phi_i = \lim_{n \to \infty} \sum_1^n \alpha_i \phi_i$$

which converges in the norm of $\mathcal{H}$. Such a basis is *complete* [AG81]: a set of vectors in $\mathcal{H}$ is said to be complete if there is no non-zero vector in $\mathcal{H}$ which is orthogonal to every vector in the set.

In a separable Hilbert space, any complete sequence of *orthonormal* vectors $\{e_i\}$ forms a basis. In addition, the cardinalities of two orthonormal bases of a separable Hilbert space are equal: they are at most countably infinite, and if there is a finite orthonormal basis $\{e_i\}_1^n$, then any other orthonormal basis has also n elements. The *dimension* of $\mathcal{H}$ is defined as the number of elements in any complete orthonormal basis. Any subspace of a separable Hilbert space is again separable; the dimension of a subspace is defined in the same way. The dimension of a linear manifold $\mathcal{L}$ is defined to be the dimension of its closure $\overline{\mathcal{L}}$.

If two Hilbert spaces $\mathcal{H}$ and $\mathcal{H}'$ have the same dimension, then they are *isomorphic* in the sense that a one-to-one correspondence between the elements of $\mathcal{H}$ and $\mathcal{H}'$ can be set up, such that, if $f, g \in \mathcal{H}$ and $f', g' \in \mathcal{H}'$ correspond to f, g, then

1. $\alpha f' + \beta g'$ corresponds to $\alpha f + \beta g$;

2. $(f', g')_{\mathcal{H}'} = (f, g)_{\mathcal{H}}$.

In fact, the isometry is defined by the transformation of a complete orthonormal basis in $\mathcal{H}$ into such a basis in $\mathcal{H}'$.

Non-orthogonal basis; Gram matrix

Let $\{f_1, \cdots, f_n\}$ be a set of n vectors in a Hilbert space $\mathcal{H}$. Consider the matrix $\Lambda_n = [(f_i, f_j)]_{i,j=1}^n$ of inner products of the f_i, *i.e.*,

$$\Lambda_n = \begin{bmatrix} (f_1,f_1) & (f_1,f_2) & \cdots & (f_1,f_n) \\ (f_2,f_1) & (f_2,f_2) & & (f_2,f_n) \\ \vdots & & \ddots & \vdots \\ (f_n,f_1) & (f_n,f_2) & \cdots & (f_n,f_n) \end{bmatrix}.$$

The set is orthonormal if $\Lambda_n = I$. It is linearly independent if and only if Λ_n is nonsingular (*i.e.*, invertible). This can readily be shown from the definition of linear independence: let $f = \alpha_1 f_1 + \alpha_2 f_2 + \cdots + \alpha_n f_n$ be a vector in the linear manifold generated by the f_i, and suppose that not all α_i are equal to zero. By definition, the set of vectors is linearly independent if $f = 0 \Rightarrow \alpha_i = 0$ $(i = 1, \cdots, n)$. Because $f = 0 \Rightarrow (f, f_i) = 0$ $(i = 1, \cdots, n)$, we obtain upon substituting the definition of f the set of linear equations

$$\begin{cases} \alpha_1 (f_1,f_1) + \alpha_2 (f_1,f_2) + \cdots + \alpha_n (f_1,f_n) = 0 \\ \vdots \qquad\qquad\qquad\qquad\qquad\qquad \vdots \\ \alpha_1 (f_n,f_1) + \alpha_2 (f_n,f_2) + \cdots + \alpha_n (f_n,f_n) = 0 \end{cases}$$

and hence $\alpha_i = 0$ $(i = 1, \cdots, n)$ follows if and only if Λ_n is invertible.

Λ_n is called the Gram matrix of the set of vectors. Gram matrices play an important role in the analysis of non-orthogonal bases, as is illustrated by the following. Let $\{f_k\}_1^\infty$ be a complete system of vectors in a Hilbert space $\mathcal{H}$, and let Λ_n be the sequence of Gram matrices $\Lambda_n = [(f_i, f_j)]_{i,j=1}^{n}$. If

$$\begin{aligned} \lim_{n\to\infty} \|\Lambda_n\| &< \infty \\ \text{and} \quad & \\ \lim_{n\to\infty} \|\Lambda_n^{-1}\| &< \infty \end{aligned}$$

(where $\|\cdot\|$ denotes the matrix 2-norm), then $\{f_k\}_1^\infty$ is a basis in $\mathcal{H}$ [AG81]. Such a basis is called a Riesz basis. It is said to be equivalent to an orthonormal basis because there is a boundedly invertible transformation (based on Λ) of $\{f_k\}$ to an orthonormal basis. We use only such bases.

Let $\{f_i\}_1^\infty$ be a non-orthogonal basis in $\mathcal{H}$, and let $\{q_i\}_1^\infty$ be an orthonormal basis of $\mathcal{H}$. Then the $\{f_i\}$ can be expressed in terms of the $\{q_i\}$ as

$$f_i = \sum_j R_{ij} q_j, \qquad \text{where} \quad R_{ij} = (f_i, q_j). \tag{A.6}$$

Define $R = [R_{ij}]_{i,j=1}^{\infty}$. The Gram matrix $\Lambda = [(f_i, f_j)]$ can be written in terms of R, using the expansion (A.6), as

$$\Lambda_{ij} = \sum_k R_{ik} (R^*)_{kj}$$

so that $\Lambda = RR^*$. Suppose that both R and R^{-1} are bounded. Then Λ and Λ^{-1} are bounded as well, so that $\{f_i\}$ is a Riesz basis, and the expression $\sum_k (R^{-1})_{ik} R_{kj} = \delta_{ij}$ shows, with (A.6), that each q_i can be written in terms of the $\{f_j\}$:

$$q_i = \sum_j (R^{-1})_{ij} f_j .$$

Hence $\{f_i\}$ can be orthonormalized by R^{-1}, where R is a boundedly invertible factor of Λ.

Bounded linear operators

Let $\mathcal{H}_1$ and $\mathcal{H}_2$ be Hilbert spaces, and let D denote a set in $\mathcal{H}_1$. A function (mapping) T which associates to each element $f \in D$ some element $g = fT$ in $\mathcal{H}_2$ is called an operator. $D = D(T)$ is called the domain of T, while $\operatorname{ran}(T) = \{fT : f \in D\}$ is its range. T is linear if D is a linear manifold and $(\alpha f + \beta g)T = \alpha fT + \beta gT$ for all $f, g \in D$ and all complex numbers α, β. The *norm* of a linear operator T is

$$\|T\| = \sup_{f \in D, f \neq 0} \frac{\|fT\|_2}{\|f\|_2}$$

and T is bounded if $\|T\| < \infty$. A bounded linear operator is continuous: for every $f_0 \in D$,

$$\lim_{f \to f_0} fT = f_0 T \qquad (f \in D).$$

If S is another bounded linear operator such that the product ST is defined, then $\|ST\| \leq \|S\| \cdot \|T\|$.

A linear operator T is finite dimensional if it is bounded and if $\mathrm{ran}(T)$ is a finite-dimensional subspace of $\mathcal{H}$. Let $\{h_k\}$ be a basis in $\mathrm{ran}(T)$, then the operator can be expressed as

$$fT = \sum_{1}^{n} (f, g_k) h_k$$

where $\{g_k\}$ is a finite system of vectors, not depending on f.

Let $T : \mathcal{H}_1 \to \mathcal{H}_2$ be a bounded linear operator defined on the whole of $\mathcal{H}_1$. The adjoint of T is the unique operator $T^* : \mathcal{H}_2 \to \mathcal{H}_1$ with the property that for all $f, g \in \mathcal{H}_1$,

$$(fT, g) = (f, gT^*).$$

T^* always exists and is unique, $(T^*)^* = T$, $(ST)^* = T^*S^*$, and if T^{-1} exists then $(T^{-1})^* = (T^*)^{-1} =: T^{-*}$. T is called self-adjoint if $T = T^*$; a self-adjoint operator is called *positive* if $(fT, f) \geq 0$ for all $f \in \mathcal{H}_1$.

Let $\{e_k\}_1^\infty$ be an orthonormal basis in $\mathcal{H}$. Suppose that the sum $\sum_{k=1}^\infty (e_kT, e_k)$ converges absolutely. Then T is said to be a *nuclear operator* whose *trace* is given by

$$\mathrm{trace}(T) := \sum_{1}^{\infty} (e_kT, e_k),$$

It can be shown that the property and the value of the trace does not depend on the basis chosen [AG81].

The null-space or kernel of a bounded linear operator $T : \mathcal{H}_1 \to \mathcal{H}_2$ is the linear manifold

$$\ker(T) = \{f \in \mathcal{H}_1 : fT = 0\}.$$

This linear manifold is actually closed, hence $\ker(T)$ is a subspace. On the other hand, the range of T is a linear manifold which is not necessarily closed; it is closed if and only if the range of its adjoint is closed. $\mathcal{H}_1$ and $\mathcal{H}_2$ satisfy an orthogonal decomposition as

$$\begin{aligned} \mathcal{H}_1 &= \ker(T) \oplus \overline{\mathrm{ran}(T^*)} \\ \mathcal{H}_2 &= \ker(T^*) \oplus \overline{\mathrm{ran}(T)}. \end{aligned} \tag{A.7}$$

T is said to be *injective* (one-to-one) if $fT = gT \Rightarrow f = g$, which reduces for linear operators to the condition $fT = 0 \Rightarrow f = 0$, *i.e.*, T is injective if and only if $\ker(T) = 0$. Hence if the range of T^* is dense in $\mathcal{H}_1$, then T is one-to-one. T is *surjective* (onto) if its range is all of $\mathcal{H}_2$. T with domain restricted to $\ker(T)^\perp$ maps one-to-one to the closure of its range, but is not necessarily surjective. If T is both injective and surjective, then (by the closed graph theorem [DS63]) it is boundedly invertible.

An operator P is a projection if it satisfies $P^2 = P$. It is called an orthogonal projection if, in addition, $P^* = P$. If $\mathcal{M}$ is a subspace in $\mathcal{H}$, then $\mathcal{H} = \mathcal{M} \oplus \mathcal{M}^\perp$. The orthogonal projector $P_\mathcal{M}$ onto the subspace $\mathcal{M}$ is unique.

The following theorem gives necessary and sufficient conditions for the range of an operator to be closed (*cf.* [Hal51, §21], [Dou66]):

Theorem A.1 *Let T be a bounded operator on a Hilbert space.*

$$\mathrm{ran}(T^*) \text{ is closed} \quad \Leftrightarrow \quad \exists\, \varepsilon > 0: \quad \|xT\| \geq \varepsilon \|x\| \quad \text{for all } x \in \overline{\mathrm{ran}(T^*)}. \tag{A.8}$$

A linear manifold (subspace) $\mathcal{M}$ is called an invariant manifold (subspace) for an operator T if $\mathcal{M}T \subset \mathcal{M}$. $\mathcal{M}$ is invariant for T if and only if $P_{\mathcal{M}}TP_{\mathcal{M}} = TP_{\mathcal{M}}$.

An operator U is called an isometry if it satisfies $UU^* = I$, a co-isometry if $U^*U = I$, and unitary if it satisfies both. If U is unitary, then it is invertible, and $U^{-1} = U^*$. Two Hilbert spaces $\mathcal{H}_1$ and $\mathcal{H}_2$ are isometrically isomorphic if there exists an invertible transformation U such that

$$(fU, gU)_2 = (f, g)_1 \qquad (\text{for all } f, g \in \mathcal{H}_1).$$

In this case, U is unitary.

Transfer function theory

In closing this chapter, let us give some ingredients of classical function theory and harmonic analysis. Consider a causal, time invariant and time discrete system with impulse response $[\cdots\ 0\ \boxed{T_0}\ T_1\ T_2\ \cdots]$, starting at time $k = 0$. We assume that all T_k are $m \times n$ matrices (for easy reading, assume them scalar). The corresponding transfer function is defined by the formal series

$$T(z) = T_0 + zT_1 + z^2T_2 + \cdots \tag{A.9}$$

where z denotes the unit delay (in engineering literature the causal unit delay is usually denoted z^{-1}, for the present discussion the definition given is more convenient.) The purely formal representation allows for formal multiplication of the transfer function with a one-sided input series. If $u = [u_0\ u_1\ u_2\ \cdots]$ is an input sequence, and $y = [y_0\ y_1\ y_2\ \cdots]$ the corresponding output sequence such that $y = uT$, then

$$y_k = \sum_{i=0}^{k} u_i T_{k-i},$$

The same would be obtained if we look at the series $U(z) = u_0 + zu_1 + z^2u_2 + \cdots$ and $Y(z) = y_1 + zy_1 + z^2y_2 + \cdots$, and formally equate $Y(z) = U(z)T(z)$.

In the linear time invariant (LTI) case, the transfer operator corresponding to $T(z)$ is actually given by the Toeplitz operator

$$\mathcal{T}(T(z)) = \begin{bmatrix} \ddots & \ddots & \ddots & \ddots & \ddots & \ddots & \ddots \\ \ddots & 0 & T_0 & T_1 & T_2 & T_3 & \ddots \\ \ddots & 0 & 0 & \boxed{T_0} & T_1 & T_2 & \ddots \\ \ddots & 0 & 0 & 0 & T_0 & T_1 & \ddots \\ \ddots & \ddots & \ddots & \ddots & \ddots & \ddots & \ddots \end{bmatrix}.$$

However, $T(z)$ can also be interpreted as a $m \times n$ matrix function of a complex variable z. The convergence of the series representation for $T(z)$ in the complex plane can then be studied and related to input/output properties of the system whose transfer function is $T(z)$. From the theory of Maclaurin series, we know that if the growth in magnitude of the series $[T_k]_0^\infty$ is sufficiently restricted, then the series will converge to an analytic function inside a disc around the origin, which is also denoted by $T(z)$ but now has the meaning of a complex matrix function. Of course, formal multiplication of series in a symbol z is consistent with the multiplication of Maclaurin series in the intersection of their domains of convergence.

For the benefit of the reader, we recall a few relevant facts from the theory of complex series. For an extensive treatise on the subject, see [WW92], a more compact account of properties is found in [Rud66].

- For a one sided series as given in equation (A.9), there exists a positive number ρ called the *convergence radius* which is such that the series converges absolutely in the open disc $\{z : |z| < \rho\}$ of the complex plane. The series will diverge outside the closed disc $\{z : |z| \leq \rho\}$ (on the circle convergence is dubious). ρ is given by the expression

$$\rho = \lim_{k\to\infty} \|T_k\|^{\frac{1}{k}}$$

  in which $\|T_k\|$ indicates the Euclidean (induced matrix 2-norm) of the matrix T_k.

- Inside the open circle of convergence, $T(z)$ is an *analytic* function of $T(z)$, meaning that the derivative $\frac{\mathrm{d}T(z)}{\mathrm{d}z}$ exists as a complex matrix at each point $\{z : |z| < \rho\}$ (a single complex matrix independent of the direction of $\mathrm{d}z$). The series has a converging termwise derivative in that region as well, *i.e.*,

$$\frac{\mathrm{d}T(z)}{\mathrm{d}z} = \sum_{k=1}^{\infty} kz^{k-1}T_k.$$

  The region of analyticity of $T(z)$ can extend much beyond the radius of convergence, thanks to analytic extension. For example, the series $1 + z + z^2 + \cdots$ has convergence radius 1 but its analytic extension is given by $1/(1-z)$, which is analytic in the whole complex plane except the point $z = 1$. The convergent series corresponding to this $T(z)$ outside the closed unit disc is given by

$$T(z) = -\sum_{k=1}^{\infty} z^{-k}$$

  which is also of the Maclaurin type, but now in the complex variable z^{-1}. None of these two series converge on the unit circle in the usual sense, but they may do so in an extended sense.

  In particular, if $T(z)$ is analytic in the open unit disc $\mathbf{D} = \{z, |z| < 1\}$, then it has a onesided series representation $T(z) = \sum_0^\infty z^k T_k$ which converges in $\mathbf{D}$.

- A one sided input sequence $U(z) = u_0 + zu_1 + z^2u_2 + \cdots$ of the ℓ_2-type is analytic in the open unit circle and is such that the integral

$$\frac{1}{2\pi}\int_{-\pi}^{\pi} \|U(\rho e^{i\theta})\|^2 \mathrm{d}\theta$$

is uniformly bounded for $\rho < 1$ (the integral is a monotonously increasing function ρ). Its limit for $\rho \to 1$ is given by

$$\frac{1}{2\pi}\int_{-\pi}^{\pi} \|U(e^{i\theta})\|^2 d\theta$$

which is known to exist and to equal the ℓ_2-norm of the sequence $[u_k]_0^\infty$. This is a special case of the celebrated Parseval theorem:

$$\|u\|_{\ell_2} = [\sum_{k=0}^{\infty} \|u_k\|^2]^{\frac{1}{2}} = [\frac{1}{2\pi}\int_{-\pi}^{\pi} \|U(e^{i\theta})\|^2 d\theta]^{\frac{1}{2}} = \|U(e^{i\theta})\|_{L_2(\mathbf{T})}$$

in which the last quantity is by definition the L_2-norm for functions on the unit circle $\mathbf{T} = \{z : z = e^{i\theta}\}$ whose squared norm is integrable in the measure $\frac{d\theta}{2\pi}$. Such one-sided functions $U(z)$ are said to belong to the *Hardy space* $H_2(\mathbf{T})$, which can be viewed as the subspace of $L_2(\mathbf{T})$ functions with vanishing Fourier coefficients of strictly negative index. Indeed, such L_2-functions have a unique analytic extension to the open unit disc $\mathbf{D}$, uniquely defined by the corresponding one-sided Fourier expansion. For more information see the introductory survey of [Hof62] and the treatment of Hardy spaces in [Rud66].

- $H_\infty(\mathbf{D})$ is the space of functions $T(z)$ which are uniformly bounded in $\mathbf{D}$:

$$\|T(z)\|_{H_\infty(\mathbf{D})} := \sup_{z\in\mathbf{D}} \|T(z)\| < \infty.$$

On the unit circle, $H_\infty(\mathbf{T})$ is a subspace of the space of essentially bounded, measurable matrix functions $L_\infty(\mathbf{T})$ on the unit circle $\mathbf{T}$ of the complex plane.

- A system is uniformly stable in the "bounded input bounded output" (BIBO) sense for the ℓ_2-norm, if

$$\|T\| = \sup_{u\in\ell_2} \frac{\|uT\|}{\|u\|} < \infty$$

By using Parsevals theorem, it follows that this operator norm is equal to the $L_\infty(\mathbf{T})$-norm.

Harmonic analysis shows that a causal system T is BIBO stable if and only if $T(z)$ is analytic inside the unit disc and uniformly bounded: $T(z)$ is in $H_\infty(\mathbf{D})$ (for the original proof, see [BC49]). In this case, $\sup_{-\pi<\theta\leq\pi} \|T(\rho e^{i\theta})\|$ is a monotonously increasing function of ρ and its norm as a transfer operator is given by

$$\sup_{\rho<1,\, -\pi<\theta\leq\pi} \|T(\rho e^{i\theta})\| = \sup_{-\pi<\theta\leq\pi} \|T(e^{i\theta})\|$$

so that

$$\|T(z)\|_{H_\infty} = \|T(e^{i\theta})\|_{L_\infty} = \|T\|.$$

We see that in this case, the norm of $T = \mathcal{T}(T(z))$ as an input-output operator over ℓ_2-spaces equals the L_∞-norm of the Fourier transform $T(e^{i\theta})$ of $T(z)$ on the unit circle, which, for causal systems described by one-sided transfer functions, is actually the supremum of the norm of $T(z)$ over the open disc $\{z : |z| < 1\}$. The fact that in this case we have $\sup_{|z|\leq 1} \|T(z)\| = \sup_\theta \|T(e^{i\theta})\|$, *i.e.*, the H_∞ norm of $T(z)$ on the unit disc is equal to the L_∞ norm of $T(e^{i\theta})$ on the unit circle actually follows from the celebrated maximum modulus theorem valid in a domain of analyticity. Much more is known about these functions, see *e.g.*, [Hof62].

The danger of misinterpretation resides in partial reversals of this result. It is *not* true that the conditions "$T(z)$ is analytic in the unit disc" and "$\sup_\theta \|T(e^{i\theta})\| < \infty$" entail that $T(z) \in H_\infty$, *i.e.*, T causal and $\|T\| < \infty$. The standard and elementary counterexample is given by

$$T(z) = \exp(\frac{1+z}{1-z}).$$

$T(e^{i\theta}) = \exp(i\cot\frac{\theta}{2})$ so that for all θ, $\|T(e^{i\theta})\| = 1$. However, for $0 < \rho < 1$, we see that $T(\rho) = \exp\frac{1+\rho}{1-\rho} \to \infty$ as $\rho \to 1$! Note that quite the opposite is true for $T(z) = \exp(\frac{z+1}{z-1})$ which does correspond to a BIBO stable system. $T(z) = \exp(\frac{1+z}{1-z})$ should be interpreted as a bounded but anticausal transfer function.

Projections of L_∞-functions of the unit circle onto their causal or anticausal parts may produce similar kinds of problems. Suppose that $T(e^{i\theta}) = \sum_{k=-\infty}^{\infty} T_k e^{ik\theta}$ is some (double sided) Fourier series and consider the "projection to causal" given by

$$\mathbf{P}(T)(e^{i\theta}) = \sum_{k=0}^{\infty} T_k e^{ik\theta}.$$

It is *not* true that $\|T(e^{i\theta})\|_{L_\infty} < \infty \Rightarrow \|\mathbf{P}(T)(e^{i\theta})\|_{L_\infty} < \infty$. This fact is fundamental to harmonic analysis and symptomatic for the relation between the time domain and the frequency or spectral domain. The most classical example is perhaps the ideal low pass filter. Assume that $T(e^{i\theta})$ is real and specified by $T(e^{i\theta}) = 1$ for $-\frac{\pi}{2} \leq \theta \leq \frac{\pi}{2}$ and zero for other values of θ. We have $T_0 = \frac{1}{2}$ and for $k \neq 0$

$$T_k = \int_{-\pi}^{\pi} T(e^{i\theta}) e^{-ik\theta} \frac{d\theta}{2\pi} = \int_{-\pi/2}^{\pi/2} e^{-ik\theta} \frac{d\theta}{2\pi} = \frac{1}{\pi k} \sin(\frac{\pi k}{2}).$$

In matrix form, the corresponding transfer operator is

$$T = \frac{1}{\pi} \begin{bmatrix} \ddots & \ddots & \ddots & \ddots & \ddots & \ddots & \ddots & \ddots & \ddots & \ddots & \ddots \\ \ddots & 0 & 1 & \frac{\pi}{2} & 1 & 0 & -\frac{1}{3} & 0 & \frac{1}{5} & 0 & \ddots \\ \ddots & -\frac{1}{3} & 0 & 1 & \frac{\pi}{2} & 1 & 0 & -\frac{1}{3} & 0 & \frac{1}{5} & \ddots \\ \ddots & 0 & -\frac{1}{3} & 0 & 1 & \boxed{\frac{\pi}{2}} & 1 & 0 & -\frac{1}{3} & 0 & \ddots \\ \ddots & \frac{1}{5} & 0 & -\frac{1}{3} & 0 & 1 & \frac{\pi}{2} & 1 & 0 & -\frac{1}{3} & \ddots \\ \ddots & 0 & \frac{1}{5} & 0 & -\frac{1}{3} & 0 & 1 & \frac{\pi}{2} & 1 & 0 & \ddots \\ \ddots & \ddots & \ddots & \ddots & \ddots & \ddots & \ddots & \ddots & \ddots & \ddots & \ddots \end{bmatrix}.$$

The projection $\mathbf{P}(T)$ is given by the series

$$\begin{aligned}\mathbf{P}(T)(z) &= \tfrac{1}{2}+\tfrac{1}{\pi}(z-\tfrac{1}{3}z^3+\tfrac{1}{5}z^5-\cdots)\\ &= \tfrac{1}{2}+\tfrac{1}{\pi}\arctan z\end{aligned}$$

since $\frac{d}{dz}(z-\frac{1}{3}z^3+\frac{1}{5}z^5+\cdots)=\frac{1}{1+z^2}$. It has an essential singularity at the points $z=\pm i$ on the unit circle ($i=\sqrt{-1}$), and hence neither belongs to L_∞ nor to H_∞, although it is analytic in the unit disc. Hence we see that $\mathbf{P}(T)$ is unbounded in the operator norm, while T is perfectly bounded.

References

[AAK71] V.M. Adamjan, D.Z. Arov, and M.G. Krein, "Analytic properties of Schmidt pairs for a Hankel operator and the generalized Schur-Takagi problem," *Math. USSR Sbornik*, **15**(1):31–73, 1971 (transl. of *Iz. Akad. Nauk Armjan. SSR Ser. Mat. 6 (1971)).*

[AD90] D. Alpay and P. Dewilde, "Time-varying signal approximation and estimation," In M.A. Kaashoek, J.H. van Schuppen, and A.C.M. Ran, editors, *Signal Processing, Scattering and Operator Theory, and Numerical Methods*, volume III of *Proc. Int. Symp. MTNS-89*, pp. 1–22. Birkhäuser Verlag, 1990.

[ADD90] D. Alpay, P. Dewilde, and H. Dym, "Lossless Inverse Scattering and reproducing kernels for upper triangular operators," In I. Gohberg, editor, *Extension and Interpolation of Linear Operators and Matrix Functions*, volume 47 of *Operator Theory, Advances and Applications*, pp. 61–135. Birkhäuser Verlag, 1990.

[ADM82] H. Ahmed, J. Delosme, and M. Morf, "Highly concurrent computing structures for matrix arithmetic and signal processing," *Computer*, pp. 65–82, January 1982.

[AG81] N.I. Akhiezer and I.M. Glazman, *"Theory of Linear Operators in Hilbert Space,"* volume I and II. Pitman Publishing Ltd, London, 1981.

[AHD74] B.D.O. Anderson, K.L. Hitz, and N.D. Diem, "Recursive algorithm for spectral factorization," *IEEE Trans. Circuits Syst.*, **21**(6):742–750, 1974.

[AK74] H.B. Aasnaes and T. Kailath, "Initial-condition robustness of Linear Least-Squares filtering algorithms," *IEEE Trans. Automat. Control*, **19**:393–397, August 1974.

[AM69] B.D.O. Anderson and J.B. Moore, "New results in linear system stability," *SIAM J. Control*, **7**(3):398–414, August 1969.

[AM79] B.D.O. Anderson and J.B. Moore, *"Optimal Filtering,"* Prentice Hall, 1979.

[AM81] B.D.O. Anderson and J.B. Moore, "Detectability and stabilizability of time-varying discrete-time linear systems," *SIAM J. Control and Optimization*, **19**(1):20–32, January 1981 Comments in *IEEE Trans. Automat. Control,* vol. 37, no. 3, 1992, pp. 409-410.

[AM92] B.D.O. Anderson and J.B. Moore, "Comments on "Stabilizability and detectability of discrete-time time-varying systems"," *IEEE Trans. Automat. Control*, **37**(3):409–410, March 1992.

[And67] B.D.O. Anderson, "An algebraic solution to the spectral factorization problem," *IEEE Trans. Automat. Contr.*, **12**:410–414, 1967.

[Arv75] W. Arveson, "Interpolation problems in nest algebras," *J. Functional Anal.*, **20**:208–233, 1975.

[AV73] B.D.O. Anderson and S. Vongpanitlerd, *"Network Analysis and Synthesis,"* Prentice Hall, 1973.

[AY89] T.Ya. Azizov and I.S. Yokhvidov, *"Linear Operators in Spaces with an Indefinite Metric,"* Pure and Applied Mathematics. John Wiley & Sons, 1989.

[BAGK94] A. Ben-Artzi, I. Gohberg, and M.A. Kaashoek, "Exponentially dominated infinite block matrices of finite Kronecker rank," *Int. Eq. Operator Th.*, **17**(4), 1994.

[BC49] S. Bochner and K. Chandrasekharan, *"Fourier Transforms,"* Princeton Univ. Press, Princeton, NJ, 1949.

[BCHM74] F. Burns, D. Carlson, E. Haynsworth, and T. Markham, "Generalized inverse formulas using Schur complements," *SIAM J. Applied Math.*, **26**:254–259, 1974.

[BCN88] S. Bittanti, P. Colerani, and G. De Nicolao, "The difference periodic Riccati equation for the periodic prediction problem," *IEEE Tr. Automat. Control*, **33**:706–712, 1988.

[BD80] A. Bultheel and P.M. Dewilde, "On the Adamjan-Arov-Krein approximation, identification, and balanced realization," In *Proc. 1980 Eur. Conf. on Circ. Th. and Design*, volume 2, pp. 186–191, 1980.

[Bel68] A.V. Belevitch, *"Classical Network Theory,"* Holden Day, San Francisco, 1968.

[Beu49] A. Beurling, "On two problems concerning linear transformations in Hilbert space," *Acta Math.*, **81**:239–255, 1949.

[BG81] A. Bunse-Gerstner, "An analysis of the HR algorithm for computing the eigenvalues of a matrix," *Lin. Alg. Appl.*, **35**:155–173, 1981.

[BGD92] A. Bojanczyk, G. Golub, and P. Van Dooren, "The periodic Schur decomposition. Algorithms and applications," In F.T. Luk, editor, *Proc. SPIE, "Advanced Signal Processing Algorithms, Architectures, and Implementations", III*, volume 1770, pp. 31–42, San Diego, July 1992.

[BGK79] H. Bart, I. Gohberg, and M.A. Kaashoek, *"Minimal Factorization of Matrix and Operator Functions,"* Birkhäuser Verlag, Basel, 1979.

[BGK92a] J.A. Ball, I. Gohberg, and M.A. Kaashoek, "Nevanlinna-Pick interpolation for time-varying input-output maps: the discrete case," In I. Gohberg, editor, *Time-Variant Systems and Interpolation*, volume 56 of *Operator Theory: Advances and Applications*, pp. 1–51. Birkhäuser Verlag, 1992.

[BGK92b] J.A. Ball, I. Gohberg, and M.A. Kaashoek, "Nevanlinna-Pick interpolation for time-varying input-output maps: the continuous time case," In I. Gohberg, editor, *Time-Variant Systems and Interpolation*, volume 56 of *Operator Theory: Advances and Applications*, pp. 52–89. Birkhäuser Verlag, 1992.

[BGKD80] H. Bart, I. Gohberg, M.A. Kaashoek, and P. Van Dooren, "Factorization of transfer functions," *SIAM J. Control and Optimization*, **18**(6):675–696, November 1980.

[BGR90] J.A. Ball, I. Gohberg, and L. Rodman, *"Interpolation of Rational Matrix Functions,"* volume 45 of *Operator Theory: Advances and Applications*. Birkhäuser Verlag, 1990.

[BH83] J.A. Ball and J.W. Helton, "A Beurling-Lax theorem for the Lie group $U(m,n)$ which contains most classical interpolation theory," *J. Operator Theory*, **9**:107–142, 1983.

[BL58] M. Brodskii and M.S. Livsic, "Spectral analysis of non-self adjoint operators and intermediate systems," *Amer. Math. Soc. Transl. Ser. 2*, **13**:265–346, 1958.

[BL93] M.D. Di Benedetto and P. Lucibello, "Inversion of nonlinear time-varying systems," *IEEE Trans. Automat. Control*, **38**(8):1259–1264, August 1993.

[BLW91] S. Bittanti, A.J. Laub, and J.C. Willems, editors, *"The Riccati Equation,"* Comm. Control Eng. Series. Springer Verlag, 1991.

[BN78] J.R. Bunch and C.P. Nielsen, "Updating the singular value decomposition," *Numerische Mathematik*, **31**:111–129, 1978.

[Bog74] J. Bognar, *"Indefinite Inner Product Spaces,"* Springer Verlag, 1974.

[BR76] F.J. Beutler and W.L Root, "The operator pseudo-inverse in control and systems identification," In M. Zuhair Nashed, editor, *Generalized Inverses and Applications*, pp. 397–494. Academic Press, 1976.

[BS92] C.H. Bischof and G.M. Shroff, "On updating signal subspaces," *IEEE Trans. Signal Proc.*, **40**(1):96–105, January 1992.

[CC92] C.K. Chui and G. Chen, *"Signal Processing and Systems Theory,"* Springer Verlag, Berlin, 1992.

[CH90] T.F. Chan and P.C. Hansen, "Computing truncated singular value decomposition least squares solutions by rank revealing QR-factorizations," *SIAM J. Sci. Stat. Comput.*, **11**(3):519–530, May 1990.

[CH92] T.F. Chan and P.C. Hansen, "Some applications of the rank revealing QR-factorization," *SIAM J. Sci. Stat. Comput.*, **13**(3):727–741, May 1992.

[Cha87] T.F. Chan, "Rank revealing QR factorizations," *Lin. Alg. Appl.*, **88/89**:67–82, 1987.

[CHM74] D. Carlson, E. Haynsworth, and T. Markham, "A generalization of the Schur complement by means of the Moore-Penrose inverse," *SIAM J. Applied Math.*, **26**:169–175, 1974.

[Cho94] C.T. Chou, *"Geometry of Linear Systems and Identification,"* PhD thesis, Trinity College, Cambridge (UK), March 1994.

[Chu89] J. Chun, *"Fast Array Algorithms for Structured Matrices,"* PhD thesis, Stanford Univ., Stanford, CA, 1989.

[CI94] S. Chandrasekaran and I.C.F. Ipsen, "On rank-revealing factorisations," *SIAM J. Matrix Anal. Appl.*, **15**(2):592–622, April 1994.

[CKL87] J. Chun, T. Kailath, and H. Lev-Ari, "Fast parallel algorithms for QR and triangular factorizations," *SIAM J. Sci. Stat. Comp.*, **8**(6):899–913, 1987.

[CSK94] T. Constantinescu, A.H. Sayed, and T. Kailath, "A recursive Schur-based solution of the Four-Block problem," *IEEE Tr. on Automat. Contr.*, **39**(7):1476–1481, July 1994.

[Dar39] S. Darlington, "Synthesis of reactance 4-poles which produce prescribed insertion loss characteristics," *J. Math. Phys.*, **18**:257–355, 1939.

[DBN71] P.M. Dewilde, A.V. Belevitch, and R. Newcomb, "On the problem of degree reduction of a scattering matrix by factorization," *J. Franklin Inst.*, **291**:387–401, May 1971.

[DD80] E. Deprettere and P. Dewilde, "Orthogonal cascade realization of real multiport digital filters," *Circuit Theory and Appl.*, **8**:245–272, 1980.

[DD81a] P. Dewilde and H. Dym, "Lossless chain scattering matrices and optimum linear prediction: The vector case," *Circuit Theory and Appl.*, **9**:135–175, 1981.

[DD81b] P. Dewilde and H. Dym, "Schur recursions, error formulas, and convergence of rational estimators for stationary stochastic sequences," *IEEE Trans. Informat. Th.*, **27**(4):446–461, July 1981.

[DD81c] P. Van Dooren and P.M. Dewilde, "Minimal cascade factorization of real and complex rational transfer matrices," *IEEE Trans. Circuits Syst.*, **28**(5):390–400, May 1981.

[DD84] P. Dewilde and H. Dym, "Lossless inverse scattering, digital filters, and estimation theory," *IEEE Trans. Informat. Th.*, **30**(4):644–662, July 1984.

[DD87] P. Dewilde and E. Deprettere, "Approximative inversion of positive matrices with applications to modeling," In *NATO ASI Series, Vol. F34 on Modeling, Robustness and Sensitivity Reduction in Control Systems*. Springer Verlag, Berlin, 1987.

[DD88] P. Dewilde and E. Deprettere, "The generalized Schur algorithm: Approximation and hierarchy," In *Operator Theory: Advances and Applications*, volume 29, pp. 97–116. Birkhäuser Verlag, 1988.

[DD92] P. Dewilde and H. Dym, "Interpolation for upper triangular operators," In I. Gohberg, editor, *Time-Variant Systems and Interpolation*, volume 56 of *Operator Theory: Advances and Applications*, pp. 153–260. Birkhäuser Verlag, 1992.

[DDN84] P. Dewilde, E.F. Deprettere, and R. Nouta, "Parallel and pipelined VLSI implementation of signal processing algorithms," In H.J. Whitehouse S.Y. Kung and T. Kailath, editors, *VLSI and Modern Signal Processing*. Prentice-Hall, Englewood Cliffs, NJ, 1984.

[DDR84] Ed. F. Deprettere, P.M. Dewilde, and P. Rao, "Orthogonal filter design and VLSI impementation," In *Proc. Int. Conf. Computers, Systems, and Signal Proc.*, pp. 779–790, Bangalore, India, 1984.

[Den75] M.J. Denham, "On the factorization of discrete-time rational spectral density matrices," *IEEE Trans. Automat. Control*, pp. 535–537, 1975.

[Dep81] E. Deprettere, "Mixed-form time-variant lattice recursions," In *Outils et Modèles Mathématiques pour l'Automatique, l'Analyse de Systèmes et le Traitement du Signal*, Paris, 1981. CNRS.

[Des91] U.B. Desai, "A state-space approach to orthogonal digital filters," *IEEE Trans. Circuits Syst.*, **38**(2):160–169, February 1991.

[Dew76] P. Dewilde, "Input-output description of roomy systems," *SIAM J. Control and Optimization*, **14**(4):712–736, July 1976.

[Dew85] P. Dewilde, "Advanced digital filters," In T. Kailath, editor, *Modern Signal Processing*, pp. 169–209. Springer Verlag, 1985.

[Dew91] P.M. Dewilde, "A course on the algebraic Schur and Nevanlinna-Pick interpolation problems," In Ed. F. Deprettere and A.J. van der Veen, editors, *Algorithms and Parallel VLSI Architectures*. Elsevier, 1991.

[Dew95] P. M. Dewilde, "On the synthesis of lossless computational circuits," *Archiv f. Elektronik u. Übertragungstechnik*, **49**(5/6):279–292, September 1995.

[Dew97] P. Dewilde, "Minimal complexity realization of structured matrices," *preprint, Delft University of Technology*, 1997.

[DF97] H. Dym and B. Freydin, "Bitangential interpolation for upper triangular operators," *preprint, Weizmann Institute*, pp. Part I, 43pp, Part II, 24pp, 1997.

[DKV92] P. Dewilde, M.A. Kaashoek, and M. Verhaegen, editors, *"Challenges of a Generalized System Theory,"* volume 40 of *Essays on Physics*. Royal Netherlands Ac. of Arts and Sciences, 1992.

[DN90] P. Dewilde and Z.-Q. Ning, *"Models for large integrated circuits,"* Kluwer, Boston, 1990.

[Dou66] R.G. Douglas, "On majorization, factorization and range inclusion of operators on Hilbert space," *Proc. Amer. Math. Soc.*, **17**:413–415, 1966.

[DS63] N. Dunford and J.T. Schwartz, *"Linear Operators,"* volume 1, 2. Interscience, New York, 1963.

[dS91] C.E. de Souza, "Periodic strong solution for the optimal filtering problem of linear discrete-time periodic systems," *IEEE Trans. Automat. Control*, **36**(3):333–338, 1991.

[DS92] W.N. Dale and M.C. Smith, "Existence of coprime factorizations for time-varying systems—an operator-theoretic approach," In H. Kimura and S. Kodama, editors, *Recent Advances in Mathematical Theory of Systems, Control, Networks and Signal Processing I (Proc. Int. Symp. MTNS-91)*, pp. 177–182. MITA Press, Japan, 1992.

[DvdV93] P.M. Dewilde and A.J. van der Veen, "On the Hankel-norm approximation of upper-triangular operators and matrices," *Integral Eq. Operator Th.*, **17**(1):1–45, 1993.

[DVK78] P. Dewilde, A.C. Vieira, and T. Kailath, "On a generalized Szegö-Levinson realization algorithm for optimal linear predictors based on a network synthesis approach," *IEEE Trans. Circuits Syst.*, **25**(9):663–675, September 1978.

[Dym89] H. Dym, *"J-Contractive Matrix Functions, Reproducing Kernel Hilbert Spaces and Interpolation,"* Number 71 in CBMS reg. conf. ser. American Math. Soc., Providence, 1989.

[e.a87] J.W. Helton e.a., *"Operator Theory, Analytic Functions, Matrices, and Electrical Engineering,"* volume 68 of *CBMS regional conference series*. American Math. Soc., Providence, 1987.

[Eva72] D.S. Evans, "Finite-dimensional realization of discrete-time weighting patterns," *SIAM J. Applied Math.*, **22**:45–67, 1972.

[Fet70] A. Fettweis, "Factorization of transfer matrices of lossless two-ports," *IEEE Trans. Circuit Th.*, **17**:86–94, 1970.

[FM96a] A. Feintuch and A. Markus, "Isometric dilations in nest algebras," *Int. Eq. Oper. Th.*, **26**:346–352, 1996.

[FM96b] A. Feintuch and A. Markus, "The lossless embedding problem for time-varying contractive systems," *Systems and Control Letters*, **28**:181–187, 1996.

[FMKL79] B. Friedlander, M. Morf, T. Kailath, and L. Ljung, "New inversion formulas for matrices classified in terms of their distance from Toeplitz matrices," *Lin. Alg. Appl.*, **23**:31–60, 1979.

[Fos86] L.V. Foster, "Rank and null space calculations using matrix decomposition without column interchanges," *Lin. Alg. Appl.*, **74**:47–71, 1986.

[FS82] A. Feintuch and R. Saeks, *"System Theory: A Hilbert Space Approach,"* Academic Press, 1982.

[Fuh74] P.A. Fuhrmann, "On realizations of linear systems and applications to some questions of stability," *Math. Systems Theory*, **8**:132–141, 1974.

[Fuh75] P.A. Fuhrmann, "Realization theory in Hilbert space for a class of transfer functions," *J. Functional Anal.*, **18**(4):338–349, April 1975.

[Fuh76] P.A. Fuhrmann, "Exact controllability and observability and realization theory in Hilbert space," *J. Math. Anal. Appl.*, **53**(2):377–392, February 1976.

[Fuh81] P.A. Fuhrmann, *"Linear Systems and Operators in Hilbert Space,"* McGraw-Hill, 1981.

[GCP88] K. Glover, R.F. Curtain, and J.R. Partington, "Realization and approximation of linear infinite-dimensional systems with error bounds," *SIAM J. Control and Optimization*, **26**(4):863–898, 1988.

[GDK+83] Y. Genin, P. Van Dooren, T. Kailath, J.M. Delosme, and M. Morf, "On Σ-lossless transfer functions and related questions," *Lin. Alg. Appl.*, **50**:251–275, 1983.

[GH77] R.P. Gilbert and G.N. Hile, "Hilbert function modules with reproducing kernels," *Non-linear Analysis, Methods and Applications*, **1**(2):135–150, 1977.

[Gil63] E.G. Gilbert, "Controllability and observability in multivariable control systems," *SIAM J. Control*, **1**:128–151, 1963.

[GK81a] Y.V. Genin and S.Y. Kung, "A two-variable approach to the model reduction problem with Hankel norm criterion," *IEEE Trans. Circuits Syst.*, **28**(9):912–924, 1981.

[GK81b] W.M. Gentleman and H.T. Kung, "Matrix triangularization by systolic arrays," *Proc. SPIE, Real Time Signal Proc. IV*, **298**:19–26, 1981.

[GKL92] I. Gohberg, M.A. Kaashoek, and L. Lerer, "Minimality and realization of discrete time-varying systems," In I. Gohberg, editor, *Time Variant Systems and Interpolation*, volume OT 56, pp. 261–296. Birkhäuser Verlag, 1992.

[GKvS84] I. Gohberg, M.A. Kaashoek, and F. van Schagen, "Non-compact integral operators with semi-separable kernels and their discrete analogues: Inversion and Fredholm properties," *Int. Eq. Operator Th.*, **7**:642–703, 1984.

[GKW89] I. Gohberg, M.J. Kaashoek, and H.J. Woerdeman, "The band method for positive and strictly contractive extension problems: an alternative version and new applications," *Integral Eq. Operator Th.*, **12**:343–382, 1989.

[GKW91] I. Gohberg, M.A. Kaashoek, and H.J. Woerdeman, "A maximum entropy principle in the general framework of the band method," *J. Functional Anal.*, **95**(2):231–254, February 1991.

[GL95] M. Green and D.J.N. Limebeer, *"Linear Robust Control,"* Prentice Hall, Englewood Cliffs, NJ, 1995.

[Glo84] K. Glover, "All optimal Hankel norm approximations of linear multi-variable systems and their L^∞-error bounds," *Int. J. Control*, **39**(6):1115–1193, 1984.

[GMW91] I. Gohberg, M.A.Kaashoek, and H.J. Woerdeman, "Time variant extension problems of Nehari type and the band method," In C. Foias,

B. Francis, and J.W. Helton, editors, *H^∞-Control Theory (lectures given at the 2nd session of C.I.M.E., Como, June 18-26, 1990)*, Lecture Notes Math. 1496, pp. 309–323. Springer Verlag, 1991.

[GS72] I. Gohberg and A. Semencul, "On the inversion of finite Toeplitz matrices and their continuous analogs," *Mat. Issled.*, **2**:201–233, 1972.

[GS84] G.C. Goodwin and K.S. Sin, *"Adaptive Filtering, Prediction and Control,"* Prentice Hall, Englewood Cliffs, NJ, 1984.

[GV89] G. Golub and C.F. Van Loan, *"Matrix Computations,"* The Johns Hopkins University Press, 1989.

[GvdV96] J. Götze and A.J. van der Veen, "On-line subspace estimation using a Schur-type method," *IEEE Trans. Signal Processing*, **44**(6):1585–1589, June 1996.

[Hal51] P.R. Halmos, *"Introduction to Hilbert Space,"* Chelsea Publ. Comp., NY, 1951.

[Hay91] S. Haykin, *"Adaptive Filter Theory,"* Prentice-Hall, 1991.

[Hel64] H. Helson, *"Lectures on Invariant Subspaces,"* Academic Press, New York, 1964.

[Hel72] J.W. Helton, "The characteristic functions of operator theory and electrical network realization," *Indiana Univ. Math. J.*, **22**(5):403–414, 1972.

[Hel74] J.W. Helton, "Discrete time systems, operator models, and scattering theory," *J. Functional Anal.*, **16**(1):15–38, May 1974.

[Hel76] J.W. Helton, "Systems with infinite-dimensional state space: The Hilbert space approach," *Proceedings of the IEEE*, **64**(1):145–160, January 1976.

[Hel78] J.W. Helton, "Orbit structure of the Möbius transformation semigroup acting on H_∞ (broadband matching)," In *Topics in Functional Analysis*, volume 3 of *Adv. in Math. Suppl. Studies*, pp. 129–133. Academic Press, 1978.

[Hel82] J.W. Helton, "Non-Euclidean functional analysis and electronics," *Bull. of the AMS*, **7**(1):1–64, 1982.

[Heu96] R. Heusdens, "Design of lapped orthogonal transforms," *IEEE Trans. on Image Processing*, **5**(8):1281–1284, August 1996.

[HI93] A. Halanay and V. Ionescu, "Generalized discrete-time Popov-Yakubovich theory," *Systems & Control Letters*, **20**(1):1–6, January 1993.

[HI94] A. Halanay and V. Ionescu, *"Time-Varying Discrete Linear Systems,"* volume 68 of *Operator Theory: Advances and Applications*. Birkhäuser, 1994.

[HJ89] R.A. Horn and C.R. Johnson, *"Topics in Matrix Analysis,"* Cambridge Univ. Press, Cambridge, NY, 1989.

[HK66] B.L. Ho and R.E. Kalman, "Effective construction of linear, state-variable models from input/output functions," *Regelungstechnik*, **14**:545–548, 1966.

[HL94] J.J. Hench and A.J. Laub, "Numerical solution of the discrete-time periodic Riccati equation," *IEEE Tr. Automat. Contr.*, **39**(6):1197–1210, June 1994.

[Hof62] K. Hoffman, *"Banach Spaces of Analytic Functions,"* Prentice-Hall, Englewood Cliffs, NJ, 1962.

[IO96] V. Ionescu and C. Oara, "The time-varying discrete four block Nehari problem: A generalized Popov-Yakubovich type approach," *Int. Eq. Oper. Th.*, **26**:404–431, 1996.

[JD89] K. Jainandunsing and Ed F. Deprettere, "A new class of parallel algorithms for solving systems of linear equations," *SIAM J. Sci. Stat. Comp.*, **10**(5):880–912, 1989.

[JM91] M.R. Jarmasz and G.O. Martens, "A simplified synthesis of lossless cascade analog and digital two-port networks," *IEEE Trans. Circuits Syst.*, **38**(12):1501–1516, December 1991.

[Kai80] T. Kailath, *"Linear Systems,"* Prentice Hall, Englewood Cliffs, NJ, 1980.

[Kai86] T. Kailath, "A theorem of I. Schur and its impact on modern signal processing," In *Operator Theory: Advances and Applications*, volume 18, pp. 9–30. Birkhäuser Verlag, Basel, 1986.

[Kal63] R.E. Kalman, "Mathematical description of linear dynamical systems," *SIAM J. Control*, **1**:152–192, 1963.

[Kal65] R.E. Kalman, "Irreducible realizations and the degree of a rational matrix," *SIAM J. Applied Math.*, **13**:520–545, 1965.

[Kam75] E.W. Kamen, "On an algebraic theory of systems defined by convolution operators," *Math. Systems Theory*, **9**(1):57–74, 1975.

[Kam76a] E.W. Kamen, "Module structure of infinite-dimensional systems with applications to controllability," *SIAM J. Control and Optimization*, **14**(3):389–408, May 1976.

[Kam76b] E.W. Kamen, "Representation and realization of operational differential equations with time-varying coefficients," *Journal of the Franklin Inst.*, **301**(6):559–571, June 1976.

[Kam79] E.W. Kamen, "New results in realization theory for linear time-varying analytic systems," *IEEE Trans. Automat. Control*, **24**(6):866–878, December 1979.

[KFA70] R.E. Kalman, P.L. Falb, and M.A. Arbib, *"Topics in Mathematical System Theory,"* Int. Series in Pure and Applied Math. McGraw-Hill, 1970.

[KH79] E.W. Kamen and K.M. Hafez, "Algebraic theory of linear time-varying systems," *SIAM J. Control and Optimization*, **17**(4):500–510, July 1979.

[KH90] R. E. Kearney and L. W. Hunter, "System identification of human joint dynamics," *Biomedical Engineering*, **18**(1):55–87, 1990.

[Kim97] H. Kimura, *"Chain Scattering Approach to H_∞-Control,"* Birkhäuser, Boston, 1997.

[KKM79] T. Kailath, S.Y. Kung, and M. Morf, "Displacement ranks of matrices and linear equations," *J. Math. Anal. Appl.*, **68**(2):395–407, 1979.

[KKMH91] R.E. Kearney, R.E. Kirsch, B. MacNeil, and I.W. Hunter, "An ensemble time-varying identification technique: Theory and biomedical applications," In *9th IFAC/IFORS Symposium on Identification and System Parameter Estimation*, pp. 191–196, 1991.

[KKP85] E.W. Kamen, P.P. Khargonekar, and K.R. Poolla, "A transfer-function approach to linear time-varying discrete-time systems," *SIAM J. Control and Optimization*, **23**(4):550–565, July 1985.

[KKY87] V.E. Katsnelson, A.Ya. Kheifets, and P.M. Yuditskii, *"An abstract interpolation problem and the theory of extensions of isometric operators,"*, volume 146, Operators in Function Spaces and Problems in Function Theory, pp. 83–96 V.E. Marchenko, 1987.

[KL81] S.Y. Kung and D.W. Lin, "Optimal Hankel norm model reductions: Multi-variable systems," *IEEE Trans. Automat. Control*, **26**(4):832–852, August 1981.

[KN79] H. Kano and T. Nishimura, "Periodic solutions of matrix Riccati equations with detectability and stabilizability," *Internat. J. Control*, **29**:471–487, 1979.

[KP86] P.P. Khargonekar and K. Poolla, "On polynomial matrix fraction representations for linear time-varying discrete-time systems," *Lin. Alg. Appl.*, **80**:1–37, 1986.

[Kre70] M.G. Krein, "Introduction to the geometry of indefinite J-spaces and to the theory of operators in those spaces," *Amer. Math. Soc. Transl.*, **93**:103–176, 1970.

[Kro90] L. Kronecker, "Algebraische Reduction der schaaren bilinearer Formen," *S.B. Akad. Berlin*, pp. 663–776, 1890.

[KS95] T. Kailath and A.H. Sayed, "Displacement structure: Theory and applications," *SIAM Review*, **37(3)**:297–386, 1995.

[Kuc72] V. Kucera, "A contribution to matrix quadratic equations," *IEEE Tr. Automat. Control*, **17**:344–347, 1972.

[Kun78] S.Y. Kung, "A new identification and model reduction algorithm via singular value decomposition," In *Twelfth Asilomar Conf. on Circuits, Systems and Comp.*, pp. 705–714, Asilomar, CA., November 1978.

[Lax59] P.D. Lax, "Translation invariant subspaces," *Acta Math.*, **101**:163–178, 1959.

[Lev47] N. Levinson, "The Wiener RMS error criterion in filter design and prediction," *J. Math. Phys.*, **25**:261–278, 1947.

[Lev83] H. Lev-Ari, *"Non-stationary Lattice-Filter Modeling,"* PhD thesis, Stanford University, 1983.

[LG90] D.J.N. Limebeer and M. Green, "Parametric interpolation, H_∞-control and model reduction," *Int. J. Control*, **52**(2):293–318, 1990.

[Liv72] M.S. Livsic, *"Operators, Oscillations, Waves (Open Systems),"* volume 34. Amer. Math. Soc. Transl. Math. Monographs, Providence, 1972.

[LK84] H. Lev-Ari and T. Kailath, "Lattice filter parametrization and modeling of non-stationary processes," *IEEE Trans. Informat. Th.*, **30**(1):2–16, January 1984.

[LK86] H. Lev-Ari and T. Kailath, "Triangular factorizations of structured Hermitian matrices," In *Operator Theory: Advances and Applications*, volume 18, pp. 301–324. Birkhäuser Verlag, 1986.

[LK91] H. Lev-Ari and T. Kailath, "Lossless arrays and fast algorithms for structured matrices," In Ed. F. Deprettere and A.J. van der Veen, editors, *Algorithms and Parallel VLSI Architectures*, volume A, pp. 97–112. Elsevier, 1991.

[LK92] H. Lev-Ari and T. Kailath, "State-space approach to factorization of lossless transfer functions and structured matrices," *Lin. Alg. Appl.*, **162**:273–295, February 1992.

[LP67] P.D. Lax and R.S. Phillips, *"Scattering Theory,"* Academic Press, New York, 1967.

[LR95] P. Lancaster and L. Rodman, *"Algebraic Riccati Equations,"* Oxford Univ. Press, Oxford, UK, 1995.

[MDCM95] E. Moulines, P. Duhamel, J.-F. Cardoso, and S. Mayrargue, "Subspace methods for the blind identification of multichannel FIR filters," *IEEE Trans. Signal Proc.*, **43**(2):516–525, February 1995.

[MDV93] M. Moonen, P. Van Dooren, and F. Vanpoucke, "On the QR algorithm and updating the SVD and URV decomposition in parallel," *Lin. Alg. Appl.*, **188/189**:549–568, July 1993.

[MMVV89] M. Moonen, B. De Moor, L. Vandenberghe, and J. Vandewalle, "On- and off-line identification of linear state-space models," *Int. J. Control*, **49**:219–232, 1989.

[Mol75] B.P. Molinari, "The stabilizing solution of the discrete algebraic Riccati equation," *IEEE Tr. Automat. Control*, **20**:396–399, June 1975.

[Moo79] B.C. Moore, "Singular value analysis of linear systems," In *Proc. IEEE Conf. Dec. Control*, pp. 66–73, 1979.

[Moo81] B.C. Moore, "Principal component analysis in linear systems: Controllability, observability and model reduction," *IEEE Trans. Automat. Control*, **26**(1):17–32, February 1981.

[Mur84] J. Murray, "Time-varying systems and crossed products," *Math. Systems Theory*, **17**:217–241, 1984.

[MVV92] M. Moonen, P. Van Dooren, and J. Vandewalle, "An SVD updating algorithm for subspace tracking," *SIAM J. Matrix Anal. Appl.*, **13**(4):1015–1038, 1992.

[ND91] H. Nelis and E.F. Deprettere, "Approximate inversion of partially specified positive definite matrices," In G. Golub and P. van Dooren, editors, *Numerical Linear Algebra, Digital Signal Processing and Parallel Algorithms*, pp. 559–568. Springer-Verlag, 1991.

[Neh57] Z. Nehari, "On bounded bilinear forms," *Ann. of Math.*, **65**(2):153–162, 1957.

[Nel89] H. Nelis, *"Sparse Approximations of Inverse Matrices,"* PhD thesis, Delft Univ. Techn., The Netherlands, 1989.

[Ner58] A. Nerode, "Linear automaton transformations," *Proc. American Mathematical Society*, **9**:541–544, 1958.

[Nic92] G. De Nicolao, "On the time-varying Riccati difference equation of optimal filtering," *SIAM J. Control and Optimization*, **30**(6):1251–1269, November 1992.

[Obe91] R. Ober, "Balanced parametrization of classes of linear systems," *SIAM J. Control Optim.*, **29**(6):1251–1287, November 1991.

[OSB91] R. Onn, A.O. Steinhardt, and A.W. Bojanczyk, "The hyperbolic singular value decomposition and applications," *IEEE Trans. Signal Proc.*, **39**(7):1575–1588, July 1991.

[Ove95] P. Van Overschee, *"Subspace identification: theory, implementation, application,"* PhD thesis, KU Leuven, Leuven, Belgium, 1995.

[OY54] Y. Oono and K. Yasuura, "Synthesis of finite passice $2n$ terminal networks with prescribed scattering matrices," *Mem. Fac. Eng. Kyushu Univ.*, **14**(2):125–177, May 1954 (French translation, *Ann. Telecomm.*, **9** (3):73-80, March 1954; (4):109-115, April 1954; (5):133-140, May 1954).

[PK87] K. Poolla and P. Khargonekar, "Stabilizability and stable-proper factorizations for linear time-varying systems," *SIAM J. Control and Optimization*, **25**(3):723–736, May 1987.

[PLS80] T. Pappas, A.J. Laub, and N.R. Sandell, "On the numerical solution of the discrete-time algebraic Riccati equation," *IEEE Tr. Automat. Control*, **25**(4):631–641, August 1980.

[Pot60] V.P. Potapov, "The multiplicative structure of J-contractive matrix functions," *Amer. Math. Soc. Transl. Ser. 2*, **15**:131–243, 1960.

[PS79] L. Pernebo and L.M. Silverman, "Balanced systems and model reduction," In *Proc. IEEE Conf. Dec. Control*, pp. 865–867, 1979.

[PS82] L. Pernebo and L.M. Silverman, "Model reduction via balanced state space representations," *IEEE Trans. Automat. Control*, **27**(2):382–387, April 1982.

[Rak92] M. Rakowski, "Minimal factorization of rational matrix functions," *IEEE Trans. Circuits and Systems–I: Fund. Th. Appl.*, **39**(6):440–445, June 1992.

[Red62] R. Redheffer, "On the relation of transmission line theory to scattering and transfer," *J.Math.Phys.*, **XLI**:1–41, 1962.

[RK84] S.K. Rao and T. Kailath, "Orthogonal digital filters for VLSI implementation," *IEEE Trans. Circuits Syst.*, **31**(11):933–945, November 1984.

[RM87] R.A. Roberts and C.T. Mullis, *"Digital Signal Processing,"* Addison-Wesley, 1987.

[RMV88] P.A. Regalia, S.K. Mitra, and P.P. Vaidyanathan, "The digital all-pass filter: A versatile signal processing building block," *Proc. IEEE*, **76**(1):19–37, January 1988.

[RPK92] R. Ravi, A.M. Pascoal, and P.P. Khargonekar, "Normalized coprime factorizations for linear time-varying systems," *Systems and Control Letters*, **18**:455–465, 1992.

[Rud66] W. Rudin, *"Real and Complex Analysis,"* McGraw-Hill, New York, 1966.

[Rug93] W.J. Rugh, *"Linear System Theory,"* Prentice Hall, Englewood Cliffs, NJ, 1993.

[SA68] L.M. Silverman and B.D.O. Anderson, "Controllability, observability and stability of linear systems," *SIAM J. Control*, **6**(1):121–130, 1968.

[SA73] L.H. Son and B.D.O. Anderson, "Design of Kalman filters using signal model output statistics," *Proc. IEE*, **120**(2):312–318, 1973.

[Saa96] Y. Saad, *"Iterative methods for sparse linear systems,"* PWS Publ. Co., Boston, 1996.

[Say92] A. Sayed, *"Displacement Structure in Signal Processing and Mathematics,"* PhD thesis, Stanford University, 1992.

[Sch17] I. Schur, "Uber Potenzreihen, die im Innern des Einheitskreises beschränkt sind, I," *J. Reine Angew. Math.*, **147**:205–232, 1917 Eng. Transl. *Operator Theory: Adv. Appl.*, vol. 18, pp. 31-59, Birkhäuser Verlag, 1986.

[SK94] A.H. Sayed and T. Kailath, "A state-space approach to adaptive RLS filtering," *IEEE Signal Proc. Mag.*, **11**(3):18–60, July 1994.

[Slo94] D. Slock, "Blind fractionally-spaced equalization, perfect-reconstruction filter banks and multichannel linear prediction," In *Proc. IEEE ICASSP*, pp. IV:585–588, 1994.

[SM66] L.M. Silverman and H.E. Meadows, "Equivalence and synthesis of time-variable linear systems," In *Proc. 4-th Allerton Conf. Circuit and Systems Theory*, pp. 776–784, 1966.

[SM69] L.M. Silverman and H.E Meadows, "Equivalent realizations of linear systems," *SIAM J. Applied Math.*, **17**:393–408, 1969.

[SNF70] B. Sz.-Nagy and C. Foias, *"Harmonic Analysis of Operators on Hilbert Space,"* North-Holland, Amsterdam, 1970.

[Ste77] G.W. Stewart, *"Introduction to Matrix Computations,"* Academic Press, 1977.

[Ste92] G.W. Stewart, "An updating algorithm for subspace tracking," *IEEE Trans. Signal Proc.*, **40**(6):1535–1541, June 1992.

[SV95] J.M.A. Scherpen and M. Verhaegen, "On the Riccati equations of the H_∞ control problem for discrete time-varying systems," In *Proc. Third European Control Conference*, pp. 1824–1829 vol.3, Rome, Italy, 1995. Eur. Union Control Assoc.

[SV96] J.M.A. Scherpen and M. Verhaegen, "H_∞ output feedback control for linear discrete time-varying systems via the bounded real lemma," *Int. J. Control*, **65**(6):963–993, 1996.

[TD95] H.G.J. Theunis and E. F. Deprettere, "Piecewise stationary perfect reconstruction filter banks," *Archiv f. Elektronik u. Übertragungstechnik*, **49**(5/6):344–361, September 1995.

[Vai85a] P.P. Vaidyanathan, "The discrete-time Bounded-Real Lemma in digital filtering," *IEEE Trans. Circuits Syst.*, **32**(9):918–924, 1985.

[Vai85b] P.P. Vaidyanathan, "A unified approach to orthogonal digital filters and wave digital filters, based on LBR two-pair extraction," *IEEE Trans. Circuits Syst.*, **32**:673–686, July 1985.

[VD77] J. Vandewalle and P.M. Dewilde, "On the irreducible cascade synthesis of a system with a real rational transfer matrix," *IEEE Trans. Circuits Syst.*, **24**(9):481–494, September 1977.

[VD92a] M. Verhaegen and P. Dewilde, "Subspace model identification. part I: The output-error state space model identification class of algorithms," *Int. J. Control*, **56**(5):1187–1210, 1992.

[VD92b] M. Verhaegen and P. Dewilde, "Subspace model identification. part II: Analysis of the elementary output-error state space model identification algorithm," *Int. J. Control*, **56**(5):1211–1241, 1992.

[vdV93a] A.J. van der Veen, "Computation of the inner-outer factorization for time-varying systems," In P. Dewilde e.a., editor, *Challenges of a Generalized System Theory*, Essays of the Royal Dutch Academy of Sciences, pp. 99–117, Amsterdam, The Netherlands, 1993. North-Holland.

[vdV93b] A.J. van der Veen, *"Time-Varying System Theory and Computational Modeling: Realization, Approximation, and Factorization,"* PhD thesis, Delft University of Technology, Delft, The Netherlands, June 1993.

[vdV95] A.J. van der Veen, "Time-varying lossless systems and the inversion of large structured matrices," *Archiv f. Elektronik u. Übertragungstechnik*, **49**(5/6):372–382, September 1995.

[vdV96] A.J. van der Veen, "A Schur method for low-rank matrix approximation," *SIAM J. Matrix Anal. Appl.*, **17**(1):139–160, January 1996.

[vdVD91] A.J. van der Veen and P.M. Dewilde, "Time-varying system theory for computational networks," In P. Quinton and Y. Robert, editors, *Algorithms and Parallel VLSI Architectures, II*, pp. 103–127. Elsevier, 1991.

[vdVD93] A.J. van der Veen and P.M. Dewilde, "Time-varying computational networks: Realization, orthogonal embedding and structural factorization," *Integration, the VLSI journal*, **16**:267–291, 1993.

[vdVD94a] A.J. van der Veen and P.M. Dewilde, "Embedding of time-varying contractive systems in lossless realizations," *Math. Control Signals Systems*, **7**:306–330, 1994.

[vdVD94b] A.J. van der Veen and P.M. Dewilde, "On low-complexity approximation of matrices," *Linear Algebra and its Applications*, **205/206**:1145–1201, July 1994.

[vdVP96] A.J. van der Veen and A. Paulraj, "An analytical constant modulus algorithm," *IEEE Trans. Signal Processing*, **44**(5):1136–1155, May 1996.

[vdVTP97] A.J. van der Veen, S. Talwar, and A. Paulraj, "A subspace approach to blind space-time signal processing for wireless communication systems," *IEEE Trans. Signal Proc.*, **45**(1):173–190, January 1997.

[vdVV96] A.J. van der Veen and M. Viberg, "Minimal continuous state-space parametrizations," In *Proc. Eusipco*, pp. 523–526, Trieste (Italy), September 1996.

[Ver94] M. Verhaegen, "Subspace model identification. part III: Analysis of the ordinary output-error state space model identification algorithm," *Int. J. Control*, **58**:555–586, 1994.

[Vib94] M. Viberg, "Subspace methods in system identification," In *Proc. 10-th IFAC Symp. on Syst. Id.*, Copenhagen, Denmark, July 1994.

[Vib95] M. Viberg, "Subspace-based methods for the identification of linear time-invariant systems," *Automatica*, **31**(12):1835–1851, December 1995.

[VY95] M. Verhaegen and X. Yu, "A class of subspace model identification algorithms to identify periodically and arbitrarily time-varying systems," *Automatica*, **31**(2):201–216, 1995.

[Wei72] L. Weiss, "Controllability, realization, and stability of discrete-time systems," *SIAM J. Control and Optimization*, **10**:230–251, 1972.

[Woe89] H. Woerdeman, *"Matrix and Operator Extensions,"* PhD thesis, Dept. Math. Comp Sci., Free University, Amsterdam, The Netherlands, 1989.

[Won68] W.M. Wonham, "On a matrix Riccati equations of stochastic control," *SIAM J. Control and Optimization*, **6**:681–697, 1968.

[WW92] E.T. Whittaker and G.N. Watson, *"A Course of Modern Analysis,"* University Press, Cambridge, 1992.

[Yan95] B. Yang, "Projection approximation subspace tracking," *IEEE Trans. Signal Proc.*, **43**(1):95–107, January 1995.

[Yos71] K. Yoshida, *"Functional Analysis,"* Springer Verlag, Berlin, 3rd edition, 1971.

[You66] D.C. Youla, "The synthesis of linear dynamical systems from prescribed weighting patterns," *SIAM J. Applied Math.*, **14**(3):527–549, May 1966.

[You86] N.J. Young, "Balanced realizations in infinite dimensions," In H. Bart, I. Gohberg, and M.A. Kaashoek, editors, *Operator Theory and Systems*, volume OT-19, pp. 449–471. Birkhäuser Verlag, 1986.

[YS91] I. Yaesh and U. Shaked, "A transfer function approach to the problems of discrete-time systems: H_∞-optimal linear control and filtering," *IEEE Trans. Automat. Control*, **36**(11):1264–1271, November 1991.

[YSvdVD96] X. Yu, J. Scherpen, A.J. van der Veen, and P.M. Dewilde, "Outer (j_1, j_2)-lossless factorizations of linear discrete time-varying systems," In *Proc. IEEE CDC*, pp. 2249–2254, December 1996.

[YT66] D.C. Youla and P. Tissi, "*n*-Port synthesis via reactance extraction–part I," *IEEE Int. Conf. Rec.*, **14**(7):183–205, 1966.

[Yu96] Xiaode Yu, *"Time-varying System Identification, J-lossless Factorization, and H_∞ Control,"* PhD thesis, Delft Univ. of Technology, Delft, The Netherlands, May 1996.

[YV93] X. Yu and M. Verhaegen, "Application of a time-varying subspace model identification scheme to the identification of the human joint dynamics," In *Proc. European Control Conf.*, volume 2, pp. 603–608, Groningen, The Netherlands, June 1993.

[Zad50] L.A. Zadeh, "Frequency analysis of variable networks," *Proc. IRE*, **38**:291–299, March 1950.

[Zad61] L.A. Zadeh, "Time-varying networks, I," *Proc. IRE*, **49**:1488–1503, October 1961.

[Zam81] G. Zames, "Feedback and optimal sensitivity: Model reference transformations, multiplicative seminorms, and approximate inverses," *IEEE Trans. Automat. Control*, **26**(2):301–320, April 1981.

[ZM74] H.P. Zeiger and A.J. McEwen, "Approximate linear realizations of given dimension via Ho's algorithm," *IEEE Trans. Automat. Control*, **19**(2):153, April 1974.

Glossary of notation

Diagonal algebra

$\mathcal{N} = \mathbb{C}^N$: space of (non-uniform) sequences with i-th entry in $\mathbb{C}^{N_i}$ (p. 20).

$N = \#\mathcal{N}$: the sequence of dimensions of $\mathcal{N}$ (p. 20).

$\ell_2^{\mathcal{N}}$: space of bounded (non-uniform) sequences in $\mathcal{N}$

$\mathcal{X}(\mathcal{M},\mathcal{N})$: space of bounded operators $\ell_2^{\mathcal{M}} \to \ell_2^{\mathcal{N}}$ and $\mathcal{X}_2^{\mathcal{M}} \to \mathcal{X}_2\mathcal{N}$ (p. 22).

$\mathcal{U},\mathcal{L},\mathcal{D}$: upper/lower/diagonal bounded operators in $\mathcal{X}$ (p. 23).

$\mathcal{X}_2,\mathcal{U}_2,\mathcal{L}_2,\mathcal{D}_2$: (Hilbert) spaces of operators in $\mathcal{X},\mathcal{U},\mathcal{L},\mathcal{D}$ with bounded HS-norm (p. 25).

π_i: sequence constructor. $A \in \mathcal{X}$ has entries $A_{ij} = \pi_i A \pi_j^*$ (p. 22).

Z: bilateral causal shift operator (p. 26).

$T^{(k)}$: diagonal shift of $T \in \mathcal{X}$ over k positions into south-west direction (p. 27).

$r(X)$: spectral radius of X (p. 24).

$\mathbf{P}_{\mathcal{H}}$: projection onto a subspace $\mathcal{H} \subset \mathcal{X}_2$ (p. 25).

$\mathbf{P},\mathbf{P}_0,\mathbf{P}'$: projection onto $\mathcal{U}_2$, $\mathcal{D}_2$, $\mathcal{L}_2 Z^{-1}$ (p. 25).

$\{A,B\}$ $= \mathbf{P}_0(AB^*)$: diagonal inner product (p. 77).

$\{A,B\}_J$ $= \mathbf{P}_0(AJB^*)$: indefinite diagonal inner product (p. 200).

$A \gg 0$: A is uniformly strictly positive definite (p. 77).

$A^{\{k\}}$ $= A^{(k)}A^{(k-1)}\cdots A^{(1)}$ (p. 27).

$A^{[k]}$ $= AA^{(1)}\cdots A^{(k-1)}$ (p. 26).

$T_{[k]}$ $= \mathbf{P}_0(Z^{-k}T)$: the k-th diagonal above the main (0-th) diagonal of T (p. 28).

$\Lambda_{\mathbf{F}}$ $= \mathbf{P}_0(\mathbf{F}\mathbf{F}^*)$: the Gram operator associated to a basis representation $\mathbf{F}$ (p. 84).

sdim $(\cdot)$: the sequence of dimensions of a left D-invariant subspace (p. 78).

$\ell_2^{\mathcal{D}}$: the space of bounded sequences with entries in $\mathcal{D}$

$(\cdot)^\dagger$: the pseudo-inverse (Moore-Penrose generalized inverse)

System theory

T: realization matrix. $\mathbf{T} = \{A, B, C, D\}$ stands for the matrix $\mathbf{T} = \begin{bmatrix} A & C \\ B & D \end{bmatrix}$ (p. 37).

ℓ_A: the spectral radius of AZ (p. 38).

$\mathcal{H}(T), \mathcal{H}_o(T), \mathcal{K}(T), \mathcal{K}_o(T)$: input state space, output state space, input null space, output null space of an operator $T \in \mathcal{X}$ (p. 89).

$\mathbf{Q}, \mathbf{F}, \mathbf{G}, \mathbf{F}_0$: typically, **Q** and **G** are orthonormal basis representations of the input and output state space. **F**, $\mathbf{F}_0$ are strong basis representations of these spaces (p. 105 *ff.*).

$\mathcal{C}, \mathcal{O}$: controllability, observability operators (p. 54).

H_T, K_T, E_T: the operator T on restricted domains and ranges. $H_T : \mathcal{L}_2 Z^{-1} \to \mathcal{U}_2$ is the Hankel operator. $K_T : \mathcal{L}_2 Z^{-1} \to \mathcal{L}_2 Z^{-1}$, $E_T : \mathcal{U}_2 \to \mathcal{U}_2$ (p. 88).

$T_\Theta[S_L]$ $= (\Theta_{11} S_L - \Theta_{12})(\Theta_{22} - \Theta_{21} S_L)^{-1}$ (p. 199).

Index

MIX
Papier aus verantwortungsvollen Quellen
Paper from responsible sources
FSC® C105338

If you have any concerns about our products,
you can contact us on
ProductSafety@springernature.com

In case Publisher is established outside the EU,
the EU authorized representative is:
Springer Nature Customer Service Center GmbH
Europaplatz 3, 69115 Heidelberg, Germany

Printed by Libri Plureos GmbH
in Hamburg, Germany